Handbook of Brewing

Edited by
Hans Michael Eßlinger

Further Reading

Ziegler, H. (Ed.)

Flavourings

Production, Composition, Applications, Regulations

Second, Completely Revised Edition
2007
ISBN: 978-3-527-31406-5

Surburg, H., Panten, J. (Eds.)

Common Fragrance and Flavor Materials

Preparation, Properties and Uses

Fifth, Completely Revised and Extended Edition
2006
ISBN: 978-3-527-31315-0

Brennan, J. G. (Ed.)

Food Processing Handbook

2006
ISBN: 978-3-527-30719-7

Handbook of Brewing

Processes, Technology, Markets

Edited by
Hans Michael Eßlinger

WILEY-VCH Verlag GmbH & Co. KGaA

The Editor

Dr. Hans Michael Eßlinger
Freiberger Brauhaus AG
Am Fürstenwald
09599 Freiberg
Germany

All books published by **Wiley-VCH** are carefully produced. Nevertheless, authors, editors, and publisher do not warrant the information contained in these books, including this book, to be free of errors. Readers are advised to keep in mind that statements, data, illustrations, procedural details or other items may inadvertently be inaccurate.

Library of Congress Card No.: applied for

British Library Cataloguing-in-Publication Data
A catalogue record for this book is available from the British Library.

Bibliographic information published by the Deutsche Nationalbibliothek
The Deutsche Nationalbibliothek lists this publication in the Deutsche Nationalbibliografie; detailed bibliographic data are available on the Internet at <http://dnb.d-nb.de>.

© 2009 WILEY-VCH Verlag GmbH & Co. KGaA, Weinheim

All rights reserved (including those of translation into other languages). No part of this book may be reproduced in any form – by photoprinting, microfilm, or any other means – nor transmitted or translated into a machine language without written permission from the publishers. Registered names, trademarks, etc. used in this book, even when not specifically marked as such, are not to be considered unprotected by law.

Composition   SNP Best-set Typesetter Ltd., Hong Kong
Printing   Betz-Druck GmbH, Darmstadt
Bookbinding   Litges & Dopf GmbH, Heppenheim

Printed in the Federal Republic of Germany
Printed on acid-free paper

ISBN: 978-3-527-31674-8

Contents

Preface *XXVII*
List of Contributors *XXIX*

1 **A Comprehensive History of Beer Brewing** *1*
Franz G. Meussdoerffer
1.1 Introduction *1*
1.2 'The Truly Happy Man Has His Mouth Full of Beer': From Prehistory to the End of the Roman Empire *2*
1.2.1 Advent of Agrarian Societies *2*
1.2.2 Mesopotamia and Egypt *5*
1.2.3 Hellenistic Period: Greeks, Romans and Their Neighbors *7*
1.2.4 Celts and the Germans *8*
1.3 'I Would Like to Have a Great Lake of Beer for Christ the King': The Christian Middle Ages *9*
1.3.1 Monasteries *9*
1.3.2 Beginnings of Professionalism *12*
1.4 'Woar Volk Is, Bint Klaantn': Hopped Beer and the Seaports *16*
1.4.1 Hanseatic League *16*
1.4.2 Rise and Decline of the Cities in Central Europe *18*
1.4.3 Rise of Dutch Brewing *20*
1.5 'For a Quart of Ale is a Dish for a King': John Bull and the Industrialization of Brewing *21*
1.6 'We Live in a Country Where Beer Constitutes Quasi the Fifth Element': Advent of Lager and the Internationalization of Brewing *25*
1.6.1 Bavaria and the Rise of Lager Beer *25*
1.6.2 Spread of Lager Brewing *27*
1.6.3 From Alchemy to Biochemistry: The Science of Brewing *30*
1.6.4 Europe, the United States and the Internationalization of Beer Brewing *33*
References *37*

Handbook of Brewing: Processes, Technology, Markets. Edited by H. M. Eßlinger
Copyright © 2009 WILEY-VCH Verlag GmbH & Co. KGaA, Weinheim
ISBN: 978-3-527-31674-8

2 Starchy Raw Materials 43
Franz Meussdoerffer and Martin Zarnkow

2.1 Introduction 43
2.2 Principles of Structure and Metabolism 44
2.2.1 Grain Structure 44
2.2.2 Basic Anabolic Processes 46
2.2.2.1 Starch Formation 47
2.2.2.2 Storage Proteins 50
2.2.2.3 Regulation of Grain Filling and Impact of Environmental Factors 52
2.2.3 Catabolic Processes 53
2.3 Major Brewing Cereals 55
2.3.1 Introductory Remarks 55
2.3.2 Selected Cereals 59
2.3.2.1 Barley (*H. vulgare* L.) 59
2.3.2.2 Oats (*Avena sativa* L.) 60
2.3.2.3 Short-Grain Millet 61
2.3.2.4 Maize (*Zea mays* L.) 63
2.3.2.5 Rice (*Oryza sativa* L.) 64
2.3.2.6 Rye (*Secale cereale* L.) 65
2.3.2.7 Sorghum (*Sorghum bicolor* L.) 65
2.3.2.8 Wheat Group (*Triticum* L.) 66
2.3.3 Pseudocereals 69
2.3.3.1 Grain Amaranth (Main Sorts: *Amaranthus cruentus, Amaranthus hypochondriacus* and *Amaranthus caudatus*) 69
2.3.3.2 Buckwheat (*Fagopyrum esculentum* Moench) 70
2.3.3.3 Quinoa (*Chenopodium quinoa* Willd.) 70
2.4 Concluding Remarks 71
References 71

3 Hops 85
Martin Krottenthaler

3.1 Introduction 85
3.2 Cultivation of Hops 85
3.3 Components of Hops 87
3.3.1 Bitter Acids 87
3.3.2 Aroma Substances 88
3.3.3 Polyphenols 89
3.4 Hop Products 90
3.5 Analytics 91
3.6 Hopping Technology 91
3.6.1 General Aspects of Hop Addition 91
3.6.2 Beer with Hop Flavor 94
3.6.3 Beer with a Hop Flavor 95
3.6.4 Beer Enriched with Xanthohumol 97

3.6.5	Yield of Bitter Principles	97
3.6.6	Foam	98
3.6.7	Microbiology	99
3.6.8	Addition of 'Downstream Products'	99
3.7	Storage of Hops	100
	References	101
4	**Brew Water**	**105**
	Martin Krottenthaler and Karl Glas	
4.1	General Requirements	105
4.2	Characteristics of Constituents Relevant for Brewing	105
4.3	Quality Criteria for Brew Water	108
4.4	Water Treatment	110
4.4.1	Removal of Problematic Inorganic Substances	110
4.4.1.1	Deferrization and Demanganization	110
4.4.1.2	Nitrate Reduction	111
4.4.2	Removal of Problematic Organic Substances	112
4.4.2.1	Aeration	112
4.4.2.2	Activated-Carbon Filtration	112
4.4.2.3	Combination Processes Using Oxidation/UV Irradiation	112
4.4.3	Common Processes for Brew Water Treatment	112
4.4.3.1	Lime Softening	112
4.4.3.2	Ion Exchange	113
4.4.3.3	Membrane Processes	115
	References	117
5	**Yeast**	**119**
	Christoph Tenge	
5.1	Brewing Yeast	119
5.1.1	History of Yeast Research	119
5.1.2	Yeast for Brewing Applications	120
5.1.2.1	Flocculation	123
5.1.3	Yeast Morphology and Chemical Composition	124
5.1.3.1	Cell Wall	124
5.1.3.2	Periplasm	125
5.1.3.3	Cell Membrane	125
5.1.3.4	Cytoplasm	126
5.1.3.5	Mitochondria	126
5.2	Yeast Management	127
5.2.1	Nutrient Requirements and Intake	127
5.2.1.1	Carbohydrates and Fermentable Sugars	127
5.2.1.2	Nitrogen Sources	128
5.2.1.3	Oxygen	129
5.2.1.4	Minerals and Trace Elements	129
5.2.1.5	Vitamins and Other Growth Factors	130

5.2.2	Metabolic Pathways during Propagation and Fermentation	*131*
5.2.2.1	Carbohydrate Metabolism for Cell Growth and Energy Generation	*131*
5.2.2.2	Formation of Vicinal Diketones	*132*
5.2.2.3	Formation of Higher Alcohols	*132*
5.2.2.4	Formation of Esters	*134*
5.2.2.5	Phenolic Compounds	*134*
5.2.2.6	Formation of Sulfur Dioxide	*135*
5.2.3	Yeast Cultivation, Propagation and Post-Fermentation Treatment	*135*
5.2.3.1	Yeast Cultivation in the Laboratory	*136*
5.2.3.2	Yeast Propagation in the Yeast Cellar	*138*
5.2.3.3	Post-Fermentation Treatment of Yeast	*140*
5.2.3.4	Yeast Crop	*140*
5.2.3.5	Yeast Treatment after Cropping and Yeast Storage	*141*
	References	*142*

6	**Malting**	*147*
	Stefan Kreisz	
6.1	Brewing Barley	*147*
6.2	Barley Intake and Storage	*149*
6.2.1	Barley Cleaning	*149*
6.2.2	Steeping	*150*
6.2.3	Germination	*152*
6.2.4	Kilning	*155*
6.2.5	Cleaning, Storage and Polishing of the Malt	*156*
6.2.6	Malt Yield	*157*
6.2.7	Malt Quality	*157*
6.2.8	Quality Criteria of Barley Malt	*158*
6.2.8.1	Cytolysis	*158*
6.2.8.2	Proteolysis	*160*
6.2.8.3	Amylolysis	*160*
6.2.8.4	Enzymes	*161*
6.2.8.5	Further Malt Quality Criteria	*161*
6.3	Special Malts	*163*
6.3.1	Dark Malt (Munich Type)	*163*
6.3.2	Caramel (Crystal) Malt	*163*
6.3.3	Roasted Malt	*164*
6.3.4	Wheat Malt and Malt Made from Other Cereals	*164*
6.3.5	Other Special Malt	*164*
	References	*164*

7	**Wort Production**	*165*
	Martin Krottenthaler, Werner Back, and Martin Zarnkow	
7.1	Introduction	*165*
7.2	Technology of Grinding	*165*

7.3	Mashing Technology	168
7.3.1	Mashing Parameters	169
7.3.2	Selected Mashing Processes	173
7.3.2.1	Step-Mashing Process	173
7.3.2.2	Maltase Process	174
7.3.2.3	Dark Beer Varieties	175
7.3.2.4	Adjunct Mashing	176
7.4	Technology of Lautering	181
7.4.1	Lauter Tun	182
7.4.2	Mash Filter	184
7.4.3	Strainmaster	186
7.5	Technological Basics of Wort Boiling	186
7.5.1	Hot Holding	186
7.5.2	Evaporation	187
7.5.3	Modern Boiling Systems	190
7.5.3.1	Internal Boilers	190
7.5.3.2	Optimized Internal Boiler 'Stromboli'	190
7.5.3.3	Optimized Internal Boiler Subjet	191
7.5.3.4	External Boiler	192
7.5.3.5	High-Temperature Wort Boiling	194
7.5.3.6	Dynamic Low-Pressure Boiling	194
7.5.3.7	Soft Boiling Method 'SchoKo'	194
7.5.3.8	Wort Stripping	196
7.5.3.9	Vacuum Evaporation	197
7.5.3.10	Flash Evaporation 'Varioboil'	197
7.5.3.11	Thin Film Evaporator 'Merlin'	199
7.5.4	Vapor Condensate	201
7.5.5	Cold Trub	202
	References	202
8	**Fermentation, Maturation and Storage**	207
	Hans Michael Eßlinger	
8.1	Pitching	207
8.2	Aeration	209
8.3	Topping-up	209
8.4	Changes during Fermentation	209
8.4.1	Changes in the Composition of Nitrogen Compounds	211
8.4.2	pH Drop	211
8.4.3	Changes in the Redox Properties of Beer	212
8.4.4	Beer Color	212
8.4.5	Precipitation of Bitter Substances and Polyphenols	212
8.4.6	CO_2 Content	212
8.4.7	Clarification and Colloidal Stabilization	213
8.5	Appearance during Fermentation	213
8.6	Fermentation Parameters	213
8.7	Control of Fermentation	214

8.8	Fermenters	*214*
8.9	Maturation	*215*
8.10	Storage	*215*
8.11	Bottom Fermentation in Practice	*216*
8.11.1	Cold Fermentation with Conventional Storage	*216*
8.11.2	Cold Fermentation with Well-Directed Maturation in a Cylindroconical Vessel (CCV)	*217*
8.11.3	Pressureless Warm Fermentation	*217*
8.11.4	Accelerated Fermentation under CO_2 Pressure	*217*
8.11.5	Cold Fermentation with Integrated Maturation at 12 °C	*218*
8.11.6	Cold Fermentation with Programmed Maturation at 20 °C	*219*
8.11.7	Accelerated Fermentation and/or Maturation	*219*
8.12	Yeast Crop and Yeast Storage	*220*
8.13	Beer Recovery from Yeast	*221*
8.14	CO_2 Recovery	*221*
8.15	Types of Bottom-Fermented Beers	*222*
8.16	Top Fermentation	*222*
8.17	Types and Production of Top-Fermented Beers	*223*
8.17.1	Wheat Beer	*223*
8.17.2	Alt Beer	*223*
8.17.3	Kölsch Beer	*224*
	References	*224*
9	**Filtration and Stabilization**	***225***
	Bernd Lindemann	
9.1	Introduction	*225*
9.2	Purpose of Filtration	*225*
9.3	Theoretical Considerations of Cake Filtration	*225*
9.4	Filtration Techniques	*226*
9.4.1	Kieselguhr Filtration	*226*
9.4.1.1	Plate and Frame Filter	*227*
9.4.1.2	Horizontal Leaf Filter	*227*
9.4.1.3	Metal Candle Leaf Filter	*229*
9.4.2	Filter Aids for Pre-coating	*230*
9.4.3	Methods in Kieselguhr Pre-coating Filtration	*230*
9.4.4	Membrane Filtration	*230*
9.5	Variables Influencing Beer Filtration	*231*
9.6	Beer Stabilization	*231*
9.7	Technical Design of a Filtration and Stabilization Plant	*232*
	References	*234*
10	**Special Production Methods**	***235***
	Felix Burberg and Martin Zarnkow	
10.1	Alcohol-Free Beers	*235*
10.1.1	Introduction	*235*

10.1.2	Techniques for the Production of Alcohol-Free Beers	*236*
10.1.2.1	Physical Techniques	*236*
10.1.2.2	Biological Methods	*238*
10.1.2.3	Combination Physical-Biological Processes	*240*
10.2	Dietetic Beer	*240*
10.2.1	Introduction	*240*
10.2.2	Methods for the Production of Dietetic Beers	*241*
10.3	'Nährbier' and 'Malzbier' ('Malztrunk')	*242*
10.3.1	Introduction	*242*
10.3.2	Methods for the Production of 'Nährbier' and 'Malzbier' ('Malztrunk')	*242*
10.4	XAN™ Wheat Beer	*243*
10.4.1	Introduction	*243*
10.4.2	Methods of Production of XAN™ Wheat Beer	*243*
10.5	Gluten-Free Beer	*244*
10.5.1	Introduction	*244*
10.5.2	Production Methods for Gluten-Free Beer	*244*
10.5.2.1	Conventional 'Gluten-Containing' Raw Material	*245*
10.5.2.2	Sources of Gluten-Free Sugars and Starch	*245*
10.6	Brewing with High Original Wort	*246*
10.6.1	Introduction	*246*
10.6.2	Methods for Beer Production with High Original Wort	*247*
10.7	Ale and Cask-Conditioned Ale	*248*
10.7.1	Introduction	*248*
10.7.2	Methods for the Production of Cask-Conditioned Ale	*249*
10.8	Lambic, Gueuze and Fruit Lambic	*250*
10.8.1	Introduction	*250*
10.8.2	Method for the Production of Lambic and Gueuze	*251*
10.8.2.1	Wort Production	*251*
10.8.2.2	Beer Production	*251*
10.8.3	Method for the Production of Fruit Lambic	*252*
10.9	Berliner Weisse	*253*
10.9.1	Introduction	*253*
10.9.2	Method for the Production of Berliner Weisse	*254*
10.10	Porter	*254*
10.10.1	Introduction	*254*
10.10.2	Method for the Production of Porter Bier	*255*
10.11	Summary	*255*
	References	*255*
11	**Beer-Based Mixed Drinks** *257*	
	Oliver Franz, Martina Gastl, and Werner Back	
11.1	Development of Beer-Based Mixed Drinks	*258*
11.2	Ingredients and Mixing Formulations	*260*
11.2.1	Constituent Beer	*260*

11.2.2	Water Quality	261
11.2.3	Sweetening	261
11.2.3.1	Sweetening Agents	261
11.2.4	Food Acids	262
11.2.5	Flavor and Juices	263
11.3	Quality Control of Beer-Based Mixed Drinks	263
11.3.1	Wet Chemical Analysis	263
11.3.2	Sensory Assessment of Beer-Based Mixed Drinks	264
11.3.3	Assessment by the Deutsche Landwirtschafts-Gesellschaft (DLG) (German Agricultural Society)	264
11.3.4	Off-flavors	266
11.4	Microbiology of Beer-Based Mixed Drinks	268
11.5	Preservation of the Final Beverage	269
11.5.1	Use of Antioxidants	269
11.5.2	Use of Preserving Agents	269
11.5.3	Thermal Processes–Pasteurization	269
11.6	Technological Aspects for the Production of Beer-Based Mixed Drinks	270
11.6.1	Mixing	270
11.6.2	Filtration	271
11.6.3	Filling	272
11.6.4	Influence of the Packaging	272
11.7	Technical Equipment for the Production of Beer-Based Mixed Drinks	273
	References	273
12	**Filling** 275	
	Susanne Blüml	
12.1	Choice of Packaging	275
12.1.1	Glass Bottles	275
12.1.2	Cans	277
12.1.3	Plastic Bottles	278
12.1.4	Kegs	280
12.2	Framework Conditions for Filling Beer	281
12.2.1	Significance of the Gases	282
12.2.1.1	Oxygen Content	282
12.2.1.2	CO_2 Content	283
12.2.2	Filling Pressure	284
12.2.3	Temperature	285
12.3	Process Steps When Filling Beer	285
12.3.1	Evacuation	285
12.3.2	Flushing with Ring-Bowl or Pure Gas	287
12.3.2.1	Flushing with Ring-Bowl Gas	288
12.3.2.2	Flushing with Pure Gas	288
12.3.3	Pressurization	288

12.3.4	Filling 288	
12.3.5	Settling and Snifting 290	
12.4	Filling Systems for Beer 290	
12.4.1	Mechanical Level-Controlled Filling Systems 290	
12.4.1.1	Mecafill VKPV – The Mechanical System for Bottling Beer 292	
12.4.2	Electronic Level-Controlled Filling Systems 292	
12.4.2.1	Sensometic VPVI – Filling With an Electronic Probe 294	
12.4.2.2	Sensometic VPL-PET – Probe-Controlled Long-Tube Filler for Single-Chamber and Multi-Chamber Operation 295	
12.4.3	Electronic Volumetric Filling Systems 296	
12.4.3.1	Volumetic VOC – The Can Filler 297	
12.4.3.2	Volumetic VODM-PET – Filling With Flow-Metered Quantitative Measurement 300	
12.4.4	Associated System Modules 301	
12.4.4.1	Fobbing 301	
12.4.4.2	Crowners 303	
12.4.4.3	Screw-Cappers for Plastic Screw Caps 304	
12.5	Constituent Parts of a Bottling Line 305	
12.5.1	Bottle Washer 305	
12.5.1.1	Treatment zones 306	
12.5.1.2	Components of a Bottle Washer 307	
12.5.1.3	Typical Bottle Treatment Sequence 311	
12.5.2	Inspection and Monitoring Units 313	
12.5.2.1	Machine Types 313	
12.5.2.2	Inspection Tasks 316	
12.5.2.3	Inspection Technology 318	
12.5.2.4	Reliability of the Inspectors 319	
	References 320	
13	**Labeling** 321	
	Jörg Bückle	
13.1	Some Basic Remarks on Machine Construction 321	
13.2	Wet-Glue Labeling 324	
13.2.1	Foiling 326	
13.3	Hotmelt Labeling 327	
13.3.1	Hotmelt Labeling with Pre-cut Labels 328	
13.3.2	Reel-Fed Hotmelt Labeling 330	
13.3.3	Roll on/Shrink on 331	
13.3.4	Tamper-Evident Seals 332	
13.4	Pressure-Sensitive Labeling 332	
13.5	Sleeving 334	
13.5.1	Stretch-Sleeve Process 335	
13.5.2	Shrink-Sleeve Process 336	
13.6	Date Coding and Identification 337	

14	**Beer Dispensing** *339*
	Reinhold Mertens
14.1	Beer Quality in the Draft Beer System *339*
14.1.1	Temperature *339*
14.1.2	Time on Tap *339*
14.1.3	CO_2 Content *339*
14.1.4	Foamhead *341*
14.1.5	Pouring the Beer *341*
14.2	Design of Draft Beer Systems *341*
14.2.1	Requirements for Rooms *341*
14.2.2	Requirements for Refrigeration *342*
14.2.2.1	Storage/Cabinet Cooling *342*
14.2.2.2	Ancillary Cooling *342*
14.2.2.3	Bar Cooling *342*
14.2.3	Requirements for Beer Lines *342*
14.2.4	Requirements for CO_2 Lines *343*
14.2.5	Requirements for Beer Bars/Bar Counter *343*
14.2.6	Requirements for Glass-Washing Equipment *343*
14.2.6.1	Two-Sink Installation *343*
14.2.6.2	Glass Cleaning System *343*
14.2.6.3	Glass Washing Machines *344*
14.2.7	Calculation of Applied Gauge Pressure *344*
14.3	Dispensing *346*
14.3.1	Types of Dispensing *346*
14.3.1.1	Dispensing from Underneath the Beer Bar *346*
14.3.1.2	Dispensing from above the Beer Bar *346*
14.3.1.3	Dispensing Direct from the Beer Bar *346*
14.3.2	Dispensing with Beer Pumps *346*
14.3.3	Dispensing Beer with Pre-mixed Gas (CO_2/N_2) *348*
14.3.4	Use of Gas Blenders *349*
14.3.5	Computerized Beer Dispensing *349*
14.3.6	Beer-Dispensing Tanks *350*
14.4	Parts of Draft Beer Systems *350*
14.4.1	Requirements for Gas-Pressurized Parts *350*
14.4.2	Requirements for Beverage Parts *351*
14.4.3	Keg-Tapping Equipment *351*
14.5	Hygiene Requirements in Draft Beer Systems *352*
14.5.1	Hygiene Target *352*
14.5.2	Cleaning and Disinfecting Procedures *353*
14.5.3	Hygiene Problem Areas *354*
14.6	Testing *355*
14.6.1	Hygiene Testing *355*
14.6.2	Leak Tests *355*
14.6.3	Temperature Tests *355*
14.7	Safety Precautions *356*

14.7.1	CO$_2$ Gas Alert Units	356
14.8	Final Remarks	356
	Reference	357
15	**Properties and Quality**	**359**
	August Gresser	
15.1	Composition of Finished, Bottom-Fermented Beer	359
15.2	Overall Qualities of Bottom-Fermented Beer	360
15.3	Redox Potential	360
15.4	Beer Color	360
15.5	Taste of Beer	361
15.6	Beer Foam	361
15.6.1	Basis of Beer Foam	362
15.6.2	Influence of Gas	362
15.6.3	Influence of Foam Stability	362
15.6.4	Influence of Brewing Liquor	362
15.6.5	Influence of Hop Products	363
15.6.6	Influence of Malt	363
15.6.7	Influence of Mash Filtration	363
15.6.8	Wort Boiling	363
15.6.9	Cold Break Removal	364
15.6.10	Main Fermentation	364
15.6.11	Storage Conditions	364
15.6.12	Beer Filtration	365
15.6.13	Precocious Indicators for the Foam Appearance	365
15.6.14	Conclusion	365
15.7	Bitter Substances in Hops	366
15.7.1	Influence of Cohumulone on the Bittering Quality	368
15.7.2	Influence of Cohumulone on Foam Stability	369
15.7.3	The Influence of Cohumulones on Beer Aging	369
15.8	Aroma Substances in Hops	371
15.9	Polyphenols in Beer Production	371
15.9.1	Definition of Polyphenols	373
15.9.2	Origin of Polyphenols	373
15.9.2.1	Malt Polyphenols	373
15.9.2.2	Hop Polyphenols	374
15.9.4	Polphenolic-Related Reactions during Brewing	379
15.9.4.1	Reaction with Proteins	379
15.9.4.2	Influence of Hops	379
15.9.4.3	Advantages and Disadvantages	379
15.9.5	Value of Anthocyanogens and Other Beer Characteristics	380
15.9.6	Reaction Path of Polyphenolic Components during the Brewing Process	380
15.9.7	Conclusions	381
15.10	N-Heterocycles	381

15.10.1	Presence of Heterocycles *381*
15.10.2	N-Heterocycles in the Malting Process *383*
15.10.3	Mashing Conditions *383*
15.10.4	Wort Boiling *386*
15.11	DMS *386*
15.11.1	Formation of DMS *386*
15.11.2	Barley and Malt *387*
15.11.3	Temperature *387*
15.11.4	Withering and Kilning *387*
15.11.5	Malt Cleaner *388*
15.11.6	Brewhouse *388*
15.12	Gushing (Uncontrolled Overflow or Overfoaming of Beer) *389*
15.12.1	General *389*
15.12.2	Determination of the Gushing Behavior of Beer Induced by Raw Materials *389*
15.12.3	Metal Ions in Bottled Beer *389*
15.12.4	Precipitation of Calcium Oxalate Crystals *390*
15.12.5	Filter Media *392*
15.12.6	Malt-Induced Gushing *392*
15.12.7	Chemical Components Causing Gushing *394*
	References *395*

16 **Stability of Beer** *399*
August Gresser

16.1	Flavor Stability *399*
16.1.1	Introduction *399*
16.1.2	Reasons for Beer Aging *401*
16.1.3	Changes of Aromatic Compounds *402*
16.1.4	Definition of Indicator Substances *403*
16.1.5	Technological Measurements to Preserve Organoleptic Stability *404*
16.1.5.1	Barley Variety *404*
16.1.5.2	Germination *405*
16.1.5.3	Wort Preparation *409*
16.1.5.4	Mash Filtration *413*
16.1.5.5	Wort Boiling *413*
16.1.5.6	Hot Break Removal *414*
16.1.5.7	Flotation (Removal of Cold Break) *414*
16.1.5.8	Yeast Handling *414*
16.1.5.9	Fermentation and Maturation *417*
16.1.5.10	Filtration *418*
16.1.5.11	Filling *420*
16.1.5.12	Analytical Control of Flavor Stability and Stale Flavor Compounds *422*
16.2	Lightstruck Flavor *423*
16.3	Colloidal Stability of Beer *428*

16.3.1	Introduction	*428*
16.3.1.1	Composition of Turbidity in Beer	*430*
16.3.2	Mechanism of Turbidity Formation	*430*
16.3.3	Influence of Raw Materials and Auxiliary Materials	*431*
16.3.4	Brewhouse	*431*
16.3.4.1	Malt Milling	*431*
16.3.4.2	Mashing Procedure	*431*
16.3.4.3	Mash Filtration	*431*
16.3.4.4	Wort Boiling	*431*
16.3.4.5	Hot Break	*431*
16.3.5	Fermentation and Storage	*432*
16.3.6	Filtration	*432*
16.4	Stabilization Systems	*432*
	References	*434*
17	**Analysis and Quality Control**	*437*
	Heinz-Michael Anger, Stefan Schildbach, Diedrich Harms, and Katrin Pankoke	
17.1	Introduction	*437*
17.2	Analyses	*439*
17.2.1	Density, Extract, Alcohol Content, Original Gravity and Degree of Fermentation	*440*
17.2.2	Photometric Measurements	*444*
17.2.2.1	Color	*445*
17.2.2.2	Free Amino Nitrogen (FAN)	*446*
17.2.2.3	Bitter Units	*446*
17.2.2.4	Photometric Iodine Reaction	*446*
17.2.2.5	Thiobarbituric Acid Index (TBI)	*446*
17.2.2.6	Total Polyphenols and Anthocyanogens	*447*
17.2.2.7	Ions	*447*
17.2.2.8	α-Amylases According to EBC/ASBC: Dextrinizing Units	*447*
17.2.2.9	Other Photometric Measurements	*447*
17.2.3	pH Measurement	*448*
17.2.4	Conductivity	*449*
17.2.5	Titration Methods	*449*
17.2.5.1	Acid–Base Titration	*449*
17.2.5.2	Complexometric Titration	*450*
17.2.5.3	Manganometric Titration	*450*
17.2.5.4	Diastatic Power: Iodometric Titration	*450*
17.2.6	Determination of Nitrogenous Compounds	*451*
17.2.6.1	Determination According to Kjeldahl	*451*
17.2.6.2	Determination According to Dumas (Combustion Method)	*451*
17.2.6.3	Near-IR Transmission Spectroscopy	*451*
17.2.6.4	Fractions of Nitrogenous Compounds	*452*
17.2.7	Carbon Dioxide	*453*

17.2.8	Measurement of Oxygen	454
17.2.9	Measurement of Chlorine Dioxide by a Sensor	455
17.2.10	Head Retention (Foam)	455
17.2.11	Turbidity and Non-Biological Stability	456
17.2.12	Viscosity	457
17.2.13	Congress Mash	458
17.2.14	Spent Grain Analysis	459
17.2.15	Friabilimeter	459
17.2.16	Grading	460
17.2.17	Hand Assessment	460
17.2.18	Homogeneity and Modification	460
17.2.19	Protein Electrophoreses	460
17.2.20	Gushing	461
17.2.21	Hop Bitter Substances	461
17.2.22	Continuous Flow Analysis (CFA)	461
17.2.23	Chromatographic Analyses	461
17.2.23.1	High Performance Liquid Chromatography (HPLC)	462
17.2.23.2	Gas Chromatography (GC)	463
17.2.24	Enzymatic Analyses	464
17.2.25	Determination of the Calorific Value of Beer	466
17.2.26	Atomic Absorption Spectroscopy (AAS) and Inductively Coupled Plasma Optical Emission Spectrometry (ICP-OES)	466
17.2.27	Sulfur Dioxide: Distillation Method	467
17.2.28	Emzyme Linked Immuno Sorbent Assay (ELISA)	468
17.2.29	Electron Spin Resonance Spectroscopy (ESR)	469
17.3	Analyses in Daily Quality Control	469
	References	473
18	**Microbiology**	**477**
	Werner Back	
18.1	Microflora in the Brewery	477
18.2	Manufacturing Cultures	479
18.3	Foreign Yeasts	480
18.4	Beer-Spoilage Bacteria	481
18.5	Detection of Beer Pests	484
	References	490
19	**Certification**	**491**
	Bernd Lindemann	
19.1	Management Systems and Business Management	491
19.2	Management Systems Standards	491
19.2.1	DIN EN ISO 9000, 9001 and 9004	492
19.2.2	DIN EN ISO 14001	492
19.2.3	DIN EN ISO 22000	492
19.2.4	Global Food Safety Initiative	493

19.2.5	IFS *493*
19.2.6	BRC *494*
19.2.7	DIN EN ISO 17025 *494*
19.3	Principles and Similarities *494*
19.4	Legal Requirements *495*
19.4.1	Regulation (EU) 178 *495*
19.4.2	Regulation (EU) 852 *495*
19.5	Certification According to ISO Standards *495*
19.6	Certification through IFS and BRC *496*
19.7	Certification through HACCP *496*

20	**World Beer Market** *497*
	Jens Christoph Riese and Hans Michael Eßlinger
20.1	Introduction *497*
20.2	Statistics *498*
20.2.1	Raw Materials for Brewing *498*
20.2.1.1	Barley and Malt Market *498*
20.2.1.2	Global Malt Production *501*
20.2.2	Beer Consumption *503*
20.2.3	Beer Styles *503*
20.2.4	Packaging *505*
20.3	Beer Markets and Their Key Players in 2004 *505*
20.3.1	Beer Markets *507*
20.3.1.1	Europe *507*
20.3.1.2	America *508*
20.3.1.3	Asia *509*
20.3.1.4	Africa *509*
20.3.1.5	Australia and Pacific Beer Markets *509*
20.3.1.6	Profitability *510*
20.3.2	World's Top Brewers *510*
20.3.3	Branding in the Global Brewing Industry *511*
	References *513*

21	**Physiology and Toxicology** *515*
	Manfred Walzl
21.1	Astounding Health Benefits of Beer *515*
21.2	Beer and Alcohol *521*
21.3	Beer and Cancer *521*
21.4	Beer Helps to Protect the Stomach and the Arteries *522*
21.5	Lower Risk of Developing Kidney Stones *522*
21.6	Ideal Sports Drink *523*
21.7	Improved Concentration, Better Performance and Quicker Reactions *523*
21.8	Against Bacteria *524*
21.9	Beer Removes Metals from the Organism *524*

21.10	Beer is 'Clean' 525
21.11	Beer Makes Beautiful 525
21.12	Beneficial Minerals 526
21.13	Legend of the Beer Belly 527
21.14	'Beer Prescription' 527
	References 528

22 Automation 531
Georg Bretthauer, Jens Uwe Müller, and Markus Ruchter

22.1	Introduction 531
22.2	Measurement Technology 533
22.2.1	Level 534
22.2.2	Temperature 534
22.2.3	pH Value 536
22.2.4	Pressure 536
22.2.5	Flow 536
22.2.6	Conductivity 537
22.2.7	Oxygen 538
22.2.8	Turbidity 539
22.2.9	Dosing 541
22.2.10	Limit Monitors 541
22.2.11	'In-Line' Measurement 541
22.3	Control Strategies 542
22.3.1	Classic Algorithms 542
22.3.2	Advanced Algorithms 543
22.3.3	Advanced Control of the Lauter Tun 545
22.3.3.1	Description of the Lauter Tun 545
22.3.3.2	Process Characteristics 547
22.3.3.3	Structure of the Controller 548
22.3.3.4	Results 548
22.4	Process Control System (PCS) 550
22.5	Information Technologies 551
22.6	Conclusions 552
	References 554

23 Malthouse and Brewery Planning 555
Walter Flad

23.1	Malthouse Planning 555
23.1.1	Introduction 555
23.1.2	Storage of Barley and Malt 555
23.1.3	Steeping 556
23.1.4	Germination 556
23.1.5	Kilning 557
23.1.6	Show Case Malting Plant 557
23.1.6.1	Design of the Steeping, Germination and Kilning Tower 557

23.1.6.2	Calculations	*560*
23.1.7	Consumption Data	*561*
23.2	Brewery Planning	*562*
23.2.1	Brewhouse	*563*
23.2.2	Wort Cooling	*564*
23.2.3	Malt Silos and Malt Treatment	*564*
23.2.4	Fermenting, Maturation and Storage Tanks	*565*
23.2.5	Yeast Management	*566*
23.2.6	Filtration	*566*
23.2.7	Bright Beer Tanks	*566*
23.2.8	Bottling Plant	*567*
23.2.9	Kegging Plant	*567*
23.2.10	Space Requirement of Full Packs and Returned Empties	*567*
23.2.11	Utilities and Power Supply	*568*
23.2.11.1	Supply with Heat	*568*
23.2.11.2	Supply with Coldness	*568*
23.2.11.3	CO_2 Recovery	*569*
23.2.11.4	Supply with Compressed Air	*569*
23.2.11.5	Supply with Electrical Power	*569*
23.2.11.6	Supply with Fresh Water	*569*
23.2.12	Key Figures for New Breweries	*569*
23.2.12.1	Required Land Area	*569*
23.2.12.2	Required Investment Costs	*573*
23.2.13	Documentation and Specifications	*575*
	References	*576*
24	**Packaging** *577*	
	Jörg Bückle	
24.1	Selecting the Suitable Machine Configuration	*577*
24.2	Packing into Packs Open at the Top	*578*
24.2.1	Classical Machine Design	*578*
24.2.2	Robot Technology	*582*
24.3	Wrap-around Packaging	*584*
24.4	Shrink-Wrap Packaging	*587*
24.5	Multipacks Made of Paperboard	*592*
24.6	Multipacks with a Plastic Carrier	*593*
25	**Cleaning and Disinfecting** *595*	
	Udo Praeckel	
25.1	Cleaning	*596*
25.1.1	Cleaning Agents	*597*
25.1.1.1	Alkaline Cleaning Agents	*597*
25.1.1.2	Acidic Cleaning Agents	*599*
25.2	Disinfecting	*600*
25.3	Cleaning Methods	*603*

25.3.1	Non-Recovery CIP Cleaning Method	603
25.3.2	Recovery Tank CIP Cleaning Method	604
25.3.3	Combined CIP Cleaning Method	605
25.4	Material Compatibility	611
25.5	Cleaning Glass Bottles	613
25.5.1	Bottle-Cleaning Machine	614
25.5.1.1	Residual Draining	614
25.5.1.2	Pre-soak with Pre-spray	615
25.5.1.3	Main Caustic Soaker Bath and Caustic Spray	615
25.5.1.4	Intermediate Spray, Hot- and Cold-Water Zones	615
25.5.2	Use of Chemicals in the Bottle-Washing Machine	616
25.6	Cleaning PET Bottles	617
25.7	Cleaning Barrels	617
25.8	Foam Cleaning	619
25.9	Work Safety and Environmental Protection	620

26 Waste Water 621
Karl Glas

26.1	Introduction	621
26.2	Characterization of Brewery Waste Water	621
26.2.1	Types of Waste Water	621
26.2.2	Waste water Constituents	622
26.2.3	Analysis of Waste Water	623
26.3	Preliminary Investigations to Determine Waste Water Pollutant Load and to Plan Waste Water Plants	625
26.4	Practical Example of a Waste Water Measurement	626
26.4.1	Determination of Concentration and Volume	626
26.4.2	Load Values	626
26.5	Specific Characteristic Parameters of Waste Water	627
26.5.1	Total Waste Water	627
26.5.2	Split Streams	629
26.6	In-house Measures	631
26.6.1	Classification of In-house Measures	631
26.6.2	Goal of In-house Measures	632
26.6.3	Practical Check-List of In-house Measures for the Production Steps	632
26.7	Waste Water Treatment	632
26.7.1	Neutralization	632
26.7.1.1	Neutralization of Carbonic Acid (CO_2)	632
26.7.1.2	Neutralization with Flue Gas	634
26.7.2	Mixing and Equalizing Tanks	634
26.7.3	Aerobic Waste Water Treatment	635
26.7.3.1	Activated-Sludge Plant	635
26.7.3.2	High-Performance Reactors	636
26.7.3.3	SBR	636

26.7.3.4	Aerated Waste Water Ponds	637
26.7.4	Anaerobic Waste Water Treatment	637
26.7.4.1	Biochemical Basics	637
26.7.5	Combination of Aerobic–Anaerobic Techniques	640
26.7.6	Comparison of Anaerobic and Aerobic Techniques	640
	References	641

27 Energy 643

Georg Schu

27.1	Introduction	643
27.2	Heat Requirements of the Brewery	643
27.2.1	Heat Consumption in the Brewery	644
27.2.1.1	Brewhouse	644
27.2.1.2	Service Warm Water	645
27.2.1.3	Keg Cleaning	646
27.2.1.4	Bottle-Rinsing Machinery	646
27.2.1.5	Others (Pasteurization, Flash Pasteurization, CIP)	646
27.2.1.6	Room Heating and System Losses	647
27.2.2	Boiler House	647
27.2.2.1	Boiler Plant and Combustion	647
27.2.2.2	Heat Carrier Systems	647
27.2.2.3	Fuels	648
27.2.3	Optimization Possibilities: Exhaust Emission Heat Exchanger, Degassing, Oxygen Regulation, Water Treatment and Blow-Off	648
27.2.4	Possibilities for Heat Recovery	648
27.2.4.1	Wort Cooling	648
27.2.4.2	Vapor Condenser	650
27.2.4.3	Waste Heat from Vapor Condensate	650
27.2.4.4	Waste Heat from Compressed Air	650
27.2.4.5	Waste Heat from the Refrigeration System	651
27.3	Power Supply	652
27.3.1	Requirement Figures	652
27.3.2	External Power Supply	652
27.3.2.1	Supply and Measurement at the Release Point	652
27.3.2.2	Power Factor Correction	653
27.3.2.3	Supply Contracts	653
27.3.3	Electric Power Consumption of the Brewery	653
27.3.3.1	The Brewhouse	653
27.3.3.2	The Filling Area	654
27.3.3.3	Drive System and Components	655
27.3.3.4	Lighting	655
27.3.3.5	Heat Supply	656
27.3.4	Optimization of the Electrical Power Supply: Load Management	656
27.3.5	Combined Heat and Power (CHP)	657

27.4	Cold Supply 657	
27.4.1	Cooling Requirements 658	
27.4.1.1	Design of Fermentation and Storage Cellar 658	
27.4.1.2	Location/Climatic Zone 658	
27.4.2	Cold Production 658	
27.4.3	Goals for an Optimal Cold Supply 661	
27.5	Compressed Air Supply 662	
	References 663	

28 Environmental Protection 665
Jochen Keilbach

28.1	Introduction 665
28.2	Environmentally Relevant Subjects in Relation to Brewing 665
28.2.1	Waste Water 665
28.2.1.1	Avoidance of Waste Water by Reduction of Water Usage 665
28.2.1.2	Composition of Brewery Waste Water 666
28.2.1.3	Waste Water Disposal 667
28.2.2	Energy 668
28.2.2.1	Renewable Energy 669
28.2.3	Brewery Emissions 670
28.2.3.1	Gaseous Emissions 670
28.2.3.2	Dust Emissions 672
28.2.3.3	Noise Emission 672
28.2.4	Waste 673
28.2.4.1	Special Brewery-Specific Production Waste 673
	References 674

29 Sensory Evaluation 675
Bill Taylor and Gregory Organ

29.1	Introduction to the Five Senses 675
29.2	How to Assess the Flavor of Beer 676
29.2.1	Technique for Flavor Assessment 677
29.2.2	Additional Points 677
29.2.3	Requirements for Attendance 678
29.2.4	Overall Assessment 678
29.3	Description of the Main Flavor Attributes 679
29.3.1	Sulfur Dioxide 679
29.3.2	Hydrogen Sulfide/Mercaptan 680
29.3.3	Dimethylsulfide/Cooked Vegetable 681
29.3.4	Solvent 682
29.3.5	Acetaldehyde 682
29.3.6	Estery/Fruity 682
29.3.7	Hoppy 683

29.3.8	Floral	684
29.3.9	Spicy	684
29.3.10	Fresh Grass	684
29.3.11	Clove/4-VG	684
29.3.12	Grainy/Straw	685
29.3.13	Malty	685
29.3.14	Caramelized	685
29.3.15	Roasted	685
29.3.16	Fatty Acid	685
29.3.17	Butyric	686
29.3.18	Cheesy	686
29.3.19	Diacetyl	686
29.3.20	Yeasty	686
29.3.21	Oxidized	687
29.3.22	Acidic/Sour	687
29.3.23	Alcoholic	688
29.3.24	Body	688
29.3.25	Sweetness	689
29.3.26	Bitterness	689
29.3.27	After-bitterness	690
29.3.28	Astringency	690
29.3.29	Metallic	690
29.4	Sensory Evaluation Environment	691
29.4.1	Panel Leader	691
29.4.2	Sensory Manual	692
29.4.3	Panel Motivation	692
29.5	Types of Sensory Tests	693
29.5.1	Paired Comparison Test	693
29.5.2	Duo–Trio Test	693
29.5.3	Triangle Test	693
29.5.4	Flavor Profile and Rating Test	693
29.5.5	Fresh and Aged Test	694
29.6	Selection, Training and Validation of Panelists	694
29.7	Building a Sensory Capability	695
29.7.1	Training Programme Overview	695
29.7.2	Level 1 – Difference Panelist	696
29.7.3	Level 2 – 'Beer Quality Acceptance' Panelist	696
29.7.4	Level 3 – 'Profile' Panelist	698
29.7.5	On-going Level 3 Performance Monitoring	699
29.8	Applications for Flavor Assessment	699
29.8.1	In-Process Sample Testing	699
29.8.2	Final Beer Testing	700
29.8.3	Consumer Research	700
	References	701

30	**Technical Approval of Equipment** *703*	
	Walter Flad and Hans Michael Eßlinger	
30.1	Generalities *703*	
30.2	Technical Approval of Brewhouses *703*	
30.3	Technical Approval of Filling Lines *704*	
30.3.1	Efficiency tests according to DIN 8782 *705*	
30.3.2	Reporting of Efficiency Tests *706*	
30.4	Other Key Figures *706*	
	References *708*	

Index *709*

Preface

Beer is one of the most fascinating products of mankind. It is produced and consumed all over the world with a passion. Beer encompasses gregariousness and communication, adds to the joy of living, and has – above all – a unique taste. The positive effects of moderate consumption of beer on health, body and the mind cannot be denied, even by the most serious of critics.

The keyword "Beer" is the most frequently requested article in the Ullmann's Encyclopedia of Industrial Chemistry of Wiley-VCH. So, it was a natural next step for the publisher to pull together the expertise on the fine art of beer-brewing not only in the compact form of the classic textbook of my dear doctoral father, Professor Dr. Ludwig Narziß (Abriss der Bierbrauerei), but also in its complete breadth for an international audience in the successful 'Handbook' series. This project continues the long-standing cooperation between Wiley-VCH and me, that started in my time at Weihenstephan, intensified during my years in Mannheim and was cultivated during my time here in Freiberg.

I am grateful to the 29 authors who contributed 30 articles to this book. I am especially indebted to the Director of the Weihenstephan Institute of Brewing Technology, Prof. Dr. Werner Back and his team for writing down the latest findings in brewing-technology. As every author has eagerly been awaiting the publication date, it is great to see that all their hard work will now be rewarded with the printed book. It is especially nice to see previously unpublished results in several chapters, so that the brewers will gain from this valuable information.

Let this book introduce you to the world of beer: from the complicated markets of procurement and retail to the many types of production of the natural and healthy cultural item, beer. Join me in the exploration of a drink, that has a tradition of thousands of years, is still an exciting flavor adventure and as refreshing as ever.

Freiberg, April 3rd 2009 *Prof. Dr. Hans Michael Eßlinger*

List of Contributors

Heinz-Michael Anger
Versuchs- und Lehranstalt für
　Brauerei
Seestraße 13
13353 Berlin
Germany

Werner Back
Lehrstuhl für Technologie der
　Brauerei I
Weihenstephaner Steig 20
85354 Freising
Germany

Susanne Blüml
KRONES AG
Böhmerwaldstraße 5
93073 Neutraubling
Germany

Georg Bretthauer
Forschungszentrum Karlsruhe
　GmbH
Institut für Angewandte Informatik
Hermann-von-Hemholtz-Platz 1
76344 Eggenstein-Leopoldshafen
Germany

Jörg Bückle
KRONES AG
Böhmerwaldstraße 5
93073 Neutraubling
Germany

Felix Burberg
KRONES AG
Werk Steinecker
Raiffeisenstraße 30
85356 Freising
Germany

Hans Michael Eßlinger
Freiberger Brauhaus GmbH
Am Fürstenwald
09599 Freiberg
Germany

Walter Flad
Technisches Büro Weihenstephan
　GmbH
Johannisstraße 6
85353 Freising
Germany

Oliver Franz
Unilever Deutschland GmbH
Werk Heilbronn
Knorrstraße 1
74074 Heilbronn
Germany

Handbook of Brewing: Processes, Technology, Markets. Edited by H. M. Eßlinger
Copyright © 2009 WILEY-VCH Verlag GmbH & Co. KGaA, Weinheim
ISBN: 978-3-527-31674-8

Martina Gastl
Lehrstuhl für Technologie der
  Brauerei I
Der TU München-Weihenstephan
Weihenstephaner Steig 20
85354 Freising
Germany

Karl Glas
CPW Competence Pool
Weihenstephan
Weihenstephaner Steig 23
85354 Freising
Germany

August Gresser
Bergner Bräu GmbH
Via J.-G. Palzer 31
39057 Eppan
Italy

Diedrich Harms
Versuchs- und Lehranstalt für
  Brauerei
Seestraße 13
13353 Berlin
Germany

Jochen Keilbach
Eichbaum-Brauereien AG
Käfertaler Straße 17
68167 Mannheim
Germany

Stefan Kreisz
Novozymes A/S
Krogshoejvej 36
2880 Bagsvaerd
Denmark

Martin Krottenthaler
Lehrstuhl für Technologie der
  Brauerei I
der TU München-Weihenstephan
Weihenstephaner Steig 20
85354 Freising
Germany

Bernd Lindemann
Fachhochschule Wiesbaden
Fachbereich Weinbau und
  Getränketechnologie
Von-Lade-Straße 1
65366 Geisenheim
Germany

Reinhold Mertens
Concept Cleanmanagement GmbH
Mescheder Straße 24
59846 Sundern
Germany

Franz G. Meussdoerffer
Universität Bayreuth
Lehrstuhl für Mikrobiologie
Universitätsstraße 30
95447 Bayreuth
Germany

Jens Uwe Müller
Forschungszentrum Karlsruhe GmbH
Institut für Angewandte Informatik
Hermann-von-Helmholtz-Platz 1
76344 Eggenstein-Leopoldshafen
Germany

Gregory Organ
Lion Nathan Australia Pyt Limited
20 Hunter Street
Sydney, NSW 2000
Australia

Udo Praeckel
Tensid Chemie GmbH
Heinkelstraße 32
76461 Muggensturm
Germany

Katrin Pankoke
Versuchs- und Lehranstalt für
  Brauerei
Seestraße 13
13353 Berlin
Germany

Jens Christoph Riese
Aspera Brauerei Riese GmbH
Rheinstraße 146-152
45478 Mülheim/Ruhr
Germany

Markus Ruchter
Forschungszentrum Karlsruhe
  GmbH
Institut für Angewandte
  Informatik
Hermann-von-Helmholtz-Platz 1
76344 Eggenstein-Leopoldshafen
Germany

Stefan Schildbach
EUWA H. H. Eumann GmbH
Daimlerstraße 2–10
71116 Gärtringen
Germany

Georg Schu
IGS Ingenieurbüro für
Energie- und Umwelttechnik
Maximilianstraße 28 b
85399 Hallbergmoos
Germany

Bill Taylor
Lion Nathan Australia Pyt Ltd.
20 Hunter Street
Sydney NSW 2000
Australia

Christoph Tenge
Spaten-Franziskaner Bräu GmbH
Marsstraße 46 + 48
80335 München
Germany

Manfred Walzl
Wagner-Jauregg-Platz 1
8053 Graz
Austria

Martin Zarnkow
TU München Weihenstephan
Lehrstuhl für Technologie der
  Brauerei I
Weihenstephaner Steig 20
85354 Freising
Germany

1
A Comprehensive History of Beer Brewing
Franz G. Meussdoerffer

1.1
Introduction

Brewing has been a human activity ever since the beginning of urbanization and civilization in the Neolithic period. Beer is a product valued by its physico-chemical properties (i.e. quality) as much as by its entanglement with religious, culinary and ethnic distinctiveness (i.e. tradition). Accordingly, the history of beer brewing is not only one of scientific and technological advancement, but also the tale of people themselves: their governance, their economy, their rites and their daily life. It encompasses grain markets as well as alchemy.

There exists a vast literature on beer and brewing. Among the most comprehensive reviews are the books by Arnold [1] and Hornsey [2]. Some aspects have recently been covered by Unger [3] and Nelson [4]. A major problem is posed by language – there is an abundance of information available in, for instance, English, German, Dutch, French, Danish or Czech, which, due to insufficient command of the various languages, is are not acknowledged by other authors. If evaluated in a broader context these publications would yield very interesting insights.

There are two fundamental limits to any history of beer brewing. First of all it is the unambiguous definition of its object, namely beer. Does 'beer' broadly refer to fermented beverages based on grain or does it designate the hopped drink obtained from liquefied starch after fermentation with specific strains of *Saccharomyces* yeasts, which is understood to be beer in our times? Although including the history of all grain-based fermented beverages would exceed the scope of this chapter, a consideration of hopped beer only would be too selective, and would ignore the fundamental roots of brewing technology and beer culture. Therefore, this chapter follows the approach taken by many previous authors – illustrating the process leading from the beginning of agriculture in the Neolithic period to today's beer. This approach necessarily limits the scope to Mediterranean, European and North-American developments. This does not imply, however, that development elsewhere is inferior. In contrast, the dissemination of European/American beer culture all over the world and the recent trend towards globalization generates new variants of beer-like beverages that follow regional traditions and preferences.

A second difficulty arises from the availability and reliability of sources. Our knowledge about brewing comes either from archaeological artifacts or written documents. Thus, grain residues and the presence of oxalate, respectively, have been used as archaeological evidence for beer. This is certainly appropriate if the surroundings indicate conditions favorable for beer production [5]; however, it is not proof in itself. Difficulties might also arise from the interpretation of written documents. As brewing consists principally of at least three distinct processes (germination/drying = malting; enzymatic hydrolysis of natural polymers = mashing; microbial transformation of amino acids, sugars or oligosaccharides = fermentation) one is tempted to apply modern technical concepts to earlier technology. This might not always be appropriate and may result in misinterpretations. Also, terms and definitions might change over the times, complicating the correct interpretation of texts. For instance, the Indo-Germanic terms *beor*, possibly etymologically related to our word 'beer', and *alu*, the likely origin for the word 'ale', might both have been used at times not for grain-derived beverages, but for mead and fermented fruit juices, respectively [4, 6].

This uncertainty arising from the meaning of terms is further complicated by the high ideological significance of beer and brewing. Ever since antiquity beer drinking has also implied a cultural, national and ideological affiliation, and publications have served not only the documentation of technological or economical facts, but also the satisfaction of cultural pride. From antiquity to modern times many authors have approached the theme with considerable bias. Accordingly, while beer and wine probably coexisted in most societies, frequently attempts have been made to mark a civilization as 'wine' or 'beer' drinking. This chapter tries to avoid valuations as far as possible, and to focus on the economic, technological and cultural developments that enabled men and women to provide their fellow men with a nutritious, healthy and joyful drink for over 6000 years. However, it follows the advice of Professor William Whewell from Cambridge (who coined the term 'science' around 1840) that new insights should not be taught until 100 years have passed to avoid scientists ridiculing themselves. It ends therefore with the advent of World War I.

1.2
'The Truly Happy Man Has His Mouth Full of Beer' [7]: From Prehistory to the End of the Roman Empire

1.2.1
Advent of Agrarian Societies

> 'Very deep is the well of the past. Should we not call it bottomless? Bottomless indeed, if-and perhaps only if-the past we mean is the past merely of the life of mankind, that riddling essence of which our own normally unsatisfied

and quite abnormally wretched existences form a part;
whose mystery, of course, includes our own and is the
alpha and omega of all our questions, lending burning
immediacy to all we say, and significance to all our striving.
For the deeper we sound, the further down into the lower
world of the past we probe and press, the more do we find
that the earliest foundations of humanity, its history and
culture, reveal themselves unfathomable. No matter to what
hazardous lengths we let out our line they still withdraw
again, and further, into the depths.' THOMAS MANN,
JOSEPH AND HIS BROTHERS, *Translated by H.T. Lowe-Porter*, ALFRED A. KNOPF, New York, 1945.

These sentences written by Thomas Mann at the beginning of his novel *Joseph and his Brothers* describe the dilemma of defining the beginnings of brewing. While humans might have used intoxicants at all times and even fermented ones in Paleolithic times [8], three prerequisites must be met for brewing: (i) the availability of suitable grains, (ii) a controllable source of energy (i.e. a fireplace) and (iii) suitable brewing vessels (i.e. pottery or metal kettles). Most probably some experience with handling and processing grain (storage, controlled germination and milling) was also to be acquired before a fermented beverage could be made from cereals. These prerequisites were met on a wider scale not earlier than 5000 BC.

The end of the ice age as result of climate change around 12 500 BC had induced changes to human societies as well. As a consequence, the small groups of 25–50 humans who dotted the country (0.1–0.2 humans/km^2) and made their living from hunting and gathering began to build up supplies. They developed simple harvesting utensils as well as the means to grind seeds. As population growth rates increased, the first permanent settlements emerged. This development occurred independently in the so-called Fertile Crescent (Israel/Jordan/Syria/Turkey/Iraq), Southern and Northern China, Sub-Saharan Africa, the Andes region, and Middle and North America. Concomitantly, humans began to adapt other species to their needs (i.e. to domesticate plants as well as animals). Among the plants selected and bred were einkorn [9], wheat [10, 11] and barley [12]. As these grains provide a useful source of carbohydrates and protein, are easy to handle, and can be stored over long periods, they quickly became a major object of interest and innovation. It is no surprise, therefore, that the first evidence of grain-based fermented beverages mainly originates from the regions where grain cultivation first flourished – the Fertile Crescent, Mesopotamia and Egypt. However, the oldest evidence for such drinks comes from China. Chemical analyses of ancient organics absorbed into pottery jars from the early Neolithic village of Jiahu in Henan province revealed that a mixed fermented beverage of rice, honey and fruit was being produced as early as the seventh millennium BC [13]. Even the first pictorial and chemical evidence for grain-based fermented beverages from the near east dates from times when the last mammoths became extinct and the iceman 'Özi' traveled the Alps, namely from the fourth millennium BC [14, 15]. Thus, at the very beginning of scientific (i.e. recorded) history, beer-like drinks were already part of human life. It has even been argued that the development of fermented cereals acted as a driving force of human development [16].

While cereals as a source for alcoholic drinks offer the great advantage of being available all year around if stored appropriately, they have two great disadvantages: (i) the starch content of the cereal must somehow be modified and converted into fermentable sugars, and (ii) the few yeasts suitable for effectively transforming carbohydrates to ethanol [17] do not settle on grain surfaces [18]. Thus, the brewing yeast must come from another source. As at the beginning of civilization and under Mediterranean climate conditions suitable airborne yeasts might have been rather scare, plants can be assumed to have been the source of the earliest brewing yeasts. Certain fruits, such as grapes or dates, host appropriate wild yeast on their skins.

Brewing beverages from starch can be performed by different processes; however, these are always composed of a few invariable modules (Table 1.1, modified from [19]).

As conversion of starch to sugars (malting) was poorly perfected initially, either additional fermentable sugars had to be added in the form of juices or honey and/or microorganisms capable of liquefying starch with exoenzymes had to be present to ensure fermentation. Moreover, contamination with harmful microor-

Table 1.1 Basic modules of brewing.

Sugar source		Yeast conservation/ starter culture	Reduction of Contamination	Fermentation	Storage
Grain starch modification	Other source				
Heating/boiling	honey/mead	unmodified plants (fruits/berries)	plants/herbs	heterofermentation	clay bottles
Soaking/ germinating/ malting	fruit syrups	leavens/bread	lactic acid bacteria	homofermentation	glass bottles
Dough/baking (beer-bread)	wine	sugar solution (wort, honey)	ethanol formation or addition (wine)	immobilized yeast	barrels
Enzymatic saccharification in solution (wort)		fermentation broth	wort boiling		
Enzymatic saccharification in solid state (mold amylases)		pure cultures	filtering		

ganisms had to be minimized by lowering the pH through lactic acid fermentation and the addition of suitable plant components. Thus, a great variety of beverages became available from grains shaped by climatic and cultural peculiarities. Mixed beverages were widespread throughout Roman–Hellenistic times [20] and the Middle Ages, and live on in some of the Belgian lambic and geuze beers [21]. The 'barley wines' mentioned by several ancient authors [22] might be such products as well. The complexity and long fermentation times of such beverages result in a wine-like drink [23].

1.2.2
Mesopotamia and Egypt

As the life of settled humans became increasingly stratified in social classes organized by administrative structures in urban centers and an exchange of goods evolved, both developments were accompanied by the development of writing in the Sumerian states and Egypt (3300–3100 BC) as well as in Akkad (2500 BC). Even the earliest written documents provide information about beer, stressing its key role in those societies. In the *Epic of Gilgamesh*, king of Ur, the savage Enkidu is 'civilized' by a woman teaching him to eat bread and to drink beer (second table), revealing brewing as one cornerstone of civilization.

Most of our scant knowledge about brewing in early Mesopotamian cultures originates from administrative and literary texts, the best known of which is the 'Hymn to Ninkasi' [24]. This poem has come to us in copies from the Old Babylonian period (about 1800 BC) and – in spite of many ambiguities – details the brewing process. It specifies as basic ingredients of the beverage: *bappir* (often translated as beer-bread, a cooked or fermented mixture of leaven and aromatic herbs), *munu* (a hulled cereal that is malted during the brewing process) and *titab* (probably a mash prepared from malted grain that is dried after cooking). Honey and wine are added to the mixture before fermenting.

Beer and other alcoholic drinks played a vital role in early societies. Ecstasy was considered an indispensable spiritual exercise and a state of unification with the gods, while banquets and ceremonial dinners served as divine service and to foster tribal allegiance. In the form of libations, beer served as an offering to the gods, believed to be an essential constituent of their divine diet as it was of that of their worshippers. As disease was believed to result from 'banes' or 'numinous anger', magic and ritual were integral parts of medical therapy. It is no wonder that beer was important as a basic element of the medicines as well as of magic evocations. Thus, to implore Nergal, the god of plague, to spare the city one had to recite: 'O lord, do not enter this beer-house, do not slay those sitting at the place of beer!', while a snake dissolved in beer was considered a remedy for jaundice.

Moreover, beer, malt and 'beer-bread' were a common element of tributes and tithes, as well as an integral part of the daily remunerations of magistrates, servants, priests or workers. Given the vast commercial and social importance of brewing intermediates and beer in early societies it is not surprising that immense efforts were made to adapt to changing consumer preferences and to optimize brewing technology to improve efficiency and profitability. There is no doubt that brewing techniques as well as ingredients changed dramatically over a period ranging from the fourth millennium BC (late Uruk period) until the fifth century BC (New Babylonian Empire) [25]. While originally at least nine different beer types had been produced from barley and barley malt [26], in Babylonian times up to 70 different types of beer were manufactured from (white) emmer, barley and intermediates thereof [27, 28]. The cheapest beverage was a 'black' beer prepared from barley only [29]. The color probably results from the grain as the wild barley varieties commonly found in Syria and Mesopotamia exhibit a purple color [30]. Unfortunately, details of technical improvements are prone to speculation, as our knowledge depends on the differing specialist interpretation of cuneiforms and sparse illustrations on seals. The latter often indicate that, in particular, in the northwest of Mesopotamia a specific brewing technique yielded a beer that had to be sipped through a reed or metal tube in order to avoid swallowing husks and other insoluble plant constituents. Xenophon, the Athenian historian and soldier, encountered such a beer in the fourth century BC in Armenia and found it quite tasty [31]. However, bone and metal straw-tip beer-strainers have also been found in the central Jordan valley [32], indicating a broader distribution of this brewing technique.

It is assumed that the Jewish elite learned to prepare beer during their exile in Babylon. Although there are controversies about brewing in ancient Israel, evidence for beer and beer consumption can be found in the scripts [29, 33–35]. The designation *shekar* (strong drink) is certainly related to the Akkadian *sikaru* and the Babylonian *shikaru*, both meaning beer, although at times *Shekar* might have also been used for other beverages. It is interesting to note that one of the earliest references to hopped beer, a list of tithes owed to the abbey of Fontanelle from the early ninth century AD, refers to this drink using the Carolingian term *sicera* [4], an analog to *Shekar* and *sikaru*. It had been postulated that *Shekar* was a hopped beverage [34].

There is better knowledge about brewing in the other great early civilization – Egypt – since the Egyptians displayed the brewing process on the walls of their tombs and supplied their dead with clay figures of laborers (*uscheptis*) to ensure provisioning with indispensable victuals in the afterlife. Beer was a crucial staple and played as dominant a role in cult and medicine in ancient Egypt as it did in Mesopotamia [36]. There were a variety of beer styles differing in raw materials, ingredients, alcohol content and taste. Some beers were brewed for religious purposes only, like the 'beer of truth' reserved for the 12 gods who guarded the shrine of Osiris or their priests [29]. The god Osiris, in Hellenistic times equated with Dionysus, was considered the one who had taught humans to brew beer. This legend lived on for centuries. As late as 1536 AD the German humanist A. Althammer informed his readers that the god Osiris and his wife Isis had taught the Thuiskon chief Gambrivius how to cook beer [37]. The legend of Osiris – a god murdered and dismembered by his brother, and then reconstructed and reanimated by his wife – served not only as a template for later theological paradigms, but also for much of the alchemistic theory. One reason for this was that the Egyptians had not only developed an alloy named electron, which resembled gold in appearance and properties, but also their mastery of fermentation.

The first evidence of beer in Egypt dates from about 3500 BC [38]. Excavations from two pre-dynastic sites at Abydos and Hierakonpolis uncovered large, fixed vats, supported by distinctive firebricks, which are considered the remains of breweries [39]. Later, in the New Kingdom times and probably earlier, brewing took place in smaller moveable pottery vessels. The majority of ancient Egyptian beer residues contained barley, whereas only some were made entirely from emmer. Occasionally, both cereals were mixed together [40]. In most Egyptian beers presumably a special type of bread [41] or sour dough was used as a source of yeast and lactic acid. Scanning electron microscopy analysis of desiccated bread loaves and beer remains suggest that bread was made not only with flour from raw grain, but sometimes also with malt and with yeast. Brewing blended cooked and uncooked malt with water; the mixture was strained free of husk before inoculation with yeast [42]. It is natural that the Egyptian brewing technology did not remain the same over the centuries, but was adapted to market trends and technological progress [19, 40, 43]. Moreover, there is evidence that different brewing processes were used in Upper and Lower Egypt [19]; this is not surprising as Egyptian texts also differentiate between the six-rowed barley from Upper and Lower Egypt [19].

Many hundreds of years of brewing resulted in a process that has been preserved in the writings of the fourth-century alchemist Zosimus of Panopolis. He described malting by soaking and germinating barley before drying it in the sun and air, and subsequently grinding it. The malt was made into leavens which were briefly heated (baked/fermented?) and crumbled into water, slightly heated for mashing, and subsequently fermented. This Egyptian beer, called *Zythos*, was generally synonymous for beer in Hellenistic times. Beverages made by the recipe of Zosimus were consumed even in Islamic Egypt and a drink named *bouza* was prepared by Fellaheen in modern times from wheat according to Zosimus' recipe [4].

1.2.3
Hellenistic Period: Greeks, Romans and Their Neighbors

The Greeks under Alexander the Great conquered Egypt in 331 BC and established the Ptolemaic dynasty there. The Ptolemeian rulers introduced wine in Egypt, which was soon to become the favorite drink of the upper classes. In contrast, beer production and sale was tightly regulated, and ultimately beer making became a state monopoly. Under the pretext of combating the public abuse of alcoholic beverages, a tax was imposed on beer – unheard of during the millennia before.

One reason for this was that Greeks (as well as Romans later on) were determined wine drinkers for cultural, not to say ideological, reasons. For once, they were surrounded by hostile beer-drinking nations. Their rivals for the trade in the Eastern Mediterranean from Egypt and Mesopotamia were beer drinkers as well as the Thracians, the Illyrians, the Scythes, the Phrygians and the Gauls, who obstructed their territorial expansion. Although the Greeks had been originally beer drinkers too, their term for beer *methou* being closely related to our word mead, the tragedian of Aeschylus in the fifth century BC let one character exclaim in the *Danaids*: 'Truly, you will find that the dwellers of this land are men indeed, not beer drinkers'. This prejudice against beer in Greek culture originates not only from the necessity to distinguish themselves as men of culture and knowledge from the uncultivated barbarians, but also from philosophical considerations [4]. First, they had developed the theory of the four elements into a universal concept, attributing to substances properties like cold, hot, dry or humid. While wine was considered a 'hot' drink suiting the 'hot' male principle, beer was thought to be 'cold', matching the female essence. Beer drinkers therefore had to be effeminate. Second, the nature of fermentation was regarded with suspicion. The excretion of yeast indicated an unnatural and impure process. Cicero called the proletariat *'faex populi'* (yeast of the people) and Theophrastus of Lesbos specified in the late fourth century: 'They even turn into drinkable juices some [products] which they have caused to depart from their nature and have somewhat rotted'. The Greek prejudice was absorbed by the Romans who encountered hostile beer drinkers in Hispania, Britain and Germany. However, the necessity to accommodate their legions in regions without viniculture with suitable drinks required the provision of beer. This they organized with their typical thoroughness and beer can be found on the shopping lists of Roman pursers in England [44]. Remains of Roman breweries at the border forts Loesnich, Xanten and Regensburg as well as at a villa in the hinterland at Namur point to a high degree of professionalism. Moreover, inscriptions from tombstones indicate that brewers and beer dealers were organized in distinct guilds(e.g. *'ars cervesaria'* or the *'cervesa'* guild) [4]. The different beer styles known to the Romans are also mentioned in the commentaries on civic law by Sabinus in the first century AD and Ulpian in the third century AD.

1.2.4
Celts and the Germans

The cultivation of grain did not spread to Northern Europe until the start of the Neolithic period, about 6000 years ago. Afterwards, a distinct beer manufacturing tradition emerged there that differed significantly from that developed in the Mediterranean. First of all, except for rare examples [45], leavened bread was not widespread in the northern regions. Moreover, air-drying of soaked cereals would not yield malt, but mold contamination. Therefore, kilning, as shown at a Celtic brewery site [46], had to be established in malting.

The Celts had been around since at least 700 BC in Central Europe. During the fifth to third centuries BC they expanded south towards Italy, west to Gaul and Iberia, and east into Hungary, Greece and Turkey. The Celts had certainly acquired great experience in brewing; however, we have only the testimony of Pliny who remarks in his Natural History (XXII, lxxxii): 'from grain they make beverage, which is called "*Zythos*" in Egypt, "*Caelia*" and "*Cerea*" in Spain, cervesia in Gaul and other countries'. He also mentions (XIV, xxix) that 'the populace of the west of Europe have a liquid with which they intoxicate themselves, made from grain and water. The manner of making this liquid is somewhat different in Gaul, Spain and other countries, and it is called by different names, but its nature and properties are everywhere the same', and states (XVIII, xii) that a Spanish beer is made from wheat and very stable. In spite of this apparently sophisticated brewing technology, the Gauls soon turned to wine. The Celtic beer culture, however, survived in Ireland and was brought back to the continent by Irish missionaries 500 years later.

Even less is known about the brewing technology of the German tribes, who slowly rolled back the Celts in Central Europe between 300 BC and 100 AD. The early Germanic tribes spoke mutually intelligible dialects, and shared a common culture and mythology, as is indicated by Beowulf and the Volsunga saga. We know that they drank beer in considerable quantities from a report by Tacitus who wrote (*Germania*, 23): 'they make a liquid from barley or wheat, which, if fermented [rotted], resembles wine'. Archaeological evidence indicates that germinating cereals was a step in Germanic brewing [47]. These were not dried, but squashed immediately and the resulting mash was fermented with airborne yeasts, and eventually contained plants and honey [48]. From the poetic Edda's tale of the allwise dwarf Alvíss we come to know that there were different names for beer or beer styles: *bjórrr, veig, oel, hreinna lögrr* or *mjöð*, probable reflecting various beer/mead drinks.

In the third and fourth century the Germanic tribes approached the Roman fortifications. Germanic preference for beer would meet with Roman technology and herald a new chapter in brewing.

1.3
'I Would Like to Have a Great Lake of Beer for Christ the King' [49]: The Christian Middle Ages

Thus ended Roman rule in the Western parts of the empire. Germanic tribes overran the Roman defenses and roamed the country. Military commanders, already of German ancestry, added to their Roman titles of magister or consul that of 'king' and parted with their Germanic kin. However, some continued to cooperate with the gentry and the administration of state and church. In continental Europe the Franks were the most successful, not least because they accepted the Roman Catholic religion, thus allying themselves with the old system. Their cavalry finally accomplished what the Roman legions had failed to enforce for so long – to subordinate the Northern German tribes.

1.3.1
Monasteries

The church acquired a predominant position in those turbulent times, and managed to conserve some of the immense medical, scientific and technical knowledge compiled in Roman times. This applied to brewing too. The early Christians, standing in the Hellenistic/Judaic tradition, scorned beer and beer drinkers. In Ireland, however, which was never under Roman rule, a quite independent church and monasticism had developed. Here, beer played a significant role as exemplified by Saint Brigid of Kildare who is reported to have worked several beer wonders. Irish missionaries like St Columbanus spread the Celtic predilection for beer as well as the Christian faith in Europe. St Columbanus' hagiographer, Jonas of Bobbio, remembered around 643 that 'as their drink they brewed beer (cerevisia) from grain (wheat) or barley-juice, which at that time was not only common at the Skordic and Dardaneic people but also in Gallia, Britannia, Ireland and Germania and their kin' [50]. St Columbanus also worked beer miracles, as did other holy men at that time. Thus, the attitude of the church towards beer changed during the sixth and seventh centuries as a consequence of the predominance of Germanic traditions and the Celtic mission among Gallic and Germanic tribes. The documented recognition of beer by the church, however, only came in the year 816 at the synod of Aachen where a standardized binding order of monastery life was deliberated. It was decided that a monk should receive daily one beaker (hemina, 0.273 l) of wine or, where no wine was available, twice as much (one sextarius, 0.546 l) of 'good beer' [4].

In the Christian culture, the aspect of 'holy drunkenness' never had the significance attributed to it in the Roman–Hellenistic and particularly the Celtic–German cultures. Therefore, missionaries and clergy had little compassion with their flock's view of Christ as highest chieftain and its church as mead hall. Ordinances of archbishop Dunstan of Canterbury, however, prove that such an association was not so far from reality. Dunstan decreed (Canon 26) 'Let no drinking be allowed in church' and (Canon 58) 'Let no priest be an ale ale-scop nor in any wise act the gleeman' [51]. Thus, consumption of alcoholic beverages was observed with utmost suspicion by the church. Accordingly, King Edward of England, instigated by Dunstan, ordered alehouses to close. Proof that this had at best a temporary effect comes from the canons of his successor Aelfric, who ordered (Canon 35) 'Nor ought men to drink or eat intemperately in God's house ... yet men often act so absurdly as to sit up by night and drink to madness within Gods house and to defile it with scandalous games and lewed discourse'. These tenth-century examples from Britain characterize a situation to be found all over Northern European Christianity.

At that time the prospering Northern monasteries had become brewing centers. Nothing exemplifies this more than an outline of a Carolingian monastery drawn around 830 and known as the plan of St Gallen. There, three different brewhouses are designated: one to provide beer for noble guests, one for the daily consumption of the monks, and another one to grant beer for the pilgrims and paupers. The brewing process apparently consisted of crushing malted and unmalted grain (a kiln is depicted) with mortars (presumably water propelled) and mills. The brewing facility consisted of a central room with a hearth, and a second room for cooling and filtering the beer. The storage room contained 14 barrels of different size with an estimated capacity of 350 hl [2, 52].

It was also the monasteries that documented the first evidence of using hops in brewing. It is assumed today that hops originated in China. The closely related hemp was in use by the bell-beaker people and the Scythes [2], and spread to

Middle Europe in the early Middle Ages. It is assumed that Caucasian people brewed hopped beer before our times, and that this technology came with their migration to Northern and Eastern Europe [53]. It is interesting to note that a hop garden in Geisenfeld near Freising mentioned in 736 AD is said to have been set up by Wendic prisoners of war. That the Wendes, a tribe living on the shores of the Baltic Sea, were especially familiar with hops confirms a customs ordinance of Lübeck from around 1220 stating that hops is carried by a Wende into the city were free of duties.

The first evidence of the use of hops in brewing in Europe originates from the ninth century. In 822 AD, Abbot Adalhard of Corbie issued a set of instructions for his abbey. Corbie counted among the most important monasteries of the empire and was the location of the imperial library, and Abbot Adalhard was a member of the royal family, a cousin to Emperor Charlemagne. In his *Consuetudines Corbeienses*, Adalhard decreed that the porter of the monastery might receive a fraction of the hops given as tithe to the abbey to make beer thereof [54]. The use of hops in Western monasteries has been further documented in Fontanelle and St. Denis during the eighth and ninth centuries AD [4]. Another monasterial statute, the Polyptychon of St Germain (Paris) from the early ninth century, names hops as tithe. Moreover, at the same time (822 AD) that Adalhard enacted his *Consuetudines*, he founded the monastery Corvey (*Corbeia nova*, near the river Weser in Central Germany) and it is safe to assume that his regulations were, if not destined for it, at least followed there also. Cultivation of hops is also documented in the diocesan records of Freising (Bavaria) between 859 and 875 [55]. Thus, it seems likely that hops were used for brewing in at least some monasteries.

Evidence of the use of hops for brewing in the ninth and tenth centuries comes also from archeobotanical evidence. The remains of Haithabu, one of the most important Viking trading places on the Schleswig coast, confirm the frequent use of hops [56]. Moreover, hop residues were detected in high numbers in the remains of a cargo vessel from the ninth century at Graveney on the English coast [54]. Thus, it can be assumed that hops were in use outside the monasteries, and even traded between the continent and the British Isles. References to hops (*hymele*) are somewhat later found in a ninth- or tenth-century English version of Pseudo-Apuleius's Herbarium, where it is recommended 'that men mix it with their usual drink' [4, 57].

Hops were also – at least at some times – part of another beer additive – *Gruit*. There is some of confusion about *Gruit* as it designates a brewing privilege, beer additives, a beer type, or even flour or something baked, respectively, depending on the context, the time and the place. *Gruit* is known first as a brewing privilege. In Germanic societies brewing was a right that every free member possessed. However, by his *Capitulare de Villis* from 811, Charlemagne made the right to brew (for more than one's own needs) a royal prerogative. The edict limited the right to brew for the surrounding territory to the royal manors and imperial homesteads. Thus, brewing changed from a common law to a privilege linked to a place. Consequently, it became a general privilege of the king who later bestowed it to his lords [58]. Thus, the German emperors Otto III and Heinrich mention a brewing right and brewing ingredient called *fermentum*, stating it is synonymous with *Gruit*. The brewing privilege was of commercial interest particularly to the cities, where a relatively large population increasingly depended on beer produced by professionals. Before the thirteenth century the majority of the few cities in Central Europe were of Roman origin and under the rule of bishops. The bishops – and probably other lords owning cities – excised their brewing privilege by compelling the municipal brewers to buy essential brewing ingredients from their administration. At least in the case of the bishop of Magdeburg, we have written evidence that he had the privilege to provide the yeast for the brewers in his entire diocese. That meant that brewers as far as Hamburg or Halle had to buy their yeast (*bärme*) from the Episcopal *Bärmamt* (Yeast Administration) [59]. *Fermentum*, always synonymous with '*Gruit*', could indicate the physical form in which yeast was kept. Pliny had already mentioned that yeast for wine making was preserved in antiquity by mixing flour and must at harvest time and baking little cakes thereof [60]. As *fermentum* originally had designated the particle of the eucharistic bread sent by the Bishop of Rome to the bishops of the other churches as a symbol of unity and intercommunion (until the Council of Laodicea forbade the custom), *fermentum* = *Gruit* could have originally meant a bread-like substance containing viable yeast. It is noteworthy that Robert Boyle writing on digestion in the seventeenth century concludes that something in a dogs stomach 'boild flesh, bread, gruit, &. to the consistence of a fluid Body', thus putting *Gruit* next to bread [61].

Later, *Gruit* designated the ready mix for brewing, to be exclusively purchased from the local authorities (bishop, count or city council) or from their representatives. This was an early form of taxation as the price of *Gruit* included a fee for the authorities. *Gruit* contained malt, herbal ingredients like sweet gale (bog myrtle), marsh (wild rosemary), coriander, yarrow and milfoil; and probably yeast, attached to straw. The use of straw as a means to concentrate and preserve yeast was widespread [62, 63]. Moreover, it had been common in monasteries to add herbal ingredients like *Myrica*, *Ficaria* and *Iris* to the yeast [64], probably as a bactericide to protect the cultures against contamination. Apart from their bactericidal and fungicidal properties, *Gruit* herbs also produces flavor and psychotropic effects. As Lobelius put it later in 1551: 'the same plants [*Myrica gale*] are added at times of lacking hops/to the beer in the Nordic countries: sometimes only in order to make the drinker happy/since gale goes up to the head and its spirit pleasures the limbs...' [65]. Thus, almost all drinks, including malt liquors and wine, had been spiced since antiquity and the medieval preference for spiced food included drinks, too. Therefore, condiments were regarded as essential constituents of beer, and their type and quantity contributed to its valuation. In many locations hops were used as a constituent of *Gruit* before it became the sole beer additive due to commercial and fiscal necessities.

1.3.2
Beginnings of Professionalism

Professionalism means not only mastering a technology, but also developing brands and markets. In the latter respect, medieval brewers differed from their predecessors in ancient times. The emergence of professional brewers can be traced back to the times of Charlemagne, when he admonished governors in his administrative instructions *Capitulare de Villis* to always have a sufficient number of experienced craftsmen at hand, among them brewers (*siceratores*). This referred probably to the profession already known as *braxatores* in Roman times. *Brassatores* and *cervisarii* are also mentioned next to bakers, cooks and tailors in the Doomsday Book of 1086 in England [51]. In cities, the presence of brewers (*brassatores*) is recorded as early as the eleventh and twelfth centuries (Aachen, Bamberg, Huy), most likely denoting a specialist in the service of the overlord. It was in the cities, however, where professional brewing began to flourish.

The cities founded in Roman times had been gradually abandoned during the early Middle Ages. However, when trade and commerce slowly resumed after the crusades, cities regained their function as intersections of trading routes and administrative centers. In the twelfth century, new words began to appear in urban records, both in Latin and in vernacular languages, to designate brewers and brewing, indicating that a change in the status of brewers had occurred [3]. This is illustrated in a thirteenth-century manuscript of the Bibliotheque de Bourgogne in Ghent that depicts typical urban craftsmen with their professional tools: butcher, blacksmith mason and the brewer with a stirring paddle [66]. This display in an expensive manuscript shows that the craftsmen had gained a position previously reserved for overlords or bishops, and indicates the economic and political significance of their cities, many of which managed to gain more or less independence from their overlords during the eleventh to thirteenth centuries. As a consequence, most cities acquired special liberties, among them the brewing privilege. This included the right to grant brewing permits and regulate beer production and distribution. Brewing was regarded as a 'civic aliment' (mode of subsidy), indicating that commercial brewing was restricted to the inhabitants of cities. Moreover, many cities obtained additionally a right of precinct (Bannmeile, Banlieux), giving them a monopoly for their products within a specified parameter around the city.

Although in the beginning every citizen was entitled to brew, this right was soon confined to houseowners. The increasing demand required specialization and professionalism, as larger production volumes required high investment in equipment and raw materials. Moreover, the ever stricter fire prevention legislation could be more easily enforced on houseowners. As cities grew and brewing turned out to be quite profitable, the group of brewing citizens had to be constricted further. This was accomplished by restricting the period of time during which each brewer was permitted to produce, establishing the chronological order by lottery and by complicating admission to brewing [67].

Division of labor and specialization is one of the most significant features of urban economies. The expert artisans and traders were the pillars of urban economy. The craftsmen were quick to organize into corporations – autonomous guilds responsible for the admission to brew, training, product quality, and pension provision and fraternities for charity. Guilds had a considerable political influence as they represented all brewers, and could decree binding regulations and impose fines. Records of brewer's guilds exist as early as 1200 for London, 1230 for

Regensburg, 1267 for Ypern and 1280 for Munich. Separate maltster's guilds existed in several cities, such as Brunswick. Slightly later, the laborers also organized. In 1447 at Bruges (Flanders) the first association of brewery workers (*fraternitas sancti vincenti*) convened to protect themselves against 'innkeeper, women and provost' [68]. The emergence of brewer's guilds is an indicator of the transformation from home brewing to professional (and profitable) brewing. By the sixteenth century, brewers guilds were found in most cities all over Europe and continued to exert considerable influence in many states until the early nineteenth century.

Concomitantly, the display of guild emblems came into use. Until the late Middle Ages, emblems (e.g. a coat of arms) were strictly confined to the nobility. Now craftsmen who were certainly not entitled to armorial bearings used the symbols of their trade to distinguish themselves from other citizens. Moreover, the display of a guild emblem soon served as a certificate that a professional was at work. The tools used by professionals were therefore commonly displayed in the guild's emblem. In addition, the signs were soon customized by serving as an early version of a trademark.

The guilds of the staple food producing professions not only had an impact on the urban economy, but also greatly influenced the health and satisfaction of the citizens through the quality and the price of their respective product. Thus, city governments enacted regulations concerning the measure and quality of food. When taxation of food – in particular, beer – had become an important source of income for the city treasury, such legislation became even more substantial. Brewing ordinances are recorded as early as 1156 from Augsburg, 1268 from Paris and 1293 from Nuremberg. Those urban 'purity laws' dealt with [69]

- Consumer protection (beer additives, brewing seasons, attenuation, technology).
- Security of supply (grain used for brewing, pricing, obligation to brew).
- Organization of brewing (guilds, training, authorization to brew).
- Implementation of regulations (beer inspection).
- Trade (sale and taxation of foreign beers).

Thus, one main concern was to ensure the supply of grain. The adoption of the undemanding rye in Central Europe between the eighth and tenth centuries and the corresponding implementation of the three-field economy had shifted human nourishment unidirectionally towards grain-based diets [70]. In the Middle Ages, rye (*Secale cereale*), bread wheat (*Triticum aestivum*) and spelt wheat (*Triticum spelta*) were the dominant winter cereals, oats (*Avena* spp.) and to a lesser extent barley (*Hordeum vulgare*) being in general the preferred spring grains; although local variations occurred [71]. As the ratio between seed and crop yield amounted to 1:3 on average, there was no chance to accumulate stocks and supplies lasted only for 1 year. Famines ravaged Europe – there were 29 major ones alone between the years 750 and 1100 and the great famine of 1315 is still remembered. Even locally restricted crop failures severely affected regions without connections to waterways, as transport was extremely poor. As the grain demand for beer accounted for some 25–45% of the grain demand in towns [3] (and hardly less in the countryside), regulating the less important consumption of beer to ensure the supply of indispensable bread appeared reasonable. During severe famines brewing was interdicted altogether, as in London during the great famine of 1315. At other times, the type of grain for brewing could be dictated (e.g. barley in 1325 in Nuremberg). The most widespread brewing grain in the Middle Ages, however, was oats–a summer cereal; wheat, barley and rye being also in use. The type of grain or the ratio of different grains could change annually depending on their availability [72]. Many urban brewing ordinances also specified the amount of grain to be used for a given volume of beer.

The other major concern of urban ordinances was beer additives. Constituents of more than 40 different plants are known to be used as beer supplements and 14 additional ones were employed in beer for medical application [55]. As the properties of medieval plants might differ from today's plants, and in most cases the part of the plant used and the point of addition (before or after boiling) it is unknown, conclusions about the reason for their use and physiological effects remain speculative. However, much of the vast ancient information about plants and the insights of Islamic scholars as well as the popular knowledge was available to the educated upper classes. On the other hand, medieval beers contained only little alcohol in normal times and were diluted even more in times of scarceness. Therefore, the temptation to enhance the psychotropic effects of beers by the addition of eventually harmful herbs was always great. The ordinances therefore constitute an early legislation to protect consumer's health. However, they also reflect attempts to confine brewing to professionals using *Gruit* (which was not harmful) and hops.

As soon as brewers became organized professionals, they tried to develop their markets. Taverns were originally part of the brewer's premises. However, producing and serving beer soon became separate occupations, although linked by manifold financial interests. During the fourteenth and fifteenth centuries taverns differentiated further, the various types recognizable on their sign. In Roman times taverns along the roads were already marked with a pole and British alehouses as well as European inns retained a pole or a broom as their emblem. As traffic increased and taverns offered food and accommodation, such hostels distinguished themselves by sophisticated signposts. In Central Europe such signs were only allowed to be exhibited as long as the innkeeper had beer in stock.

Accordingly, a foreigner could get all of the necessary information about a place, like the types of beverages served, the price category, and whether food and accommodation were available, by looking at the sign outside.

The typical emblems of beer-serving premises had in part developed from the trade signs. This was exemplified by the drawings in the books of the Mendel and Landauer charities in Nuremberg [73]. The earliest known depiction of a professional brewer can be found in the Mendel charity book. It dates from 1425 and shows Herttel, the brewer, together with his brewing utensils. Interestingly, a hexagram is depicted above the brewing scene, indicating the use of this symbol in connection with beer brewing. It has been assumed that the use of the hexagram as a icon for beer brewing and selling originates from its ancient meaning as a protective symbol and has little in common with the same symbol used by the alchemists or the Magen David used for the first time by the Jewish militia in neighboring Bohemia in 1350 [73]. Brewers' emblems others than the hexagram are also depicted in the charity books, for example, together with Jorg Prewmaister in 1437 (gate) and Hans Franck in 1506 (star). These symbols were soon used not only to indicate a professional brewer, but also the availability of beer.

This trend was strengthened further when Guilds built drinking parlors of their own and city councils furnished cellars in the town hall. This process favored the professional brewers who could offer beers of much superior quality in comparison to country brewers or brewing victuallers. Hence, professional brewing emerged in the medieval cities. It was the cities along the shores of the Baltic and North Sea where brewing technology was perfected further.

1.4
'Woar Volk Is, Bint Klaantn' [74]: Hopped Beer and the Seaports

1.4.1
Hanseatic League

After the downfall of the Roman Empire, several seaborne empires arose successively on the shores of the North Sea, based on powerful navies and international trade. The first was that of the Vikings/Normans followed by the German Hansa, the Dutch and finally the Britons. All had in common the fact that beer was an indispensable prerequisite for their seaborne ventures and constituted a trading commodity that contributed much to their wealth. The beer of course had to be particularly nutritious and stable in order to suit the necessities of sea voyages. Strong hopped beers proved to satisfy these requirements.

It is not certain that the inhabitants of the Viking settlement of Haithabu (Schleswig) already brewed beer with hops. The Vikings at Haithabu were neighbors of the Wendes, who certainly knew of hops. Moreover, large quantities of hops found at Haithabu can only be explained by invoking brewing [56]. Beer had been recorded as a trading good in the North as early as the mid-eleventh century [66]. Certainly, the citizens of Bremen brewed a hopped beer, about 100 years after Haithabu had been destroyed by the Wendes in 1066, and exported it overseas [75]. Beer from Bremen is mentioned at Holland in 1252 and Groningen (The Netherlands) in 1278. Soon, however, nearby Hamburg overtook Bremen as a brewing center, eventually because beer in Hamburg was brewed with wheat and barley malt instead of the common oat malt. The commerce in beer flourished and Hamburg was soon known as 'the Hanseatic League's Brewhouse'. In 1376, 475 of Hamburg's 1075 manufacturers were brewers who had exported some 133 000 hl in 1369 [75].

A confederation of Hamburg and Lübeck in 1241 served as a starting point for a much broader League of up to 170 cities, which was formalized in the first general Diet of the Hansa held in 1356. There were now other important brewing centers associated with the League: Wismar, Rostock, Lübeck and Danzig. Commanding at the height of its power about 1000 ships, beer consumption by its own fleet might have amounted to over 250 000 hl annually if the quantities consumed at the time are considered [76]. However, most of the beer was exported; at its height the Hansa sold abroad well above 500 000 hl annually [75]. The League traded its famous hopped beer from Hamburg mainly to The Netherlands, where the little village of Amsterdam had been named as one of two exclusive custom places for beer from Hamburg in 1323. Another trade was established from the ports at the Baltic Sea (Lübeck, Wismar or Danzig) that exported what was often referred to as German or Prussian beer to Denmark, Norway and Sweden. In the period from 1562 to 1596, when the peak of the beer trade had already passed, Hansa ships still carried an average of 40 000 hl of beer annually through the Danish Straits [77].

Apart from its large trading network for beer exports, the League offered safe access to raw materials. Hops were no longer collected in the thirteenth century but planted in hop gardens in and around cities. Such hop gardens are reported from Wismar in 1250 and many other Hansa cities thereafter. Moreover, Hamburg had developed into the leading trading center for hops in Europe where hops were shipped to on the Elbe River even from Bohemia. This far-reaching trade allowed compensating for regional crop failures. Even more significant was the League's trade in malt. As it patronized the eastern granaries of Prussia, Pomerania and Mecklenburg, Hansa's brewers were independent of local grain supply, and could afford to always use choice grain and brew a strong beer of constant quality. Moreover, cities like Stralsund had specialized in malting and exported large amounts of malt [77].

The large-scale production of hopped 'export' beers of course required special production methods. Oats as the traditional brewing cereal were replaced during the fourteenth and fifteenth century by barley [72]. Some beers like that from Hamburg contained considerable amounts of wheat malt. Malt was air dried or kilned, resulting in 'white' or 'red' beers, the latter constituting the majority. Moreover, the amount of malt per brew varied considerable. Thus, taking the variation in brewing waters and technological variants into account, an astonishing diversity of beer styles emerged in Central and Northern Europe during the fifteenth century [75].

The manufacturing technology also changed. This started with brewing water, which was pumped and delivered by water mills as early as 1291 in Lübeck. As hops had to be boiled to benefit from their use, boiling vessels and heating technology had to be optimized. Until the early fourteenth century wooden vats had prevailed in rural brewing and iron pots with a capacity never exceeding 600 l were in use in the cities [75]. Wort preparation and fermentation were often conducted in the same vessels. Soon, however, larger copper kettles came into use with capacities of several thousand liters. This resulted not only in a significant increase in quality, but also in higher profitability. The urban kettles outclassed in their capacity particularly those of the monasteries – the greatest competitors of urban brewers. The corresponding mashing technology ensured yields of 1.05–1.80 l of beer per liter of grain, while aristocratic household in fourteenth-century England got 0.77–1.16 l of ale per liter of grain [72], which can be assumed as pretty much the average for European brewing at that time. Of course, malting, fermenting and storage facilities had to be adapted to such volumes, which consumed 800–3000 kg of malt per brew [75]. Moreover, copper brewing kettles required significant capital expenditures. Therefore, the large kettles were first bought by the community and carried from house to house upon demand. Later, when the revenues justified the investment, brewers bought kettles themselves. While at first hung from the ceiling fixed to huge timbers, they were later integrated in furnaces to enhance fuel efficiency. Bricked ovens under copper kettles with fixed equipment to move the liquids in and out had had become the standard of urban brewing by the sixteenth century [72].

The expansion of Russia and the loss of privileges in London initiated the downfall of the Hansa. After the Thirty Years War, the grain and malting region of Pomerania was lost to the Swedes. Their successors became the Dutch seaborne empire, which was better organized and commanded superior financial instruments. However, the Hansa brewers had been the first to develop a beer that could be transported over great distances overseas. Thereby they went far beyond the magnitudes handled by rural home brewers and urban victuallers until then. The limits they set were only surpassed by the London breweries of the eighteenth century.

1.4.2
Rise and Decline of the Cities in Central Europe

The overall conditions for brewing and selling beer changed dramatically at the end of the fifteenth century. Several factors contributed to this change.

First of all, the overall climate changed. Average temperatures dropped [78], although the actual levels varied regionally [79]. A decline in wine yields and a consecutive price rise resulted in an increased demand for beer [80]. The transition of Bavaria from a wine to a beer country must be seen in this context. This move had wider consequences, however. Next to grain, wine was the most important cash crop and was taxed heavily on the level of trade as well as on the level of consumption. As state revenue from wine started to fall, beer taxation and regulation of the brewing process increased.

Second, an increase in the population caused what has been termed the 'sixteenth-century price revolution'. Population growth, currency debasement and growth of the money supply, combined with climatic effects, resulted in an accelerated inflation, particularly of grain prices [81, 82]. This had an impact on nutrition in general. From the sixteenth century meat consumption started to drop from about 100 kg/head in the fourteenth and fifteenth centuries to a minimum of 14 kg/head in the eighteenth and nineteenth centuries. Therefore, bread became the most important staple food and the daily amount considered as the minimum to survive almost doubled [83]. As 1 l of good beer provided only 400–800 calories in the fifteenth century, while 1 kg of bread yielded 2500 calories [3], brewing was restricted by legislation with respect to the types of grains used and malting/brewing periods (beginning only well after harvest until spring). Also during that time, the important fodder grain oats was increasingly substituted for barley. Urban and state purity laws prescribing barley as the sole brewing cereal perhaps also meant that the summer grain barley might be sewn only after yields in winter varieties (wheat) appeared sufficient for bread supply.

The competition between bread and brewing in turn accelerated the rise in grain prices even more; the latter increased faster than beer prices in most regions [84]. This required restructuring and process optimization from brewers. The numerous ordinances interdicting malting and brewing during the summertime might have been motivated by concerns over the quality of malt and beer as well as the price situation before harvest time [3]. It might also, but by no means exclusively, be an indicator of bottom fermentation.

By the sixteenth century the infrastructure in Central Europe had improved so far that traveling and transport became possible at speeds not encountered since the times of the Roman legions [85]. Thus, overland transport became feasible even for commodities like beer, although at low volumes and high costs. A chart carrying 6 barrels of 1.75 hl (1 Rostock Ton) moved at a speed of 20–30 km/day, hampered by the numerous toll stations and bad roads. In contrast, a Hansa Kogge carried 600 such barrels much faster to distant customers. Accordingly, land transport added about 50–70% to the price of beer for every 100 km transported overland [84], and only very famous and stable beers were transported over longer distances. Strong hopped beers suitable for overland transport were developed, for instance in Einbeck, Hannover (Broyhan) and Braunschweig (Mumme), their increasing reputation being successfully marketed.

After an unprecedented increase in beer consumption during the fifteenth and sixteenth centuries the situation changed during the seventeenth and eighteenth century. The Thirty Years War had ravaged Central Europe, destroying capital and knowledge. In particular, the cities, centers of commercial brewing, were hit. The political structures had changed too, abating the significance of the cities. The territorial states emerged, and taxed and regulated professional brewing. Nobility as well as the monasteries and the rural populace increasingly ignored the privileges of the cities, particularly the precinct. As beer sales within the monopolized zone could make up to 80% of overall sales [3], theses tendencies had a severe impact. After a war the Bohemian cities, for instance, lost their precincts in the treaty of St Wenzeslau in 1517 and the nobility began to erect breweries on their domains. Being producers of brewing raw materials themselves, the domain breweries could produce cheap beer of high quality. As in Bavaria, however, where the nobility had held the brewing privilege since the Middle Ages, the greater competition secured in the long run the supply of wholesome beer. More pressing, however, was increased taxation. No less than 18 different state and local taxes applied to beer production and beer selling in the renowned brewing city of Freiberg in Saxony [86]. The most decisive development, however, was the pricing of beer and its raw materials. As the raw materials made up about 60% of the costs

[84] (fuel and labor together made up 20%, and taxes 20%), the ratio of raw material cost to beer price was the most important parameter in brewing and the only one a brewer could govern to a degree. In the North, costs of raw materials increased at a faster rate than beer prices, inducing the brewers to compromise on quality. Beer consumption in these regions was therefore replaced by coffee, tea and particularly boozes. On the other hand, in the southern regions of Central Europe like Bohemia, Württemberg and Bavaria, lower raw material costs ensured profitability of brewing and consequently a constant beer quality. Therefore, these regions saw a constant rise in beer consumption over the seventeenth and eighteenth centuries [84].

1.4.3
Rise of Dutch Brewing

In the sixteenth and seventeenth the Dutch commanded the largest fleet on the seven seas and established a trading empire stretching from New York to Jakarta. The rise of The Netherlands to this height was accompanied by the formation of a brewing industry unmatched in its time.

The Low Countries had been a stronghold of *Gruit* beer until the fifteenth century. However, in the fourteenth century the Hansa had exported its hopped beer along the shores of the North Sea and Holland became the preferred market. While the textile industry was the major economic factor in Holland and Flanders, the displacement of the Hansa town of Lübeck from the lucrative herring and salt trade with Scandinavia was a starting point for the Dutch seaborne trade in the late fourteenth century. Concomitantly, an impressive brewing industry emerged in the Dutch cities, producing a stable hopped beer. This product first replaced the Hansa beers from the domestic markets, and consequently supplanted them in the markets of Scandinavia and Northern France [87]. Apart from beer, the export of textiles and herring and the import of grain and timber fuelled the Dutch economy. From 1400 to 1700, Dutch per capita income growth was the fastest in Europe.

The Dutch brewing industry resembled that of the Hansa towns in many ways: it was based on an urban economy and shaped by the peculiarities of the respective city. Although the Low Countries established the most advanced and efficient agriculture of its time, grains had to be imported to a considerable degree from Prussia and Pomerania. Barley, wheat, oats and in some cases rye were used for brewing depending on local and temporary trends [72]. Altogether, Dutch brewing did not differ significantly technologically from the European standards, but did differ in the fuels used. While wood was the standard heating fuel for brewing in Germany until the middle of the nineteenth century, Dutch brewers used peat and coal. Wood, which was rare and expensive, had to be imported from Scandinavia and Russia; however, peat was a domestic product. Beginning in the fifteenth century, Scottish coal was used for brewing. Fourteen tonnes of peat were required for a brew, requiring handling huge volumes. That volume was reduced by 75% if coal was used [87].

During the seventeenth and eighteenth centuries brewing in the Netherlands declined. For once, beer export had lost its significance. Moreover, from 1600 to the 1800s Dutch incomes were the highest in Europe, permitting the consumption of tea, coffee and French wine. Thus, the domestic market shrunk, too. The most serious threat, however, came from distilled spirits like brandy, gin and genever [87].

The Dutch brewers had contributed significant improvements to fuel efficiency in brewing, and they had spread their brewing technology to other continents, erecting the first breweries in New York, the Cape Province Indonesia.

1.5
'For a Quart of Ale is a Dish for a King' [88]: John Bull and the Industrialization of Brewing

Fermented malt liquors had been native to the British Isles since primeval times. In Roman times the shopping lists of the garrison document that beer was an important constituent of the legion's diet. In the eighth and ninth centuries different types of malt liquors were already in common use: clear ale, mild ale and Welsh (*wylisc*) ale [2, 51], the latter probably being an originally Celtic drink made from ale and honey [89]. At the same time, the *Historia Brittonum* speaks of '*vinum et siceram*', using the Carolingian expression '*sicera*' (see above) for alcoholic drinks. On the other hand, a lease dated 901 AD mentions 'beer' (*beor*) besides 'sweet Welsh ale' and 'clear ale', indicating that 'beer' was already common [51]. The term beer vanishes thereafter for 500 years to be used later for a hopped malt liquor. Moreover, 'bragot' (braggot), a drink mentioned again in the fourteenth century in Geoffrey Chaucer's *The Canterbury Tales* was probably a honey-containing malt liquor of Celtic origin.

As early as the ninth century the numerous alehouses could be identified by a long pole (and a bush of evergreens if they also sold wine). Later, the bush also indicated the serving of ale [90]. The pronounced affinity of his subjects towards ale compelled King Edgar, the reformer, on the advice of St Dunstan, to issue several laws against abuse and decreed that the large drinking vessels passed round at meals should be marked with pins at intervals to show the amount to be drunk. The distinct fondness of the English people for ale intrigued other governments too. The Magna Charta included a provision for the standardization of measures, mentioning explicitly ale in this context. However, it was the prospect of sharing profits with the government that soon fired the imagination of various administrations. The first tax on all movables, the 'Saladin tithe', was imposed on ale in 1188 and Henry III levied the 'Assize of Bread and Ale' on these basic victuals in 1267.

At the end of the fourteenth century their trading partners from the Low Countries confronted the Englishmen with hopped beer. Up to then, ale was the characteristic British drink of the time. It was probably not fermented throughout and the omission of hops gave it a heavy, sweet taste [91]. It was, as all British beers were until the advent of lager from the continent in the nineteenth century, a top-fermented beverage. The new drink, soon called beer (bière), was adopted at a few places at the southern coast, but was at first vehemently opposed in the rest of the country. Although the brewers of beer were officially recognized as a guild in 1493, their product was viewed with utmost suspicion until well into the sixteenth century. 'Hops, Reformation, Bays, and Beer Came into England all in one year' goes a rhyme from the early sixteenth century based on Henry Buttes' *Dyets Dry Dinner* and Andrew Boorde states in his *Breuiary of Helthe* that ale is the natural drink for an Englishman. In contrast, beer 'is a natural drink for a Dutchman. And now of late days it is much used in England to the detriment of many Englishmen; for the drink is a cold drink; yet it doth make a man fat and doth inflate the belly, as it doth appear by the Dutchmen's faces and bellies' [92]. Nevertheless, hop growing was legalized by an Act of Parliament in 1554 and hopped beer slowly established itself alongside ale in England during the first half of the sixteenth century.

At this time producing ale and beer and selling it had become separate activities in Britain, much like on the continent. Thus, in 1522 the licensing of alehouses was made mandatory by law. The census of 1577 lists 14 202 alehouses, 1631 inns and 329 taverns (which sold wine in addition to ale) – one license per 183 people [51]. These licenses were an important revenue for the state and a century later the desperate need for money by successive governments resulted in the first excise ever on ale, beer, cider and perry in Britain early in 1643.

However, apart from tax revenues there was yet another reason for the various governments to be interested in the production of high-quality ales and beers. The Tudors laid the foundations for British sea power and beer soon proved to be a prerequisite for British maritime expansion, as indispensable as had it been for the Vikings, the Hansa and the Dutch. Thus, ale was a standard article of the sea ration as early as the fourteenth century. Henry VII established a naval brewery at Portsmouth in 1492 to supply his ships with beer [93] and Henry VIII granted his sailors 10 pints (5.7 l) of beer daily. Later other naval brewhouses followed at East Smithfield (1683), London and Portsmouth [91]. At the latest after a series of stunning military misfortunes due to lack of beer and near catastrophes like the one caused by tainted beer during the battle with the Armada, the significance of beer in building an empire became evident to everybody. Beer was not only a rather stable beverage during sea voyages, but was also indispensable for the sailors health. Accordingly, Shakespeare's son-in-law, Dr John Hall, cured scurvy by brewing a beer or ale of 'ascorbutic herbs, viz.: scurvy grass, watercresses and brook lime' [94]. As Samuel Pepys, Secretary to the Admiralty, observed: 'Englishmen, and more especially seamen, love their bellies above anything else, and therefore it must always be remembered in the management of the victualling of the Navy that to make any abatement from them in the quantity or agreeableness of the victuals is to discourage and provoke them in the tenderest point and will sooner render them disgusted with the King's service than any one other hardship that can be put upon them'. It was Pepys who laid down that every sailor was entitled to 1 gallon of beer daily – a standard upheld until after Nelson's time.

The provision of the fleet with victuals drew heavily on national resources. Between September 1651 and December 1652, for instance, the victualling contracts for the fleet came to 332 000 pounds sterling [95]. The building of the British maritime empire had two important ramifications for brewing: (i) the necessity to cater the ships for wholesome beer at all stations abroad resulted in the spread of British brewing techniques all over the world and (ii) the need to supply beer in large quantities at home contributed to the industrialization of beer production.

Attempts to discourage small-scale brewing had been made by local administrations since the fifteenth century. This trend was accelerated during the seventeenth century, when commercial constraints and an increased capital requirement for brewing technology added to the disadvantage of small-scale brewing. Large commercial ('Common') brewers were noted first in London in the sixteenth century and emerged soon in other large cities all over the kingdom.

Thus, at the dawn of the eighteenth century a number of powerful commercial breweries had already been concentrated in the London area. At the end of the century commercial large-scale brewing had been firmly established over the entire country with its center still being London. Nothing indicates better than these numbers the tremendous changes that took place in brewing during the eighteenth century.

This change was associated with one name: 'porter'. There are numerous legends around the invention of this black, bitter and thick beer. One has it that it was brewed for the first time at the Bellin brewhouse at Shoreditch in 1722, although an earlier version called 'Entire' might have consisted of a ready-made blend of three different styles previously available: 'ale', 'beer' and 'twopenny'. A sharp rise in beer taxes in 1692 would have encouraged brewers to blend the differently taxed beers. Most likely, however, 'porter's beer' was a popular synonym for 'brown butt-beer', a long established London drink [96]. Porter soon became known as a distinct beer style and dominated the British Empire for about 100 years. It defied the gin fever in the 1730s, and the competition of tea and coffee as well. John Bull with a glass of porter and a barley ear on his hat as sketched by Hogarth came to symbolize the Englishman and British way of life.

Three factors triggered the emergence of industrial brewing [96]. The first was the British exchequer, which depended heavily on the excise duties from beer. The authorities granted the large London brewers higher rates of tax relief than their smaller competitors, officially on the grounds that large-scale operations suffered more wastage [91]. The second factor was the London market. At times when beer was very susceptible to contamination, and transport was difficult and expensive, a booming home market in the world's most industrialized city was a distinctive incentive for scale-up. The third factor was porter itself. Made from dark brown malts, a specialty of the Hertfordshire maltsters, and large quantity of hops, it required extended maturation over several months to attain its characteristic flavor. Storage of large quantities made it easy to blend different batches of beer, smoothing out the inevitable variations to create a standard, reliable product [96]. Porter seems to have been the first beer technically suited to mass production at contemporary standards of control [91].

Porter brewing triggered a series of groundbreaking technical, organizational and commercial innovations to facilitate industrial-scale brewing. The porter brewers were affluent men, ready to command any innovation they thought potentially useful. The new breweries combined aesthetics with practicability. They comprised several floors to facilitate transport of intermediates and reduce unwanted contamination through compartmentalization. Industrial (i.e. mechanized) brewing reverted to cast iron equipment like malt mills, pumps and tubing. The scale-up of pans (kettles), fermentation vessels and storage vats proved challenging. The first sizable vats installed in 1736 were capable of holding 1500 barrels each; a vat mounted in 1790 held 10 000 barrels [2]. When such a vat collapsed in 1814, 7600 barrels of porter flooded the brewery and its neighborhood, destroying several houses and killing eight people. At that time, walled storage facilities holding up to 20 000 barrels were in use at Whitbread's brewery. Other important inventions concerned the advancement of the brewing process itself like the cooling of large quantities of wort or introduction of isinglass (collagen) for fining and stabilization.

In 1784, Henry Goodwin and Samuel Whitbread were the first brewers ever to establish a steam engine at a brewery. Needless to say that by then coal was the fuel in breweries. Coal had been used in manufacturing in England as far back as 1307 AD [97] and by the end of the seventeenth century some 1 million tons of coal were consumed in Britain in manufacturing [2]. The soaring price of wood had soon enticed many brewers to use coal for heating and it took only a little more time for the adoption of coal in malting. Thus, indirect kilning was devised and around 1800 the first experiments with malting drums were conducted. The use of the thermometer – described by Michael Combrune in 1762 [98] – permitted a fairly exact adjustment of malt colors. Although many more obstacles had to be overcome, the resulting 'English' malting process of indirect kilning proved to be far superior to anything known up to then.

Equally important, the acquisition, storage and transport of huge quantities of raw materials and providing appropriate financing had to be organized. Raw materials made up to more than half of the costs; duties constituting the biggest fraction of the rest [91]. This resulted in sophisticated malt and hops trading networks as well as in the spread of commercial malting, either at one's own risk or commissioned by the brewers who provided the barley. Finally, marketing strategies also changed. The industrialization of brewing started in the urban centers, where a mass market existed, and there new patterns in the distribution of beer emerged first too. The brewing victualler yielded to the retailer and the independent inns became tied public houses. The percentage of publicans tied by London breweries around 1850 exceeded 60% of all served and the loans made to them valued up to 1.5–2.7 pounds per barrel [91]. Thus, most of the fundamentals of manufacturing and selling beer in our times had been put in place in London by 1850.

The technology of industrialized brewing did not remain confined to England for long, but spread first to Scotland and Ireland, and consequently to other regions of the British Empire. In particular, in Ireland, the development of a brewing industry linked to the name of Guinness is noteworthy. The London porter brewers had established a prospering market for their product in Ireland and sold considerable quantities of their beer there. However, high import duties rendered English beer rather expensive and tilted the market in favor of ardent spirits. The war with Napoleonic France increased the price of exports drastically and the determination of the government to curb hard liquor consumption at the same time opened a unique opportunity for Irish brewers [91]. Arthur Guinness of Dublin was most successful in adapting London brewing technology to Irish conditions. By 1840, over half of Guinness's annual production of 80 000 barrels was sold to England [91].

In the mid nineteenth century porter lost public favor and had to yield to new ale types. This change was initiated by modifications in porter manufacture itself. During the 1780s research had established that pale malts contained more fermentable sugars than the dark malts used to make traditional porter [99]. Porter breweries therefore substituted dark malts by pale ones and shortened maturation times. The traditional dark color was now obtained by the addition of charred 'color malt'. Only by increasingly sophisticated temperature management techniques could the stability of the product be maintained, but its flavor changed, resembling that of dark-colored ale [91]. In the meantime the brewers of paler ales and beers in the northern provinces had adopted the London technologies and adjusted them to their product. With the advent of fast and cheap freight, provincial ale brewers gained access to urban markets and began to scale-up their production. While London porter slowly sunk into oblivion, the different ale styles dominated the British market until the 1960s.

1.6
'We Live in a Country Where Beer Constitutes Quasi the Fifth Element' [100]: Advent of Lager and the Internationalization of Brewing

1.6.1
Bavaria and the Rise of Lager Beer

Bavaria is a medium-sized state stretching between the rim of the Alps and the Danube. After the Napoleonic wars it absorbed the Franconian countries up to the central German mountains and with them a long brewing tradition, a fine barley region and the then most important hop market in Germany. Until the twentieth century Bavaria was an agrarian state, with only few urban centers. Wine had constituted the preferred beverage of its inhabitants until the first half of the sixteenth century, and there was widespread consensus in the fifteenth and early sixteenth centuries that the qualitatively best beers originated from the Hansa towns followed by beers from Saxony, Thuringia and Bohemia. In the south, the brews from the northern parts (Franconia, Upper Palatinate) were considered superior to those from Bavaria. The reasons for this are manifold: the competition of cheap wine, smaller production volumes as well as the use of wooden vessels and instruments which facilitated infections.

In the sixteenth century the situation changed dramatically and subsequently Bavaria became synonymous with beer. Several factors, among them climate change and state intervention, precipitated this development. Nothing exemplifies the role of the state in shaping the drinking habits of its subjects and in regulating brewing better than the famous Bavarian Purity Law ('Reinheitsgebot'; literally 'purity order') of 23 April 1516 [69]. The purity law was certainly not the most pressing of the topics negotiated at the diet of Ingolstadt in this spring of 1516 and hardly any of the participants would have thought that the small paragraph dealing with brewing in the new governance passed then would be of any particular importance. However, the decree limiting brewing ingredients to hops, barley, malt and water turned out to be one of the few legislative measures lasting half a millennium and granted the name of the sovereign Wilhelm (the fourth) some status of immortality.

Purity laws had already been part of urban legislation for more than four centuries. However, these were issued by cities and never attained more than local significance. In contrast, the Bavarian Purity Law and aligned later ordinances regulated brewing in an entire state. It laid down fundamental standards for beer production at a time when rapid changes in brewing technology and beer varieties alienated authorities and consumers alike. The Bavarian Purity Law in the first place defined unambiguously the term 'beer', it stated the price of this indispensable staple food, and it established administrative measures to ensure a sufficient supply and a satisfactory quality. Its best known provisions are the instructions concerning quality: the exclusive use of barley malt, water and hops, the omission of other herbs and inorganic ingredients (with the exception of low quantities of salt, juniper berries and caraway, as specified later in 1553 and 1616), and the bottom fermentation, interdicting brewing activities during summertime.

Bottom fermentation had probably been known in Bavaria since the beginning of the fifteenth century as brewing ordinances from Landshut (1409) and Regensburg (1454) indicate [69]. In Munich, bottom fermentation was presumably first practiced by a 'Bohemian' brewer from Eger (Cheb, at the Bohemian–Franconian border) [101]. While in Bohemia proper top fermentation was universally in use until the nineteenth century [1], the brewing ordinances of Nuremberg of 1303–1325 indicate that a dark, hopped, bottom-fermented beer was brewed in Franconia at the beginning of the fourteenth century [102, 103]. In Nuremberg, a special guild, the 'Hefner' (Yeasters), was responsible for maintaining a stock of (dried) yeast over the summer (when brewing was not permitted) and providing sufficient yeast when brewing was resumed in September. From Franconia, bottom fermentation might have spread to Bavaria where it was firmly established by 1600.

While the Bavarian Purity Law set a binding norm for what beer is and how it is manufactured, it left sufficient room for competition between the three emancipated brewing classes: the burghers, the nobility and the church. Unlike other German regions, where such competition was curtailed by precincts and privileges, contributing to a decline in beer quality, an acceptable excellence of beer was maintained in Bavaria. The great public concern in all brewing matters and the strict governmental scrutiny of producing and serving beer encouraged the widespread idolization of beer. There were three styles of beer available. The first was the white (top-fermented) wheat beer, which only the sovereign was permitted to manufacture [104], resembling an earlier Bohemian type. The second, available only in some regions, was a 'white', top-fermented barley beer. The ubiquitous, typical Bavarian beer, however, was the bottom-fermented brew of dark brown color.

After the Napoleonic Wars, Bavaria was the first German State to enact a constitution, liberalize trade regulations, and harmonize weights and measures, thus paving the way for the industrialization of brewing. However, as elsewhere it was the taxation (the 'Bierregulativ') that favored larger production units by setting a maximum price [105]. Although slowly, industrial-scale brewing emerged, always following British examples. 'So much is certain that the English are the very masters of brewing. They had been the first ones who wrested brewing from empirical handicraft and practiced it with reliability and precision by applying chemistry, mathematics and physics' wrote a compendium in 1811 [106].

However, it was not possible just to copy British technology and processes as Bavarian lager brewing significantly differed in some respects from porter and ale brewing (Table 1.2).

The outstanding property of Bavarian lager was its stability as it was intended to last over the summer. Therefore, enough nutrients had to be left after the main fermentation to allow residual yeast cells to remain viable and continue fermentation at the low temperatures in the storage cellars. A low degree of attenuation was therefore desired, and this was accomplished by special bottom-fermenting yeasts and low temperatures. This type of fermentation of course results in a distinct flavor pattern [111], which differed significantly from the other beers of the time. Differences existed also in the malting process, where British technology was unsurpassed at the time.

1.6.2
Spread of Lager Brewing

Until about 1860 there were two different developments proceeding in Munich and Franconia, respectively. In Munich, which attracted increasing numbers of inhabitants, rapidly expanding breweries served the home market. As Munich brewers commanded respectable financial resources, they were able to implement the newest technology. The first 'English-type' malting kiln with indirect heating was installed in 1818, the first steam engine in 1840 and wooden equipment replaced by metal appliances as far as possible. After the completion of the important railways connecting Bavaria with the rest of Germany the most severe limitation of Munich breweries – cooling and supply of coal – could be overcome. Wood remained the principle fuel for brewing and malting in Bavaria until the middle of the nineteenth century. Even more pressing was the development of technical means for cooling wort, fermentation vats and maturing vessels – an essential prerequisite to overcome the legal restriction of (bottom-fermenting) brewing during summertime. The construction of underground cellars in the vicinity of

Table 1.2 The three primary beer styles (about 1875–1914 [107–109]).

Process	Brown Bavarian lager	Pale American lager	British pale ale
Malt	summer barley, germination 4–10 days, air-drying and subsequent direct and indirect kilning, close at 85–87 °C	six-row barley malt, drying/kilning for 26–30 h, close at 85–87 °C	germination 10–15 days, indirect kiln, air drying for 3–4 days, close 5 h at 75 °C; pale, excellent solution
Mashing process	decoction, two- or three-mashing process	malt, corn, rice, syrups; 'double' mashing technology: unmalted materials, mashed at 40–70 °C, boiled; malt doughed at 38 °C, boiled adjuncts added, subsequently infusion mashing at 67 °C	infusion process, high starting temperature (64–69 °C)
Main fermentation	bottom fermentation, 6–12 days, 6–10 °C, limited attenuation	bottom fermentation, closed tubs, CO_2 collected, one tub, 2–5 days, 8–12 °C	top fermentation, two tubs, 49 h, 17–22 °C, high degree of attenuation
Microorganism	slow-fermenting yeasts (*S. uvarum*) with uniform genomes [110],	bottom-fermenting yeasts, slight flocculation	fast-fermenting yeasts (*S. cerevisiae*), heterogeneous genomes, several species
Second fermentation, storage	after saccharification for 6–8 days, storage up to 6 month	2–12 weeks at ≤3 °C; fining (chip tanks), carbonating	>3 months
Beer (W: wort; A: alcohol)	W: 11.0–13.5%; A: 2.7–4.0%	W: 12–13%; A: 3.8%	W: 13.5–14.0%; A: 4.1–5.0%

Munich and the availability of natural ice were decisive in the abolition of the time-honored proscription of summer brewing in 1850. Soon the enormous quantities of natural ice required by the brewers and other food industries could only be provided by an extensive international trading network [112]. Only the advent of the Linde ammonia refrigeration in 1875 finally permitted large-scale lager brewing anywhere at any season. From 1806 to 1871, beer production in Bavaria rose from 3 700 000 to 8 600 000 hl; 97% of it being the characteristic bottom-fermented brown beer. By 1897, it had almost doubled again to 17 000 000 hl; 2 600 000 hl of which were exported. More than half, 1 500 000 hl, of the exports were produced in Munich [105].

This success of Bavarian lager beer can be attributed to a large extent to the Franconian breweries that had previously popularized this beer type in Saxony and

Table 1.3 Beer export of selected Bavarian cities (1000 hl).

	1843	1869	1879	1889	1896
Kitzingen	10.8	13.1	7.9	1.7	0.6
Erlangen	0.1	42.9	100.4	112.3	99.3
Kulmbach	3.8	60.6	143.7	379.8	626.1
Nürnberg	8.6	78.5	186.9	236.7	151.7
Munich	0.01	28.3	254.5	ca. 1200.0	1450.1

the Prussian countries. Much smaller and technically less advanced than their Munich counterparts, Franconian breweries nevertheless succeeded in developing a beer type that matched the taste of the northern German public. This beer, dark and bottom fermented like the products from Munich, contained about 1% more alcohol [107, 113] and exhibited an excellent stability. In the adjacent northern countries, top-fermented beers still prevailed until the last quarter of the nineteenth century, which were inferior to the Franconian products with respect of price and quality.

As indicated in Table 1.3, three distinct phases in the spread of Bavarian beer can be roughly distinguished. In a first phase until about 1850, beer exports were driven by the abolition of trading barriers as a consequence of the Germen customs accord of 1834 and the expanding railway network. A second phase spanning roughly between 1850 and 1890 was characterized by great technical improvements, the final liberalization of trade in 1869 and the German unification of 1871. The latter resulted in a dramatic acceleration in the pace of industrialization with a corresponding increase in beer consumption. During this period, however, the well-established market for Bavarian beer was threatened by the large newly built lager breweries in other German states.

During the industrialization of Germany, lager beer played the role that porter had played in eighteenth-century London. Accordingly, lager breweries came into being all over Germany. Particularly dramatic was the development in the centers of industrialization. In Berlin, lager beer production rose from 150 000 hl in 1860 to 1 940 000 hl in 1890, while the output of the traditional top-fermented beer increased over the same time from 370 000 to only 1 060 000 hl [114]. In Dortmund, where a strong, stable lager beer called 'Export' was brewed, production soared form 120 000 hl in 1866 to 1 400 000 hl in 1900 [115].

As the scale of production necessary to compete in the larger markets required considerable working capital, only companies able to raise equity carried the day. During this time Munich breweries rushed into the markets developed previously by Franconian brewers (Table 1.3). The final phase, from the last decade of the nineteenth century until World War I, is characterized by a saturated market where tied houses, extensive advertising and marketing as well as new sales channels like bottled beer become widespread.

Moreover, new beer styles came into vogue. In particular, Pilsner, a pale, heavily hopped beer styled by a head brewer at the civic brewhouse at Plzen (Pilsen), Josef Groll in 1842, quickly attained attention throughout Europe. Between 1880 and 1897, Pilsner exports to Germany multiplied 6-fold and in the Austrian empire the Civic Brewery in Plzen came only second to Anton Dreher's brewery in Schwechat [113]. Soon other breweries tried to copy the Pilsner brew. As the soft water of Plzen was a decisive factor in Pilsner brewing, at first only a few breweries that had exceptional soft water at their disposal succeeded in producing Pilsner style beers. One of them was the brewery in Freiberg (Saxony) in 1863. Successively, however, the Pilsner style was reproduced in many places. However, at first only places where a particular soft water was available and in 1899, when breweries generally had recognized the marketing value of a 'place of origin' labeling, a court decided that 'Pilsner' did not designate a regional provenance, but a distinct beer style.

As indicated in Table 1.3, only Kulmbach, a small city of about 8000 inhabitants at the time, and Munich, numbering about 400 000 inhabitants, were major centers of Bavarian beer export in 1900. The disproportionate ranking of Kulmbach can be explained by three factors: its distinct dark beer, its brand name relying on the high recognition of Kulmbach as a brewing and malting center, and, most important, the affiliation to the Dresden–Berlin brewery holding. The latter was structured by the brothers Max and Georg Arnhold who concentrated their Dresdner private bank business early in brewing. During the 1880s they acquired malt factories, breweries in the industrial centers (Berlin, Dresden and Frankfurt) as well as famous brands (Radeberg, Kulmbach, Kiel), and set up a special bank (Braubank) to manage their holdings. By cost cutting and optimizing marketing and sales channels, they became an all-round provider that could offer different products and services through a standardized distribution network [116].

The British brewing industry had experienced exceptional growth too, using newly raised equity for expansion, technology and – a specialty of the British market – the acquisition of tied houses. This increasingly costly engagement in estates provoked a concentration process in the brewing industry and dangerously exposed it to the hazards of the property markets [117]. However, as in Germany, dividends of incorporated breweries and malt factories at that time never fell short of 5%.

After 1900, the economic boom of the 1880s and 1890s came to an end all over Europe, resulting in unprecedented restructuring and concentration of processes in Germany and Britain alike. By 1914, in Britain, the big breweries, in particular Guinness, had weathered the storm and experienced prosperity again [117]. In Germany, the Arnholds were in a comfortable position to adapt to the new realities. The structures that they had created around 1900 were to last until the 1990s when globalization reshaped German brewing.

1.6.3
From Alchemy to Biochemistry: The Science of Brewing

The quantum leap in brewing during the eighteenth and nineteenth century was for the most part the result of the implementation of scientific principles that facilitated the development of appropriate equipment and technologies.

The first step towards scientific brewing was the adaptation of measuring devices to the specific needs of the brewers and maltsters, and the development of new instruments. Here, British brewers led the way. In 1762, Michael Combrune published his *Essay on Brewing* stressing the importance of using a thermometer in brewing and malting. The book was almost immediately translated into German and French, and a subsequent edition found its place on the bookshelf of Thomas Jefferson. A little later, in 1769, James Baverstock used a hydrometer to measure hop extract. However, it was John Richardson's treatise on *Statistical Estimates of the Materials of Brewing* that precipitated the search for means to measure the 'attenuation' (i.e. the decrease of wort density and concomitant formation of ethanol with increasing precision). After considerable progress linked to the names of T. Thomson, S. Hermbstädt, J. Long, C. Steinheil and J. Fuchs, it was Carl Josef Napoleon Balling from Prague who finally established a reliable method to measure and control the conversion of wort sugars into ethanol by fermentation in the brewery in 1843 [118].

Even though it was possible from the 1830s on to quantify satisfactory the wort sugar content and its conversion to alcohol (the Spaten brewery in Munich had used a Long saccharometer since 1834), the process of fermentation itself still remained a mystery. In 1803, the Institut de France had offered a medal worth 1 kg of gold for an answer to the question: 'What are the characteristics which distinguish vegetable and animal substances acting as ferments [substrates] from those that undergo fermentation [enzymes]'? [119] The Frenchmen A. Lavoisier (1789) and J. Gay-Lussac (1815) had established the chemical formula of the oxidation of glucose to CO_2 and ethanol. The Swede J. Berzelius had formulated the principles of catalysis (1835) and the German J. Liebig had applied these to sugar fermentation (1839), giving a fairly accurate description of the overall process. The snag with the chemical theories was that they were of little use for brewing practice.

Meanwhile, the first enzymological studies were published: amylase (diastase) was discovered in barley by the Frenchmen A. Payen and J. Perzoz in 1833, invertase in yeast by P. Berthelot in 1860, and the acid protease pepsin by the German T. Schwann in 1836 in the mucosa. On the other hand, progress in optics permitted Giovanni Battista Amici from Italy to build a microscope in 1837 that had a numerical aperture and a resolution not much inferior to modern microscopes [119, 120] These improvements opened up the possibility of detecting and characterizing microbes for the first time, and laid the foundations of a new science-- microbiology. Soon, in 1837, three scientists, C. Cagniard-Latour of Paris, F. Kützing from Halle and T. Schwann of Berlin, discovered independently that yeasts are living organisms. The fascinating aspect of the theories of the 'microscopers' was that the catalyst, yeast, was not only well known to brewers, but could now be observed directly under the microscope. The Spaten brewery in Munich bought a microscope for 7 Gulden in 1855.

It was another Frenchman, L. Pasteur, who subsequently laid the fundaments of scientific microbiology and established unequivocally between 1855 and 1875 the role of yeast in alcoholic fermentation, the physiological nature of fermentation, and the differences between aerobic and anaerobic metabolism. The immense interest awakened in microbes sensitized the scientists to the differences of microbial genera and resulted in the characterization of different yeasts [121]. In 1876, in *Études sur la Bière*, Pasteur had already recommended ensuring that brewing yeast was free of 'disease-causing ferments' by purifying it from bacteria. However, it was C. E. Hansen who in 1883 separated three different species from the yeast used the Carlsberg brewery. The original strain had been obtained from the Spaten brewery in Munich in 1845 and used ever since. It turned out that only one clone resembled the original strain, while the two others were 'wild yeasts' from the Carlsberg orchards just outside the brewery [122]. Although some time passed until it was generally accepted, the theory of pure yeasts prevailed at the end of the century.

The newly flourishing science of chemistry had its impact on brewing too. In 1820, the German-born London-based Professor Frederic Accum published *A Treatise on the Adulteration of Food and Culinary Poisons*. In his manuscript, Accum not only revealed a long list of unwholesome additives in beer, but also named those brought to the courts for beer adulteration. His book caused an outcry among food manufacturers in London and after much harassment Accum fled to Germany.

Ten years later, Arthur Hill Hassall, John Postgate and Thomas Wakley resumed the fight against food adulteration. Their analyses revealed that industrialization had by no means improved beer quality. Among other poisonous substances, the presence of picric acid, *Cocculus indicus* (fishberry), Levant nut (containing the alkaloid picrotoxin) and strychnine were detected in beer. Moreover, beer purchased directly from the brewers was shown to contain 4.53–7.15% alcohol, while the same beer bought from a publican varied in its ethanol content from 3.23 to 4.87%. Thus, at the sales level beers were adulterated too and spiked with poisonous bittering substances to delude the consumer. In Germany, the situation was by no means better; work by G. Hopf (1846), H. Klencke (1858) and M. Bauer (1877) detected strychnine, picric acid and opium among many detrimental substances in beer. In spite of these findings, powerful brewing lobbies delayed legislation for many years. Only in 1860 was an 'Adulteration of Foods Act' adopted and in 1885 a bill regulating the content of beer passed in Britain. In Germany, a law regulating food safety was enacted in 1879 and the Bavarian Purity Law became federal German law in 1919.

The most lasting consequence of the developments outlined above, however, was the emergence of brewing as a scientific discipline. Certainly, by the middle of the nineteenth century a vast literature on brewing existed, accumulated ever since movable letters had been used to print books. However, these were pure descriptions of individual knowledge on how to fabricate beer. Their use for others was rather limited, as the publications lacked the most important prerequisites: standardized equipment and objective measurements. Only after thermometers and saccharometry had been established in brewing, weights and measures had been aligned, and brewing equipment became comparable, could scientific theories be conceptualized. Brewing science commences with small groups of apprentices being taught the basics of physics and chemistry, and their application in brewing. In Munich, Professor Cajetan Kaiser started teaching in his parlor in 1836. Of the scholars, only about at third were from Bavaria – the majority came from other German states or abroad. In 1865, a brewing curriculum was established at the Technical University at Weihenstephan near Munich and in 1888, brewing classes started at the University of Berlin [122]. The Institute of Brewing was formally established in London in 1890. In the United States, John E. Siebel opened the Zymotechnic Institute in 1872 [123] and Anton Schwarz founded the Brewers' Academy of the United States in 1880 in New York. The establishment of an academic brewing science was accompanied by the publication of scientific journals and a multitude of books on brewing technology. By 1900, brewing had become an accepted branch of the natural sciences.

1.6.4
Europe, the United States and the Internationalization of Beer Brewing

At the turn of the twentieth century beer production and consumption had spread all over the world (Table 1.4; the numbers are estimates due to large fluctuations in beer output at the time and inaccuracy of records).

While at that time 79% of the world's beer still was of European origin, today the corresponding figure is rated at 34%, as the Americas and Asia have developed their own production and beer markets. Moreover, grouping of beer output according to countries was reasonable in 1900, when national habits formed beer styles and markets. However, such a classification is of less significance today, when companies rather the than states have the main impact on brewing. Currently, the world's leading brewing conglomerate produces more beer than all the breweries in Germany and Britain – in 1900 the leading brewing nations – combined. The basis for this development was laid down at the end of the nineteenth century with the concentration of breweries within Europe and the emergence of a powerful brewing industry in the United States.

American brewing mirrors the cultural diversity of its immigrants and the variety of brewing materials available as much as the ingenuity of its inhabitants. Legend has it that the Pilgrim Fathers debarked in Massachusetts because their beer supply was exhausted. Following the advice of established settlers it was recommended to newcomers in 1620 to bring a sufficient supply of malt for brewing with them in order 'to avoid the sudden transit to water drinking which can cause great transformations of their body' [126]. The first authentic record of the existence of a public brewery dates back to 1637, as far as Massachusetts Bay, and to 1638, as far as Rhode Island is concerned [127]. Later, William Penn built a brewhouse at Pennsbury in 1683 and the brewery Leonhard Baretz established in 1744 at Baltimore was the largest in the states at its time. However, it was at 'New Amsterdam' (Manhattan) where brewing flourished first. New Netherlands established public breweries in 1632, an excise tax on private brewers in 1644, and laws regulating brewing and the quality of beer in 1664. However, home brewing prevailed in the sparsely populated country and resorted to suitable materials and methods. As only the rich brewers in the cities could afford to buy malt from

1 A Comprehensive History of Beer Brewing

Table 1.4 World beer production around 1900 and 2003 [113, 124, 125].

Country	Output (1000 hl)	
	1897–1901	2004
Germany	69,000	106,190
British Empire	64,000	
United Kingdom	60,500	58,911
Ireland		8,142
British Colonies	3,500	
Australia	1600	16,910
Canada	1100	23,130
India	245	7,000
Cape Province/South Africa	100	25,000
United States	46,400	233,300
Austria–Hungary	21,500	
Austria	4,800	8,670
Bohemia/Czech Republic	9,400	18,753
Moravia/Slovakia	2,000	4,218
Galicia/Poland	1,200	27,700
Hungary	1,400	6,872
Rest Austro-Hungarian Empire	2,700	
Italy		13,125
Belgium	14,100	17,409
France	9,600	16,801
Russia (Commonwealth of Independent States)	5,200	85,200
Sweden	3,300	3,788
Denmark	3,000	8,550
Switzerland	2,500	3,561
Netherlands	2,200	23,828
Middle and South America	1,500	233,626
Mexico	300	68,482
Chile	400	4,200
Argentina	290	12,800
Brazil	220	85,600
Peru	15	6,100
Norway	500	
Japan	270	65,490
Smaller states of Europe	230	73,000
Serbia	70	6,000
Romania	65	14,000
China		291,000
Spain		30,677

Britain, poorer people used molasses, corn or bran for brewing, or even resorted to a kind of beer-bread [128]. John Winthrop was awarded a prize for succeeding in preparing an acceptable beer from corn. The following years saw a decrease in beer consumption, as Robert Proud in his history of Pennsylvania notes, because 'the consumption of tea, coffee and chocolate spread more and more and supplanted beer' [123]. The beverages made from domestic grain or other native raw materials were considered of inferior quality even before the War of Independence discontinued the malt supply from Britain and caused further deterioration in malt liquor quality. As a measure for business development as well as to curb hard liquor consumption, Massachusetts passed an act encouraging the manufacture and consumption of ale and beer in 1789. Around 1810, Americans still consumed 18 quarts of distilled drinks and 5 quarts of beer per head; 218 000 hl of beer were produced by 140 breweries – Pennsylvania and New York being the leading brewing centers.

In 1840, a German born brewer, John Wagner, produced the first lager beer in the United States in a tiny hut next to his house, using yeast he had brought from Bavaria [123, 124]. Increased immigration and a resulting change in the ethnic composition as well as the industrialization in the big cities had a dramatic impact on beer consumption and production. Brewing spread along the waterways and, later on, along the railways. Due to the deep-rooted ale tradition, it would take almost 40 years for lager to outsell ale. In 1881, John Ewald Siebel published an article in the *Western Brewer* concerning the use of unmalted cereals in brewing. This was a milestone in the development of beers specifically adapted to the United States taste for lighter colors, flavors and texture, and marked the beginning of a distinct American brewing technique [129]. Thus, the ensuing famous pale American lager differed significantly from European beers (Table 1.2). First, it was made not only from malt but contained corn (grits, flakes and flour), rice and corn syrups as well, since the native six-rowed barley did not yield malts of the desired properties. Enzymes were applied in brewing from 1910 on to improve raw material utilization and enhance beer stability Mashing, fermenting and storage were perfected to yield a clear, pale beer of about 3.8% alcohol by weight. The typical European decoction technique was replaced by most brewers by the American infusion mashing method to yield more 'fruity' beers. Finally, the new lager styles dominated the market; ale and porter accounting for less than 5% of beer production between 1910 and 1915 [130]. Thus, American breweries at the dawn of the twentieth century developed the technologies to make a distinct beer from any starch-containing material, that is, even at places where the traditional brewing grains cannot be grown.

Secondly, America was a relatively sparsely populated country. To reach consumers outside of the big cities made long-range transport necessary, and required very stable beers, adequate packing and sophisticated logistic networks. Thus, refrigeration technology was not only optimized for production, but also for (rail) transport. Moreover, bottled beer played an important role, accounting for approximately 30% of the total output during the period from 1910 to 1915. Standardized bottles and barrels were the norm, and the crown cork was introduced by W.

Painter in 1892. To enhance stability, elaborate filter systems, pasteurization and CO_2 management (artificial carbonization) became routine. Thus, beer production and distribution became standardized and independent of seasonal and climatic variations.

Nothing changed the American brewing scene as much as prohibition, which became general law (Volstead Act) in 1920. Although prohibition laws had been enacted in some states before and Canada had prohibited alcohol consumption since 1917, now the whole of America was affected. While 1243 breweries existed there in 1916, only 31 were left in 1933 and 756 in 1934. At first brewers tried to make a living from 'near beer' or find alternative uses for their factories, but only a few managed to diversify and bring innovative products on the market to a degree to ensure their survival. The Anheuser-Bush brewery, for instance, spent US$11.3 million on equipment for alcohol removal, diversification in yeast production, and the manufacture and marketing of alternative beverages like ginger ale [130]. However, the restraints of prohibition also resulted in innovations (e.g. in accelerating fermentation or home brewing equipment) and it had forged a powerful, consolidated brewing industry.

In summary, industrialization created the prerequisites for beer production all over the world, wherever a market had developed. Accordingly, it was not the availability of malt or suitable grains or favorable climatic conditions that were preconditions for brewing anymore, but demand and capital. Lead by the British colonies, brewing industries developed in many countries. In Japan, brewing started in 1853. By 1900, Japanese breweries manufactured bottled draft beer and Japanese beer received a Grand Prix at the Paris World Exhibition. In Rio de Janeiro, the first brewery operated in 1836 and over 900 000 hl of beer were produced in Brazil by 1913 [131]. Kulmbacher and Pilsner style beers were made at almost 4000 m above sea level in the Bolivian Andes in 1900, while Colonial ales were brewed at the same time in Tasmania [124]. This development is still going on and opens a wide scope for innovations, as native raw materials and regional preferences have increasingly to be considered.

We have encountered now almost 10 000 years history of fermented malt liquors. Based on the fundamentals established in antiquity, the cities generated the technology and organization for professional brewing. The necessity to provide a wholesome and stable beverage for sea voyages resulted in the creation of hopped beer and a flourishing beer trade along the shores of Northern Europe. The British Isles led the way from manufacture to industry, and the industrialized production of lager beer in Central Europe and North America set the stage for a global brewing industry.

As soon as man managed to pass on his most important experiences to further generations in writing, he immediately used this ability to describe brewing and the enjoyment of drinking beer. After so many years and regardless of numerous attempts to spoil this pleasure, the esteem of this sociable and wholesome beverage is the same as 10 000 years ago. Therefore, although beer has had a very eventful history so far, it will continue to be an indispensable attendant of humanity and there will always be the necessity to rewrite its history in the centuries to come.

References

1 Arnold, J.P. (1911) *Origin and History of Beer and Brewing*, Alumni Association of the Wahl-Henius Institute of Fermentology, Chicago, IL.
2 Hornsey, J.S. (2003) *A History of Beer and Brewing*, The Royal Society of Chemistry, Cambridge.
3 Unger, R. (2004) *Beer in the Middle Ages and The Renaissance*, University of Pennsylvania Press, Philadelphia, PA.
4 Nelson, M. (2005) *The Barbarian's Beverage*, Routledge Taylor & Francis, London.
5 Michel, R.H., McGovern, P.E. and Badler, V.R. (1993) The first wine and beer. *Analytical Chemistry*, **65**, A408–413.
6 Thunaeus, H. (1965) Sprachliches vom Bier, in *Jahrbuch der Gesellschaft für Geschichte und Bibliographie des Brauwesens 1965*, Institut für Gärungsgewerbe, Berlin, pp. 169–97, p. 174.
7 Egyptian inscription 2200 BC.
8 Maurizio, A. (1933) *Geschichte der gegorenen Getränke*, Verlagsbuchhandlung Paul Parey, Berlin.
9 Heun, M., Schäfer-Pregl, R., Klawan, D., Castagna, R., Accerbi, M., Borghi, B. and Salamini, F. (1997) Site of einkorn wheat domestication identified by DNA fingerprinting. *Science*, **278**, 1312–14.
10 Feldman, M. (2001) Origin of cultivated wheat, in *The World Wheat Book, A History of Wheat Breeding* (eds A.P. Bonjean and W.J. Angus), Lavoisier Publishing. Paris, pp. 1–56.
11 Simons, K., Fellers, J., Trick, H., Zhang, Z., Tai, Y., Gill, B. and Faris, J. (2006) Molecular characterization of the major wheat domestication gene Q. *Genetics*, **172**, 547–55.
12 Badr, A., Müller, K., Schäfer-Pregl, R., El Rabey, H., Effgen, S., Ibrahim, H.H., Pozzi, C., Rohde, W. and Salamini, F. (2000) On the origin and domestication history of barley (*Hordeum vulgare*). *Molecular Biology and Evolution*, **17**, 499–510.
13 McGovern, P. *et al.* (2004) Fermented beverages of pre- and proto-historic China. *Proceedings of the National Academy of Sciences of the United States of America*, **101**, 17593–8.
14 Michel, R. and McGovern, P. (1992) Chemical evidence for ancient beer. *Nature*, **360**, 24.
15 A seal from Tepe Gawra in Northern Iraq, now in the University of Pennsylvania Museum of Archaeology and Anthropology, shows two people drinking from a jar. They use bent straws characteristically for the consumption of grain-based fermented beverages.
16 Katz, S. and Voigt, M. (1986) Bread and beer. *Expedition*, **28**, 23–34.
17 Benitez, T. and Codon, A. (2003) Ethanol-tolerance and production by yeast, in *Handbook of Fungal Biotechnology*, 2nd edn (ed. D.K. Aurora), Marcel Dekker, New York, pp. 249–65.
18 Rabie, C. *et al.* (1997) Enumeration of fungi in barley. *International Journal of Food Microbiology*, **35**, 117–27.
19 Ishida, H. (2005) Two different brewing processes revealed from two ancient Egyptian mural paintings. *MBAA Technique Quarterly*, **42**, 273–82.
20 McGovern, P.E. *et al.* (1999) A funerary feast fit for King Midas. *Nature*, **402**, 863–4.
21 De Keersmaecker, J. (1996) The mystery of Lambic beer. *Scientific American*, **275** (2), 74–80.
22 'Oinos krithios' or 'oinos krithown' (types of barley wine) are mentioned by Herodotus, Aeschylos and Xenophon.
23 Van Oevelen, D., Spaepen, M., Timmermans, P., Geens, L. and Verachtert, H. (1977) Microbial aspects of spontaneous wort fermentation in the production of lambic and geuze. *Journal of the Institute of Brewing*, **83**, 356–60.
24 Civil, M. (1963) *A Hymn to the Beer Goddess and a Drinking Song: Studies Presented to A. Leo Oppenheim June 7, 1964*, Oriental Institute Press, Chicago, IL, pp. 67–89.
25 Hartman, L. and Oppenheim, A. (1950) On beer and brewing techniques in

ancient Mesopotamia. *Supplement to the Journal of the American Oriental Society*, **70** (4), 10.
26 Damerow, P. (2001) Sumerian beer: the origins of brewing technology in ancient Mesopotamia, *Proceedings of the 16th ICAF Conference 1, Andechs*.
27 Huber, E. (1926) Bier und Bierbereitung bei den Babyloniern, in *Bier und Bierbereitung bei den Völkern der Urzeit, Veröffentlichungen der Gesellschaft für die Geschichte und Bibliographie des Brauwesens*, Institut für Gärungsgewerbe, Berlin pp. 9–28.
28 Röllig, W. (1971) Das Bier im Alten Mesopotamien, in *Jahrbuch der Gesellschaft für Geschichte und Bibliographie des Brauwesens 1971*, Institut für Gärungsgewerbe, Berlin, pp. 9–104.
29 Lutz, H. (1922) *Viticulture and Brewing in The Ancient Orient*, J.C. Hinrich'sche Buchhandlung, Leipzig.
30 Zarnkow, M., Spieleder, E., Back, W., Sacher, B., Otto, A. and Einwag, B. (2006) Interdisziplinäre Untersuchungen zum altorientalischen Bierbrauen in der Siedlung von Tall Bazi/Nordsyrien vor rund 3200 Jahren. *Technikgeschichte*, **73**, 3–25.
31 *Anabasis*, **IV**, 26–27.
32 Maeir, A. and Garfinkel, Y. (1992) Bone and metal straw tip beer-strainers from the ancient near east. *Levant*, **XXIV**, 218–23.
33 Homan, M. (2002) Beer production by throwing bread into water: a new interpretation of QOH. XI, 1–2. *Vetus Testamentum* **52**, 275–8.
34 Huber, E. (1927) Die Völker unter babylonischem Kultureinfluss. Auftreten des gehopften Bieres, in *Bier und Bierbereitung bei den Völkern der Urzeit (II)*, Gesellschaft für die Geschichte und Bibliographie des Brauwesens, Berlin.
35 Death, J. (1887) *The Beer of the Bible*, Trübner, London.
36 Helck, W. (1972) Das Bier im Alten Ägypten, in *Jahrbuch der Gesellschaft für Geschichte und Bibliographie des Brauwesens 1972*, Institut für Gärungsgewerbe, Berlin, pp. 9–120.
37 Mechow, M. (1977) Wer war Gambrinus? in *Jahrbuch der Gesellschaft für Geschichte und Bibliographie des Brauwesens 1977*, Institut für Gärungsgewerbe, Berlin, pp. 120–9.
38 Maksoud, S. *et al.* (1994) Beer from the early dynasties (3500–3400 ca. BC) of upper Egypt, detected by archeochemical methods. *Vegetation History and Archeobotany*, **3**, 219–24.
39 Geller, J. (1992) From prehistory to history: beer in Egypt, in *The Followers of Horus (Egyptian Studies Association Publication 2)* (eds R. Friedman and B. Adams), Oxbow Books, Oxford, pp. 19–26.
40 Samuel, D. (2000) Brewing and baking, in *Ancient Egyptian Materials and Technology* (eds P. Nicholson and I. Shaw), Cambridge University Press, Cambridge, pp. 537–76.
41 Faltings, D. (1998) *Die Keramik der Lebensmittelproduktion im Alten Reich*, Heidelberger Orientverlag, Heidelberg.
42 Samuel, D. (1996) Investigation of ancient Egyptian baking and brewing methods by correlative microscopy. *Science*, **273**, 488–90.
43 Ishida, H. (2002) Insight into ancient Egyptian beer brewing using current folkloristic methods. *Master Brewers Association of the Americas Technical Quarterly*, **39**, 81–8.
44 Bowman, A. (1974) Roman military records from Vindolanda. *Britannia*, **5**, 360–73.
45 Audouze, F. and Büchsenschütz, O. (1992) *Villages and Countryside of Celtic Europe*, Indiana University Press, Bloomington, IN.
46 Stika, H. (1996) Traces of a possible Celtic brewery in Eberdingen-Hochdorf, Kreis Ludwigsburg, southwest Germany. *Vegetation History and Archeobotany*, **5**, 81–8.
47 Grüss, J. (1932) Zwei altgermanische Trinkhörner mit Bier- und Metresten. *Forschung und Fortschritte*, **8** (23/24), 289–91.
48 Grüss, J. (1931) Zwei altgermanische Trinkhörner mit Bier- und Metresten. *Praehistorische Zeitschrift*, **XII**, 180–7.

49 From 'St Brighid's alefeast', Irish poem from the 11th century [2].
50 Jonas: Vita Columbani I 16a, according to Poll, I. (1928) Beiträge zur Geschichte des Klosterbrauwesens, in *Gesellschaft für die Geschichte und Bibliographie des Brauwesens*, Institut für Gärungsgewerbe, Berlin, pp. 9–33.
51 Monckton, H.A. (1966) *A History of English Ale & Beer*, The Bodley Head, London.
52 Riedesel zu Eisenbach, M. (1992) 'Fortis ab invicta cruce celia sit benedicta'. Das Klosterbrauwesen in St. Gallen, in *Jahrbuch der Gesellschaft für die Geschichte und Bibliographie des Brauwesens 1991.1992*, Institut für Gärungsgewerbe, Berlin, pp. 101–21.
53 Huber, E. (1927) Bier und Bierbereitung bei den indogermanischen Völkern in Persien und am Kaukasus, in *Bier und Bierbereitung bei Völkern der Urzeit II*, Veröffentlichungen der Gesellschaft für die Geschichte und Bibliographie des Brauwesens, Institut für Braugerwerbe, Berlin.
54 Wilson, D. (1975) Plant remains from the Graveney boat and the early history of *Humulus Lupus* in W. Europe, *New Phytology*, **75**, 627–48.
55 Behre, K. (1999) The history of beer additives in Europe–a review. *Vegetation History and Archaeobotany*, **8**, 35–48.
56 Behre, K., (1983) Ernährung und Umwelt der wikingerzeitlichen Siedlung Haithabu: die Ergebnisse der Untersuchungen der Pflanzenreste. Ausgr. Haithabu 8, Neumünster.
57 Eifler, O. (1940) Das Schrifttum über die physiologischen Wirkungen des Hopfens, in *Jahrbuch für die Geschichte und Bibliographie des Brauwesens 1940*, Institut für Gärungsgewerbe, Berlin.
58 Lutterbeck, W. (1940) Das Bier am Niederrhein, in *Veröffentlichungen der Gesellschaft für die Geschichte und Bibliographie des Brauwesens*, Institut für Gärungsgewerbe, Berlin.
59 Bartscherer, A. (1967) Vom Magdeburger Bärm-Amt und dem Kampf um dessen Ende, in *Jahrbuch der Gesellschaft für die Geschichte und Bibliographie des Brauwesens 1967*, Institut für Gärungsgewerbe, Berlin, pp. 67–79.
60 Plinius, *Historia Naturalis*, **XVIII**, 26, quoted according to Arnold [1], p. 125.
61 Boyle, R. (1661) *Tentamina quaedam Physiologica, Cum Historia Fluiditatis et Firmitaris. Accessit de novo Tractatus de Absoluta Quiete in Corporibus*, 9th edn, Samuel de Tournes, Geneve.
62 Bartscherer, A. (1952) Beiträge zur Lösung der Gruitbier-Rätsel, in *Jahrbuch der Gesellschaft für die Geschichte und Bibliographie des Brauwesens 1952*, Institut für Gärungsgewerbe, Berlin, pp. 109–32.
63 Anonymous (1936) Die anlässlich des 50-jährigen Jubiläums des schwedischen Brauerbundes im Nordischen Museum zu Stockholm veranstaltete Ausstellung, in *Jahrbuch der Gesellschaft für die Geschichte und Bibliographie des Brauwesens 1936*, Institut für Gärungsgewerbe, Berlin, pp. 166–73.
64 Poll, I. (1936) Das Brauwesen des Klosters Prüfening, in *Veröffentlichungen der Gesellschaft für die Geschichte und Bibliographie des Brauwesens: Beiträge zur Geschichte des Klosterbrauwesens 1*, Institut für Gärungsgewerbe, Berlin.
65 Anonymus (1992) Lodewijk van Gruuthuse und sein Bier, in *Jahrbuch der Gesellschaft für die Geschichte und Bibliographie des Brauwesens 1991/1992*, Institut für Gärungsgewerbe, Berlin, pp. 227–8.
66 Huber, E. (1934) Bier und Brauerei im Bereich der Hanse, in *Jahrbuch der Gesellschaft für die Geschichte und Bibliographie des Brauwesens 1934*, Institut für Gärungsgewerbe, Berlin, pp. 50–66.
67 Herzog, B. (1928) *Die Konkurrenzverhältnisse im Deutschen Braugerwerbe*, Verlag Hans Carl, Nürnberg.
68 Meister, H. (1970) Ein Beitrag zur Geschichte des Hamburger Bieres, in *Jahrbuch der Gesellschaft für die Geschichte und Bibliographie des Brauwesens 1970*, Institut für Gärungsgewerbe, Berlin, pp. 92–9.
69 Hackel-Stehr, K. (1987) Das Brauwesen in Bayern vom 14. bis 16. Jahrhundert, in

Gesellschaft für Öffentlichkeitsarbeit der Deutschen Brauwirtschaft e.V., Deutschen Brauwirtschaft, Bonn.

70 Behre, K. (1986) Die Ernährung im Mittelalter, in *Mensch und Umwelt im Mittelalter* (ed. B. Herrmann), Deutsche Verlags-Anstalt, Stuttgart, pp. 74–87.

71 Rösch, M., Jacomet, S. and Karg, S. (1992) The history of cereals in the region of the former dutchy of Swabia (Herzogtum Schwaben) from the Roman to the post-medieval period. *Vegetation History and Archaeobotany*, **1**, 193–231.

72 Unger, R. (2001) *A History of Brewing in Holland*, Brill, Leiden.

73 Trum, M. (2002) Historische Darstellungen, Zunftzeichen und Symbole des Brauer- und Mälzerhandwerks, Diplomarbeit, Lehrstuhl für Technologie der Brauerei I, Technischen Universität München-Weihenstephan.

74 Dutch proverb: 'Where there are people there are customers'.

75 von Blanckenburg, C. (2001) *Die Hanse und ihr Bier*, Böhlau Verlag, Köln.

76 Pilgrim, K. (1969) Der Durst auf den Weltmeeren, in *Jahrbuch der Gesellschaft für die Geschichte und Bibliographie des Brauwesens 1969*, Institut für Gärungsgewerbe, Berlin, pp. 70–123.

77 Gaessner, H. (1938) Bier, Malz und Hopfen im deutschen und besonders hansischen Überseeverkehr von 1562bis 1657, in *Jahrbuch der Gesellschaft für die Geschichte und Bibliographie des Brauwesens 1938*, Institut für Gärungsgewerbe, Berlin, pp. 80–100.

78 Moberg, A., Sonechkin, D., Holmgren, K., Datsenko, N., Karlen. W.and Lauritzen and S. (2005) Highly variable Northern Hemisphere temperatures reconstructed from low- and high-resolution proxy data. *Nature*, **433**, 613–17.

79 Osborn, T. and Briffa, K. (2006) The spatial extent of 20th-century warmth in the context of the past 1200 years. *Science*, **311**, 841–4.

80 Landsteiner, E. (1999) The crisis of wine production in late sixteenth-century central Europe: climatic causes and economical consequences. *Climatic Change*, **43**, 323–34.

81 Braudel, F. and Spooner, F. (1967) Prices in Europe from 1450–1750, in *The Cambridge Economic History of Europe*, Vol. **IV** (eds E.H. Rich and C.H. Wilson), Cambridge University Press, Cambridge, pp. 374–486.

82 Bauernfeind, W. and Woitek, U. (1999) The influence of climatic change on price fluctuations in Germany during the 16th century price revolution. *Climatic Change*, **43**, 303–21.

83 Montarini, M. (1999) *Der Hunger und der Überfluss*, Verlag C.H. Beck, Munich, pp. 126–30.

84 Huntemann, H. (1970) Bierproduktion und Bierverbrauch in Deutschland vom 15. bis zum Beginn des 19. Jahrhunderts, Dissertation, Wirtschafts- und Sozialwissenschaftlichen Fakultät, Georg-August-Universität zu Göttingen.

85 Behringer, W. (1990) *Thurn und Taxis: die Geschichte ihrer Post und ihrer Unternehmen*, Piper-Verlag, München.

86 Thiel, U. (1998) Das Brauwesen im 16. und 17. Jahrhundert, *Mitteilungen des Freiberger Altertumsvereins*, Freiberg/SN, **81**, 4–96.

87 Unger, R. (1995) The scale of Dutch brewing, 1350–1600. *Research in Economic History*, **15**, 261–92.

88 Shakespeare, W. (1611) A Winters Tale, *Act 4, Scene III*.

89 Breeze, A. (2004) What was 'Whelsh Ale' in Anglo-Saxon England? *Neophilologus*, **88**, 299–301.

90 Anonymus (1880) A bush is a sign of an ale-house. *Notes and Queries*, **s6-II**, 467.

91 Mathias, P. (1959) *The Brewing Industry in England 1700–1830*, Cambridge University Press, Cambridge.

92 Scarlett, E. (1938) A Tudor worthy: Master Andrew Boorde of Physycke Doctour. *The Canadian Medical Association Journal*, June 1938, 588–95.

93 Stubbs, B. (2003) Captain cook's beer: the antiscorbutic use of malt and beer in late 18th century sea voyages. *Asia Pacific Journal of Clinical Nutrition*, **12**, 129–37.

94 Waife, S. (1953) 1753 lind, lemons and limeys. *The American Journal of Clinical Nutrition*, **1**, 471–3.

95 Oppenheim, M. (1896) The navy of the commonwealth, 1649–1660. *English Historical Review*, **XI**, 20–81.
96 Sumner, J. (2005) Powering the porter brewery. *Endeavour*, **29**, 72–7.
97 Nef, J. (1932) *The Rise of the British Coal Industry*, George Routledge & Sons, London.
98 Combrune, M. (1762) *The Theory and Practice of Brewing*, J. Haberkorn, London.
99 Sumner, J. (2001) John Richardson, saccharometry and the pounds per-barrel extract. *British Journal for the History of Science*, **34**, 255–74.
100 Freiherr, W. (1756) Rechtsregeln und Sprüche/ Herausgezogen aus des Wiguläus Xaver Alois Freiherrn von Kreittmayr Anmerkungen zu den bayerischen Gesetzbüchern, Durch den Magistrat München gedruckt bei J. G. Weiss 1848, chapter 9, paragraph 1752, p. 164.
101 Sedlmayr, F. (1941) Die Irrungen der Münchner Brauer und Bäcker wegen der Hefe, in *Jahrbuch der Gesellschaft für die Geschichte und Bibliographie des Brauwesens 1941*, Institut für Gärungsgewerbe, Berlin, pp. 7–36.
102 Schultheiss, W. (1978) *Brauwesen und Braurechte in Nürnberg bis zum Beginn des 19. Jahrhunderts*, Nürnberger Werkstücke zur Stadt- und Landesgeschichte, Nürnberg.
103 Sprotte, J. (2005) Von 1303/1305 bis zum Jahre 2005. 700 Jahre Nürnberger Bier, in *Jahrbuch der Gesellschaft für die Geschichte und Bibliographie des Brauwesens 2005*, Institut für Gärungsgewerbe, Berlin, pp. 87–131.
104 Letzing, H. (1995) *Die Geschichte des Bierbrauwesens der Wittelsbacher*, Verlag Dr. Bernd Wissner, Augsburg.
105 Struve, E. (1893) *Die Entwicklung des Bayerischen Braugewerbes*, Duncker & Humblot, Leipzig.
106 Anonymus (1811) Geschichte der Künste und Wissenschaften seit der Wiederherstellung derselben bis an das Ende des achtzehnten Jahrhunderts. Von einer Gesellschaft gelehrter Männer ausgearbeitet. 3rd volume, Göttingen (quoted according to [105]).
107 Heiss, Ph. (1875) *Die Bierbrauerei*, 6. Auflage, edited by V. Griessmayer, Lampart & Co., Augsburg.
108 Schönfeld, F. (1938) *Obergärige Biere und ihre Herstellung*, Verlag Paul Parey, Berlin.
109 Hartung, J. (1932) Die Brautechnik in den Vereinigten Staaten vor und nach der Einführung der Prohibition, in *Die Brauindustrie in den Vereinigten Staaten in ihrer technischen und wirtschaftlichen Entwicklung, Gesellschaft für die Geschichte und Bibliographie des Brauwesens*, Institut für Gärungsgewerbe, Berlin, pp. 101–90.
110 Hutter, K. (2001) Flußzytometrische Prozesskontrolle obergäriger Betriebshefen. *Monatszeitschrift für Brauwissenschaft*, 3/4, 48–54.
111 Fix, G. and Fix, L. (1997) *An Analysis of Brewing Techniques*, Brewers Publications, Boulder, CO.
112 Teich, M. (2000) *Bier, Wissenschaft und Wirtschaft in Deutschland 1800–1914*, Böhlau Verlag, Wien.
113 Michel, C. (1899) *Geschichte des Bieres von der ältesten Zeit bis zum Jahre 1899*, Verlagsbuchhandlung Gebrüder Reichel, Augsburg.
114 Stresemann, G. (1900) Die Entwicklung des Berliner Flaschenbiergeschäfts, in *Inauguraldissertation der Philosophischen Fakultät der Universität Leipzig*, R.F. Funke, Berlin.
115 Tappe, H. (2005) Dortmunder Bierproduktion und –ausfuhr im 19. und 20. Jahrhundert, *Jahrbuch der Gesellschaft für die Geschichte und Bibliographie des Brauwesens 2005*, Institut für Gärungsgewerbe, Berlin, pp. 49–84.
116 Starke, H. (2005) *Vom Brauerhandwerk zur Brauindustrie*, Böhlau Verlag, Wien.
117 Vaizey, J. (1960) *The Brewing Industry 1886–1951*, Sir Isaac Pitman & Sons, London.
118 Balling, C. (1843) *Die sacharometrische Bierprobe*, Borrosch & André, Prag.
119 Barnett, J. (2003) Beginnings of microbiology and biochemistry. *Microbiology*, **149**, 557–67.
120 van Cittert, P.H. and van Cittert-Eymers, J.G. (1951) Some remarks on the development of the compound microscopes in the 19th century.

121 Barnett, J. (2000) A history of research on yeasts 2: Louis Pasteur and his contemporaries, 1850 ± 1880. *Yeast*, **16**, 755–71.
122 Luers, H. and Weinfurtner, F. (1931) Die Einführung der Hefereinzucht im Brauereibetrieb, in Die Hefereinzucht in der Entwicklungsgeschichte der Brauerei, in *Gesellschaft für die Geschichte und Bibliographie des Brauwesens*, Institut für Gärungsgewerbe, Berlin, pp. 23–105.
123 Baron, S. (1962) *Brewed in America, The History of Beer and Ale in the United States*, Little, Brown and Company, Boston, MA.
124 Western Brewer (1903) *One Hundred Years of Brewing*, H. S. Rich, Chicago, IL.
125 Johann Barth & Son (2005) The Barth Report 2004/2005.
126 Schmölders, G. (1932) Die Brauindustrie in den Vereinigten Staaten von der Kolonialzeit bis zur Gegenwart, in *Die Brauindustrie in den Vereinigten Staaten in ihrer technischen und wirtschaftlichen Entwicklung*, Proceedings of the Koninklijke Nederlandse Akademie Van Wetenschappen, **54**, 73–80.

Gesellschaft für die Geschichte und Bibliographie des Brauwesens, Institut für Gärungsgewerbe, Berlin, pp. 3–100.
127 Thomann, G. (1909) *American Beer: Glimpses of its History and Description of Its Manufacture*, United States Brewers Association.
128 Acrelius, I. (1759) *History of New Sweden: Or the Settlements on the River Delaware*, Translated from the Swedish original by William M. Reynolds, Historical Society of Pennsylvania, Philadelphia, PA, 1874.
129 Ryder, S. (2000) Editorial–The World Brewing Congress, *ASBC Newsletter*, **60**, (2).
130 Hartung, J. (1932) Die Brautechnik in den Vereinigten Staaten vor und nach der Einführung der Prohibition, in Die Brauindustrie in den Vereinigten Staaten in ihrer technischen und wirtschaftlichen Entwicklung, in *Gesellschaft für die Geschichte und Bibliographie des Brauwesens*, Institut für Gärungsgewerbe, Berlin, pp. 101–90.
131 Köb, E. (1999) Wie das Bier brasilianisch wurde, in *Jahrbuch der Gesellschaft für die Geschichte und Bibliographie des Brauwesens 1998*, Institut für Gärungsgewerbe, Berlin, pp. 85–118.

2
Starchy Raw Materials

Franz Meussdoerffer and Martin Zarnkow

2.1
Introduction

Unlike most other organisms, plants are generally immobile. Therefore, they have developed several elaborate mechanisms for their propagation. Two of those are of special interest with respect to the production of fermented beverages: grass seeds and berries. While the sweet taste of sugar-containing berries fools mammals and birds into assimilating the berry and transporting its indigestible seed until shedding it in their excretions, grass seeds are disseminated by the wind. The nutrients required for germination are resorbed by the berry seeds from the decaying excrement, while the grass seeds carry storage polymers (starch and proteins) with them. The controlled conversion of those storage materials into dextrins, sugars, peptides and amino acids during malting and mashing constitutes the first steps of brewing.

The most important families (Pooideae, e.g. wheat and barley; Erhartoideae, e.g. rice; and Panicoideae, e.g. maize and sorghum) diverged from a common ancestor about 50 million years ago [1–3]. A wheat–rice comparative genomics analysis indicated that gene evolution occurs preferentially at the ends of chromosomes, driven by duplication and divergence associated with high rates of recombination [4]. The domestication of the major cereal species commenced around 7800 BC in the so-called Fertile Crescent (Israel/Jordan/Syria/Turkey/Iraq) [5]. This enabled our Neolithic ancestors to cultivate the cereals and store their grains for various use. At least at that time man devised processes to prepare more or less intoxicating beverages from cereals, and grains still constitute the principal raw material for beer, providing starch (sugars), proteins (amino acids), vitamins and trace elements for the fermenting microorganism and subsequently to the consumer. However, different cultures employ different grains for brewing and thus vary in the processes for deriving fermentable fragments from the respective grain. Exploiting inducible endogenous kernel enzymes for the breakdown of starch and proteins during malting of Pooideae might be the most common process; however, the application of mold exoenzymes in the fermentation of rice-based beverages

Handbook of Brewing: Processes, Technology, Markets. Edited by H. M. Eßlinger
Copyright © 2009 WILEY-VCH Verlag GmbH & Co. KGaA, Weinheim
ISBN: 978-3-527-31674-8

like sake or amazake or the use of technical bacterial exoenzymes for the degradation of corn polymers are also widespread. Thus, the multitude of available technologies permits us to use a variety of starchy plants for brewing. To summarize all alternatives would significantly exceed the scope of this chapter, all the more as genetically modified plants and enzymes also deserve to be considered. Therefore, this chapter is confined to a general overview of the metabolic processes during kernel formation that determine its technological properties and a special section summarizing the most common brewing cereals.

2.2
Principles of Structure and Metabolism

Due to their great importance to the survival of the human species, cereals have been extensively characterized by biological, biochemical and chemical methods. Recently, molecular biology has contributed new insights into the processes generating seeds and determining their properties. However, cereal crop species differ greatly in their DNA content, from below 400 Mb in the small rice and foxtail millet genomes to 5000 Mb in barley [6] and 17000 Mb in bread wheat [7], and the frequent genomic variations might result in different properties within a species. Thus, considering all known alternatives of genomic and biochemical processes affecting grain properties is beyond the scope of this chapter. On the other hand, cereal crop genes tend to be highly conserved at the DNA sequence level, permitting us to characterize some fundamental principles.

Three factors therefore determine the suitability of a cereal for brewing: (i) its basic anatomy as determined by the species (genome); (ii) its content of useful and undesired metabolites including storage polymers, as determined by its species-specific metabolic pathways, nutrient availability and climate; and (iii) – at least for malting grains – its capacity to degrade the storage polymers. Thus, structural and general aspects of anabolic and catabolic metabolism will be briefly summarized.

2.2.1
Grain Structure

Cereal grains are complex organs. The mature grain, sometimes surrounded by husks, contains the caryopsis, a fruit characterized by a fusion between the pericarp and seed coat. Enclosed by the seed coat (testa) are the embryo and endosperm, the latter differentiated into the outer aleurone/sub-aleurone layers and the inner starchy endosperm. The embryonic tissues are located in the basal part of the grain, and comprise the scutellum, a nursing tissue and the embryo proper (Figure 2.1). Thus, the essential parts are detailed in the following paragraphs.

(i) A more or less structured hull, including the pericarp (fruit coat) and the seed coat (primarily testa). The (outer layer) pericarp is a tough skin that protects

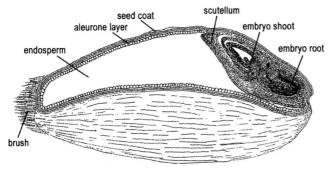

Figure 2.1 Simple diagram of the dry wheat grain showing its principle parts (from [8]).

the inner seed from soil organisms that may attack it. The inner seed coats control the intake of water by the seed. The barley hull usually constitutes about 100–130 g/kg of the grain dry weight, and consists mainly of cellulose, hemicelluloses (xylans), lignin and a smaller quantity of protein [9]. Moreover, arabinoxylans constitute a major component of the outer layers of the endosperm and the husk in barley [10].

(ii) During the development of cereal grain, endosperm cells differentiate into two different tissues: the starchy endosperm and the aleurone layer. Observations from wheat [11] and maize [12–14] have provided indications that positional signaling is involved as aleurone cells are converted to become starchy endosperm cells upon internalization.

Although they share a common origin, only the aleurone cells remain alive in the mature grain, whereas the starchy endosperm cells accumulate storage compounds and subsequently die [15]. The aleurone layer comprises a layer of specialized cells, which in the mature kernel exhibit quite distinct inner and outer walls. It covers most of the perimeter of the endosperm. In maize and wheat, there is one cell layer; in rice, one to several layers; and in barley, three layers [16]. In maize, the aleurone layer consists of an estimated 250 000 cells [17], whereas barley grains include around 100 000 aleurone cells [18]. The aleurone cell cytoplasm is densely packed with small vacuoles with inclusion bodies and grains that are completely surrounded by lipid droplets and interspersed with occasional plastids and numerous mitochondria [11, 19]. Aleurone cells may contain two major types of inclusion bodies [20]: the globoid bodies, which contain a crystalline matrix of phytin, protein and lipid, and protein–carbohydrate bodies. Toward the end of seed maturation, a specialized developmental program confers desiccation tolerance to the aleurone cells [21, 22]. After moistening and subsequent onset of germination the aleurone cells release hydrolytic enzymes to convert the storage proteins and starch granules of the starchy endosperm into sugars and amino acids for the growing embryo [23–25].

(iii) The starchy endosperm contains most of the storage biopolymers. It represents the largest body of cell mass in the kernel, and consists of an estimated 80 000–90 000 cells in barley and 60 000 cells in wheat [16, 26]. The bulk of the starchy endosperm is starch, synthesized within amyloplasts. The starchy endosperm cell walls of barley contain considerable amounts of β-glucans (80%) and arabinoxylans (20%) [27], while the cell walls of wheat have been shown to consist predominantly of arabinoxylans (70% of wall polysaccharides) and (1–3, 1–4)-β-D-glucans (20% of wall polysaccharides) [28]. Barley therefore differs from other grains in its β-glucan content in the starchy endosperm [29, 30]. This, in turn, has a markedly effect on malting quality. Recently, quantitative trait locus mapping [31] has been applied to identify gene regions in barley conferring favorable polygenic traits during malting [32], including those involving β-glucan metabolism [33].

(iv) The germ is an embryo plant, consisting of a radicle, which can grow into a root system, and a plumule, which can develop into stems, leaves and ears. The embryo contains all of the necessary modules to initiate plant growth. Like the cells of the aleurone layer, the embryo remains viable during the ripening phase. The survival has been attributed to a range of protective proteins that allow cells to tolerate desiccation. Simultaneously, developmental arrest and a state of dormancy in the embryo and aleurone layer are induced to prevent premature germination. There is good evidence that the plant hormone abscisic acid (ABA) is involved in the regulation of embryo dormancy. Transcription factors and an ABA-responsive kinase affecting dormancy have been described [34, 35]. When the embryo has grown to about half its final size it digests the endosperm cells nearest to it. These cells consequently fail to develop as normal endosperm and will become the 'crushed cell layer'. This layer has important technological functions during wheat milling (separation of 'germ' and starchy endosperm) and malting (restricting water uptake during the steeping). Part of the embryo is (v) the scutellum (seed leaf), which plays an important role in dormancy and germination. In the context of starchy endosperm utilization, the scutellar epithelium, located at the interface of the embryo and the starchy endosperm, participates in both enzyme secretion and the absorption of degradation products. The scutellar epithelium in resting grains is, like aleurone cells, packed with specialized protein bodies surrounded by lipid bodies [15]. Proteomic studies have shown that germ and aleurone exhibit distinct patterns of functional proteins in wheat [36] and barley [37].

The different parts of the kernel differ with respect to their weight, reflecting their diverse content of protein, carbohydrates and glycerides (Table 2.1). Differences between the various parts of the kernel also exist with respect to compounds of nutritional value (Table 2.2). Different grains vary of course in their content of these constituents, reflecting varying properties. Thus, the starch content of barley comprises about 65%, while maize contains up to 72%, wheat about 64% and oats around 45% [41].

2.2.2
Basic Anabolic Processes

Cereal grain development can be divided into three main stages: grain enlargement, grain filling and desiccation. Grain enlargement involves early, rapid

Table 2.1 Weight of kernel fractions and distribution of protein and starch in the wheat kernel (based on [38–40]).

Fraction of kernel	Maize	Rice	Wheat		
	% of kernel weight	% of kernel weight	% of kernel weight	% of total starch	% of total protein
Hull, pericarp, testa	5–6	1–2	8.0	0	4.5
Aleurone	2–3	5	7.0	0	15.5
Endosperm	80–85	89–91	82.5	100	72.0
Scutellum	10–12	2–3	1.5	0	4.5
Embryo			1.0	0	3.5

Table 2.2 Physiologically important nutrients in the wheat kernel (from [42]).

Fraction	Kernel (% weight)	Key nutrients
Bran	8	dietary fiber minerals: potassium, phosphorus, magnesium, iron and zinc
Aleurone layer	7	vitamins: niacin, thiamine, folate minerals: phosphorus (mainly as phytate), potassium, magnesium, iron and zinc
Endosperm	82	starch, protein, minerals
Germ: embryo	1	fats and lipids, protein and sugars
Scutellum	2	B vitamins (especially thiamine), phosphorus

division of the zygote and triploid nucleus. Cell division is followed by the influx of water, which drives cell extension. This stage spans approximately 3–20 days post-anthesis [43]. During the second stage (grain filling), cell division slows and subsequently ceases. Simultaneously, storage biopolymers are accumulated, beginning at around 10 days post-anthesis until maturity [6, 16, 44]. Grain filling is a process highly coordinated at the level of gene expression [45, 46] and enzymatic activity [47]. Figure 2.2 gives an overview over the developments.

2.2.2.1 Starch Formation
The formation of starch granules is a complex process [39, 48–53], with species-specific peculiarities resulting in distinct starch granules. When grain filling

48 | *2 Starchy Raw Materials*

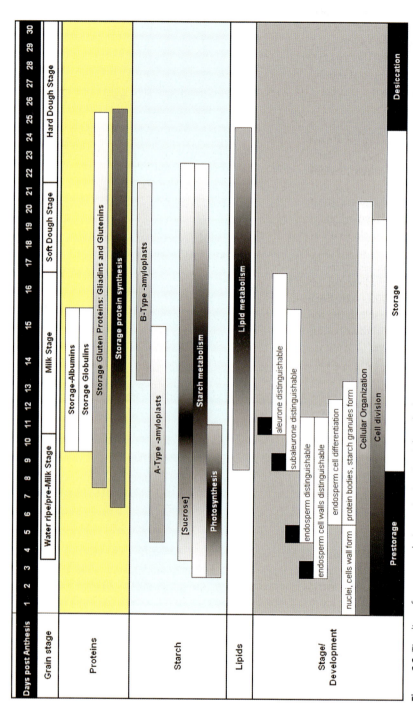

Figure 2.2 Timeline of processes during grain ripening (wheat/barley) [6, 16, 48]. Shaded areas indicate the intensity of the process from dark = high intensity to pale = low intensity.

begins, sucrose produced by leaf chloroplasts is transported in phloem to the immature grain. Sucrose is been converted in the endosperm cells to adenosine diphosphate (ADP)-glucose and transported into the starch storage organelles, the amyloplasts. The ADP-glucose units are there polymerized into the α-1–4-glucosyl chains of starch by two basic classes of starch synthetic enzymes: granule-bound starch synthases, which synthesize amylose, and soluble starch synthases, which synthesize a linear amylopectin. This linear amylopectin chain is then modified by other enzymes including branching enzymes that transfer short α-1–4-glucosyl chains of amylopectin to an α-1–6 configuration, creating the branched amylopectins. The starch-synthesizing enzymes have recently attracted much interest for their feasibility to generate 'tailored' starches by genetic engineering [54–56].

Grain starches typically consist of linear or lightly branched amylose with a degree of polymerization (DP) of 1000–5000 glucose units and the much larger branched amylopectin (DP 105–106) containing frequent α-1–6 branch points (5–6% α-1–6 linkages). In most common types of cereal endosperm starches, the relative weight percentages of amylopectin and amylose range between 72 and 82% amylopectin and 18 and 33% amylose [49, 57].

The varying activities of one or several isoform(s) of the starch biosynthetic enzymes result in the formation of starches differing with respect to amylopectin chain length distribution. These structural differences become evident from the 'growth rings' that are visible by light microscopy, and that reflect amorphous and semicrystalline regions (or domains) within the granule. Thus, starches with varying physical behavior and functionality are obtained, which differ in important aspects like starch crystallinity (A-, B- and C-types), gelatinization or enzyme digestibility [58], (Table 2.3).

At least 10 major proteins exhibiting molecular weights in the range of approximately 5–149 kDa are associated with most starches, among them starch-synthesizing enzymes. A substantial number of these proteins are located at the starch granule surface, where their presence in association with that of other minor granule components (such as lipids) appears to significantly influence the overall properties of both starch granules and starchy products [60].

In some grains, like wheat, barley rye and triticale, two granule species exist designated A- and B-type, respectively, which differ in size and physico-chemical properties [61]. While A-type amyloplasts are synthesized first in proliferating endosperm cells, B-type granules arise later in grain development (Figure 2.2) by evagination or constriction of the envelopes of existing amyloplasts in wheat [48], triticale [62] and barley [63]. The smaller B-type particles comprise about 80–97% of the total number of granules, but contribute 50–80% of the total weight of endosperm starch. The A- and B-types might also differ in their amylose content [64, 65], although environmental factors exert a strong influence on their starch qualities [66]. The difference has been attributed to preferential incorporation of starch-branching isoenzymes into A-type starch granules of grains like wheat, barley, rye and triticale that exhibit a bimodal starch granule size distribution [67].

Table 2.3 Properties of starch granules [59].

Starch species	Number of granule types	Granule size range (µm)	Amylose (%)	Gelatinization temperature (°C)
Triticale	2		0	
High amylose maize		4–22	70	
Maize		5–25	28	70–75
Sorghum		3–27	0	70–75
Wheat	2	<10 20–45	26	52–54
Barley	2	<6 10–30	28	61–62
Rice				68–75
Potato		10–70	20	56–69

The cell wall of barley endosperms differs from that of other grains in its content of glucuronoarabinoxylans carrying ester-linked ferulic acid residues and (1–3, 1–4)-β-D-glucans as the major non-cellulosic polysaccharides, and low levels of pectic polysaccharides. The synthesis proceeds via (1–3)-β-D-glucan (callose) and (1–4)-β-D-glucan (cellulose), formed between day 3 and day 4 after angiothesis. (1–3, 1–4)-β-Glucan emerges in the developing cell walls approximately at 5 days post-anthesis, while arabinoxylan deposition begins at 7 and 8 days post-anthesis [68]. The synthesis of the (1–3, 1–4)-β-glucans in the endosperm cell walls of barley and other cereals is probable mediated by cellulase-like glucan synthases encoded by members of the Csl gene family [69].

2.2.2.2 Storage Proteins

Grain proteins have been classified either according to their physico-chemical properties as albumins, globulins and prolamines or according to their function as catalysts (enzymes), structural, storage and defense proteins, respectively. The process of storage protein deposition in maturing seeds appears to be highly specialized, involving compartmentalized special membrane-enveloped protein vesicles that either directly bud from the endoplasmic reticulum as in maize or from the protein storage vacuoles as in the majority of cereals [70]. The sorting of storage proteins takes place on the messenger RNA [71, 72], vesicular [73, 74] and protein [75] levels. In particular, specific proteolytic processing plays an important role [76, 77].

The resulting protein bodies are located in the endosperm and embryo, and serve the purpose of supplying the embryo with amino acids and sulfur during germination. They account for about 50–80% of the total protein in mature cereal grains [78].

The embryo and outer aleurone layer of the endosperm of at least some cereal grains contain globulin and albumin storage proteins. In oats and rice, globulins form the major endosperm storage protein fraction, accounting for about 70–80% of the total protein. They are related to the widely distributed 'legumin'-type globulins that occur in most dicotyledonous species. However, similar proteins (triticins) are also present in wheat [29, 79], where they account for about 5% of the total grain protein.

The majority of the albumin proteins of mature grain endosperm of wheat and barley belong to a family of α-amylase/trypsin inhibitors [80, 81]. Beyond their storage function, these proteins have the capability to inhibit certain enzymes, thereby exerting protection against, for example, premature seed sprouting or pathogens [82].

Another type of protease inhibitors in grain endosperm belongs to the serpin superfamily. The corresponding barley endosperm Z protein inhibits chymotrypsin and is immunochemically related to the major protein component in beer [83]. Serpins have also been isolated from wheat, rye and oats [84].

Except for oats and rice, the major endosperm storage proteins of all cereal grains are prolamins [29]. The prolamins vary from about 10 000 to almost 100 000 in their molecular masses, and are much more variable in structure than the globulins [85]. They are characterized by unusual amino acid compositions, being particularly rich in proline and glutamine, thus affecting the overall amino acid composition of grains (Table 2.4). Furthermore, many prolamins are complex multidomain proteins, with one or more domains comprising repeated sequences based on the reiteration of one or more short peptide motifs [87]. It is likely that

Table 2.4 Content of important amino acid in different grains (% of total; values from [86]).

	Barley	Wheat	Maize	Millet	Rye
Lysine	3.0	2.5	0.0	0.10	3.0–3.5
Leucine	6.5	6.5	14.0	7.5	5.5–6.0
Isoleucine	3.5	3.5	3.0	2.5	3.5–6.0
Valine	5.0	4.0	3.5	1.5	4.0–4.5
Phenylalanine	5.0	5.0	4.0	3.0	4.5–6.0
Tyrosine	3.0	2.5	3.0	2.0	2.0–3.0
Threonine	3.5	2.5	2.0	2.0	2.5–3.0
Methionine	2.5	2.0	0.5	2.5	2.0–2.5
Glutamic acid	26.0	33.0	15.5	13.5	27.0–35.0
Proline	14.5	11.5	8.0	6.5	10.0–13.0

the major groups of prolamins in the Triticeae (wheat, barley, rye) and the Panicoideae (maize, sorghum, millets) might have separate evolutionary origins [88]. However, the prolamins from the Pooideae subfamily, such as barley (hordeins), wheat (glutenins, gliadins) and rye (secalins, gliadins), are closely related to each other.

The close evolutionary relationship is also manifested by the conservation of a common cis-acting element in the promoter region of many genes expressed during endosperm development of cereal seeds – the prolamin box (P-box) [89, 90]. A factor called P-box binding factor is assumed to modulate transcription by binding at the P-box [91].

A database compilation indicates that approximately 65% of plant food allergens originate from only four protein families. Among them the prolamine superfamily (including the albuminious inhibitors and the grain softness proteins) plays a predominant role [92, 93]. Moreover, celiac sprue, also known as celiac disease or gluten-sensitive enteropathy, is an autoimmune disease of the small intestine caused by the intake of gliadins from foods containing, for example, wheat, rye, barley and possibly oats, which affects about 0.1–1% of the world population [94]. In contrast to other foodborne proteins, due to their high proline content these gliadins release long, proteolytically stable, multivalent peptides as a result of duodenal proteolysis, which trigger the autoimmune reaction in genetically predetermined patients [95]. As neither genetically unmodified barley nor yeast has so far been shown to express a suitable prolyl-peptidase, other approaches have been taken to produce a gluten-free beer [96, 97].

2.2.2.3 Regulation of Grain Filling and Impact of Environmental Factors

As grain filling is of outmost importance for the adequacy of a grain for processing, multiple studies have been devoted to elucidate the processes involved. Moreover, the impact of environmental factors like moisture, temperature or oxygen availability has been extensively examined.

The cereal plants have an immense metabolic capability, resulting in a great flexibility to environmental stress. Thus, multiple genes encoding tissue-specific variants of a given class of enzymes might exist [46, 98, 99]. Furthermore, gene expression for the starch-synthesizing machinery is highly regulated with respect to tissue and developmental stage, as has been shown for rice [100], barley [101, 102] and wheat [103, 104]. The same holds true for the synthesis of storage proteins [80, 105]. Finally, the enzymatic activity might be modulated by the availability of energy [106], post-translational modification [107] or the presence of inhibitors [108, 109].

Apart from the plant species, environmental factors like heat or drought have a significant impact on grain properties. Multiple studies have demonstrated a key role of sugars and the plant hormone ABA in the stress response of cereals [34, 110–114]. As ABA levels increase during the seed development program or in response to environmental stress, a number of ABA-induced genes are expressed. These include the genes coding for LEA ('late embryogenesis abundant' [115]) and dehydrin [116], which may serve to protect the developing seed from desiccation, or proteins resembling known antifreeze proteins [117]. It has been shown that elevated temperatures during grain filling also have an impact on malt produced from stressed barley [118].

Another external factor influencing grain properties results from human activity. Cultivation and selection of cereals over millennia has had a significant impact on the properties and genetic diversity of grains. Maize molecular diversity, for instance, is roughly 2- to 6-fold higher than that of other domesticated grass crops. However, the diversity of maize starch-biosynthesizing enzymes has been reduced by breeding to the magnitude of wheat [119]. The advent of genetic engineering has delivered tools for selectively altering specific properties of grains (e.g. starch composition) [54, 120, 121]. Transgenic plants have also been created with alterations in storage proteins [122, 123] or insect resistance [124]. Another approach has been taken by the production of recombinant proteins in transgenic grains to enhance feed or malting properties [125]. However, so far such varieties have not yet arrived on the market for brewing cereals.

2.2.3
Catabolic Processes

The germination of grains has little technological significance for milling or if starches and proteins are utilized by means of microbial enzymes. However, the metabolic reactions and enzymes involved in germination are of great consequence for malting and mashing in the traditional brewing process. Therefore, many studies have been dedicated to the elucidation of the underlying reactions and catalysts, particularly in barley. Proteome analysis has indeed shown significant differences in the protein spectrum of mature and germinated seeds [88, 126, 127]. As the disposition for the degradation of storage polymers is predisposed by the properties of the mature seed, the relevant processes and enzymes will be briefly summarized.

By definition, germination incorporates those events that commence with the uptake of water by the quiescent dry seed and terminate with the elongation of the embryonic axis [128]. This process can be separated in two distinct phases: germination and seeding growth. Germination might be further divided as a first stage of activation for resuming metabolism including onset of respiration, amino acid metabolism and RNA synthesis, and a second stage encompassing the preparation of plant elongation. The latter stage comprises the mobilization of storage biopolymers [129].

Water initiates germination and the controlled soaking of grains is an important tool to control the progress of malting. Uptake of water by a dry seed occurs in a triphasic manner. A rapid initial uptake (imbibition) results in temporary structural perturbations, which lead to an immediate and rapid leakage of solutes and low-molecular-weight metabolites into the surrounding imbibition solution. This phase is followed by a plateau phase and further water uptake after germination is completed. Several factors influence both the uptake of water in the grain as well as its distribution within, among them grain size, and starch, β-glucan and protein contents [130]. Moreover, it has been demonstrated in the case of tobacco seeds that different seed tissues and organs hydrate to different extents, and that the endosperm of tobacco acts as a water reservoir for the embryo [131]. The loss of barley viability during storage has also been attributed to premature germination due to preferential hydration of the scutellum region [132].

One of the first changes upon imbibition is the recommencement of respiratory activity, which can be detected within minutes. It is an indicator of the resumption of metabolic processes, which in malting are subsequently controlled by the provision of oxygen (e.g. by the resting/aeration frequency).

As has been shown in a number of studies, gibberellic acid (GA) increases in cereal grains after imbibition [133]. Although it is still disputed whether GA is a necessary prerequisite for germination, it is beyond doubt that the synthesis of GA is required for the expression of hydrolytic enzymes. The precursors of GA accumulate rapidly in the scutellum and the embryo after 24 h of imbibition, and diffuse in the endosperm and the aleurone cells [134, 135]. Studies with barley suggest that aleurone cells exhibit a differential sensitivity towards GA, depending on the position within the grain and developmental factors [136].

Due to its technological significance, the processes governing GA induction of hydrolytic enzymes are being studied extensively on the genetic and molecular level [134, 137]. Common sequences in GA-responsive genes have been identified and a complex regulation by proteinacous factors suppressing or inducing transcription is being elucidated [138–141]. ABA, the hormone controlling seed dormancy, has been shown to inhibit the responses to GA as well as causing the upregulation of several ABA-responsive genes [128, 142–144]. Apart from GA and ABA, the availability of sugars modulates aleurone gene expression [145, 146].

Although GA induces enzyme formation in sensitive cells, it also elicits the programmed cell death of aleurone cells [147–150]. Thus, the application of GA in malting is, irrespective of any legal aspects, not without problems [151].

Mobilization of reserves from endosperm cells is accomplished by hydrolytic enzymes (glucanases, proteases, lipases and others) formed and secreted by the aleurone and the scutellum, and by enzymes released from starchy endosperm cells prior to grain maturation. An acidic pH (4.9–5.1) prevails in the endosperm of barley and wheat during the first stages of germination, being a prerequisite for many physiological processes and the activity of hydrolytic enzymes (as it is in mashing). However, while starchy endosperm acidification already occurs during ripening in barley, resulting from an accumulation of malic acid at late stages of grain development [152], it is initiated by GA only after imbibition in wheat and advances from the embryo to the distal part of the grain [153].

There are three major factors delimiting the extent of hydrolysis: (i) the ordered process of water uptake and distribution within the kernel, (ii) the type and amount of hydrolases formed, and (iii) the accessibility of the endosperm vesicles. All three factors vary widely with cereal type, plant variant and environmental conditions. The first step in the degradation of storage biopolymers is the perforation of the β-glucan sheath of the starchy endosperm cells, to enable the amylases, peptidases and other hydrolases to access substrates within the cell [154]. The barley glucanase enzymes have been well characterized on the protein and genomic level [155, 156]. Starch is subsequently degraded by the well-characterized families of amylases and glucosidases [157–160]. It is noteworthy, however, that in spite of structural similarities the pattern of gene expression might differ between species, as shown in the case of rice and barley [161]. Storage proteins are degraded by peptidases already present in the protein storage bodies [162] and by *de novo* synthesized proteinases [163, 164]. It is noteworthy that the cysteine and metalloproteinases are able to hydrolyze hordeins, and contribute predominantly to protein solubilization during mashing [165].

The multitude of variables influencing the composition of malt and wort, and the complexity of the processes, reveals that the most important factor for beer quality is the knowledge and experience of maltsters and brewers accumulated over centuries.

2.3
Major Brewing Cereals

2.3.1
Introductory Remarks

Presently, only two cereals are used in malted form for beer production and another three are applied unmodified worldwide. The most important consideration for brewing purposes is the carbohydrate content, which accounts for the extract as well as for coloring and flavoring.

Barley (*Hordeum vulgare*) and barley malt have been shown to be exceptionally suited for beer brewing purposes, above all due to the high content in hydrolytic enzymes in the malted form. Moreover, barley hulls have turned out to be a convenient filtering adjuvant. The advantage of barley for brewing was realized very early in human history, as convincingly documented by archeobotanical findings from Egypt and Mesopotamia [166, 167]. This early insight resulted in a specific selection over the centuries to mobilize the full potential of this brewing grain. The incentive to improve barley varieties had been further accelerated by the Bavarian Purity Law, which approved malted barley as the sole cereal source for the production of beer. Successive progress in both breeding and malting expertise optimized the starch content and enzymatic potential of the grains as well as the technological capabilities to utilize them for brewing. Malted grain is thus the single most important raw material for beer production. It is therefore useful when making predictions about the suitability of a cereal for brewing purposes to compare its malt properties with others, in particular barley malt. A compilation of the properties of different cereals malted under standard conditions is given in Table 2.5.

Today, brewing in Germany is regulated by the 'Provisional Beer Law', which is a direct successor to the Bavarian Purity Law and still maintains that bottom-fermented beer has to be brewed exclusively with barley malt. Top-fermented

Table 2.5 Properties of selected cereals and pseudocereals using a standard malting regime[a].

Attribute	Barley	Oats	Proso millet	Blue maize	Black rice	Rye	Sorghum	Spelt	Einkorn	Emmer	Kamut	Triticale	Tritordeum	Wheat	Amaranth	Buckwheat	Quinoa
Extract in d.m. dry matter (%)	82.0	64.4	63.6	60.7	87.4	89.2	73.9	84.3	85.1	88.9	74.8	88.0	81.4	86.2	79.7	52.9	83.2
Saccharification (min)	<10	<10	<25	no	no	<10	no	<10	<10	10–15	10–15	10–15	10–15	10–15	No	No	No
Color (EBC)	2.9	3.7	2.1	3.9	7.6	7.7	7.5	3.1	3.5	2.8	2.0	6.1	4.4	3.2	5.6	2.5	5.0
Protein (dry basis, %)	10.5	12.6	13.6	14.6	8.4	10.4	7.8	13.6	14.3	12.3	15.9	10.6	15.1	12.1	15.2	15.4	13.7
Viscosity (mPas)	1453	1511	1404	1407	1842	6467	1959	2034	2920	1704	1732	2219	1587	1756	1969	3507	1520
Soluble nitrogen (mg/100 g d.m.)	682	681	604	481	283	964	426	654	756	843	797	840	898	725	1022	713	888

Degree of modification (%)	40.7	33.8	27.8	20.6	21.1	57.9	34.1	32.7	33.0	42.3	31.3	49.5	37.2	37.4	42.0	29.7	40.5
Free amino nitrogen (mg/100 g d.m.)	140	145	75	119	58	121	119	97	119	114	58	123	143	110	187	111	206
Diastatic power	311	269	78	72	82	177	83	367	304	346	461	430	466	405	88	77	81
α-amylase activity (ASBC) dry basis	55	24	11	7	6	18	8	18	27	24	21	21	50	20	1	6	2
Final attenuation (%)	82.1	77.3	80.9	57.5	51.4	68.4	79.7	79.3	71.3	69.1	69.7	75.9	79.2	80.2	22.4	46.4	63.5

a The malts were prepared according the following standard procedure. Steeping: 5 h submerged at 14.5 °C; 19 h dry, 4 h submerged at 14.5 °C, 20 h dry, final degree of steeping 45%. Germination: 6 days at 15 °C. Drying/kilning: 16 h at 50 °C, 1 h at 60 °C, 1 h at 70 °C, 5 h at 80 °C. Analyses were performed with standard methods for barley malt.

The respective malts were fermented with top-fermenting yeast (yeast straim W 68; 10^7 cells/ml) at 20 °C for 1 week. Afterwards the yeast was removed and the resulting beer filled into bottles without further filtering or stabilization.

beers, however, might be produced with malts made of other cereals [168]. However, legislation maintains that rice, maize and dari are not regarded as cereals within the scope of this law. Furthermore, an annotation to the law equals dari with milo [169], thus excluding the entire category of sorghum [170, 171] from brewing. Beyond German legislation, however, various alternative starch/sugar sources are used for brewing.

There are three – mostly interconnected – incentives to use raw materials other than barley malt for brewing: (i) favorable price, (ii) availability in regions not suited for barley production or (iii) special coloring and flavoring characteristics for the creation of innovative beer types [172]. Alternative brewing materials are available in many forms. Grains in the form of grits or flakes have been well known to brewers for a long time, as these are byproducts of wet or dry milling technologies. However, the requirements on baking and brewing cereals differ significantly with regard to protein content, being in the range of 9–12% for brewing grains [173]. Thus, dehulled and germless preparations are preferentially utilized in this technology. Refined corn and rice grits are today the most important sources of brewing adjuncts worldwide (for a comparison of price/wort properties for several common adjuncts, see [174]). The recent surge in maize syrup applications has made these products economically attractive. Molasses or other products of sugar production have also been used in brewing, but these are not derived from starchy plants and therefore are outside the scope of this chapter.

The use of alternative carbohydrate sources requires appropriate enzymes for the mashing process. These can be provided by malting the respective grain or the simultaneous application of suitable malts, setting an upper limit to the ratio of unmalted adjunct added. Alternatively, exogenous enzymes can be added. It should be noted, however, that malted grains (and even malt meal) contain a selection of appropriate complementary enzymes in sufficient quantity closely attached to the corresponding carbohydrate and protein superstructure, while the use of external added proteases and glucohydrolases requires easily accessible (i.e. prepurified) starches and proteins. As grits consist predominantly of unmodified endosperm, containing native (i.e. densely packed) granules not immediately accessible to hydrolytic enzymes, they must be cooked prior to mashing to gelatinize the starch. This might prove energy consuming. Moreover, using starchy adjuncts always requires a complementation with suitable nitrogen sources to ensure proper fermentation [175]. Also, in all-malt brewer's worts, maltose accounts for around 60% of the total fermentable sugars, and glucose and maltotriose each account for roughly 20%. Adjunct carbohydrates often contain relatively more glucose. As yeast fermentation of maltose is repressed by glucose [176], fermentation patterns of all-malt and malt/adjunct mixtures might differ from each other. Another aspect of the utilization of alternative grains or adjuncts is the appropriate mash separation technology for unhulled adjuncts [177].

The Food and Agriculture Organization (FAO) of the United Nations defines 17 cereals of which 11 (wheat, barley, maize, rye, oats, millets, sorghum, buckwheat, quinoa, fonio and triticale) have a significance for brewing. Table 2.6 gives an

Table 2.6 Acreage and production of important brewing cereals 2004 (source: http://faostat.fao.org).

Cereal	Area planted (1000 ha)	Production (1000 tons)
Wheat	213 571	626 679
Maize	146 823	472 157
Rice	149 919	362 611
Barley	56 073	152 387
Sorghum	42 214	58 553
Millets	33 862	27 747
Oats	11 730	26 047
Rye	6 908	17 645

overview over the planted area and world production of the most important brewing cereals.

The ultimate criterion for the relevance of any brewing cereal and the ensuing beverage is the quality of the drink and its acceptance by the consumer. The grains described in the following section have met these requirements. It will provide an overview of the multitude of cultivated, carbohydrate-rich cereals suited for beer production from a technological point of view. In principle, any alternative to barley malt must meet its specifications. Moreover, such cereals should be economic, available in adequate quantity and quality, and readily malted with standard equipment. The malted product should comply with the common specifications for brewing materials.

2.3.2
Selected Cereals

2.3.2.1 Barley (*H. vulgare* L.)

Culturally and historically barley belongs, besides einkorn, wheat and emmer, to those varieties of grains first cultivated by humans in Europe and Asia. The oldest prehistoric residuals of barley were found on the Peloponnesus and the strata of the Pre Pottery Neolithic period at the upper Euphrates valley (between 10 000 and 7000 BC) [178]. Barley is one of the most ubiquitous cereals worldwide; Russia, Canada and the European Countries being the biggest producers. Basically, two-rowed and multirowed varieties, whose sowing and crop periods as well as their usage vary, must be differentiated. In Europe, the two-rowed variety is mostly employed for malting. Barley prospers particularly in moderate climates on fertile and profound loamy soils with good water diffusion. Of all cultivated grain varieties, barley exhibits the best adaptability and also grows well in cold, rainy climate zones with long daylight periods as at the edge of hot, dry steppes.

Barley is the major source for brewing malts, which constitute the single most important raw material for beer production. The physico-chemical properties of barley and barley malt starches and the biochemistry of its synthesis and degradation are well known [179]. In many countries outside Germany unmalted barley is used as a source of carbohydrates for beer production. Due to the abrasiveness of the kernels, fine grists must be produced, in this case by means of a hammer mill or a dispersion mill to ensure an optimal extract yield. Raw grain mashes (throw ratio usually 1:4.5) are first solubilized at boiling temperatures [180]. It could be demonstrated, however, that it is cost-effective (less energy consumption and shorter mashing times) and advantageous under quality aspects to determine the gelatinization temperature in advance [181]. Varieties with low gelatinization temperatures, which can be found among almost any cereal, should be preferred, since they require less malt or exogenous enzymes to be added. In mixtures of malted and unmalted barley, the portion of unmalted grains should not exceed 50%, as higher ratios exceed the capacity (quantity and activity) of the endogenous hydrolases in malt during the mashing process [182, 183]. Moreover, to avoid lautering problems an infusion process starting at 50 °C is typically applied and exogenous β-glucanases are added, even if only small fractions of raw grain are processed. If the ratio of malted to unmalted barley exceeds of 50%, the addition of exogenous hydrolases, such as protease, α-amylase and β-glucanase, as well as separate mashing conditions are mandatory. By increasing the addition of barley, the β-glucan concentration rises arithmetically and the viscosity increases even exponentially [182]. This necessitates the addition of enzyme adjuncts (β-glucanases) to avoid imminent lautering and filtration difficulties. Barley can also be added to the throw in a gelatinized form; however, the raw grain must be treated with IR light or microwave radiation in advance and then be milled or flaked. Extrusion (impact extrusion) is another alternative treatment, whereby the raw grain is gelatinized at pressures of 45–75 bar and temperatures of 120–160 °C. In any case, only brewing-quality barley should be used.

Another important, although often unnoticed, application of barley is its processing into color and aroma malts. On the one hand, barley can be roasted in a raw state, as happens particularly in the case of Irish stout beers. On the other hand, barley black malt or barley black malt beers can be used conformable to the Bavarian Purity Law for color adjustment. It is possible to add barley black malt during the entire brewing process until shortly before the filling. The broad variety of aroma and specialty malts has been described elsewhere [184, 185].

2.3.2.2 Oats (Avena sativa L.)
Oats probably developed from the wild form *A. fatua* [186]. First findings in Central Europe can be dated back to the Bronze Age. Oats belong to the group of secondary cultivated plants, which pre-existed as weeds in the primary cultivated plants [187]. While oats served as the predominant brewing cereal during the Middle Ages [188] and was long afterwards used to brew inferior beers [189], they have now lost their significance for brewing. Today, oats play only a minor role in world grain production; Russia, Canada and the United States being the largest producers. As a panicled grain, oats differs phenotypically from the other major grain varieties. Cultivation is possible in areas of humid and moderate climate [186].

During steeping and germinating the slim oat grain absorbs water rather quickly [190], so that a short rest and a subsequent sputter in the germination box are sufficient to achieve the desired degree of steeping [191]. The germination conditions resemble those of barley; the germs, however, are looser because of the voluptuous husks.

Oats are known for their high contents of protein, fat and β-glucan [192]. Moreover, oats exhibit significantly reduced hydrolytic activities in comparison to barley. Thus, malted oats from most contemporary varieties are only of little use for beer production. However, special oat types with a reduced β-glucan content (e.g. Duffy) are perfectly suited for brewing.

Worts from 100% oat malt are comparable with respect to their physicochemical properties with barley malt worts. Due to the high husk content, a very fast lautering is possible. The worts differ mainly by increased contents of zinc (to 0.6 mg/l), β-glucan and tryptophan [193, 194]. The beers differ, however, with respect to their sensory properties, exhibiting a distinct, oat-typical taste and good reduction properties.

Beers with a high oat ratio exhibit a typical, stable turbidity [195]. It is impossible to produce a bright-fine oat beer offhand. However, the use of oat malt might improve the turbidity stability of top-fermented beers in accordance the purity laws.

2.3.2.3 Short-Grain Millet

As a counterpart to sorghum millet (sorghum), which is botanically rather related to maize, the collective term 'millets' is used by the FAO for short-grain millet [196]. India, Nigeria and Niger produce over 60% of the world's harvest. A multitude of millets is counted in this group; however, only the most important for beer production are treated in the following sections. The properties of malted proso millet, the predominant millet in Germany, are specified in Table 2.5.

2.3.2.3.1 Pearl Millet (*Pennisetum glaucum* (L.) R. Br.)

Pearl millet evolved originally in West Africa, from where the oldest dated findings also originate (1000 BC in Mauritania). With a cultivated area of 26 million hectares, it is the economically most important short-grain type of millet. It is grown primarily in Africa south of the Sahara. The dry climate with yearly precipitations between 200 and 600 mm is favored by pearl millet and makes it the most drought-resistant grain [197].

In some African states, pearl millet malt is used for traditional opaque beers. For this purpose only short germination times are required and germination is completed mostly after 24 h. In Malawi, even raw, unmalted pearl millet is the basis for a local beer [198]. Optimal malting conditions were determined with different varieties at 25–30 °C and 3–5 days germination [199]. Different trials have established β-amylase activity levels in pearl millet malt similar to that of sorghum malt [200].

2.3.2.3.2 Foxtail Millet (*Setaria italica* (L.) P. Beauv.)

Foxtail millet originates in China. Local findings indicate its cultivation 7000–8000 years ago. It was grown for a time in Germany too (e.g. in the Rhineland and Eastern Germany), but it was completely replaced by other cereals till the beginning of the 20th century. Today, the most important foxtail millet producer is China [196].

During malting the germination period is remarkably long compared with other sorts, which frequently results in inhomogeneous germs. In terms of other properties, foxtail millet resembles proso millet (see Section 2.3.2.3.6).

2.3.2.3.3 Fonio Millet (*Digitaria exiliz*)

Fonio has a very small seed grain with 0.5 g thousand corn weight. It is the oldest of the cereals originating from Africa. Its presumed origin and today's major cultivation area is the dry African savannah. Fonio millet tolerates a wide range of soils and climate zones with yearly precipitation from 400 to 3000 mm. *Opaque beers* are brewed in Togo and Nigeria by using raw fonio millet. The low yields during fonio malt production are acknowledged. However, when mixed with malts from other millet, pleasing beers resulted [200]. After widespread sort trials performed by Nzelibe *et al.*, the optimal enzymatic activity was obtained after 4 days germination at temperatures around 30 °C. In order to reduce mold growth, soft water was mixed with formaldehyde. As with all millets, kiln temperature should never exceed 50 °C [201].

2.3.2.3.4 Teff (*Eragrostis tef* (Zucc.) Trotter)

The oldest references to teff are about 5650 years old. This cereal might have been domesticated in Ethiopia for the first time. Its name in Amharic, the official language of Ethiopia, means 'to lose' and is derived from the short sized seeds (0.3–0.4 g thousand corn weight). Today, its main cultivation area span from Ethiopia to Eritrea covering highlands areas between 1700 and 2800 m in height and a annual precipitation of 300–2500 mm. Apart from the malted grain, raw teff and unleavened teff bread is used for the production of opaque beers [202, 203].

2.3.2.3.5 Finger Millet (*Eleusine coracana* (L.) Gaertn.)

Finger millet was domesticated about 3000 BC ago in the present Central Sudan. Today, India is the world's biggest producer of finger millet. In Kenya, opaque beers are brewed using finger millet malt and unmalted maize. Sort trials aimed to select finger millet with low contents of tannin and high amylolytic activities [202]. According to these studies, some types of finger millet could reach levels of β-amylase activities that match those of barley malt.

2.3.2.3.6 Proso Millet (*Panicum miliaceum* L.)

The origin of this cereal is supposed to be in Northern China [204]. There it dominated grain cultivation before the advent of barley and wheat. In the Middle Ages, Proso millet spread out to Middle and West Europe, where it lost its importance only after the introduction of the potato [205]. The sparse cultivation areas in Germany disappeared almost completely before the beginning of World War I. Recently, attempts were made to reintroduce this cereal in Germany. Proso millet is cultivated in many arid and semiarid zones around the world. It is a short-day plant, with a preference for hot climates and exhibiting a high drought resistance. In comparison to barley and wheat, it is less demanding on the soil quality [206].

Proso millet has a very small seed diameter, so that the endosperm content is quickly consumed when germination is initiated. The absorption of water proceeds very slowly because of its relatively thick husks.

A degree of steeping over 46% and temperatures of 27 °C proved favorable for germination. The malting process is completed after 7 days. Proso millet malt showed complete saccharification as well as an acceptable final attenuation, even under standard malting conditions (see Table 2.5). The unusual but pleasant sweet odor during mashing could not be transferred into the finished beer so far. The beer was evaluated as a particular top fermentation specialty which clearly differs from the well-known beer types.

2.3.2.4 Maize (*Zea mays* L.)

Maize is the largest of all grain varieties with respect to plant height and seed size. Maize kernels exhibit a thousand corn weight up to 240 g. The oldest relics of maize were dated to 5000 BC; however, the archetype of maize has not been identified so far. Maize is grown in many countries, the United States, China and Brazil being the largest producers. There are many varieties, which are classified in the following subspecies: dent corn, flint corn, sweet corn, popcorn, waxy corn, flour corn and pod corn [207].

Maize (corn) starch is used as a substitute for malt. However, the starch must be purified beforehand (dry and wet milling). Of particular importance is the removal of the germ. The latter, which constitutes 5–14% of the maize seed weight [208], contains approximately 85% oil [209]. The separation of the sprouts takes place during milling. The resulting maize grits can be added to mashing in a similar way as described below for raw rice. Gelatinization takes place during mashing. Pre-gelatinized Maize flakes (as well as raw grains of rice and barley) are produced through steaming, rolling and drying. Apart from its competitive price [172], the application of maize flakes might shorten the mashing time significantly. Maize flakes can be added to throw in similar fractions as the other starchy grain, rice. A comparison of raw grains of maize, sorghum and barley showed no differences with respect to wort properties if the ratio of raw cereal did not exceed 5%. If higher ratios are applied, protein wort concentration decreases proportional; slightly less in the case of maize, however [210]. In principle, free amino nitrogen (FAN) should not drop under 150 mg/l (referring to 12 wt%) in raw grain worts, to ensure proper fermentation. If higher ratios of unmalted grains are to be applied, additional nitrogen sources must be added. A 10% malt addition is sufficient to liquefy maize starch at 78 °C. At higher temperatures near the boiling temperature, the hydrolytic enzymes are inactivated quickly and a retrogradation of starch takes place after gelatinization, resulting in incomplete liquefaction. Subsequently, starch is less accessible to enzymatic degradation [211]. It is noteworthy that the addition of maize derivatives has a significant impact on the sensory properties of the beer [212].

Malting maize requires temperatures around 28 °C and, compared with barley, an extended vegetation period of up to 2 days. Apart from lowering diastatic power and subsequently extract yield, the extended vegetation period also results in higher malting losses and an increased risk of mold growth. The numbers in Table 2.5 for standard malting are based on blue maize. Except for free amino acid content, the values compare unfavorably with barley malt, as also observed elsewhere [213].

2.3.2.5 Rice (*Oryza sativa* L.)

Rice is one of the world's most important commercial crop plants, ranking second only to wheat. The earliest evidence for rice cultivation originates from Southern China, about 7000 years ago. Today, the major producers are above all China followed by India and Indonesia, together growing more than half of the world's production. Rice (*Oryza*), like barley, wheat and millet, belongs to the Poaceae. *Oryza* comprises 25 different varieties, from which only two types were cultivated successfully and permanently: the African species *O. glaberrima* Steud, which is restricted to the Niger zone, and the Asian species *O. sativa* L., of which 120 000 varieties are known [214]. Rice is a very adaptable plant. Some varieties are well suited to dry-land cultivation (mountain rice), but also prosper in wet-land cultivation. Wet-land varieties, however, do prosper on dry land.

Before it is suitable for brewing, rice has to be processed. Apart from grain polishing, repeated washing of rice can decrease the fat content. This results in a decrease of the fat-derived metabolites γ-nonalactone and 1-hexanol. Both are species-specific flavor components, which become perceptible only after fermentation (for yet unknown reasons) [215]. Strongly swelling varieties like *O. japonica* are rich in amylopectin. These varieties are preferentially used unmalted as a carbohydrate source in brewing as they bring about a quick and thorough gelatinization of starch.

Due to its higher gelatinization temperatures, the wort production in this case resembles a 'boiling of raw grain'. On the other hand, lower mashing temperatures can be applied for low gelatinizing rice varieties, similar to those described for barley and triticale. Beers with a high content of unmalted rice exhibit a higher colloidal stability.

Malted rice, obtained after 3 days wet steeping, 3–7 days germination and subsequent air-drying, is used for a local beer in Nagaland/north-east India [216]. Due to the great variety the trial (malt results, see Table 2.5) series were initially limited to one sort with a special characteristic. In this case black rice was chosen, as it shows a red coloration only achievable by berries because of its enormous fraction of anthocyanin in a sour medium (50-fold in cold extraction in comparison with Pilsner barley malt). The malting of the peeled material is relatively trouble-free. However, the germination temperature and the growth of fungi must be coordinated, a situation which can be achieved at 19 °C and 5 vegetation days. Standard malt is characterized by its low protein content; however, the extract can be transferred to a large extent into wort and brings along a relatively high ratio of fermentable sugar. The beers have a particular and pleasing flavor, which cannot be compared with the usual rice flavors. They have also an intensive red color.

2.3 Major Brewing Cereals

2.3.2.6 Rye (*Secale cereale* L.)

Rye is a very undemanding grain of the northern hemisphere and, culturally and historically, a young cereal. The oldest precursors of rye originate in the fertile crescent from stone-aged settlements in Northern Syria and Turkey (approximately 6600 BC) [217]. It was the most abundant grain in Northern Europe for almost 1000 years, before it was replaced first by wheat than by barley during the twentieth century. Today, the world's largest producers are Poland, Germany and Russia.

Rye malt as produced by standard methods is characterized by very high viscosity, which can be attributed to the high content of pentosane. Therefore, oxygenation must be strictly avoided to ensure unproblematic lautering [218]. Moreover, it is extremely difficult to filter rye beers bright-fine. The deep color of rye malts is carried on to the final product. Rye malt results in a pleasing, unfiltered, dark, top-fermented specialty beer.

Recently, roasted rye specialty malts have become available. Due to their slightly elevated colors [500–800 European Brewery Convention (EBC) units], they have a similar application range as roasted spelt and barley malts.

2.3.2.7 Sorghum (*Sorghum bicolor* L.)

Sorghum probably originates from Ethiopia, and is grown today in many areas in Africa, India, South-East Asia, Australia and the United States. The date of its first cultivation is unclear (approximately 3000–5000 years old). It is an aridity-resistant plant, which even grows in areas that are too dry for maize [219].

The United States, Nigeria, India and Mexico are the biggest producers of sorghum. Sorghum has traditionally been used for brewing opaque and lager beers in Africa as a source of both diastatic malt and adjuncts. In Mexico, where sorghum is second only to maize with respect to area planted, sorghum grits have been considered and substituted for corn and rice grits.

Favorable malting parameters are achieved at temperatures between 24 and 26 °C [220], which facilitate the growth of molds, however. Addition of a 0.375% mixture of borax and boric acid to soft water reduced the germ number of mold significantly [221].

Sorghum malts differ from barley malts in their higher gelatinization temperatures and reduced diastatic power, which can be attributed to lower β-amylase activities, while α-amylase activity at least equals that of barley malt [222]. They are characterized by very low protein contents that might even decrease further during storage [223] and result in diminished concentrations of soluble protein in the wort. Although the raw protein content of sorghum compares to that of barley, it can be only partially degraded into smaller fragments during mashing (as the prolamines of the Panicoideae subfamily differ from that of the Pooideae subfamily, see Section 2.2.2). Accordingly, the amino acid composition of wort polypeptides differs between barley and sorghum malts [224]. However, FAN levels are in the normal range. A minimum of 5 germination days is required during malting of sorghum to ensure sufficient soluble nitrogen concentrations in the wort [221]. This extension of the germination combined with a significantly increased degree of steeping results in elevated levels of dissolved pentosans. Therefore, the viscosity of sorghum-derived worts is slightly elevated as compared with barley malts. Moreover, sorghum malts exhibit only low diastatic activities, resulting in long saccharification times [225].

Unmalted sorghum ranks third among barley malt surrogates behind maize and rice. Sorghum grits are usually obtained by a dry-milling procedure, in which the cleaned grain is first decorticated, degermed and reduced in particle size. The color of worts containing unmalted sorghum is bright. Sorghum grains vary widely with respect to secondary plant metabolites. Thus, it has been reported that tannin and polyphenol-rich sorts are not suited for the beer production [226]. However, amylopectin-rich types are used for the manufacture of alcohol-free beers, since the worts derived thereof contain a considerable fraction of non-fermentable dextrins [227]. Such varieties have relatively high gelatinization temperatures, making the use of exogenous enzymes mandatory [228, 229].

Sorghum grits are particularly used for brewing in Mexico [230] and Nigeria. Owing to an import ban for malt to Nigeria in 1988, the local breweries had to adapt to raw sorghum. This 'bondage' resulted in intensified research on sorghum-based brewing technologies [231]. Consequently, sorghum beers have been firmly established in the markets, even after malt imports resumed. Brewing with sorghum requires a 'raw grain boiler' and a mash cooler. Moreover, the lack of husk necessitates a mash filter for lautering.

2.3.2.8 Wheat Group (*Triticum* L.)

It is now established that the evolution of wheat varieties was not a continuous and directed process, but occurred at several places, at different times in a similar manner (spontaneous crossing, mutations). The cultivation began in the Neolithic period. In the course of approximately 10 000 years of adjustment processes ('domestication') cultivated plants were developed from manifold wild forms through selection by humans. Wheats are by far the most important cereal worldwide; China, India and the United States being the biggest producers.

2.3.2.8.1 Spelt (*Triticum aestivum* ssp. *spelta* Thell. = *Triticum spelta* L.)
Spelt, a hexaploid spelt wheat, is the next relative to our bread wheat [232]. Its cultivation differs from wheat, since spelt is undemanding with respect to soil and climate, and grows at altitudes where wheat no longer prospers. Spelt originates from Asia, where it was grown at least 3000 years ago. In the Middle Ages it was cultivated in Switzerland, Tyrol, Baden-Württemberg and Middle Franconia. Spelt was then called 'Schwabenkorn' (Swabia grain) because Swabia was the center of spelt cultivation.

The standard malting dates in Table 2.5 show that spelt malt resembles wheat malt. This applies also for the entire production process of wort and beer. Adding 70% spelt malt to throw proved to be technologically unproblematic and resulted in pleasing top-fermented beers, full of character.

Roasted spelt malt is not as color intensive (450–650 EBC) as the standard roasted barley malt and results in somewhat lighter beers. Its addition (maximum

5%) should take place during the final mash pumping, so further thermal stress can be avoided. Dehusking of the grain adds to smooth flavor of the beer.

2.3.2.8.2 Einkorn (*Triticum monococcum* L.)
Einkorn is a delicate cereal with only one grain on each side of the ear. This diploid husked grain is related to bread wheat and emmer [233]. The oldest findings originate from the Pre-Pottery Neolithic period in Northern Syria. Today, cultivation of einkorn is restricted to south-west Germany and nearby Switzerland [234].

Handling and malting of einkorn is difficult as great caution must be taken to preserve the very sensitive embryo as a prerequisite for maltibility. Representative results for a malt obtained from a German variety of einkorn are provided in Table 2.5. It is remarkable that soluble proteins as well as amino nitrogen are below average, although the protein content of the raw grain is within a normal range. The final attenuation is very low relative to the extract. These disadvantages can be corrected by higher germination temperatures at 20 °C under isothermal conditions. Worts prepared from einkorn malt exhibit standard properties, and the resulting beers show excellent foam stability and a distinct pleasing carbonation taste.

2.3.2.8.3 Emmer (*Triticum dicoccum* Schübl.)
The oldest findings of emmer wheat originate from the Near East about 8000 BC. A ubiquitous cereal in Neolithic times, it begun slowly to disappear in Europe during the Bronze Age. Emmer survived in south-west Germany and has recently come of age again [235]. It is a tetraploid spelt wheat with slim but compact ears. The diverse coloration (white, red and black) still characterizes the different rural types today. Unhusked glumes are sowed here too. The germination power according to Schönfeld showed satisfying values in the case of dehusked grains also.

The dehusked emmer malt shown in Table 2.5 is characterized by a very high extract yield and good saccharification times, but weak final attenuation. The protein content is within the normal range. Wort and beer production show no peculiarities. The slightly sour, tangy note of this pleasing top-fermented beer type was praised in the sensory evaluation.

2.3.2.8.4 Tetraploid Durum (e.g. *Triticum durum* L., *Triticum turgidum* L. and *Triticum polonicum* L.)
The tetraploid wheat types of the emmer series have 14 chromosome pairs and an alloploid genome – a trait revealing their hybrid character. Among the varieties cultivated, the mature grains remain husked only in the case of emmer, all others classify as durum. It is not clear yet whether the sporadic wheat findings from stone-aged strata in Near-East and Middle Asia are to be assigned to *T. durum* or *T. aestivum*. Today's cultivation areas comprise the Mediterranean, Middle Asia and the United States [236].

The tetraploid kamut, which originates from the Nile region, was used for comparative studies. It has exceptional long grains, which exhibit an irregular solution when malted. Therefore, the extract value is low as well as the degree of modification, color and α-amino nitrogen, although the protein content in the unmalted grain is high. The diastatic power is particularly high. The beers have a distinct sensory character.

2.3.2.8.5 Triticale (×*T. riticosecale* Wittmack) Triticale is a new grain variety, which was bred in the last century. It is a second-generation hybrid of tetraploid and/or hexaploid wheat as female crossing partner and diploid rye as male crossing partner. Since the resulting offspring is sterile, the chromosome sets of the plants are treated with colchicines, an alkaloid of the autumn crocus. This treatment results in an artificial chromosome doubling, thus rendering the embryos fertile. Triticale combines the yields and qualities of wheat with the hardiness, modesty and resistance to diseases of rye. Winter sorts are cultivated almost exclusively in the moderate climate zones, while the summer varieties are preferred in the tropics and subtropics. Poland, France, countries of the former USSR, Australia, Portugal, the United States, Brazil and Germany are major areas of cultivation [237].

The malting qualities of triticale, which is preferentially used for feed, have been extensively evaluated [238, 239]. The basic properties of Triticale malt resemble those of rye malt. Color and especially viscosity were not too high (see Table 2.5). It is noteworthy that the diastatic power was high, whereas α-amylase activity was below average. Unlike rye malt, triticale malt causes a permanent turbidity in the finished beer. The top-fermented beers exhibit a distinct flavor.

The application of unmalted triticale for wort production was described recently [240, 241]. If not boiled, the unmalted triticale exhibits an exceptional high diastatic power that can be exploited in mashing. Moreover, some varieties have such a low gelatinization temperature that a mashing is possible without prior boiling.

2.3.2.8.6 Tritordeum (Hexaploid) At the end of the 1970s, tritordeum was crossed from the wild barley *Hordeum chilense* and durum wheat. With respect to morphological, agronomic and bread characteristics, this cereal resembles closely wheat [242].

The standard malt excels through its enormous high diastatic power and the highest α-amylase activity among all cereals tested (see Table 2.5). In addition, the relationship to wheat is noticeable here, although the viscosity is 3% lower.

2.3.2.8.7 Wheat (*Triticum aestivum* L.) Wheat has a hexaploid chromosome set such as spelt. However, it differs from that grain in being unhulled. Today, it is the most important cereal worldwide. Wheat can be traced back to the beginnings of agriculture, as findings in the Near-East and Middle Asia (approximately 7800 BC) show. In Germany, wheat has displaced rye as the preferred bread grain since the 1940s [236].

Wheat malt is the raw material for the production of wheat beer in Southern Germany, in particular Bavaria. Standard properties of wheat malts are given in Table 2.5.

Unmalted wheat has a long tradition as a brewing material in Belgium. The Belgian wheat beers (witbier or bière blanche) as well as the spontaneous fermented beers of the lambic type contain raw wheat. Up to 50% unmalted wheat is added to the throw of these beers. Since wheat has a relatively low gelatinization temperature, barley malt hydrolases can degrade wheat biopolymers without prior boiling [243]. In the case of lambic beers it is even desired that dextrins and starch are still available for the overlapping distinct fermentations of a wild microflora (e.g. *Brettanomyces bruxellensis*, *Brettanomyces lambicus*, *Saccharomyces* spp., *Kloeckera* spp., *Pedioccocus* spp., *Lactobacillus* spp.) [244].

Even standard wheat bread flour can be applied to the mash. Pre-gelatinized wheat can also be added to throw in a popped (at temperatures over 260 °C) and milled form. While this adjunct served primarily as a flavor enhancer in the past, it is applied today to elevate wort extract [245].

Recently, roasted wheat malts have received attention as alternative coloring and flavoring components. With a color of 800–1200 EBC, up to 5% can be added to the throw of dark top-fermented specialty beers.

2.3.3
Pseudocereals

2.3.3.1 Grain Amaranth (Main Sorts: *Amaranthus cruenteus*, *Amaranthus hypochondriacus* and *Amaranthus caudatus*)

Amaranth belongs to the genus of *Celosia*. *This* cereal originates in the tropics and prefers warm climates. The first findings date it back to 6700 BC. Some types are also cultivated in Europe. Centers of breeding are South America, China, Africa, India and the United States. *A. hypochondriacus*, the variety with the best flavor characteristics, originates from Mexico, and is late ripening and frost-sensitive [246].

A. hypochondriacus is further noted for its high lysine content, which is 2–3 times as high as that of barley and wheat. It increases even more during germination [247]. The exceedingly low thousand corn weight of 0.6 g is a distinctive feature that should be particularly kept in mind if used for malting. Germination should take place on a germination floor and sieve floors from steep tank to kiln must be adjusted to the very small grain diameter. Owing to of the large specific corn surface, the quick and high water absorption must also be considered. The amount of throw is 80% lower than in the case of barley. The temperatures during steeping and germination can exceed 30 °C; however, the germs must be washed sufficiently in order to prevent molding.

The malted grains show a slight red coloration, which contributes to a coloration if applied in fermented foods [248]. Consequently, Congress wort is colored slightly pink. There is sufficient soluble protein and FAN in the malt (see Table 2.5), but it lacks any detectable amylolytic activity and consequently exhibits the lowest final attenuation of all on record.

In a pre-gelatinized form, starch from amaranth cereal can be added up to 20% to the barley malt throw without any problem [249].

2.3.3.2 Buckwheat (*Fagopyrum esculentum* Moench)

Buckwheat is a Polygonum plant and is in no way related to either the beech tree or wheat. Its name is derived from its husked seeds, which resemble the beech nut, and from its bright flour, which resembles wheat flour. Despite an annual production of 3.2 million tons, Zeller maintains that buckwheat ranks among the most neglected plants worldwide [250]. Buckwheat presumably originated from southern China and Tibet, from where it spread along the trade routes and reached Japan very early (approximately 5000 BC). In Germany, it is first mentioned in 1396 in documents of a Nuremberg archive and came into wider use during the nineteenth century, and it is sporadically planted as interlining or green fodder. Today's major cultivation areas are China, Russia, Ukraine and Poland. During steeping of buckwheat care must be taken to avoid the frequent bursting of the unhulled buckwheat grains and the resulting sliming of starch. Short steeping times are recommended as water is absorbed very quickly. Maximal β-amylase activity is achieved at a germination temperature of 20 °C [251]. During malting this cereal develops a particular nutty flavor, which partially resembles that of pistachio. This flavor is carried over to the finished beer. The green malt as well as the kilned malt must be transported with utmost care to avoid smutting and/or crumbling of the grain. Due to their very high viscosity, buckwheat congress wort and mash cannot be easily filtered [252]; however, this problem is overcome by an upstream separation. Despite these technological obstacles and the mediocre characteristics of the standard malt, the absence of gluten and the presence of valuable secondary metabolites make buckwheat potentially an attractive brewing cereal. For instance, its content of rutin, a glycoside of the quercetin group with distinctive antioxidative effects, as well as other tanning agents could be increased 7-fold by altering of the malting parameters [253, 254].

2.3.3.3 Quinoa (*Chenopodium quinoa* Willd.)

Quinoa, a minor cereal, belongs to the Goosefoots (genus Chenopodium). It originates from the Andean region in South America, where it was first cultivated in Peru 7000 years ago; quinoa played an important role in the nutrition of the Inca civilization. It is currently used for food and to make *chicha* (a fermented beverage). Quinoa is frost-resistant and gives satisfactory yields even in regions of low rainfall. Moreover, quinoa tolerates high altitudes and cultivation areas extend to altitudes of 4000 m above sea.

For handling, a gentle peeling of the seeds is very important, so that the saponins, which are located in the coating, can be separated without hurting the germ. By draining repeatedly the steeping water, further saponins are removed. Quinoa germination is initiated very quickly and intensively even at temperatures of about 8 °C. Germination can therefore be completed after 3–4 days vegetation time. The standard malt shows good extract yields and protein solution properties. The red color of the mash can be attributed to the β-cyanin pigment (like amaranth) [255], as well as to the high content of zinc [256]. Quinoa malt has the highest final attenuation of all pseudocereals.

2.4
Concluding Remarks

Fermenting (i.e. modification by microbial transformation) has been used by humans for millennia to produce wholesome and stable foods from cereals. However, while only two cereals – wheat and rye – are suited to the preparation of leavened bread, many more are used for producing alcoholic beverages. A recent survey [257] lists 10 different such beverages produced from cereals in Africa, 13 from Asia and 22 from Latin America. This list might be complemented by the many beer styles produced in Europe and the United States. This indicates that the full potential of cereals for the production of beverages has not been tapped so far. As health considerations (gluten-free, vitamin content, trace elements) become a decisive aspect for the consumer and 'ethnic food' comes of age, even conservative brewers and malsters should become interested in the options presented by common and hitherto unknown cereals.

References

1 Kellogg, E.A. (2001) Evolutionary history of the grasses. *Plant Physiology*, **125**, 1198–205.

2 Guo, H. and Moose, S.P. (2003) Conserved noncoding sequences among cultivated cereal genomes identify candidate regulatory sequence elements and patterns of promoter evolution. *Plant Cell*, **15**, 1143–58.

3 Kellogg, E.A. and Bennetzen, J.L. (2004) The evolution of nuclear genome structure in seed plants. *American Journal of Botany*, **91**, 1709–25.

4 See, D.R., Brooks, S., Nelson, J.C., Brown-Guedira, G., Friebe, B. and Gill, B.S. (2006) Gene evolution at the ends of wheat chromosomes. *Proceedings of the National Academy of Sciences of the United States of America*, **103**, 4162–7.

5 Heun, M., Schaefer, R., Klawan, D., Castagna, R., Accerbi, M., Borgi, B. and Salamini, F. (1997) Site of einkorn wheat domestication identified by DNA fingerprinting. *Science*, **278**, 1312–14.

6 Wobus, U., Sreenivasulu, N., Borisjuk, L., Rolletschek, H., Panitz, R., Gubatz, S. and Weschke, W. (2005) Molecular physiology and genomics of developing barley grains. *Recent Research Developments in Plant Molecular Biology*, **2**, 1–29.

7 Devos, K.M. and Gale, M.D. (2000) Genome relationships: the grass model in current research. *Plant Cell*, **12**, 637–46.

8 Edwards, K., Jellis, G., Lenton, J., Shepherd, S., Barker, G. and Parker, J. Wheat: the big picture, www.wheatbp.net (accessed 23. 01. 2009).

9 Andersson, A.A., Elfverson, C., Andersson, R. and Aman, P. (1999) Chemical and physical characteristics of different barley samples. *Journal of the Science of Food and Agriculture*, **79**, 979–86.

10 Egi, A., Speers, R.A. and Schwarz, P.B. (2004) Arabinoxylans and their behavior during malting and brewing. *Master Brewers Association of the Americas Technical Quarterly*, **41**, 248–67.

11 Morrison, I.N., Kuo, J. and O'Brien, T.P. (1975) Histochemistry and fine structure of developing wheat aleurone cells. *Planta*, **123**, 105–16.

12 Becraft, P.W. and Asuncion-Crabb, Y. (2000) Positional cues specify and maintain aleurone cell fate in maize endosperm development. *Development*, **127**, 4039–48.

13 Shen, B., Li, C., Min, Z., Meeley, R.B., Tarczynski, M.C. and Olsen, O.A. (2003) sal1 determines the number of aleurone

13. cell layers in maize endosperm and encodes a class E vacuolar sorting protein. *Proceedings of the National Academy of Sciences of the United States of America*, **100**, 6552–7.
14. Gruis, D., Guo, H., Selinger, D., Tian, Q. and Olsen, O.A. (2006) Surface position, not signaling from surrounding maternal tissues, specifies aleurone epidermal cell fate in maize. *Plant Physiology*, **141**, 898–909.
15. Fincher, G.B. (1989) Molecular and cellular biology associated with endosperm mobilization in germinating cereal grains. *Annual Review of Plant Physiology and Plant Molecular Biology*, **40**, 305–46.
16. Olsen, O.-A. (2001) Endosperm development: cellularization and cell fate specification. *Annual Review of Plant Physiology and Plant Molecular Biology*, **52**, 233–67.
17. Walbot, V. (1994) Overview of key steps in aleurone development, in *The Maize Handbook* (eds M. Freeling and V. Walbot), Springer, New York, pp. 78–80.
18. Kvaale, A. and Olsen, O.A. (1986) Rates of cell division in developing barley (*Hordeum vulgare* var. *distichum* cultivar Bomi) endosperms. *Annals of Botany*, **57**, 829–34.
19. Buttrose, M. (1963) Ultrastructure of the developing aleurone cells of wheat grain. *Australian Journal of Biological Sciences*, **16**, 768–74.
20. Jakobsen, J.V., Knox, R.B. and Pyliotis, N.A. (1971) The structure and composition of aleurone grains in the barley aleurone layer. *Planta*, **101**, 189–209.
21. Hoecker, U., Vasil, I.K. and McCarty, D.R. (1995) Integrated control of seed maturation and germination programs by activator and repressor functions of Viviparous-1 of maize. *Genes and Development*, **9**, 2459–69.
22. Lane, B.G. (1991) Cellular desiccation and hydration: developmentally regulated proteins, and the maturation and germination of seed embryos. *FASEB Journal*, **5**, 2893–901.
23. Russell, L.J. (1971) Fractionation of the enzymes of the barley aleurone layer: evidence for a soluble mode of enzyme release. *Planta*, **103**, 95–109.
24. Jacobsen, J.V. and Varner, J.E. (1967) Gibberellic acid-induced synthesis of protease by isolated aleurone layers of barley. *Plant Physiology*, **42**, 1596–600.
25. Jones, R.L. and Jacobsen, J.V. (1991) Regulation of synthesis and transport of secreted proteins in cereal aleurone. *International Review of Cytology*, **126**, 49–88.
26. Berger, F. (2003) Endosperm: the crossroad of seed development. *Current Opinion in Plant Biology*, **6**, 42–50.
27. Hrmova, M., Banik, M., Harvey, A.J., Garrett, T.P., Varghese, J.N., Hoj, P.B. and Fincher, G.B. (1997) Polysaccharide hydrolases in germinated barley and their role in the depolymerization of plant and fungal cell walls. *International Journal of Biological Macromolecules*, **21**, 67–72.
28. Guillon, F., Tranquet, O., Quillien, L., Utille, J.-P., Ordaz Ortiz, J.J. and Saulnier, L. (2004) Generation of polyclonal and monoclonal antibodies against arabinoxylans and their use for immunocytochemical location of arabinoxylans in cell walls of endosperm of wheat. *Journal of Cereal Science*, **40**, 167–82.
29. Miller, S.S. and Fulche, R.G. (1974) Distribution of $(1-3),(1-4)$-β-D-glucan in kernels of oats and barley using microspectrofluorometry. *Cereal Chemistry*, **71**, 64–8.
30. Chandra, G.S., Proudlove, M.O. and Baxter, E.D. (1999) The structure of barley endosperm. An important determinant of malt modification. *Journal of the Science of Food and Agriculture*, **79**, 37–46.
31. Korstanje, R. and Paigen, B. (2002) From QTL to gene: the harvest begins. *Nature Genetics*, **31**, 235–6.
32. Gao, W., Clancy, J.A., Han, F., Jones, B.L., Budde, A., Wesenberg, D.M., Kleinhofs, A. and Ullrich, S.E. (2004) Fine mapping of a malting-quality QTL complex near the chromosome 4H S telomere in barley. *Theoretical and Applied Genetics*, **109**, 750–60.

33 Han, F., Ullrich, S.E., Chirat, S., Menteur, S., Jestin, L., Sarrafi, A., Hayes, P.M., Jones, B.L., Blake, T.K., Wesenberg, D.M., Kleinhofs, A. and Kilian, A. (1995) Mapping of β-glucan content and β-glucanase activity loci in barley grain and malt. *Theoretical and Applied Genetics*, **91**, 921–7.

34 Johnson, R.R., Wagner, R.L., Verhey, S.D. and Walker-Simmons, M.K. (2002) The abscisic acid-responsive kinase PKABA1 interacts with a seed-specific abscisic acid response element-binding factor, TaABF, and phosphorylates TaABF peptide sequences. *Plant Physiology*, **130**, 837–46.

35 Bailey, P.C., McKibbin, R.S., Lenton, J.R., Holdsworth, M.J., Flintham, J.E. and Gale, M.D. (1999) Genetic map locations for orthologous Vp1 genes in wheat and rice. *Theoretical and Applied Genetics*, **98**, 281–4.

36 Mak, Y., Skylas, D.J., Willows, R., Connolly, A., Cordwell, S.J., Wrigley, C.W., Sharp, P.J. and Copeland, L. (2006) A proteomic approach to the identification and characterisation of protein composition in wheat germ. *Functional and Integrative Genomics*, **6**, 322–37.

37 Finnie, C. and Svensson, B. (2003) Feasibility study of a tissue-specific approach to barley proteome analysis: aleurone layer, endosperm, embryo and single seeds. *Journal of Cereal Science*, **38**, 217–27.

38 Food and Agriculture Organization of the United Nations (1992) *Maize in Human Nutrition, FAO Food and Nutrition Series, No. 25*, FAO, Rome.

39 Spurway, R.A. (1988) How is grain protein formed? in *The Impact of Change – The Need to Adapt: Riverina Outlook Conference 1988*, The Regional Institute, Gosford, NSW.

40 Wadsworth, J.I. (1994) Degree of milling, in *Rice Science and Technology* (eds W.E. Marshall and J.I. Wadsworth), Marcel Dekker, New York, pp. 139–79.

41 Waldo, D.R. (1973) Extent and partition of cereal grain starch digestion in ruminants. *Journal of Animal Sciences*, **37**, 1062–74.

42 Orth, R.A. and Shellenberger, J.A. (1988) Origin, production, and utilization of wheat, in *Wheat Chemistry and Technology*, 3rd edn (ed. Y. Pomeranz), American Association of Cereal Chemists, St Paul, MN, pp. 1–14.

43 Briarty, L.G., Hughes, C.E. and Evers, A.D. (1979) The developing endosperm of wheat – a stereological analysis. *Annals of Botany*, **44**, 641–58.

44 Bewley, J.D. and Black, M. (1994) *Seeds: Physiology of Development and Germination*, Plenum Press, New York.

45 Zhu, T., Budworth, P., Chen, W., Provart, N., Chang, H.-S., Guimil, S., Su, W., Estes, B., Zou, G. and Wang, X. (2003) Transcriptional control of nutrient partitioning during rice grain filling. *Plant Biotechnology Journal*, **1**, 59–70.

46 Drea, S., Leader, D.J., Arnolda, B.C., Shawa, P., Dolana, L. and Doonana, J.H. (2005) Systematic spatial analysis of gene expression during wheat caryopsis development. *Plant Cell*, **17**, 2172–85.

47 James, M.G., Denyer, K. and Myers, A.M. (2003) Starch synthesis in the cereal endosperm. *Current Opinion in Plant Biology*, **6**, 215–22.

48 Jenner, C.F., Ugalde, T.D. and Aspinall, D. (1991) The physiology of starch and protein deposition in the endosperm of wheat. *Australian Journal of Plant Physiology*, **18**, 211–26.

49 Buleon, A., Colonna, P., Planchot, V. and Ball, S. (1998) Starch granules: structure and biosynthesis. *International Journal of Biological Macromolecules*, **23**, 85–112.

50 Guan, H.P. and Keeling, P.L. (1998) Starch biosynthesis: understanding the functions and interactions of multiple isozymes of starch synthase and branching enzyme. *Trends in Glycoscience and Glycotechnology*, **10**, 307–19.

51 Smith, A. (2001) The biosynthesis of starch granules. *Biomacromolecules*, **2**, 335–41.

52 Ball, S.G. and Morell, M.K. (2003) From bacterial glycogen to starch: understanding the biogenesis of the plant starch granule. *Annual Review of Plant Biology*, **54**, 207–33.

53 Tetlow, I.J., Morell, M.K. and Emes, M.J. (2004) Recent developments in

understanding the regulation of starch metabolism in higher plants. *Journal of Experimental Botany*, **55**, 2131–45.
54 Slattery, C.J., Kavakli, I.H. and Okita, T.W. (2000) Engineering starch for increased quantity and quality. *Trends in Plant Science*, **5**, 291–8.
55 Fujita, N., Kubo, A., Suh, D.-S., Wong, K.-S., Jane, J.-L., Ozawa, K., Takaiwa, F., Inaba, Y. and Nakamura, Y. (2003) Antisense inhibition of isoamylase alters the structure of amylopectin and the physicochemical properties of starch in rice endosperm. *Plant and Cell Physiology*, **44**, 607–18.
56 Li, Z., Sun, F., Xu, S., Chu, X., Mukai, X.C., Yamamoto, M., Ali, S., Lynette Rampling, L., Kosar-Hashemi, B., Rahman, S. and Morell, M.K. (2003) The structural organization of the gene encoding class II starch synthase of wheat and barley and the evolution of the genes encoding starch synthases in plants. *Functional and Integrative Genomics*, **3**, 76–85.
57 Moorthy, S.N., Andersson, L., Eliasson, A.-C., Santacruz, S. and Ruales, J. (2006) Determination of amylose content in different starches using modulated differential scanning calorimetry. *Starch/Stärke*, **58**, 209–14.
58 Vandeputte, G.E. and Delcour, J.A. (2004) From sucrose to starch granule to starch physical behavior: a focus on rice starch. *Carbohydrate Polymers*, **58**, 245–66.
59 Satin, M. (1999) *Tropical Starches. Functional Properties of Starches*, FAO, Rome.
60 Baldwin, P.M. (2001) Starch granule-associated proteins and polypeptides: a review. *Starch/Stärke*, **53**, 475–503.
61 Lindeboom, N., Chang, P.R. and Tyler, R.T. (2004) Analytical, biochemical and physicochemical aspects of starch granule size, with emphasis on small granule starches: a review. *Starch/Stärke*, **56**, 89–99.
62 Perez, G.T., Ribotta, P.D., Aguirre, A.V., Rubiolo, O.J. and Leon, A.E. (2004) Changes in proteins and starch granule size distribution during grain filling of triticale. *Agriscientia*, **21**, 13–21.
63 MacGregor, A.W. and Fincher, G.B. (1993) Carbohydrates of the barley grain, in *Barley: Chemistry and Technology*, American Association of Cereal Chemists, St Paul, MN, pp. 73–130.
64 Kang, M.Y., Sugimoto, Y., Sakamoto, S. and Fuwa, H. (1985) Developmental changes in the amylose content of endosperm starch of barley (*Hordeum vulgare* L.) during the grain filling period after anthesis. *Agricultural and Biological Chemistry*, **49**, 3463–6.
65 McDonald, A.M.L., Stark, J.R., Morrison, W.R. and Ellis, R.P. (1991) The composition of starch granules from developing barley genotypes. *Journal of Cereal Science*, **13**, 93–112.
66 Tester, R.F. and Karkalas, J. (2001) The effects of environmental conditions on the structural features and physico-chemical properties of starches. *Starch/Stärke*, **53**, 513–19.
67 Peng, M., Gao, M., Baga, M., Hucl, P. and Chibbar, R.N. (2000) Starch-branching enzymes preferentially associated with A-type starch granules in wheat endosperm. *Plant Physiology*, **124**, 265–72.
68 Wilson, S.M., Burton, R.A., Doblin, M.S., Stone, B.A., Newbigin, E.J., Fincher, G.B. and Bacic, A. (2006) Temporal and spatial appearance of wall polysaccharides during cellularization of barley (*Hordeum vulgare*) endosperm. *Planta*, **224**, 655–67.
69 Burton, R.A., Wilson, S.M., Hrmova, M., Harvey, A.J., Shirley, N.J., Medhurst, A., Stone, B.A., Newbigin, E.J., Bacic, A. and Fincher, G.B. (2006) Cellulose synthase-like CslF genes mediate the synthesis of cell wall (1–3, 1–4)-beta-D-glucans. *Science*, **311**, 1940–2.
70 Herman, E.M. and Larkins, B.A. (1999) Protein storage bodies and vacuoles. *Plant Cell*, **11**, 601–14.
71 Crofts, A.J., Washida, H., Okita, T.W., Ogawa, M., Kumamaru, T. and Satoh, H. (2004) Targeting of proteins to endoplasmic reticulum-derived compartments in plants: the importance of RNA localization. *Plant Physiology*, **136**, 3414–19.
72 Crofts, A.J., Washida, H., Okita, T.W., Satoh, M., Ogawa, M., Kumamaru, T.

and Satoh, H. (2005) The role of mRNA and protein sorting in seed storage protein synthesis, transport, and deposition. *Biochemistry and Cell Biology*, **83**, 728–37.

73 Vitale, A. and Ceriotti, A. (2004) Protein quality control mechanisms and protein storage in the endoplasmic reticulum. A conflict of interests? *Plant Physiology*, **136**, 3420–6.

74 Galili, G. (2004) ER-derived compartments are formed by highly regulated processes and have special functions in plants. *Plant Physiology*, **136**, 3411–13.

75 Kawagoe, Y., Suzuki, K., Tasaki, M., Yasuda, H., Akagi, K., Katoh, E., Nishizawa, N.K., Ogawa, M. and Takaiwa, F. (2005) The critical role of disulfide bond formation in protein sorting in the endosperm of rice. *Plant Cell*, **17**, 1141–53.

76 Muentz, K. (1998) Deposition of storage proteins. *Plant Molecular Biology*, **38**, 77–99.

77 Gruis, D.F., Schulze, J. and Jung, R. (2004) Storage protein accumulation in the absence of the vacuolar processing enzyme family of cysteine proteases. *Plant Cell*, **16**, 270–90.

78 Shewry, P.R. and Halford, N.G. (2002) Cereal seed storage proteins: structures, properties and role in grain utilization. *Journal of Experimental Botany*, **53**, 947–58.

79 Singh, N.K., Donovan, G.R., Carpenter, H.C., Skerritt, J.H. and Langridge, P. (1993) Isolation and characterization of wheat triticin cDNA revealing a unique lysine-rich repetitive domain. *Plant Molecular Biology*, **22**, 227–37.

80 Triboi, E., Martre, P. and Triboi-Blondel, A.-M. (2003) Environmentally-induced changes in protein composition in developing grains of wheat are related to changes in total protein content. *Journal of Experimental Botany*, **54**, 1731–42.

81 Carbonero, P. and Garcia-Olmedo, F. (1999) A multigene family of trypsin/α-amylase inhibitors from cereals, in *Seed Proteins* (eds P.R. Shewry and R. Casey), Kluwer, Dordrecht, pp. 617–33.

82 Nielsen, P.K., Bønsager, B.C., Fukuda, K. and Svensson, B. (2004) Barley α-amylase/subtilisin inhibitor: structure, biophysics and protein engineering. *Biochim Biophys Acta*, **1696**, 157–64.

83 Lundgard, R. and Svensson, B. (1989) A 39 kD barley seed protein of the serpin superfamily inhibits alpha-chymotrypsin. *Carlsberg Research Communications*, **54**, 173–80.

84 Roberts, T.H., Marttila, S., Rasmussen, S.K. and Hejgaard, J. (2003) Differential gene expression for suicide-substrate serine proteinase inhibitors (serpins) in vegetative and grain tissues of barley. *Journal of Experimental Botany*, **54**, 2251–63.

85 Krüger, E. and Anger, H.-M. (1990) *Kennzeichen zur Betriebskontrolle und Qualitätsbeschreibung in der Brauwirtschaft*, Behr's Verlag, Hamburg.

86 Shewry, P.R. and Tatham, A.S. (1990) The prolamin storage proteins of cereal seeds: structure and evolution. *Biochemical Journal*, **267**, 1–12.

87 Shewry, P.R., Beaudoin, F., Jenkins, J., Griffiths-Jones, S. and Mills, E.N.C. (2002) Plant protein families and their relationships to food allergy. *Biochemical Society Transactions*, **30**, 906–10.

88 Shewry, P.R. and Tatham, A.S. (1999) The characteristics, structures and evolutionary relationships of prolamins, in *Seed Proteins* (eds P.R. Shewry and R. Casey), Kluwer, Dordrecht, pp. 11–33.

89 Kreis, M., Forde, B.G., Rahman, S., Miflin, B.J. and Shewry, P.R. (1985) Molecular evolution of the seed storage proteins of barley, rye and wheat. *Journal of Molecular Biology*, **183**, 499–502.

90 Wang, Z. and Messing, J. (1998) Modulation of gene expression by DNA–protein and protein–protein interactions in the promoter region of the zein multigene family. *Gene*, **223**, 333–45.

91 Mena, M., Vicente-Carbajosa, J., Schmidt, R.J. and Carbonero, P. (1998) An endosperm-specific DOF protein from barley, highly conserved in wheat, binds to and activates transcription from the prolamine-box of a native B-hordein promoter in barley endosperm. *Plant Journal*, **16**, 53–62.

92 Koller, A., Washburn, M.P., Lange, B.M., Andon, N.L., Deciu, C., Haynes, P.A., Hays, L., Schieltz, D., Ulaszek, R., Wei, J., Wolters, D. and Yates, J.R. (2002) Proteomic survey of metabolic pathways in rice. *Proceedings of the National Academy of Sciences of the United States of America*, **99**, 11969–74.

93 Jenkins, J.A., Griffiths-Jones, S., Shewry, P.R., Breiteneder, H. and Mills, E.N. (2005) Structural relatedness of plant food allergens with specific reference to cross-reactive allergens: an in silico analysis. *Journal of Allergy and Clinical Immunology*, **115**, 163–70.

94 van Heel, D.A. and West, J. (2006) Recent advances in celiac disease. *Gut*, **55**, 1037–46.

95 Shan, L., Qiao, S.W., Arentz-Hansen, H., Molberg, Ø., Gray, G.M., Sollid, L.M. and Khosla, C. (2005) Identification and analysis of multivalent proteolytically resistant peptides from gluten: implications for celiac sprue. *Journal of Proteome Research*, **4**, 1732–41.

96 Zarnkow, M., Kessler, M., Burberg, F., Kreisz, S. and Back, W. (2005) Gluten free beer from malted cereals and pseudocereals, in *Proceedings of the 30th EBC Congress, Prague, 14–19 May* (ed. J. Vesely), Fachverlag Hans Carl, Nürnberg, pp. 104/1–104/8.

97 Zarnkow, M., Kreisz, S. and Back, W. (2005) Glutenfreie Biere – Verschiedene Herstellungsverfahren im Fokus. *Brauindustrie*, **90**, 26–8.

98 Sun, C., Sathish, P., Ahlandsberg, S. and Jansson, C. (1998) The two genes encoding starch-branching enzymes IIa and IIb are differentially expressed in barley. *Plant Physiology*, **118**, 37–49.

99 Edwards, A., Vincken, J.P., Suurs, L.C., Visser, R.G., Zeeman, S., Smith, A. and Martin, C. (2002) Discrete forms of amylose are synthesized by isoforms of GBSSI in pea. *Plant Cell*, **14**, 1767–85.

100 Ohdan, T., Francisco, P.B., Sawada, T., Hirose, T., Terao, T., Satoh, H. and Nakamura, Y. (2005) Expression profiling of genes involved in starch synthesis in sink and source organs of rice. *Journal of Experimental Botany*, **56**, 3229–44.

101 Finnie, C., Melchior, S., Roepstorff, P. and Svensson, B. (2002) Proteome analysis of grain filling and seed maturation in barley. *Plant Physiology*, **129**, 1308–19.

102 Finnie, C., Maeda, K., Stergaard, O.O., Bak-Jensen, K.S., Larsen, J. and Svensson, B. (2004) Aspects of the barley seed proteome during development and germination. *Biochemical Society Transactions*, **32**, 517–19.

103 Skylas, D.J., Mackintosh, J.A., Cordwell, S.J., Basseal, D.J., Walsh, B.J., Harry, J., Blumenthal, C., Copeland, L., Wrigley, C.W. and Rathmell, W. (2000) Proteome approach to the characterisation of protein composition in the developing and mature wheat-grain endosperm. *Journal of Cereal Science*, **32**, 169–88.

104 Balmer, Y., Vensel, W.H., DuPont, F.M., Buchanan, B.B. and Hurkman, W.J. (2006) Proteome of amyloplasts isolated from developing wheat endosperm presents evidence of broad metabolic capability. *Journal of Experimental Botany*, **57**, 1591–602.

105 Panozzo, J.F., Eagles, H.A. and Wootton, M. (2001) Changes in protein composition during grain development in wheat. *Australian Journal of Agricultural Research*, **52**, 485–93.

106 van Dongen, J.T., Roeb, G.W., Dautzenberg, M., Froehlich, A., Vigeolas, H., Minchin, P.E. and Geigenberger, P. (2004) Phloem import and storage metabolism are highly coordinated by the low oxygen concentrations within developing wheat seeds. *Plant Physiology*, **135**, 1809–21.

107 Tetlow, I.J., Wait, R., Lu, Z., Akkasaeng, R., Bowsher, C.G., Esposito, S., Kosar-Hashemi, B., Morell, M.K. and Emes, M.J. (2004) Protein phosphorylation in amyloplasts regulates starch branching enzyme activity and protein–protein interactions. *Plant Cell*, **16**, 694–708.

108 Bønsager, B.C., Nielsen, P.K., Abou Hachem, M., Kenji Fukuda, K., Prætorius-Ibba, M. and Svensson, B. (2005) Mutational analysis of target enzyme recognition of the β-trefoil fold

barley α-amylase/subtilisin inhibitor. *Journal of Biological Chemistry*, **280**, 14855–64.
109 Jones, B. (2005) The endogenous endoprotease inhibitors of barley and malt and their roles in malting and brewing. *Journal of Cereal Science*, **42**, 271–80.
110 Saini, H.S. and Westgate, M.E. (2000) Reproductive development in grain crops during drought. *Advances in Agronomy*, **68**, 59–96.
111 Rabbani, M.A., Maruyama, K., Abe, H., Khan, M.A., Katsura, K., Ito, Y., Yoshiwara, K., Seki, M., Shinozaki, K. and Yamaguchi-Shinozaki, K. (2003) Monitoring expression profiles of rice genes under cold, drought, and high-salinity stresses and abscisic acid application using cDNA microarray and RNA gel-blot analyses. *Plant Physiology*, **133**, 1755–7.
112 Yang, J., Zhang, J., Wang, Z., Xu, G. and Zhu, Q. (2004) Activities of key enzymes in sucrose-to-starch conversion in wheat grains subjected to water deficit during grain filling. *Plant Physiology*, **135**, 1621–9.
113 Yu, L.-X. and Setter, T.L. (2003) Comparative transcriptional profiling of placenta and endosperm in developing maize kernels in response to water deficit. *Plant Physiology*, **131**, 568–82.
114 Boyer, J.S. and Westgate, M.E. (2004) Grain yields with limited water. *Journal of Experimental Botany*, **55**, 2385–94.
115 Ried, J.L. and Walker-Simmons, M.K. (1993) Group 3 late embryogenesis abundant proteins in desiccation-tolerant seedlings of wheat (*Triticum aestivum* L.). *Plant Physiology*, **102**, 125–31.
116 Close, T.J. (1997) Dehydrins: a commonality in the response of plants to dehydration and low temperature. *Physiologia Plantarum*, **100**, 291–6.
117 Griffith, M., Ala, P., Yang, D.S., Hon, W.C. and Moffatt, B.A. (1992) Antifreeze protein produced endogenously in winter rye leaves. *Plant Physiology*, **100**, 593–6.
118 Wallwork, M.A.B., Jenner, C.F., Logue, S.J. and Sedgley, M. (1998) Effect of high temperature during grain-filling on the structure of developing and malted barley grains. *Annals of Botany*, **82**, 587–99.
119 Whitt, S.R., Wilson, L.M., Tenaillon, M.I., Gaut, B.S. and Buckler, E.S. (2002) Genetic diversity and selection in the maize starch pathway. *Proceedings of the National Academy of Sciences of the United States of America*, **99**, 12959–62.
120 Davisa, J.P., Supatchareea, N., Khandelwalc, R.L. and Chibbara, R.N. (2003) Synthesis of novel starches in planta: opportunities and challenges. *Starch/Stärke*, **55**, 107–20.
121 Yanase, Y., Takaha, T. and Kuriki, T. (2006) α-Glucan phosphorylase and its use in carbohydrate engineering. *Journal of the Science of Food and Agriculture*, **86**, 1631–5.
122 He, G.Y., Rooke, L., Steele, S., Bekes, F., Gras, P., Tatham, A.S., Fido, R., Barcelo, P., Shewry, P.R. and Lazzeri, P.A. (1999) Transformation of pasta wheat (*Triticum turgidum* L. var *durum*) with high-molecular weight glutenin subunit genes and modification of dough functionality. *Molecular Breeding*, **5**, 377–86.
123 Alvarez, M.L., Guelman, S., Halford, N.G., Lustig, S., Reggiardo, M.I., Ryabushkina, N., Schewry, P., Stein, J. and Vallejos, R.H. (2000) Silencing of HMW glutenins in transgenic wheat expressing extra HMW subunits. *Theoretical and Applied Genetics*, **100**, 319–27.
124 Altpeter, F., Diaz, I., McAuslane, H., Gaddour, K., Carbonero, P. and Vasil, I.K. (1999) Increased insect resistance in transgenic wheat stably expressing trypsin inhibitor CMe. *Molecular Breeding*, **5**, 53–63.
125 Horvath, H., Huang, J., Wong, O., Kohl, E., Okita, T., Kannangara, C.G. and von Wettstein, D. (2000) The production of recombinant proteins in transgenic barley grains. *Proceedings of the National Academy of Sciences of the United States of America*, **97**, 1914–19.
126 Ostergaard, O., Melchior, S., Roepstorff, P. and Svensson, B. (2002) Initial

proteome analysis of mature barley seeds and malt. *Proteomics*, **2**, 733–9.

127 Bak-Jensen, K.S., Laugesen, S., Roepstorff, P. and Svensson, B. (2004) Two-dimensional gel electrophoresis pattern (pH 6–11) and identification of water-soluble barley seed and malt proteins by mass spectrometry. *Proteomics*, **4**, 728–42.

128 Bewley, J.D. (1997) Seed germination and dormancy. *Plant Cell*, **9**, 1055–66.

129 Bove, J., Jullien, M. and Grappin, P. (2001) Functional genomics in the study of seed germination. *Genome Biology*, **3** (1), R10021–R10025.

130 Molina-Cano, J.-L., Sopena, A., Polo, J.P., Bergareche, C., Moralejo, M.A., Swanston, J.S. and Glidewell, S.M. (2002) Relationships between barley hordeins and malting quality in a mutant of cv. triumph. II. Genetic and environmental effects on water uptake. *Journal of Cereal Science*, **36**, 39–50.

131 Manz, B., Müller, K., Kucera, B., Volke, F. and Leubner-Metzger, G. (2005) Water uptake and distribution in germinating tobacco seeds investigated *in vivo* by nuclear magnetic resonance imaging. *Plant Physiology*, **138**, 1538–51.

132 Yin, X.S., Foster, J.E., Browers, M., Schroeder, S., Izydorczyk, W., MacGregor, M.A., Gruwel, M.L.H. and Abrams, S. (2006) An investigation on the detection of potential loss of germinative capacity during storage of malting barley. *Journal of the American Society of Brewing Chemists*, **64**, 94–9.

133 Thomas, S.G. and Sun, T.P. (2004) Update on gibberellin signaling. A tale of the tall and the short. *Plant Physiology*, **135**, 668–76.

134 Olszewski, N., Sun, T.P. and Gubler, F. (2002) Gibberellin signaling: biosynthesis, catabolism, and response pathways. *Plant Cell*, **14** (Suppl.), S61–80.

135 Kaneko, M., Itoh, H., Ueguchi-Tanaka, M., Ashikari, M. and Matsuoka, M. (2002) The α-amylase induction in endosperm during rice seed germination is caused by gibberellin synthesized in epithelium. *Plant Physiology*, **128**, 1264–70.

136 Ritchie, S., McCubbin, A., Ambrose, G., Kao, T.-H. and Gilroy, S. (1999) The sensitivity of barley aleurone tissue to gibberellin is heterogeneous and may be spatially determined. *Plant Physiology*, **120**, 361–70.

137 Ritchie, S. and Gilroy, S. (1998) Calcium-dependent protein phosphorylation may mediate the gibberellic acid response in barley aleurone. *Plant Physiology*, **116**, 765–76.

138 Gubler, F., Raventos, D., Keys, M., Watts, R., Mundy, J. and Jacobsen, J.V. (1999) Target genes and regulatory domains of the GAMYB transcriptional activator in cereal aleurone. *Plant Journal*, **17**, 1–9.

139 Cercós, M., Gómez-Cadenas, A. and Ho, T.-H.D. (1999) Hormonal regulation of a cysteine proteinase gene, EPB1, in barley aleurone layers: *cis*- and *trans*-acting elements involved in the coordinated gene expression regulated by gibberellins and abscisic acid. *Plant Journal*, **19**, 107–18.

140 Gubler, F., Chandler, P.M., White, R.G., Llewellyn, D.J. and Jacobsen, J.V. (2002) Gibberellin signaling in barley aleurone cells: control of SLN1 and GAMYB expression. *Plant Physiology*, **129**, 191–200.

141 Woodger, F.J., Jacobsen, J.V. and Gubler, F. (2004) GMPOZ, a BTB/POZ domain nuclear protein, is a regulator of hormone responsive gene expression in barley aleurone. *Plant and Cell Physiology*, **45**, 945–50.

142 Zentella, R., Yamauchi, D. and Tuan-hua, T.-H.D. (2002) Molecular dissection of the gibberellin/abscisic acid signaling pathways by transiently expressed RNA interference in barley aleurone cells. *Plant Cell*, **14**, 2289–301.

143 Jacobsen, J.V., Pearce, D.W., Poole, A.T., Pharis, R.P. and Mander, L.N. (2002) Abscisic acid, phaseic acid and gibberellin contents associated with dormancy and germination in barley. *Physiologia Plantarum*, **115**, 428–41.

144 Finkelstein, R.R., Gampala, S.S. and Rock, C.D. (2002) Abscisic acid signaling in seeds and seedlings. *Plant Cell*, **14** (Suppl.), s15–45.

145 Lu, C.A., Ho, T.H., Ho, S.L. and Yu, S.M. (2002) Three novel MYB proteins with one DNA binding repeat mediate sugar and hormone regulation of α-amylase gene expression. *Plant Cell*, **14**, 1963–80.

146 Loreti, E., Alpi, A. and Perata, P. (2000) Sugar antagonistic GA: glucose and disaccharide-sensing mechanisms modulate the expression of α-amylase in barley embryos. *Plant Physiology*, **123**, 939–48.

147 Dominguez, F., Moreno, J. and Cejudo, F.J. (2004) A gibberellin-induced nuclease is localized in the nucleus of wheat aleurone cells undergoing programmed cell death. *Journal of Biological Chemistry*, **279**, 11530–6.

148 Hwang, Y.S., Bethke, P.C., Cheong, Y.H., Chang, H.S., Zhu, T. and Jones, R.L. (2005) A gibberellin-regulated calcineurin B in rice localizes to the tonoplast and is implicated in vacuole function. *Plant Physiology*, **138**, 1347–58.

149 Mrva, K., Wallwork, M. and Mares, D.J. (2006) α-Amylase and programmed cell death in aleurone of ripening wheat grains. *Journal of Experimental Botany*, **57**, 877–85.

150 Bethke, P.C., Swanson, S.J., Hillmer, S. and Jones, R.L. (1998) From storage compartment to lytic organelle: the metamorphosis of the aleurone protein storage vacuole. *Annals of Botany*, **82**, 399–412.

151 Li, Y., Rehmanji, M., Abrams, S.R. and Gusta, L.V. (1991) Effect of abscisic acid analogs on extract yield, α-amylase, and diastatic power during malting of barley. *Journal of the American Society of Brewing Chemists*, **49**, 135–9.

152 Macnicol, P.K. and Jacobsen, J.V. (1992) Endosperm acidification and related metabolic changes in the developing barley grain. *Plant Physiology*, **98**, 1098–104.

153 Dominguez, F. and Cejudo, F.J. (1999) Patterns of starchy endosperm acidification and protease gene expression in wheat grains following germination. *Plant Physiology*, **119**, 81–8.

154 Brennan, C.S., Harris, N., Smith, D. and Shewry, P.R. (1996) Structural differences in the mature endosperms of good and poor malting barley cultivars. *Journal of Cereal Science*, **24**, 171–7.

155 Fincher, G.B., Lock, P.A., Morgan, M.M., Lingelbach, K., Wettenhall, R.E.H., Mercer, J.F.B., Brandt, A. and Thomsen, K.K. (1986) Primary Structure of the $(1 \rightarrow 3, 1 \rightarrow 4)$-β-D-glucan 4-glucohydrolase from barley aleurone. *Proceedings of the National Academy of Sciences of the United States of America*, **83**, 2081–5.

156 Chen, L., Fincher, G.B. and Hoj, P.B. (1993) Evolution of polysaccharide hydrolase substrate specificity. Catalytic amino acids are conserved in barley 1-3-1-4- and 1-3-beta-glucanases. *Journal of Biological Chemistry*, **268**, 13318–26.

157 MacGregor, A.W. (1987) α-Amylase, limit dextrinase, and α-glucosidase enzymes in barley and malt. *CRC Critical Reviews in Biotechnology*, **5**, 117–28.

158 Huang, N., Stebbins, G.L. and Rodriguez, R.L. (1992) Classification and evolution of alpha-amylase genes in plants. *Proceedings of the National Academy of Sciences of the United States of America*, **89**, 7526–30.

159 Sissons, M.J. and MacGregor, A.W. (1994) Hydrolysis of barley starch granules by α-glucosidases from malt. *Journal of Cereal Science*, **19**, 161–9.

160 Henrissat, B., Callebaut, I., Fabrega, S., Lehn, P., Mornon, J.P. and Davies, G. (1995) Conserved catalytic machinery and the prediction of a common fold for several families of glycosyl hydrolases. *Proceedings of the National Academy of Sciences of the United States of America*, **92**, 7090–4.

161 Sugimoto, N., Takeda, G., Nagato, Y. and Yamaguchi, J. (1998) Temporal and spatial expression of the α-amylase gene during seed germination in rice and barley. *Plant and Cell Physiology*, **39**, 323–33.

162 Müntz, K., Belozersky, M.A., Dunaevsky, Y.E., Schlereth, A. and Tiedemann, J. (2001) Stored proteinases and the initiation of storage protein mobilization in seeds during germination and seedling

growth. *Journal of Experimental Botany*, **52**, 1741–52.
163 Koehler, S.M. and Ho, T.-H.D. (1990) Hormonal regulation, processing, and secretion of cysteine proteinases in barley aleurone layers. *Plant Cell*, **2**, 769–83.
164 Jones, B.L. (2005) Endoproteases of barley and malt. *Journal of Cereal Science*, **42**, 139–56.
165 Jones, B.L. and Budde, A.D. (2005) How various malt endoproteinase classes affect wort soluble protein levels. *Journal of Cereal Science*, **41**, 95–106.
166 Röllig, W. (1970) Das Bier im Alten Mesopotamien, in *Jahrbuch der Gesellschaft für Geschichte und Bibliographie des Brauwesen 1970*, Institut für Gärungsgewerbe, Berlin, pp. 10–18.
167 Zarnkow, M. (2005) Brautechnologische Erkenntnisse experimenteller Archäologie am Beispiel einer bronzezeitlichen Stadt (Tall Bazi), in *Lehrstuhl für Technologie der Brauerei I: 3. Rohstoffseminar*, Weihenstephan.
168 Ziptel, W., Rathke, K.-D. A410B, Biersteuer Verordnung (in der Fassung vom 29. 07. 1993) § 9 Abs. 1–8.
169 Gerstenberg, H. (1999) Vorläufiges Biergesetz, in *Lebensmittelrecht* (eds W. Zipfel and K.-D. Rathke), Verlag C. H. Beck, München.
170 Rehm, S. (1971) Hirse, in *Handbuch der Landwirtschaft und Ernährung in Entwicklungsländern*, Verlag Eugen Ulmer, Stuttgart, pp. 281–98.
171 Hansen, H.J. (1912) *Mitteilung der Deutschen Landwirtschafts Gesellschaft*, Deutsche Landwirtschafts Gesellschaft, Frankfurt am Main, p. 342.
172 O'Rourke, T. (1999) Adjuncts and their use in the brewing process. *Brewers' Guardian*, **128**, 32–6.
173 Narziss, L. (2005) *Abriss der Bierbrauerei*, Wiley-VCH Verlag GmbH, Weinheim, p. 4
174 Glatthar, J., Heinisch, J.J. and Senn, T. (2003) The use of unmalted triticale in brewing and its effect on wort and beer quality. *Journal of the American Society of Brewing Chemists*, **61**, 182–90.
175 O'Connor-Cox, E.S. and Ingledew, W.M. (1989) Wort nitrogenous sources – their use by brewing yeasts: a review. *Journal of the American Society of Brewing Chemists*, **47**, 102–8.
176 Vidgren, V., Ruohonen, L. and Londesborough, J. (2005) Characterization and functional analysis of the MAL and MPH loci for maltose utilization in some ale and lager yeast strains. *Applied and Environmental Microbiology*, **71**, 7846–57.
177 Andrews, J. (2004) A review of progress in mash separation technology. *Master Brewers Association of the Americas Technical Quarterly*, **41**, 45–9.
178 Körber-Grohne, U. (1995) *Nutzpflanzen in Deutschland*, Nikol Verlagsgesellschaft, Hamburg.
179 Bamforth, C.W. (2003) Barley and malt starch in brewing: a general review. *Master Brewers Association of the Americas Technical Quarterly*, **40**, 89–97.
180 Narziss, L. (2005) *Abriss der Bierbrauerei*, Wiley-VCH Verlag GmbH, Weinheim, p. 13.
181 Keßler, M., Zarnkow, M., Kreisz, S. and Back, W. (2005) Gelatinisation properties of different cereals and pseudocereals. *Monatsschrift für Brauwissenschaft*, **58** (5), 75–81.
182 Meilgaard, M.C. (1976) Wort composition: with special reference to the use of adjuncts. *Technical Quarterly*, **13**, 78–90.
183 Müller, K.-P. (2005) Brauversuche unter Einsatz unvermälzter Gerste, Diplomarbeit, Technische Universität München, Lehrstuhl für Technologie der Brauerei I, Freising.
184 Narziss, L. (1976) Die Technologie der Malzbereitung, in *Schuster/Weinfurtner/ Narziss Die Bierbrauerei*, 6. Auflage, Erster Band: Die Technologie der Malzbereitung, Ferdinand Enke Verlag, Stuttgart, pp. 342–8.
185 Kunze, W. (1998) *Technologie Brauer und Mälzer*, 8th edn, VLB, Berlin.
186 Zade, A. (1918) *Der Hafer: Eine Monographie auf wissenschaftlicher und praktischer Grundlage*, Gustav Fischer Verlag, Jena.

187 Ranhotra, G.S. and Gelroth, J.A. (1995) Food uses of oats, in *The Oat Crop. Production and Utilization* (ed. R.W. Welch), Chapman & Hall, London, pp. 409–30.

188 Both, F. (1998) *Gerstensaft und Hirsebier, 5000 Jahre Biergenuss*, Verlag Isensee, Oldenburg, 11–38.

189 Unger, R.W. (2001) *A History of Brewing in Holland 900–1900, Economy, Technology and the State*, Brill, Leiden.

190 Briggs, D.E. (1998) *Malting and Brewing*, Blackie Academic & Professional, London.

191 Weichherz, J. (1928) *Die Malzextrakte*, Springer Verlag, Berlin.

192 Belitz, H.-D. and Grosch, W. (1992) *Lehrbuch der Lebensmittelchemie*, 4th rev. edn, Springer-Lehrbuch, Berlin.

193 Hanke, S., Zarnkow, M., Kreisz, S. and Back, W. (2005) The use of oats in brewing. *Monatsschrift Für Brauwissenschaft*, **58**, 11–17.

194 Hanke, S., Zarnkow, M., Kreisz, S. and Back, W. (2005) Verwendung von Hafer für die Malz- und Bierbereitung. *Brauwelt*, **145**, 216–19.

195 Habich, G.E., Schneider, C. and Behrend, G. (1883) *Die Praxis der Bierbraukunde, Illustriertes Hand- und Hülfsbuch für Brauer*, Wilhelm Knapp, Halle a. d. Saale.

196 Zeller, F.J. (2000) Nutzung, Genetik und Züchtung kleinkörniger Hirsegräser: 2. Kolbenhirse [*Setaria italica* (L.) P. Beauv.]. *Journal of Applied Botany/Angewandte Botanik*, **47**, 50–4.

197 Zeller, F.J. (2000) Nutzung, Genetik und Züchtung kleinkörniger Hirsegräser: 1. Perlhirse [*Pennisetum glaucum* (L.) R. Br.]. *Journal of Applied Botany/Angewandte Botanik*, **47**, 42–9.

198 Rao, S.A., Mengesha, M.H., Sibale, P.K. and Reddy, C.R. (1986) Collection and evaluation of pearl-millet (*Pennisetum*) germplasm from Malawi. *Economic Botany*, **40**, 27–37.

199 Pelembe, L.A.M., Dewar, J. and Taylor, J.R.N. (2002) Effect of malting conditions on pearl millet malt quality. *Journal of the Institute of Brewing*, **108**, 7–12.

200 Obilana, A.B. (2002) Millets, in *Pseudocereals and Less Common Cereals* (eds P.S. Belton and J.R.N. Taylor), Springer Verlag, Berlin, pp. 204–6.

201 Nzelibe, H.C., Obaleye, S. and Onyenekwe, P.C. (2000) Malting characteristics of different varieties of fonio millet (*Digitaria exilis*). *European Food Research and Technology*, **211**, 126–9.

202 Obilana, A.B. (2002) Millets, in *Pseudocereals and Less Common Cereals* (eds P.S. Belton and J.R.N. Taylor), Springer Verlag, Berlin, p. 210.

203 Zeller, F.J. (2003) Nutzung, Genetik und Züchtung kleinkörniger Hirsegräser: 5. Teff [*Eragrostis tef* (Zucc.) Trotter]. *Journal of Applied Botany/Angewandte Botanik*, **77**, 47–52.

204 Li, H. (1970) Origin of cultivated plants in Southeast Asia. *Economic Botany*, **8**, 3–19.

205 Böckler, W. (1954) Relikte unter den Kulturpflanzen. *Zeitschrift für Agrargeschichte und Agrarsoziologie*, **2**, 22–40.

206 Hoffmann-Bahnsen, R. and Plessow, J. (2003) Alte Kulturpflanzen neu entdeckt, Rispenhirse (*Panicum miliaceum*) eine ideale Sommerung für den ökologischen Landbau. *Mitteilungen der Gesellschaft für Pflanzenbauwissenschaften*, **15**, 31–3.

207 Körber-Grohne, U. (1995) *Nutzpflanzen in Deutschland*, Nikol Verlagsgesellschaft, Hamburg, pp. 86–97.

208 Lásztity, R. (1984) *The Chemistry of Cereal Proteins*, CRC Press, Boca Raton, FL.

209 Mounts, T.L. and Anderson, R.A. (1983) Corn oil production, processing and use, in *Lipids in Cereal Technology* (ed. P.J. Barnes), Academic Press, New York, pp. 373–87.

210 Agu, R.C. (2002) A comparison of maize, sorghum and barley as brewing adjuncts. *Journal of the Institute of Brewing*, **108**, 19–22.

211 Tegge, G. (2004) Stärke und Stärkederivate. 3. vollständig überarbeitete Auflage.

212 Einsiedler Bierspezialitäten, http://www.beer.ch/ros_biere.html#maisgold (accessed 22 June 2005).

213 Eneje, L.O., Ogu, E.O., Aloh, C.U., Odibo, F.J.C., Agu, R.C. and Palmer, G.H. (2004)

Effect of steeping and germination time on malting performance of Nigerian white and yellow maize varieties. *Process Biochemistry*, **39**, 1013–16.
214 Ramseyer, U. (1988) *Reis. Konsequenzen des Geschmacks*, Verlag Edition diá, St Gallen.
215 Zarnkow, M. and Back, W. (2004) Probleme bei der Herstellung von Rohfruchtbieren mit hohen Reisgaben. *Brauwelt*, **144**, 391–6.
216 Teramoto, Y., Yoshida, S. and Ueda, S. (2000) An indigenous rice beer of Nagaland, India. *Ferment*, **13**, 39–41.
217 Körber-Grohne, U. (1995) *Nutzpflanzen in Deutschland*, Nikol Verlagsgesellschaft, Hamburg, pp. 40–6.
218 Braun, F. (1998) Mälzungs- und Maischversuche zu Möglichkeiten der technologsichen Einflußnahme auf ungünstige Fließeigenschaften bei Roggenmalzwürzen unter besonderer Berücksichtigung des Reinheitsgebotes, Diplomarbeit, Lehrstuhl für Technologie der Brauerei I, Technische Universität München, Freising.
219 Taylor, J.R.N. and Belton, P.S. (2002) Sorghum, in *Pseudocereals and Less Common Cereals* (eds J.R.N. Taylor and P.S. Belton), Springer Verlag, Berlin, pp. 25–81.
220 Palmer, G. (1993) Aspects of the malting of sorghum. *Ferment*, **6**, 339–41.
221 Ajerio, K.O., Booer, C.D. and Proudlove, M.O. (1993) Aspects of the malting of sorghum. *Ferment*, **6**, 339–41.
222 Back, W. (2005) *Ausgewählte Kapitel der Brauereitechnologie*, Fachverlag Hans Carl, Nürnberg.
223 Jani, M., Annemüller, G. and Schildbach, R. (1999) Untersuchunge über den Einfluß der Hirsemalzqualitäten auf die Extraktgewinnung. *Brauwelt*, **139**, 1062–5.
224 Dufour, J.P. and Mélotte, L. (1992) Sorghum malts for the production of a lager beer. *Journal of the American Society of Brewing Chemists*, **50**, 110–19.
225 Muoria, J.K., Linden, J.C. and Bechtel, P.J. (1998) Diastatic power and alpha-amylase activity in millet, sorghum, and barley grains and malts. *Journal of the American Society of Brewing Chemists*, **56**, 131–5.
226 Canales, A.M. and Sierra, J.A. (1976) Use of sorghum. *Master Brewers Association of the Americas Technical Quartely*, **13**, 114–16.
227 Figueroa, J.D.C., Martinez, B.F. and Rios, E. (1995) Effect of sorghum endosperm type on the quality of adjuncts for the brewing industry. *Journal of the American Society of Brewing Chemists*, **53**, 5–9.
228 Goode, D.L., Halbert, C. and Arendt, E.K. (2003) Optimization of mashing conditions when mashing with unmalted sorghum and commercial enzymes. *Journal of the American Society of Brewing Chemists*, **61**, 69–78.
229 Gorinstein, S., Kitov, S., Sarel, S., Berman, O., Berliner, M., Popovich, G. and Vermus, Y. (1980) Changes in the chemical composition of beer during the brewing process as a result of added enzymes. *Journal of the American Society of Brewing Chemists*, **38**, 23–6.
230 Osorio-Morales, S., Serna Saldivar, S.O., Contreras, J.C., Almeida-Dominguez, H.D. and Rooney, L.W. (2000) Production of brewing adjuncts and sweet worts from different types of sorghum. *Journal of the American Society of Brewing Chemists*, **58**, 21–5.
231 O'Rourke, T. (1999) Adjuncts and their use in the brewing process. *Brewers' Guardian*, **128**, 32–6.
232 Körber-Grohne, U. (1995) *Nutzpflanzen in Deutschland*, Nikol Verlagsgesellschaft, Hamburg, pp. 68–86.
233 Körber-Grohne, U. (1995) *Nutzpflanzen in Deutschland*, Nikol Verlagsgesellschaft, Hamburg, p. 324.
234 Bänninger, A., Jenny, M., Gehring, B. and Bartha, B. (2005) Einkorn/Emmer-Projekt, http://www.sab.ch/v2/marketing/projekte/klettgau.htm (accessed 21 June 2005).
235 Körber-Grohne, U. (1995) *Nutzpflanzen in Deutschland*, Nikol Verlagsgesellschaft, Hamburg, pp. 326–30.
236 Körber-Grohne, U. (1995) *Nutzpflanzen in Deutschland*, Nikol Verlagsgesellschaft, Hamburg, pp. 28–39.

237 Schuchert, W. Triticale, http://www2.mpiz-koeln.mpg.de/pr/garten/schau/Triticale/Triticale(d).html (accessed 24 June 2005).
238 Malter, R. (1992) Untersuchungen der brautechnischen Einsatzmöglichkeiten von Triticale, Doctoral Thesis (Dissertation), Technical University Köthen, Köthen.
239 Creydt, G., Mietla, B., Rath, F., Annemuller, G., Schildbach, R. and Tuszynski, T. (1999) Triticale and triticale malt: part II: oriented malting of triticale. *Monatsschrift Für Brauwissenschaft*, **52**, 126–30.
240 Glatthar, J., Heinisch, J. and Senn, T. (2002) A study on the suitability of unmalted triticale as a brewing adjunct. *Journal of the American Society of Brewing Chemists*, **60**, 181–7.
241 Glatthar, J., Heinisch, J.J. and Senn, T. (2005) Unmalted triticale cultivars as brewing adjuncts: effects of enzyme activities and composition on beer wort quality. *Journal of the Science of Food and Agriculture*, **85**, 647–54.
242 Martin, A., Alvarez, J.B., Martin, L.M., Barro, F. and Ballesteros, J. (1999) The development of tritordeum: a novel cereal for food processing. *Journal of Cereal Science*, **30**, 85–95.
243 Little, B.T. (1994) Alternative cereals for beer production. *Ferment*, **7**, 163–8.
244 Zarnkow, M. (2005) Biereinteilung, http://www.roempp.com/prod/index1.html (accessed 02. 2005).
245 Britnell, J. (1973) Torrified cereals. *Technical Quarterly*, **10**, 176–9.
246 Aufhammer, W. (1998) *Getreide- und andere Körnerfruchtarten*, Eugen Ulmer Verlag, Stuttgart, p. 162.
247 Paredeslopez, O. and Moraescobedo, R. (1989) Germination of amaranth seeds – effects on nutrient composition and color. *Journal of Food Science*, **54**, 761–2.
248 Yue, S. and Sun, H. (1997) The development of food products of grain amaranth, in *Chinese Cereals and Oils: Proceedings of the International Symposium on New Approaches to Functional Cereals and Oils*, Bejing, pp. 188–91.
249 Fenzl, G., Berghofer, E., Silberhummer, H. and Schwarz, H. (1997) Einsatzmöglichkeiten extrudierter, stärkereicher Rohstoffe zur Bierherstellung. *Tagungsband Österreichisches Brauforum*, **1997**, 1–6.
250 Zeller, F.J. (2001) Buchweizen (*Fagopyrum esculentum* Moench): Nutzung, Genetik, Züchtung. *Bodenkultur*, **52**, 259–76.
251 Wijngaard, H.H., Ulmer, H.M. and Arendt, E.K. (2005) The effect of germination temperature on malt quality of buckwheat. *Journal of the American Society of Brewing Chemists*, **63**, 31–6.
252 Wijngaard, H.H. and Arendt, E.K. (2006) Optimisation of a mashing program for 100% malted buckwheat. *Journal of the Institute of Brewing*, **112**, 57–65.
253 Dadic, M. and van Gheluwe, J.E.A. (1971) Potential antioxidants in brewing. *Master Brewers Association of the Americas Technical Quarterly*, **8**, 182–90.
254 Zarnkow, M., Kessler, M., Burberg, F., Kreisz, S. and Back, W. (2005) Glutenfree beer from malted cereals and pseudocereals, in *Proceedings of the European Brewery Convention Congress, Prague, 2005*, Fachverlag Hans Carl, Nürnberg, CD ROM, contribution 104.
255 Mabry, T.J., Taylor, A. and Turner, B.L. (1963) The betacyanins and their distribution. *Phytochemistry*, **2**, 61–4.
256 Whali, C. (1990) *Quinua hacia su cultivo commercial*, Latinreco, Quito.
257 Haard, N.F., Odunfa, S.A., Lee, C-H., Quintero-Ramírez, R., Lorence-Quiñones, A. and Wacher-Radarte, C. (1999) *Fermented Cereals: Global Perspective*, FAO Agricultural Services Bulletin No. 138, FAO, Rome.

3
Hops

Martin Krottenthaler

3.1
Introduction

The common hop (*Humulus lupulus*) belongs within the botanical classification scheme of urticales (nettle family) to the hemp family (*Cannabaceae*) [1].

Hops were cultivated in more than 50 countries on about 50 500 ha in 2005. The development of the cultivation can be seen in the increasing α-acid values of the bittering hop varieties. Despite increasing world beer production, the area of cultivation has decreased. On the one hand, this is due to an increased cultivation of high-α-hop varieties. On the other hand, the average α-dosage per hectoliter of beer has been steadily reduced. From a globally standpoint, the bittering units of beer are decreasing. Considerably less than 5 g α-acids/hl are currently added.

Detailed information on German hop varieties, such as concentration of the bitter resins, composition of the essential oils including gas chromatography chromatograms, sensory characteristics, concentration of polyphenols, resistances as well as further information on agronomy and brew quality [2, 3] can be found in the CMA varieties map [4]. Data and description on German and international hop varieties are available on the internet [5–8].

3.2
Cultivation of Hops

The hop is a hardy, dioecious vine and is fertilized by wind-borne pollen. In hop cultures only unfertilized female plants are cultivated, which develop cones from the flowers. The cultivation of hops is work and capital intensive.

The aerial part of the hop plant (vine, side shoots, leaves and umbels) is herbaceous and annual. The root stock is perennial and reaches up to 4 m into the ground. Every year the root stock develops adventitious roots (summer roots) in the upper soil layer. These sprout from the earth-covered part of the shoot and serve additionally for feeding during the main growth periods. The nutrients move from the vines to the rhizomes in autumn so that the young shoots can feed in spring (April) from these reserves from up to 1 m height. In spring, the hop plant is cut back on the upper surface to remove old shoots. Two to three new shoots are trained on a string. These shoots wind clockwise in the northern hemisphere and attach themselves to the guiding wire using stiff hairs. The hop reaches the 5.5–8.0 m (7 m in Germany) high scaffolding at the end of June, which means that the average growth rate is 10 cm/day. The hop flowers ('blooms') from the beginning to mid July. The umbel (cone) gradually develops from the unfertilized flowers [5]. The harvest takes place at the end of August until mid September. Then the umbel is closed on the tops so that the lupulin seeds, valuable for brewing, cannot be lost. Hop plants are propagated from runners that are cut at the start of vegetation.

Due to the expansive, deep-reaching roots system, hop needs recondite, well-rootable, light to medium soils. Heavy soils should be avoided, as these tend to cause waterlogging. Hops grow only between 35° and 55° latitude since the long days in summer fulfill the prerequisite of the flowers [9]. The hops' need for light is accommodated by angular training and appropriate spacing between the hop plants. The hop needs a great deal of warmth during growth, flowering and development of the umbels. Rainfall, especially during development of the umbels, is beneficial for the harvest and for the concentration of the bitter resins.

Fertilization has the function to keep and increase soil fertility so that the hop plants have all the necessary nutrients available at the right time in the right form and quantity, and can absorb them. Soil analysis, carried out on a regular basis, is a prerequisite for a tailored fertilization. In addition, this should also cover trace elements (zinc, boron, etc.). Undersown crops in the hop garden serve to avoid erosion. This green fertilization is dug-in during spring. It prevents the washing out of nitrate over the winter and serves as an organic fertilizer in spring.

The sensitivity of hops for diseases and pests is dependent on the prevalent force of infection and the resistance capacity of the hop varieties. Climatic circumstances (damp, cool weather promotes mycosis) play an equally important role as the presence of infested hop plants (clearing of virus-infested plants and replacement with healthy cuttings).

Legal requirements in different countries in which hops or the hop product are regulate the use of pesticides. The effectiveness and innocuousness of the harvest is thereby checked. Due to ecological and economic reasons, correct pest management needs to fulfill these requirements. The goal is to reduce the amount of pesticide used, and hence protect the environment and consumer. By using early warning systems (*Peronospora* warning service) and information centers for hop cultivation, the choice of the right supplement and optimal application point can be determined. Every product has specifications about what period has to be observed between the last spraying and the hop harvest.

Depending on the varieties and weather trend, the hop harvest begins in August and lasts until about mid September. The right time for harvest is a compromise of yield, concentration of bitter resins, aroma quality and respective disease infestation. After the harvest in the hop garden, the umbels are separated from the vines in a picker. With mechanical harvest a weight loss of about 10% can be expected through the loss of umbels and lupulin. The hop umbels have a water concentration of 75–80 wt% at the point of harvest. They are dried to 10–11 wt% at most to obtain storable hops. Hot air kilns are available in hop cultivation farms for this purpose. The drying needs to be carried out gently to avoid the loss of bitter resins and aroma. The maximum drying temperature is between 62 and 65 °C depending on the variety and water concentration of the crude hop. Sulfuring, for the prevention of infestation with microorganisms, has been abandoned, since hop umbels are now rarely stored over long periods of time. Conditioning is carried out after drying, in which an adjustment of moisture takes place within the charge and between the spindle and umbel, respectively. Finally, the dried raw hop is packed for further transportation into a rectangular bundle.

3.3
Components of Hops

The yellow-green, sticky in fresh conditions, cup-like glands (lupulin glands) located on the inner and outer bracteoles of the hop umbel are of greatest interest to the brewer since these contain the technically important bitter resins and aroma substances. The composition of the hops [10] depends on variety, vintage, provenance, time of harvest, drying and storage [11] (Figure 3.1).

3.3.1
Bitter Acids

The total resins represent the sum of all bitter resins, and can be divided into hard and soft resins. The main constituents in soft resins are the bitter acids, which can be divided into α-acids (3–17%) and β-acids (2–7%). The five homologs of the α-acids are humulone and co-, ad-, pre- and posthumulone. Five homologs can also be found in the β-acid (lupulones), which only differ from humulone in the side-chain at C-3 (a dimethyl allyl group instead of an OH group).

The bitter resins give the beer the fine pleasant hop bitterness, aid the salubriousness, stabilize the foam and increase the biological shelf-life with their antibacterial properties (especially against Gram-negative, but also against Gram-positive bacteria) [12]. Studies have shown a positive influence on diabetes type 2 [13].

The bitterness potential and hence the value of brewing differs for the different substances. In addition to the solubility in the wort and beer, the bitter quality is of great importance. The iso-α-acids and derivatives show the greatest bitterness potential. Due to its insolubility in wort and beer, the β-acids do not contribute to the bitterness.

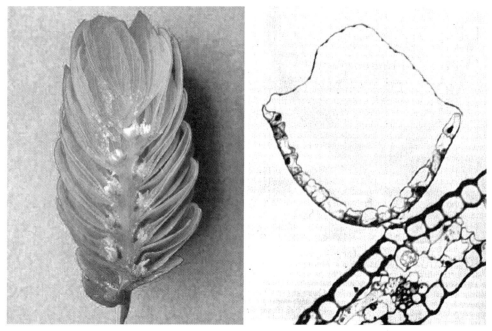

Figure 3.1 Hop umbel with lupulin gland and microscopic presentation of a lupulin gland.

During thermal isomerization, five-ring structures are formed from the optical active hexa-cyclic substances. As a result of the high oxygen sensitivity of hops, numerous oxidized derivates are formed. These non-volatile, water-soluble, often complex substances dissolve in the wort and can reach the beer [9, 14].

3.3.2
Aroma Substances

So far, more than 300 volatile substances have been identified in hops [15]. The monoterpene myrcene is the quantitatively most important substance with 17–37% mass fraction of the total oils. It is responsible for the typical taste if hops are added to the beer during storage (dry hopping). The sesquiterpene farnesene [16] is typical with the varieties Saaz (Spalt, Tettnang) and can be found in fractions up to 30% of total oils. It does not have any sensory contribution.

The oxidized terpenes linalool and geraniol have a floral- and a rose-like aroma, respectively.

Linalool is a chiral compound and is predominantly found in the notably more aroma-active (*R*)-form (92–94%) [17–19].

It was discovered that the chiral distribution of linalool in aroma hops is approximately 94% (*R*)-linalool [20] and is constant starting from raw hops up to all conventional hop products.

Esters can contribute to a fruity aroma impression. Fatty acids are mainly responsible for a cheese-like aroma, especially with unfavorably stored hops (warm).

Oxygen heterocycles (epoxides) are formed by autoxidation of sesquiterpene hydrocarbons. Hops also contain sulfur containing aroma substances such as thioester, sulfides and sulfur heterocycles [21, 22].

The composition of the hop oils is dependent on the variety and can be strongly influenced by the drying conditions as well as processing procedures. Oxidation is widely suppressed by the use of a cold chain starting with receipt of the raw hops via refinement and to the hop product.

3.3.3
Polyphenols

Between 20 and 30% of polyphenols in wort come from hops; the rest is introduced by malt. About 100 different phenolic single components can be determined analytically. Polyphenols consist of about 80% condensable and about 20% hydrolysable compounds. The first are monomeric polyphenols and their glycosides. These are able to polymerize to higher molecular products.

The group of phenolic carboxylic acids such as hydroxybenzoic acid and hydrocinnamic acid is also included in polyphenols. The principle component of the prenylflavonoids is xanthohumol [1].

Aroma hops have higher amounts of low molecular polyphenols than bittering hops [23–26]. Differences arise depending on the place of cultivation, the variety and the progress of vegetation.

In particular, proanthocyanidins composed of catechin or leucoanthocyanidin units are the cause of protein precipitation and turbidity. Proanthocyanidins are called anthocyanogens in the brewing business [16]. Low molecular polyphenols are natural antioxidants and increase the reduction power of the beer. They protect the beer from oxidation and thus enhance the taste stability indirectly. Moreover, it is postulated that they act as radical scavengers in the human body. The prenylflavonoid xanthohumol is located in the lupulin gland and shows an anticarcinogenic effect both *in vitro* and in animal tests. [24, 27–31].

High molecular polyphenols can increase the beer color, especially after increased boiling times, and can cause an astringent bitterness. They reduce the colloidal stability and cause turbidity of the beer. The entry of polyphenols is reduced or partly removed to achieve a high physical stability of the beer. Polyphenol-free hop extract as well as proanthocyanidin-free barley varieties can be used to achieve high physical stability of the beer. Alternatively, turbidity causing polyphenols can be removed during beer filtration by adsorption on polyvinylpolypyrrolidone (PVPP) [24]. Polyphenols are partly removed by separation of hot and cold trub, yeast sludge and filtration. Phenolic acids are precursors to certain beer aromas. From ferulic acid develops 4-vinyl guaiacol (with an aroma impression like a clove) which is responsible for the typical aroma of wheat beer [1, 32]. Hops also contains 4-vinyl guaiacol [22].

3.4
Hop Products

Figure 3.2 shows possibilities how to classify hop products. The key feature of the hop products is the reduced volume as compared to raw hops. In particular, enriched hop products (pellets type 45 or extract) combine many advantages. These facilitate good storage in cold rooms, simplified logistics and a reduction in the packing materials needed. Due to the homogenous distribution of α-acids in a product batch, an exact dosage of the hop is possible in the brewery. Furthermore, hop products are protected by special packaging from oxidation and, consequently, a longer preservation of the quality of the raw material is provided. Residues of pesticides lie below the maximum permissible value regulated by law and are reduced in hop products to a different extent [1, 34–36]. The products can be differentiated on the basis of their application and their point of dosage (bitterness, aromatization, improvement of foam, light stability, etc.). If dosing occurs after the wort boil (i.e. to the beer), these are then called downstream products [37]. The particular hop product the brewer decides to use is, besides company-specific factors, dependent on the hop products allowed by law within the European Union or national regulations, such as the Bavarian Purity Law [38]. Within the Bavarian Purity Law only conventional hop products are allowed to be used in the hot domain.

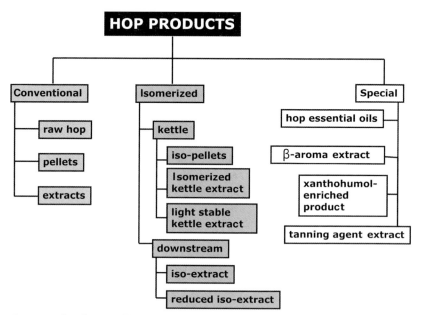

Figure 3.2 Classification of hop products [33].

3.5
Analytics

Conventional hop addition in a brewery is mostly adjusted by conductometric values (CVs) using pellets and by high-performance liquid chromatography (HPLC) for CO_2 extracts. It has been shown that in this way a constant bitter level of the beer can be achieved. In ethanol extracts, the conductometric bittering value (CBV) is used as it shows a high correlation to the sensory bitterness perceived in beer [41].

The comparison of the CVs with the sum of α-acids from HPLC analysis results to some considerable differences. Apart from the α-acids, unspecific bitter resins are detected with the determination of the sum. Therefore, the CV is higher than the sum of α-acids determined by HPLC. Furthermore, time-dependent fluctuations can be detected.

The total contents of oils are determined by distillation. The individual hop oils can be determined by capillary gas chromatography (GC) which is partly coupled to mass spectroscopy (GC-MS) or multidimensional GC-MS [22]. Sample preparation can be carried out by steam distillation or solid-phase extraction [4]. The total polyphenol concentration can be determined with wet chemical methods. The identification of single components is carried out using HPLC with diode array detection [23]. See Table 3.1.

3.6
Hopping Technology

3.6.1
General Aspects of Hop Addition

The goal of hop addition is the manipulation of the beer aroma and the beer taste. The aroma of the beer is influenced by the essential oils of hops. These are volatile substances that contribute to the characteristic hop aroma of the beer. The beer taste is determined by non-volatile substances. The biggest contribution to the beer bitterness is from iso-α-acids [42].

The hop dosage needs to be adjusted to the taste profile of the respective beer. The taste and aroma of the hops need to be integrated harmonically into the beer matrix. The following parameters should be considered:

- Composition of the raw material (water, malt, malt substitute, additives).
- Hop variety.
- Hop product.
- Amount of hop addition.
- Point of hop addition (boiling time of hops).
- Wort boil system.
- Whirlpool standing time and temperature.

Table 3.1 Methods for analyzing hop and hop products [33].

Method	EBC [39]	American Society of Brewing Chemists	Mitteleuropäische Brautechnische Analysenkommission [40]
Sampling	7.1	Hops-1	5.1.1/5.2.1/5.3.1
Evaluation by hand	–	Hops-2	5.1.2
Water content	7.2	Hops-4	5.1.4/5.2.3
Concentration of seeds	7.3	Hops-2	–
CV	7.4	Hops-6B	–
CV modified by Wöllmer for powder products	7.5	Hops-8B(II)	5.1.5.1
CV modified by Wöllmer for extracts	7.6	–	5.2.4.1
α-Acids (spectral photometric)	–	Hops-6A, -8B(I)	–
α-Acids (HPLC)	7.7	Hops-14	5.1.5.2
α-, β- and Iso-α-acids (HPLC)	7.8	Hops-9C, -9D, -15	5.3.3.1
CBV (with ethanol extract)	CBV = CV (EBC 7.6) + 50% iso-α-acid (HPLC EBC 7.8)		5.2.4.1; 5.3.3.1
Iso-α-acid and reduced iso-α-acid in hop products (HPLC)	7.9	–	–
Hop oil	7.10	Hops-13	–
Hop storage index	–	Hops-12	–

- Quality of trub separation (hot trub, cold trub).
- Recycling of hot trub.
- Fermentation (beer aroma spectra).
- Original extract.
- Attenuation limit.
- Filtration and stabilization (polyphenols).

- Container (white glass, e.g. light taste).
- Consumer expectance (regional and cultural differences).
- Duty of declaration (legal regulations).

A harmonic taste profile directly influences the taste value. Independent of this, different beer types that can be clearly placed onto the beer market can be developed by targeted use of hopping.

According to the Preliminary Purity Law, only conventional hop products (see Figure 3.2) can be added to the wort. It is possible to use raw hops, pellets and extracts [43]. Raw hops necessitate a hop strainer to remove the insoluble parts of the hop from the wort. Insoluble hop components from pellets can be eliminated with the hot trub (e.g. whirlpool). It should be kept in mind that an overdosage influences the whirlpool function in a negative way. If greater volumes of the hot trub reach the fermentation, a raspy bitterness can result. The trub cone in the whirlpool should not touch the border of the whirlpool after drainage.

It is possible that the hop matrix introduces nitrate to the wort and consequently into the beer. The introduction of nitrate can be considerably reduced by using extract and enriched pellets. A nitrate load of 100 mg/100 g hops (9% α-acids lftr) and a raw hop dosage of 8 g α-acids/hl wort would result in an increase of the nitrate concentration of 9 mg/l. Using an extract of the aforementioned raw hops results in a nitrate dosage of 0.05–1 mg/l [44]. The nitrate concentration is reduced in the hop extract by at least 98% By means of CO_2 or ethanol extraction [45].

The employment of enriched hop products (pellets type 45, extract) facilitates the cool storage in the brewery and the dosage in the brewhouse.

For international beers, outside the German purity law, there are a variety of different hop products that have undergone further processing steps (see Figure 3.2). On the one hand, isomerized hop products, including iso-pellets and isomerized kettle extracts, can be charged into the hot wort ('kettle'). On the other hand, the addition can occur before the filtration of the beer. These are called 'downstream products'. Iso-extracts are used to correct the beer bitterness. Reduced iso-extracts increase the stability of the foam and protect against light-induced flavoring of the beer. Since the isomerization of the hops' α-acids occurs at higher temperatures, an influence of the aroma and flavor profile of the beer cannot be excluded [46]. In particular, when using iso-pellets, attention should be paid to gentle processing during the isomerization in the calometric chamber. Hop products enriched with different hop fractions are available to take special requirements into account.

In particular, the reduced iso-extracts have different effects on the sensory perceived intensity of the bitterness. The determination of the sensory bitterness intensity showed that with equal amounts of tetrahydroiso-α-acids and iso-α-acids in the beer, tetrahydroiso-α-acids showed 1.0–1.7 times the bitterness potentials.

In comparison to iso-α-acid, rho-iso-α-acids have 0.6–0.8 times the bitterness potential [47–49]. Indeed, with an increased addition of reduced iso-α-acids the impression of bitterness is stronger, but the quality of the bitters is reduced at the same rate.

Furthermore, a high dosage of tetrahydroiso-α-acids leads to saturation with regard to the sensory bitterness and an improvement of the foam. If the beer possesses an intensive 'flavor matrix' or if the beer is drunk in extreme cold temperature, masking the bitters will lead to a reduction of the bitterness potential. This explains why tetrahydroiso-α-acid has a bitterness potential in water of about 1.7, which can fall to a value of 1.0 depending on the beer type [49].

There are many reasons to add hop products in different steps during the brewing process. Apart from the legal aspects (e.g. Provisional Beer Law), it is the composition of the hop products that is responsible. With a dosage into the final beer, the hop product gets directly into the drink. The sensory properties of the hop products, the solubility in beer, and much more, need to comply with certain requirements. The dosage in the brewhouse ensures that several more process steps are passed. This guarantees the elimination of, for example, beer-insoluble hop components and undesirable aroma of the bittering hops. The use of isomerized products is carried out primarily due to economic reasons and hardly due to qualitative reasons. Bittering hops should be added at the beginning of the boil to preferably achieve an optimal isomerization and steaming out of undesirable aroma notes. Aroma hops can be added after the addition of the bitter resins to obtain a mild bitterness [50].

The later the dosage occurs, the less flavor is steamed out and the more of the hop oil that reaches the beer. Addition of aroma hops for hop-flavored beer should occur towards the end of the boil or as an dosage to the whirlpool to obtain an optimal transfer of flavor into the beer.

As seen in Figure 3.3, early addition can preferably influence the bitterness and the flavor. Aromatization is achieved with the addition at the end of the boil.

3.6.2
Beer with Hop Flavor

A 'weakly hopped beer' is, for example, a beer with about 15–20 bitterness units without any noticeable nasal hop aroma. The original extracts of the bottom-fermented/top-fermented beers are about 11%. The beer should possess a basic bitterness which from a sensory point of view should not come to the fore.

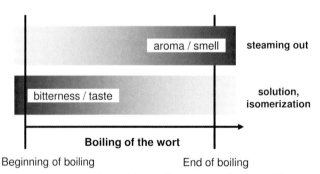

Figure 3.3 Influence of the addition of hops on the character of beer [33].

A 'strongly hopped beer' is, for example, a beer with about 20–40 bitterness units without and noticeable nasal hop aroma. The original extracts of the bottom fermented beers are about 12%. The beer should possess a noticeable bitterness.

The following application examples can be recommended.

- Beers from 100% malt have an intensive flavor matrix due to the exclusive use of malt.
  About 20 bitterness units with weakly hopped beer
  About 30–40 bitterness units with strongly hopped beer, especially since the higher original extracts additionally mask the bitterness

- With the use of adjuncts, the flavor matrix of the beer is not as pronounced and the bitterness is hence perceived sensorically more strongly.
  About 15 bitterness units with weakly hopped beer
  About 20–27 bitterness units with strongly hopped beer

- Stouts and heavy beers should be hopped more intensively to neutralize the sweet and excessive malty aftertaste. A balance needs to be found between alcohol, residual sugar and bitterness to increase the taste value.

- Addition of hops at the beginning of the boil to increase the yield of bitter resins and to relieve the whirl pool.

- Preferably, aroma hops should be added during the first half of the cooking time as this gives the drink an enjoyable bitterness.

- Alternatively, bittering hops should be added if this has proven of value in brewing experiments for one's own beer varieties.

- An increased use of aroma hops in relation to bitter hops leads to a 'milder' and more 'harmonic' bitterness

- Iso-pellets to increase the yield of bitter resins and to introduce besides iso-α-acids also further flavoring

- Isomerized kettle extract during wort boiling to increase the yield of bitter resins and microbiological stability.

- Iso-extract before filtration to correct the bitterness.

- Reduced iso-extract to stabilize the foam (tetra-hydro- or hexahydroiso-α-acids products).

- Combinations of the mentioned products.

3.6.3
Beer with a Hop Flavor

A 'hop-flavored beer', for example, a beer with about 20–40 European Brewery Convention (EBC) bitterness units possessing a noticeable nasal hop aroma will be described. The original extracts of the bottom-fermenting beer are about

10–12%. The beer should have a noticeable hop flavor. This should be perceived as a pleasant hop note during smelling and tasting [51]. The following application examples can be recommended.

- About 30–40 bitterness unit with beers from 100% malt, as the flavoring matrix is very intensive due to the exclusive use of malt.
- About 20–30 bitterness unit with the use of adjuncts, as the flavoring matrix is not as pronounced and hence the bitterness is perceived sensorically more strongly
- The first hop addition at, but not before, the beginning of the boil to increase the yield of bitter resins.
- Preferably, part of the first hop addition as aroma hops as this gives a mild bitterness to the drink.
- Alternatively, as a first addition only bittering hops proven of value for one's own beer varieties in brewing experiments.
- Aromatization by a second hop dosage which, if possible, should not be boiled anymore; thus, the steaming out loss is minimized. This guarantees maximum aroma transfer from the raw material hop to the beer.
- The second hop dosage can occur at the end of the boil in the wort kettle or as an dosage in the whirlpool
- The aroma transfer is favored by cooling the wort before addition to the whirlpool 'pre-cooling' and thus minimizing the steaming-out loss [18, 52, 53]
- Brewing experiments need to be carried out to make the decisions about which aroma hop variety is best suited. Depending on the variety, there are different influences on the flavor and the aroma of the beer.
- The intensity and the quality of the hop flavor can be individually optimized with the point of addition. The time frame for the last hop addition starts about 10 min before the end of the boil and spans to a underback in the whirlpool
- The contribution on the taste of the beer increases as more of the hop matrix (e.g. pellets type 90) is charged. This is applicable for aroma as well as bittering hops.
- Addition of iso-pellets of aroma hops at the end of the boil increases the flavor and bitter resins yield.
- Addition of isomerized kettle extract during wort boiling increases the bitter resins yield and microbiological stability
- Iso-extract before filtration to correct the bitters
- Hop oil or 'hop oil fractions' before filtration for flavoring

- Reduced iso-extracts to stabilize the foam (tetrahydroiso- or hexahydroiso-α-acids products).

- Combinations of the mentioned products

The dosage quantity of aroma hops in breweries is mostly determined by the concentration of α-acids. More logical is the dosage depending on the concentration of the hop oil or, even better, depending on the linalool concentration [54, 55] or depending on the leading component of the hop aroma [22, 56], thus ensuring that the same amounts of aroma substances are always added. Time-dependent fluctuation in the oil concentration can be adjusted. In particular, with subsequent aroma dosage the α-acid dosage is irrelevant as only a very small part isomerizes. A goal of hop addition is maximum aroma transfer. Linalool was found as the character impact compound for hop aromatic beers [20, 57, 58]. The taste threshold value in beer is 2–80 µg/l [18, 59, 60]. A sensoric noticeable hop aroma in beer can be found at a linalool concentration of about 20 µg in lager beers. This value depends on the particular taste profile of the beer. The hop aroma can be described as 'flowery perfumed' from about 50 µg/l. Apart from linalool, other hop aroma substances are involved in the hop aroma. These are in parts below the taste threshold value of beer, but can contribute to the taste profile of the beer via synergistic effects. The taste stability of the beer can be positively influenced by late aroma addition. The stale flavor of beer is masked by a discreet but noticeable hop aroma [24, 61, 62].

3.6.4
Beer Enriched with Xanthohumol

The prenylflavonoid xanthohumol has gained increasing interest in medicine and brewing technology. The enrichment of xanthohumol in beer can be achieved by using certain hop products and adapted technology [63–71].

Xanthohumol can be found in raw hops, pellets, ethanol extracts as well as xanthohumol-enriched hop products. With special brewing technologies, non-isomerized xanthohumol can be conserved in higher concentrations up to beer within the frame of the Bavarian Purity Law. Outside the Bavarian Purity Law, as well as in mixed beer and non-alcoholic beverages, xanthohumol can be added in desired concentrations by special hop products. A stabilization of the beer with PVPP should be avoided with light, xanthohumol-enriched beer, as PVPP removes polyphenol to a large extent from the beer. With dark beer brewed with a fraction of dark specialty malt, PVPP stabilization is possible with a minimal loss of xanthohumol.

3.6.5
Yield of Bitter resins

The following criteria are crucial for the yield of hop components (e.g. bitter resins, aroma, polyphenols): hop variety as well as the type of hop product, amount of hop dosage, original extract, intensity of boiling the wort, wort boil system, pH of the wort, time of the hop in the hot wort, intensity of the trub separation (hot trub, cold trub), re-use of hot trub, yeast technology, temperature of fermentation, construction of the fermentation vessel, and type of filtration and stabilization [38, 72].

Two different methods are commonly used to calculate the bitter resins yield. (i) It is defined as the ratio of the resulting amount of bitter resins in beer (EBC bitterness units) to the used amount of hop calculated as the CBV of ethanol extract and the CV in pellets. (ii) It is also defined as the ratio of iso-α-acids in beer to the α-acids in the CO_2 extract used. The α- and iso-α-acids are determined by HPLC.

The efficiency of the use of isomerized products (outside the Baverian Purity Law) should always be properly inspected in the brewery. The low required quantities of isomerized hops are in contrast to the higher purchase costs. A possibility to calculate the 'break-even' point is shown in the following [1, 73]:

$$\text{EUR/kg } \alpha\text{-acid} = \text{EUR/kg iso-}\alpha\text{-acid} \times \frac{IR_k}{IR_I}$$

$$IR_k = \frac{\text{iso-}\alpha\text{-acid beer}}{\alpha\text{-acid input}}$$

$$IR_I = \frac{\text{iso-}\alpha\text{-acid beer}}{\text{iso-}\alpha\text{-acid input}}$$

where IR_k is the rate of isomerization of conventional hop products and IR_I is the rate of isomerization of the isomerized products. If, for example, the rate of isomerization as well as the price of the iso-products is applied, then the threshold prices of the conventional products are calculated. If the actual market price of the hop products is smaller than the calculated threshold price, then the conventional product is cheaper.

3.6.6
Foam

Hops have a positive influence on the foam stability of beer. In the frame of the Bavarian Purity Law it is possible to increase the concentration of non-isomerized α-acids by the late addition of hops. These are only slightly soluble in beer (around 2 mg/l), but can increase foam stability [74].

Reduced hop products can be used in international beers. The addition of 3–5 mg tetrahydroiso-α-acids per liter of beer enhances the beer foam significantly. Excessive addition can lead to an unnatural, untypical foam structure. Furthermore, attention needs to be paid to the sensory perceived bitterness as explained in Section 3.6.1 [1, 75]. Increased hydration of the iso-α-acids increases the hydrophobicity and hence the positive influence on the beer foam. Because of a lower solubility it can lead to technical problems with the dosage (elimination).

3.6.7
Microbiology

Hops increase the microbiological stability of beer. Non-isomerized α-acids have a 3–4 times stronger antimicrobial effect on beer-damaging microorganisms of the species *Lactobacillus* and *Pediococcus* than iso-α-acids. With late hop addition, the concentration of α-acids can be increased in the beer. Similar to the foam, increasing hydration of the iso-α-acids has a positive effect on the microbiological stability of the beer.

Hop bitter resins can be listed beginning with the worst to the best microbiological beer stability [74] as follows: unhopped beer < iso-α-acids < rho-iso-α-acids < tetrahydroiso-α-acids < 50% tetrahydroiso-α-acids +plus 50% hexahydroiso-α-acids < α-acids.

During the operation of a biological acidification facility attention should be paid such that no hop bitter resins comes into contact with the lactobacillus cultures (*Lactobacillus amylolyticus*), as these do not possess hop tolerance. For the biological acidification facility, re-feeding of the hot trub into the lauter tun may be started only after the first wort is withdrawn.

3.6.8
Addition of 'Downstream Products'

The addition of downstream products to the beer is carried out mostly before filtration. An addition after filtration would increase the yield; however, the risk of turbidity would also increase. The solubility of hop products is in part strongly dependent on pH and temperature. The higher the degree of hydration of the iso-α-acids, the better the stability of the foam and its adhesion to the beer glass (lacing). The solubility in the beer deteriorates simultaneously.

Hence, there are special dosage devices as well as dosage instructions. A mass proportional dosage of partly pre-dissolved hop products to the beer is widely used [1].

The following equation can be applied for the calculation the dosage of iso-extract to the beer:

$$E[\text{mg/l}] = \frac{BE[\text{EBC BU}] * 10000}{K[\%] * Y[\%] * C},$$

where E is the amount of iso-extract to be added, BE is the desired bitterness of the beer express as EBC bitterness units, K is the concentration of iso-extracts, Y is the yield factor and C is the coefficient of the sensory bitterness of the product in relation to the bitterness of iso-α-acids.

The following equation can be applied to correct the bitterness of the beer with iso-extract:

$$E[g] = \frac{(BE-AB)[\text{EBC BU}]*1000*V[\text{hl}]}{K[\%]*Y[\%]*C},$$

where AB is the actual beer bitterness before the addition, V is the volume of beer to be corrected, Y is the yield factor and C is the coefficient of the sensory bitterness of the product in relation to the bitterness of iso-α-acids.

3.7
Storage of Hops

Hops are sensitive to oxidation and chemical alteration of their components. The bittering strength can decrease with poor storage conditions by oxidation of the α-acids. In contrast, at an optimal storage temperature (see Table 3.2), the original given value for the addition of hops can still be assumed since even the developed degradation products in turn contribute to a flavorful acceptable bitterness.

The aroma becomes displeasing and cheese-like with incorrect storage of the hops before and after refining. This applies particularly to raw hops and pellets. Therefore, hops should be stored protected from light, at cold temperature and protected from external influences (odor, water etc.). α-Acids oxidize in the presence of oxygen and isomerize under an inert atmosphere. This explains the loss of α-acids during storage. Hard resin components are already present in raw hops. Furthermore, oxidation products of lupolone can develop in the presence of oxygen (also called polar bittering substances). As a consequence, cold storage of raw hops and pellets is a necessity. Pure resin extracts are considerably more stable than hop umbel and pellets. Table 3.2 shows the advisable storage temperature for hops and hop products.

Table 3.2 Advisable storage temperatures (°C) for hops and hops products [33].

	Storage stability of hops and hops products				
	1 year	2 years	3 years	5 years	8 years
Hop umbel	<5	<−20	not feasible		
Pellets	<5	<5	<5	<−20	not feasible
CO_2 extract	<25	<25	<5	<5	<5
Ethanol extract	<10	<10	<10	<10	<5

References

1. Krottenthaler, M. (2000) Hopfen, in *Praxishandbuch der Brauerei* (ed. K.-U. Heyse), Behr's Verlag, Hamburg, pp. 1–55.
2. Krottenthaler, M., Back, W. and Engelhard, B. (2000) Großversuche mit der neuen Hüller Zuchtsorte 'Hallertauer Merkur' (MR) [Large-scale brewing trials with the new Hüll variety 'Hallertauer Merkur' (MR)], *Hopfenrundschau International*, 43–6.
3. Krottenthaler, M., Zürcher, J., Zarnkow, M. and Back, W. (2001/2002) Großversuche mit der neuen Hüller Zuchtsorte 'Saphir' [Large-scale trials with the new Hüll variety 'Saphir'], *Hopfenrundschau International*, 60–7.
4. CMA (2005) *The spirit of beer, Hops from Germany*, 5th edn, Verband Deutscher Hopfenpflanzer, Wolnzach.
5. Steiner Hopfen GmbH, www.hopsteiner.de (accessed 2009).
6. Verband Deutscher Hopfenpflanzer eV, www.deutscher-hopfen.de (accessed 2009).
7. Joh. Barth & Sohn, www.johbarth.com (accessed 2009).
8. Yakimachief Inc., www.yakimachief.com (accessed 2009).
9. Verzele, M. and De Keukeleire, D. (1991) *Chemistry and Analysis of Hop and Beer Bitter Acids 27*, Elsevier, Amsterdam.
10. Moir, M. (2000) Hops – a millennium review. *Journal of the American Society of Brewing Chemists*, **58** (4), 131–46.
11. Forster, A., Schmidt, R. and Biendl, M. (2004) Wie bestimmt sich der durchschnittliche alpha-Säuregehalt einer Hopfensorte? *Brauwelt*, **22**, 654–60.
12. Back, W. (1994) *Farbatlas und Handbuch der Getränkebiologie 1*, Fachverlag Hans Carl, Nürnberg.
13. Kondo, K. (2003) Preventive effects of dietary beer on lifestyle, in *Proceedings of the European Brewery Convention Congress 2003*, Fachverlag Hans Carl, Nürnberg.
14. Hartl, A. and Reininger, W. (1975) Über die Umwandlung von Lupulonen in bierlösliche Bitterstoffe, in *Proceedings of the European Brewery Convention Congress 1975*, Elsevier, Amsterdam.
15. Kammhuber, K. and Hagl, S. (2001) Statistische Untersuchungen zur Korrelation von Hopfenölkomponente. *Monatsschrift für Brauwissenschaft*, **54** (5/6), 100–3.
16. Benitez, J.L., Forster, A., De Keukeleire, D., Moir, M., Sharpe, F.R., Verhagen, L.C. and Westwood, K.T. (1997) Hops and hop products, in *Manual of Good Practice – European Brewery Convention*, Fachverlag Hans Carl, Nürnberg, p. 30.
17. Steinhaus, M. and Schieberle, P. (2000) Comparison of the most odor-active compounds in fresh and dried hop cones (Humulus lupulus L. variety spalter select) based on GC-olfactometry and odor dilution techniques. *Journal of Agricultural and Food Chemistry*, **48** (5), 1776–83.
18. Kaltner, D. (2000) Untersuchungen zur Ausbildung des Hopfenaromas und technologisch Maßnahmen zur Erzeugung hopfenaromatischer Biere, Dissertation, Technische Universität, München.
19. Steinhaus, M., Fritsch, H.T. and Schieberle, P. (2003) Quantitation of (R)- and (S)-linalool in beer using solid phase microextraction (SPME) in combination with a stable isotope dilution assay (SIDA). *Journal of Agricultural and Food Chemistry*, **51** (24), 7100–5.
20. Kaltner, D., Steinhaus, M., Mitter, W., Biendl, M. and Schieberle, P. (2003) (R)-Linalool as key flavor for hop aroma in beer and its behavior [(R)-Linalool als Schlüsselaromastoff für das Hopfenaroma in Bier und sein Verhalten während der Bieralterung]. *Monatsschrift für Brauwissenschaft*, **56** (11–12), 192–6.
21. Lermusieau, G., Bulens, M. and Collin, S. (2001) Use of GC-olfactometry to identify the hop aromatic compounds in beer. *Journal of Agricultural and Food Chemistry*, **49** (8), 3867–74.
22. Steinhaus, M. (2000) *Wichtige Aromastoffe im Hopfen (Humulus Lupulus L.)*, Verlag Dr Hut, München.
23. Forster, A., Beck, B. and Schmidt, R. (1995) Untersuchungen zu Hopfenpolyphenolen, in *Proceedings of the*

European Brewery Convention Congress 1995, Oxford University Press, New York.

24 Forster, A., Beck, B. and Schmidt, R. (1999) Hopfenpolyphenole – mehr als nur Trübungsbildner in Bier [Hop polyphenols do more than just cause turbidity in beer], *Hopfenrundschau International*, 68–74.

25 Forster, A., Beck, B., Schmidt, R., Jansen, C. and Mellenthin, A. (2002) Über die Zusammensetzung von niedermolekularen Polyphenolen in verschiedenen hopfensorten und zwei Anbaugebieten. *Monatsschrift für Brauwissenschaft*, 55 (5/6), 98–108.

26 Forster, A., Beck, B., Massinger, S. and Schmidt, R. (2004) Niedermolekulare Polyphenole beim Hopfenwachstum. *Brauwelt*, 144 (46–7), 1556–61.

27 Forster, A. and Köberlein, A. (1998) Der Verbleib von Xanthohumol aus Hopfen während der Bierbereitung. *Brauwelt*, 138 (37), 1677–9.

28 Kammhuber, K., Zeidler, C., Seigner, E. and Engelhard, B. (1998) Stand der Erkenntnisse zum Hopfeninhaltsstoff Xanthohumol. *Brauwelt*, 138 (36), 1633–6.

29 Biendl, M. (1999) Grünes Gold. Anticancerogene Aktivität – ein neuer Aspekt bei Hopfeninhaltsstoffen. *Brauindustrie*, 84 (9), 502–7.

30 Biendl, M., Mitter, W., Peters, U. and Methner, F.-J. (2000) Einsatz eines xanthohumolreichen Hopfenproduktes bei der Bierherstellung. *Brauwelt*, 140 (46–47), 2006–11.

31 Biendl, M., Methner, F.-J., Stettner, G. and Walker, C.J. (2004) Brauversuche mit einem xanthohumolreichen Hopfenprodukt. *Brauwelt*, 144 (9/10), 236–41.

32 Back, W., Diener, C. and Sacher, B. (1998) Hefeweizenbier – Geschmacksvarianten und Technologie. *Brauwelt*, 138 (28/29), 1279–84.

33 Back, W., Krottenthaler, M., Bohak, I., Dickel, T., Franz, O., Hanke, S., Hartmann, K., Herrmann, M., Kaltner, D., Kessler, M., Kreisz, S., Kühbeck, F., Mezger, R., Narziss, L., Schneeberger, M., Schütz, M., Schönberger, C., Spieleder, E., Thiele, F., Vetterlein, K., Wunderlich, S., Wurzbacher, M., Zarnkow, M. and Zürcher, J. (2008) *Ausgewählte Kapitel der Brauereitechnologie 2* (ed. W. Back), Fachverlag Hans Carl, Nürnberg, pp. 1–393.

34 Forster, A., Beck, B., Gehrig, M. (1991) Der Verbleib von Pestizidrückständen des Hopfens bei der Herstellung von Hopfenprodukten, in *Proceedings of the European Brewery Convention Congress 1991*, Oxford University Press, New York.

35 Schmidt, R., Anderegg, P. and Biendl, M. (2003) Umweltkontaminanten in Hopfen. *Brauwelt*, 143 (45), 1493–7.

36 Hartl, A. (1992) Kaum Pestizide in Hopfenextrakten. Beitrag zur Frage der Schadstoffreduzierung mit Hopfenprodukten. *Brauwelt*, 133 (7/8).

37 Bradley, L. (2001) Extracting success. *Brewers' Guardian*, 130 (10), 24–26.

38 Kaltner, D. (2005) Moderne Aspekte zur Hopfung des Bieres. *Brauindustrie*, 90 (2), 22–5.

39 European Brewey Convention (1998) *Analytika-EBC (Grundwerk) Aktualisierungslieferung Stand Dezember 2004, IbtEA Committee*, Verlag Hans Carl, Nürnberg.

40 Anger, H.-M. (2005) *Gerste, Rohfrucht, Malz, Hopfen. Analysensammlung der Mitteleuropäischen Brautechnischen Analysenkommission (MEBAK)*, Selbstverlag der MEBAK, Freising-Weihenstephan.

41 Mitter, W. (1989) Analytik und Bewertung von Hopfen und Hopfenprodukten im Hinblick auf die Bierbittere. *Schweizer Brauerei-Rundschau*, 100 (11), 5.

42 Narziss, L. (1992) *Die Bierbrauerei Band II: Die Technologie der Würzebereitung*, 7th edn, Ferdinand Enke Verlag, Stuttgart.

43 Mitter, W., Kessler, H. and Biendl, M. (1999) Geprüfte Variation – Bitterhopfengabe in Form von Pellets und Extrakten und deren Einfluss auf Würze und Bier. *Brauindustrie*, 84 (10), 560–4.

44 Forster, A. (1988) Zur Nitratdosage durch Hopfen und Hopfenprodukte. *Brauwelt*, 128 (6), 188–90.

45 Ertlmaier, S. (1991) Reduzierung der Nitratdosage bei Einsatz von Ethanolextrakten. *Brauwelt*, **131** (42), 1864–5.
46 Ketterer, M., Forster, A., Gahr, A., Beck, B., Massinger, S. and Schmidt, R. (2003) The influence of isomerized hop pellets on beer quality, in *Proceedings of the European Brewery Convention Congress 2003*, Fachverlag Hans Carl, Nürnberg.
47 Guzinski, J.A. (1994) Practical considerations of reduced hop extracts, in *Monograph XXII of the European Brewery Convention Symposium of Hops*, Fachverlag Hans Carl, Nürnberg.
48 Seldeslachts, D., De Winter, D., Mélotte, L., De Bock, A. and Haerinck, R. (1999) The use of high tech hopping in practice, in *Proceedings of the European Brewery Convention Congress 1999*, Fachverlag Hans Carl, Nürnberg.
49 Weiss, A., Schönberger, C., Mitter, W., Biendl, M., Back, W. and Krottenthaler, M. (2002) Sensory and analytical characterization of reduced, isomerized hop extracts and their influence and use in beer. *Journal of the Institute of Brewing*, **108** (2), 236–42.
50 Krottenthaler, M., Kummert, F., Back, W., Kaltner, D. and Mitter, W. (2005/2006) Das Geheimnis der Hopfenblume [The secret of the hop bouquet], *Hopfenrundschau International*, 16–20.
51 Kaltner, D. and Mitter, W. (2003) The influence of hops upon the aroma and taste of beer. *Scandinavian Brewers' Review*, **60** (3), 8–16.
52 Coors, G., Krottenthaler, M. and Back, W. (2000) Auswirkungen einer Würzevorkühlung beim Ausschlagen. *Brauwelt*, **140** (42–43), 1696–9.
53 Coors, G., Krottenthaler, M. and Back, W. (2003) Wort pre-cooling and its influence on casting. *Brauwelt International*, **21** (1/03), 40–1.
54 Mitter, W., Biendl, M. and Kaltner, D. (2001) Behavior of hop derived aroma substances during wort boiling, in *Monograph XXXI of the European Brewery Convention Symposium Flavor and Flavor Stability*, Fachverlag Hans Carl, Nürnberg.
55 Kaltner, D., Thum, B., Forster, C. and Back, W. (2001) Untersuchungen zum Hopfenaroma in Pilsner Bieren bei Variation technologischer Parameter. *Monatsschrift für Brauwissenschaft*, **54** (9/10), 199–205.
56 Fritsch, H.T., Kaltner, D., Schieberle, P. and Back, W. (2004) Entschlüsselung des Hopfenaromas in Bier. *Brauwelt*, **144** (39/40), 1206–7.
57 Fritsch, H.T. and Schieberle, P. (2005) Bewertung von Hopfenaromastoffen in Pilsner Bieren. *Brauerei Forum*, **20** (3), 66–7.
58 Fritsch, H.T. and Schieberle, P. (2003) Changes in key aroma compounds during boiling of unhopped and hopped wort, in *Proceedings of the European Brewery Convention Congress 2003*, Fachverlag Hans Carl, Nürnberg.
59 Fritsch, H.T. (2001) *Einfluss des Hopfens auf wertgebende Aromastoffe in Pilsener Bieren sowie in Zwischenstufen des Brauprozesses*, Institut für Lebensmittelchemie, Technische Universität München, München.
60 Moir, M. (1994) Hop aromatic compounds, in *Monograph of the European Brewery Convention Symposium of Hops*, Fachverlag Hans Carl, Nürnberg.
61 Back, W., Forster, C., Krottenthaler, M., Lehmann, J., Sacher, B. and Thum, B. (1997) Neue Forschungserkenntnisse zur Verbesserung der Geschmacksstabilität. *Brauwelt*, **137** (38), 1677–92.
62 Kaltner, D., Forster, C., Thum, B. and Back, W. (1999) Untersuchungen zur Ausbildung des Hopfenaromas während des Brauprozesses, in *Proceedings of the European Brewery Convention Congress 2003*, Fachverlag Hans Carl, Nürnberg.
63 Biendl, M. (2002/2003) Research on the xanthohumol content in hops [Untersuchungen zum Xanthohumol-Gehalt in Hopfen], *Hopfenrundschau International*, 72–75.
64 Mitter, W., Biendl, M. and Kaltner, D. (2004) Possible effects of prenylated flavonoids and their presence in beer, in

Proceedings of the 28th Convention of the Institute and Guild of Brewing – Asian Pacific Section, Hanoi.

65 Wunderlich, S., Zürcher, J., Back, W., Frank, N., Hussong, R., Gerhäuser, C., Berwanger, S. and Becker, H. (2004) Mehr Xanthohumol im Bier für mehr Gesundheit? in *33. Deutscher Lebensmittelchemikertag*, Gesellschaft Deutscher Chemiker, Lebensmittelchemische Gesellschaft, Bonn.

66 Forster, A., Ketterer, M. and Gahr, A. (2003/2004) A new xanthohumol-enriched hop-extract and its use for unfiltered beers [Ein neuartiger Xanthohumol-angereicherter Hopfenextrakt und sein Einsatz für unfiltrierte Biere], *Hopfenrundschau International*, 65–71.

67 Forster, A., Gahr, A. and Ketterer, M. (2002) Xanthohumol in Bier – Möglichkeiten und Grenzen einer Anreicherung. *Monatsschrift für Brauwissenschaft*, **55** (9/10), 184–94.

68 Walker, C.J., Biendl, M. and Lence, C.F., (2004) Investigation into the high levels of xanthohumol found in stout and porter style beers. *Brauwelt International*, **22** (2), 100ff.

69 Biendl, M., Methner, F.-J., Stettner, G. and Walker, C.J. (2004) Brauversuche mit einem xanthohumolreichen Hopfenprodukt. *Brauwelt*, (9/10), 236ff.

70 Back, W., Zürcher, J. and Wunderlich, S. (2004) Verfahren zur Herstellung eines Xanthohumolhaltigen Getränkes aus Malz- und/oder Rohfruchtwürze sowie derart hergestelltes Getränk, Patent, DE 10256166 A1.

71 Wunderlich, S., Zürcher, J. and Back, W. (2005) Enrichment of xanthohumol in the brewing process, *Molecular Nutrition and Food Research*, **49**, 874–81.

72 Kaltner, D., Mitter, W., Binkert, J., Preis, F., Zimmermann, R. and Biendl, M. (2004) Würzekochsystem SCHOKO und Hopfenkomponenten. *Brauwelt*, **144** (46/47), 1562–7.

73 Forster, A. (1998) Hopfen – ein natürlicher Rohstoff oder Basis für maßgeschneiderte Moleküle? *Mitteilungen Österreichisches Getränke Institut*, (3/4), 28–34.

74 Kaltner, D., Bohak, I., Forster, A., Gahr, A. and Back, W. (2001) Investigation of the influence of hop products on the microbiological stability of beer, in *Proceedings of the European Brewery Convention Congress 2001*, Fachverlag Hans Carl, Nürnberg.

75 Smith, R.J., Davidson, D. and Wilson, R.J.H. (1998) Natural foam stabilizing and bittering compounds derived from hops. *Journal of the American Society of Brewing Chemists*, **56** (2), 52–7.

4
Brew Water
Martin Krottenthaler and Karl Glas

4.1
General Requirements

Water, in terms of quantity, is the most important raw material of beer. The chemical and biological composition of water therefore has a significant relevance in beer production, and there is no step in the brewing process that is not influenced by the constituents of water. Consequently, treatment of water is in many cases necessary. Since the breweries and beverage industry obtain their process water either from public mains and/or from their own wells, water treatment needs to be addressed in two aspects:

- Treatment of crude water to fulfill legal criteria.
- Treatment of drinking water due to technological brewing requirements.

4.2
Characteristics of Constituents Relevant for Brewing

The total hardness is defined as the sum of all alkaline-earth ions (i.e. calcium, magnesium, strontium, and barium). In practice, strontium and barium ions are neglected, so that the total hardness is determined from the calcium and the magnesium hardness. In turn, the total hardness is divided in carbonate and non-carbonate hardness. Common counter ions for non-carbonate hardness are sulfate, nitrate and chloride, and hydrogen carbonate for the carbonate hardness. See Figure 4.1

If the carbonate hardness is greater than the total hardness, the carbonate hardness will be put equal to the total hardness. In these cases, water contains raised sodium concentrations. This is known as soda alkalinity. The total hardness is determined either volumetrically or in single analysis of calcium and magnesium ions with stoichiometric calculations. The carbonate hardness corresponds to the part of the alkaline-earth ions equivalent to hydrogen carbonate and carbonate ions present in water. These ions are measured by determining the m value, which is

Handbook of Brewing: Processes, Technology, Markets. Edited by H. M. Eßlinger
Copyright © 2009 WILEY-VCH Verlag GmbH & Co. KGaA, Weinheim
ISBN: 978-3-527-31674-8

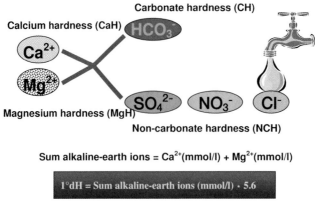

Figure 4.1 Hardness proportions of water.

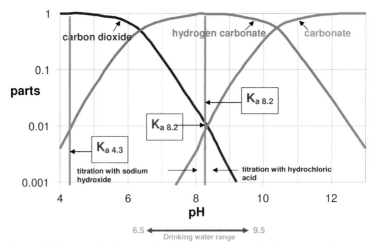

Figure 4.2 Forms of carbonic acid at 25 °C (according to Sontheimer et al.).

defined as the acid consumption up to pH 4.3 (see Figure 4.2, acid capacity $K_{a\,4.3}$) in which methyl orange is used as indicator [1].

Since only half of the carbonates are captured when titrated with phenolphthalein (p value, $K_{a\,8.2}$), it is possible to calculate the carbonate and hydrogen carbonate concentration as well as the hydroxide concentration from the p and m values as shown in Table 4.1.

The non-carbonate hardness is a pure calculation value and is calculated from the difference of the total hardness and the carbonate hardness. Consequently, a particular concentration of the alkaline-earth ions exists in the presence of an equivalent concentration of counter ions (e.g. sulfate, chloride, nitrate, etc.) as shown in Figure 4.1.

Table 4.1 p and m values of water (mval/l) [1].

Titration results	Hydroxide OH⁻	Carbonate CO_3^{2-}	Hydrogen carbonate HCO_3^-
$p = 0$	0	0	m
$p < 0.5\,m$	0	$2p$	$m - 2p$
$p = 0.5\,m$	0	$2p$	0
$p > 0.5\,m$	$2p - m$	$2(m - p)$	0
$p = m$	m	0	0

Table 4.2 Conversion of water hardness units [2].

Hardness	$c\,(Ca^{2+} + Mg^{2+})$ (mmol/l)	$c\,(Ca^{2+} + Mg^{2+})$ (mval/l)	$CaCO_3$ (ppm)	°D	°GB	°F
$c\,(Ca^{2+} + Mg^{2+})$ (mmol/l)	1	2	100	5.6	7.0	10.0
$c\,(Ca^{2+} + Mg^{2+})$ (mval/l)	0.5	1	50	2.8	3.51	5.0
$CaCO_3$ (ppm)	0.01	0.02	1	0.056	0.07	0.10
°D	0.1786	0.357	17.85	1	1.25	1.786
°GB	0.1458	0.285	14.29	0.7999	1	1.429
°F	0.10	0.20	10.00	0.5599	0.70	1

In Germany, the degree of hardness of water is defined as 10 mg CaO/l. The most important expressions based on the German degree of hardness are [1].

Calcium hardness (°d)   = mg/l Ca^{2+} × 0.1399
Magnesium hardness (°d)   = mg/l Mg^{2+} × 0.2306
Total hardness (°d)   = (mg/l Ca^{2+} × 0.1399) + (mg/l Mg^{2+} × 0.2306)
Total hardness (mval/l)   = (mg/l Ca^{2+} × 0.0499) + (mg/l Mg^{2+} × 0.0822)
Carbonate hardness (°d)   = m value × 2.804
Non-carbonate hardness (°d) = total hardness (TH) − carbonate hardness (CH)
If CH > TH   CH = TH (soda alkalinity)

The definition of an English or French degree of hardness is based on similar considerations, whereupon calcium carbonate is the reference parameter. The conversion factors for German (D), French (F) or English (GB) degrees of hardness are presented in Table 4.2.

Numerous chromatographic and spectroscopic methods for the analysis of anions and cations are available in modern water analysis. Ion balance is carried out as a plausibility check, where it is checked if the concentration of cations corresponds to the equivalent concentration of anions. The difference of the ion balance should be smaller than 5%.

4.3
Quality Criteria for Brew Water

Apart from the legal requirements, additional quality criteria for brew water need to be complied with, since water ions influence the pH value of mash, wort and beer, and thus enzymatic and non-enzymatic reactions (see Chapter 7). Consequently, these have a considerable influence on the acidity. Acidity is understood as the overall available dissociated and non-dissociated acid former. Hydrogen carbonate ions count as acid destroying since they lead to an increase in the pH value. Calcium (Ca^{2+}) and magnesium ions (Mg^{2+}) are acidity supporting and lead to a pH decrease of the mash. The chemistry of water ions on the acidity of the mash is shown in the following:

Acidity diminishing: hydrogen carbonate ions:

$$HCO_3^- + H^+ \quad H_2O + CO_2\uparrow$$

Hydrogen carbonate ions are acidity diminishing as they remove protons from the wort. Carbonic acid is formed, which disappears during heating.

Acidity diminishing: e.g. calcium hydrogen carbonate:

$$H_2PO_4^- \rightarrow HPO_4^{2-} + H^+$$
$$\underline{HCO_3^- + H^+ \rightarrow CO_2\uparrow + H_2O}$$
$$H_2PO_4^- + HCO_3^- \rightarrow HPO_4^{2-} + CO_2\uparrow + H_2O$$

Hydrogen carbonate bound to calcium, magnesium or sodium decreases the acidity since acidic-acting primary phosphates (KH_2PO_4) are converted to basic-acting secondary phosphates (K_2HPO_4). Carbonic acid is formed in the process, which is then expelled.

Calcium hydrogen carbonate:

$$2\,KH_2PO_4 + Ca(HCO_3)_2 \rightarrow CaHPO_4\downarrow + K_2HPO_4 + 2\,H_2O + 2\,CO_2\uparrow$$
$$4\,KH_2PO_4 + 3\,Ca(HCO_3)_2 \rightarrow Ca_3(PO_4)_2\downarrow + 2\,K_2HPO_4 + 6\,H_2O + 6\,CO_2\uparrow$$

Magnesium hydrogen carbonate:

$$2\,KH_2PO_4 + Mg(HCO_3)_2 \rightarrow MgHPO_4 + K_2HPO_4 + 2\,H_2O + 2\,CO_2\uparrow$$

Sodium hydrogen carbonate:

$$2\,KH_2PO_4 + 2\,NaHCO_3 \rightarrow Na_2HPO_4 + K_2HPO_4 + 2\,H_2O + 2\,CO_2\uparrow$$

The calcium ion supports the acidity. It reacts with the secondary basic phosphate (as the main constituent of the wort pH), resulting in the insoluble tertiary calcium phosphate, thus releasing a proton. Reaction with magnesium ions follows the same process. However, the resulting magnesium phosphate is only insoluble at high temperatures.

Acidity supporting: Calcium and magnesium ions:

$$3\,CaSO_4 + 4\,K_2HPO_4 \leftrightarrow Ca_3(PO_4)_2 \downarrow + 2\,KH_2PO_4 + 3\,K_2SO_4$$

$$3\,MgSO_4 + 4\,K_2HPO_4 \leftrightarrow Mg_3(PO_4)_2 + 2\,KH_2PO_4 + 3\,K_2SO_4$$

Calcium may release protons even from weak organic acids, which are present in the wort in the undissociated form, by forming insoluble salts [3].

To compare the acidity-diminishing and acidity-supporting ions, Kolbach [4–6] introduced the term residual alkalinity:

Residual alkalinity = total alkalinity − balanced alkalinity.
Residual alkalinity = carbonate hardness − (calcium hardness + 0.5 magnesium hardness)/3.5.

The total alkalinity represents the hydrogen carbonate ions, which are determined as carbonate hardness using the acid capacity. Balanced alkalinity is the acidity-supporting calcium and magnesium ions, expressed as calcium and magnesium hardness. The magnesium hardness has only half the impact of the calcium hardness. Kolbach postulated that the acidity-destroying effect of 1° of carbonate hardness is balanced by 3.5° of calcium hardness.

A residual alkalinity of zero means that the acidic-supporting and acidic-diminishing properties of the brew water ions cancel each other out. This water acts like distilled water, as it does not influence the pH of the mash. Thus, the pH value of the mash/wort is exclusively dependent on the malt used. An increase of the residual alkalinity by 10°dH results in an increase in pH of the mash by 0.3 units.

A residual alkalinity between −2 and 2°dH for Pilsner beers, <5°dH for light beers, and <10°dH for dark beers should be aimed for.

Table 4.3 Molar mass and conversion factors at the calcium dosage.

Ion	Molar mass (g/mol)	40 g calcium ions (Ca^{2+}) correspond to:	
Ca^{2+}	40.078	$CaCl_2$	111 g
S^{2-}	32.066	$CaCl_2 \cdot 4\,H_2O$	183 g
O^{2+}	15.999	$CaSO_4$	136 g
Cl^-	35.453	$CaSO_4 \cdot 2\,H_2O$	172 g
H^+	1.0079	CaO	56 g

For sensory reasons as well as for gushing prophylaxis, the ratio of non-carbonate hardness to carbonate hardness should be 3.1. To increase the non-carbonate hardness, calcium is added as calcium chloride or calcium sulfate.

In the following example the calcium hardness of brew water should be increased by 10 °dH by the addition of calcium sulfate [the required amount of calcium sulfate (gypsum) in grams is calculated for 1 hl brew water]:

$$\begin{aligned}
1°\text{dH} \quad &\text{corresponds to } 10\,\text{mg CaO/l} \\
&= 1000\,\text{mg CaO/hl} \\
&= 1000:56 \times 40 \\
&= 714\,\text{mg Ca}^{2+}/\text{hl} \\
10°\text{dH} &= 7140\,\text{mg Ca}^{2+}/\text{hl} \\
&= 7140:40 \times 136 \\
&= 24.3\,\text{g CaSO}_4/\text{hl} \\
&= 30.7\,\text{g CaSO}_4 \cdot 2\text{H}_2\text{O}/\text{hl}
\end{aligned}$$

Additional reference values and quality criteria for brew water are presented in Table 4.4.

4.4
Water Treatment

There are a number of chemical and physical alternatives available for water treatment. In the scope of this chapter, only the most common methods will be discussed. In recent years the following compounds have had to be disposed of as the most common problematic materials: iron, manganese, nitrates, halogenated hydrocarbons as well as pesticides.

4.4.1
Removal of Problematic Inorganic Substances

4.4.1.1 Deferrization and Demanganization

This process is necessary – even from a brew technological point of view – if the iron concentration is above 0.12 mg/l and the manganese concentration is above 0.05 mg/l. Through oxidation with adequate aeration devices, water-insoluble compounds are formed and removed from the water by filtration.

Through aeration, the soluble iron(II) compounds are transferred to insoluble iron(III) compounds. Organically bound iron is precipitated by the addition of oxidants (e.g. cupper permanganate and flocculant Fe(II) salts). The reaction is pH dependent, where the oxidation rate increases with higher pH.

Demanganization is a considerably more sensitive process. Manganese(II) is oxidized to manganese(IV). A precipitation of manganese oxide is achieved at a pH above 9 by the addition of oxidants (e.g. potassium permanganate). A bio-

Table 4.4 Quality criteria of brew water [7, 8].

Characteristic	Value	Reason
pH	7–8	too acidic: danger of corrosion; too basic: inhibition of enzymes
p	0–0.3 mval/l	water does not contain aggressive CO_2, but only low fraction of CO_3^{2-} and OH^- ions
m	0.7–1.2 mval/l	only low residue of acid destroying HCO_3^-; low fraction but positive for palateful taste
Non-carbonate hardness	at least twice, better three times the carbonate hardness	balanced alkalinity
Residual alkalinity	−2 to 2 °dH <5 °dH <10 °dH	for Pilsner beers for light beers for dark beers
Sulfate	100–150 mg/l	dry bitterness, tendency to a hop aroma
Chloride	<100 mg/l	salty taste, corrosion
Nitrate	<25 mg/l	fermentation disturbances are avoided; low value is better as nitrate is also introduced into the beer by hop and malt
Iron	<0.1 mg/l	flaws in taste, danger of gushing, danger of turbidity, beer taste in stability
Free aggressive CO_2	–	danger of corrosion

chemical oxidation by manganese bacteria can occur simultaneously on the filter bed. Consequently, a longer adjustment period during demanganization is to be expected as compared to the deferrization filter.

Two separated filter units are normally advantageous for the removal of iron and manganese. Deferrization occurs in the first filter (saving cleaning water, iron compounds are easily rinsed off). In the second filter demanganization is carried out (increased holding time, gradual coarsening of the filter material through black manganese dioxide).

4.4.1.2 Nitrate Reduction

Biological as well as chemical methods are used for the removal of nitrated from crude water (see Sections 4.4.3.2 and 4.4.3.3).

4.4.2
Removal of Problematic Organic Substances

4.4.2.1 Aeration
Volatile substances (e.g. halogenated hydrocarbons) can be expelled to a large extent by 'stripping' the water with air (e.g. intensive aeration).

4.4.2.2 Activated-Carbon Filtration
Activated-carbon filtration is useful for polishing and removal of problematic substances in the water. There, contaminants adsorb onto the surface of the activated carbon and are thus fixed. In this repeatedly recommended and used process, the problem is shifted as this alternative provides no elimination and mineralization (see Section 4.4.2.3) of these problematic substances. The activated-carbon filter serves, among other things, for the removal of:

- Odorous substances and flavorings.
- Discoloration.
- Organic water constituents (e.g. pesticides).
- Halogenated hydrocarbons.
- Chlorine and chlorine derivatives.
- Microorganisms.

In order to enable sterilization of the filters and thus avoid microbiological contamination, the activated-carbon filter should be made out of stainless steel

4.4.2.3 Combination Processes Using Oxidation/UV Irradiation
Both sterilization and mineralization of organic contaminants in the water are possible through the combination of chemical and physical reaction mechanisms. The chemical oxidation occurs with hydrogen peroxide or ozone. Thin-walled radiators with high application rates are used in the subsequent UV irradiation.

4.4.3
Common Processes for Brew Water Treatment

The composition of the brew water is of great importance for the quality as well as the character of the beers. Apart from the change in hardness former (softening), the reduction of nitrate concentration in the brew water plays an ever-increasing role.

4.4.3.1 Lime Softening
Lime softening is a classical process. Through the addition of saturated lime water, the calcium concentration and at sufficient basicity (pH > 10) even the magnesium concentration are reduced. The added lime slurry is quantitatively deposited as calcium carbonate or magnesium hydroxide. The mode of action is clarified in the following equations:

$$CO_2 + Ca(OH)_2 \rightarrow CaCO_3\downarrow + H_2O$$

$$Ca(HCO_3)_2 + Ca(OH)_2 \rightarrow 2\,CaCO_3\downarrow + 2\,H_2O$$

$$Mg(HCO_3)_2 + Ca(OH)_2 \rightarrow CaCO_3\downarrow + MgCO_3 + 2\,H_2O$$

$$MgCO_3 + Ca(OH)_2 \rightarrow CaCO_3\downarrow + Mg(OH)_2\downarrow\ (pH > 10)$$

$$2\,Na(HCO_3) + Ca(OH)_2 \rightarrow CaCO_3\downarrow + Na_2CO_3 + 2\,H_2O$$

The efficiency of this softening process is influenced by the following factors:

- High temperatures: above 12 °C.
- Strong convection: stirring, pumping, fluidized-bed reactor.
- Crystallization seeds: contact sludge, sand.
- Pressure or rapid decarbonization.

The pH of the brew water after softening is relatively high at about 8.5. This has no influence on the mash pH as the buffering capacity of the water is low. To control the quality of the brew water, the p and m values are determined. The m value serves to determine the softening effect; the p value gives an indication of the correct dosage of lime. See Figure 4.3.

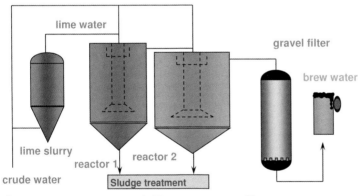

Figure 4.3 Lime precipitation unit: Morgenstern system [9].

4.4.3.2 Ion Exchange

The use of ion exchangers is widespread in the beverage industry for demineralization of crude waters, since this process allows selective removal of cations as well as anions. The hardness former of the water (calcium, magnesium, sodium) can be removed by using weak and/or strong acidic ion-exchange resins. In the subsequent use of weak basic resins, the water can be completely demineralized. Nitrate-selective (strong basic) resins, on the other hand, only reduce the nitrate concentration.

Ion exchangers are solid materials that are able to adsorb cations and anions from an electrolytic solution, and replace these with an equivalent of amount ions of the same charge. In a cation exchanger, sulfonic acid residues are fixed in the matrix of the resin. The corresponding sodium ions can move freely within the structure. Cations, such as calcium ions, enter without problem; anions, such as chloride ions, are repulsed by the negatively attached ions. See Figure 4.4.

A differentiation of ion-exchangers is presented in Table 4.5. Only legally licensed regenerating materials are allowed to be used to regenerate the ion exchanger. Furthermore, the resins need to be safe for all foods – no flavorful, olfactorial or health-damaging materials are allowed to be given off.

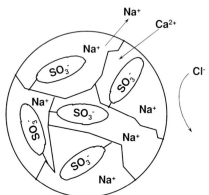

Figure 4.4 Incision through a cation-exchanger particle.

Table 4.5 Differentiation of ion exchangers.

Type		Removal of	Anchoring groups	Counter ion	Regenerant: aqueous solution of
Cation exchanger					
	strong acidic	Ca, Mg, Na	Cl, SO$_4$, NO$_3$, HCO$_3$	H$^+$ Na$^+$	HCl, H$_2$SO$_4$ NaCl
	weak acidic	Ca, Mg	HCO$_3$, CO$_3$	H$^+$ H$^+$ Na$^+$	HCl, H$_2$SO$_4$ CO$_2$ NaOH
Anion exchanger					
	strong basic	all anions Cl, SO$_4$, NO$_3$, HCO$_3$	all cations Na, K, Ca, Mg	Cl$^-$ SO$_4^{2-}$	NaCl, HCl Na$_2$SO$_4$, H$_2$SO$_4$
	weak basic	anions of strong acids Cl, SO$_4$, NO$_3$	all cations Na, K, Ca, Mg	Cl$^-$ OH$^-$ (free basic form) HCO$_3^-$	NaCl NaOH CO$_2$

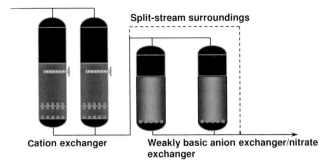

Figure 4.5 Schematic presentation of an ion exchanger used for brew water treatment [9].

The following points should be considered for the planning of an ion-exchanger plant:

- Free from floating materials.
- Polishing filter in the case of organically polluted waters.
- Deferrization and demanganization.
- Free of chlorine and chlorine derivates.
- Activated-carbon filter prior to softening to protect the ion-exchange material.

A common process combination of a cation and anion exchanger is presented in Figure 4.5. The removal of the hardness former is carried out through cation exchange. The optimal following anion exchanger can be designed either to serve as a weak basic resin for complete demineralization or as a strong basic resin to selectively remove nitrate. Depending on the choice of the combination processes, a special regeneration scheme is applied.

4.4.3.3 Membrane Processes

Reversed osmosis is normally used for brew water treatment. In this process a high pressure is administrated onto the crude water. This must be greater than the osmotic pressure, so that the water molecules are forced to migrate through a semipermeable membrane, whereas the salts are mostly retained. Thereby, the water is completely demineralized. Adjustment of the desired brew water quality (see Table 4.4) can be achieved by, for example, a blend with crude water, an increase of hardness with lime slurry and so on

Generally, filtration processes can be classified into their separation spectra, as shown in Figure 4.6.

For a better overview, the brewery-relevant processes and necessary definitions are explained by referring to Figure 4.7.

- *Semipermeable membrane.* A semipermeable membrane is partly permeable (i.e. only a certain part of a substance or mixture is able to pass). However, it is permeable towards the solvent.

- *Permeate.* The permeate is the cleaned solution of the membrane process (i.e. the part of the feeding water that migrates through the membrane).

4 Brew Water

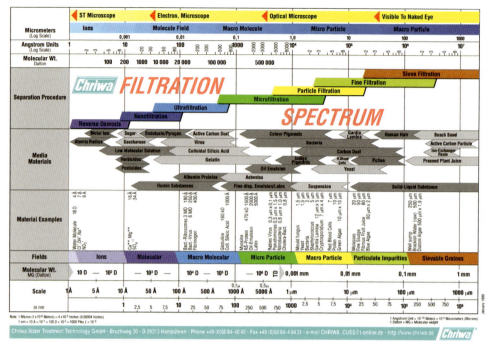

Figure 4.6 Classification of the filtration processes [10].

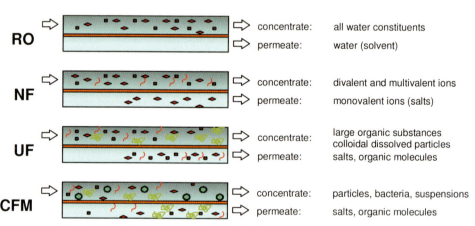

Figure 4.7 Filtration processes applied in the brewery.
RO = reverse osmosis; NF = nanofiltration; UF = ultrafiltration;
CFM = cross-flow microfiltration.

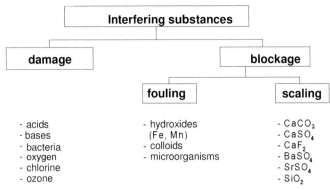

Figure 4.8 Fouling and scaling (blockage of the membranes).

- *Concentrate (retentate).* In reversed osmosis the concentrate is the concentrated water that, corresponding to the volume of obtained permeate, contains the dissolved substances of the feed in concentrated form.
- *Yield.* The yield is the rate of permeate volume to the feed volume: yield (%) = permeate/crude water × 100%. It is important to note that with increasing yield the salt concentration in the permeate also increases. An optimum between quality of the permeate and the yield of the permeate needs to be established for every use. The yield of permeate (i.e. the ratio of treated water to introduced crude water) can be arbitrarily adjusted through variations of pressure and flow rate. A brew water treatment plant commonly works with a yield of about 80%.

For an economical operation mode of membranes, there are a number of limiting factors, which either damage the membranes or lead to a blockage as shown in Figure 4.8. For example, the barium concentration of the crude water should be below 0.1 mg/l to avoid irreversible precipitation of barium sulfate onto the membrane.

References

1 Anger, H.-M. (2005) *Wasser. Analysensammlung der Mitteleuropäischen Brautechnischen Analysenkommission (MEBAK).* Selbstverlag der MEBAK, Freising-Weihenstephan.

2 Brüchler, F., Carter, D., Fandrey, P., Fischer, J., Mann, R., Born, R. and Knapp, P. (2000) *Grundlagen für die Planung von Kreiselpumpenanlagen,* Vol. 7, Sterling SIHI, Itzehoe, p. 381.

3 Narziss, L. (1992) *Die Bierbrauerei Band II: Die Technologie der Würzebereitung,* 7th edn, Ferdinand Enke Verlag, Stuttgart.

4 Kolbach, P. (1941) Über den Ausgleich der Carbonatwirkung durch den Gips des Brauwassers. *Wochenschrift für Brauerei,* **58** (36), 195–8.

5 Kolbach, P. (1941) Zur Beurteilung gipshaltiger Brauwässer. *Wochenschrift für Brauerei,* **44** (58), 231–3.

6 Kolbach, P. (1953) Der Einfluss des Brauwassers auf den pH von Würze und Bier. *Monatsschrift für Brauerei,* **6** (5), 49–52.

7 Bückle, J., (1996) Umweltrelevante Aspekte der Brauwasseraufbereitung mittels

Ionentauscher, Dissertation, Technische Universität München, München.
8 Kolbach, P. (1964) Der Einfluss des Magnesiumsulfats im Brauwasser auf die Zusammensetzung und Qualität des Bieres. *Monatszeitschrift für Brauerei*, 206–9.
9 H. H. Eumann GmbH, www.euwa.de (accessed 2009).
10 Chriwa Water Treatment GmbH, www.chriwa.de (accessed 2009).

5
Yeast

Christoph Tenge

5.1
Brewing Yeast

Yeast is known more for what it can do than what it is [1].

5.1.1
History of Yeast Research

Yeast was employed to carry out fermentation more or less from the early beginnings of brewing. Although microorganisms or even yeast were not specifically known at that time, it was a well-known secret in the Middle Ages that the best beers were produced next to bakeries. The Bavarian Purity Law (Reinheitsgebot), one of the oldest restrictions on food, which regulates German beer production, was also proclaimed without any knowledge of yeast.

First progress was made with the development of the microscope by Antonie van Leeuwenhoek (1652–1725) [2]. Gay-Lussac set up the first complete formula for the fermentation of sugar; however, he still believed in fermentation as an oxidative process [3].

An important advance was made in 1837 by Theodor Schwan through his experiments involving sterilized grape juice. Under the microscope he found small proliferating items and concluded that these items were living organisms. He called them sugar fungi. Meyen continued Schwan's studies and named these microorganisms *Saccharomyces* (sugar fungi) [4]. Although the evidence for the relationship between yeast and fermentation had been found, leading chemists at that time, like Liebig and Wöhler, still criticized these results.

The emerging brewing industry in the nineteenth century already knew about the importance of yeast and many brew masters who set up new businesses brought along their own yeast. The first differentiations between top- and bottom-fermenting strains were introduced.

Between 1855 and 1876, Luis Pasteur published his fermentation theory and distinguished between aerobic and anaerobic utilization of sugars by yeast [5]. In

1897, Eduard Buchner achieved fermentation by cell-free extracts, making it practical to study the biochemistry of fermentation *in vitro*. It took almost 100 years to elucidate the total biochemical pathways of fermentation. Again, it was Louis Pasteur who discovered the production of fermentation byproducts (glycerol) in 1860. Arthur Harden and William Young showed that phosphate is required for glucose fermentation. Furthermore, they set up the fermentation formula: $C_6H_{12}O_6 \rightarrow 2\ C_2H_5OH + CO_2$.

The importance and fate of pyruvate was examined by Neubauer, Fromherz and Neuberg in 1911. Much progress in this field was made between 1910 and 1940. A number of scientists were involved, such as Embden, Meyerhof, Parnas and Warburg. They discovered important reactions, the impact of coenzymes, leading finally to the enlightenment of the whole fermentation pathway [6].

In 1929, the Crabtree effect was identified by the homonymous scientist. The effect describes a limited respiratory activity in yeast in the presence of a certain sugar concentration. It impacts almost the whole yeast metabolism under brewing conditions [7].

Probably the most important steps in yeast research concerning the brewing industry were made by Hansen. In 1883, he cultivated the first pure cultures of brewing yeast and implemented them in breweries. The first cultivated cultures were named 'Carlsberg 1 and Carlsberg 2' because he carried out his studies in the Carlsberg Brewery Laboratories. Hansen adapted methods invented by the famous physician Robert Koch, isolating single cells on plate agar cultures. His work was the first that allowed working with pure cultures, not only free from beer spoilage bacteria, but also void of wild yeasts [8]. Lindner elaborated Hansen's method and cultivated single cells in small wort drops on an object slide under the microscope. He also isolated two different bottom-fermenting yeast strains in 1930: type 'Frohberg' (strong fermenting) and type 'Saatz' (weak fermenting) [9]. Hansen and Lindner also took great care about transferring the isolated cultures from the laboratory into the brewery. Both developed simple propagation devices in order to grow greater yeast amounts and pitch larger fermentation vessels. In 1913, Coblitz and Stockhausen introduced a method of growing yeast by transferring it during the high kräusen stage [10]. This process is still the basis for yeast propagation today.

Since then, much progress has been made in the field of yeast research. The possibility of modern molecular biology has provided major insights into the yeast cell (e.g. into yeast genomics and yeast physiology). These results have more or less affected modern yeast management in today's breweries. The first beers have now been brewed with molecular biologically modified yeasts. In addition, yeast has become the model microorganism in cell research and its genome was the first of any eukaryotic cell to be completely decoded.

5.1.2
Yeast for Brewing Applications

Apart from some yeasts used in specialty beers, the yeasts used for brewing all belong to the genus *Saccharomyces*.

Table 5.1 Overview of yeast classification [12].

Fungi (kingdom)
   Eumycota (division)
     Ascomycotina (subdivision)
       Hemiascomycetes (class)
         Endomycetales (order)
           Saccharomycetaceae (family)
             Saccharomyces (genus)
     Basiomycotina (subdivision)
       Tremellales (order)

Yeasts are fungi that are mainly unicellular and reproduce vegetatively by budding. *Saccharomyces* is one of about 40 ascosporogenous yeasts [11]. An outline of yeast classification is given in Table 5.1. Currently 10 species of *Saccharomyces* are recognized: *S. bayanus, S. castelli, S. cerevisisae, S. dairensis, S. exiguous, S. kluyveri, S. paradoxus, S. pastorianus, S. servazii* and *S. unisporus*. Two types of *Saccharomyces* yeasts are involved in beer fermentation, top-fermenting (or ale) yeasts and bottom-fermenting (or lager) yeasts. Since the last century, ale yeasts have been classified as *S. cerevisiae*, whereas lager yeasts have been known under several names such as *S. carlsbergensis, S. uvarum* and *S. cerevisiae*. Both yeast species belong to the closely related *Saccharomyces sensu stricto* species [13]. Although closely related, employing molecular typing techniques has revealed several genetic differences between top- and bottom-fermenting yeasts [14]. Ale yeasts are still classified as *S. cerevisiae*, but for lager yeasts the classification is more complicated. The evolution of lager yeast seems to be the result of hybridization between *S. cerevisiae* and another *Saccharomyces* yeast (*S. monacensis* or *S. bayanus* may be the other progenitor). To make things more confusing, *S. monacensis* is used as a synonym for *S. pastorianus* and lager yeast is currently classified as *S. pastorianus* [15]. This classification is taxonomically most precise, but is less helpful for brewers. In breweries, *S. pastorianus* is well known as a beer-spoiling wild yeast. The culture yeast can hardly be compared with it. Since the yeast's fermentation properties are more important in breweries than taxonomy, it is recommended to continue referring to lager yeast as *S. carlsbergensis*. In addition, *S. carlsbergensis* is sometimes used as a synonym for *S. pastorianus*, as well.

From a brewer's point of view, there are several differences between ale and lager yeast with vital consequences for the brewing process. Thus, it is still sensible to maintain these two kinds of yeasts as separate types. Since a lot of these top-fermenting yeasts are employed in cylindroconical tanks nowadays and are also cropped from the cone like bottom-fermenting yeast, the segregation becomes blurred when using modern fermentation technology. Most differences between these two types of yeast are in their fermentative ability, rate of sugar utilization, tolerance to temperature, flocculation characteristics and profile of volatiles. Top

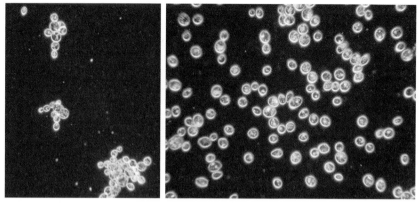

Figure 5.1 Bottom- (left) and top-fermenting yeasts (right) under the microscope (×640 magnification).

fermentation produces beers that are more fruity and estery, whereas lager fermentation provides beers with a purer and partly sulfurous aroma [16].

Morphological differences between ale and lager yeasts are small. Shape and size of the cells cannot be used for differentiation. Under the microscope both types of yeast can only be distinguished by their budding characteristics. Lager yeasts separate very soon after budding, and mother and daughter cells then bud again. This results in single or pairs of cells in the microscope image. The cells of top-fermenting yeasts still stick together when they bud again. As a result of multilateral budding, a small complex cell cluster is formed. Side branches break away and build new clusters. A differentiation can be made under the microscope (see Figure 5.1). At the end of fermentation the disintegration of the clusters leads to complications in differentiation.

In terms of physiological aspects, the distinctive features are more numerous. The main characteristic is the utilization of raffinose. Raffinose is a trisaccharide composed of fructose, glucose and galactose. The enzyme endowment of the lager yeast allows it to cleave both glycosidic bonds between the monosaccharides. All sugars can be utilized and fermented. The ale yeast is missing one enzyme – an α-galactosidase (melibiase). Galactose and glucose can neither be cleaved nor utilized; only fructose is fermented. Thus, the fermentation rate is only one-third of the rate of lager yeast fermentation. This context is used for a differentiation test – the so-called raffinose test. A raffinose solution is fermented and the resulting amount of CO_2 is evaluated to determine the yeast type. The test is not totally precise because some seldom-used lager yeasts also lack melibiase. Sporulation can also be used for differentiation. Cultivated on suitable media, ale yeasts already produce ascospores after 48 h whereas lager yeast show spores after 72 h, at the earliest. Other physiological differences are a better maltotriose utilization by lager yeast compared to some ale yeasts [17] and distinctions in the fructose-transport systems [18].

The most important differences for the brewer are visible in fermentation management. The temperature profile differs significantly. Lager yeast naturally ferments between 7 and 15 °C. Ale yeast is employed at higher temperatures between 18 and 25 °C. This leads to a 2-fold greater yeast crop for ale yeast. Although lager yeast prefers higher temperatures (optimum growing temperature around 28 °C), it is able to maintain the metabolism under much colder conditions. This characteristic makes lager yeast suitable for cold fermentation with its typical flavor profile. Ale and lager yeast can also be separated according to temperature requirements. Only ale yeast still grows at cultivation temperatures over 34 °C.

Despite the widespread use of top- and bottom-fermenting yeast of the genus *Saccharomyces*, other yeasts are involved in the brewing of some specialty beers. Most of these beers are fermented spontaneously; at later stages of fermentation yeasts such as *Dekkera* spp. (perfect form of *Brettanomyces* spp.) are involved [19]. *B. bruxellensis* and *B. lambicus* are the most common species. These *Brettanomyces* yeasts are weak fermenting, but they are able to utilize dextrins. A slow growth during conditioning is characteristic. In most of these special beers the *Brettanomyces* yeast is socialized with brewing yeast or acid-producing bacteria. This results in totally different flavor profiles because of high amounts of ethyl acetate and acetic or lactic acid [16].

5.1.2.1 Flocculation

A phenomenon shown by brewing yeast is very important for the fermentation performance and following beer production steps: flocculation. The ability of yeast to build cell clusters and settle to the bottom of the tank can be positive or negative for the process. If the yeast settles down too early, not enough yeast cells remain active to reduce diacetyl. If the flocculation is too weak and too many yeast cells stay in suspension, filtration and haze problems may occur. At present, the exact mechanism behind flocculation remains uncertain. Several theories have been proposed and some factors influencing the process have been determined (e.g. zinc and calcium, see below).

The cell wall seems to play a major role in flocculation; its composition is discussed below. Some of the proteins found in the cell wall are coded by genes (Flo1p, Flo5p, Flo9p, Flo10p) and an influence of these genes in flocculation has been shown [20].

Carbohydrate-binding domains were also found on the cell wall. The domains are cell wall-linked proteins, which are called zymolectins. These zymolectins seem to bind to carbohydrate receptors of adjacent cells [21]. Based on these results, the lectin hypothesis for flocculation has been widely accepted. The zymolectins bind to the terminal non-reducing mannose, glucose and fructose residues of saccharides of the respective receptors. Calcium is suspected of being an essential part of the sugar-binding center, keeping up its structure. The zymolectins are first synthesized at the end of the exponential growth, activated by calcium [22]. Although the mechanism of flocculation is becoming clearer, strategies for directly influencing flocculation are few and of empirical character.

5.1.3
Yeast Morphology and Chemical Composition

Brewing yeast is mostly round or ovoid. Partly elliptical or cylindrical cells can also be observed. They are of very regular shape and size. Cells measure 5–10 μm in diameter, 3–10 μm in width, and 4–14 μm in length. The values are imprecise because the size of the cell depends greatly on the physiological state (e.g. prior to budding, yeast cells can reach 3 times the volume of regular cells).

Like other living cells, yeast is composed mainly of water. The non-aqueous cell material is polymeric. Carbohydrates, proteins and nucleic acids made up of the six elements carbon, hydrogen, oxygen, nitrogen, phosphorous and sulfur form the bulk of the material with a large range of low-molecular-weight organic compounds and inorganic ions making up the remainder [23]. Table 5.2 shows the main chemical composition of a yeast cell.

Considerable information about a yeast cell can be obtained under the microscope. For brewing purposes, a magnification of ×640 with phase contrast is sufficient. Young cells show a clear interior and a thin cell wall. Deeper insights into the cell require higher resolutions, which can be achieved with an electron microscope or through confocal microscopy.

The yeast cell contains the typical organelles of other eukaryotic cells. A cross-section in Figure 5.2 illustrates the most important ones.

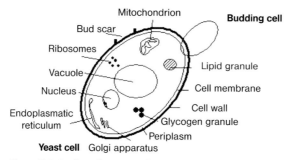

Figure 5.2 Outline of a yeast cell in cross-section.

5.1.3.1 Cell Wall

On the one hand, the cell wall of *Saccharomyces* yeasts is responsible for the resistance of the cell against mechanical forces. On the other hand, it protects the

Table 5.2 Chemical composition (% dry weight) of yeast [24].

Ash/non-organic compounds	8–9
Nitrogen compounds	45–60
Total fat	1.2–12
Total carbohydrates	10–30
Glycogen	28–43

cytoplasm against external effects. The cell wall is built up in a two-layer structure. The inner layer is responsible for the stability and resilience, consisting mainly of β-1-3-glucane, which is linked covalent to chitin. In contrast, the outer layer consists of densely packed glycosylated mannoproteins. This decreases the permeability and builds a barrier for dissolved substances. In addition, hydrophilic attributes of the cell wall are achieved by phosphorylation of the mannoproteins. They also regulate the signal transfer by cell–cell contact and flocculation.

The connection between the inner and outer layer consists of β-1–6-glucan or the proteins are directly linked to the β-1–3-glucan chitin net [25].

5.1.3.2 Periplasm
The so-called periplasm is located between the cell wall and the cell membrane. Some enzymes that start the first degradation steps can be found here. For example, an invertase already cleaves sucrose in the periplasm. Also, it seems to be a protection system of the cell membrane [26].

5.1.3.3 Cell Membrane
Although the cell wall is the stabilizing factor, the main barrier for substrate exchange in and out of the cell is the cell membrane. The two main components are phospholipids and proteins. Next to those, sterols are important membrane components, mostly ergosterol and zymosterol. Phospholipids, molecules with polar heads and aliphatic tails (fatty acids), build a bilayer up to 7.5 nm thick. The heads are oriented towards the surface and hydrophobic areas of the molecules fill out the space in between the two layers; proteins are integrated within the construction.

The bilayer structure shows a certain fluidity that changes under varying physiological conditions of the cell. Maintaining this fluidity is essential for nutrient uptake and substrate exchange. By varying the fatty acid composition of the phospholipids, the cell can adapt to changing conditions via regulating the membrane fluidity within certain limits. For example, cold temperatures reduce membrane fluidity; the cell reacts with the integration of more unsaturated fatty acids, resulting in a regaining of fluidity [27].

The membrane-building phospholipids vary from the outer to the inner layer [28]. Sphingolipids are mainly found on the outside and their task is to give the membrane a structure. They play an important role in cell growth and cell regulation, and have an influence on the stress reactions of the cell [29].

Proteins that are built into the membrane are more or less mobile. They are spread asymmetrically over the bilayer. Intrinsic proteins are stretched through the whole membrane, whereas extrinsic membrane proteins can only be found on the inner or the outer layer. Most of them are specialized to different tasks [30]:

- Bracing the cytoskeleton.
- Enzymes for cell wall synthesis and repair.
- Signal transfer.
- Primary transport.
- Secondary transport.

As is apparent, the substrate transport systems of the cell are all located in the membrane. Some are protein dependent, others are not. Four different transport systems can be differentiated [31]:

- Free diffusion is the slowest mode of transport into yeast. Lipid-soluble solutes move from a high extracellular concentration to a lower intracellular concentration. Undissociated organic acids and long-chain fatty acids enter the cell in this way. Ethanol and gaseous compounds are exported in this manner.

- Facilitated diffusion is a faster method, since solutes are translocated, in an enzyme-mediated manner, down a transmembrane concentration gradient. Transport of nutrients stops when the intracellular concentration equals that of the extracellular medium. Notably, glucose is transported like this.

- Diffusion channels exits for some ions. These voltage-dependent gates move ions down concentration gradients. Normally the channels are closed in resting cells and open when the membrane potential becomes positive.

- Active transport is responsible for the uptake of the majority of nutrients into yeast cells. It is an energy-dependent mechanism. This ATP-consuming proton motive force enables nutrients to either enter with influxed protons or against effluxed protons. Concentrative sugar uptake and expulsion of metabolites works in such a fashion.

5.1.3.4 Cytoplasm

The cytoplasm of yeasts is an aqueous, slightly acidic fluid. Several organelles are located here: vacuoles, ribosomes, the endoplasmic reticulum, the Golgi apparatus and the nucleus. In addition, solved proteins, lipid particles and glycogen can also be found. The lipid particles play a role in sterol biosynthesis and serve as lipid storage for membrane synthesis. As storage carbohydrates, glycogen granules are built up at the end of fermentation. They are used as an energy reserve during storage and as starting energy for new cell cycles.

The chromosomes are located in the nucleus. The endoplasmic reticulum and the Golgi apparatus serve as an intracellular transport net; mainly proteins are moved to membranes, vacuoles or even excreted out of the cell. Under the microscope, vacuoles are the organelles that can be recognized best. Enclosed by their own membrane they are a dynamic storage of nutrition (e.g. amino acids or inorganic phosphate are stored here). The vacuoles also create a space for intercellular protein modifications [32].

5.1.3.5 Mitochondria

The main contribution of mitochondria to the cellular economy is the oxidative transformation of potential chemical energy into biochemically useful forms. Under brewing conditions, this function is not used very much because the aforementioned Crabtree effect inhibits some of the enzymes involved. Thus, it was long believed that mitochondria only play a secondary role in fermentation performance. Studies of O'Connor cox gave evidence to the contrary. She showed that these organelles are involved in the following contexts [33]:

- Synthesis of unsaturated fatty acids and membrane lipids.
- Mitochondrial cytochromes participate in ergosterol synthesis.
- Physiological adaptation to stress (i.e. ethanol stress).
- Modification of the cell surface (important for flocculation and cell division).
- Mobilization of glycogen.

Mitochondria are double-wall structures with internal membranes. They appear round or elongated. However, yeast mitochondria are dynamic. Their number and structure undergo extensive modification depending on changes in the life cycle and physiological state of the yeast. The mitochondrial structure differs drastically at high sugar concentration or under anaerobic conditions [34].

5.2 Yeast Management

5.2.1 Nutrient Requirements and Intake

Wort represents a rich source of nutrients for yeast. It contains fermentable sugars, assimilable nitrogen, minerals and vitamins, as well as minor growth factors. The important oxygen must be supplied by aeration. By careful brewhouse work all malt worts can be created containing all essential nutrients in ample amounts. Only zinc and, rarely, biotin can be critical. Different conditions may occur by applying high-gravity brewing under utilization of large amounts of unmalted cereals or sugar, where a deficiency in nutrients can be observed. Commercially available, so-called yeast nutrients, are employed to cover the requirements. During yeast propagation and fermentation, the concentration of various nutrients changes. The yeast has to respond immediately; therefore, the nutrient requirements and intake are of tremendous importance.

5.2.1.1 Carbohydrates and Fermentable Sugars

Regular wort contains mainly the following sugars: fructose, glucose, sucrose, maltose, maltotriose and dextrins. Except for the dextrins, all sugars can be utilized for the generation of energy and biosynthesis. These fermentable sugars are the main carbon source for the yeast. After being pitched, the yeast (in an appropriate physiological condition) immediately injests the monosaccharides. They enter the cell via the membrane by facilitated diffusion [35]. The uptake of sucrose proceeds simultaneously. It is cleaved by an invertase in the periplasm; the resulting monosaccharides are introduced into the cell as mentioned above [36]. Following the monosaccharides, maltose and maltotriose are injested via active transport. Specific energy-consuming transport systems import the saccharides into the cytoplasm, where they are split into glucose and metabolized.

The initial glucose concentration in the wort plays a key role in the order of the sugar consumption. As long as glucose is still present in the fermentation media the yeast will take in no other sugar. This is due to a glucose repression. The transcription of enzyme-coding genes (e.g. maltase) is repressed [37]. This phenomenon can be important when employing high amounts of glucose (crystals or liquid) as additional extract. The yeast enzymatic system becomes adapted to the high glucose amounts and reduces or even halts the ensuing maltose intake. Insufficient fermentation results occur because of the inadequate maltose metabolism.

Another important carbohydrate source for yeast is glycogen. It serves as a reserve carbohydrate and is generated in the cell during the later anaerobic stages of fermentation. After repitching the yeast to aerated wort, glycogen is utilized for initializing the enzyme activities and yeast metabolism. It is important to preserve the glycogen content during yeast storage to ensure a fast fermentation start.

5.2.1.2 Nitrogen Sources

In terms of nitrogen sources, wort also provides yeasts with readily utilizable sources of nitrogen (amino acids and low-molecular-weight peptides) essential for cellular biosyntheses, enzyme and nucleic acid function. They are present in abundance. *S. cerevisiae* is non-diazotrophic (cannot fix nitrogen) and non-proteolytic (being unable to utilize proteins as nitrogen sources) [38]. Without a certain yeast growth, sufficient fermentation performance is not possible. Therefore, nitrogen sources are essential. For all-malt wort (12 °P) an amount of 900–1200 mg/l total soluble nitrogen and 200–240 mg/l free amino nitrogen are considered adequate. It should be taken into account that the required amount is also strain dependent – some strains show perfect fermentation results despite a much lower nitrogen supply.

Amino acids are removed from wort in a certain order. According to the intake order, several groups can be differentiated (group A, immediate intake; group D; hardly absorbed):

Group A: glutamine, glutamate, asparagines, aspartate, serine, threonine, lysine.
Group B (intake after A): valine, methionine, leucine, isoleucine, histidine.
Group C (slow consumption): glycine, phenylalanine, tyrosine, tryptophan, alanine.
Group D: proline.

Amino acids in groups A and B are transported into the cell by specific permeases; the group C amino acids are injested by the general amino acid permease. Once amino acids and peptides are consumed, they are not assimilated intact, but instead pass through a transaminase system, which removes the amino group. The separated amino group can be used to synthesize new amino acids so that the necessary supply of amino acids can be ensured independently of the intake [39].

During fermentation, the order of amino acid intake has instantaneous consequences. Valine is needed immediately for yeast growth, but can be metabolized only after the group A amino acids have been consumed. Therefore, the cell synthesizes it. During synthesis a byproduct is produced, acetolactate, which is converted extracellularly to diacetyl. Therefore, the diacetyl concentration does not decline until the yeast starts taking in valine. Thus, the amino acid metabolism directly influences beer quality.

This is underlined in the creation of the main fermentation byproducts such as higher alcohols and aldehydes, whose precursors are also generated in these pathways.

5.2.1.3 Oxygen

Under brewing conditions, oxygen can be considered as a nutrient for yeast. Although yeast goes through long anaerobic phases during fermentation and most of the respiratory pathways are blocked due to the Crabtree effect, yeast needs oxygen for a sufficient yeast growth. Furthermore, oxygen seems to be the limiting factor for yeast growth during propagation and fermentation. An amount of 7–9 mg/l dissolved oxygen is considered sufficient for proper fermentation, but it must be mentioned that the oxygen demand of the yeast is very strain dependent [40]. The oxygen enters the cell via facilitated diffusion and at the beginning of fermentation the oxygen dissolved in wort is consumed within hours. At the same time, the glycogen reserves are depleted.

As mentioned earlier, only very poor respiratory metabolism takes place. The yeast uses the oxygen for certain growth-maintaining hydroxylations. It is involved in the synthesis of sterols and unsaturated fatty acids. Both components are scarcely represented in wort, but are essential for yeast growth. Therefore, a synthesis by the yeast is imperative. Studies have shown that the addition of these substances to wort can replace aeration [41].

In detail, the oxygen is needed for the cyclization of squalene to lanosterol, which finally results in ergosterol – the main yeast sterol. Saturated fatty acyl-CoA components are unsaturated with oxygen. In addition, the main regulatory enzymes in the sterol synthetic pathway are induced [42].

The importance of the synthesized components becomes visible if their disposition is determined. Both sterols and unsaturated fatty acids are essential parts of the cell membrane. As mentioned earlier, they are responsible for the fluidity and integrity of membranes. If there is a lack of sterols or unsaturated fatty acids, no functioning cell membrane exists and, for example, no nutrient intake is possible.

After the synthesis of sterols they are rapidly esterized and a sterol pool is built. When the oxygen is used up, these sterol pools are employed for yeast growth. They limit the yeast growth.

Other influences of oxygen can also be observed. It regulates several genes and influences the building of mitochondria [33].

5.2.1.4 Minerals and Trace Elements

Metals are very important for yeast cell physiology. They are needed to maintain the cell's structural integrity, flocculation, gene expression, cell division, nutrient intake, enzyme activity and more. The most important metals that influence yeast fermentation are potassium, magnesium, calcium, manganese, iron, copper, and zinc.

Potassium is the most abundant cation in wort. It is the main electrolyte essential for osmoregulation, and additionally acts as cofactor for enzymes involved in oxidative phosphorylation, protein and carbohydrate metabolism.

Magnesium is absolutely essential for yeast growth. It acts as an essential cofactor for over 300 enzymes (e.g. synthesis of DNA, glycolytic enzymes). Magnesium-deficient cells will not complete cell division; generally, it maintains cell viability and vitality. Magnesium is also involved in the stress response of the cell and is important for high-gravity brewing where the ion is involved in protection against ethanol stress.

Calcium's main role for brewing purposes is in flocculation. It binds to yeast cell walls and stabilizes the lectin-binding center of other yeast cells. On the other hand, calcium can affect yeast physiology by suppressing magnesium-dependent enzymes.

Zinc is one of the nutrients that can be deficient even in carefully prepared wort. Zinc is found in abundance in the mash, but most of it remains in the spent grains after lautering. A critical limit for yeast physiology is reached below 0.12 mg zinc per liter wort. More important is the zinc pool in the yeast; if it is maintained, even zinc-deficient worts can be fermented well. Zinc plays a major role in yeast's fermentative metabolism because it is essential for alcohol dehydrogenase activity, and also stimulates the intake of maltose and maltotriose. In addition, it promotes yeast flocculation and maintains protein structures. A zinc deficiency in yeast definitely leads to slower fermentation rates with poor results.

Other trace elements that influence fermentation include manganese, copper and iron. These are required for yeast metabolism as enzyme cofactors (e.g. manganese) and in yeast respiratory pathways as components of redox pigments (iron and copper). In conclusion, minerals and trace elements play an important role in yeast propagation and fermentation. Currently, research is still being conducted. Very detailed information on this topic can be found in Walker [38].

5.2.1.5 Vitamins and Other Growth Factors

This group contains mainly organic compounds that are needed in very low concentrations. For yeast, these are vitamins, purines, pyrimidines and fatty acids. Vitamins are used as components of cofactors. Biotin is an absolute necessity for most yeast strains; pantothenate and inositol are sometimes required, whereas pyridoxine and thiamin seem only to be needed by top-fermenting brewer's yeast [43].

Purines and pyrimidines are used for DNA and RNA synthesis; the fatty acids are assimilated to construct lipids.

Sulfur and phosphorus are needed as inorganic compounds. Sulfur participates in the synthesis of sulfur-containing amino acids. Methionine and inorganic sulfur are the main sources for yeast. Phosphorus is essential for phospholipids, in the phosphorous bonds of the nucleic acids and for the numerous phosphorylations in the yeast metabolism [44].

Due to the low amounts in which the yeast needs these substances, they do not play a critical role in regularly prepared worts.

5.2.2
Metabolic Pathways during Propagation and Fermentation

5.2.2.1 Carbohydrate Metabolism for Cell Growth and Energy Generation

The dominating metabolic pathway of brewer's yeast during beer production is the formation of ethanol by consumption of wort carbohydrates. In general, alcoholic fermentation is an energy generation under anaerobic conditions, whereby glucose is metabolized to ethanol and CO_2. The balance of energy is 2 mol ATP/mol glucose. Acetaldehyde serves as the hydrogen acceptor. In detail, glucose is transferred via glycolysis to pyruvate. The enzyme pyruvate decarboxylase secedes CO_2 and acetaldehyde remains. Then alcohol dehydrogenase transforms the aldehyde to ethanol.

Under strictly aerobic conditions the same pathways are taken up to pyruvate, but pyruvate is then changed to acetyl-CoA by the pyruvate dehydrogenase. In the following tricarboxylic acid cycle the acetyl-CoA is used for energy generation. The hydrogen acceptor here is oxygen and 38 mol ATP/mol glucose can be gained (see Figure 5.3).

A very important regulatory phenomenon interacts in this pathway. The Crabtree effect refers to the occurrence of alcoholic fermentation (rather than respiration) of glucose under aerobic conditions because of higher sugar concentrations [45]. These sugar concentrations are always found under brewing conditions (e.g. working with wort). Thus, the Crabtree effect occurs constantly during propagation and the start of fermentation, and is totally independent of aeration.

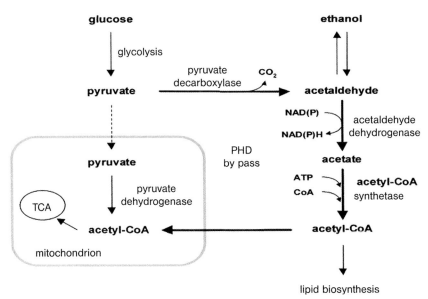

Figure 5.3 Aerobic and anaerobic fate of glucose [49].
PDH = pyruvate dehydrogenase; TCA = tricarboxylic acid.

Despite the low level of aerobic metabolism, yeast should also grow under brewing conditions. Acetyl-CoA as a precursor for biomass production has to be generated in a different way. Yeast metabolism takes a devious route. The acetyl-CoA is built via the so-called pyruvate dehydrogenase bypass. Here, pyruvate is transferred to acetaldehyde. This is metabolized to acetate and then to acetyl-CoA via two enzymes: (i) acetaldehyde dehydrogenase and (ii) acetyl-CoA synthetase [46] (see Figure 5.3).

The limiting effect of the Crabtree phenomenon is counteracted by the generation of the essential byproducts of the aerobic metabolism via the pyruvate dehydrogenase bypass. Yeast manages a sufficient yeast growth with a mainly anaerobic metabolism. Oxygen still is needed, but it plays a role as a nutrient (see above). The idea of switching from aerobic to anaerobic metabolism is not very precise. Rather, it seems that yeast exists in a growing or in a fermenting physiological state when employed in the brewery [47].

5.2.2.2 Formation of Vicinal Diketones

Vicinal diketones like 2,3-butanedione (diacetyl) and 2,3-pentanedione are produced by yeast during fermentation. If they are transferred into finished beer they cause an undesirable buttery flavor note. Diacetyl is the more important substance because of its lower flavor threshold. During conditioning, yeast should consume the vicinal diketones and lower the concentration under the threshold. This reaction is used as the major parameter for control of conditioning [23].

Both vicinal diketones are formed from intermediates of the amino acid biosynthesis. Diacetyl relates to valine and 2,3-pentandione relates to isoleucine. The first intermediates in this metabolism are α-acetolactate and α-aceto-hydroxybutyrate. These components are discharged from the cell and undergo an oxidative decarboxylation to form diacetyl and 2,3-pentanedione. Yeast takes in these substances again and reduces them to 2,3-butanediol and 2,3-pentanediol, respectively. Owing to their high threshold, both resulting components show little influence on flavor [48]. The formation of the diketones is illustrated in Figure 5.4.

5.2.2.3 Formation of Higher Alcohols

Higher alcohols represent the majority of the volatiles in beer. The flavor impressions reach from flowery to solvent-like and alcoholic. The main representatives are n-propanol, iso-butanol, iso-amylalcohol, 2-methylbutanol, phenylethanol and tyrosol.

At the end of primary fermentation over 90% of the higher alcohols have been built, the rest arise during conditioning. The formation of higher alcohols is linked with the amino acid metabolism of yeast. Two pathways for generation can be taken. First, the catabolic pathway (Ehrlich mechanism), whereby the original amino acid is desaminated following a decarboxylation. The resulting ketoacid (2-oxoacid) is then decarboxylated to an aldehyde and finally reduced to the corresponding alcohol (Figure 5.5). The second possibility for the formation of higher alcohols results from the amino acid synthesis. The necessary ketoacids are byproducts of the carbohydrate mechanism (Figure 5.5).

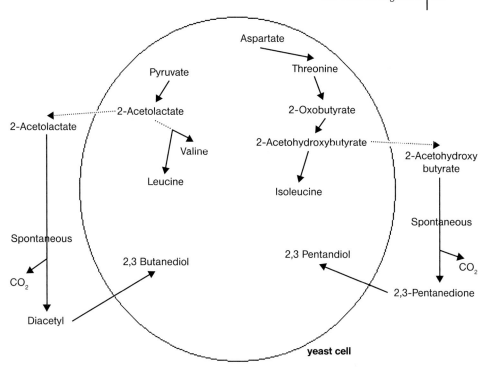

Figure 5.4 Formation and reduction of vicinal diketones [23].

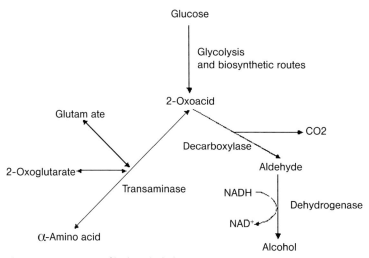

Figure 5.5 Formation of higher alcohols [23].

5.2.2.4 Formation of Esters

Volatile aroma-active esters form the largest group of flavor compounds in beer. They are responsible for the fruity-like aromas. Esters are only trace compounds, but they are extremely important for the flavor profile. Over 100 different esters have been identified, which can be divided into two groups: acetate esters and medium-chain fatty acid esters. In general, esters are constructed out of alcohol and an organic acid. For the acetate esters, acetate (acyl-CoA) is the acid residue and ethanol or higher alcohols serve as the alcohol part. The medium-chain fatty acid esters are composed of a medium-chain fatty acid as well as ethanol or higher alcohols [50].

As mentioned above, most of the higher alcohols are formed during primary fermentation and, since these compounds are required for ester synthesis, ester synthesis is delayed. A significant amount of esters is built when yeast growth declines. Ester synthesis is an intracellular process and is catalyzed by enzymes (e.g. alcohol acyltransferase). Their formation during fermentation depends on the enzyme activity and the amount of available acetyl-CoA. Both parameters play opposing roles during yeast growth. The enzyme synthesis is high during proper yeast growth, whereas the amount of acetyl-CoA can be the limiting factor. A maximum level of esters can be achieved through intermediate yeast growth, which is hard to achieve in practical terms [16].

The other main influencing factor for ester synthesis is temperature; an increase in temperature also increases ester synthesis. Table 5.3 displays the main esters in beer and their flavor impression.

It should be taken into account that the flavor impression in beer is a composition of all these aromas and synergistic effects occur which results partly in different flavor impressions.

Table 5.3 Esters in beer.

Ester	Flavor impression
Ethyl acetate	fruity, like a solvent
Iso-butyl acetate	fruity
Iso-amyl acetate	fruity, banana
Ethyl butyrate	apple, papaya
Ethyl hexanoate	soapy, estery
Ethyl dodecanoate	fruity, strawberry

5.2.2.5 Phenolic Compounds

A special flavor attribute of some German wheat beers is reminiscent of the aroma of cloves. The responsible substances behind this aroma are 4-vinylguaiacol and 4-vinylphenol. For this type of beer the flavor of 4-vinylguaiacol and 4-vinylphenol is of utmost important.

The substances are no esters (also responsible for the flavor of German wheat beers); they belong to the group of phenols. The synthesis is regulated by the *POF* (phenolic off-flavor) gene. This gene can also be found in lager yeasts and even in wild yeasts, but it is only expressed in ale and wild yeast. A contamination with wild yeast results in a phenolic off-flavor in lager beers. For wheat beers, this flavor attribute is desired to a certain degree. It is synthesized by yeast out of the precursors ferulic acid and coumaric acid. These acids are extracted out of the malt during mashing.

5.2.2.6 Formation of Sulfur Dioxide

Sulfur dioxide is generated during primary fermentation by yeast. It is very important as a natural antioxidant and enhances flavor stability of the finished beer.

The component seems to be formed from intermediates in the pathway that reduces sulfate taken out of the wort. The sulfate enters the yeast by active transport and is reduced enzymatically to sulfite. After another reduction, the resulting sulfide is used for the synthesis of sulfur-containing amino acids. When yeast growth ceases, no further sulfur-containing amino acids are used and the sulfite is excreted out of the cell. Four phases of sulfur dioxide formation during primary fermentation can be distinguished.

Phase 1: no sulfur dioxide excretion because sulfite metabolism is repressed due to a high level of sulfur-containing amino acids.
Phase 2: sulfur dioxide is used for the synthesis of amino acids.
Phase 3: yeast growth ceases, the synthesized sulfur dioxide is excreted.
Phase 4: the total metabolism and the sulfur dioxide production come to a standstill.

The sulfur dioxide synthesis is directly linked to yeast growth – more yeast growth means less SO_2; however, reduction of yeast growth to enhance flavor stability has to be assessed carefully.

5.2.3
Yeast Cultivation, Propagation and Post-Fermentation Treatment

After having discussed yeast nutrition and the most important pathways, the following section is focused on aspects of practical yeast management. Adapted yeast management is the basis for reproducible beer production with high qualitative standards. The availability of the required amounts at the right time in the right physiological state is the main goal of yeast management. Yeast goes through various conditions (anaerobiosis, pressure, nutrient depletion) and it is the brewer's task to keep it in a state that guarantees proper fermentations with the desired flavor and stability properties. An overview of the main aspects of yeast management is given in Figure 5.6.

The main stages are yeast strain cultivation, yeast propagation in the laboratory, propagation in the yeast cellar and post-fermentation treatment.

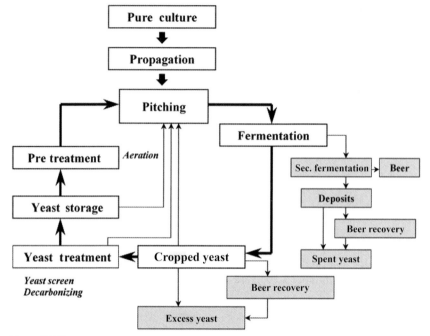

Figure 5.6 Yeast management.

5.2.3.1 Yeast Cultivation in the Laboratory

Selecting the suitable strain is the first step. Several points have to be taken into account:

- Wort composition and concentrations (e.g. high gravity).
- Fermentation temperature, temperature management.
- Pressure management.
- Fermentation speed/extract utilization.
- Yeast growth.
- Diacetyl production and reduction.
- Production of volatiles and other flavor components.
- Attenuation limit.
- Flocculation.
- Sensory attributes of the finished beer.
- Foam quality.

In order to achieve and maintain these desired attributes and the cell morphology, careful yeast cultivation is necessary. It could be carried out in the brewery's own laboratory or, for smaller breweries, by an external yeast bank.

Different techniques for cultivation are possible. Yeast can be cultivated in wort. The wort can be prepared as a wort slant agar culture, making it necessary to recultivate the yeast every 4 weeks to prevent mutations and retain the desired attributes. In order to survive longer periods of storage, yeast can be frozen in liquid nitrogen at −196 °C or at −80 °C with glycerol as antifreeze. The rapid freezing at −196 °C or the glycerol prevents the formation of ice crystals that destroy the cells. By employing these methods, yeast can be stored for at least 1 year. After that it should be cultivated once in liquid wort and frozen again.

Before cultivating the yeast, it has to be ensured that it is a pure culture. This is done by growing the culture out of a single cell. Two techniques are common: for both, the yeast is isolated during the logarithmic growth phase. The yeast solution is then diluted until small drops with single cells are found under the microscope. These drops can be picked up and cultivated in wort. Through this method, it can be guaranteed that the resulting culture has been grown out of one cell, but it is very complex because of the many dilution steps. The second method is easier to handle. The cells are thinned out on a plate agar culture. The thinning out is repeated up to 3 three times with the small colonies grown after 2–3 days. There is only a statistical certainty that the culture is grown out of one cell, but the method is widespread because of its convenience.

For a brewery's own culture collections it is recommended to check the purity of the cultures periodically: (i) for microbiological contaminants and (ii) for the consistency of the strains. Modern microbiological methods like karyotype analyses or DNA fingerprinting are employed [51].

An alternative for small breweries or specialty beers is the use of dried yeasts. Formerly known for bad fermentation performance and microbiological contaminants, they are now available in excellent quality. The sealed packages can be stored for a long period. The critical step is the rehydration; the cells are very sensitive at that point. The yeast should only be spread onto the surface of the culture medium and kept there at the ambient temperature. Not stirring prevents the cells from being damaged. After that step, the yeast can be transferred to the fermentation medium.

The cultivated strains are the basis for new yeast generations in the brewery. Old yeast that has been employed for several fermentations is discharged and replaced for microbiological reasons or because of degeneration. In the brewing field, many different opinions can be found concerning how often the harvested yeast is repitched. Traditionally, from the top of the fermenter, cropped ale yeast can be used for years because it is always cropped in the best physiological state, whereas modern processing in conical vessels makes it necessary to replace the yeast every three to six generations. Some breweries even pitch every batch with new yeast. This is necessary especially under extreme brewing conditions (high gravity, pressure) where yeast is exposed to high stress factors.

To cultivate a new generation, a yeast sample is taken from the slant agar and transferred to liquid wort. The step-by-step cultivation starts in 50 ml of wort and is than transferred to 1 l of sterile wort. In the next step, the volume is increased by a ratio between 1:5 and 1:10. The cultivation temperature can be room temperature and should be reduced little by little to the pitching temperature in the yeast cellar. So-called Carlsberg flasks are often used for transferring the yeast from the lab to the breweries. These can be sterilized and yeast can be transferred with sterile air.

5.2.3.2 Yeast Propagation in the Yeast Cellar

Conventional yeast propagations in open vessels or yeast tanks are still widespread in smaller breweries. These conventional propagations all abide by the method of Coblitz and Stockhausen [10]. It is a consequent continuation of the cultivation in the lab. In the logarithmic phase, yeast is transferred to larger wort volumes until enough yeast for pitching has been grown. The temperature is decreased carefully to pitching temperature. It is a time-consuming process, taking between 10 and 14 days. No special equipment is necessary for this kind of propagation and it is feasible in simple tanks; however, it is time intensive, as mentioned above, and microbiological problems can easily occur. The biggest disadvantage is that the yeast cannot be sufficiently supplied with oxygen, which results in slow growth rates and occasionally poor yeast physiology.

To fulfill demands of efficient beer production processes, conventional propagation is no longer applicable. Automated propagation plants with aeration are state of the art. These plants should have the following features:

- Possibility of sterile inoculation.
- Aeration equipment.
- Equipment for homogenization.
- Sampling device for representative samples.
- Temperature control.
- Oxygen control (helpful).
- Hygienic design.

The time for propagation can be reduced from several to a single day depending on the temperature.

The systems have additional advantages, since the enclosed design makes them less susceptible to microbiological contaminations. The aeration and agitation result in ideal oxygen and nutrition supply for the yeast. An excellent yeast quality showing high and reproducible fermentation performance are the consequences. The beer and foam quality are also affected positively.

The available systems are based on one or two tanks. Differences can be found in the methods of homogenization and aeration. Some systems are equipped with an aeration device fixed in the tank (with or without agitator), others inject the oxygen via an aeration jet installed in a recirculation pipe. The pipe is also used for homogenization. Without recirculation or agitation, an inadequate dispersion of the air (the oxygen) may occur. In addition, aeration and homogenization cannot be separated; non-aerated homogenization is not possible. So if only a part of the yeast is removed for pitching, the remainder stays aerated in the tank. To avoid a loss in quality, new wort has to be filled up immediately. These systems have a deficiency in flexibility. The aeration jets installed in some systems ensure a good supply of oxygen to the yeast. Combined aeration and agitation systems are also available, and although viewed critically (but controllable) from a microbiological point of view, these systems show perfect oxygen supplication with corresponding growth rates.

No matter what system is used, the demands of the yeast have to be fulfilled to guarantee sufficient yeast growth and proper physiological conditions. The main parameters during the propagation process are the temperature, oxygen supply and wort composition. In addition, agitation and inoculum play a role.

Early approaches in yeast propagation had the goal of maximizing cell count. However, it has to be taken into account that the physiological condition during yeast propagation is not constant and a high number of cells may result in poor yeast quality. Therefore, it is necessary to determine the optimum moment to crop the yeast for pitching. The yeast should be cropped in the logarithmic growth phase.

Later an increase in cell count can still be observed, but the physiological conditions will already have declined due to a lack of nutrients [52]. On the one hand, some systems are not able to deliver enough oxygen for greater yeast growth and, on the other hand, the propagation is a nutrient-intensive process. During fermentation a maximum of 50% of the amino acids are metabolized by the yeast. During propagation the yeast may completely deplete the medium of single amino acids. The yeast has to rely on internal proteins (enzyme systems) that as a result might be missing in the following fermentation. Also, minerals and trace elements can be limiting factors for yeast growth. Greater aeration and subsequent depletion of minerals worsens the physiological condition of the yeast [53].

Oxygen supply is the basis for yeast growth during propagation. It is needed for the synthesis of essential sterols and lipids (see above). During propagation, yeast can be kept in the logarithmic phase via oxygen supplication. The concentration of dissolved oxygen is not the critical factor, but it must be ensured that the current demand of the yeast is covered. This can be achieved through low concentrations of dissolved oxygen. Monitoring the oxygen control with an oxygen probe could be helpful. Over-aeration leads to massive foam production and a change in yeast behavior (e.g. less flocculation) [54].

In addition to oxygen, temperature is the main factor for influencing the propagation process. The ideal growth temperature of lager yeast is around 26 °C. At this temperature the ratio between biomass building and degradation of extract reaches a maximum. However, it is recommended to propagate at a temperature maximum 2–4 °C over pitching temperature, otherwise the yeast might experience a cold shock upon being pitched. Rather, temperature is a perfect tool to control the speed of propagation.

If the yeast is needed in a short time, raising the temperature can speed up the process, while a prolonging of the propagation can be achieved with a decrease in temperature. Both do not affect the physiological condition.

Propagations can be carried out as batch processes, where the necessary amount of yeast is grown in one batch. Another possibility is a sequencing batch process: after cropping the required yeast, the remainder is topped up with fresh wort and propagated again. A continuous supply of yeast is possible. Figure 5.7 shows a sequencing batch process for a one-tank system. In panel (I), the propagator is

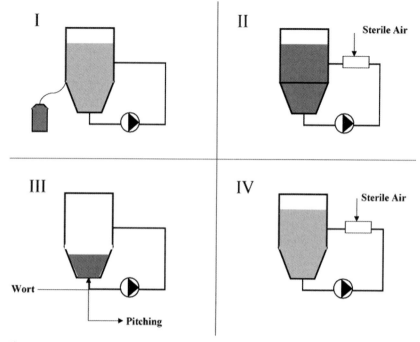

Figure 5.7 Yeast propagation as a sequencing batch process.

filled with sterile wort and inoculated with sterile yeast from a Carlsberg flask. After inoculation the yeast propagation starts with circulation and aeration (panel II). Yielding a certain cell count, the propagated yeast is pitched and the remainder in the propagator is filled up with fresh wort (panel III). To continue the process the new batch is circulated and aerated again to achieve the yeast growth (panel IV).

The latest methods of controlling a propagation process are based on mathematical models. The control calculates the growth rates to achieve the required cropping time and amounts [55].

5.2.3.3 Post-Fermentation Treatment of Yeast

Repitching the yeast, instead of a single use, is the common method in the brewery. Therefore, it is one important aspect of yeast management to maintain yeast quality from one fermentation to the next. The process starts with harvesting the yeast, following an optional washing or sieving, and finally yeast storage and treatment before repitching.

5.2.3.4 Yeast Crop

In classical open vessels or horizontal tanks, the yeast sediments in three layers. The bottom layer consists of yeast cells and fast-sedimenting particles like trub. The core yeast in the middle layer are mainly cells that show a good fermentation performance. Finally, the top layer also consists of high-quality yeast, cells but is polluted with hop resins that stick on the yeast surface. This differentiation between the three layers becomes blurred when cropping yeast from cylindroconical vessels. The yeast settles down in the cone and, if it is possible, only the core yeast should be reused. For a two-tank fermentation, the yeast should be cropped after primary fermentation and the spent yeast that remains in the tank following the secondary fermentation should be discarded. If fermentation and conditioning is carried out in one tank, the yeast crop should start as soon as possible. In general, it is not recommended to crop the yeast in one batch just prior to transferring, but rather to crop the yeast as soon as possible, in several small crops. Since that would cause the green beer to remain in contact with the yeast too long at fermentation temperature [56]. The yeast then starts to metabolize its energy reserve resulting in a lack for the next fermentation.

Furthermore, the yeast excretes fatty acids and proteases. These components diminish the flavor and foam stability. Thus, the yeast should be harvested in divisional steps. The main advantage of conical tanks is the possibility for immediate yeast cropping, which should be utilized.

Despite the latest discoveries about the lotus effect (versus smooth surfaces), it is still state of the art to enhance the sliding of the yeast on the inner surface by using electrically polished cones. The cone angel should range between 60° and 70° [57]. The yeast crop in cylindroconical vessels can be easily controlled with haze analysis. Main technological aspects for a sufficient yeast harvest are: [58]

- Wort composition (mainly calcium and zinc, see above).
- Precisely controllable cooling (intensive cooling at the end of fermentation including cone cooling).
- Degree of attenuation near final degree of attenuation.
- Cropping under pressure to prevent vigorous foam building.

A sufficient yeast crop cannot be achieved by employing non-flocculating yeast strains. In this case, a centrifuge is used for harvesting the yeast. Any oxygen uptake during centrifuging has to be avoided. Mainly, there are two ways to use a separator – the whole yeast is segregated at once or the yeast is separated after an initial partial cropping. The centrifuge can also be used to fix a certain yeast count for the production of specialty beers with a natural haze.

5.2.3.5 Yeast Treatment after Cropping and Yeast Storage

The best way to use cropped yeast is to repitch it immediately after cropping. In this case, the yeast should be homogenized and aerated to take in oxygen and release CO_2. However, in most cases, the production schedule makes it necessary to store the yeast. Thus, the yeast has to be treated and stored in way that maintains the yeast quality.

If the yeast was exposed to pressure during fermentation (pressure fermentation or high cylindroconical vessels) the yeast slurry is enriched with CO_2, which is toxic for the cell. The pressure can be released in a cooled tank in approximately 4 h, during which time most of the CO_2 is also removed [59]. If the CO_2 release is achieved via an aeration, it has to be taken into account that the sterile air can raise the temperature of the yeast slurry significantly [60].

After that, a sieving of yeast has shown advantages. An additional degassing of the CO_2 is achieved and the yeast can be supplied with air via aeration equipment in the sieve. In addition, trub, hop resins and other impurities are removed mechanically. Modern sieves are encapsulated, thus no contact to the surrounding atmosphere is possible. This reduces the risk of microbiological contaminations [61]. Washing yeast with water or acids remains an exception, for example in breweries employing and reusing top-fermenting yeasts over very long periods. Here, washing yeast with acid reduces microbiological risks. In modern yeast management, washing yeast is less helpful.

To maintain the nutrient reserves during storage and save them for repitching, it is necessary to reduce yeast metabolism. Therefore, a temperature between 2 and 4 °C is necessary. Achieving this temperature homogenously over the total tank volume using only cooling jackets is hard to accomplish [62]. One solution for this problem is an agitation of the tank volume (microbiologically precarious), another is the employment of a yeast chiller (modified plate cooler). Hereby, a careful treatment can be guaranteed [63].

The main parameters to ensure yeast quality during storage are:

- Sufficient trub removal.
- Degassing.
- Depressurizing.
- Cool storage.
- Short storage.
- No aeration during storage.

The storage should only last a few days. The storage vessel ought to be equipped with cooling jackets (also at the cone) and a possibility for aeration and homogenization. However, homogenization and aeration is only carried out right before repitching.

Taking care of the yeast will definitely enhance beer quality. Yeast quality has to be maintained during a proper and adapted yeast management. Based on a good nutrition, the yeast then will take the discussed pathways in the right order to rapidly ferment the wort and produce the desired flavor attributes.

References

1 Robinow, C. and Johnson, B. (1991) Yeast cytology: an overview, in *The Yeasts*, 2nd edn, Vol. 4 (eds H. Rose and J. Harrison), Academic Press, London, pp. 8–120.

2 De Kruif, P. (1935) *Mikrobenjäger*, Orell Füssli, Zurich.

3 Gay-Lussac, J. (1810) *Annales des Chimie*, 76, S. 247.

4 Schwann, T. (1837) Vorläufige Mitteilung, betreffend Versuche über die Weingärung und Fäulnis. *Poggendorfs Annalen*, 41, S. 184.

5 Pasteur, L. (1861) *Researches on the Molecular Asymmetry of Natural*

Organic Products, Alembic Club, Edinburgh.
6 Barnett, J. (2003) A history of research on yeasts 5: the fermentation pathway. *Yeast*, **20**, 509–43.
7 Annemüller, G. et al. (2005) *Die Hefe in der Brauerei*, 1 edn, VLB, Berlin, pp. 156 and 187.
8 Hansen, E. (1887) Über Hefe und Hefereinzucht. *Allgemeine Zeitschrift für Brauerei und Malzfabrikation*, 518.
9 Lindner, P. (1930) *Mikroskopische und biologische Betriebskontrolle in den Gärungsgewerben*, Paul Parey, Berlin.
10 Coblitz, W. and Stockhausen, F. (1913) Ein neues Verfahren zur Herführung reiner Anstellhefe für Großbetriebe. *Wochenschrift für Brauerei*, 518.
11 Barnett, J. (1992) The taxonomy of the genus *Saccharomyces* Meyen *ex* Reess: a short review for non-taxonomists. *Yeast*, **8**, 1–2.
12 Boekhout, T. and Pfaff, H. (2003) Yeast biodiversity, in *Yeasts in Food* (eds T. Boekhout and V. Robert), Behr's Verlag, Hamburg, pp. 7–11.
13 Hansen, J. and Piskur, J. (2003) Fungi in brewing: biodiversity and biotechnology perspectives, in *Handbook of Fungal Biotechnology* (ed. D. Arora), Dekker, New York, pp. 233–48.
14 Jespersen, L. et al. (2000) Phenotypic and genetic diversity of *Saccharomyces* contaminants isolated from lager breweries and their phylogenetic relationship with brewing yeast. *International Journal of Microbiology*, **60**, 43–53.
15 Kodama, Y. et al. (2006) Lager brewing yeast. *Topics in Current Genetics*, **15**, 4–5.
16 Dufour, J. et al. (2003) Brewing yeasts, in *Yeasts in Food* (eds T. Boekhout and V. Robert), Behr's Verlag, Hamburg, pp. 347–88.
17 Stewart, G. et al. (1995) Wort sugar uptake and metabolism – the influence of genetic and environmental factors, in *Proceedings of the 25th European Brewery Convention Congress, Brussels, Belgium*, Oxford University Press, Oxford, pp. 403–10.
18 Rodriguez de Sousa, H. et al. (1995) The significance of active fructose transport and maximum temperature for growth in the taxonomy of *Saccharomyces sensu stricto*. *Systematic and Applied Microbiology*, **18**, 44–51.
19 Romano, P. et al. (2006) Taxonomic and ecological diversity of food and beverage yeasts, in *Yeasts in Food and Beverages* (eds A. Querol and G. Fleet), Springer Verlag, Berlin, pp. 13–54.
20 Hagen, I. (2002) Untersuchungen zur Zellwandbiogenese der Bäckerhefe *Saccharomyces cerevisiae*, Dissertation, Universität Regensburg.
21 Stratford, M. (1992) Yeast flocculation: a new perspective. *Advances in Microbiol Physiol*, **33**, 1–71.
22 Masy, C. et al. (1992) Fluorescence study of lectin-like receptors involved in the flocculation of yeast *Saccharomyces cerevisiae*. *Canadian Journal of Microbiology*, **38**, 405–9.
23 Slaughter, J. (2003) Biochemistry and physiology of yeast growth, in *Brewing Microbiology* (eds F. Priest and I. Campell), Kluwer, New York, pp. 20–66.
24 Krüger, E. and Anger, H. (1990) Hefe, in *Kennzahlen zur Betriebskontrolle und Qualitätsbeschreibung in der Brauwirtschaft*, Behr's Verlag, Hamburg, pp. 3–8.
25 Lipke, N. and Ovalle, R. (1998) Cell wall architecture in yeast: new structure and new challenges. *Journal of Bacteriology*, **180**, 3735–40.
26 Arnold, W. (1991) Periplasmic space, in *The Yeasts*, 2nd edn, Vol. 4 (eds H. Rose and J. Harrison), Academic Press, London, pp. 279–95.
27 Shinitzky, M. (1984) Membrane fluidity and cellular functions, in *Physiology of Membrane Fluidity* (ed. M. Shinitzky), CRC Press, Boca Raton, FL, pp. 1–51.
28 Zinser, E. et al. (1991) Phospholipid synthesis and lipid composition of subcellular membranes in the unicellular eukaryote *Saccharomyces cerevisiae*. *Journal of Bacteriology*, **173**, 2026–34.
29 Hannun, Y. and Obeid, L. (1997) Ceramide and the eukaryotic stress response. *Biochemical Society Transactions*, **25**, 1171–5.
30 Van der Rest, M. et al. (1995) The plasma membrane of *Saccharomyces cerevisiae*: structure, function and biogenesis, *Microbiolgical Reviews*, **59**, 304–22.

31 Walker, G. (2000) Yeast nutrition, in *Yeast Physiology and Biotechnology*, John Wiley & Sons Ltd, Chichester, UK, pp. 51–100.

32 Boulton, C. and Quain, D. (2001) *Brewing Yeast and Fermentation*, Blackwell Science, Oxford, pp. 155–8.

33 O'Connor Cox, E. et al. (1996) Mitochondrial relevance to yeast fermentative performance: a review. *Journal of the Institute of Brewing*, **103**, 19–25.

34 Guerin, B. (1991) Mitochondria, in *The Yeasts*, 2nd edn, Vol. 4 (eds H. Rose and J. Harrison) Academic Press, London, pp. 541–600.

35 Reifenberger, E. et al. (1997) Kinetic characterization of individual hexose transporters of *S. cerevisiae* and their relation to the triggering mechanism of glucose repression. *European Journal of Biochemistry*, **245**, 324–33.

36 Meneses, F. and Jiranek, V. (1008) Expression patterns of genes and enzymes involved in sugar metabolism in industrial *Saccharomyces cerevisiae* strains displaying novel fermentation characteristics. *Journal of the Institute of Brewing*, **3**, 322–35.

37 Klein, C. et al. (1998) Glucose control in *Saccharomyces cerevisiae*: the role of MIG1 in metabolic function. *Microbiology*, **144**, 13–24.

38 Walker, G. (2004) Metals in yeast fermentation processes. *Advances in Applied Microbiology*, **54**, 197–229.

39 Hammond, J. (1993) Brewer's yeast, in *The Yeasts, Vol. 5, Yeast Technology* (eds A. Rose and J. Harrison), Academic Press, London, pp. 27–9.

40 Jakobsen, M. and Thorne, R. (1980) Oxygen requirements of brewing strains of *Saccharomyces carlsbergensis*, *Journal of the Institute of Brewing*, **102**, 284–7.

41 Grützmacher, J. (1991) Bedeutung des Sauerstoffs für die Bierhefe. *Brauwelt*, **23**, 958.

42 Boll, M. et al. (1975) Sterol biosynthesis in yeast. 3-Hydroxy-3-methylglutaryl-coenzyme A reductase as a regulatory enzyme. *European Journal of Biochemistry*, **54**, 435–44.

43 Hammond, J. (1993) Brewer's yeast, in *The Yeasts, Vol. 5, Yeast Technology* (eds A. Rose and J. Harrison), Academic Press, London, p. 31.

44 Walker, G. (2000) Yeast nutrition, in *Yeast Physiology and Biotechnology*, John Wiley & Sons Ltd, Chichester, UK, pp. 55–8.

45 Postma, E. et al. (1989) Enzymatic analysis of the Crabtree effect in glucose-limited chemostat cultures of *Saccharomyces cerevisiae*. *Applied and Environmental Microbiology*, 468–77.55

46 Flikweert, M. et al. (1996) Pyruvate decarboxylase: an indispensable enzyme for growth of *Saccharomyces cerevisiae* on glucose. *Yeast*, **12**, 247–57.

47 Wellhoener, U. (2006) Beurteilung des physiologischen Zustandes von Brauereihefen mittels Aktivitätsmessungen von Schlüsselenzymen bei der Propagation und Gärungb, Dissertation, Technische Universität München, München.

48 Debourg, A. (1999) Yeast flavor metabolites. *EBC Monograph*, **28**, 60–73.

49 Remize, F. et al. (2000) Engineering of the pyruvate dehydrogenase bypass in *Saccharomyces cerevisiae*: Role of cytosolic Mg^{2+} and mitochondrial K^+ acetaldehyde dehydrogneases Ald6p and Ald4p in acetate formation during alcoholic fermentation. *Applied and Environmental Microbiology*, **66** Nr. 8, S. 3151–9.

50 Verstrepen, K. et al. (2003) Esters in beer–part 1: the fermentation process: more than ethanol formation. *Cerevisiae*, **28** (3), 41–7.

51 Scherer, A. (2003) Entwicklung von PCR-Methoden zur Klassifizierung industriell genutzter Hefen, Dissertation, Technische Universität München, München.

52 Tenge, C. and Kurz, T. (2002) Yeast propagation: technology and modeling, Bulletin of the 4th technical Meeting EBC Brewing Science Group, Oporto, pp. 113–19.

53 Briem, F. (2005) Modernes Hefemanagement. *Brauwelt*, **145** (33), 965–6.

54 Annemüller, G. and Manger, H. (1999) Die Belüftung der Hefereinzucht – maximal ist nicht gleich optimal. *Brauwelt*, **139** (21/22), 993–1007.

55 Kurz, T. (2002) Mathematically based management of *Saccharomyces* sp. batch propagations and fermentations, Dissertation, Technische Universität München, München.
56 Schmidt, H. (1996) Modernes Hefemanagement. *Brauwelt*, **135** (14), 653–4.
57 Schuttelwood, J. (1984) How to engineer cylindroconicals correctly. *Brewing and Distilling International*, **8**, 22–30.
58 Litzenburger, K. (1996) Hefereinzucht, Hefegabe, Hefeernte. *Brauwelt*, **136** (45), 444–8 and 690–8.
59 Geiger, E. and Tenge, C. (2004) Optimierungsaspekte der Hefetechnologie. *Brauwelt*, **144** (8), 185–7.
60 Lustig, S. and Eidtmann, A. (1999) Technological factors to improve fermentation performance and beer quality. *EBC Monograph*, **28**, 128–38.
61 Tenge, C. *et al.* (1996) Untersuchungen an einem neuentwickeltem Hefesieb. *Brauwelt*, **136** (45), 2195–9.
62 Cahill, G. *et al.* (1999) A study of thermal gradient development in yeast crop, in *Proceedings of the 27th European Brewery Convention Congress, Cannes*, IRL Press, Oxford, pp. 695–702.
63 Eils, H. *et al.* (2000) Maßnahmen zur Verbesserung der Hefetechnologie und Umsetzung in der Praxis. *Brauwelt*, **140** (15), 590–5.

6
Malting
Stefan Kreisz

6.1
Brewing Barley

Barley malt is the main raw material and the main starch source for brewing worldwide. Even if the use of unmalted starch sources like barley, rice, corn or sorghum as an adjunct become more and more popular, most beers produced contain a least 70% barley malt. Brewing barley is a highly specialized cereal with a long breeding, malting and brewing tradition. New research results from archaeological excavations in Tall Bazi (Syria) show that 3200 years ago not only barley, but malted barley was used for beer production [1].

The barley grain (*Hordeum vulgare*) has a complex structure described by different authors [2–4]. Brewing barley contains mainly starch (approximately 62–65% dry weight), protein, cell wall polysaccharides, and a smaller amount of fat and minerals. It has a typical husk, which is needed for the filter bed in the lauter tun.

The varieties are divided by their seeding season and the number of rows. In Europe, the most common brewing barley is two-rowed spring barley (seeding in March and April) as well as in some regions (United Kingdom and France) two-rowed winter barley (seeding in autumn). Some six-rowed winter varieties like Esterel may have brewing quality, but they are more often used outside Europe (Canada and United States).

Quality specifications for brewing barley are the most challenging specifications in comparison to other cereals in the food industry. The high quality demanded is specified by several quality parameters like germinative capacity, protein content, sorting (size of the kernels), water content, kernel abnormalities and infestation. Therefore, an effective quality check before barley intake is essential. The evaluation of the grain involves both visual and laboratory assessments. Each delivery should be checked before it is unloaded. The necessary analytical methods are published in several collections [5, 6]. Retaining representative samples is very important for a successful evaluation.

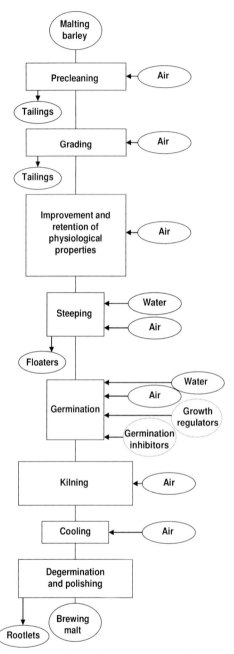

Figure 6.1 Flowchart of malt production. Additives that are shown in dotted ellipses are not necessary.

6.2
Barley Intake and Storage

Delivery may be done by railway, ship or truck. The grain will be inspected for moisture content and protein content by means of rapid analytic methods like near-IR spectroscopy. Pre-germination, infesting insects, kernel abnormities and signs of fungal attack are checked visually. If the barley does not match the declared specification, the load is likely to be rejected. In the laboratory the samples will be graded and checked for the most important parameter – the germinative capacity (above 96%) – with tetrazolium staining.

After accepting the barley the grain should be handled and stored in batches, separated by variety and protein content. The following handling depends on the intake conditions of the barley. The storage conditions must be controlled to preserve the quality. Barley has to be stored under dry conditions with a maximum water content of 14% and moderate temperature. If possible, the barley may be cooled to low temperatures (below 12 °C). However, this treatment can conserve dormancy and water sensitivity, and therefore barley should not be cooled with existing dormancy or water sensitivity. The barley has to be pre-cleaned and dried if the water content is too high. The pre-cleaning includes the removal of coarse impurities like straw and sticks, large stones and clods of earth as well as nuts, bolts and other metallic contaminations that can cause dust explosions. Chaff and dust, small stones, foreign seeds, and broken or thin barley kernels have to be removed. To conserve the quality over longer periods, barley is normally stored in concrete or steel silos with aeration, drying and cooling facilities, temperature control, and pest control. After cleaning, drying and storage, the barley has to be analyzed again before the steeping-in process. The levels of residues of insecticides, fungicides, plant growth regulators or herbicides have to be checked occasionally. Germinative energy, water sensitivity and micro malting trials are needed to predict the malting performance. On the basis of these results, the maltster is able to choose the right date for steeping-in and the right malting regime.

6.2.1
Barley Cleaning

After the definition of the right steeping and malting regime for each batch, barley has to be transported, cleaned, graded and weighed. The barley transport may be done by screw conveyers, chain conveyers (Redler), elevators and pneumatic conveyers. According to the local conditions and regulations, different combinations of systems may be of advantage. Generally, the transport system should prevent damage to the barley. It should combine the ideal speed for the planned capacity with a minimum risk of dust explosion and the possibility to empty the whole system. Dust must be cleaned away, not only because it can form explosive mixtures when mixed with air, but it can also endanger the working staff as it may cause allergies and lung infections. It is necessary to ground all equipment to prevent sparks that might trigger an explosion, and all conveyors, ducts and so on should have explosion vents. Therefore, the transport lane must be planned carefully, and it should be as short and smooth as possible. On the way to the steep the barley has to be cleaned, graded and weighed. The grading is necessary because the water uptake by smaller kernels during steeping is faster and they have higher protein levels. Both may cause inhomogeneities in modification. Barley must be graded into first grade (kernels larger than 2.5 mm), second grade (2.2–2.5 mm) and screenings (smaller than 2.2 mm; only feed quality). Cleaning and grading is done by means of several machines (e.g. classifier, winnower, trieur or plansifter). Nowadays, cleaning and grading is normally done by one machine with combined techniques. After cleaning the barley is weighed by an automatic balance to separate the right batch size for one steep.

6.2.2
Steeping

Steeping-in has two major functions:

- Raise the moisture content from ±12% in different steps to ±40% to initiate germination.
- Wash the grain and remove germination inhibitors as well as all floating material by skimming.

The increase of moisture over 30% starts germination. The water uptake has to be done under controlled conditions of:

- Temperature.
- Vegetation time.
- Alternating between dry and wet periods.
- Oxygen supply.

See Figure 6.2.

The water quality must be at least as high as drinking water quality. Dependent on the system the water consumption for steeping is around 2–3.5 m^3/ton. The water uptake is not evenly distributed over the whole kernel. The spelt and the region around the germ show much higher water contents at the beginning than the endosperm.

The water uptake mainly depends on the steep water temperature and the duration of the wet periods. Higher steeping water temperatures and long wet periods do increase the water level faster, but the risk of drowning water-sensitive barley is higher and microbial growth is accelerated. Common steeping temperatures are about 12–18 °C. As the grain hydrates it swells to 1.4 times its original volume. By blowing air into the base of the steeping vessel it is possible to prevent packing and the barley is mixed. This also adds oxygen. The oxygen is needed by the kernels for respiration. A lack of oxygen may provoke CO_2 accumulation followed by fermentation and therefore a poisoning of the germ. Infestations of microbes are undesirable. They compete with the grain for oxygen and reduce the percentage

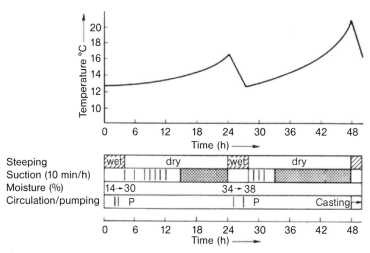

Figure 6.2 Steeping.

germination. Some microorganisms like *Fusarium* spp. produce mycotoxins, which are regulated by limit values for human nutrition and thought to cause gushing. It is possible to use agents like mineral acids, potassium and sodium hydroxides, potassium permanganate, sodium metabisulfite or formaldehyde, but they are subject to local regulations and add costs to the process.

The water uptake and the germination performance are heavily influenced by barley quality. Barley may show a high germination capacity, but still not the full germinative energy at the moment of steeping-in. The phenomenon is called dormancy, which naturally prevents the grain from germinating in a too early state and is caused by inhibitors of the grain. Therefore, the germinative energy has to be checked before steeping-in with a Schönfeld or Aubry test. The result should be above 95%. European barley shows about 30% water content by using a wet period of about 4–6 h. The wet period is followed by an air rest. In this period the barley should be streamed by humid tempered air by downward ventilation. This provides the grain with oxygen, and removes CO_2 and heat generated by the metabolizing grain. Common systems work with two wet and dry periods up to water contents about 38–40%. Sometimes a third wet period is executed to a steeping-out level about 43%. The water content (degree of steeping) should be checked after every period. In addition to the change of the volume, the kernel changes visually by developing a small white root at the base of the kernel. This occurrence is called chitting. Homogenous chitting of all kernels can be checked visually and is essential for a homogenous malt quality. There are different steeping and steeping-in systems. Steeps are built cylindroconically or with a flat bottom. Wet steeping-in systems are screw conveyors or washing drums.

6.2.3
Germination

According to the system, the chitting grains are transferred to the germination box or vessel. There are different types like floors (older systems), Saladin boxes, Wanderhaufen (System Lausmann) or circular germination vessels. The objectives of germination are:

- Controlled breakdown of cell walls and matrix proteins.
- Produce optimal level of hydrolytic enzymes.
- Hydrolyze certain barley reserves [e.g. protein to form free amino nitrogen (FAN)].
- Minimize loss of potential extract from growth and respiration while achieving optimal modification.
- Produce balanced, well-modified green malt for kilning.

These objectives can be achieved by using the following parameters:

- Time.
- Maximum degree of steeping.
- Temperature.
- $O_2:CO_2$ ratio.
- Adjuncts (most commonly gibberellic acid).

See Figure 6.3.

As described above, the appearance of the kernel is changing. After chitting, the grain grows, producing several rootlets at the base of the grain. Simultaneously, the acrospire grows along the dorsal side of the grain, beneath the husk. The length

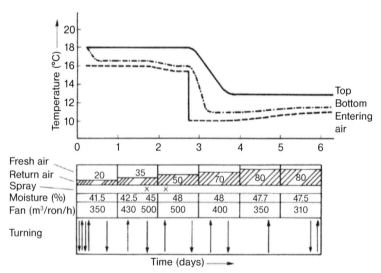

Figure 6.3 Germination with falling temperatures.

of acrospire growth in relation to the length of the grain may be used to evaluate the homogeneity of the batch. Different acrospire lengths and very long acrospires (acrospires which are longer than the grain itself are called 'Husaren') indicate heterogeneity in growth and therefore a non-homogenous malt quality. The visual detectable growth reflects the internal change in the grain.

Around 12 h after adding the vegetation water the acrospire shows first activities. As shown in Figure 6.4, the barley kernel consist of three main regions: the germ, the endosperm and the covering. Modification starts at the scutellum. The scutellum and the epithelium are thin layers between the germ and the endosperm. The enzyme induction is caused by gibberellin hormones. Gibberellic acid induces the production of different hydrolysing enzymes in the aleurone layer which covers the whole endosperm. The products of endosperm breakdown (sugars, amino acids, etc.) together with materials from the aleuronic layer (phosphate, metal ions, etc.), are needed for the growing germ.

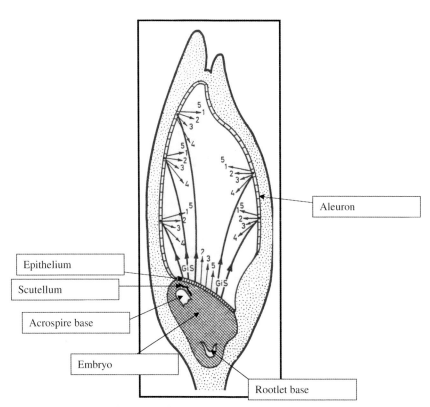

Figure 6.4 Enzyme development in the barley kernel during germination: (1) endo-β-glucanase, (2) α-amylase, (3) protease, (4) phosphatase and (5) β-amylase. GiS = gibberellic acid.

It is now the challenge for the maltster to control the hydrolysis of proteins (proteolysis), cell walls (cytolysis) and starch (amylolysis). The breakdown of the cell walls and the hydrolysis of the proteins, as well as the production of enzymes needed for mashing, has to be advanced to match the desired malt specification. On the other hand, extensive growth and respiration provoke malting loses and over-modification. Therefore, the right combination of temperature, moisture level, vegetation time and (if allowed) the addition of gibberellic acid has to be chosen carefully according to the barley variety and quality (and if available the micro malting results). Different varieties react with varying behavior to changes in malting parameters. General reactions can be described as followed.

Vegetation time (steeping + germination = 120–168 h):

- Enzyme development continues from day 1 through to the beginning of kilning, although different enzyme groups peak at different times.
- Extract development peaks at day 6, then reaches a plateau.
- Soluble nitrogen continues to increase until kilning.

Moisture content (42–48%):

- Modification is faster at higher moisture levels.
- Enzyme development is faster at higher moisture levels.
- Extract development is faster at higher moisture levels.
- Malt yield is lower at higher moisture levels.
- Energy needed for kilning is higher at higher moisture levels.

Temperature (12–20 °C):

- Cooler temperatures favor enzyme development of most enzymes.
- Cooler temperatures favor soluble nitrogen development.
- Higher temperatures favor cytolysis.
- Increased temperature increases respiration and malting loss, and lowers yield.
- The most germination programmes are not isothermal. So-called falling temperature programmes (18 °C at the beginning to 13 °C at the end) may improve quality (see Figure 6.3).

Addition of gibberellic acid (0.01–0.5 ppm applied shortly after steep-out):

- May help to break dormancy or stimulate grain with reduced vigor.
- Speeds up modification.
- Increased enzyme activity generally leads to greater color formation.
- Helps to improve cytolysis, but may provoke very high proteolysis.
- Intensify inhomogeneity when added to inhomogeneous chitting batches.

As already mentioned, the barley quality is strongly influenced by the variety and the local vegetation conditions. Thus, in combination with the installed steeping and germination technique determines the used malting technology.

6.2.4
Kilning

The green malt is transferred to the kiln. Kilns have a false bottom where large quantities of hot air can pass through to dry the green malt. There are different types of kilns. Common modern kilns are high-performance kilns with heat recovery systems and different numbers of floors. The objectives of kilning are:

- Terminate the modification process and the growth of the plant.
- Reduce moisture to levels suitable for grain storage.
- Conserve enzyme complexes developed during malting.
- Develop color and flavor (both taste and aroma) characteristics as required by the brewer.

The process can be managed by the temperature of fresh air and performance of the fan. It is controlled by the temperature of the air below and above the bed as well as air humidity of the exhaust air and the moisture level of the upper layer. The heating has to be performed carefully at the beginning. Enzymes can be destroyed by wet heat, and case-hardened malt may occur if the wet starch gelatinizes and becomes glassy after drying at high temperatures. The choice of the initial temperature depends on the desired malt quality. For darker malts, a continuing hydrolysing process to produce reactants for the Maillard reaction and therefore more melanoidins is desired. For any pale malts like Pilsner malt with a color of around 3 European Brewery Convention (EBC) units, exaggerated hydrolysis has to be avoided. Common temperatures of the fresh air for pre-drying (withering) of green malt start at around 50 °C and end at around 65 °C with full fan power. The green malt bed can be divided in three (upper, middle and lower) layers. The incoming hot air is able to absorb moisture from the green malt. At the beginning of the drying process the exhaust air above the bed is saturated and has a significantly lower temperature. The drying process starts at the lower layer and constantly develops through the bed. After 10–12 h (single-floor, high-performance kiln) the temperature above the bed is rising and relative humidity of the exhaust air is decreasing. This point is called the 'breakthrough' and marks the earliest point (the moisture level of the upper layer should not be above 20%) when a slow increase of the fresh air temperature to curing temperature (around 85 °C is advisable. At this point the fan power can be reduced. As mentioned above, major changes that are essential for the malt quality take place during withering and curing. Color and flavor are produced mainly through the Maillard reaction. The most important Maillard products are melanoidins. They have a major impact on the color and aroma as well as on the pH and taste stability of the beer. Dimethylsulfide (DMS) precursor (*S*-methylmethionine) is converted into DMS and driven out. As higher the temperature, DMS precursor is better reduced, but higher temperatures also increase the color and the thiobarbituric acid number (TBN). Enzymes are partly inactivated. Lower temperatures at higher

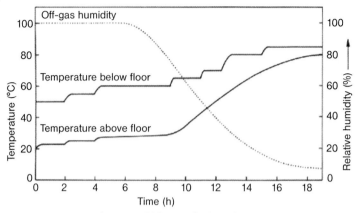

Figure 6.5 Drying (withering and kilning) of pale malt.

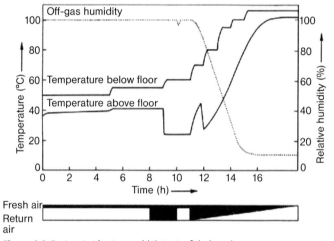

Figure 6.6 Drying (withering and kilning) of dark malt.

moisture contents help to conserve enzymes. Carcinogenic nitrosamines such as N-nitrosodimethylamine (NDMA) may be produced. They are built from hordine (a phenolic alkaloid) fractions of the green malt and nitrogen oxide from the air; therefore, they are mainly formed by direct heating of the kiln (which is not common) and may be prevented by sulfurizing while withering. The limit value for NDMA in the European Union is 2.5 µg/kg.

See Figures 6.5 and 6.6.

6.2.5
Cleaning, Storage and Polishing of the Malt

After kilning the malt should be cooled and cleaned. The rootlets, which are 3–5% of the malt, cannot be used in the further process and have to be removed because they may provoke uncontrolled water uptake of the malt. Cleaning is performed in a deculming machine, mostly by pressing the kernels again sieving cylinders. The rootlets break off and can be removed by conveyors. They are useful for feeding and are sold separately. Malt should be stored for at least 2 weeks after kilning. The enzymatic abilities are recovering during this time. The malt should be stored cold and dry to avoid excessive water uptake. Before dispatch the different malt batches are blended to the costumers needs and polished. Polishing again includes dust removal, magnetic control and, according to the specification, removal of broken kernels/split husk fragments.

6.2.6
Malt Yield

Approximately 800 kg malt can be produced from 1 ton of barley. The loss is called malting loss and is composed of different losses through the individual production steps (Table 6.1). The major part occurs from the different water content of barley (12–14%) and malt (4–5%). Losses of dry weight are cause by steeping, respiration and growth of rootlets.

Table 6.1 Volume and mass change during malting.

	Moisture content (w/%)	Volume (hl)	Mass (kg)
Malting barley	14	100	100
Steeped barley	41	145	145
Green malt	48	220	147
Kilned malt	3.5	118	79
Stored malt	4.7	120	80

6.2.7
Malt Quality

Malt quality has a basic influence on beer production and on the quality of the final beer. Each production step (e.g. lautering, fermentation and filtration) as well as important characteristics of beer (e.g. flavor, color, foam and stability,) are strongly influenced by malt quality. The brewer determines the necessary quality of raw material for each beer type by choosing a barley variety and the malt quality, and therewith the specification of normal and limit values for the malt analytical features. The precision of the analyses as well as the interaction of different parameters should be considered when choosing the analytical criteria. An accurate execution must be guaranteed when determining the results. The Mitteleuropäische Brautechnische Analysenkommission (MEBAK) and the EBC publish the analyses specifications commonly used in Europe. [1, 2] Before treating each analyses criterion, it must be pointed out that the quality of the malt analyses and therewith the quality of the evaluation of the malt delivery crucially depend on a representative sampling. Correspondingly, there are published rules that must be followed during sampling and sample preparation. [1, 3].

6.2.8
Quality Criteria of Barley Malt

Above all, barley malt analytics describe the three modification procedures of cytolysis, proteolysis and amylolysis, divided on the basis of the main components of the grain. The preferably smooth and complete degradation of the cell walls and the correct degree of protein modification are no longer adjusted in the brewhouse due to the use of modern brewing procedures with mashing-in temperatures above 60 °C. For this reason the importance of malt quality, especially for the processability, increases. In large breweries with up to 12 brews/day a correction of the modification procedures through an individual adjustment of temperatures and breaks is no longer planned due to time requirements of the brewhouse procedure. In such cases, the mashing work is mainly restricted to amylolysis and, with it, to the degradation of amylose and amylopectin to the desired degree.

6.2.8.1 Cytolysis
Cytolysis describes the degradation of the supporting and structural cell wall substances in the coating of the starch-bearing cells of the endosperm. Structural proteins and cell wall polysaccharides, especially β-glucan, are degraded. A wide degradation of structural substances during malting enables an easier enzymatic attack of the starch in the course of mashing and, consequently, a better yield in the brewhouse. Reciprocally, undermodified all walls causes, on the one hand, yield losses and, on the other hand a transfer of a large amount of high-molecular β-glucans in dissolved form.

Older sources certify that high molecular β-glucan has a favorable effect on foam and taste. As long as β-glucan is not available as gel, an amount up to 350 mg/l is technologically unproblematic. However, recent research papers could not confirm that β-glucan has a reproducible influence on beer foam and body in a technological justifiable range. In small quantities of 10–15 mg/l, which are only just above the reliable detection limit, β-glucan gel can cause filtration problems. In this context, mashing methods with high mashing-in temperatures above 60 °C turned out to be very susceptible. Due to the enzyme β-glucan solubilase a release of β-glucan still bound in the cell walls can be observed at mashing temperatures of 60–65 °C. However, this high molecular β-glucan cannot be decomposed anymore, since the endo-β-glucanase, which is responsible for it, is already inactivated at temperatures above 52 °C. Thus, brewing methods with high mashing-in temperatures at identical malt qualities always result in higher total β-glucan contents in wort and beer, and therefore demand raw material with particular strong cell wall modification. Different key data are used to describe the extent of cytolysis. The friability turned out to be simple and quick to determine. The cell wall modification is determined through the friability of the malt grains and a statement about the homogeneous processing of barley in the malthouse can be made by means of the quantity of the whole unmodified grains. In opposition to earlier opinions, detailed research by Sacher, which incorporated many practical observations, showed that a friabilimeter value of at least 85% is necessary at high mashing-in temperatures and should not be limited upwards. High friability values are not a general indicator for over-modification as long as only cell wall modification is high without an over-modified protein modification at the same time. Viscosity of the congress wort, viscosity of isotherm 65 °C mash, β-glucan contents of the congress wort and the isotherm 65 °C mash as well as homogeneity and modification of malt are further criteria for the evaluation of cytolysis. Here, the iso 65 °C values show better results than the values of the congress mash.

It must be pointed out here that both MEBAK and EBC canceled the coarse-grist difference (extract difference) because of the imprecision of the analysis and replaced it with the aforementioned analyses. On the basis of statistical evaluations, the analysis criteria and limit values at high mashing-in temperatures given in Table 6.2 proved to be reasonable in combination with practical tests.

In practice, the limits of the viscosity of iso 65 °C mash and the friabilimeter values (including whole unmodified grains) as well as the homogeneity should be realized as routine analyses, and only in the case of doubt (exceeding as well as undershooting of the limits values) to widen the analysis through the remaining analyses. The results should be interpreted as follows. If the values of the above described analyses were exceeded or below target in at least two cases, lautering and filtration difficulties would be expected because of malt usage at high mashing-in temperatures. By contrast, in case that the limit values are kept and filtration disturbances nevertheless occur, these are probably due to mistakes in the brewing technology. If only one value exceeds the limit, the analysis has to be repeated in any case. The measurement of the whole unmodified grains as well as of the β-glucan partially shows very poor comparability. The corresponding statistical evaluation basics can be seen in the Methodensammlung of MEBAK. Low homogeneity values and a strong increase of viscosity as well as of the β-glucan of the

Table 6.2 Cytolytic malt analyses criteria for mashing methods with high mashing-in temperatures.

Friabilimeter	> 85%
Whole unmodified grains	< 2%
Viscosity (8.6%)	< 1.56 mPas
Viscosity 65 °C (8.6%)	< 1.6 mPas
β-Glucan (8.6%)	< 200 mg/l
β-Glucan 65 °C (8.6%)	< 350 mg/l
Homogeneity	> 75%

congress wort, which is disproportionate to wort of iso 65 °C mash, indicate a blending of well and bad modified malts [7].

6.2.8.2 Proteolysis

Proteolysis describes the degradation of the grain protein, and its transfer in low, middle and high molecular soluble form. While high values behave neutral with regard to cell wall modification, improving the processability, excessive as well as too low protein modifications should be regarded as disadvantageous. Too low protein modifications may not provide yeast with sufficient assimilable nitrogen compounds. Low yeast propagation and the development of undesired fermentation byproducts (e.g. diacetyl) may result. However, excessive protein modification results in too strong degradation of high molecular protein. The lack of sufficient quantities of high molecular proteins as well as the backlog of mid molecular compounds and certain amino acids (lysine, arginine and histidine) has a negative influence on foam stability. These malts, worts and beers tend to high colors, and a surplus of certain amino acids could cause an off-flavor and poorer flavor stability. In addition, beers with higher amino acids contents are very susceptible to beer-spoiling microorganisms.

The Kolbach index, content of soluble nitrogen and FAN are used as parameters of proteolysis. In practice, the Kolbach index is the most frequently used parameter for the evaluation of proteolysis. It measures the percentage of raw protein of the entire grain protein, which was dissolved during malting and the congress mashing method. The desired range for all malt beers lies between 38 and 42%. It restricts the combination possibilities which are mathematically calculative feasible on the basis of the absolute statements for the protein content of malt (norm values 9.5–11%) and e soluble protein (norm value 3.9–4.7%). Thus, a balanced composition of the soluble protein in the high molecular area (head retention) as well as and in the low molecular area (yeast nutrition) should be guaranteed despite the raw protein content of malt. The soluble protein is mostly determined through the evaluation of the soluble nitrogen (conversion factor 6.25). From the abovementioned data, the following values are reached: for soluble nitrogen 630–750 mg/100 g dry weight and for FAN content, which should amount approximately 21% of the soluble nitrogen, 130–160 mg/100 g dry weight. When applying a too close mashing method or the addition of raw grain (e.g. rice or maize), higher values are to be used.

6.2.8.3 Amylolysis

Among the amylolytic parameters, the extract, final attenuation degree, diastatic power as an estimation of α-amylase activity should usually be acquired in malt. The extract content shows the percentage of malt dry matter that can be dissolved through the congress mashing method with fine grist and indicates the yield during brewing procedure. In the case of barley malt, the values range between 78 and 83.5%. Even at higher raw protein values modern summer barley varieties should have extract contents above 81%. The final attenuation degree states a summary parameter for the evaluation of (congress) wort quality and gives information about the processability of the extract by yeast. The quantity of the fermentable sugar and its relative fraction are very important here. The role of trace elements and nitrogen composition should not be under-rated. As a quality feature for the congress mash, it should be considered that the final attenuation degree should be as high as possible (above 81%). Owing to too high attenuation degrees in the brewery, which are hardly controllable, it is always being discussed whether the desired final attenuation degree should be limited already in the malt. Discussing the amylolytic enzyme supply, especially the diastatic power of malt is paid regard to, since high amounts of raw grain are applied without noteworthy enzymatic activities of their own. Activity values above 200 WK should be sufficient for all malt beers. In extreme cases too low α-amylase activity can cause a displacement of the sugar spectrum and lead to an atypical fermentation developing (diauxie).

α-Amylase is the pacemaker enzyme of the starch degradation. It forms starch fragments through the degradation of amylopectin and amylose, and thus contact points for the β-amylasis. An α-amylasis activity above 40 American Society of Brewing Chemists units can be regarded as sufficient. If mash acidification takes place, the α-amylasis activity is affected. Increased iodine figures with negative consequences, such as turbidity in the filtered beer, can be caused. In addition, the lack of contact points for β-amylasis can be disadvantageous for the final degree of attenuation.

6.2.8.4 Enzymes

Table 6.3 shows the great variety of enzymes formed during malt production.

6.2.8.5 Further Malt Quality Criteria

6.2.8.5.1 DMS Precursor

The DMS precursor is an amino acid that occurs in free form in the malt, although not yet in the barley. DMS is formed from DMS precursor at approximately 70 °C, and thus at every thermal treatment step in the course of malt and wort preparation. During malt kilning and wort boiling the DMS-P should be widely decomposed and expelled. It must be considered that free DMS is reproduced in the whirlpool. Kilning temperature and period allows the most effective control of the quantity of DMS precursor. Basically, the higher and the longer both are selected, the lower the DMS precursor content. However, there are economic aspects as well as a too strong thermal stress of the pale malt that argue against a too intensive kilning.

In the finished beer DMS can raise an odor and taste like vegetables or celery. The sensory swell for free DMS in beer is specified as 130 µg/l. Depending on the intensity of boiling the DMS precursor content in malt should not exceed 5–7 ppm.

6.2.8.5.2 Thiobarbituric Acid Number (TBN)

The TBN gives a summary parameter for the thermal stress of malt during kilning. This thermal stress causes Maillard products, which increases the color and may negatively influence the flavor stability. In case of pale malt, the TBN should not exceed 18 in the congress

Table 6.3 Malt enzymes.

Enzyme	CAS registry number	EC number	Optimum conditions in mash		Inactivation temperature (°C)
			pH	t (°C)	
Oxidoreductases					
peroxidase	[9003-99-0]	1.11.1.7		40–50	65
lipoxygenase	[9029-60-1]	1.13.11.12	6.5	40	70
polyphenoloxidase		1.14.18.1		60–65	80
Hydrolases					
lipase	[9001-62-1]	3.1.1.3	6.8	35–40	60
acid phosphatase	[9001-77-8]	3.1.3.2	4.5–5.0	50–53	70
α-amylase	[9000-90-2]	3.2.1.1	5.6–5.8	70–75	80
β-amylase	[9000-91-3]	3.2.1.2	5.4–5.6	60–65	70
endo-β-(1–4)-glucanase	[9074-99-1]	3.2.1.8	4.7–5.0	40–50	55
cellulase	[9012-54-8]	3.2.1.4	4.5–5.0	20	20
laminarinase	[9025-37-0]	3.2.1.6	5.0	37	50
limit dextrinase	[9025-70-1]	3.2.1.11	5.1	55–60	65
maltase	[9001-42-7]	3.2.1.20	6.0	35–40	40
β-mannosidase	[9025-43-8]	3.2.1.25	3–6	55	70
invertase	[9001-57-4]	3.2.1.26	5.5	50	55
exo- and endo-xylanases		3.2.1.37	5.0	45	
endo-β-(1–3)-glucanase	[9044-93-3]	3.2.1.39	4.7–5.0	40–45	55
exo-β-glucanases		3.2.1	4.5	40	40
pullulanase	[9012-47-9]	3.2.1.41	5.0–5.2	40	70
arabinosidase	[9067-74-7]	3.2.1.55	4.6–4.7	40	60
β-glucan solubilase					
with esterase activity		3.2	6.6–7.0	62	73
with carboxypeptidase activity		3.4	4.6–4.9	62	73
aminopeptidases		3.4.1	7.2	40–45	55
carboxypeptidases		3.4.2	5.2	50–60	70
dipeptidases		3.4.3	7.2–8.2	40–45	55
endopeptidases		3.4.4	5.0–5.2	50–60	70

wort, whereas the demands must be adjusted with the maximum for DMS precursor. As described before, a compromise between the decomposition of DMS precursor and evaporation of DMS and a preferably low thermal stress must be made.

6.2.8.5.3 **Color and Boiled Wort Color** The color of malt influences the color of the finished beer significantly. It is determined photometrical and visually from the congress wort and the boiled congress wort. The method applied should be considered when comparing different analysis results. The color should range between 2.5 and 3.5 EBC in the pale malt and provide information about adequate malt production. On the basis of the boiled wort color with the color of the finished beer can be conditionally predicted. The thermal processes in combination with possible oxidative stresses (especially during mashing) and the displacement of the pH during fermentation, both taking place during the brewing procedure, have a considerable influence (independent from malt color) on the beer color. Pale beers can occur because of gentle boiling methods and too low thermal stresses of malt as well as wort. Adjustment of the desired color can be carried out by means of different special malts.

6.2.8.5.4 **pH** The pH of malt is measured in the congress wort. It ranges between 5.80 and 5.95 in case of pale barley malt. Dark malt shows lower pH values of 5.5–5.8 because of a larger amount of Maillard products. Too low pH values of pale malt indicate a strong modification or possibly a too intensive sulfuring of malt. Low pH values reveal low mashing pH.

6.3 Special Malts

Special malts are malts that are used mainly in a minor percentage of the grist to add individual color, flavor and aroma to different beer types. They are mainly made out of barley (but also wheat, rye, sorghum or other cereals), and by normal steeping and germination procedures, but different kilning or even roasting techniques. As the kilning and the roasting temperatures are significantly higher than the temperatures for pale malt they have lower extract yields and a lack of enzymes (except some special malts made for enzyme addition like diastatic malt).

6.3.1 Dark Malt (Munich Type)

Dark malt has a color between 15 and 25 EBC. The barley has a higher protein level (11–13%) and hydrolysis while germination is more intensive. Moist and hot conditions are desired during withering, and the curing temperature is about 105 °C.

6.3.2 Caramel (Crystal) Malt

The enzymatic breakdown in caramel malt is increased by high temperatures at the last 2 days of germination. Therefore, low molecular nitrogen and sugars are formed extensively. The malt is saccharified in a kiln or roasting drum and dried according to the desired color at temperatures between 80 °C (kiln, color up to 5 EBC like Carapils) and 180 °C (roasting drum color up to 150 EBC).

6.3.3
Roasted Malt

Roasted malt is made out of well-modified, already kilned pale malt. The pale malt is conditioned (5–10% increase of the water content) and then heated to 180–220 °C in a roasting drum until the desired color appears.

6.3.4
Wheat Malt and Malt Made from Other Cereals

In Germany and Belgium, wheat malt is a popular malt to brew special wheat beers. German wheat beer contains at least 50% wheat malt. The malting technology is quite similar to barley malt, but the lack of a husk (fast water uptake) and different enzymatic abilities has to be taken in account. Roasted or dark wheat malt is also available. Other cereals like oat or rye are malted by some malteries. They are added to special brews by mostly small or craft breweries all over the world. Sorghum malt may be solution for a gluten-free beer, but research is still in progress.

6.3.5
Other Special Malt

Other special malts exist such as smoked malt (for special taste and flavor), acid malt (to adjust the pH of the mash), diastatic malt (low-calorie beers) or chit malt (to enhance the foam properties of the beer).

References

1 Zarnkow, M. et al. (2006) Kaltmaischverfahren – eine mögliche Technologie im Alten Orient. *Brauwelt*, **146** (10), 272–5.
2 Narziss, L. (1999) *Die Bierbrauerei: Band I: Die Technologie der Malzbereitung*, 7th edn, Ferdinand Enke Verlag, Stuttgart, pp. 21–65.
3 Briggs, D.E. et al. (2004) *Brewing Science and Practice*, Woodhead Publishing, Cambridge, pp. 11–14.
4 Kunze, W. (1996) *Technology Brewing and Malting*, 7th edn, VLB, Berlin, pp. 28–9.
5 Anger, H. (2006) *Brautechnische Analysenmethoden: Band Rohstoffe*, 1st edn, Selbstverlag der MEBAK, Freising.
6 van Erde, P. (1998) *Analytica-EBC*, Verlag Hans Carl, Nürnberg.

7
Wort Production

Martin Krottenthaler, Werner Back, and Martin Zarnkow

7.1
Introduction

An overview on the processing steps needed for beer production in a brewery is shown in Figure 7.1.

7.2
Technology of Grinding

The mechanical decomposition of the malt grain happens with grinding. Since the composition of the grist influences the lautering properties of the spent grains, the mill type and mill setting need to be adjusted to the lautering system. The sorting of the grist is dependent on the milling system and can vary depending on the malt solution even with the same mill settings.

On the one hand, grinding is of importance for extracting the inside of the malt or cereal grain as much as possible out of the endosperm to facilitate the following extraction and dilution steps. On the other hand, the husks should be kept intact with the application of a lauter tun to obtain a permeable filter cake. With excessive milling, the pores of the filter cake clog up very rapidly during lautering so that the permeability is also reduced due to decreased porosity. The consequences are longer run-off times of the filtrate and the need to loosen the filter cake mechanically. This is accomplished in a lauter tun with raking knives. The husk should also be ground finely when using a mash filter; however, the flour dust part should not be too high. If milled too little, the result is yield losses as the starch adsorbed to the grain sheath is discarded with it.

Currently, hammer mills and roller mills are mostly employed. Furthermore, there is a differentiation between wet milling and dry milling. The most used process is where the dried malt is ground by counter-rotating roller pairs. Depending if the rollers rotate at the same or at different speed, the grinding of the dry grist is carried out by applying pressure or by pressure and shearing. The efficiency

Handbook of Brewing: Processes, Technology, Markets. Edited by H. M. Eßlinger
Copyright © 2009 WILEY-VCH Verlag GmbH & Co. KGaA, Weinheim
ISBN: 978-3-527-31674-8

7 Wort Production

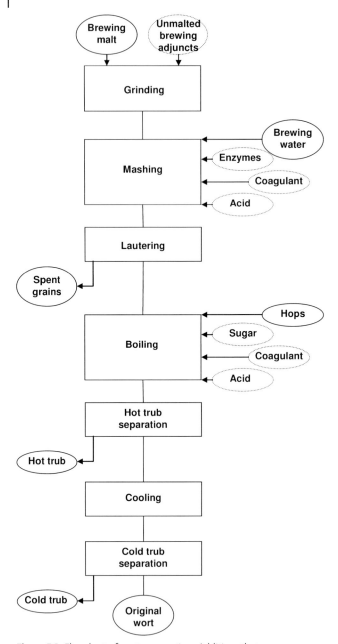

Figure 7.1 Flowchart of wort preparation. Additives that are shown in dotted ellipses are not necessary.

of the mill (i.e. how much malt can be ground per time unit) depends on the length of the roller, time of revolutions, riffle and possibly differential speed of the rollers. Depending on the numbers of milling steps needed, dry grist mills have one to three roller pairs, and are hence differentiated between two-, four-, five- and six-roller mills. After every single milling step shaking or vibration screens are arranged to remove the sufficiently ground parts. Consequently, only the coarsely ground malt particles pass through the milling passage again. The setup of a six-roller mill is shown in Figure 7.2.

An overview of grist specifications of different milling systems is given in Figure 7.3.

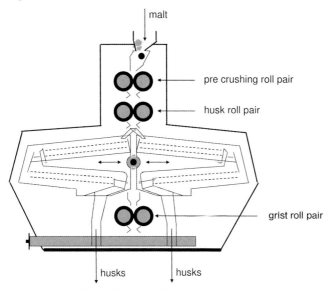

Figure 7.2 Six-roller mill for dry milling (Bühler, www.buhlergroup.com).

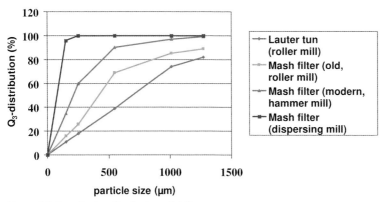

Figure 7.3 Particle size distribution of different grist (cumulative distribution) (Meura, www.meura.be; Ziemann, www.ziemann.com) [1].

Since the brittle endosperm can easily break during dry milling with roller mills, the grist used for the lauter tun is nowadays steeped before the milling process (conditioning chute before dry milling, water uptake 1–2%) or milled totally wet (steeping chute before wet milling, water uptake 15–20%) [2]. Dispersing mills, hammer mills and disk mills are mostly used for the mash filter.

7.3
Mashing Technology

Modern mashing needs to fulfill both economic as well as qualitative demands. Considering the malt quality, the technical equipment in the brewhouse and the choice of the mashing, a wort composition needs to be achieved guaranteeing a high beer quality. The proteolytic breakdown processes need to be controlled in such a way that sufficient assimilating nitrogen is available for an optimal fermentation as well as high molecular nitrogen reactions remain for foam stability. The cytolysis needs to be controlled so that a smooth lautering and filtration is ensured. Amylolysis provides the required fermentable extracts. Furthermore, the brewer has the possibility to influence the aroma and taste profile of the beer with the degradation products developed during mashing.

Mashing is the solubilization of malt components by enzymatic, physical and chemical solution processes. Starch, proteins and cell wall substances are the most important classes of substances for malting, which are dissolved by hydrolysis. Furthermore, during mashing organic phosphate is transformed into primary phosphates by phosphatases (phosphorolysis). This increases the buffering capacity of the mash and wort, and can influence the pH fall during fermentation. Lipids are decomposed by autoxidation as well as enzymatically to numerous products. Polyphenols undergo oxidation and polymerization processes, which result in the decrease of valuable antioxidants and consequently in a reduction of the taste stability [1].

Therefore, many aspects need to be considered for the mashing technology. The process needs to be adjusted to the desired beer variety considering the raw materials (malt and brew water) and the available brewhouse equipment. The used malt significantly affects the color and aroma. By enriching low molecular nitrogen compounds there is a possibility to introduce additional starting materials for the Maillard reaction into the mash. These can be transformed at high temperature (e.g. during wort boiling) into color- and aroma-rich compounds. Not only the equipment of the brewhouse needs to be considered, but also aspects such as the heating rate and stirring speed. A high heating rate is particularly required during mashing if certain enzymatic reactions should be inhibited, such as maltose development during the fabrication of alcohol-reduced beers. Alternatively, with the addition of hot brew water the mash temperature can be escalated. The adjustment of stirrer speed is of importance in view of the low oxygen introduction during mashing and homogenization of the mash during heating up, mashing in or mashing out.

7.3.1
Mashing Parameters

The enzymatic breakdown during mashing can be controlled by the parameters of temperature, viscosity, pH value and time [3]. Over the course of time many different mashing processes have been developed, which can in general be divided into infusion and decoction methods. The decoction method supports the enzymatic breakdown by a physical–thermal decomposition of starch with the aid of the boiling parts of the mash [4]. The partial mashes lead already during the mashing process to an increased thermal stress and an increased protein precipitation. It needs to be taken into account that during transferring the partial mash into a boiler additional oxygen introduction and mechanical stress of the husks and colloids occurs. With the breeding progress to date of brewing barley varieties, good cytolysis and proteolysis take place during the malting process so that the mashing for light beers needs only to be adjusted for amylolysis. Thus, the two-mash procedure (compare Figures 7.4 and 7.5) and infusion methods with high mashing-in temperatures of 60–64 °C are possible and common.

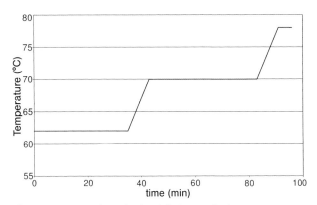

Figure 7.4 Two-mash method as infusion method.

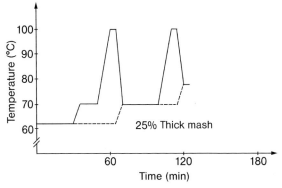

Figure 7.5 Two-mash method as decoction method [1].

Apart from saving energy, which is nowadays of great importance, the two-mash method offers a technological advantage. By inhibiting the phosphatase activity possessing a temperature optimum between 50 and 53 °C (compare Table 7.1) a lower buffering capacity is formed which leads to a stronger pH fall during fermentation. The consequences are also a lighter color and a higher colloidal stability. Furthermore, the proteolytic activity is reduced and thus the high molecular proteins possessing a positive influence on the foam are not further broken down. Nevertheless, it needs to be ensured that the malt used delivers the required free amino nitrogen (FAN) concentration. Additional substances in malt with a positive influence on the foam are the glycoproteins. These are partly dissolved during malting and at low mashing temperatures, and then also broken down to low molecular compounds. A rest at 70 °C assists the solubility of these glycoproteins. However, they are not broken down further and hence aid the foam stability [5].

The major task of mashing is the breakdown of starch to fermentable sugar. Starch, at 55–65% of the dry weight, is the main constituent of barley, and is present in the form of large (A-type, 10–25 µm) and small (B-type, 1–5 µm) starch grains. The large starch grains make up only 10% of the total starch grains, but represent 90% of the starch mass. Normal barley starch is composed of 20–30% amylase and 70–80% amylopectin. The most important property of starch for brewing is gelatinization – the water adsorption-induced swelling of the starch grains and the resulting irreversible loss of the crystalline structure of the starch [7, 8]. The temperature at which gelatinization starts is the gelatinization temperature (GT), which for large starch grains is 61–62 °C and for small grains is 75–80 °C [9]. The GT as well as the gelatinization behavior of barley or barley malt can be determined with rotation (see Table 7.2), amongst other methods [7]. It should be pointed out that the determination of the GT can differ to some degree depending on the measuring method used. The height of the GT is dependent on the variety, place and vintage as shown in Figure 7.6 [1, 7]. In particular, an increased temperature is of importance.

Figure 7.7 shows that the gelatinization did not start until 65 °C in 2005 and that despite a weakened β-amylase this temperature needed to be kept during rest to obtain the desired attenuation limit.

Below the GT, only those starch grains are hydrolysed that were enzymatically attacked during malting [10] or that were mechanically damaged during milling [11]. Only those grains can also take up water below the GT and are thus enzymatically hydrolysable [8, 12].

The GT is hence of importance during mashing as in malt the bulk of starch grains are present in native form and consequently are only hydrolysed at temperatures above the GT. The breakdown of starch during mashing is accordingly influenced by the activity of amylolytic enzymes and the corrodibility of the starch. A fast gelatinization and a high amylolytic activity guarantee a rapid saccharification and a high degree of attenuation.

Relevant enzymes for mashing are α-amylase and β-amylase; both cleave α-(1–4)-D-glycosidic linkages of the starch [13]. β-Amylase cleaves maltose from the non-reducing end of the starch molecule and is thus significantly responsible for

Table 7.1 pH and temperature optima of some barley malt enzymes in mash [6].

	Enzyme	Temperature optimum in the mash (°C)	pH optimum in the mash	Substrate	Product
Cytolysis	β-glucan solubilase	62–65	6.8	matrix bound β-glucan	soluble high molecular β-glucan
	endo-1–3-β-glucanase	<60	4.6	soluble high molecular β-glucan	low molecular β-glucan, cellobiose, laminaribiose
	endo-1–4-β-glucanase	40–45	4.5–4.8	soluble high molecular β-glucan	low molecular β-glucan, cellobiose, laminaribiose
	exo-β-glucanase	<40	4.5	cellobiose, laminaribiose	glucose
Proteolysis	endopeptidase	45–50	3.9–5.5	proteins	peptides, free amino acids
	carboxypeptidase	50	4.8–4.6	proteins, peptides	free amino acids
	aminopeptidase	45	7.0–7.2	proteins, peptides	free amino acids
	dipeptidase	45	8.8	dipeptides	free amino acids
Amylolysis	α-amylase	65–75	5.6–5.8	high molecular and low molecular α-glucans	melagosaccharides, oligosaccharides
Other enzymes	β-amylase	60–65	5.4–5.6	α-glucans	maltose
	maltase	35–40	6.0	maltose	glucose
	limit dextrinase	55–65	5.1	limit dextrin	dextrins
	lipase	55–65	6.8–7.0	lipids, lipid hydroperoxides	glycerine plus free long-chain fatty acids, fatty acids, hydroperoxides
	lipoxygenase	45–55	6.5–7.0	free long-chain fatty acids	fatty acids hydroperoxides
	polyphenol oxidase	60–65	6.5–7.0	polyphenols	oxidized polyphenols
	peroxidase	>60	6.2	organic and inorganic substrates	free radicals
	phosphatases	50–53	5.0	organic-bound phosphate	inorganic phosphate

7 Wort Production

Table 7.2 GTs.

Cereal	Number of samples	GT (°C)			Span (°C)
		minimum	maximum	average	
Unmalted cereal					
barley adjunct	16	65	69	67	4
corn adjunct	9	73	79	76	7
rice adjunct	25	67	91	81	24
Malt					
barley malt	48	58	65	62	7

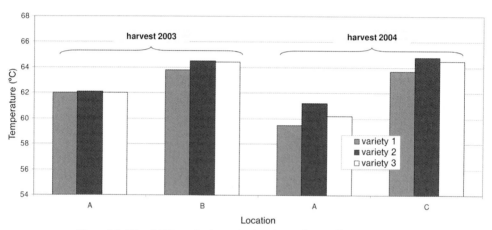

Figure 7.6 GTs of different barley malts, location and year of harvest.

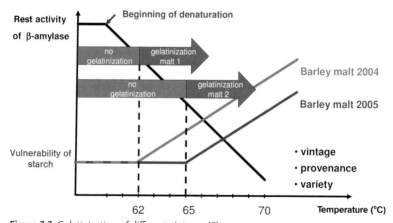

Figure 7.7 Gelatinization of different vintages [6].

the degree of attenuation of the wort [14]. β-Amylase is a thermolabile enzyme that is 60% inactive after only 10 min at 65 °C [15]. Therefore, during resting at 62 °C the maltose concentration increases only slightly after 20 min. An extension of the maltose resting to increase the degree of attenuation is consequently only beneficial to a certain degree. Using the parameter 'viscosity', the inactivation can be slowed down as β-amylase is more thermostable in the presence of proteins [1]. A low viscosity leads to a slower inactivation. This aspect is particularly of interest in high-gravity brewing. A maltose rest at 60 °C leads indeed to a slower inactivation, but with the reduced vulnerability of starch also to a slower breakdown.

In contrast to α-amylase, β-amylase is not able to attack the starch grains; consequently α-amylase is necessary as a preceding enzyme [1, 16]. α-Amylase is by far more thermostable and still shows a 30% reactivity after 10 min at 80 °C. As an endoenzyme it increases the fraction of non-reducing ends and delivers the substrate for β-amylase. This synergetic effect is highest between α- and β-amylase when gelatinization has advanced so far that the starch grains have a high accessibility for α-amylase and β-amylase is at the same time sufficiently active. This effect occurs above the GT and is thus dependent on the malt used.

Limit dextrinase cleaves α-(1–6)-D-glycosidic linkages of amylopectins and is essential for a complete degradation of starch [13, 14, 17]. Due to the low activity and near inactivation at 65 °C, the contribution of limit dextrinase in barley malt is very small in terms of starch breakdown.

A wrongly controlled amylase can lead to too high or too low degree of attenuation, high iodine values in wort, α-glucan tarnishing (gelatinization tarnishing) of beer as well as difficulties with filtration caused by high viscosity.

Even if the mash shows iodine normality it can happen that the later wort shows high iodine values. This can be attributed to the starch grains that do not gelatinize during mashing and reach the kettle-full-wort [9] and are dissolved there during further heating. This results in soluble high molecular α-glucans causing tarnishing, problems with filtration and high iodine values.

7.3.2
Selected Mashing Processes

Depending on the demand, different mashing processes can be applied. Examples of specific processes are described in the following sections.

7.3.2.1 Step-Mashing Process
The step-mashing process is particularly suited for the production of alcohol-free beers [1, 18]. The principle of this process is based on the mash temperature being increased from below 50 °C to over 70 °C by the 'jumping in' in boiling water over a short period of time and hence a rapid inactivation of maltose-producing β-amylase is achieved. The results are worts with a low attenuation limit. Apart from β-amylase, α-amylase also is affected so that worts produced by the step-mashing process do not always reach the required values for the photometric iodine

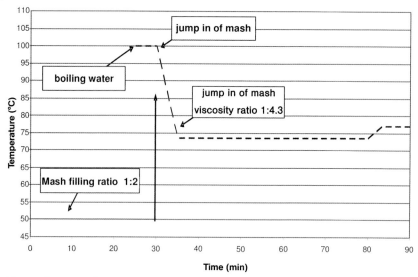

Figure 7.8 Step-mashing process.

sampling. Careful attention should be paid that the temperature does not drop below 73 °C. The mash-in temperature and the rest need to be adapted such that the wort contains sufficient assimilable nitrogen (20–22% of the total nitrogen). Worts mashed isothermically at 45 °C show a degree of attenuation of 65–75%. As already mentioned, starch is also hydrolysed at low temperatures. Malts need to be chosen that are proteolytically and cytolytically well dissolved, but posses a low attenuation limit (Figure 7.8).

7.3.2.2 Maltase Process

Particularly with wheat beers, but also with other beer varieties, higher ester concentrations can be desirable. These can be adjusted in the beer by the glucose concentration. High glucose concentrations can be achieved by promoting maltase. This enzyme cleaves maltose to two glucose units at a temperature optimum of 45 °C. However, with common infusion processes only a low concentration of maltose is present at this point in time as these only develop during the maltose rest. Therefore, a process with maltase rest is appropriate [19]. This process combines two steps (Figure 7.9). Initially, 60% of the total filling volume is mashed-in with 45% of the main cast at 62 °C and after a 40-min rest it is heated up to 70 °C for the saccharification rest. After the saccharification rest, cold water is added to the wort. The remaining malt husks are added to the cooled mash. The maltase of the added malt hydrolyses the maltose of the first filling to glucose. The mashing process can also be used to reduce high β-glucan values from extremely low dissolved malts [20].

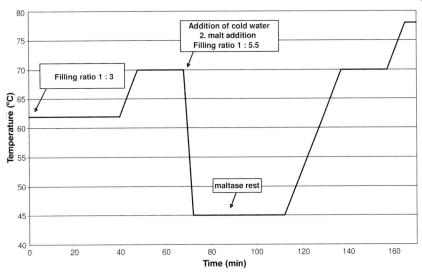

Figure 7.9 Maltase process.

7.3.2.3 Dark Beer Varieties

The most noticeable criteria of dark beers are the color and the malty aroma. The color of the darker beers lies between 50 and 80 European Brewery Convention units. To achieve these characteristic dark malts, dark caramel malts (3–5%), color malts (maximum 5%) or roast malt beers can be utilized. Color malt should be used as little as possible as too many constituents can be introduced with a ticklish aroma (e.g. pyrazine). To ensure sufficient enzyme concentration, the addition of 10–20% of light malt is advantageous. If hard water with high rest alkalinity is used, a darker beer color also develops; in contrast, soft water leads to a stronger developed malt character [1]. Another important aspect is the concentration of FAN. During the production of darker malt FAN is depleted during the formation of melanoidins in the course of the Maillard reaction as well as enzymes increasingly deactivated due to the higher kiln temperature. Therefore, an advanced proteolysis is important during mashing to avoid a deficiency in feed of the yeast with amino acids during the later fermentation. To produce a malt-aromatic dark beer the mashing needs to be adapted to the modified conditions. Decoction processes are particularly suitable for the production of dark beer varieties, whereas two- (Figure 7.10) and three-mash processes with mash-in temperatures of 35 °C are not uncommon. Due to the weaker enzyme activity of darker malt, a thermal–physical decomposition of starch with boil mash is beneficial. Since protein-rich barley parts are often used for the production of dark malt, the precipitation of proteins with the boil mash is also advantageous in view of the high colloidal stability.

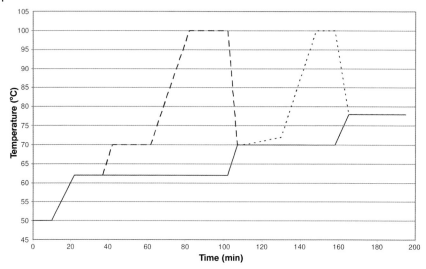

Figure 7.10 Mashing process to produce dark beer varieties.

7.3.2.4 Adjunct Mashing

Malt replacements are classified in different categories. In general, their main purpose is to produce extract. Secondarily, the malt replacements are used to provide yeast nutrition. They also provide a distinct taste and flavor profile. The different areas of application are shown in the following scheme [21]:

- Mash tun adjuncts:
  - if GT ≤ $T_{opt.}$ β-amylase → infusion mashing (e.g. barley adjunct)
  - if GT > $T_{opt.}$ β-amylase:
    - pre-gelatinized starch (e.g. corn flakes, or micronized or torrified cereal)
    - gelatinization of adjuncts with adjunct boiler.
- Copper adjuncts:
  - sugar, syrup.

The first category of raw materials comprises malt replacements, which consist of starch and must pass through a mashing process. At this point, this material must be again classified depending on whether the GT of the non-gelatinized raw material lies under or is equal to the optimal temperature of the β-amylase (e.g. barley raw grain) or not (e.g. corn and sorghum). This category includes also the raw material that is already pre-gelatinized outside the brewhouse, such as flaked, popped or micronized cereals, which finally makes up the third group. GT is greatly influenced by the type of cereal and pseudo-cereal, the varieties, the provenience, and The climatic circumstances.

Valuable information is obtained from the GT which is, however, not obvious from the congress mashing. With the use of adjuncts the GT can be of importance from an energy management aspect. For example, the GT of rice, depending on the variety, lies in the range of 67–91 °C, the temperature rest of the adjunct mash does not necessarily need to be kept above 90 °C [22]. With adjunct malt mixtures with corresponding low GT, the use of exogenous enzymes can even be abandoned or an infusion process can be carried out.

Figure 7.11 clarifies the possibilities considering the analytical requirements [23].

7.3.2.4.1 Mashing with Barley Adjunct

In beer preparation, barley adjunct is used in many countries outside Germany as extract provider. Due to the grain hardness and optimal extract yield, the fine grist needs to be produced with a hammer or dispersion mill. The adjunct mashes (addition ratio often 1:4.5) are initially decomposed at boiling temperatures [24]. Some experiments and new evidence from the literature [22] have, however, shown that it is economically (less use of energy and shorter mashing times) and qualitatively sensible to determine the GT [23]. Of all adjunct varieties, the types with low GT should be favored as these can be worked with a reduced use of malt or exogenous enzymes. Below about 40% (maximum 50%) use of adjunct is still possible to accomplish the extract production and protein breakdown with malt enzymes [25].

A typical mashing diagram is shown in Figure 7.12. Exogenous amylolytic enzymes as well as a separated mash run are mandatory above 50% of adjunct. With the increase of barley adjunct, the β-glucan concentration and the viscosity increase even exponentially [26], which necessitates further addition of enzymes (β-glucanases) due to the impending difficulties with lautering and filtration.

Barely flakes can make up to 30% of the addition in stout beers as in pale ale malt sufficient amylase should be present to saccharify its starch. About 15% barley flakes are common in the practical brewing process. A high concentration of β-glucan is introduced with barley. Too much β-glucan can lead to turbidity, gel-like precipitations and increased viscosity, which cause problems with filtration. This can be counteracted by the addition of β-glucanase or by the use of centrifuges instead of filters [27].

7.3.2.4.2 Mashing with Corn Adjuncts

Corn adjunct is commonly used as malt replacement whereby care needs to be taken that the seedlings are sufficiently separated to introduce as little as possible fat into the beer brewing process. The mainly used corn grits can be mixed in similar ratios to the additions as rice. Figure 7.13 shows that corn adjuncts mashes are sufficiently liquefied at 78 °C with 10% malt addition. If almost boiling temperatures are chosen instead, the enzymes are inactivated too early and a retrogradation of starch takes place after gelatinization and uncompleted liquefaction. This is harder to attack during the further course of mashing [6]. It should be mentioned that a high corn addition changes the beer character [28]. Possible aroma contributors are close to 'popcorn' aroma, such as 2-acetylpyrroline and 2-acetyltetrahydropyridine [29].

7.3.2.4.3 Mashing with Very High Rice Adjunct Addition (Up to 90%)

The mashing program is a kind of decoction (Figure 7.14). The rice mash is treated separately.

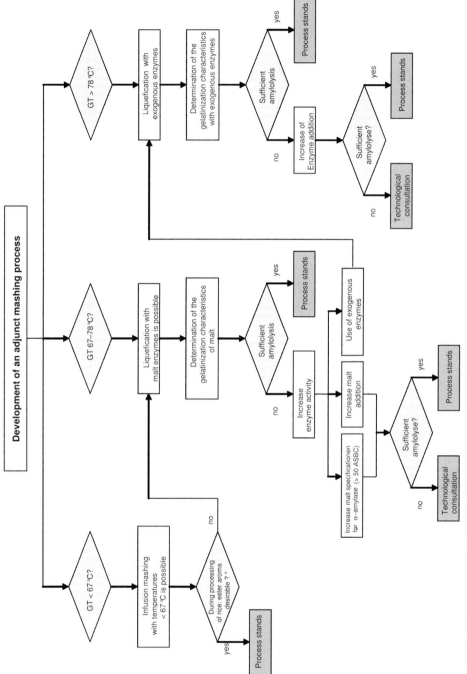

Figure 7.11 Determination of the GT of adjunct. *Rice contains thermostable α-glucosidase, giving an increased glucose fraction; consequently, during fermentation a higher content of esters is developed.

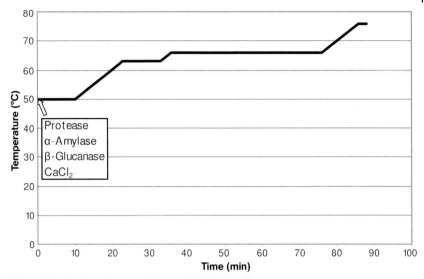

Figure 7.12 Mashing diagram with up to 50% barely adjunct.

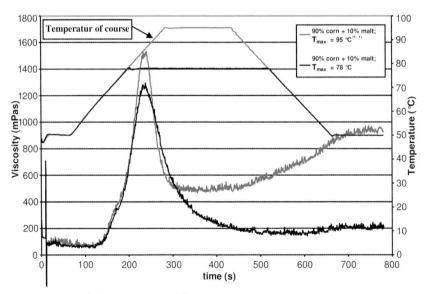

Figure 7.13 GT of 90% corn mash at different maximum temperatures (measured with a rotation viscosimeter from Newport Scientific, Australia).

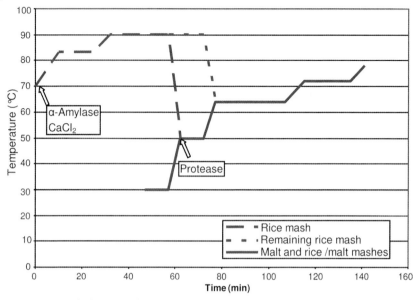

Figure 7.14 Mash diagram with very high rice adjunct addition (up to 90%).

The total hammer milled or dispersed rice grist is mashed-in in a so-called adjunct boiler just above the GT of rice. The grist sorting with hammer mills should be stressed on fractions fine grist I (27.2%), fine grist II (38.6%) and dusting flour (10.5%).

Exogenous thermostable α-amylases and a corresponding amount of calcium ions, depending on the enzyme provider's instructions, are immediately added to the mash, and compulsory pH adjustment carried out with technical lactic acid. Then the temperature is raised to 90 °C and held for 30 min. Following iodine normality, the first part of this rice mash is brewed with the 29 °C malt mash to 52 °C. Shortly afterwards, in order not to inactivate them prematurely, proteases are added according to the manufacturer's instructions in order to dissolve further protein from at least the barley malt.

The rice proteins, actually being present (Boisen et al. reported 6.2–10.9% [30]), can only be insufficiently brought into solution. This is independent if the proteases originate from barley malt or given as exogenous enzymes. The remaining rice mash is then mashed to 64 °C including a short rest of 5 min since the attenuation limit is already sufficiently high and then raised to the saccharification temperature of 72 °C. This rest is kept for 20 min, then heated to 78 °C and mashed-off over a mashing filter.

7.3.2.4.4 Mashing with Extremely High Rice Addition (Up to 98%) With such a high addition of adjunct it is no longer possible to use adjunct mash to brew malt mash. Now, a mash cooler needs to be used that works in a similar way as described under mashing with very high rice addition and cools the rice mash to 50 °C. Proteases are added additionally and operated stepwise via the β-amylase

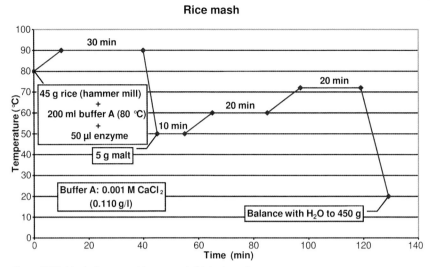

Figure 7.15 Mash diagram with extremely high rice adjunct addition (up to 98%) for laboratory scale.

rest at 62 °C to saccharification rest at 72 °C. With only 2% addition ratio malt is demoted only to β-amylases deliverer. Since only exogenous thermostable α-amylases are active, the lautering temperature can be chosen considerably higher than 78 °C. In any case, only a mash filter can filter this mash so that the choice of filter cloth needs to take the higher temperatures into consideration. The color and aroma can only be affected with roast malt beer or caramel products, which are primarily used during wort boiling. See Figure 7.15.

7.3.2.4.5 Mashing with Sorghum Adjunct Together with corn and rice, sorghum is the third main malt surrogate. The sorghum varieties used have relatively high GTs that necessitate the use of exogenous enzymes. Interestingly, the raw protein content, being comparable with that of barley, can only be brought into solution to a low degree or broken down during mashing into corresponding small fragments. The wort has a very light color.

Due to the import ban on malt to Nigeria in 1988, the local breweries had to change nearly totally to sorghum adjunct. Some elementary research results derive from this constraint [31]. In Figure 7.16 such a mash program is shown which works with many adjuncts, and requires an 'adjunct boiler' and a mash cooler. Due to the missing husks, it inevitably needs to be lautered over a mash filter.

7.4
Technology of Lautering

Lautering is a solid–liquid separation and serves to separate the compounds of malt dissolved during mashing from the insoluble parts (e.g. husk). The lauter

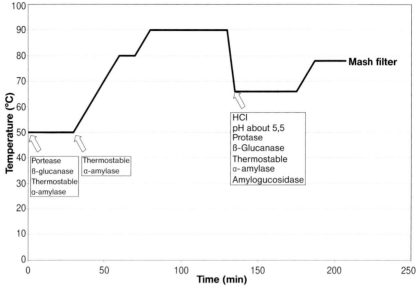

Figure 7.16 Mash diagram with very high parts of sorghum adjunct (up to 100%).

tun (Figure 7.17) and mash filter (Figure 7.18) are widely used systems. The Strainmaster as well as membrane separation processes [32–34] are hardly found in practice.

7.4.1
Lauter Tun

Commonly used lauter tuns have a total capacity of 8 hl/100 kg and spent grain depths of 30–65 cm. They have a perforated false bottom with an open surface representing about 6–30% of the plate area. Rotating, height-adjustable raking machines loosen up the spent grain bed and a spray device delivers the sparging water. Aeration of the wort should be avoided during lautering. The mash should remain homogeneous during pumping in, so that a loose, even filter cake can form. Finally, the wort that runs off should be as clear as possible, so that no particles that could disintegrate further during wort boiling (filterability) and only small amounts of the long-chain fatty acids, which destroy foam, can get into the kettle.

Owing to the danger of washing out iodine-reactive β-glucans, the sparge water should have a maximum temperature of 78 °C. In order to avoid channeling, it should be delivered evenly. The ideal performance of a lautering process in a modern lauter tun is shown in Figure 7.19.

7.4 Technology of Lautering

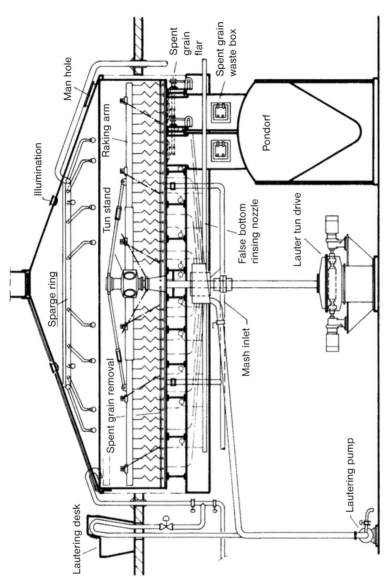

Figure 7.17 Lauter tun [1].

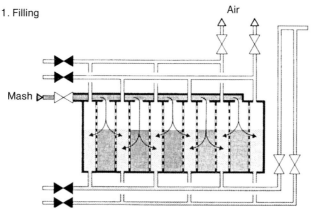

Figure 7.18 Mash filter during filling with mash [1].

7.4.2
Mash Filter

In place of the lauter tun, several brewers employ mash filters to separate the mash into solids and liquid. The total mash is transferred into a vertically arranged filter press. The frames are covered on both sides with filter cloth made of synthetic material. At the same time, the air must escape quickly when the homogeneous mash is pumped into the chambers. After yielding the first wort, water is pumped in and the filter compartments of the filter press are pressed together with corrugated steel plates. Since very low volumes of sparging water (0.5 hl/100 kg) are necessary, this facility is very suited for high-gravity brewing.

Characteristic differences between well-automated mash filters and lauter tuns are: (i) independence from the quality of the malt and from the proportion of adjunct, (ii) quicker lautering of the more highly concentrated first wort, (iii) higher yields, and (iv) mostly a hazier filtrate.

In contrast to the mash filter, the husks of the malt should largely be obtained intact in the lauter tun as these are needed to form the filter cake (see Section 7.2). While the spent grain cake lies horizontal on the false bottom in the lauter tun, it is arranged vertically in the mash filter.

The following points should be considered and differ in their construction depending on the manufacturer:

- Raking machine in the lauter tun (rake design, number of knifes per square meter).
- Headwater in the lauter tun (number per square meter, central lautering or separate regulated lauter zones, design of the lauter tubes).
- False bottom in the lauter tun (free transition area, distance between false bottom and base of lauter tun).
- Composition of the cloths and mode of the incoming flow of the mash filter.

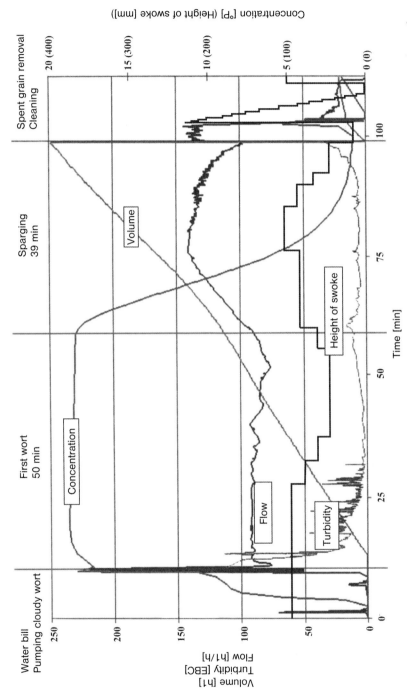

Figure 7.19 Idealized lautering diagram using a lauter tun: ---, extract concentration (wt%); —··—, pressure under false bottom (mm); – – –, volume of wort (hl); –■–, position of raking machine (mm) (full view).

- Mash filter without or with membranes for pressing the spent grain cake.
- Specific filling per square meter and height of spent grains. While the filling in the lauter tun can be varied, it is constant in the mash filter.
- Mash storage (oxygen depleted and homogenic)

Apart from the construction differences, there are diverse regulatory systems. The goal is to preferably reach a homogenic spent grain leaching with a high yield of extract with moderate wort tarnishing [35–38] in a certain amount of time. For this purpose, different lautering algorithms, artificial neural networks as well as fuzzy logic [39] are used.

The control of the yield can be achieved by determining the obtained extract volume (original extract, volume of the wort) or, even better, from the loss of spent grains [40].

7.4.3
Strainmaster

The Strainmaster consists of a rectangular vessel, the bottom half of which is conically shaped. In the lower part of the vessel perforated sieve pipes of triangular cross-section, which have an open surface area of 10%, are arranged on top of each other. After pumping off the first wort the sparging water can be pumped in from the top and/or the bottom. A very high first wort concentration of 20–23% is necessary in order to achieve a homogeneous mash. Even with large amounts of sparging water it is not possible to achieve yields as high as those obtained with the lautering systems previously discussed, primarily because the spent grains must be withdrawn while very wet. The major advantages of the Strainmaster are its large capacity, rapid action (lautering is finished in 70–80 min) and simple, automatic operation.

7.5
Technological Basics of Wort Boiling

The necessary steps of wort boiling can roughly be divided in two processes: hot holding and evaporation.

7.5.1
Hot Holding

During hot holding, different chemical reactions take place such as hop isomerization, development of aroma substances, development of color and dissolution processes, as well as inactivation of enzymes and sterilization. The inactivation of the malt enzymes is necessary as otherwise the result would be atypical taste profiles (e.g. uncontrolled over-fermented beers). Furthermore, during wort boiling the proteins and protein tannin complexes (break) need to be eliminated to obtain clarified wort. If the protein coagulation is too strong foam-positive high molecular proteins (10 and 40 kDa) are also precipitated that worsen the stability of the foam. This is particularly important if no foam-aiding additives are added (e.g. in accordance with the Bavarian Purity Law). Insufficient protein coagulation leads to colloidal unstable beers. For the customer turbid beer is a reason for reclamation. Turbidity-causing substances can, however, be partially removed by absorption before or during beer filtration using filtering aids.

The Maillard reactions that occur at about 80 °C generate new aroma substances. Of particular importance are Strecker aldehydes developed from amino acids that influence the taste stability of the beer. These are primary and secondary products of the Maillard reaction, and can be reduced depending on volatility, boiling system used and evaporation (i.e. boiled wort contains more of the aroma substances than unboiled wort).

7.5.2
Evaporation

Evaporation serves to remove undesired aroma substance such as myrcene from hops and different carbonyl as well as sulfur substances, especially dimethylsulfide (DMS). Aroma substances from lipid metabolism are also reduced by evaporation during wort boiling [41]. Some of these substances can be used as an analytical indicator of the evaporation efficiency of the boiling process as these are not reformed during boiling. Moreover, the original extract is adjusted. This is necessary to ensure the product constancy and legal requirements/marketability of the beer.

A high evaporation allows an effective leaching of the spent grain and thus saves malt. At the same time more energy is needed for the evaporation. In this process the thermal stress of the wort is increased, which has a negative affect on the taste stability of the beer. A high evaporation efficiency of the boiling system results in a good quality as the basic analytic markers of the boiling (Figure 7.20) such as free DMS, thiobarbituric acid number (TBN) and coagulable nitrogen can be ideally adjusted. Consequently, the total evaporation can be reduced, so that the heat and time demand for the boiling process is reduced.

Practical experiments have consistently shown that comparable total evaporation can cause varying concentration decreases of the evaporation indicators. The

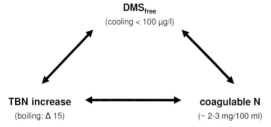

Figure 7.20 Interdependent analytical markers of wort boiling [6].

concentration decrease can be described using a dimensionless number, the evaporation efficiency, which is dependent on the measurable indicator substance, method of evaporation and the wort boiling system.

The evaporation efficiency is defined as:

$$\text{Evaporation efficiency} = \frac{\text{Concentration decrease}_{evaporation\ indicator}\ (\%)}{\text{Total evaporation}\ (\%)}$$

In classical wort boiling systems, such as in an internal or external boiler, the hot holding and evaporation goes on simultaneously. After the boiling phase, the hot trub separation takes place in, for example, a whirlpool.

In modern boiling systems, it is possible to carry out the hot holding and evaporation at the same time or delayed. Modern wort boiling offers more possibilities and a greater flexibility. Flexibility means that time lapses, energy input and quality characteristics can be consciously altered and coordinated.

Evaporation can be carried out, for example, before or after hot trub separation. In addition, wort cooling before hot trub separation is possible.

During wort pre-cooling the wort is cooled from 98–99 °C to temperatures below 90 °C before the hot trub separation. This cooling can either be achieved with a heat exchanger or a flash evaporator. In a flash evaporator the wort is pump after boiling so long in circulation until the desired temperature is reached, adjusted by the corresponding vacuum in the flash evaporator. With wort pre-cooling, the thermal load of the wort is reduced [42]. This has a positive effect on the taste stability and results in a lower development of free DMS from the precursor substance S-methylmethionine (also know as DMS precursor).

The post-evaporation occurs after hot trub separation directly before wort cooling and can be carried out in two ways: under atmospheric conditions with a postboiling or under a vacuum. The principle advantage of post-evaporation after boiling and hot trub separation is the evaporation of undesired and so far reproduced wort aroma substances that can lead to negative taste interference.

Table 7.3 shows the markers for free DMS, TBN and coagulable nitrogen for light beer wort. In addition, further criteria also play a role in the taste profile. For example, a low boiling temperature or low total evaporation often results in more characteristic and richer beers.

TBN is a dimensionless number and provides information on the thermal load of the wort. Maillard products are detected, which are formed from amino acids

Table 7.3 Marker of wort boiling with wort out of 100% malt [6, 45].

	TBN	Total DMS (µg/l)	Coagulable nitrogen relating to 12 wt-% (mg/l)
Hot wort	<45	<100	15–25 (practical experience: 20–35)
Whirlpool wort (drainage)	<60	<100 (free DMS)	–

and reducing sugars during boiling. Values from guidelines for brewhouse control [43] serve as basics in the approval of the brewhouse. These indicate that the value for the cool medial worts should be less then 60. A low thermal stress is positive with respect for the taste stability of the beers since fewer deterioration-relevant carbonyl compounds (Strecker aldehydes) are formed [44, 45]. These values are not applicable for dark worts, as due to the entirely different taste profiles the Maillard products and Strecker aldehydes are even desirable here. Since the absolute values of TBN are dependent on the type of malt, the TBN increase needs to be considered during evaluation of a boiling system.

DMS precursor is transformed into free DMS during wort boiling and should be present at concentrations below 100 µg/l (taste threshold in beer) [46]. Depending on the beer matrix, concentrations in the original wort above 150 µg/l occasionally result in a vegetable-like off-taste. It needs to be considered that during hot trub separation in the whirlpool additional free DMS is formed that, however, cannot be evaporated anymore in this process step. This has to be considered during classical boiling. In contrast, post-evaporation after hot trub separation permits a final reduction to the desired values in the original wort as shown in Figure 7.21 (adequate sampling place: middle of the cooling wort).

The necessary equipment for post-evaporation or wort pre-cooling can be retrofitted into already existing systems. Hence, it is possible to achieve the desired wort composition within the brewhouse while saving energy and responding directly to quality fluctuations of the raw materials.

Protein coagulation occurs during boiling in two steps: (i) the proteins denature and (ii) then the actual coagulation takes place. This process is dependent on the boiling temperature, mixing (homogeneity), duration of boiling and wort pH. A beneficial pH lies below pH 5.2 [47] for the protein coagulation and can be achieved by acidification of the wort. The concentration of still coagulable nitrogen in the

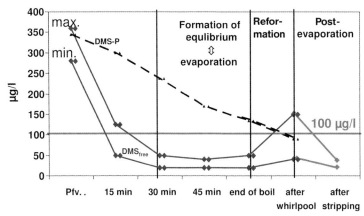

Figure 7.21 Minimal and maximal course of concentration for free DMS and DMS precursor (DMS-P) during wort boiling including post-evaporation [6].

cooled wort has an effect on colloidal stability, beer foam, taste and palate. Guide values for still coagulable nitrogen [43] lies in the range from 15 to 25 mg/l. In practice, it has been constantly shown that a value up to 35 mg/l does not cause a problem, but rather can be of advantage for the taste and foam stability.

7.5.3
Modern Boiling Systems

Due to the improved and henceforth excellent evaporation efficiencies, nowadays total evaporations of 4–6% are possible and common with modern boiling systems. In the following, common wort boiling systems and new developments over the last 10 years are presented [48]. Flawless worts can be produced and the MEBAK guidelines can be kept with all boiling systems. This applies also for DIN 8777 [49] which is often employed to guarantee declarations in brewhouse certifications.

7.5.3.1 Internal Boilers

Internal boilers were used for the first time in the middle to the end of the 1970s. These boilers are mostly vertical-standing tube-bundle exchangers and function according to the natural circulation principle. This means that the wort is heated up with steam or hot water inside the tube bundle. A temperature difference as well as evaporation due to development of bubbles in the tubes develops between the boiler inlet and boiler outlet. The resulting differences in temperature and density result in an upward pressure that ensures natural circulation. One problem with internal boilers is the pulsation occurring during heating-up that should normally end once the boiling temperature is reached. This disadvantage can be reduced or even eliminated by using a circulation pump [50]. The smaller the difference in temperature between the internal boiler outlet and the surrounding wort or boiler inlet, the smaller the tendency of the internal boiler to pulsate (Figure 7.22).

7.5.3.2 Optimized Internal Boiler 'Stromboli'

The internal boiler system Stromboli is a development and optimization of the boiler system Ecotherm. Wort is passed through a central tube inside the internal boiler by force using a frequency-regulated pump whereupon an undertow develops above the tube bundle. Due to the undertow, suction develops at the place marked with 'jet pump'. Consequently, a branch current of the wort inside the internal boiler tube is transported upwards and distributed together with the wort pumped through the central tube onto the wort surface.

With the forced incoming flow during the heating-up phase, pulsing of the wort is avoided. Furthermore, the thermal load can be reduced with existing circulations by the reduction of the heating temperature and a preservation of the remaining coagulable nitrogen is achieved. The forced incoming flow and the increased number of attached outlets in the bottom of the kettle ensure a good mixing of the kettle content during the heating-up phase and the total boiling time. Thus,

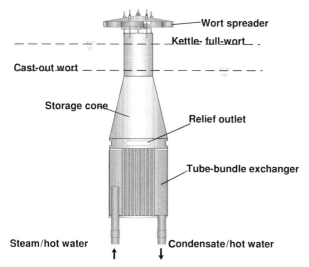

Figure 7.22 Layout of an internal boiler (source: GEA-Huppmann GmbH, Kitzingen, Germany).

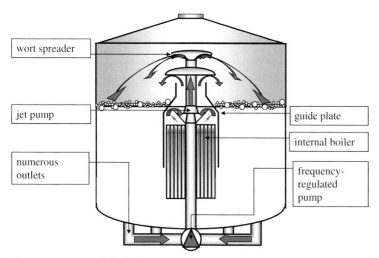

Figure 7.23 Layout of the boiler system Stromboli (source: Krones AG, Werk Steinecker, Freising, Germany).

despite a low total evaporation, a sufficient steaming out of undesired aroma substance is achieved and a beneficial energy balance is additionally obtained [51] (Figure 7.23).

7.5.3.3 Optimized Internal Boiler Subjet

During discharging the wort over the classical deflector a great deal of energy is converted into turbulence during the immersion of the liquid stream. Hence, the

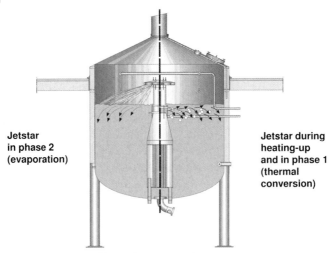

Figure 7.24 Layout of the boiler system Subjet (source: GEA-Huppmann GmbH, Kitzingen, Germany).

wort flows mainly along the outside wall back to the boiler inlet (see Figure 7.24, left side). Part of the wort is kept standing for a long time without mixing. With the new subjet wort boiling, particularly in the heating-up phase, the wort is introduced below the wort level in the kettle. With the height-adjustable subjet spreader, the wort is discharged below the wort level.

Velocity measurements along the inner wall of the vessel indicate an improved impulse exchange with the surrounding wort by the subjet. A short-circuit flow along the kettle wall is avoided. The result is a considerably better convection and an even treatment of the whole wort volumes. The wort is more homogeneous. The internal boiler with subjet spreader can be retrofitted to existing internal boilers.

7.5.3.4 External Boiler

External boilers are also mostly conducted as tube-bundle exchangers. Here, the wort is constantly removed from the wort kettle and the kettle content is pumped through an external boiler 6–8 times/h. The wort temperature is adjusted after the external boiler by an expansion valve. In the kettle, the pumped wort hits the spreader through which a cone-shaped spray pattern emerges, and the surface is increased for the evaporation of undesirable aroma substances and water. The greatest disadvantage of this boiler design can be the increased space required in the brewhouse as well as the relative large plant-specific complexity. The technological advantages, such as good mixing of the kettle contents, large variability of heating area size during installation as well as a defined hot-keeping time in the boiler by adjusting the wort volume flow over the pump capacity allow targeted wort treatment [50]. Additionally, up to three wort kettles can be fed with one external boiler (Figure 7.25).

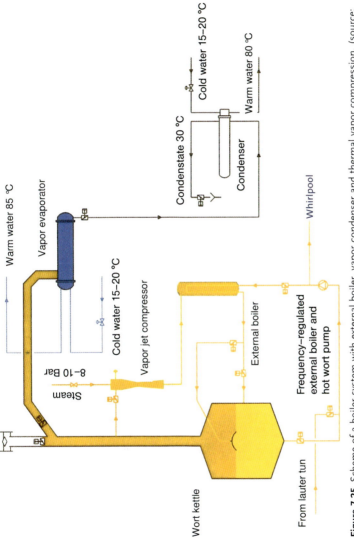

Figure 7.25 Scheme of a boiler system with external boiler, vapor condenser and thermal vapor compression. (source: Krones AG, Werk Steinecker, Freising, Germany).

7.5.3.5 High-Temperature Wort Boiling

During high-temperature wort boiling the wort is often heated to temperatures of 128–135 °C by three wort heaters. Subsequently, two resting steps take place to lower the wort temperature. During the resting of the wort the necessary evaporation of the aroma substance occurs. The energy of the arising vapors is recovered and serves for the stepwise heating-up of the wort.

In the first two wort boilers the wort is heated with these vapors to 87 and 107 °C, respectively. The heating to 130 °C occurs with primary energy. This temperature level is held for 2.5–3.0 min and then cooled in two steaming-out vessels to 117 °C and finally to 100 °C by resting. The hot trub can be separated in a whirlpool. The evaporation rate here is 6–8% [47]. During this process the main part of the energy is recovered so that this boiling system belongs to the most energy-efficient systems. According to Narziss et al. [52] the best beers were produced at 130 °C and a hot-keeping time of 3 min followed by 140 °C for 2 min. For darker beers, a displeasing boiling note can occur by the various Millard products generated. In the beginning, fouling developed prematurely on the heating areas of heating step 3 during the brewing week in the plant, resulting in an uneven thermal load of the wort. This could be resolved by a second parallel heating system allowing a cleaning every 3–6 h without stopping the system (Figure 7.26).

7.5.3.6 Dynamic Low-Pressure Boiling

Conventional low-pressure boiling has been available worldwide since 1976. Here, the wort is boiled with slight over-pressure whereby boiling temperatures of 102–105 °C can be reached. A development of this boiling system is dynamic low-pressure boiling, in which pressure build-up and pressure reduction occurs repeatedly in alternation. These systems are used with an internal boiler and a vapor condenser. The total evaporation can be reduced to values around 4.5% (related to the hot wort volume) and a boiling time of 40–50 min [53, 54]. An effective steaming out of aroma substances is achieved by repeated flash evaporation, that is pressure relaxation (e.g. from 103 to 101 °C) results in an intense boiling action as vapor bubbles develop in the wort and ascend suddenly similar as during boiling delay. To carry out this kind of wort boiling the wort kettles need to be constructed to be pressure-resistant to an absolute pressure of 1.5 bar (Figure 7.27).

7.5.3.7 Soft Boiling Method 'SchoKo'

In the soft boiling method (SchoKo) the boiling is divided into two phases: (i) the hot-keeping of the wort, and (ii) the evaporation between whirlpool and staple cooler. During the first phase the wort is kept hot at 97–99 °C for 40–60 min, resulting in practice in an evaporation of below 1%. With the appropriate equipment of the hot-keeping tank (e.g. with an internal boiler), it could also be boiled conventionally. During the second phase about 5% is also evaporated in a flash evaporator with an absolute pressure of about 400 mbar. In order to carry out this vacuum evaporation, the wort is tangentially introduced into an evacuated specially constructed vessel, resulting in a rotation of the liquid as well as a thin film of wort

7.5 Technological Basics of Wort Boiling | 195

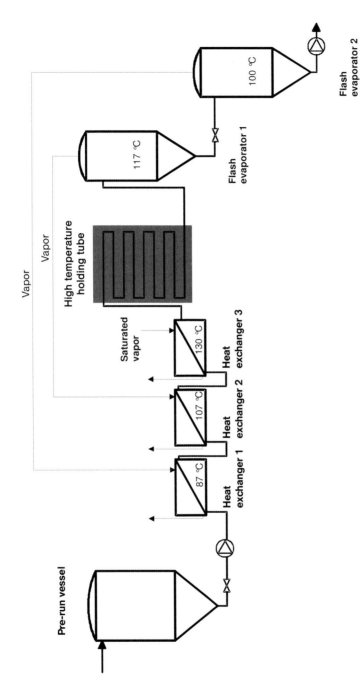

Figure 7.26 Layout of high-temperature wort boiling [6].

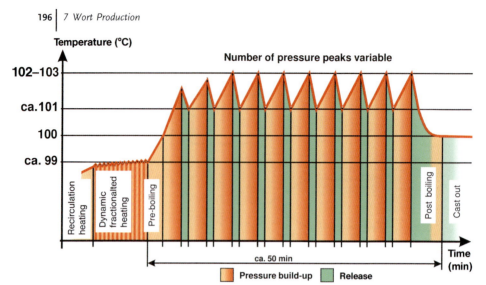

Figure 7.27 Course of dynamic low-pressure boiling (source: GEA-Huppmann GmbH, Kitzingen, Germany).

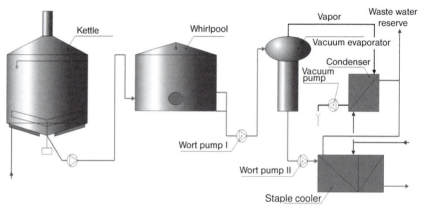

Figure 7.28 Process scheme of the wort boil system SchoKo (source: Kaspar Schulz Brauereimaschinenfabrik, Bamberg, Germany).

along the side. Consequently, a large surface area is produced for the evaporation of aroma substances and water. The special build-up of the system results in a great flexibility in view of the parameterization and layout of the boiling process. The flash evaporator is easily retrofitted [55–57] (Figure 7.28).

7.5.3.8 Wort Stripping

Furthermore, a system is available with which post-evaporation is possible on a stripping column placed between hot trub separation and wort cooling. The system is composed of a wort kettle, hot trub settling tank or whirlpool, stripping column

7.5 Technological Basics of Wort Boiling | 197

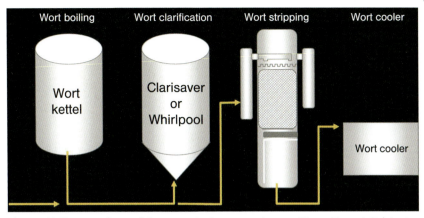

Figure 7.29 Process scheme of the wort stripping system (source: Meura, Tournai, Belgium).

and wort cooler. After reaching the boiling temperature the wort is kept for 30–50 min at 100 °C while being slightly agitated. After that the contents of the kettle are pumped into a settling tank or whirlpool. As soon as the settling time or whirlpool rest is finished, the stripping process starts. The wort is directed downwards through a packed column while steam is injected in counter-current flow. For this purpose the wort needs to be heated up to boiling temperature before the stripping. The condensate generated in the upper part of the column is discharged over a side aggregate [58] (Figure 7.29).

7.5.3.9 Vacuum Evaporation

Another possibility is the combination of conventional boiling and post-evaporation after the whirlpool. First, the wort is boiled conventionally, followed by hot trub separation in the whirlpool with subsequent post-evaporation in a flash evaporator with an absolute pressure of about 600 mbar. The conventional boiling normally lasts for about 40–50 min with a total evaporation of about 4%. About 2% is additionally vaporized in the flash evaporator. This system also allows flexible parameterization and is easily retrofitted or can be combined [55, 59, 60] (Figure 7.30).

7.5.3.10 Flash Evaporation 'Varioboil'

The main components of this system are external boilers, a flash evaporator and a batch tank. In the Varioboil system the wort is also boiled conventionally in the first phase (wort boiling). The vaporization of water and aroma substances occurs in the flash evaporator. The developed vapors are condensed on a vapor condenser. In the second phase, the flash evaporator is set under vacuum, resulting in an evaporation below 100 °C and hence a cooling of the wort (wort cooling). After this second step of boiling the hot trub is discharged conventionally in a whirlpool. During vacuum evaporation care should be taken that the wort does not cool below 80 °C as this can result in divergence of the trub body. Furthermore, there is the

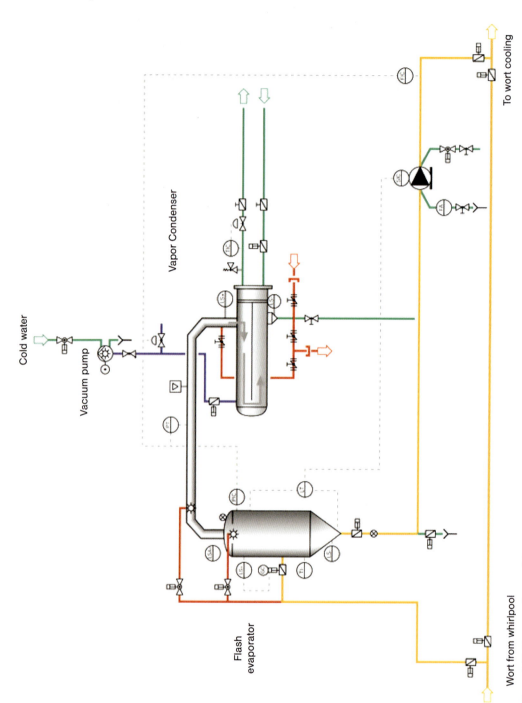

Figure 7.30 Process scheme of the vacuum evaporation system (source: A. Ziemann, Ludwigsburg, Germany).

7.5 Technological Basics of Wort Boiling | 199

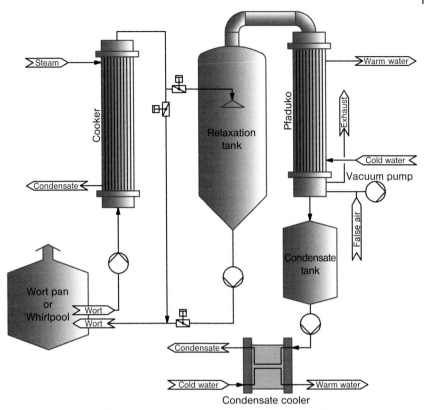

Figure 7.31 Process scheme of the Varioboil system (source: Nerb, Attaching, Germany).

possibility to produce brew water with a temperature of at least 80 °C during wort cooling.

Re- or retrofitting is also easily done with this system so that existing vessels can be made further use of. In addition, the boiling parameterization can also be flexibly varied similar to the already described systems [61, 62] (Figure 7.31).

7.5.3.11 Thin Film Evaporator 'Merlin'

The boiling system Merlin consists of a vessel with a conical heating base and a whirlpool that can be used both as a storage vessel for the wort and also in the actual sense as a vessel for hot trub separation. The wort is lautered in the whirlpool and at the end of lautering passed over a heated cone area using a rotating pump where it is heated to boiling temperature. The thin and turbulent layer yields good heat exchange. During the 35–40-min boiling the total content is circulated 4–6 times. The hot trub separation takes place already during circulation since it is tangentially pumped into the whirlpool. Owing to the early onset of hot trub separation, the whirlpool rest can be reduced to about 10 min. Subsequently, the wort is once again passed over the cone to ensure a post-evaporation under atmospheric conditions before transferring to the staple cooler [63, 64] (Figure 7.32).

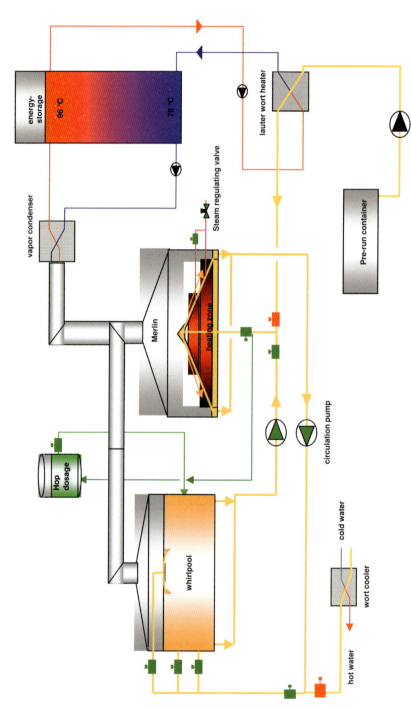

Figure 7.32 Process scheme of the Merlin system with an external boiler as a thin film evaporator (source: Krones AG, Werk Steinecker, Freising, Germany).

7.5.4
Vapor Condensate

To avoid odor emissions and to save energy, the vapors from the wort boiler are mostly condensed in a vapor condenser or, after thermal or mechanical vapor compression, in external or internal boilers. The vapor condensate accumulates in quantities corresponding to the total evaporation. For the reduction of waste water volumes in the brewery (and the corresponding costs) it was investigated whether it was possible to recycle the vapor condensate with or without processing.

The vapor condensate consists of over 99% water. It is salt-free, and contains only organic substances from hops and malt. To date, 161 constituents have been identified. These belong to the classes of aldehydes, alkanes, alcohols, esters, furanes, ketones and terpenes. The concentration of the different substances is in the range of 5–338 µg/l [65]. If the vapor condensate is colored yellow and if at the same time extract is found, a carry-over of the wort by aerosol development can be assumed.

Reverse osmosis and active charcoal filtration have been investigated as processes to purify the vapor condensate. In brewing experiments, a vapor condensate purified with reverse osmosis was used as main and post casts. From a sensory point of view, the quality of the fresh beers was comparable with beers from conventional production. Only with forced, aged beers could impairment be detected if purified vapor condensate was used. For this reason vapor condensate should not reach the product in large quantities [66, 67].

With active charcoal filtration the chemical oxygen demand (COD) of vapor condensate can be reduced by a maximum of 40%. Since drinking, brewing or boiler water quality cannot be achieved, this process is not applicable to purify vapor condensate.

Some possibilities were investigated in which vapor condensate could be used without treatment (Table 7.4). Vapor condensate can still be used for more applications as it is very soft. Only the load with organic substances (measurable with the COD) presents a problem.

Table 7.4 Possibilities to re-use untreated vapor condensate [6]

Process	Usable part (%)
Spouting of lauter tun perforated false bottom	33
Last addition during lautering	28
Cooling water for evaporation condenser	100
Preparation of alkaline detergents	
cleaning of bottles	100
cleaning in place and lost cleaning	100

For spouting the lauter tun, only 33% of the arising vapor condensate is needed. For the preparation of alkaline detergents, the total vapor condensate can be utilized if the lye is freshly prepared often enough. The vapor condensate has a lower COD than detergent so that the cleaning power is not weakened. The vapor condensate is normally cooled to 25–30 °C and 6 ppm chlorine dioxide added for disinfection. Gelatinization and silting of the evaporation condensers and of potential intermediate piling containers result without the addition of disinfectant. Furthermore, the addition of chloride dioxide inhibits unpleasant odors. The use of H_2O_2 was shown to be unsuitable as H_2O_2 has no storage properties and involves higher costs [68].

7.5.5
Cold Trub

The processing step of cold trub separation during beer wort preparation incurs costs and represents a microbiological risk. Its need is controversial in the literature. Brewing experiments with flotation processes, cold wort filtration and cold wort centrifugation showed that cold trub separation was not necessary if the following conditions are fulfilled in the respective brewery [69]:

- Hot trub separation functions properly.
- Beers are constant in their qualities.
- Yeast management corresponds with the state of the art.
- The taste profile of beer produced is as desired by the brewery.

This is based on the fact that the taste of the beers can indeed be influenced by different cold trub entries, but the quality parameters such as taste stability, chemical–physical stability, reduction ability (lag time) and foam stability do not experience a negative influence, but tend to be improved. In detail, the taste changes brewery-specifically with respect to odor, bitterness and palate. The beers obtain a characteristic taste by the cold trub. The aged beer with cold trub tends to be preferred. Both the reductive potential and the chemical–physical stability increase by leaving the cold trub in the wort. The foam stability does not change. The oxygen feed of the yeast can be enhanced by flotation, so that this method of cold trub removal is especially applicable if the physiological condition of the pitching yeast is not optimal.

References

1 Narziss, L. (1992) *Die Bierbrauerei. Band II: Die Technologie der Würzebereitung*, 7th edn, Ferdinand Enke Verlag, Stuttgart.

2 Ferstl, F., Banke, F., Kain, J. and Krottenthaler, M. (2002) Examination of the Factors Influencing the Operational Limits of a Wet Mill. *Chemical Engineering & Technology*, **25** (8), 220–5.

3 Kühbeck, F., Dickel, T., Krottenthaler, M., Back, W., Mitzscherling, M., Delgado, A. and Becker, T. (2005) Effects of Mashing Parameters on Mash ß-Glucan, FAN and

Soluble Extract Levels. *Journal of the Institute of Brewing*, **111** (3), 316–27.
4 Köhler, P., Krottenthaler, M., Herrmann, M., Keßler, M. and Kühbeck, F. (2006) Tests on Infusion and Decoction Method. *Brauwelt International*, **24** (1), 33–7.
5 Narziss, L., Reicheneder, E. and Barth, D. (1982) Untersuchungen über den Einfluss der Glykoproteide auf die Schaumeigenschaften des Bieres. *Brauwissenschaft*, **35** (11), 275–83.
6 Back, W., Krottenthaler, M., Bohak, I., Dickel, T., Franz, O., Hanke, S., Hartmann, K., Herrmann, M., Kaltner, D., Keßler, M., Kreisz, S., Kühbeck, F., Mezger, R., Narziss, L., Schneeberger, M., Schütz, M., Schönberger, C., Spieleder, E., Thiele, F., Vetterlein, K., Wunderlich, S., Wurzbacher, M., Zarnkow, M. and Zürcher, J. (2008) *Ausgewählte Kapitel der Brauereitechnologie*. Nürnberg: Fachverlag Hans Carl GmbH.
7 Tegge, G. (2004) *Stärke und Stärkederivate*, Vol. 3, Behr's Verlag, Hamburg.39
8 French, D. (1984) Organization of starch granules, in *Starch: Chemistry and Technology* (ed. E.F. Paschall), Academic Press, London, pp. 184–247.
9 Palmer, G.H. (1989) Cereals in malting and brewing, in *Cereal Science and Technology* (ed. G.H. Palmer), Aberdeen University Press, Aberdeen, pp. 61–242.
10 Bertoft, E. and Henriksnas, H. (1983) Starch hydrolysis in malting and mashing. *Journal of the Institute of Brewing*, **89** (4), 279–82.
11 Gibson, T.S., Alqalla, H. and McCleary, B.V. (1992) An improved enzymatic method for the measurement of starch damage in wheat-flour. *Journal of Cereal Science*, **15** (1), 15–27.
12 Bathgate, G.N. and Palmer, G.H. (1973) The in-vivo and in-vitro degradation of barley and malt starch granules. *Journal of the Institute of Brewing*, **79** (5), 402–6.
13 Kruger, J.E. and Lineback, D.R. (1987) Carbohydrate-degrading enzymes in cereals, in *Enzymes and their Role in Cereal Technology* (ed. C.E. Stauffler), American Association of Cereal Chemists, St Paul, MN, pp. 117–35.
14 MacGregor, A.W., Bazin, S.L., Macri, L.J. and Babb, J.C. (1999) Modeling the contribution of alpha-amylase, beta-amylase and limit dextrinase to starch degradation during mashing. *Journal of Cereal Science*, **29** (2), 161–9.
15 Eglinton, J.K., Langridge, P. and Evans, D.E. (1998) Thermostability variation in alleles of barley beta-amylase. *Journal of Cereal Science*, **28** (3), 301–9.
16 Ziegler, P. (1999) Cereal beta-amylases. *Journal of Cereal Science*, **29** (3), 195–204.
17 Stenholm, K. and Home, S. (1999) A new approach to limit dextrinase and its role in mashing. *Journal of the Institute of Brewing*, **105** (4), 205–10.
18 Narziss, L., Miedaner, H. and Krottenthaler, M. (1995) Development of a selection method suitable for serial application for improving the quality of the beer and the quality parameters of barley which influence the taste of beer. *Monatsschrift für Brauwissenschaft*, **48** (3/4), 108–11.
19 Herrmann, M., Back, W., Sacher, B. and Krottenthaler, M. (2003) Options to technologically influence flavor compounds in beer. *Monatsschrift für Brauwissenschaft*, **56** (5/6), 99–106.
20 Krottenthaler, M., Zürcher, J., Back, W., Schneider, J. and Weisser, H. (1999) Grist Fractionating and Adjusted Mashing to Improve the Reduction of β-Glucan. In: *Proceedings of the Congress – European Brewery Convention*. Cannes, 603–10.
21 Zarnkow, M. (2006) Perspektiven beim Brauen mit Rohfrucht, Seminario Tecnologico Cervecero Internacional, San Salvador, El Salvador.
22 Zarnkow, M. *et al.* (2007) Influence of cereal adjuncts on beer flavor and flavor stability in consideration of rice adjunct. *Cerevisia*, **32** (2), 110–19.
23 Kessler, M. *et al.* (2005) Gelatinization properties of different cereals and pseudocereals. *Monatsschrift für Brauwissenschaft*, **9/10**, 82–8.
24 Narziss, L. (2004) *Abriss der Bierbrauerei*, 7th edn, Wiley-VCH Verlag GmbH, Weinheim.
25 Meilgaard, M.C. (1976) Wort composition: with special reference to the use of adjuncts. *Technical Quarterly*, **13** (2), 78–90.

26 Müller, K.-P. (2005) Brauversuche unter Einsatz unvermälzter Gerste, in *Lehrstuhl für Technologie der Brauerei I*, Technische Universität München, Freising.
27 Lewis, M. (1995) *Stout*, Brewers Publications, Boulder, CO.
28 Einsiedler Bierspezialitäten, http://www.beer.ch/ros_biere.html#maisgold (accessed 22 June 2005).
29 Lebensmittelchemie, D.F.A. (1991) Identifizierung von Aromastoffen mit Röstgeruch, Jahresbericht, http://dfa.leb.chemie.tu-muenchen.de/DJahr1991, html#Index (accessed 15. 01. 2009).
30 Boisen, S. et al. (2001) Comparative protein digestibility in growing rats of cooked rice and protein properties of indica and japonica milled rice. *Journal of Cereal Science*, 33, 183–91.
31 O'Rourke, T. (1999) Adjuncts and their use in the brewing process. *Brewers' Guardian*, 128 (3), 32–6.
32 Lotz, M., Schneider, J., Weisser, H., Krottenthaler, M. and Back, W. (1997) New Mash Filtration Technique for Processing of Powder Grist. In: *Proceedings of the Congress – European Brewery Convention*. Maastricht, 299–305.
33 Schneider, J., Krottenthaler, M., Back, W. and Weisser, H. (2001) Vibrating Membrane Filtration of Mash for the Beer Production. In: *Proceedings of the Congress – European Brewery Convention*. Budapest, 217–25 (Poster no. 222).
34 Schneider, J., Krottenthaler, M., Back, W. and Weisser, H. (2005) Study on the Membrane Filtration of Mash with Particular Respect to the Quality of Wort and Beer. *Journal of the Institute of Brewing*, 111 (4), 380–7.
35 Kühbeck, F., Back, W. and Krottenthaler, M. (2006) Influence of Lauter Turbidity on Wort Composition, Fermentation Performance and Beer Quality – A Review. *Journal of the Institute of Brewing*, 112 (3), 215–21.
36 Kühbeck, F., Back, W. and Krottenthaler, M. (2006) Release of long-chain fatty acids and zinc from hot trub to wort. *Monatsschrift für Brauwissenschaft*, 59, 67–77.
37 Kühbeck, F., Back, W. and Krottenthaler, M. (2006) Influence of lauter turbidity on wort composition, fermentation performance and beer quality – a review. *Journal of the Institute of Brewing*, 112, 215.
38 Kühbeck, F., Back, W. and Krottenthaler, M. (2006) Influence of lauter turbidity on wort composition, fermentation performance and beer quality in large-scale trials. *Journal of the Institute of Brewing*, 112, 222–31.
39 Back, W., Krottenthaler, M., Voigt, T., Hege, U., Van de Braak, B. and Stippler, K. (2000) New Control System for the Lautering Process in the Lauter Tun. In: *Brauwelt International*, 18 (6), 475–7.
40 Anger, H.-M. (2006) *Gerste, Rohfrucht, Malz, Hopfen*, Selbstverlag der MEBAK, Freising-Weihenstephan.
41 Yamashita, H., Kühbeck, F., Hohrein, A., Herrmann, M., Back, W. and Krottenthaler, M. (2006) Fractionated boiling technology: wort boiling of different lauter fractions. *Monatsschrift für Brauwissenschaft*, 59 (7/8), 130–47.
42 Coors, G., Krottenthaler, M. and Back, W. (2003) wort pre-cooling and its influence on casting. *Brauwelt International*, 21 (1/3), 40–1.
43 Miedaner, H. (ed.) (2002) *Brautechnische Analysenmethoden (MEBAK)*, Band 2.4, Selbstverlag der MEBAK, Freising-Weihenstephan, pp. 19, 20.
44 Schwill-Miedaner, A. and Krottenthaler, M. (1999) Neue Kochsysteme im Überblick. *Der Weihenstephaner*, 67 (1), 69–73.
45 Back, W., Forster, C., Krottenthaler, M., Lehmann, J., Sacher, B. and Thum, B. (1999) New research findings on improving taste stability. *Brauwelt International*, 17 (5), 394–405.
46 Anness, B.J. and Bamforth, C.W. (1982) Dimethyl sulphide – a review. *Journal of the Institute of Brewing*, 88, 244–52.
47 Narziss, L. (2005) *Abriss der Bierbrauerei*, 7th edn, Wiley-VCH Verlag GmbH, Weinheim.
48 Mezger, R., Krottenthaler, M. and Back, W. (2003) Modern wort boiling systems – an overview. *Brauwelt International*, 21 (1), 34–9.
49 DIN 8777 (1996) Sudhausanlagen in Brauereien.
50 Kunze, W. (1998) *Technologie Brauer und Mälzer*, Vol. 8, VLB, Berlin.
51 Wasmuht, K., Weinzierl, M., Gattermeier, P. and Baumgärtner, Y. (2004) Stromboli–

ein 'Vulkan' auf der Überholspur. *Brauwelt*, **30/31**, 925–7.
52 Narziss, L., Miedaner, H. and Schneider, F.P. (1991) Weiterführende Untersuchungen zur Technologie der Würzekochung unter besonderer Berücksichtigung energiesparender Maßnahmen (Teil 3). *Monatsschrift für Brauwissenschaft Heft*, **9**, 304.
53 Hackensellner, T. (2001) Würzebereitung mit dynamischer Niederdruckkochung: Energie- und Anlagentechnik–Teil 1. *Brauindustrie*, **3**, 14–16.
54 Bühler, T., Michel, R., Kantelberg, B. and Baumgärtner, Y. (2003) Die dynamische Niederdruckkochung–systematisch qualitätsoptimiert. *Brauwelt*, **38**, 1173–8.
55 Energietechnisches Seminar (2002) Thema: Würzekochung, Veranstalter: IGS Ingenieurbüro Dr Georg F. Schu, Bayrischer Brauerbund, Schirmherrschaft.
56 Binkert, J. and Haertl, D. (2001) Neues Würzkochsystem mittels Expansionsverdampfung. *Brauwelt*, **37**, 1494–503.
57 Mezger, R., Krottenthaler, M. and Back, W. (2004) Dividing wort boiling into two phases. Using the 'SchoKo' boiling process as an example. *Brauwelt International*, **22** (1), 29–32.
58 Meura (2001) Meura–Traditionally Pioneers Since 1845. Meura, Tournai, CD-ROM.
59 Fohr, M. (2000) Höhepunkte der Sudhaustechnik auf der Brau 2000. *Brauwelt*, **48**, 2090–2.
60 Krottenthaler, M., Lehmann, J. and Mieth, R. (2003) Use of a vacuum evaporator unit in gilde Brauerei AG. *Brauwelt International*, **21** (6), 382–7.
61 Krottenthaler, M., Hartmann, K. and Back, W. (2001) Use of a flash evaporator for wort treatment. *Brauwelt International*, **19** (6), 457–9.
62 Mezger, R., Krottenthaler, M. and Back, W. (2006) Vacuum boiling–a new alternative for gentle wort processing in the brewhouse. *Brauwelt International*, **24** (1), 22–5.
63 Weinzierl, M., Stippler, K., Wasmuht, K., Miedaner, H. and Englmann, J. (1999) Ein neues Würzekochsystem. *Brauwelt*, **5**, 185–9.
64 Wasmuht, K., Weinzierl, M. and Vasel, B. (2000) Neue Ergebnisse zum Kochsystem Merlin®. *Brauwelt*, **26**, 1057–9.
65 Fohr, M., Meyer-Pittroff, R., Krottenthaler, M. and Back, W. (1999) State-of-the-art in utilisation of vapor condensate. *Brauwelt International*, **17**, 360–5.
66 Back, W., Krottenthaler, M. and Vetterlein, K. (1998) Zurück zum Ursprung. Aufbereitung von Brüdenkondensat und Weiterverarbeitung im Sudhaus. *Brauindustrie*, **2**, 81–6.
67 Back, W., Krottenthaler, M. and Vetterlein, K. (1996) Technologische Möglichkeiten zur Wiederverwendung von Brüdenkondensat. *Der Weihenstephaner*, **64** (4), 235–41.
68 Fohr, M., Meyer-Pittroff, R., Krottenthaler, M. and Back, W. (1998) Verwertung von Brüdenkondensat ohne Aufbereitung. *Der Weihenstephaner*, **1**, 25–8.
69 Dickel, T., Krottenthaler, M. and Back, W. (2002) Investigations into the influence of residual cold break on beer quality. *Brauwelt International*, **20** (1), 23–5.
70 Qian, J.Y. and Kuhn, M. (1999) Evaluation on gelatinization of buckwheat starch: a comparative study of brabender viscoamylography, rapid visco-analysis, and differential scanning calorimetry. *European Food Research and Technology*, **209** (3/4), 277–80.
71 Kessler, M., Kreisz, S., Zarnkow, M. and Back, W. (2005) Investigations about the relative hartong extract at 45 °C. *Monatsschrift Für Brauwissenschaft*, in press.
72 Schur, F., Anderegg, P. and Pfenninger, H. (1978) Charakterisierung filtrationshemmender Stoffe. *Brauerei Rundschau*, **89** (8), 129–44.
73 Hermann, M. (2005) Technologische Möglichkeiten zu Beeinflussung des Aromaprofils von Weizenbieren, Dissertation, Lehrstuhl für Technologie der Brauerei I, Technische Universität München.
74 Kessler, M. et al. (2005) Investigations about the relative hartong extract at 45 °C. *Monatsschrift für Brauwissenschaft*, **58** (4), 56–62.

8
Fermentation, Maturation and Storage

Hans Michael Eßlinger

Transferring wort into beer is the third main step in brewing (Figure 8.1). Fermentation means to metabolize substrates into products by the activity of microorganisms and simultaneously to gain energy. In our case yeast transfers sugars to ethanol and CO_2. During this process we have also the formation of fermentation byproducts, which have a considerable effect on the aroma profile and the taste of the resulting beer. Fermentation is started by adding yeast to the wort – a process called pitching [1].

Brewery yeasts are mainly two types, called top-fermenting (*Saccharomyces cerevisiae*) and bottom-fermenting yeast (*Saccharomyces uvarum* var. *carlsbergensis* and *Saccharomyces bayanus*). The *Saccharomyces* yeasts are facultative anaerobes, which means that they can easily adjust their metabolism from aerobic to anaerobic conditions [2].

Yeast doubles or triples its mass during fermentation. For the build-up of cell substances yeast needs mostly amino acids, which it either takes from the fermenting substrate or must synthesize by itself. Apart from proteins, lipids have to be synthesized for yeast propagation because they are important components of the cell wall and are also needed for the uptake of nutrients. Molecular oxygen is necessary for the synthesis of these lipids from acetyl coenzyme A. Wort itself contains only few lipids. Finally, yeast also requires minerals for the stabilization of its enzyme systems [3].

8.1
Pitching

The fermentation process is initiated by the addition of 0.5–0.7 l of a heavy yeast slurry per hectoliter of wort, corresponding to 15–20 million yeast cells per milliliter of cold and aerated wort. After the addition of yeast, the wort is called young beer or simply beer.

The single yeast cells must quickly come in contact with the nutrients of the wort. Consequently, the yeast is injected continuously into the flow of the cold

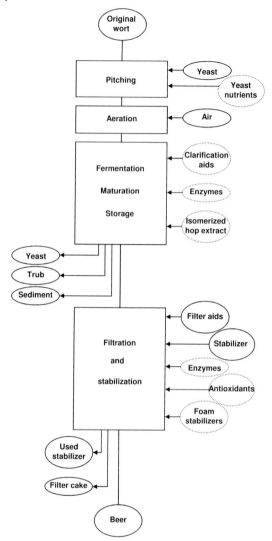

Figure 8.1 Flowchart of beer fermentation from pitching to final product. Additives that are shown in dotted ellipses are not necessary.

wort. Pitching can either be performed by blending fermenting young beer (kräusen) or by adding a slurry of yeast from a yeast buffer tank. In the latter case it is necessary to measure the turbidity of the wort before and after pitching in-line. The difference in turbidity is directly proportional to the concentration of the yeast cells.

8.2
Aeration

In addition to the yeast cell concentration, a satisfying fermentation depends on a sufficient supply of the cells with oxygen. This is done by aerating the wort with sterile air (or exceptionally with oxygen itself). An intensive distribution with fine bubbles is achieved by aeration with sintered candles, venturi tubes, special nozzles or static mixers. By using air for aeration, no problems of over-aeration can occur. A value of 8–10 mg O_2 per liter wort is optimal.

8.3
Topping-up

Topping-up means the addition of wort to an already fermenting tank. By means of pitching and aeration, the time to fill a fermenter can be prolonged up to 24 h. Care has to be taken that the temperature of the freshly added wort is not below the temperature of the fermenting beer.

8.4
Changes during Fermentation

The dissolved extract substances of the wort are fermented to ethanol and CO_2 by the activity of yeast enzymes. This metabolic pathway is exothermic and is called glycolysis (see Chapter 5). Yeast hydrolyses hexose and sucrose first (starting sugar), then takes maltose (main fermentation sugar) and afterwards maltotriose (secondary sugar).

Fermentation byproducts are formed on metabolic sideways (see Chapter 5). They can be characterized into six groups:

- Higher aliphatic and aromatic alcohols.
- Multivalent alcohols.
- Esters.
- Carbonyl compounds.
- Sulfur-containing compounds.
- Organic acids.

All these compounds have different taste and odor thresholds. Their combined contributions make up the flavor or off-flavor of the beer; the amounts produced can be influenced to some degree by brewing technology.

During the main fermentation, the pH decreases by one unit because volatile (acetic, formic) and non-volatile organic acids (pyruvic, malic, citric, lactic) are formed from amino acids by deamination. The final pH of the beer ranges from 4.3 to 4.6. The intensity and speed of acid formation is determined by the buffering capacity of the wort, amount of easily assimilated nitrogen, yeast strain and fermentation method used. The pH has a direct influence on the flavor and the sparkle of the beer.

Short-chain fatty acids are formed at the beginning of the main fermentation process: butyric, isovaleric, hexanoic, octaoic and decanoic acids. Their amounts can be controlled by the wort composition, aeration, yeast strain and general fermentation conditions. During pressure fermentation, increased levels of these compounds can be expected. They cause a yeasty odor and impair head retention.

The higher aliphatic alcohols (1-propanol, 2-methyl-1-propanol, 2-methyl-1-butanol and 3-methyl-1-butanol) and the aromatic alcohols (especially 2-phenyl-1-ethanol) represent the largest fraction of the compounds responsible for the aroma of the beer. They are formed during the first 2–3 days of the fermentation; later on there is only a slight increase. In concentrations above 100 mg/l, however, they will adversely affect taste and quality. Their levels can be controlled to some extent by the content of free amino nitrogen, wort concentration, pitching rate and yeast strain, but mainly by the pitching temperature and the fermentation temperature. The formation of higher alcohols is decreased by cold pitching temperatures, colder fermentation temperatures and the use of pressure as early as a degree of attenuation of about 50% is reached.

Owing to their low threshold values, esters strongly influence the organoleptic properties of beer to interesting fruity notes. Esters are the products of the enzymatic catalysis of organic acids and alcohol (mainly ethanol, but also higher alcohols). Their formation is closely related to yeast propagation and lipid metabolism. Beer contains more than 50 different esters, from which six are of greater importance for the beer flavor:

- Ethylacetate.
- Iso-amylacetate.
- Iso-butylacetate.
- β-Phenylacetate.
- Ethylcaproate.
- Ethylcaprylate.

Bottom-fermenting beers contain up to 60 mg esters/l, top-fermenting beers up to 80 mg esters/l. Ester production is increased by:

- Increasing the original gravity above 13%.
- Restricting wort aeration.
- Higher fermentation temperatures.
- Increased movement during fermentation and maturation.

Pressure during fermentation decreases ester formation in the same way as the higher alcohols.

Higher alcohols and esters in defined concentrations are necessary for the aroma profile of a high-quality beer. Once formed, they stay in beer and are not removed in later fermentation stages.

Volatile sulfur compounds [hydrogen sulfide, dimethylsulfide (DMS), 3-methylthio-1-propanol and thiols] are not desirable in beer because of their strong vegetable odor and taste even in very low concentrations. Apart from efficient trub removal, the most important factor in the formation and reduction of these flavor-active substances is the yeast strain and the absence of spoiling microorganisms in beer. Referring sulfur dioxide see page 214.

Mercaptans are thioalcohols, compounds in which the OH group of the alcohol is replaced by a SH group. They can strongly impair beer flavor and are also responsible for the so-called lightstruck flavor in beer.

DMS is not affected by yeast. DMS in beer depends on the amount present in wort.

Glycerol, a multivalent alcohol, is formed during glycolysis; its concentration depends on the amount of fermented sugars (1300–2000 mg/l) and is therefore proportional to the gravity of the wort.

Aldehydes and ketones are responsible for the aroma of the green beer and for the stale flavor. Acetaldehyde is formed in the green beer during the first 3 days and gives beer an unripe, unbalanced taste. Since it vanishes in later fermentation stages, it presents no technological difficulties. In the green beer phase the acetaldehyde content is about 20–40 mg/l; in the finished beer values of 8 to 10 mg/l are found.

Off-flavors in beer are usually caused by high levels of the vicinal diketones diacetyl and 2,3-pentandione. They impart an unfavorable, rancid and cheesy/buttery odor and taste to beer. The taste threshold of diacetyl depends on beer type and ranges from 0.08 to 0.2 mg/l; that of 2.3-pentandione from 0.5 to 0.6 mg/l. The vicinal diketones are transferred to the corresponding multivalent alcohols and they are seen to be indicators of the stage of maturation.

In addition to the formation of byproducts, a number of other reactions and changes take place during the fermentation and these are outlined in the following sections.

8.4.1
Changes in the Composition of Nitrogen Compounds

Yeast converts nitrogen compounds from wort to synthesize its own cellular substances. The free amino nitrogen (FAN) content is reduced from 200–250 to 100–120 mg/l. High-molecular-weight polypeptides become insoluble and are later filtered out of the beer.

8.4.2
pH Drop

The pH decreases from 5.4–5.6 in wort (biological acidified wort 5.2–5.0) to a value of 4.3–4.6 and then remains constant. A rapid fermentation is advantageous for the precipitation of protein–polyphenol complexes. The beer matures quicker, can be filtered without problems and has an excellent non-biological stability. Lower pH values, for instance below 4.2, must be avoided because they impair an acidic beer taste. An increase in pH after fermentation indicates yeast autolysis.

8.4.3
Changes in the Redox Properties of Beer

The redox properties indicate the relationship between the reducing and oxidizing power in a solution. The increase of the reducing potential is closely related to the uptake of dissolved oxygen by the yeast at the beginning of the fermentation.

The redox potential in beer can be measured by the rH value, the Indicator Time Test or the oxygen content in beer. The rH value of wort ranges between 20 and 30; in green beer between 8 and 12.

The oxygen of the aerated wort is absorbed by the yeast some hours after pitching and then reaches values of 0.0 mg O_2/l beer. The oxygen content must be kept low during filtration and filling.

8.4.4
Beer Color

In the first days of fermentation the color of the beer becomes 2–3 European Brewery Convention (EBC) color units lighter. Some substances change their color according to the pH drop; some are adsorbed on the surface of the yeast and are removed with the settling yeast.

8.4.5
Precipitation of Bitter Substances and Polyphenols

As a further result of the pH drop a number of colloidally dissolved bitter substances and polyphenols reach their isoelectric point (pI) and then precipitate. The rest of the non-isomerized α-acids and some isohumulones are captured from ascending CO_2 bubbles and carried ahead to the foam. The loss of bitter substances ranges from 25 to 40% during fermentation.

8.4.6
CO_2 Content

According to the desired CO_2 content, CO_2 is enriched in fermenting beer. The CO_2 content can range from 4.3 to 5.5 g CO_2/l for bottom-fermented beers and 6 to 10 g CO_2/l for top-fermented beers. The solubility of CO_2 in beer depends on temperature and pressure. Most of the CO_2 produced during fermentation is, however, recovered.

8.4.7
Clarification and Colloidal Stabilization

After maturation, beer is lagered at temperatures of 0 to −2 °C to clarify for 1–2 weeks. During this process its filterability and its colloidal stability are improved. The yeast cell count should be below 2 million cells/ml after conditioning beer in this way.

8.5
Appearance during Fermentation

During fermentation the young beer passes through various stages, which can be recognized by their appearance. First, the young beer becomes covered by a thin white layer of foam – the sign that fermentation has begun. Then, the fine bubble foam becomes tougher and brown particles appear. In the following stage of high kräusen, fermentation has entered in its most intensive part. The curls and crests of the kräusen become higher and the CO_2 bubbles bigger. At the end of fermentation the kräusen collapses slowly and a thin brown layer indicates that the yeast has done its work. A 'kräusenring' of brown residuals is left at the borders of the fermenter.

8.6
Fermentation Parameters

Fermentation is a decisive step to obtain a well-balanced and high-quality beer. The main goals of the fermentation are to reach:

- Constant fermentation times.
- Vigorous extract degradation and pH drop.
- A desired degree of fermentation.
- A constant beer quality.
- A long shelf-life.
- Maintain the analysis parameters.

Numerous parameters influence the result. They can be divided in the following groups:

- Composition of the wort:
  - original gravity
  - FAN > 230 mg/l
  - Ca 10–20 mg/l
  - Mg > 40 mg/l
  - Zn 0.10–0.15 mg/l

- Temperature course:
  - speed of fermentation start influences the total fermentation time
  - fermentation byproducts
  - CO_2 counter-pressure on top of the fermenter
- Yeast cell count:
  - yeast viability
  - yeast vitality
  - yeast strain
- Aeration:
  - rate of reproduction
  - sulfur dioxide formation

Sulfur dioxide is a highly antioxidative substance that is positive to the shelf-life of the beer. SO_2 has come into focus as an ingredient that can be an allergen to some people. Therefore, if a concentration above 10 mg/l is reached, it has to be declared on the label. Formation during fermentation depends mostly on the yeast strain. In addition, it can also be influenced by technological parameters during pitching. Lower aeration, high original gravity and poor yeast vitality lead to higher sulfur dioxide contents; however, yeast count at pitching is also important, as higher extract and lower oxygen per cell can increase sulfur dioxide [4].

- Topping-up method:
  - pitching and aerating each brew
  - pitching and aerating the first brew and aerating only the following brews
- Geometry of the fermenter:
  - open or closed fermentation
  - hydrostatic pressure

8.7
Control of Fermentation

Control of fermentation and maturation needs to measure temperature, extract and diacetyl. In addition, pH drop, cell count during fermentation, turbidity, decrease in color, redox potential or CO_2 content also need to be checked once per fermentation or from time to time [5].

8.8
Fermenters

Fermentation and maturation are carried out in open or closed fermenters, horizontal tanks with a manhole or vertical tanks with conical bottoms. In a two-tank process, fermentation, maturation and storage occur separately.

The capacity of the cooling equipment and the heat exchange surface on the tanks should be designed for maximum heat development during fermentation or, in the case of single-tank processes, for a maximum cooling rate (see also Chapter 27).

8.9
Maturation

The total diacetyl concentration is used to judge the maturity of fermenting beer and must be decreased below the flavor threshold by means of brewing technology. The diacetyl precursor 2-acetolactate is also called 'potential diacetyl', because it transforms into free diacetyl only in the filtered, yeast-free beer and then cannot be broken down any further. In calculating the total diacetyl concentration, 2-acetolactate must be added to the amount of free diacetyl.

During the propagation phase the yeast cells need numerous nitrogen compounds for the formation of yeast protein. Therefore, the yeast assimilates amino acids and peptides, and transfers NH_2 groups on α-ketoacids coming from carbohydrates. As described in Chapter 5, diacetyl is formed during valine synthesis. This highly temperature-dependent step is catalyzed by yeast enzymes and occurs very slowly below 10 °C. Diacetyl itself is present in very small quantities in fermentation samples and in green beer, because its reduction to acetoin is much faster than its formation. Bacteria, which may occur in the brewery as infections, are also likely to promote the formation of diacetyl.

8.10
Storage

Beer must reach its desired CO_2 level during the phase of cold storage (up to 4.8 g/l for draft beer, 5.0 g/l for canned beer and up to 0.55 g/l for bottled bottom-fermented beer). This can be achieved in the conventional procedure by using a definite bunging over-pressure of 0.2–0.6 bar, depending also on the hydrostatic pressure and temperature. Beers produced with warm maturation require either higher pressure or CO_2 to be added during transfer from warm to cold storage tanks. During storage, the beer must clarify by allowing the yeast and other haze-causing materials to settle, and its taste must refine and round-off. In order to achieve these requirements, the beer must be stored at 0 to −2 °C for 1–2 weeks.

Frequently, fermentation, maturation and conditioning take place in the same vessel (one-tank process). By using separate tanks for the sedimentation of the yeast after fermentation there is a chance to add kräusen or stabilizers such as bentonite during the last production period. Continuous fermentation or maturation systems use through-flow or tributary-flow systems or a bioreactor; however, these methods are rarely found in large production facilities.

8.11
Bottom Fermentation in Practice

Fermentation processes for bottom-fermented beers can be divided in three schematic groups [6]:

- Cold fermentation and cold maturation.
- Warm fermentation and warm maturation.
- Cold fermentation and warm maturation.

In Figures 8.2–8.7 these processes are shown by depicting temperature (———), apparent extract (– . – . –), diacetyl (- - - -) and CO_2 over-pressure (– .. – .. –). Transfer and yeast crop are indicated by large downward and small upward arrows, respectively.

8.11.1
Cold Fermentation with Conventional Storage

This is the common production process in open fermenters by simple means. Pitching is performed at 6–7 °C, then the temperature is allowed to rise to 8–9 °C. Four days later the beer is cooled down slowly to 3–4 °C. Afterwards it is transferred in a lager tank still containing a residual fermentable extract of 1.1–1.3%. However, it is also possible to add 10–12% kräusen with a degree of attenuation of 25% to a totally fermented young beer to depict the residual extract. In the lager tank the beer is cooled slowly to –1 °C. In this time the extract is fermented completely and the diacetyl is reduced below the human threshold. At –1 °C the beer is lagered for a further or 2 weeks. See Figure 8.2.

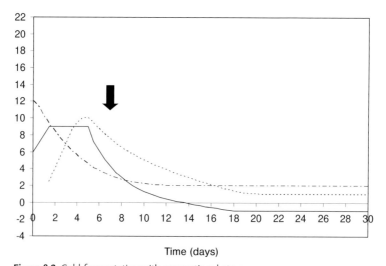

Figure 8.2 Cold fermentation with conventional storage.

8.11.2
Cold Fermentation with Well-Directed Maturation in a Cylindroconical Vessel (CCV)

The first stages are the same as before. For maturation, the maximum fermenting temperature is maintained until diacetyl is below 0.1 mg/l. Then it is cooled down to a lagering temperature of −1 °C and kept at this temperature for 1 week. Kräusening as described above is also possible. Yeast crop is done 3 or 4 times, the last time shortly before filtration. The yeast cropped during the maximum fermentation temperature is suitable for repitching; the yeast removed later is not. See Figure 8.3.

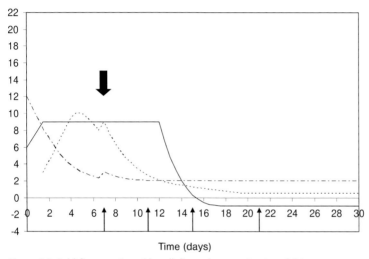

Figure 8.3 Cold fermentation with well-directed maturation in a CCV.

8.11.3
Pressureless Warm Fermentation

Higher temperatures are necessary when looking for ways to shorten fermentation and maturation. As known from experiments, the formation of byproducts is then increased and the character of the beer changed to a more yeasty flavor. Pitching is performed at 8–9 °C and temperature is allowed to rise to 12–14 °C. Much more diacetyl is produced, but it is also more rapidly removed. When in addition the attenuation limit is reached, the beer is cooled down and lagered cold for 1 week. Thus, the production takes 17–20 days. See Figure 8.4.

8.11.4
Accelerated Fermentation under CO_2 Pressure

The higher the fermentation temperatures, the more fermentation byproducts are formed. A remedy is CO_2 as counter-pressure on the top of the fermenter. Using

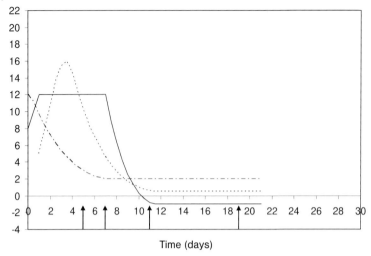

Figure 8.4 Pressureless warm fermentation.

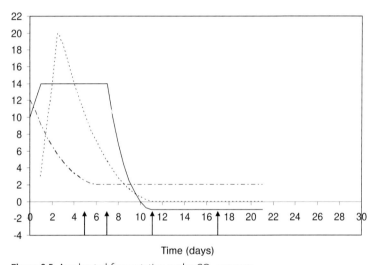

Figure 8.5 Accelerated fermentation under CO_2 pressure.

temperatures above 14 °C, pressure has to be increased at an attenuation degree of about 50%. The desired counter-pressure is then kept at least until to the end of the maturation phase. After cooling to −1 °C, this temperature is maintained for 1 week. The production time is then 12–15 days per batch. See Figure 8.5.

8.11.5
Cold Fermentation with Integrated Maturation at 12 °C

The combination of cold fermentation (less byproducts) and warm maturation (quicker decrease of diacetyl) avoids all disadvantages and leads to beers with constant high quality. Temperature control is optimally adjusted to the metabolism of yeast. In the case of cold fermentation with integrated maturation, fermentation is performed at 8–9 °C to an attenuation degree of about 50%, then the refrigeration is switched off and temperature control set at 12–13 °C. At this temperature diacetyl removal and final attenuation are reached, then the process proceeds as usual. This method is also possible in a conventional lager cellar. The total batch time takes 20 days. See Figure 8.6.

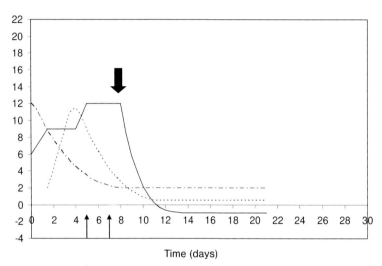

Figure 8.6 Cold fermentation with integrated maturation.

8.11.6
Cold Fermentation with Programmed Maturation at 20 °C

In the so-called 9/20 procedure, heat exchangers are used in order to increase the temperature to 20 °C. The removal of the primary fermentation yeast by centrifuging and the addition of 10% kräusen with an apparent degree of attenuation of 20–30% is practiced at the beginning of the maturation phase. The temperature of 20 °C is kept for 1.5–2 days and the diacetyl removal is monitored several times. The total production time takes 18–20 days. See Figure 8.7.

8.11.7
Accelerated Fermentation and/or Maturation

Further methods to achieve a faster fermenting or maturing of the beer are

- Fermenting with stirring and purging the green beer with CO_2.
- Using immobilized yeast or enzymes for diacetyl removal.

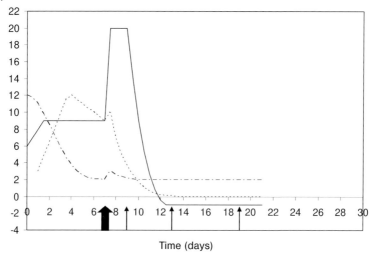

Figure 8.7 Cold fermentation with programmed maturation at 20 °C.

8.12
Yeast Crop and Yeast Storage

The yeast sedimented after fermentation has metabolized all the nutrients of the wort and is then exposed to CO_2 and ethanol, as well as to the hydrostatic pressure and eventually higher temperatures in the cone of the CCV. Thus, a separation of the yeast and the beer is necessary. Yeast in a constant good condition can be cropped – as mentioned – up to reaching of the final degree of attenuation before cooling the beer down to −1 °C.

Yeast crop is 2–3 l of thick yeast slurry per hectoliter of beer. During cropping care has to be taken regarding pressure release so that CO_2, ethanol and undesired fermentation products can leave the yeast cell. In no case may yeast be aerated, because then it would metabolize its reserve fund, and this would weaken and autolyse the yeast.

It is of great importance to cool the yeast for further storage to 0–3 °C. At higher temperatures yeast suffers a loss of vitality. The use of a external yeast cooler is necessary. The stream of volume should not exceed 10 hl/h from a CCV, so each crop lasts several hours.

If the yeast is not repitched immediately it is often washed and sieved. Therefore, it is mixed with water and passed through a fine sieve which retains large impurities such as trub from fermentation and hop resin particles.

For a good pitching yeast, the following requirements are necessary:

- It must not contain beer spoiling microorganisms.
- The viability should exceed 95%.
- It should reach the final degree of attenuation.
- The viability should be high (intracellular pH > 6.0).
- It should be possible to use it for eight to 10 cold fermentations or four to five warm fermentations.
- It should be clean (white color), and smell and taste typically fresh.

8.13
Beer Recovery from Yeast

The cropped yeast has a dry matter content of 10–12%. Several pieces of process equipment allow for yields up to 50% of the included beer. Afterwards, the yeast has a consistency of 18–22% dry matter and can be diluted with water to enable pumping.

The amount of recovered beer (%) can be calculated as: [(100 − dry matter before recovery) × 100]/dry matter after recovering. About 2% of the produced beer can be recovered from surplus yeast.

Equipment used for recovering can be a:

- Filter chamber press.
- Separator.
- Decanter.
- Cross-flow membrane filter.

The quality of recovered beer depends on the equipment used and the condition of the yeast. Recovered beer has a slightly higher pH value, more protein and nucleic acid constituents, and a yeasty and fruity smell/taste. Therefore, it is prudent to add it before fermentation again.

8.14
CO_2 Recovery

The CO_2 formed during fermentation amounts to 2–2.5 kg/hl of beer depending on original gravity. From a 12% original gravity wort, about 2.2–2.4 kg CO_2/hl can be recovered. This corresponds approximately to the brewer's own needs.

A CO_2 recovery plant consists of a foam separator and a gas container, which serves as a buffer for the collected CO_2. The CO_2 is then freed from water-soluble impurities in the gas scrubber. The gas is compressed to 18–22 bar in a compressor. The CO_2 is led and then to a condense over the drying tower and an active carbon filter, where it is liquefied and then stored at −20 °C.

In the brewery, CO_2 is used to prevent oxygen from affecting beer after fermentation, and it is used as pressure gas on tanks, pipes, vessels, filters, fillers or road tankers.

Table 8.1 Composition of bottom-fermented pale lager beer.

Extract of original wort	12.0 wt% (125.6 g/l)
Attenuation limit (apparent)	78–85%
Real degree of attenuation	65–80%
Apparent residual extract	1.7–3.0 wt%
Real residual extract	2.0–3.5 wt%
Alcohol concentration	3.5–4.5 wt%
Total nitrogen	700–900 mg/l
Coagulable nitrogen	8–28 mg/l
High molecular nitrogen	150–250 mg/l
Low molecular nitrogen	300–600 mg/l
FAN	80–160 mg/l
Bitter substances	16–25 EBC bitterness units
Color	8–15 EBC color units
Total polyphenols	130–180 mg/l
Anthocyanidines	40–80 mg/l
pH	4.3–4.6
Viscosity	1.4–1.7 mPas

8.15
Types of Bottom-Fermented Beers

In many countries, bottom-fermented beer is designated as lager beer. Its extract of original wort varies according to the local laws (tax classification) from 7 to 14%. Lager beers are the most popular, with an average bitter substance content of 20 EBC bitter units (Table 8.1).

Bottom-fermented beer with an extract of original wort of 10–14% comprises an extraordinarily large variety of beer types, including pale and dark beers, export beers (more than 12% extract of original wort), Märzen beers, and special beers and festival beers (13–14% extract of original wort). Within these limits there are such different beer types as Pilsener, Dortmunder, Munich, as well as smoky-flavor beer and cellar beers; these are, however, restricted to certain localities. The upper gravity limits for special beers differ from country to country between 13 and 15%. Strong beers range from 15/16% extract of original wort up to a maximum of 28%.

8.16
Top Fermentation

Top fermentation is the oldest method of beer production and was the only one used until about the middle of the nineteenth century. Top-fermented beers differ from bottom-fermented beers in their ingredients (more than 50% wheat malt or other malted cereals) and by their special aroma, which is primarily induced by the top-fermenting yeast strains of *S. cerevisiae*. The particular yeast strain employed has a higher optimum fermentation temperature, and therefore the fermentation proceeds between 15 and 24 °C. During fermentation, the yeast rises and can be skimmed off the top. However, in large vessels, especially CCVs, the yeast can be cropped in the same manner as bottom-fermenting yeast. The number of yeast generations is – as mentioned – considerably greater. At the higher fermentation temperatures, the diacetyl is easily decreased. Owing to the fast rate at which fermentation proceeds, a relatively low pH value of 4.1–4.3 results.

8.17
Types and Production of Top-Fermented Beers

8.17.1
Wheat Beer

The extract of original wort in wheat beers is 11–14%; the wheat malt portion can range from at least 50 to 100%. An intensive two-mash decoction procedure in the brewhouse is advantageous in order to ensure a satisfactory protein modification, because an increase in wheat malt proportionally decreases the concentration of assimilable nitrogen compounds in the wort. Owing to the higher pitching temperature (15–18 °C), the pitching rate required will be lower (0.3–0.5 l yeast/hl wort, corresponding to 4–10 million yeast cells/ml wort). The aeration should ensure an oxygen concentration of 6–8 mg/l. The initial fermentation stage is characterized by the rise of trub particles and hop resins to the surface. After their removal, yeast rises to the top and can be cropped; this continues until the end of fermentation. Wheat beer yeast can be repitched as often as 200–500 times.

Wheat beer is characterized by a typical spectrum of fermentation byproducts, such as 4-vinylguajacol and 4-vinylphenol. These two compounds are, in addition to esters, responsible for the typical aroma of wheat beer. A more rapid and extensive pH drop, an increased formation of higher alcohols and esters, together with a greater decrease in nitrogen and bitter compounds, mark the course of wheat beer fermentation as compared with lager beer procedures. A higher bunging pressure during storage ensures a CO_2 concentration in wheat beer of 6–10 g CO_2/l. If wheat beer is marketed as 'naturally hazy', the secondary fermentation can be accomplished in the bottle by adding unfermented wort and bottom-fermenting yeast or bottom-fermenting kräusen. Crystal-clear wheat beer remains in the tank until mature and is subsequently filtered and bottled.

Top-fermented strong beers include pale and dark Weizenbock as well as different stout beers, especially Imperial stout.

8.17.2
Alt Beer

Alt beer has an extract of original wort of 11–14%. Methods vary widely for the production of this dark beer which gives readings of 25–40 EBC color units; the wort may be produced from pale malt with the addition of roasted malt beer, 100% dark malt or 90% pale malt and 10% dark caramel malt. A portion of 10–15% of wheat malt sometimes is used to round off the taste. The dark malt which is used as a substitute for caramel may also be produced from wheat. The bitter substance concentration of Alt beer amounts to 28–40 EBC bitter units. The fermentation temperature is 12–22 °C.

8.17.3
Kölsch Beer

This beer may be produced only in the town of Cologne. It is brewed with pale barley malt by adding up to 10–20% wheat malt and has a original gravity of 11–14%. The wort is fermented at 12–22 °C, but sometimes up to 28 °C. The character of Kölsch – a top-fermented bitter beer – is strongly determined by the properties of the specific yeast.

For further special beers, see Chapter 10.

References

1 Eßlinger, H.M. and Narziß, L. (2002) Beer, in *Ullmann's Enzyclopedia of Industrial Chemistry*, 6th edn, Vol. 4, Wiley-VCH Verlag GmbH, Weinheim, pp. 657–700.
2 Geiger, E. (2002) Gärung, in *Praxishandbuch der Brauerei* (ed. K.U. Heyse), Gehr, Hamburg, Chapter 3.3, pp. 1–30.
3 Kunze, W. (2004) *Technologie Brewing and Malting*, 3rd edn, VLB, Berlin.
4 Back, W. (ed.) (2005) *Ausgewählte Kapitel der Brauereitechnologie*, Fachverlag Hans Carl, Nürnberg.
5 Priest, F.G. and Stewart, G.G. (2006) *Handbook of Brewing*, 2nd edn, Taylor & Francis, Boca Raton, FL.
6 Miedaner, H. (1982) Gärverfahren in der Praxis, in: *Technologisches Seminar des Lehrstuhles*, Technologie der Brauerei I, Weihenstephan.

9
Filtration and Stabilization
Bernd Lindemann

9.1
Introduction

An important characteristic of beer is its brilliant clarity. After storage, in which besides maturing the sedimentation of solid materials also takes place, the ready-to-discharge beer is filtered. The originally used paper and pulp filters have been replaced by kieselguhr (a light soil consisting of siliceous diatom remains) filters since the end of World War II. Nowadays, kieselguhr is nearly exclusively used as a filter aid for beer clarification. Test installations for membrane filtration of beer show encouraging results. Research for alternatives to kieselguhr by using regenerative filter aids has been prompted due to the increasing difficulties of disposing of the waste guhr. The first membrane filtration plants are still in industrial trials.

9.2
Purpose of Filtration

The purpose of filtration is to preserve the beer so that no visible changes occur in the long run and the beer keeps its original appearance. Generally, the filtration steps fulfill two roles:

- To remove suspended materials from the green beer (the real filtration).
- To unhinge potential turbidity formers (stabilization).

9.3
Theoretical Considerations of Cake Filtration

Filtration processes can be described as a flow-through of layers. The layer is formed of filter materials, separated solid materials and, in addition, in a pre-coating of pre-coat filter materials.

Handbook of Brewing: Processes, Technology, Markets. Edited by H. M. Eßlinger
Copyright © 2009 WILEY-VCH Verlag GmbH & Co. KGaA, Weinheim
ISBN: 978-3-527-31674-8

Filtrations can be classified as surface and depth filtration depending on the place of solid separation. In surface filtration, the particles to be separated are retained on the surface of the active media (filter material). By contrast, in depth filtration, the separation process takes place inside (in the depth) of the filter material. The process is called a cake filtration if a filter cake is built up by the separated solid materials during surface filtration and the outer layer of the filter cake takes over the separation.

Cake filtration can be improved if a filter aid (e.g. kieselguhr) supports the cake build-up. This procedure is applied if the solid material is not able to form a rigid and (for liquids) permeable solid matrix. This is called pre-coating filtration since this filter aid pre-coats the filter surface specifically. Solid materials in combination with the filter aid build up the filter cake. Solid materials are often gel-like and amorphous. The goal of pre-coating is to build up a permeable, but still effective filter cake.

There has been no lack of attempts to describe the filtration processes by mathematical equations; however, the relation to practical filtration could only be conditionally established. More recent experiments in simulating filtration processes via mathematical models seems to be more promising [1].

9.4
Filtration Techniques

In general, cake filtration and surface filtration are the main techniques available for beer clearing. An important representative of cake filtration is kieselguhr filtration. Surface filtration has not yet prevailed in beer clearing. The first large-scale plant with membrane filters seems to be working promisingly.

9.4.1
Kieselguhr Filtration

Kieselguhr filtration is a type of cake filtration in which the eliminated solids form the filter aid. The suspended solids in beer, however, show unfavorable properties with regard to cake formation. These are fine, soft, compressible and, in part, gel-like. For this reason kieselguhr is used as a filter aid, which, on the one hand, forms a layer permeable for the filtrate and, on the other hand, functions as a loose filter cake. The ratio of turbid matter of the beer and the kieselguhr to be added lies strongly on the side of the filter aid. Kieselguhr addition of 50 g/hl can be found in practice. The dosage is generally carried out by practical experience of the master brewer. A selective choice of guhr based on analytical criteria cannot occur in practice. This is due to the fact that it is very problematic to measure the suitability of the respective kieselguhr as filter aid for beer filtration.

Three technical possibilities are available for the use of filters:

- Kieselguhr frame filter.
- Horizontal leaf filter.
- Candle leaf filter.

9.4.1.1 Plate and Frame Filter

The kieselguhr plate and frame filter has an arbitrary number of filter plates that are fixed on a frame in an alternating manner. On the filter plate is a filter layer, mostly made up of cellulose and kieselguhr, which forms the actual filter material. The filter element, filter layer and filter frame are pressed closely together [2]. A schematic presentation of the flow-through of the beer in the frame filter is shown in Figure 9.1.

The unfiltered beer arrives via an inlet into the frame and is dispensed onto the filter cake surface. The total filtration active layer is passed through in the order: secondary layer, primary layer, filter layer and filter material. The filtrate passes through the filter plate and is then ejected from the filter. The filter cake grows on the filter material and in the filter layer in the space between two filter plates, respectively. The maximum height of the filter cake is restricted by the width of the frame.

Figure 9.1 shows that the flow of the turbid material is initially tangential to the surface of the filter cake before it flows perpendicularly [4].

9.4.1.2 Horizontal Leaf Filter

Figure 9.2 shows the design of a horizontal leaf filter. It consists of horizontal, circular plates covered on one side with filter material and arranged one upon the

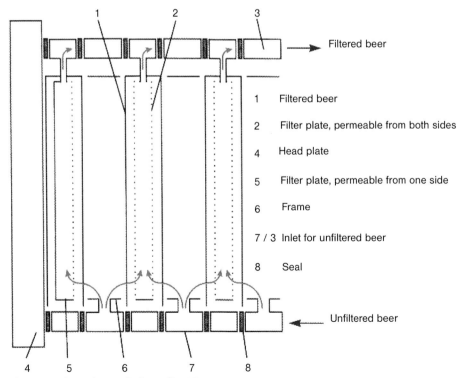

1 Filtered beer
2 Filter plate, permeable from both sides
4 Head plate
5 Filter plate, permeable from one side
6 Frame
7 / 3 Inlet for unfiltered beer
8 Seal

Figure 9.1 Flow scheme of a frame filter [3].

other in a pressure vessel. The plates are made up of a bottom plate, a drainage member arranged above and filter material. In the center of the plates is a rotating, perforated quill connected to the plates through which the filtrate flows. The filtration occurs on the upper surface of the plates. Plate diameters up to 1.50 m are common and a filter area of up to 150 m² can be realized in a filter. The filter elements are dimensionally stably tightened in order that the leaf does not crinkle and a smooth surface can be formed. Otherwise interference can occur through the build-up of the filter cake. The filter cake can tear and the solid materials can pass through.

The beer with added filter aid arrives in the pressure vessel via a pump, passes through the filter layers in the order secondary layer, primary layer and leaf, and flows via the quill off the leaf element and out of the filter. Here, two different flow directions are generally differentiated – either the leaf element is overflown from the outside (overflow of the filter element from the element edge) or from the inside (overflow of the filter element form the element's center).

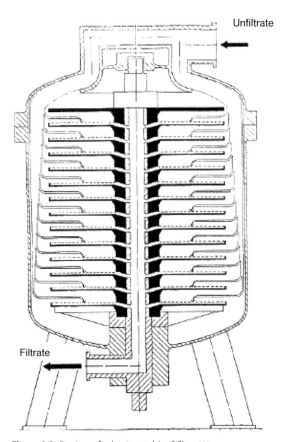

Figure 9.2 Design of a horizontal leaf filter [3].

The advantage of the horizontal leaf fitter is that the vessel can be emptied with CO_2 without changing the filter cake. In this way it is possible to filter at the beginning and the end of a charge with a low water–beer blend. The maintenance requirements of the movable parts and stress of the seals can be seen as disadvantages.

9.4.1.3 Metal Candle Leaf Filter

Figure 9.3 shows a schematic presentation of a candle leaf filter. Candles of porous materials are arranged in the filter vessel. In beer filtration, hanging candles are always used. These hang on a horizontal plate (head plate) that also forms the closure of the non-filtrate area. The candles have permeable materials on their surface and are traversed through from outside to inside. The filtrate is drained off through the inside of the candle and horizontal plate into the filtering space. Initially, the pre-coat is applied to the candle surface followed by the secondary layer. The flow through occurs in the order filter vessel (non-filtrate area), secondary layer, primary layer, candle surfaces, candle inside, perforated plate and filtrate area. From there the filtrate is discharged off the filter.

Filter candles have a diameter of 20–35 mm and are of various lengths up to a maximum of 2.5 m. A filter surface area of up to 180 m^2 can be achieved in the pressure vessel by a parallel arrangement of candles. The candles are made up of a trapezoidal wire which is spirally welded around a support. Between the coils is a space of 50–80 μm. Due to the round form it is difficult to calculate the filter area as the radius of the cake increases with the continuous addition of filter aid. Generally, the average cake thickness is taken for the calculation of the filter area. Thus, the filter area given by the plant constructors is greater than the metallic filter area of the support.

The advantage of the candle leaf filter is that no moving parts need to be maintained. A pre- and post-run free filtration is, however, not possible.

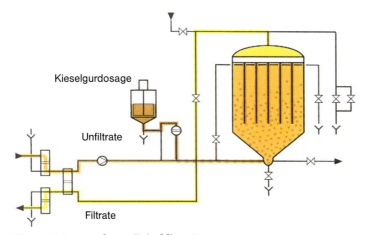

Figure 9.3 Design of a candle leaf filter [5].

9.4.2
Filter Aids for Pre-coating

The amorphous beer constituents to be filtered off show very unfavorable properties with regard to the filtration processes. For this reason rigid materials are added that form mainly the filter cake. Such filter aids are primarily kieselguhr; furthermore, perlites and cellulose are added. The addition of the materials occurs generally by experience of the plant constructer or master brewer in charge. So far, attempts to automate the addition of filter aids has met with only limited success. This is due to the fact that the filterability of the beer can only be determined and predicted with difficulty.

9.4.3
Methods in Kieselguhr Pre-coating Filtration

Fundamentally the method is the same for all pre-coatings:

- A one- to two-step pre-coating with perlites, cellulose, kieselguhr or a mixture of the three filter materials.
- The actual filtration with continuous filter material addition.
- Emptying of the filter.
- Removal of the filter cake.
- Cleaning and disinfection.

A consistent filter cake build-up is essential for the filtration result. Thereby, the flow speed is of particular importance. The filter capacity for frame filters is given as 3.5–4.0 hl/m²h and for vessel filters as 4.0–7.0 hl/m²h. For primary pre-coating, the double capacity is recommended.

The primary pre-coating takes place with a total of 1000 g/m²; the ongoing dosage lies between 50 g/hl for well-filterable beers and 300 g/hl for very poorly filterable beers.

Primary pre-coat is added together with degassed water; the water is pressed off the filter with beer. At the end of a filtration the beer is pressed off the filter with water. In this case pre- and post-runs accumulate which are used to dilute stronger brewed beers.

9.4.4
Membrane Filtration

The membrane filtration technique is divided into dead-end filtration and cross-flow filtration. Dead-end filtration is often applied as a safety filter after the pre-coat filter in which membrane modules are used with a large mesh size (up to 8 μm). Cross-flow filtration is used in on a large scale only in a few breweries. In cross-flow filtration the filtrate flow is perpendicular to the flow direction and the product to be filtered flows continuously tangentially over the membrane with a high flow rate and plant-specific working pressure. The direction of filtration and the flow direction are not identical. Only part of the liquid is filtered through the membrane; the accumulated non-filtrate (concentrate or retentate) is circulated again to achieve a good filtration result. The concentration of separated material is thus slowly increased. Turbid material is continuously added and permeates discharged. A turbulent flow is created by the overflow of the membrane surface, which results in self-cleaning of the membrane. The flow has the role to keep solid materials, microorganisms and colloids suspended, and to avoid their deposition on the membrane surface. Due to the self-cleaning effect in dynamic filtration, the filtrate capacity is considerably increased and stays constant at the same level. The endurance of the membrane is increased considerably. The flow rate in the filter modules is 2 m/s and thus considerably higher than in pre-coating filtration. The cross-flow filtration is operated to a certain transmembrane pressure, then rinsed back with filtered beer and subsequently filtered further. Cleaning is carried out after about five back flashes [6].

The concentrate accumulated during filtration can be returned to the beer tank and virtually constantly filtered. It is, however, also possible to discard the concentrate as a residual of a filtration charge. The methods used need to be adapted in the brewery.

9.5
Variables Influencing Beer Filtration

Beer filtration is influenced by the filter technique (pre-coating filtration or cross-flow filtration), filter aids used (kieselguhr), substances suspended in beer (yeast, proteins), substances dissolved in beer (glucans) and colloids (glucan gels).

All dissolved materials influence the dynamic viscosity and thus the filter capacity. Colloid gels such as β-glucan gel rearrange the filter channels in the filter cake and block the filter very quickly. The concentration of yeast and other non-dissolved substances also has a very negative influence on kieselguhr filtration.

9.6
Beer Stabilization

The removal of turbidity-forming materials from beer can be carried out in various ways:

- Adsorptive during kieselguhr filtration.
- Adsorptive in a separate filtration step.
- Sedimentative.
- Enzymatic.

Two major groups of substances count as potential turbidity formers: proteins and tannins. There are also indications of other substances, but only methods for the removal of proteins and tannins have prevailed in the stabilization techniques.

Proteins are removed by adsorption using colloidal silica. This is done by adding silica to the kieselguhr dosage. Kieselsols are also added to the cold wort for primary treatment. Bentonite is hardly used due to the unspecific protein adsorption, since this affects the froth stability of the beer.

Tannins are nearly exclusively removed by adsorption using polyvinlylpyrrolidone (PVPP). PVPP can both be added to the kieselguhr and applied in a separated filtration step. The filters used mostly are horizontal leaf filters. The method is carried out analogously to kieselguhr pre-coating.

Proteins can also be decomposed enzymatically (e.g. by the addition of papain); however, the beers need to be pasteurized very quickly to avoid a too strong interference with the froth stability.

A newer method for stabilization is based on the usage of agarose gels, which show an effect on the protein side as well as the tannin side.

9.7
Technical Design of a Filtration and Stabilization Plant

Clarification of beer and beer stabilization are directly related processes. Possible filtration solutions are show in Figure 9.4.

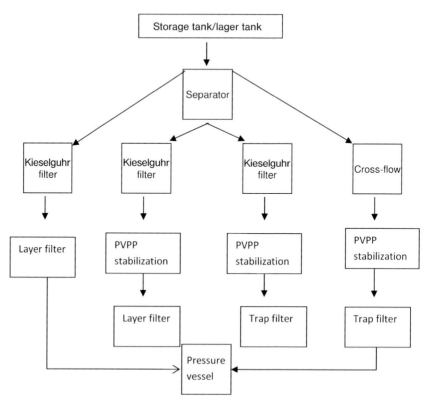

Figure 9.4 Possible filtration solutions.

The simplest type of filtration and stabilization is the kieselguhr filtration with simultaneous addition of silica for the stabilization from the protein side. This is the most cost-effective technique for smaller breweries. For bigger breweries, separated stabilization with PVPP after kieselguhr filtration is the most appropriate technique. If cross-flow membrane filtration is used the stabilization should

Table 9.1 Record of a beer filtration in a filter line for 400 hl/h.

Time (h:min)	Pressure inlet (bar)	Pressure outlet (bar)	ΔP (bar)	Turbidity			Oxygen after trap filter (mg/l)	Flow (hl/h)
				after kieselguhr filter 90°EBC	after PVPP filter 90°EBC	after PVPP filter 25°EBC		
0:00	1.70	1.50	0.20	0.96	0.43	0.31	0.09	380
0:30	1.80	1.50	0.30	1.61	0.76	0.51	0.05	380
1:00	2.00	1.50	0.50	1.55	0.71	0.46	0.03	380
1:30	2.15	1.50	0.65	1.27	0.65	0.36	0.03	380
2:00	2.30	1.50	0.80	1.15	0.61	0.33	0.03	380
2:30	2.50	1.50	1.00	1.09	0.58	0.31	0.03	380
3:00	2.70	1.50	1.20	1.03	0.53	0.29	0.03	380
3:30	2.80	1.50	1.30	0.93	0.49	0.27	0.03	380
4:00	3.00	1.50	1.50	0.86	0.46	0.27	0.03	380
4:30	3.20	1.50	1.70	0.81	0.44	0.23	0.03	380
5:00	3.40	1.50	1.90	0.78	0.41	0.21	0.03	380
5:30	3.65	1.50	2.15	0.68	0.38	0.21	0.03	380
6:00	3.80	1.50	2.30	0.65	0.38	0.19	0.03	380
6:30	3.85	1.50	2.35	0.70	0.38	0.21	0.02	380
7:00	3.90	1.50	2.40	0.63	0.37	0.19	0.02	380
7:30	4.15	1.50	2.65	0.64	0.37	0.17	0.02	350
8:00	4.25	1.50	2.75	0.55	0.34	0.17	0.02	350
8:30	4.45	1.50	2.95	0.55	0.32	0.14	0.02	320
9:00	4.70	1.50	3.20	0.53	0.32	0.17	0.02	320
9:30	4.90	1.50	3.40	0.51	0.31	0.17	0.01	300
10:00	5.40	1.50	3.90	0.51	0.31	0.17	0.02	300
10:15	5.95	1.50	4.45	0.51	0.31	0.18	0.02	300

Candle filter with 490 filter candles (66 m² metal area) and a specific flow of 4 hl/m²h followed by a PVPP filter of 36 m² area and 11 hl/m²h and a trap filter of 5.4 m² area and 74 hl/m²h.
First pre-coat   50 kg coarse guhr (750 hl/h)
Second pre-coat   30 kg middle and 10 kg fine guhr (750 hl/h)
Second body feed   150 kg middle and 50 kg fine guhr
Second body feed   100 kg middle guhr
Third body feed   60 kg middle guhr
Fourth body feed   60 kg middle guhr
Fifth body feet   40 kg middle guhr
Total filtered volume   3.829 hl (374 hl/h on average)
Kieselguhr consumption   550 kg (according to 144 g/hl).
°EBC = degrees European Brewery Convention.

be employed in a second process. Here, the application of PVPP or agarose gels would be a good choice. Pre-sedimentation using a separator improves the filtration capacity. In particular, a separation is advisable with regard to beers with a short maturation time.

Table 9.1 shows the most important results recorded for large-scale filtration. The following system is used: separator–kieselguhr candle filter–PVPP stabilization filter and trap filter.

References

1 Velten, K., Günther, M., Lindemann, B. and Loser, W. (2004) Optimization of candle filters using three-dimensional flow simulations. *Filtration–Transactions of the Filtration Society*, **4** (4), 276–80.
2 Hebmüller, F. (2003) Einflussfaktoren auf die Kieselgurfiltration des Bieres, Dissertation, TU Freiberg.
3 Autorenkollektiv (1999) *Beer Filtration, Stabilisation and Sterilisation. EBC Manual of Good Practice*, Fachverlag Hans Carl, Nürnberg.
4 Oliver-Daumen, B. (1999) Kieselgurrahmenfilter–Relikte aus vergangener Zeit oder Klassiker im Aufwärtstrend. *Brauwelt*, **139**, S246–9.
5 Kunze, W. (1998) *Technologie Brauer und Mälzer*, Vol. **8**, VLB, Berlin.
6 Henseler, R. (2004) Vergleichende Betrachtung von Membranfiltrationsanlagen in der Brauerei, Diplomarbeit, FH-Wiesbaden, Geisenheim.

10
Special Production Methods

Felix Burberg and Martin Zarnkow

Since beer was 'invented', different sorts of beer have been developed at various production locations worldwide. This could happen because of the diversity of raw materials and technologies. However, the brewing technologist has kept active and created new sorts or improved the production methods in the course of the millennia, in order to obtain a stable and pleasing product. The discovery of malting technology, usage of hops to flavor and knowledge of the different fermentation microorganisms were definitely important for the exertion of this influence. Relating to the entire development of this young biotechnology, only recently have new beer sorts appeared, such as alcohol-free beers and 'Nährbier'. This could only take place through the development of modern equipment and processes.

The original wort or the alcoholic content necessarily enables the classification of the different sorts. Also, a classification in the fermentation types is possible (i.e. top-fermented and bottom-fermented beers). The beers discussed here can even be divided into homofermentative and heterofermentative. This means that the alcohol-free beers, the dietetic beers, XAN™, wheat beer, 'Nährbier' and beers produced by means of brewing techniques with high original gravity can be regarded as homofermentative. By contrast, lambic, Berliner Weisse and porter can be regarded as heterofermentative. Note that this statement does not claim to be exhaustive, since, for example, wheat beer can be categorized as heterofermentative when the secondary fermentation takes place with an addition of bottom-fermenting yeast.

10.1
Alcohol-Free Beers

10.1.1
Introduction

Alcohol-free beers enjoy great, actually worldwide increasing popularity due to their beer-typical characteristics and their low alcoholic content. Quite different aspects are important for the consumers, including the changing concept of a healthy lifestyle, the abdication of alcohol based on it and the awareness of calorific content. Also, the stricter legal requirements for safety at work and the traffics laws, as well as the considerably increasing demands on people's mobility in labor and leisure (and the associated loss in the case of withdrawal of the driving license) influence the increasing consumption of alcohol-free beers. In addition, isotonic properties could be proved in alcohol-free beers, so that an interesting alternative beverage (e.g. for athletes) could be produced [1]. It is also an interesting alternative for breast-feeding mothers, since alcohol-free beer was proved to stimulate the production of breast milk [2]. People who cannot consume alcohol for religious reasons [3] also represent a certain market potential.

As the reasons for the consumption of alcohol-free beers differ, so the national laws also differ. Based on these laws, a beer with less than 0.5 vol% alcohol is considered alcohol-free in Germany. By contrast, the Arabic region needs less than 0.05 vol% if a beer wants to be referred to as alcohol-free. Strictly interpreted, this could mean that the wort could not come into contact with the yeast. Methods using lactic acid bacteria are suitable here.

Alcohol-free beers can be produced by means of different technologies. However, it must be pointed out that all techniques imply that the alcohol has an essential effect on the flavor of a beer. Therefore, beers produced with the technologies applied today will reach only approximately the high sensory quality of a 'normal' beer. Nevertheless, these beers make up independent, qualitatively high-class products, which have earned significance in the market [1].

10.1.2
Techniques for the Production of Alcohol-Free Beers

The techniques for the production of alcohol-free beers can be classified in two groups: physical techniques and biological techniques. The main difference between them is that the alcohol content developed is removed later when physical techniques are applied, whereas, in the case of biological techniques, the development of alcohol is inhibited and kept within a limit [1].

10.1.2.1   Physical Techniques
For the production of alcohol-free beers by physical techniques the alcoholic content is removed later from normally fermented original beer. This happens by means of thermal methods, such as vacuum distillation or membrane separation methods (i.e. reverse osmosis or dialysis) [1].

10.1.2.1.1   **Thermal Dealcoholization**   Distillation and rectification of alcohol-water mixtures are thermal separation methods with the procedural steps of evaporation and condensation. Thereby, easier boiling components in the headspace and lower boiling components in the liquid become enriched. The use of under-pressure causes a decrease of the boiling point of the liquid mixture and the thermal stress of the beer is strongly diminished. Vacuum distillation equipments, which work at absolute pressures of 60–200 mbar at 37–60 °C, are almost solely employed for thermal dealcoholization. The smaller the remaining alcoholic content, the stronger the evaporation of the beer. However, the separation of substances does not only depend on the boiling balance, but also on the kinetics of the substance transportation to the phase boundary. Thin-layer evaporators with mechanically produced thin-films or falling-stream evaporators are mainly used for vacuum evaporation. The construction of an evaporator affects the composition of the resulting exhaust vapors and therefore also the necessary evaporation rate. In the case of simple distillation by means of thin-layers or falling-stream evaporators, evaporation rates between 40 and 60% are usually demanded to lower the alcoholic content of the beer to under 0.5 vol%. If the evaporator is combined with a rectification column, an alcoholic content of less than 0.05 vol% can be obtained in the finished product when the evaporation rates ranges between 6 and 10%.

Compared to the original beer, beers resulting from vacuum evaporation show a slight color increase of about 0.5–1.5 European Brewery Convention (EBC) units and a decline of bitter substances of about 1–2 EBC units. The real extract increases according to the evaporation rate. The pH rises by approximately 0.1–0.2 units due to the loss of volatile organic acids and CO_2. The foam qualities of the beer hardly change. The turbidity increase is usually below 1 EBC. To achieve these requirements, however, it is necessary to have a well-stabilized original beer and a very oxygen-poor working method.

The sensory qualities of the beer clearly change over the course of the thermal dealcoholization. A loss of aroma, body and carbonation flavor can be seen. Apart from the loss of aroma, the aroma profile also changes. Less desirable aroma impressions, such as bready, worty and caramel-like, can come to the fore. Also, the appearance of a fatty acid taste, reminiscent of a 'billy goat' (caprylic acid), can arise with vacuum evaporation.

To achieve sensory improvements, dealcoholized beers can be blended up to the limit of the maximum allowable alcoholic content with a part of the untreated beer or a part of the aroma extract obtained in the alcohol extraction. In addition, a separately fermented aromatic beer can be added to the dealcoholized beer. Also, the addition of fresh yeast or kräusen of the dealcoholized beer influences smell and taste in a positive way. The adaptation of the recipe of the original beer through the application of special malts and the execution of a mashing method to reduce the degree of attenuation can also contribute to quality improvement [1, 4]. A fermented Bock or high-gravity beer is often dealcoholized because these beers are comparatively richer in esters, which will also be removed, although only in part, by the vacuum distillation. Thus, definitely more esters remain in the alcohol-free beer than if a 'normal' beer had been used.

10.1.2.1.2 Reverse Osmosis Diffusion by means of appropriate membranes is suitable for dealcoholization. The principle of reverse osmosis is one of these methods. Reverse osmosis makes use of an effect that takes place when a special membrane separates two liquids with different salt concentrations. A carriage of the solvent (in this case water) takes place through the membrane in the solution with the higher salt concentration (osmosis). The direction of the flow of water can be turned around when, a sufficiently large hydrostatic pressure operates on the side with higher salt concentration (reverse osmosis). Thereby, parts of the water of the beer, the alcohol and other beer substances migrate through the membrane into the permeate.

Compared to thermally dealcoholized beers, beer dealcoholized by reverse osmosis shows a somewhat fuller taste. However, aroma impressions such as worty and bready also appear here. In addition, a slightly sour character can be noticed compared to the original beers. In comparison with beers produced in the thermal process, the losses of bitter substances is somewhat higher. The increase of turbidity is in general less than 1 EBC unit. However, the quality of the original beer needs to be high when using this process [1, 4].

10.1.2.1.3 **Dialysis** The dialysis process is another membrane separation process, in which, in contrast to reverse osmosis, no pressure is applied, because the transportation of substances takes place as a spontaneous diffusion.

The moving power is the concentration difference of the substances in the retentate and dialysate. When beer is dialysed against water, all the beer ingredients except water tend to diffuse in the dialysate. The molecule exclusion takes place by the pore structure of the membrane. As in case of reverse osmosis, in the course of dialysis beer ingredients with a similar molecule size and polarity, such as ethanol, are separated by the membrane. Although dialysis is in principle a procedure without pressure, a certain pressure must be maintained on the beer as well as on the dialysate side (CO_2 release). On the basis of the small concentration differences, dialysis is less suited to reach very low alcoholic contents.

The taste evaluation of the beer produced by dialysis is generally considered as good, but also here a somewhat sour aroma appears as well as losses of harmony and body are registered. An adaptation of the recipe of the original beer by application of special malts and the execution of a mashing method for the reduction of the degree of attenuation contribute to the quality improvement. When employing this procedure, however, the quality of the beer produced also corresponds to the claims of an alcohol-free beer.

The composition of the beer dealcoholized by the membrane separation process depends on the selectivity of the membranes for the beer ingredients as well as from procedural parameters. Membrane separation processes distinguish themselves particularly by the fact that the product is not exposed to thermal stress [1, 4].

10.1.2.2 Biological Methods

In contrast to the physical processes, the beer is not entirely fermented during the biological procedure, but fermentation is interrupted when the desired alcoholic content is reached. This is referred to as interrupted fermentation. If the fermentation takes place at particularly low temperatures, the method used is called the yeast cold-contact process. The production of interrupted alcohol-free beers takes place mostly in the batch procedure; in addition, continuous procedures (bioreactors) are also applied. Moreover, procedures are nowadays used in which special microorganisms ferment selective parts of the fermentable extract.

As fermentation is interrupted at a desired alcoholic content, the beers receive a high fraction of non-fermented residual extract – a fact that is reflected in their taste and flavor. These beers can show a sweet, worty character as well as a bright body.

The advantage of the biological procedures particularly consists in the fact that, at least in the batch procedure, no additional (plant-specific) expenditure is necessary for the extraction of alcohol [1, 4].

10.1.2.2.1 Interrupted Fermentation (Batch Processing) When the method of interrupted fermentation is applied, technological measures are used in the brewhouse in order to produce a degree of attenuation as low as possible. The maltose rest at 60–64 °C is, for example, skipped over in a so-called jump-mash process; worts with a low final attenuation limit result this way. In addition, these worts show little difference between the final attenuation limit and the apparent attenuation limit, which leads to less strongly sweet, worty tasting beers. The original gravity is reduced to 5–8%, by which the degree of attenuation can be increased and with it the sensory qualities can be improved. Wort boiling should take place carefully with an extensive expulsion of unwanted aromatic substances. Part of the hopping should be carried out at the end of wort boiling with aromatic hops. A very long contact between wort and yeast is desirable during fermentation. On the one hand, this enhances the clarification of the beer by the adsorptive qualities of the yeast, as it has a clearing effect on its surface. On the other hand, the yeast possesses reducing qualities, by which wort carbonyls are converted into corresponding alcohols and a decline of the Strecker aldehydes already sets in after a short contact time [1]. The contact time between wort and yeast can be extended with decreasing temperature (yeast cold-contact process), by which a stronger decline of the Strecker aldehydes sets in with continuous alcohol development. Furthermore, the beers show a more balanced, beer-typical character when colder fermenting methods are applied. The addition of washed yeast has the advantage that, among other things, lower quantities of yeast-beer and with it alcohol are brought in, which in turn raises the contact times. A CO_2 washing of the green beer can also intensify the yeast contact, and remove a part of the wort flavor and green beer flavor. Fermentation is interrupted when the yeast is removed by means of a centrifuge at a concentration of below 0.1 million yeast cells/ml. The pre-clarified green beer is stored at a temperature of 0 °C for more than 1–3 weeks, which causes an improvement of the colloidal stability and a lowering of the gushing risk.

For the production of alcohol-free beers, special microorganisms are also used as well as industrial yeast. Yeasts that are not able to convert maltose or maltotriose to ethanol are suitable for this purpose, but also yeast mutants that have a defect in the citrate cycle and therefore produce large quantities of organic acid. Furthermore, cultures of *Lactobacillus* are also used [1, 4].

10.1.2.2.2 **Bioreactors (Immobilized Yeast Fermentation)** Alcohol-free beer can also be produced in a continuously working process in so-called bioreactors, when the yeast is bound to a carrier material, such as DEAE-cellulose, calcium alginate, calcium pectate or sintered glass. Carrier material and yeast are brought into reactors, which are streamed by wort. The immobilization of yeast induces a larger biomass per unit of reactor volume and thereby a higher fermenting performance with a shorter dwell time. The reactors divide themselves as fixed-bed reactors, in which the wort flows through the carrier material bottom-up relatively slowly, and fluidized-bed reactors like loop reactors.

In the case of loop reactors, a forced oncoming flow is applied, which leads to a good mixing and a lower chance of blockage. The parameters of flow speed, temperature, substrate composition and strain of the carrier material can be used to adjust the degree of attenuation and the alcoholic content.

Owing to the high content of residual extract, the alcohol-free beer produced in the biological process distinguishes itself by a high microbiological sensitivity. Filtration and stabilization of the beer with silica gel preparations and polyvinylpyrrolidone as well as a pasteurization of the bottled beer (at least 30 PU = pasteurisation units) is thereby inevitable [1].

10.1.2.3 Combination Physical-Biological Processes

Combining physical and biological processes is a good possibility to take advantage of both processes. When beers are blended, the wort character can be diluted, and body and softness can be brought into the beer at the same time; the sour character of the dealcoholized beers can also be partially covered. However, the input regarding equipment and costs is very high, because dealcoholization equipment as well as pasteurization are needed [1, 4].

10.2
Dietetic Beer

10.2.1
Introduction

The denomination 'dietetic beer' causes different associations in the consumer's mind. The designation induces associations with a healthy, conscious lifestyle. Accordingly this beer type is specifically marketed in some countries. Particularly in the United States, dietetic beer has achieved great market success in recent years. This can particularly be ascribed to the appearance of new forms of diets, such as the Atkins Diet, in which carbohydrates are strictly rejected [5]. Therefore, so-called low-carb beers with a decreased amount of carbohydrates were produced, which satisfy the pretended need of the consumers for beers that fit their 'diet lifestyle'. 'Low-carb' beers are not to be mistaken for 'light' (lite) beers, which also enjoy great popularity in the United States [6]. Low-carb beers show a low content of alcohol and calories, but a constant content of carbohydrates [7]. In contrast to the

American market, dietetic beers have been produced elsewhere (e.g. in Germany) for a long time with the main idea of a decreased carbohydrate content. However, compared to the United States, this is not based on a sudden increase in consciousness for health and diet, but on the need to produce beers for diabetics. Due to this different viewpoint, dietetic beer has only a low significance in the German market. Different national regulations are in force with regard to the nutritionally relevant composition of these beers. According to US Alcohol and Tobacco Tax and Trade Bureau, American 'low-carb' beers may not have more than 7 g carbohydrates/12 ounce serving (approximately 350 ml). In addition to the general regulations, the German Verordnung über diätische Lebensmittel (Regulations about Dietetic Food) must be followed when producing and labeling diet beer in Germany. According to this regulation, D-glucose, inverted sugar, disaccharide, maltodextrin or glucose syrup are not permitted to be added to dietetic beer.

It is only allowed to be produced and put in circulation with a content of no more than 0.75 g of the mentioned carbohydrates per 100 ml. Also, in Switzerland, a beer poor in carbohydrates may show at the most 7.5 g carbohydrates per liter. In Austria, however, production and marketing are forbidden [8].

10.2.2
Methods for the Production of Dietetic Beers

For the compliance of the obligatory minimum requirements for a beer poor in carbohydrates, technological peculiarities must be considered during the production of such beers. A very high degree of attenuation, which lies at more than 100% apparent degree of attenuation, is to be aimed for. Due to these high degrees of attenuation the alcoholic content of these beers increases, so that these must be submitted to an alcohol reduction or blended with brewing water after the brewing process. Another possibility is to brew with a lower original gravity, by which an alcoholic content comparable with 'normal' beers is reached [4, 9]. The mashing procedure is executed very intensely in order to unfold the activity of both amylases to a large extent. In this case, mashing-in mostly takes place at very low temperatures and intense rests are observed while mashing. To reach a high degree of attenuation, a part of the mash is additionally taken at 55–62 °C and is added in the pre-run vessel or the copper, where the remaining enzymes work till deactivitation. The further course of the wort preparation takes place analogously to the production of 'normal' worts, in which a somewhat stronger hopping is mostly chosen for a dietetic beer. As the degree of attenuation would still be too low, a malt extract (approximately 3%) is mostly added to the beer during the main fermentation. This malt extract is taken from a 'normal brew'. In addition to malt extract, malt flour, which is also given to the main fermentation [4], can also be worked with. The procedure described before conforms to the German Vorläufiges Biergesetz (Temporary Beer Law), which is based on the Bavarian Purity Law (Reinheitsgebot). In other cases, a high degree of attenuation can also be achieved by addition of technical enzymes during mashing and/or the main fermentation [7]. A stabilization of the beer should be considered, because the malt extracts contain proportionately large quantities of high molecular, coagulable nitrogen. In addition, a heat treatment of the beer should take place to inactivate the remaining enzymes and to avoid contamination of the beer [4].

10.3
'Nährbier' and 'Malzbier' ('Malztrunk')

10.3.1
Introduction

Beers that are exclusively produced from malt and may, respectively, be bottom fermented are classified as 'Nährbier'. These beers are very dark and are alcohol-free or have a low alcoholic content. The original gravity is often lower than 12% by weight, the alcoholic content ranges, according to the type, between less than 0.5 and 1.5% by volume, and the degrees of attenuation are, respectively, lower. Unlike 'Nährbier', 'Malzbier' is brewed with lower original wort and, in the filtered condition, is mixed with so much sugar shade or sugar syrup that a beer with an original gravity of approximately 12% by weight results [4]. According to the German Temporary Beer Law, this procedure is only allowed to be realized if the wort is fermented with top-fermenting yeast. In Bavaria and Baden-Württemberg, Germany, no malt substitutes are permitted even during top-fermenting brewing [10]. Also, 'Malzbiers' are produced as beers low in alcohol or alcohol-free. Initially, expectant or breast-feeding mothers, in particular, were the targeted group of consumers. Nowadays, these beers find a much wider basis among sportsmen and active people who do not want to resign the pleasure and the taking of valuable malt substances with the simultaneous absence of alcohol [11].

10.3.2
Methods for the Production of 'Nährbier' and 'Malzbier' ('Malztrunk')

For the most part the throw consists of malt mixtures of darker malt sorts. Apart from pale malt, particularly dark malt, pale caramel malt, dark caramel malt and black malt are used. Furthermore, acid malt is given in order to lower the pH. For 'Nährbier', a two-mash process is mostly chosen for wort preparation, whereas 'Malzbier' is also partially produced by means of the infusion process. The maltose formation rest is skipped to reach low degrees of final attenuation. During the boiling time, dependent on the system, hopping takes place at 15–20 BU (= EBC bitter units). In both cases the pitching wort is fermented up to the desired degree of attenuation at temperatures of approximately 10 °C; afterwards it is cooled down to 2–0 °C and, if necessary, kept fermenting as long as the final attenuation limit is reached, so that CO_2 is enriched. Afterwards the separation of the (remaining) yeast takes place by means of filtration or centrifugation, or a combination of both. The bottled beer must definitely be submitted to a heat treatment because of the high nutrient offer for microorganisms [4].

10.4
XAN™ Wheat Beer

10.4.1
Introduction

XAN™ wheat beer is proof that the brewers' creativity has no limits. This beer was been developed in 2001 at the Lehrstuhl für Technologie der Brauerei I of the Technische Universität München, Germany, and features a clearly higher content of xanthohumol. Xanthohumol is a prenylflavonoid (polyphenol), which is only found in the lupulin glands of the hop and shows a high variety of health-positive qualities in continuous *in vitro* examinations. Cancer-preventive activities and growth-inhibiting effects towards certain tumor cells were reported [12]. Furthermore, xanthohumol shows a strong antioxidative effect towards hydroxyl and peroxyl radicals, which are able to produce cell damage. As xanthohumol is relatively unstable and is lightly converted and eliminated again during the brewing process, a special brewing procedure had to be developed, with which the concentration of the hop substances can be 10-fold enriched in the beer. Usually, the content of xanthohumol in standard beers is clearly below 0.2 mg/l.

10.4.2
Methods of Production of XAN™ Wheat Beer

According to the Bavarian Purity Law, it is only possible to enrich xanthohumol through a hop addition with hop sorts rich in xanthohumol and variations of the brewing process. Xanthohumol is a non-polar substance, which solubilizes itself in polar media only in low quantities and isomerizes to iso-xanthohumol during boiling. When applying the so-called 'XAN™ technology' the resulting losses are reduced by brewing with higher original gravity, by late and high hop addition as well as by addition of cold brewing water. A rapid cooling of the wort down to 80 °C is an important influencing factor to largely prevent any isomerization of xanthohumol. The hop addition for the enrichment of xanthohumol can take place with standard ethanol extracts or products enriched with xanthohumol and is based upon the desired content of xanthohumol in the beer. Another important step is the use of roasted malt. The xanthohumol recovery clearly increases when the fraction of roasted malt increases. Roasted substances, which keep xanthohumol in solution in the course of the beer production, are held responsible [13]. Further technological influencing factors, such as reutilization of yeast or the waiving of stabilizers, have a clearly positive effect on yield in the case of xanthohumol beer production without roasted malt. Xanthohumol contents of 1–3 mg/l are possible in unfiltered beers without roasted products. Through the application of roasted malt combined with 'XAN™ technology', contents of over 10 mg xanthohumol/l can be reached in the filtered beer [1, 14].

10.5
Gluten-Free Beer

10.5.1
Introduction

A relatively large amount of people cannot drink normal beer. They suffer from an incompatibility to gluten, a protein found in grain sorts like wheat and all cultivars of the Triticeae, rye and barley. It is still unclear to what extent this incompatibility also concerns oats (in some European Union countries it is not seen in persons under observation, but this is not the case in Germany!). However, other vegetable and animal proteins can be consumed. This disease is called celiac disease or endemic sprue, if it appears at the adult age. The frequency of endemic sprue strongly fluctuates depending on the population. Thus, in Ireland it amounts to approximately 1:300; in Berlin, however, it amounts to only 1:2700. The number of unreported cases is very high. Many have unspecific, subjectively light symptoms and are never diagnosed as patients suffering from celiac disease. The ingestion of gluten-containing food leads to an inflammatory reaction of the lining of the small intestine, so that nutrients cannot be absorbed. This can lead to serious deficiency symptoms or also to psychological disturbances. More than half of all affected persons frequently complain of diarrhea, fatigue, lack of energy, weight loss and flatulence. Other important symptoms are abdominal pain, nausea and vomiting, as well as bone pains and aching muscles. Due to malnutrition in infancy or infant growth delays, vitamin deficiency appearances, and anemia and even mental aberrations can result. There is also the skin disease based on gluten sensitivity, dermatitis herpetiformis (duhring), which can be treated by gluten-free nutrition (however, in combination with a medicament). In Great Britain this rare disease occurs at an incidence of 1:15000. Patients with celiac disease and/or Duhring must keep on a strict lifelong diet, which should be free of products made of wheat, rye, barley and oats, except for their pure starch. This is often difficult, because many food products are made of these cereals, such as noodles, bread and pasta. Products with 'hidden' gluten (e.g. chocolate and canned meat) are even much more dangerous, because gluten-containing additives are used in products here, which are normally harmless [15].

10.5.2
Production Methods for Gluten-Free Beer

In principle there are five different strategies for the production of gluten-free beers. Each technology, which is briefly described below, can also be applied in combined form. However, the complete production process of each technology should be executed with great care in order to avoid contamination with gluten-containing material. It should be pointed out that the yeast also must be cultivated in gluten-free wort. Therefore, products supplied by farmers, trade as well as the malthouse must be genuine. This is based on the fact that the limit of tolerance for the daily absorption of gluten is only 10 mg for adults. A wheat beer possibly has 10-fold that content per liter and a bottom-fermenting beer possibly half or less.

10.5.2.1 Conventional 'Gluten-Containing' Raw Material

10.5.2.1.1 Selectively Bred Grains Conventional raw material could be used for a gluten-free beer production, if genetically modified grains do not have the possibility to develop gluten. This is targeted by the systematic deactivation of all gliadin genes in wheat [16, 17]. However, such a grain is not yet available. It would be advantageous if common technology was also applicable to these products. However, with regard to proteins, modification is unavoidable and certain quality attributes, like foam, aroma, turbidity tendency as well as taste stability, would certainly have to be reconsidered. The fact that these transgenic grains are not allowed for food preparation in many countries represents a clear disadvantage in terms of consumer acceptance.

10.5.2.1.2 Enzymatic Modification Two enzymatic methods have been tried to reduce gluten brought in by the raw materials (barley malt, wheat malt, etc.) to a 'harmless' degree. On the one hand, genetic engineering is the inspiration here again, where genetically modified yeast is enabled to express a specific enzyme. On the other hand, transglutaminase can also be added exogenously to modify the gluten fraction. In both cases these procedures have the advantage that virtually all beer types can be produced without loss of their specific characteristics. The additional expense factor, a lower consumer acceptance and the fact that these beers do not correspond to the German Temporary Beer Law are certainly unfavorable.

10.5.2.2 Sources of Gluten-Free Sugars and Starch

10.5.2.2.1 Sugars Sugary raw materials can be used in beer production that do not originate from gluten-containing sources, such as honey, sugar, syrup. This is a fundamental mixture of the definitions beer and wine, since beer has starch as a carbohydrate source, whereas wine has sugar. In the United States, however, such a product is known as beer [18]. The worldwide availability of these raw materials and a good consumer acceptance are favorable. Whether it is advantageous to get the raw material from the sugar factory, for example, instead of the malthouse is to be judged differently in each country. With respect to fermentation, it is particularly unfavorable that many necessary yeast growth substances must be added, and that aroma and foam can only be shaped satisfyingly by using expensive additives. It is also unfavorable that only a beer-similar product would be expected, which could not be put in circulation, for example in Germany, as a beer, because it complies with the Wine Law. Therefore, the special claim is also lost in this country, because the aim is not to provide patients suffering with celiac or duhring disease with any alcoholic beverage (wine and hard liquor do not contain gluten), but just with a special beer.

10.5.2.2.2 Grains Rich in Carbohydrates Another way to produce gluten-free beers is by using grains rich in carbohydrates, which do not contain allergenic gluten and are therefore accepted by patients suffering from celiac or duhring disease. These are the pseudocereals amaranth (e.g. *Amaranthus hypochondriacus*), buckwheat (*Fagopyrum esculentum*) and quinoa (*Chenopodium quinoa*), as well as the grain sorts sorghum (e.g. *Sorghum bicolour*), millets (e.g. *Panicum miliaceum, Setaria italica*, etc.), corn (*Zea corn*) and rice (*Oryza sativa*). Other sources are also being discussed, for example peas [19], which are afflicted, beside starch, with a very rich fraction of proteins. The necessity of proteins for beer production is undoubted, but in less valuable proportion.

10.5.2.2.3 As Raw Grains The possibility to use grains rich in carbohydrates for beer preparation is as an unmalted raw grain. This is absolutely possible, but the lack of sufficient native, amylolytic enzymes must be compensated for by the use of exogenous ones. This is an additional expense factor, which will even be increased insofar as aroma and color substances are missing and would have to be added when malting does not take place. The raw materials are comparatively low priced, available in many countries and do not demand additional equipment (sugar factory, malthouse). The product does not correspond to the German Temporary Beer Law and, in many respects, comes – at most – near 'real' beer.

10.5.2.2.4 As Malt It is possible to use these grains rich in carbohydrates in malted form. As barley was already selected brewing-specifically for millennia, these malts differ greatly from the usual barley malt data [20]. The philosophy of the Bavarian Purity Law is nevertheless evident when completely followed, since further additives must not be used. A challenge is the very different thousand corn weight of the grains, because it can vary between 0.3 g, as in case of teff (millet type with small grain sizes), and approximately 240 g, as in case of corn. Particularly in the malthouse, corresponding effort has to be invested, since the storage bins, vessels and routes of transportation must be adapted to the grain sizes. Regarding the malthouse and brewery, further capacities could be favorably balanced. Agronomic advantages are the promotion of biodiversity and the fact that some fruits still produce sufficient yield on poor ground.

Extensive tests were presented for the first time at the EBC Congress in 2005 (Prague), where beers made of 100% proso millet malt with very pleasing data on malt, wort and beer were shown [20].

10.6
Brewing with High Original Wort

10.6.1
Introduction

Brewing with high original wort is based on the idea to take better advantage of the existing equipment and capacities. This method enables one to work economically despite a lower exploitation of the raw materials. In addition, energy can be saved. The higher brewed-in brews are diluted during further production either before fermentation or before filtration onto the 'normal' original gravity. If the wort is already diluted before fermentation, an increase in capacity arises from it in the brewhouse area. If the worts are diluted shortly before filtration, capacity savings are still possible in the fermentation and storage cellar areas. During attenuation, special demands are put on the water to which the original wort is set. The blending water must be microbiologically perfect. Furthermore, the water must be as free of oxygen as possible, it should be very cold (0 °C) and must show the same CO_2 content as beer after carbonization. If the method of operation is carried out correctly, the quality of the beers is equivalent to 'normally' produced beers. In case unwanted deviations of the bitter substances content or the hop aroma occur in the finished product, addition of isomerized hop extracts is possible in the fermentation/maturation cellar according to the legal regulations.

10.6.2
Methods for Beer Production with High Original Wort

In order to achieve a higher wort concentration the mashing-in takes place at a somewhat higher throw proportion of approximately 1 : 3. Attention must be paid on the saccharification, because the activity of α-amylase can necessarily be slightly limited by the high throw proportion. If the brewhouse was not designed for high-gravity brewing from the beginning, the lauter tun would represent the limiting factor, because it cannot be overloaded intensely. Attention has to be paid that the knives of the raking machine are long enough. The number of the knives should be between 2 and 2.5 knives/m². Furthermore, the removal of the spent grains must be adapted to the larger throw, so that too long set-up times can be avoided. In addition, attention must be paid that the spent grains bunker or the spent grains buffer tank can take the larger amount of spent grains. With the increasing concentration of the finished wort, the concentration of the last runnings increase and the yield accordingly decreases.

In case of wort boiling it must be pointed out that the hop addition must correspond to the desired bitterness units after the reverse dilution. Furthermore, higher losses occur due to the higher amounts of coagulable nitrogen. A wort acidification down to pH 5.0 should take place approximately 10 min before the end of boiling in order to enhance the fermentation of the higher-proof worts. If the legal basis allows, an essentially simpler working is given if it added approximately 10 min before the end of boiling starch syrup or saccharose syrup to raise the original wort concentration. The hot break separation in the whirlpool can be somewhat problematic. This is caused, on the one hand, by the higher viscosity of the worts, and, on the other hand, by the increase of broken and spent hops. In this case, the application of a centrifuge can correct the situation. This indicates that a hot break recirculation is unnecessary for the extract production.

During cooling of the finished wort there is the possibility to dilute wort to the desired concentration. The demands on quality made on the blending water have already been described above. Within this segment, however, a complete oxygen release is not required.

If the beer is diluted only later, the fermentation is executed with high original wort. This induces a longer fermentation time. To guarantee sufficient yeast propagation and a possible shortening of the fermentation time, the supply of oxygen to the yeast must be ensured. In addition, the pH value of the pitching wort should be adjusted to 4.9–5.0. If the yeast is drawn off early enough, less higher alcohols and esters are developed. Higher temperatures are less suitable for the acceleration of fermentation, because they also increase unwanted fermentation metabolites.

The high-gravity beers are blended on the way to the filter. A treatment of the beer must take place in the unfiltered beer anyway, because after the filter 'fining mechanisms' no longer exist and the generated balance could be damaged by such an intervention. This could influence the stabilities (e.g. colloidal stability, foam stability, gushing, etc.). With respect to the blending water, make sure that the water is completely free of oxygen [4].

10.7
Ale and Cask-Conditioned Ale

10.7.1
Introduction

Before describing cask-conditioned ale, the several ale types should be characterized first. Ale is an old-established beer, which particularly enjoys great popularity in England and is also well-known beyond the borders of the United Kingdom. Ale beers are top-fermented beers, which differ themselves in color and hop enhancement. The transitions among the relative types are often smooth. The beers are mostly produced by means of an infusion mashing method. A classic English infusion mashing method for ale beers provides for a saccharification rest [21] as a single temperature step. Different sorts of malt, raw grains (e.g. roasted barley) [22] and hops are used for the different ale types. English ales can be classified in the following types:

- *Mild ale.* The denomination mild refers to the hop character and describes a slightly hopped beer. The mild ale is brewed as pale as well as dark ale and, therefore, can feature a strong, malty character.

- *Bitter ale.* In contrast to mild ale, bitter ale describes the stronger hopped ale that is brewed in extensive varieties beyond the hop character.

- *Pale ale.* The characteristics of pale ale and bitter ale strongly overlap. This beer also shows a stronger hop note, which can be somewhat milder than in the case of bitter ale. The color can be described as amber; the original wort can be slightly higher than in the case of bitter ale.

- *India pale ale.* The name India pale ale has its origins in the Victorian period, as England ale beers were exported to India by sea. These beers were often very strongly brewed and hopped to avoid ageing during the long sea voyages. Nowadays, India pale ale is no longer being brewed and hopped as strongly as its historic forerunner.

- *Brown ale.* Brown ale designates a dark, very malt-aromatic ale.

- *Old ale.* Old ale is mostly a dark type of ale, which is not completely fermented, because the sweetness and taste of the maltose should be preserved in the beer. A historic interpretation of the name is based on the fact that these beers were already brewed in the wintertime for the summer months. According to this, the beer was already some months old before it was drunk.

- *Scottish ale.* These are very malt-aromatic beers, with a higher amount of residual extract, so that the full character is brought out. The color of these beers is described as golden brown to dark brown. The name derives from the original geographic position of the brewing region (i.e. Scotland).

- *Irish ale.* Irish ale is characterized as malt-aromatic with a typically red color [23].

The ale types described are a part of the classic English ale types. Further ale beers are brewed in the United States and Europe. It must be pointed out that the borders between the respective ale types often become indistinct or merge fluently and an exact correlation to a certain type is therefore almost impossible.

The different types of ale can be divided in two categories again, as shown in Figure 10.1. On the one hand, there are the 'normal' ale beers, whose brewing process is finished in the brewery; on the other hand, there are the 'real' ale beers. 'Real ale' beers can be defined as follows: 'Real ale is the name for draft or bottled beer brewed from traditional ingredients, matured by secondary fermentation in the container, from which it is dispensed and served without the use of exogenous carbon dioxide' [22].

10.7.2
Methods for the Production of Cask-Conditioned Ale

As the term 'cask-conditioned' already describes, the secondary fermentation of these beers takes place in the barrel. The ale types described above or other ale

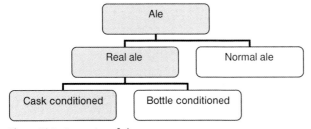

Figure 10.1 Categories of ale types.

types can be used as the basis of a cask-conditioned ale. After main fermentation the beer is transferred, if necessary under addition of finings, secondary fermentation sugar and hops, to barrels in which the secondary fermentation is executed. The secondary fermentation is executed under control of the brewery or of an expert and trained pub staff. The execution of the secondary fermentation in the pub represents the more traditional version. After delivery the barrel is stored on a rack, and should no longer be moved in order to allow clarification and sedimentation. The CO_2 developed during the secondary fermentation produces, beside carbonization of the beer, pressure in the barrel that must be reduced in a controlled manner. The release of pressure takes place by means of a soft spile (porous, gas-permeable 'wooden plug'), which is carefully pressed through the shive and can be seen as a kind of pressure release valve. This soft spile should be frequently changed during the secondary fermentation. After 12–24 h only few or even no fermentation CO_2 is developed anymore, so that the soft spile can be replaced by a hard spile (gas-impermeable 'wooden plug'). For the execution of the secondary fermentation, the knowledge of the 'cellarmaster' is required, who has to make sure that the beer is sufficiently carbonized (approximately 1.2–1.8 g CO_2/l) and clarified, and does not show any unwanted green beer character. During plugging the hard spile should be replaced by a soft spile again, since this takes over the function of a primitive air filter. At 12–24 h before serving the beer, the barrel should be tapped to guarantee another clarification of the beer. The tapping of the beer takes place traditionally through a beer pump (beer engine) without use of 'dispensing gas', while the beer tapped from the barrel is replaced by air through the soft spile. However, in modern times a gas mixture of CO_2 and nitrogen is used for tapping, by which the system draws near to the keg system and finds only little approval of traditionally minded purists. The beer should be enjoyed at a temperature of 10–15 °C [7].

10.8
Lambic, Gueuze and Fruit Lambic

10.8.1
Introduction

Lambic is a spontaneously fermented Belgian beer, which is strongly affected by regional characteristics.

The real lambic is only brewed in a certain area within a radius of approximately 15 km of Brussels, mainly in the so-called Payottenland in the Sennevalley, Belgium. This is explained with the fact that only there are the desired special microorganisms and yeast found. As the main fermentation can hardly be controlled in summer, October and March are traditionally the main brewing months [24]. By royal decree, lambic may be fermented only spontaneously; a transferring of cultures is not permitted. The throw should consist of at least 30% of unmalted wheat [25], the original wort should be at least 11% by weight and the acidity at 30 mmol

NaOH [26]. Lambic is submitted to a kind of chain reaction of spontaneous fermentations in wooden barrels [23]. The fermenting time ranges up to 3 years. The beers of the lambic family are highly complex, spontaneously fermented beers, which feature very different qualities according to the method of working. Mostly, lambic serves as the basis. Today, lambic is rather rarely enjoyed in pure form in Belgium, but it is processed to gueuze, fruit lambic and other variants. The nearly uncarbonated, sour-bitter lambic shows a character that reminds one of sherry or dry vermouth [23]. Gueuze, blended from lambic beers stored for different lengths of and submitted to fermentation in the bottle, is often called the champagne of beer [24]. It is characterized as a fresh, mild, somewhat sour, aromatic dry and sparkling beer [25]. The fruit lambics show a very complex character, which varies depending on the type of fruit used. The addition of the fruit type also plays a role. Fruit lambics traditionally produced with whole fruits are considered more bitter than fruit lambics industrially produced with fruit syrup. The most famous fruit lambics are kriek (cherry) and frambozen/framboise (raspberry) [23].

10.8.2
Method for the Production of Lambic and Gueuze

10.8.2.1 Wort Production
The throw consists of barley malt and up to 40% of unmalted wheat. A high well-modified enzyme-strong malt is used as a barley malt. Partially, wheat chaff is mixed to the mash to improve the lauter qualities. The throw ratio is usually high (about 1:8 to 1:9). A combination of mash and lauter tun serves as a mash vessel. Normally, the decoction method is applied for mashing; the heating-up to the relative temperature rests takes place by scalding of boiling mash or hot water. A mash method can be carried out as follows: mashing-in at 45 °C, the next rests take place by scalding of boiling mash/hot water at 58, 65, 72 and 76 °C. The infusion method with subsequent mash boiling is a further procedure. The respective temperature rests are reached by a scalding of hot water. The separation of solid and liquid takes place through a lauter tun. Due to the high throw ratio the wort is boiled for a long time (up to 6 h) until the desired original wort concentration is reached. Hopping takes place with aged hop cones at the beginning of boiling. The hop can be 3 years old or even older. Therefore, the application of aged hops in not primarily to adjust the bitterness when hops are added, but to make use of the conserving force of the hops. The boiled wort is pumped through a hop sieve on a coolship and left overnight [24]. The wort is thereby not only cooled, but also inoculated with the available microflora of the brewery and the ambient air led through the wort, so that a spontaneous fermentation can start. Afterwards the wort is homogenized and bottled in wood barrels, in which the fermentation begins after some time [26].

10.8.2.2 Beer Production
The production of lambic wort can take up to 3 years, and its fermentation can be divided in four and/or five stages (in the case of gueuze production). The *Entero-*

bacter phase is considered to be the first stage. It already begins during wort cooling on coolship. In this stage especially *Enterobacter cloacae* and *Klebsiella aerogenes*, but also *Hafnia alvei, Enterobacter aerogenes* and *Citrobacter freundii*, can be selected. In addition, yeasts, such as *Kloeckera apiculata, Saccharomyces globosus* and *Saccharomyces dairensis*, can also be determined. During this stage a degree of attenuation of about 15% is achieved. At this time the pH drops around approximately 1 unit, and low amounts of alcohol as well as acetic acid, lactic acid and formic acid are developed. The diacetyl content is approximately 1 g/l and the dimethylsulfide (DMS) content approximately 500 ppb. After about 1 month, depending on the nutrient supply and pH value, these microorganisms disappear again.

Afterwards, the second phase, the so-called *Saccharomyces* phase, follows. In this phase mainly *Saccharomyces* yeasts of the types *Saccharomyces cerevisiae, Saccharomyces bayanus, Saccharomyces uvarum* and *Saccharomyces inusitatus* are to be found, which are responsible for the majority of the alcoholic fermentation. The DMS content drops during this phase.

The so-called acidification phase follows after approximately 4 months. This phase is characterized by a strong increase of lactic acid and ethyl lactate, and a slight increase of acetic acid and ethyl acetate. The diacetyl values increases to 200 ppm and the pH drops to 4 or lower. The degree of attenuation continues to rise. During this phase the *Saccharomyces* yeasts are inactivated to a large extent due to the pH decrease and the lack of nutrients. During acidification, yeasts of *Pediococcus, Lactobacillus* and *Brettanomyces* are mainly observed and slime layers can be developed.

After 10 months the maturation phase starts, in which the lactic acid bacteria at first and then, finally, also the *Brettanomyces* yeast decrease. During this phase the degree of attenuation increases further, and DMS and diacetyl continue to decrease to values of 100 or 80 ppb.

If the lambic is processed to gueuze, a fifth phase, the 'refermentation phase', follows. This phase is executed as bottle fermentation. For the bottle fermentation, lambic beers of different ages are blended, roughly filtered and bottled. Due to the entry of dextrins through green lambic while blending, bottle fermentation takes place and CO_2 contents of 6–7 g/l are reached. During the refermentation phase film-developing yeasts like *Candida, Torulopsis, Hansenula, Pichia* and *Cryptococcus* can be seen. These yeasts originate from a thick yeast film that was developed in the barrel and disappears again after approximately 10 months. The percentage of *Brettanomyces* yeast and lactic acid bacteria increases during this phase [26].

Figure 10.2 shows a possible microbiological profile of a lambic fermentation. The abscissa shows the duration of the fermentation and the ordinate shows the yeast cell amount.

10.8.3
Method for the Production of Fruit Lambic

The most famous variants of fruit lambic are probably kriek (cherry) and frambozen (raspberry); in addition, however, other sorts made with peaches, blackcur-

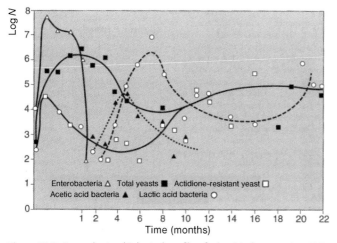
Figure 10.2 General microbiological profile of a lambic fermentation [26].

rants or grapes are produced. The production of fruit lambic takes place as follows. The desired amount of fruit is added to green lambic, which is fermented and aged for some months. After the maturation. the finished product is blended with green lambic, bottled and stored again for up to 1 year; a carbonization of the fruit lambic then takes place. Nowadays, juices or syrup are increasingly added instead of whole fruits [24, 25].

10.9
Berliner Weisse

10.9.1
Introduction

Berliner Weisse is a trademarked top-fermented beer specialty, which is only allowed to be brewed under this designation in Berlin, Germany [4]. It distinguishes itself by its typical sour but soft taste, the high CO_2 content and flavor richness [27]. Napoleon's troops called Berliner Weisse the 'champagne of the north' due to its sour character and the high CO_2 content [23]. The sour character results from the fact that during fermentation not only the industrial yeasts known in breweries, but also lactic acid bacteria are used; a pH of 3.2–3.4 is reached in this way. Since this special beer character is too sour to some people, Berliner Weisse is often mixed with syrup of woodruffs or raspberries [23]. Nowadays, Berliner Weisse is mostly brewed as a pale (5–8 EBC) draught beer (7–8% of original wort) with low bitter substances content (4–6 EBC). The percentage of wheat in the total throw is approximately two-thirds to three-quarters and the rest of the throw consists of barley malt [4].

10.9.2
Method for the Production of Berliner Weisse

The decoction as well as the infusion processes can be applied as mashing methods. The hops are already added to the mash. Thus, only a slight hop aroma is reached in the wort. Originally, the wort was not boiled, but was cooled directly after the lauter tun. This is a cold mashing method, which represented an original method of production [28, 29]. Meanwhile, the wort is heated up for reasons of sterilization.

The cooled wort is pitched with a microorganism mixture of top fermentation yeast and lactic acid rods in a relation of 4–6:1. According to the fermentation conditions, the increase of lactic acid rods and with it the acid development can be controlled. Thus, the increase of lactic acid bacteria is forced when the temperatures rises. As the yeast is restrained by the lactic acid bacteria, the main fermentation can last up to 4 days despite the low original wort. The secondary fermentation mostly takes place as bottle fermentation. Before the green beer is bottled, it is mixed with approximately 10% Kräusen. The desired content of CO_2 of 0.6–0.8% and lactic acid of 0.25–0.8% is reached in the secondary fermentation phase. The beer is subsequently stored. The storage time can vary greatly from 3 weeks to 3 years.

Difficulties during the production of Berliner Weisse can occur in the case of contaminations. A contamination with slime-forming *Pediococcus* can cause a 'stringing' of the beer. These are problems that cause a strong viscosity increase and the development of a slimy texture [4, 27].

10.10
Porter

10.10.1
Introduction

Porter is a dark, almost black beer that is distinguished by its malty character [23], which results from the usage of special malts. Apart from pale malt, roasted malts like crystal malt, chocolate malt, black malt and partially roasted barley are also used. As well as the barley malts, wheat malt can also be used to improve the foam stability [30]. Porter beers show different flavors depending on their origin. These beers are referred to as roasted malt-aromatic, toffee-like, malt-sweet, bitter or with a 'horse blanket aroma' [23]. The fermentation conditions represent another peculiarity in the production of classic porter beers. Another variety of porter is stout, which has a stronger taste. Stout is also a dark, roasted malt-aromatic beer, which can be strongly hopped.

10.10.2
Method for the Production of Porter Bier

Wort preparation does not play an important role in the production of porter. The wort is produced with the malt sorts discussed above. Furthermore, the porter can be produced as a surrogate beer by applying, for example starch, sugar or sucrose. The addition of hops takes place partly as dry hopping. The original wort ranges between 12 and 16% by weight [27]. In contrast to porter, there is a peculiarity in the wort preparation of stout beers. Stout worts are traditionally produced according to the so-called Partigyle system. When applying this fractioning method, the first wort and the corresponding spargings are boiled separately, and only blended when fermentation takes place [31].

The main fermentation at porter is a classical top fermentation, which is run warm and where fruity ale yeasts are applied; however, industrially produced porters are partially made of bottom-fermenting yeasts, too [30]. The secondary fermentation, where not only industrial yeasts but yeasts of the *Brettanomyces* type are used, plays an important role in the production of traditional porter beers. These yeasts develop distinctive aroma substances (e.g. a smoky taste similar to a 'horse blanket') [23, 27].

10.11
Summary

Only a short overview of the special production methods could be described here. This shows, however, where the creativity of brewing technologists can lead to, partly due to the raw materials and also due to consumer demand. Thus, the range of the different production methods is reflected in the beer types: low to high alcoholic contents, low to high original wort, pale to dark, clear to turbid, poor to intensive aroma and even aromas with a fruity character are possible.

References

1 Back, W. (2005) *Ausgewählte Kapitel der Brauereitechnologie*, Fachverlag Hans Carl, Nürnberg.
2 Deutscher-Brauerbund (2006) http://www.brauer-bund.de/presse/pressetexte.php32id=25 (accessed 29. 01. 2009).
3 Piendl, A. (1991) Zielgruppen für alkkoholfreies Bier. *Brauwelt*, **131**, 914–17.
4 Narziss, L. (2004) *Abriss der Bierbrauerei*, 7th edn, Wiley-VCH Verlag GmbH, Weinheim.
5 Bamforth, C.W. (2005) Beer, carbohydrates and diet. *Journal of the Institute of Brewing*, **111**, 259–64.
6 Lagorce, A. (2003) Low-Carb Beer Fattens Anheuser Busch. http://www.forbes.com/2003/09/25/cx_al_0925bud.html (accessed 25. 09. 2006).
7 Briggs, D., Boulton, A., Brooks, P. and Stevens, R. (2004) *Brewing, Science and Practise*, CRC Press, Boca Raton, FL.
8 Miedaner, H. (2002) *Brautechnische Analysenmethode Band II*, Selbstverlag der MEBAK, Freising.

9 Annemüller, G. and Schöber, J. (1999) Diätbier – die ungeliebte biersorte der deutschen brauindustrie? *Brauwelt*, **139**, 862–7.
10 Zipfel, W. (1999) *Lebensmittelrecht – Vorläufiges Biergesetz i. d. Bek. der Neufassung vom 29.07.1993 (BGBL. I S. 1399), geändert durch Art. 17 der Verordnung zur Neuordnung der Vorschriften über Zusatzstoffe vom 29.01.1998 (BGBL. I S. 230, 298)*, Verlag C. H. Beck, München.
11 Deutscher-Brauerbund (2006) http://www.brauer-bund.de/index1.html (accessed 2006).
12 Gerhauser, C., Alt, A., Heiss, E., Gamal-Eldeen, A., Klimo, K., Knauft, J., Neumann, I., Scherf, H.-R., Frank, N., Bartsch, H. and Becker, H. (2002) Cancer chemopreventive activity of Xanthohumol, a natural product derived from hop. *Molecular Cancer Therapeutics*, **1**, 959 69.
13 Walker, C.J., Fernández Lence, C. and Biendl, M. (2003) Untersuchung zum hohen Xanthohumolgehalt in Bieren vom Typ Stout/Porter. *Brauwelt*, **143**, 1709–12.
14 Wunderlich, S., Zurcher, A. and Back, W. (2005) Enrichment of xanthohumol in the brewing process. *Molecular Nutrition and Food Research*, **49**, 874–81.
15 Zeitz, M. (1999) *Thiemes Innere Medizin*, Georg Thieme, Stuttgart.
16 HN., HN. (1999) Grüne Gentechnik soll Zöliakie-Kranken helfen. *Lebensmittel Zeitung*, **8**, 64.
17 Lörz, H. and Becker, D. (2005) Gentechnische Erzeugung von Weizen ohne Zöliakie-Toxizität, http://www.vvgvg.org/projekt3.htm (accessed 2005).
18 Ramapo Valley Brewery (2006) http://www.ramapovalleybrewery.com/.
19 Brauer, J., Walker, C. and Booer, C. (2005) Of pseudocereals and roasted rice. *The Brewer & Distiller*, 24–6.
20 Zarnkow, M., Kessler, M., Burberg, F., Kreisz, S. and Back, W. (2005) Gluten free beer from malted cereals and pseudocereals, in *Proceedings of the 30th European Brewery Convention Congress in Prague*, Hans Carl Verlag, pp. 104/1–104/8.
21 Foster, T. (1990) *Pale Ale*, Brewers Publications, Boulder, CO.
22 Sutula, D. (1999) *Boulder*, Brewers Publications, Boulder, CO.
23 Jackson, M. (1999) *Bier International*, Hallwag, Bern.
24 Guinard, J.-X. (1990) *Lambic*, Brewers Publication, Boulder, CO.
25 Piendl, A. (1995) Belgische lambic-fruchtbiere. *Brauwelt*, **135**, 942–52.
26 Verachtert, H. and Debourg, A. (1995) Properties of Belgian acid beers and their microflora – 1. The production of Gueuze and related refreshing acid beers. *Cerevisia – Belgian Journal of Brewing und Biotechnology*, **1**, 37–41.
27 Hayduck, F., Bode, G., Gesell, H. and Haehn, H. (1925) *Illustriertes Brauerei-Lexikon*, Verlag für Landwirtschaft, Gartenbau und Forstwesen, Berlin.
28 Zarnkow, M., Spieleder, E., Back, W., Sacher, B., Otto, A. and Einwag, B. (2006) Interdisziplinäre Untersuchungen zum altorientalischen Bierbrauen in der Siedlung von Tall Bazi/Nordsyrien vor rund 3200 Jahren. *Technikgeschichte*, **73**, 3–25.
29 Zarnkow, M., Spieleder, E., Sacher, B., Otto, A. and Einwag, B. (2006) Kaltmaischverfahren – eine mögliche Technologie im Alten Orient. *Brauwelt*, **146**, 272–5.
30 Foster, T. (1992) *Porter*, Brewers Publications, Boulder, CO.
31 Lewis, S. (1995) *Stout*, Brewers Publications, Boulder, CO.

11
Beer-Based Mixed Drinks

Oliver Franz, Martina Gastl, and Werner Back

The market for beer-based mixed drinks has strongly increased in recent years. Beer-based mixed drinks, in particular, find a high acceptance in different consumer groups. Innovative beverages on the basis of beer give the brewer a great chance to selectively address a modern and unconventional consumer group and conquer new markets. Fruity additives to beer have a long tradition, mainly in Belgium and northern France. There you can find, for example, kriek (dark beer fermented with sour cherries), pêche (with peach), framboise (with raspberry), cassis (with cranberries), banana, strawberry, pineapple, mirabelle.

Beer-based mixed drinks have long been established in gastronomy and represent a rapidly growing segment for 'ready drinks' available in cans and bottles. A young, 'trendy' audience is selectively targeted for these products where the packaging also has an important influence on the taste value of a drink. However, there are only few limitations to the variety of packing and taste. Beer-based mixed drinks have a great potential due to this variety of possibilities for product diversification. The basis is always beer, which is mixed with, for example, lemonade, cola, fruit juices or other additives. The drinks can be alcoholic or non-alcoholic, with one or more additives, with or without functional ingredients, isotonic, containing juice, clear or turbid, with mineral additions, with one or more flavor added, or sweetened with sugar or artificial sweetener. Beer-based mixed drinks can also be produced using all kinds of beers. Thus, there are numerous variations possible and the formulation concept is very versatile. Newer products contain flavors such as lemon, orange, grapefruit or pomegranate. Whether peach or cassis, cherry or lemon, raspberry, woodruff or pitaya, caffeine, guarana or mineral – there are hardly any limits to the imagination. The traditionally flavored beers with lemonade and cola still continue to dominate the beer-based mixed drinks market.

11.1
Development of Beer-Based Mixed Drinks

Beer-based mixed drinks are beverages that consist of beer and a soft drink, often in equal fractions. The alcohol content is around 2.5% in most beer-based mixed drink.

It is actually not a new idea to mix beer with lemonade or cola. Depending on the region, 'Radler', 'Russ'n Mass', 'Alsterwasser' or 'Diesel' belong to the fixed repertoire in pubs and beer gardens. Berliner Weisse with a shot of raspberry or woodruff syrup is also a traditional beer-based mixed drink.

The 'father' of all beer-based mixed drinks surely is 'Radler' or 'Alsterwasser' composed of beer and lemonade. Originally this drink was mixed from lager beer and lemonade in a ratio of 1:1 and was supposed to be invented by a Bavarian landlord out of necessity. One nice summer day he was hard pushed with his beer storage due to the arrival of many cyclists and diluted his lager beer with lemonade without further ado. The original from about 1922 was thus called 'Radler' (cyclist) and originated at the 'Kugler-Alm' in the South of Munich. However, a mixture of beer and lemonade with the name 'shandy', appears to have already been popular in England in the late nineteenth century.

In North Germany, 'Alster' and 'Alsterwasser' or 'Potsdamer' were developed which consist of Pils mixed with either lemonade or orangeade.

Another variant is 'Russ'n Mass' in which lemonade and wheat beer are mixed equally. Legend has it that it was developed in Munich during the revolution of 1918. The communistic supporters of the republic of councils—called 'Russ'n' in the vernacular—met preferably in Munich's 'Mattäser-Keller'. It might be that the revolutionists wanted to keep a clear head or that the wheat beer was running short. In any case, it was there that 'Russ'n Mass' was supposed to be mixed for the first time.

The mixture of beer and cola, often called 'Diesel', is also popular with many consumers. The name developed due to the color of the mixture that reminds one of 'fuel'. In the French-speaking part of Swiss it is called 'Mazout'.

With these extended offers and new ideas, new priorities can be set to the range of products in the beverage industry. Many of the following mixed drinks or their names are only known regionally, but are shown in brief to demonstrate the diversity.

- *Altbierbowle (Altbier punch)*. Altbier, fruit syrup and fruits. Traditionally, raspberries and strawberries have been used, but in many places simply mixed canned fruits are used. Instead of syrup, tinned strawberries together with the juice are also used.

- *Alt-Schuss (Alt-lacing)*. Altbier with a 'lacing' Malzbier (malt beer). In some regions a mixture of Altbier and wheat beer is also called that, although it is also is called 'Krefelder' elsewhere. In some pubs in Munster 'Alt-Schuss' relates to a mixture of Altbier and raspberry syrup or fruit liqueur. In Eastern Westphalia, a mixture of Altbier and cola is also called Alt-Schuss.

- *Berliner Weisse.* A light draught beer and not a beer-based mixed drink. Many use these terms equally because it is nearly always served as Weisse with an added shot of raspberry or woodruff syrup.

- *Bismarck.* Equal fractions of a dark beer (preferably Guinness) and sparkling wine. It is mixed in a big glass a great deal of foam can develop during mixing.

- *Bockbierbowle (Bock punch).* Similar to Altbier punch. Consists of Bock, fruit syrup and fruits. Traditionally, raspberries and strawberries are used, but in many places simply mixed canned fruits are used.

- *Clara.* In Spain, this is diluted light-colored beer with the lemonade Gaseosa, comparable to Radler.

- *Colabier (cola-beer).* A mixture of beer and cola, regionally called Colabier (cola-beer), Moorwasser (swamp water) or Krefelder (often with Altbier; in Emsland), Schmutz (dirt) or Schmutiges (dirty) (in Eastern Westphalia), Neger (negro) (in Bavaria and Tirol), Dreckiges Bier (dirty beer), Gestreiftes (striped) (in Wetterau), Gespritztes (sprayed) (Saarland), Drecksack (scumbag) (mixture of cola and Kölsch in the Rhineland), Schussbier (shot beer), or Schweinebier (pig's beer). The amount of cola varies regionally and from pub to pub and goes from a small fraction of cola up to a 50:50 mixture.

- *Dr Pepper.* Beer with amaretto, depending on preference in a ratio from 3:1 to 5:1. It is named after the American lemonade Dr Pepper as is very similar in taste.

- *Flieger (flyer).* A mixture of Pils and Emsgold, known in the area around Rheda-Wiedenbrück in the southern district Gutersloh (Westphalia). Emsgold is a redish fizzy powder of the beverage company Paul Berlage in Wiedenbrück. Comparable with the drink Tango known in the neighboring Munsterland, which is mixed by using the red fizzy powder Regina.

- *Wheat beer mixtures.* Russ (see above) and also banana, cherry, cranberry or peach wheat beer. These consist of wheat beer and the corresponding fruit juice. The mixture with banana juice is often called 'weiba'. Colaweizen (cola wheat beer), Colahefe (cola yeast), neger (negro), Cab (cola and beer) and Mohren (blackmoor) consist of equal fractions of wheat beer and cola. The correct pouring out order in northern Germany is first cola then the wheat beer, otherwise the drink will froth. In Southern Germany, as well as in Salzburger Flachgau and bordering districts, the order is the other way round, as a head is favored and the cola is mixed better instead of remaining in the lower level of the wheat beer glass; furthermore, less cola is generally used (about three-quarters wheat to one-quarter cola, or even less cola). In addition, the very favorable sweet head on top of the wheat beer develops here.

11.2
Ingredients and Mixing Formulations

Beer-based mixed drink are generally composed of the following constituents:

- Beer.
- Water.
- Sweetening.
- Food acids (citric acid).
- Antioxidants.
- Flavor (base note and possible top note).
- coloring/caramel color.
- Turbidity constituent.

The different constituents will now be explained in more detail.

11.2.1
Constituent Beer

Beer has a contribution of 50% in 'classical' beer-based mixed drinks. Especially in Germany, beer is brewed after the 'Reinheitsgebot' (Bavarian Purity Law). Thus, there is the regulation for this fraction of the drink that no other ingredients are allowed to be used except water, barley and/or wheat malt, and hops. No soluble adsorbents are allowed to be used for stabilization and polishing.

The overall impression of the beer-based mixed drink should be harmonic. The basic beer component already has a significant influence on the taste, color, alcoholic content and physico-chemical stability of the beverage. Accordingly, the soft drink component in the formulation needs to be adjusted individually and specifically to the beer used. This requires that the beer is produced with a constant quality. Thus, there must not be any serious fluctuation in the beer parameters. The following beer parameters need hence to be check at regular intervals:

- Color.
- Original extract.
- pH.
- Flavor profile.
- Bitter units.
- Constituents positive for foam formation.

Most parameters belong to standard analysis. The flavor profile should be well documented for the beer used since, for example, a change in fermentation, yeast or even storage phases due to seasonal capacity load can result in a fluctuation in the quality of the product.

11.2.2
Water Quality

The water used for the soft drink production needs be of drinking water quality, and accordingly be free of pathogens and low on germs and comply with the legal regulations. Furthermore, the water should be as low as possible on oxygen. Otherwise, in addition to the aging reactions of beer, oxidation processes take place on the added flavors and a loss or change occurs in the flavor profile.

Likewise, demineralization of too hard water is important as a neutralization of the food acid can take place. Thus, about 10°dH carbonate hardness neutralizes 228 mg/l citric acid. This can be compensated for with adequate adjustments of the formulation [1].

11.2.3
Sweetening

Often a sweetening is necessary to round off the taste of the ready drink. Sugar (saccharose and invert sugar, respectively) give the beverage not only sweetness, but also palate. However, sugar has a microbiological weakness and can be attacked by yeast. In this case the ready drinks need to be pasteurized. The necessary procedure for preservation, however, needs be adjusted with regard to the desired packaging.

Artificial sweeteners display a microbiologically lower risk. However, due to their natural flavor, they can have an influence on the overall sensory impression. In particular, the development of the sweetness perception which clearly differs from sugar, the lack of palate and possible bitter aftertaste needs to be contrasted with the aforementioned advantages. Nowadays, very good taste quality can be achieved using a mixed sweetening. Attention should be paid here that the legally specified upper limit value is not exceeded. If sweeteners are used, only a low concentration of fermentable sugar should be brought from the beer into the beverage from a microbiological point of view. In this case a high attenuation limit should be aimed for in the beer used [2].

11.2.3.1 Sweetening Agents
Sweetening agents play a central role in soft drinks and thus in beer-mixed drinks for the sweetness and palate of the product. They are thus characterized in the following in more detail.

11.2.3.1.1 **Sugar** The commonly used sugar, saccharose, is a disaccharide and is composed of glucose and fructose. Saccharose is hydrolyzed by acidic or enzymatic reactions. If the two monosaccharides are present in equal parts it is called invert sugar (invertose) [3]. A solution consisting mainly of glucose is called glucose syrup, iso-glucose, corn syrup or starch syrup.

The following products are used in crystalline form, where they need to be solubilized prior to addition or as prefabricated syrup:

- Liquid sugar (saccharose).
- Invert sugar syrup (contain at least 50% inverse sugar according to the sugar regulation).
- Invert liquid sugar (invert sugar not the predominant part).
- Glucose syrup.

Inverse sugar syrup has proven of value in the beverage industry. It is composed of one-third saccharose and two-thirds invert sugar (ratio 33 : 67). The content of the dry substance is standardized at 72.7%. The specifications of invert sugar are given as 72.7/67 in this case [3].

11.2.3.1.2 Artificial Sweeteners Artificial sweeteners possess a higher sweetening strength than sugar. Thus, they can be applied in lower concentrations. In addition, they have no or little calories and are neutral to the mouth flora. Table 11.1 shows artificial sweeteners usable in beer-based mixed drinks and their sweetening strength factors as compared to sugar.

11.2.3.1.3 Multiple Sweetening The issue of the synergetic potential of the substances has already been addressed for the several artificial sweeteners. Artificial sweeteners often show a metallic, bitter after taste and the dynamic characteristics do not concur with those of sugars. The palate is also lower. Mixtures of several artificial sweeteners are used to achieve a higher similarity to sugars and additionally a building block is added (e.g. from the area of fruit sweeteners). The adjustment of several artificial sweetener results in a low-calorie sweetening similar to that of sugar, which is especially advantageous for beer-based mixed drinks.

11.2.4
Food Acids

The addition of fruit acids is important for a balanced sweet/sour ratio. In contrast to beer, this leads to a decrease of the pH in the ready beer-based mixed drink. This needs to be considered in view of precipitations especially caused by the

Table 11.1 Sweetening strength of individual sweeteners in comparison with sugar [4].

Kind of artificial sweetener	Sweetening strength
Acesulfame (E950)	130–200
Aspartame (E951)	200
Aspartame acesulfame salt	350
Cyclamate (E952)	30–50
Neohesperidine DC (E959)	400–600
Saccharin (E954)	300–500
Sucralose (E955)	600

protein fraction of the beer [5]. Care should be taken during mixing that the pH gradient is not too great during contact with beer.

A rapid fall in pH does not only affect the beer constituents, but also the acid-sensitive components of the soft drink recipe. Thus, flavorings and colorings as well as sweetening can be affected, and it can lead to a lightening, flavor loss or even flocculation.

These problems should already be considered during the development of the mixing plant and the mixing procedure. If the acids can only be pumped into the line during an in-line mixing at a certain moment wild fluctuations can be expected for the pH value. Care should be taken that the acid is either previously diluted in the syrup preparation or added constantly in low quantities during the mixing phase. Another possibility would be the late addition to the last rinsing water fed into the mixing vessel. However, the precision of the dosage should not be affected [1].

11.2.5
Flavor and Juices

Flavorings can already influence the taste at low concentrations. It is thus essential to turn one's attention to the added flavors or flavor mixtures to obtain a harmonic texture. They can be used as a top note or as subliminal taste contribution. Furthermore, constituents in flavors and essences can lead to turbidity or precipitation depending on the beer used. This needs to be tested in advanced during development of the formulation and possibly adjusted [5].

If turbid juices are used (e.g. of the 'lemon' variant) it needs to be considered that precipitations can also occur due to reactions with beer constituents. In addition, undesirable clearing of the drink can arise and dregs can form. Moreover, turbid juices represent a microbiological risk, since yeasts can grow in them very well. Thus, one should switch to clear and polished juices [2].

11.3
Quality Control of Beer-Based Mixed Drinks

11.3.1
Wet Chemical Analysis

Mitteleuropäische Brautechnische Analysenkommision (MEBAK) suggests a range of analysis to judge the quality of beer-based mixed drinks [6]. Analogous to the analysis of beer, the following are determined:

- Original extract.
- Apparent extract.
- Alcohol.

However, it is not considered here that part of the extract is introduced via the lemonade fraction, which can contain sugar, artificial sweetener or multiple sweetening. Thus, in a beer-based mixed drink sweetened with sugar the lemonade fraction is included in taxation, while this is not the case for a lemonade sweetened with artificial sweetener.

Analogous to the analysis of beer, the following are also measured:

- Color.
- pH.
- Turbidity.
- CO_2.

All the specific analyses for the soft drink fraction, such as sugar determination, detection of artificial sweeteners and preservatives, can be found in the method compiled by MEBAK.

11.3.2
Sensory Assessment of Beer-Based Mixed Drinks

Besides the physico-chemical investigation, sensory assessments play also a major role in beer-based mixed drinks. Generally, it is suggested to take to following survey criteria into account [6].

- Harmony between the beer and fruity taste.
- Flavor impression.
- Sweet/sour/bitterness ratio.
- Texture.
- Off-flavor.
- Foam formation and foam stability (beer dependent!).

There is no uniform official scheme for pure sensory assessment and relevant rating attributes need to be worked out for the beverage-type-specific profile analysis. MEBAK provides the assessment procedure shown in Table 11.2 for an 'Apple Radler' (apple Shandy), for example. It needs to be taken into account that this may result in a subjective presentation of the total character which does not necessarily comply with the consumer-oriented profile. It is advisable to have these kinds of sensory assessments conducted by a professional institute in the form of a consumer or even expert panel [1]. The result can then be presented as a 'spider-web' diagram. This allows a comparable illustration to clearly show changes in quality.

11.3.3
Assessment by the Deutsche Landwirtschafts-Gesellschaft (DLG) (German Agricultural Society)

At present, the German Agricultural Society approves only beer-based mixed drinks with lemonade and cola (both normal and low in calories) for assessment (see Figure 11.1). Other mixed beverages are not approved. The samples are examined with regard to the following properties:

Table 11.2 Attributes for a profile assessment [6].

Properties	1	2	3	4	5
Refreshing	very				not at all
Sweetness	too much				too little
Carbonic acid	too high				too low
Thirst-quenching	very good				not at all
Liking	very good				not at all
Drinking enjoyment	great				small
Pleasantness to the taste	very				to heavy
Complete in taste	very complete				too little
Lightness	too light				too heavy
Bitter impressions	too bitter				too little
Acids	too sour				too little
Beery	very beery				not at all
Taste of lemonade	very				not at all
Apple flavor	too much				too little
Harmonic	very harmonic				not at all
Overall impression	very good				bad

- Beer analysis:
  - original extract.
  - alcohol content.
  - apparent extract.
  - apparent degree of fermentation.
  - color.
  - pH value.
  - CO_2 concentration.
  - organic acids.
- Turbidity (less than 2 European Brewery Convention units).
- Biological stability (free from viable organisms).
- Sensory characteristics.
- Flavor stability.

In the frame of the sensory analysis, smell, taste, carbonation taste and harmony are assessed on a five-point scheme. If one aspect is given 3 points or less, a description of the off-flavor is necessary.

An over-flavored impression can be present for odor and taste. This assessment is accompanied with the evaluation of the harmony. Hence, the beverage is usually perceived as unbalanced. A deficiency in flavor can also be criticized in the context of harmony. An unbalanced bitterness leads to a disharmonic overall impression of the mixed beverage. This shows how important it is to adjust the soft drink fraction to the beer, which in turn implies a reproducible quality of the beer [1].

DLG-Qualitätswettbewerb für Bier
Prüfbeleg für Bier-Mischgetränke

Ort:	Journal-Nr. Probe A:
Datum:	Journal-Nr. Probe B:
Prüfer:	Prüf-Nr. DLG:

Bei der Vergabe von 3 und weniger Punkten in den Prüfmerkmalen Geruch, Geschmack und Harmonie muß eine Produkteigenschaft bzw. ein Fehler genannt werden.

Prüfmerkmale	Erreichbare Punkte	Beschreibung		Erreichte Punkte	
				Linke Probe ( )	Rechte Probe ( )
				Nr.	Nr. ....
Geruch	5	rein	aromatisch, fruchtig, reintönig, typisch		
	4	noch rein			
	3	leichte Geruchsfehler	überaromatisiert, oxidiert, dumpf	....	....
	2	deutliche Geruchsfehler	Sonstiges:		
	1	starke Geruchsfehler	....		
Reinheit des Geschmacks	5	rein	aromatisch, fruchtig, reintönig, typisch, angenehm sauer, süß		
	4	noch rein			
	3	leichte Geschmacksfehler	überaromatisiert, oxidiert, dumpf, aufdringlich süß, übersäuert, bitter	....	....
	2	deutliche Geschmacksfehler	Sonstiges:		
	1	starke Geschmacksfehler	....		
Rezenz	5	angenehm rezent			
	4	rezent			
	3	wenig rezent		....	....
	2	schal			
	1	sehr schal			
Harmonie	5	volle Harmonie			
	4	sehr harmonisch			
	3	harmonisch	Unausgewogenes süß-sauer Verhalten, nicht aromatisiert, überaromatisiert, unausgewogene Bittere	....	....
	2	noch harmonisch	Sonstiges:		
	1	wenig harmonisch	....		

© 2000 Deutsche Landwirtschaftsgesellschaft e.V. (DLG), Eschborner Landstraße 122, D-60489 Frankfurt am Main

Figure 11.1 Assessment scheme for beer-based mixed drinks according to the Germany Agricultural Society, Deutsche Landwirtschaftsgesellschaft (DLG).

11.3.4
Off-flavors

Beer-based mixed drinks are generally composed of two independent kinds of beverages. During an assessment it thus needs to be considered that the single components are evaluated without any flaws. In addition to the classical off-flavors in beer, such as diacetyl, dimethylsulfide, lightstruck and oxidation flavors, the off-flavors also need to be known for the soft drink component. Flavor, fruit flavor, sweet/sour ratio, body, flavor development and off-flavor need to be considered. The most important attributes for the evaluation of the non-alcoholic components are:

- Typical in taste and color.
- Intrusive or missing flavor.
- Too much or missing sweetness (palate).
- Too much or missing acid (harmony with beer/stability).
- Lightstruck flavor/oxidation flavor.
- Other off-flavor notes.

Serious off-flavors for soft drinks are listed in Table 11.3. The corresponding reference substances are also isted. Tasters can be trained using these substances

Table 11.3 Presentation of different off-flavors and their reference substances [6].

Chemical name of the odor/taste substances	Description of the odor/taste	Possible causes
Citric acid	sour	malfunction during addition
Acetaldehyde	unripe apple, pungent, fruity	carbonic acid, fermentation onset of the beverage
Iron	metallic, rusty water, ink-like, iron-like, blood like	production water, contact with iron surfaces
Hydrosulfide	rotten, rotten eggs, like fecal, sulfurous, like a stink bomb	carbonic acid, hydrogen sulfide-forming microorganisms
Ethyl mercaptan	Drain link odor, rain gutter drain like, leek-like, like rotten vegetables	carbonic acid
Acetic acid	vinegar-pungent, sour, fermented, pungent	beginning fermentation of the beverage
Ethyl fenchol	earthy, damp soil, humus-like	production water, microorganisms
2,6-dichlorophenol	pharmacy/medicine taste, like dentist, phenolic, like hospital	
Trichloroanisole	cork note, moldy, musty, like a damp basement	production water, detergent residues, bottle cleaning machine

and thus be sensitized for the occurrence of these flaws. Furthermore, possible causes are mentioned for the occurrence of off-flavor notes. Hence, it is possible to evaluate the quality of the production process via the flavor profile.

In order to judge the deterioration stability of beer-based mixed drinks, methods are used either according to the DLG or those of other institutes and companies to simulate the changes during distribution to the costumers.

11.4
Microbiology of Beer-Based Mixed Drinks

Beer-based mixed drinks need to be examined more extensively than beer. Microorganisms that do not present a threat in beer alone can be of importance in this case. Since the production consists primarily of addition and mixing, and takes place in the syrup area, pressure tank and filling area, it is particularly important to look out for secondary infections here.

Yeasts that have no opportunity to grow in the nearly end-fermented beer can find very good growth medium in the sugar-containing soft drink fraction and also in the finished mixed drink. An excellent medium consisting of carbohydrates and amino acids is available. Thus, a pasteurization step is needed [2]. In badly cleaned niches, lactic acid bacteria can adapt to the new medium possessing a lower pH than beer and this presents a great risk. In this case it is important to maintain hygiene and cleanness [2]. Due to carbonization, aerobic microorganisms have hardly any chance to grow even in beer-based mixed drinks.

Different microbiological risk potentials arise in the ready drink depending on the beverage formulation [2]:

- *Beer + sugar + acid + essence*. Due to the sugar, very high sensitivity towards yeast, especially at pH above 3.5. Thus, pasteurization!

- *Beer + fruit juice + sugar/artificial sweetener*. Fruit juice is an ideal growth medium for microorganisms. Thus, the ready drink needs to undergo sterilization.

- *Beer + artificial sweetener + acid + essence*. Without any fermentable carbohydrates only lactic acid bacteria play a role in this medium. Thus, keep appropriate high hygiene standards.

Generally, a microbiological control step is advisable for the production of beer-based mixed drinks. Attention should be paid to microbiologicals belonging to the following categories [7]:

- Obligate (e.g. *Lactobacillus, Pediococcus, Megasphaera*).
- Potential (e.g. Lactobacillus, *Lactococcus, Leuconostoc*).
- Indirect (e.g. *Enterobacter*).
- Indicator organisms (e.g. *Klebsiella, Acetobacter, Gluconobacter*).

11.5
Preservation of the Final Beverage

11.5.1
Use of Antioxidants

Addition of ascorbic acid (vitamin C) or tocopherol as antioxidant is possible for preservation. Primarily, they are used to protect the color. Antioxidants can be used according to legal regulations.

A certain concentration of vitamin C gets lost even during the storage phase due to oxygen inside the packaging and dissolved in the product. This loss needs be considered in the formulation. Higher additions are thus necessary, which need to be determined in advanced in stability and load tests. The *quantum satis* ('the amount which is needed') is the necessary dose to achieve a detectable antioxidative effect [1].

11.5.2
Use of Preserving Agents

The corresponding statutory threshold values for permitted acids are in force for the preservation of food stuffs. In particular, the preservation of beer is prohibited in Germany; it is only allowed to preserve the soft drink according to the German regulations on additives (Zusatzstoff-Zulassungsverordnung) and the permitted threshold value must not be exceeded.

Effective conservation of sugar-sweetened beer-based mixed drinks with preserving agents may not be sufficient and should be verified accurately. The conservation of the raw material is advisable since the microbiological stability can just be ensured during fractional removal in the bottling factory.

The most effective method for stabilizing the final beverage is generally pasteurization of the filled bottles. It needs to be taken into account that with high sugar concentrations, higher pasteurization unit (PU) values need to be applied than common in beer production. Due to the increased thermal load, the CO_2 concentration of the beverage needs to be adjusted because of high pressure in the bottles. This can also have an influence on the filling level [5]. It might be necessary to change the parameters usually used during beer filling.

11.5.3
Thermal Processes – Pasteurization

The necessary PUs for beer-based mixed drinks can be taken from Table 11.4 [8]. The PUs are calculated depending on the beverage with following equations:

- For beer [8]: PU = time [min] $\times\ 1.393^{(\text{temp}[°C] - 80[°C])}$.
- For non alcoholic, CO_2-containing beverages [9]:
  PU = time [min] $\times\ 1.393^{(\text{temp}[°C] - 80[°C])}$.

Table 11.4 Sterilization and pasteurization processes in the beverage industry: beer and beer-based mixed [8].

Beverage/product/medium	Process	Temperature (°C)	Incubation time	Comments
Beer	flash pasteurization, (flow pasteurization with heat storage)		vegetative cells, beer pests	ascospores from wild yeast can survive
Beer	hot filling (high-pressure filler)		vegetative cells, beer pests	cooling down necessary; ascospores from wild yeast can survive
Beer	pasteurization ('pasteurization' in bottles or cans)		vegetative cells, beer pests	cooling down necessary; ascospores from wild yeast can survive
Beer-based mixed drinks, malt beer	flash pasteurization, (flow pasteurization with heat storage)		vegetative cells, beverage pests	stability limit for ascospores of yeasts
Beer-based mixed drinks, malt beer	hot filling (high-pressure filler)		vegetative cells, beverage pests	cooling down advantageous; stability limit for ascospores of yeasts
Beer-based mixed drinks, malt beer	pasteurization ('pasteurization' in bottles or cans)	(core temperature)	vegetative cells, beverage pests	cooling down advantageous; stability limit for ascospores of yeasts (below 68 °C)

- For non-CO_2-containing beverages (fruit juices, vegetable juices and other fruit-containing products) [8]: PU = time [min] × $1.2586^{(temp[°C]-80[°C])}$. ...

The temperature/time frame are coordinated and differ for the respective beverages, such as beer, beer-based mixed drinks or soft drinks. The determined and coordinated temperature/time frame can be taken from Figure 11.2.

11.6
Technological Aspects for the Production of Beer-Based Mixed Drinks

11.6.1
Mixing

Homogenization of the syrup as well as of the beer-based mixed drink is very important for a consistent quality. Contact times of the beer with components

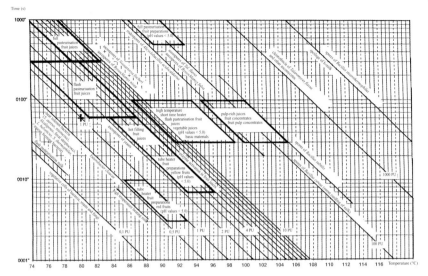

Figure 11.2 Effect of temperature and time on PUs [8].

causing physicochemical disadvantages need to be minimized if present in high concentrations.

In charge processes it is recommended to put in the syrup for the drink first and then add the beer. A homogenous mixing should be aimed for during syrup production to avoid local pH gradients. This can be achieved by pumping or CO_2 pressure blows.

As mentioned above, a well-established order of the components needs to be chosen during in-line mixing in order to avoid pH shocks for beer constituents. Additionally, it should be pointed out that even a too high pH value can cause a negative effect [1].

11.6.2
Filtration

Since beer-based mixed drinks are prone to physicochemical reactions and associated turbidities, they should be stabilized from the protein side (bentonite, silica) as well as the tannin side (polyvinylpyrrolidone). In addition, a sharp filtration is necessary so that no yeasts get into the sugar-sweetened ready drinks in particular [5]. The protein concentration of the beer has a direct influence on possible turbidity after mixing with the basic components or flavors. Thus, clear physicochemical boundaries are set [5] and not all mixing options can simply be carried out for every beer.

Since 'Radler' and cola beer are being displaced more and more by exotic mixing variations, the classical polyphenol–protein turbidity is becoming more and more relevant. In this case, particular care should be taken which fraction (e.g. containing phenolic substances) is added with the soft drink. However, this can already be investigated in stability tests carried out by the supplier during formulation preparation.

On the other hand, beer has also a potential to form these kinds of turbidities in the final beverage. The brewery needs to use sharp filtered and stabilized beers to be able to use the variety of new beverage types [1].

11.6.3
Filling

The filling of beer-based mixed drinks in glass bottles using the machines commonly found in breweries does generally not present a problem. The products are well protected from oxidation by pre-evacuation and pre-loading of the bottles.

The modified microbiological sensitivity, in particular for those sweetened with sugar, is again important in the context of beer-based mixed drinks. Thus, in order to avoid secondary infections [2] attention should especially be paid towards hygiene in the bottle washer and in the filling area. An aseptic filling is safest if polyethylene terephthalate (PET) bottles are used.

11.6.4
Influence of the Packaging

It is already know from beer that cans and brown bottles guard best against the effects of light. Light-induced taste can occur in green glass. This is also a influencing factor in beer-based mixed drink. Since mixed beverages increasingly represent fashionable beverages, the mixtures are likely to be filled in white glass bottles. Light-induced oxidation reactions are extremely high in white glass; however, these can be suppressed by covering the bottles. This is reflected in higher costs.

The increased sensitivity towards oxidation is increased in PET bottles due to the migration of oxygen. Even in glass bottles the flavors of the soft drink are strongly affected besides the classical oxidation reactions of the beer constituents. The whole flavor looses intensity. A know reaction is, for example, the breakdown of citral to cresol which causes an off-flavor. This needs to be taken into account for mixed beverages with a citrus note such as 'Radler'. Apart from the loss of flavor, oxidative-induced lightening of the color is a problem which, as already mentioned, can be avoided by the addition of antioxidants.

It is important to know the influence of the packaging on the product quality, since changing packing (e.g. from brown to white glass) can cause unwanted quality losses.

In addition to the mentioned aspects, it needs to be considered that new mixture compositions, in particular for exotic variants, can cause problems. Furthermore, non-optimally adjusted mixture systems can lead to problems.

In order to avoid undesirable precipitation reactions the formulation needs to be individually checked with respect to mixing order and hot spots due to high concentrations of single constituents [1].

11.7
Technical Equipment for the Production of Beer-Based Mixed Drinks

Important steps during the production of beer-based mixed drinks are [1]:

- Beer production (in Germany according to the Reinheitsgebot).
- Degassing of water for the soft dinks.
- Possible pre-dissolving of sugars.
- Mixing of the components: (i) preparation of the syrup for the soft drink fraction and the mixing of beer in a charge process or in-line mixing, or (ii) continuous in-line mixing;
- Carbonization/impregnation.
- Sterilization.
- Filling.
- Pasteurization (tunnel Pasteur).

References

1 Franz, O. (2006) Biermischgetränke–technologie, in *Praxishandbuch der Brauerei* (ed. K.-U. Heyse), Behr's, Hamburg, pp. 1–20.
2 Tretzel, J. (1997) Technology and microbiology of beer mix beverages. *Brewing and Beverage Industry International*, 2, 91–6.
3 (2000) *Handbuch Erfrischungsgetränke*, Südzucker AG, Mannheim.
4 http://www.suessstoff-verband.de/verwendung/suesskraft_und_dosierung/ (accessed 01. 12. 2007).
5 Schwarzenberger, M.J. (2002) Die Abfüllung von Biermischgetränken. *Brauwelt*, 18/19, 647–9.
6 Pfenninger, H.B. (2002) *Brautechnische Analysenmethoden, Band II*, Selbstverlag der MEBAK, Freising-Weihenstephan.
7 Back, W. (1994) *Farbatlas und Handbuch der Getränkebiologie, Teil 1. Kultivierung/ Methoden, Brauerei, Winzerei*, Fachverlag Hans Carl, Nürnberg.
8 Back, W. (2000) *Farbatlas und Handbuch der Getränkebiologie, Teil 2. Fruchtsaft- und Limonadenbetriebe, Wasser, Betriebshygiene, Milch und Molkereiprodukte, Begleitorganismen der Getränkeindustrie*, Fachverlag Hans Carl, Nürnberg.
9 Back, W. (ed.) (2008) Haltbarmachung von Getränken, *Mikrobiologie der Lebensmittel*, Ch. 8, Behr's Verlag, Hamburg, pp. 275–319.

12
Filling
Susanne Blüml

12.1
Choice of Packaging

In addition to marketing aspects, the properties of the packaging material, in conjunction with the product-specific idiosyncrasies of the beer concerned, have to be given due consideration by the beer bottler or canner when choosing the appropriate container.

For filling beer, there are four main categories of packaging in use worldwide:

- Glass bottles.
- Cans made of aluminum or tinplate.
- Plastic bottles made of polyethylene terephthalate (PET) and polyethylene naphthalate (PEN).
- Kegs.

The following factors are important for assessing whether and to what extent these packaging materials are suitable for packaging beer: light-proofing characteristics, barrier properties for preventing the escape of CO_2 or the entry of oxygen, and inertness in terms of mass transfer between the packaging material and the product, plus ability to withstand mechanical stresses and breakage. In the case of a returnable concept, it will be necessary to clarify whether the container can be cleaned without leaving behind any residues. In addition, although more from a marketing viewpoint, the weight and the reclosability of the container are also important.

12.1.1
Glass Bottles

By reason of the wall thickness (averaging 3–4 mm for returnable bottles and 1.5 mm for non-returnables) and a weight of approximately 140 g for non-returnable beer bottles (0.33 l), glass is the heaviest of the packaging materials presented here in terms of handling and transport (Figure 12.1). This is of

Handbook of Brewing: Processes, Technology, Markets. Edited by H. M. Eßlinger
Copyright © 2009 WILEY-VCH Verlag GmbH & Co. KGaA, Weinheim
ISBN: 978-3-527-31674-8

Figure 12.1 Glass bottle.

relevance when (i) the topic of convenience for consumers is addressed and (ii) transportation costs have to be calculated.

Glass bottles excel in terms of high mechanical strength when exposed to mechanical stresses and axial influences. This has to be offset against their high susceptibility to breakage, due to the physical brittleness of glass as a material, which may have adverse effects both during transport and at the consumer's home. The Special Technical Conditions of Delivery and Purchase of the German Brewers' Confederation lay down limit values for the quality of glass bottles, specifying a lower tolerance limit for a random sample. For internal pressure resistance, this specification demands a minimum value of 10 bar for non-returnable bottles and not less than 12 bar for overseas-export non-returnables (with a volume of more than 0.6 l). The duty internal pressure resistance for returnable bottles is specified as not less than 10 bar. In addition, the specification defines an impact resistance of greater than 35 ips for non-returnable bottles, of greater than 50 ips for overseas-export bottles and of greater than 60 ips for returnable bottles. Thanks to its high mechanical strength, glass provides excellent protection against the diffusion of gases – oxygen and CO_2 – from inside to outside and vice versa. This barrier property is particularly important in the case of beer, by reason of its tendency to oxidation reactions by some of its constituents following a lengthy period of storage. Since glass is absolutely inert in its behavior and does not absorb any constituents of the contents, nor release any substances into the bottle's contents, interaction between the glass and the product can be safely ruled out.

However, protection against light is a different matter. On the one hand, the transparency of the glass is a desirable aspect for the marketing people and the consumer; on the other hand, it is precisely this aspect that causes the risk of a taste change in the beer caused by possible exposure to UV radiation, also known as 'lightstruck flavor'. One counter-measure employed here is the customary brown coloration of the bottles, which in the UV wavelength range of 330–380 nm produces a transmission of less than 10% in comparison to white glass, which exhibits a transmission of up to 90% in the same waveband. By reason of its high mechanical strength and resistance to chemicals, glass is ideally suited for returnable applications. Mineral-water and beer bottles in a returnables circuit, for instance, achieve up to 40 turnarounds.

For certain markets, the thermal and dimensional stability of glass is a further significant factor; since the filled bottles can be pasteurized at temperatures above 60 °C, avoiding any substantial and abrupt temperature changes, thus enabling shelf-life to be extended.

12.1.2
Cans

Beverage cans (Figure 12.2) are made of aluminum or tinplate, and excel in terms of their low empty weight of 14 g, for example, in the case of a 50-cl aluminum can and 30 g for its tinplate counterpart. The cans comprise two constituent parts – the can and the lid. For the can and its specification, the determinant parameters are the diameter, flange width, inner diameter, internal pressure stability, axial strength and porosity of the internal coating. For the lid, they are the

Figure 12.2 Can.

diameter, flange height and opening, bead depth, compound placing, internal pressure resistance, porosity of the internal coating, and opening function. Their resistance to axial stresses has been reduced by a downsized body wall thickness as can design has progressed. One specification (in conformity with the Special Technical Conditions of Delivery and Purchase issued by the German Brewers' Confederation) provides for a capability to withstand stress of 800–1000 N. The stability of the cans is assured by the container's internal pressure after being filled with carbonated beer. The can's pressure resistance during pasteurization and sterilization is stated as 6.2 bar. By reason of their unbreakability, cans are a very safe packaging route.

Beverage cans are basically a non-returnable package, so that the question of their cleanability need not be pursued. For the canning facility, the metal's thermal properties mean that the filled product is simple to pasteurize, which in turn means a long shelf-life (e.g. for export). The can's good barrier characteristics, permitting no measurable exchange between product and surroundings, are based on the seamed lid and the precise, practically gas-tight flange of the lid's seam. Possible migration phenomena from the container's wall into the beverage are, in view of the food-grade paint applied to the interior – and given an undamaged can – minimal, and within the limits laid down in the European Union's migration directive. The cans offer absolute protection against light, since no light whatsoever can penetrate through the material. The protective effect of the packaging material is particularly optimal for beer. One key factor that should be mentioned when dealing with cans as packages is the can's unprotected opening for drinking or pouring. In contrast to that of glass or plastic bottles, the can's closure is not protected against soiling. If the can's head area is dirty, this dirt may be passed into the beverage when pouring or drinking it.

One special feature involved when handling beverage cans is what are called widgets, particularly well-known on the English market. A widget is something that is inserted in the empty can; after the can has been filled, liquid nitrogen is admitted and the can is then seamed. The nitrogen evaporates immediately and creates an internal pressure inside the can, also penetrating into the widget. After the can has been opened, the overpressure escapes from the can, and the nitrogen flows out of the widget and into the beer. This creates the fine-pored foam for which British and Irish beers are famed.

12.1.3
Plastic Bottles

Since the mid-1980s, plastic has been accounting for a steadily increasing proportion of beverage packaging (Figure 12.3). When it comes to beer packaging, bottles made from the most widely used plastics PET and PEN still contribute only a comparatively small proportion – a fact attributable to certain properties of the material involved. In terms of PET bottle variants, a distinction is made between monolayer and multilayer bottles. While monolayer bottles consist of a single layer of PET, their multilayer counterparts are laminates comprising several layers of

Figure 12.3 Plastic bottle.

PET, with layers of polyamide or ethylene vinyl alcohol polymer (EVOH) inside. The monolayer category also includes the PET blends, which are processed with admixtures of PEN or polyamide. Moreover, in some cases scavenger materials are added to the PET bottles, enabling oxygen molecules to be intercepted and thus providing additional protection for the product. In this way, an oxygen barrier can thus be created.

A high degree of moldability, low weight, unbreakability and extensive flexibility with regard to the producible sizes (150 ml to 3 l) are the advantages of this material when packaging alternatives are being assessed.

The internal pressure resistance of plastic bottles depends on their shape, the process used for handling them and the material involved, and is in each case evidenced in burst tests – an internal pressure resistance of more than 7 bar (as required for filling carbonated beverages) is reached by the plastic bottles. Owing to the lower axial pressure resistance of plastic bottles in comparison to their glass counterparts (between 850 and 1000 N), appropriately dimensioned forces are required for handling these bottles under the filling valve and at the sealing head.

When describing the properties of plastics, in general, and of PET, in particular, the material's permeability in relation to gases is an important criterion when it comes to beverage filling. In this context, the gas permeation in the bottle's head space and in the liquid area must be taken into due account. The value of the oxygen permeation coefficient for PET at 23 °C and 0% relative humidity is 55–65 ml · 100 μm/(m² · day · bar) [1]. Through oxygen diffusing into the product, fatty acids, vitamins (A, C and E), amino acids and aromatic substances may be affected – a phenomenon manifested by discolorations, changes in the aroma or also by product spoilage. This is why oxygen permeability is regarded as the paramount criterion for assessing the packaging material. These characteristics of PET are also one reason why PET bottles have so far not found widespread acceptance among beer bottlers, since oxidative reactions may occur here that may damage the beer inside the bottle concerned.

The diffusion of CO_2 through PET is around 1.12 g/day in the case of a monolayer bottle, 0.5-l non-returnable, weighing 28 g, at an ambient temperature of 38 °C and a CO_2 pressure of 5 bar. The quality-impairing CO_2 loss can in individual cases be offset by appropriately higher carbonation. It should also be noted that the gas permeability of plastics is temperature-dependent, which means that when the bottles are stored in tropical temperatures the gas permeability will rise in comparison to storage in temperate climates.

In addition, interactions (migration) between the plastic and the product have to be considered, which are particularly relevant when the bottles are being used in a returnables system. In addition to the migration processes caused by the consumer (e.g. by using the empty returnable bottles for different substances), the primary factor here is the migration of beer constituents into the wall of the bottle. The inadequate lightproofing effect of PET bottles must also be mentioned, which in the beer-spoilage spectral range of 300–500 nm is comparatively slight. In comparison to a brown non-returnable glass bottle, a brown-colored plastic bottle with a wall thickness of 0.3 mm lets through about 3 times as much radiation.

PEN, by contrast, with barrier properties against oxygen that are 2–3 times higher and significantly better migration characteristics, constitutes an alternative for bottling beer that is being blocked only by the high price of the raw material, which is approximately 5–6 times as high as that of PET. In Scandinavian countries, returnable PEN containers are already in use for bottling beer.

12.1.4
Kegs

The keg (Figure 12.4) is employed for bulk use and open tapping in the catering trade and at events. This alternative to the wooden barrel and the aluminum cask is a metal cask that is always pressurized, and is filled and cleaned through an integrated fitting, and a riser connected to it (located inside the keg). The riser inside the keg extends from the fitting connection to a short distance above the bottom keg base, which permits the keg to be emptied almost completely. The dispensing head integrated at the fitting is used to connect the keg to the tapping system.

The fitting serves for filling and emptying the keg, with a flat-type or well-type fitting being used in most cases. In the flat-type fitting variant, only one valve is

Figure 12.4 Kegs.

provided for beer and pressurization gas; in the well-type fitting, two separate valves are integrated for beer and pressurization gas. A combi-fitting marries the features of both valve types; the dispensing head is pushed on as with the flat-type fitting, but there are two separate valves for beer and pressurization gas. The aluminum construction, with a synthetic resin or epoxy resin coating, renders the kegs absolutely impermeable to light and gas, offering optimum conditions for storing beer. In the combined-keg version, sheathed in polyurethane, advantages can be obtained in terms of handling and sturdiness.

12.2
Framework Conditions for Filling Beer

Before opting for a particular filling process, the product's characteristics have to be assessed plus the framework conditions required for the filling process involved. In the case of beer, these are:

- Influence of gases, particularly the oxygen and CO_2 content.
- Filling pressure.
- Temperature of the product.

By meticulously examining these variables and modifying any of them whenever necessary, the performance data can be optimized, both for lines already up and running and for those still in the design phase: the number of bottles filled per hour can be maximized, for example, and product quality during filling preserved to optimum effect.

12.2.1
Significance of the Gases

Gas is encountered at various points during filling: CO_2, for example, is a known constituent of various beverages and is, moreover, also used as a 'pressurization gas' for the ring bowl or as flushing gas for the containers prior to filling. Oxygen is part of the air inside the bottles being filled or also picked up by the beverage upstream of the filler infeed. Nitrogen is present in the liquid as an absorbed gas and is usually removed from the liquid during degassing. In specific cases, it can also be used for flushing or pressurization.

12.2.1.1 Oxygen Content

The oxygen content in the beverage after filling is an aspect of major importance for preserving the product's quality. Beer is the most oxygen-sensitive of all beverages. The oxygen pick-up in beer causes aging processes or a downsized content of antioxidants, so that the beer's taste stability may be reduced, its bitterness more pronounced, and cold and continuous cloudiness, as well as changes in color, may occur.

Oxygen is contained in the beverage in a dissolved state (Figure 12.5) and is also present as a proportion of the air in the bottle prior to filling. It is precisely this latter point that can be significantly influenced by selecting an appropriate filling process. Oxygen pick-up in the product can thus be considered at three points:

- Oxygen loading in the beer at the filler inlet – initial loading.
- Oxygen pick-up during the filling process.
- Oxygen quantity in the residual air content in the bottle's head space after filling.

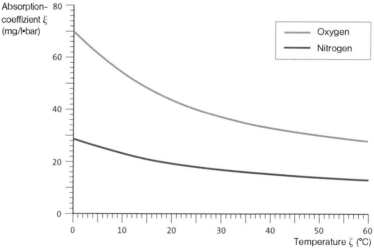

Figure 12.5 Solubility diagram (absorption coefficient oxygen and nitrogen) depending on temperature.

Oxygen pick-up before the filler inlet can, for example, be caused by air inclusions in pipes or tanks, or by defective seals. Oxygen pick-up by the beverage during filling is directly correlated to the purity of the CO_2 as pressurization gas in the ring bowl and residual air in the bottle prior to filling. The third factor contributing to oxygen pick-up, accounting for the main proportion of oxygen in the filled bottle, is the residual air in the bottle's head space. At first, the filling operation produces bubbles that rise from the beverage into the bottle's head space, thus displacing some of the air in this head space; nonetheless, the residual amount of oxygen in the bottle's head space accounts for around 75% of the oxygen present in the filled bottle.

Measures for reducing the influence of oxygen when bottling beer are, accordingly:

- Examining the product feed to the filler for unwanted oxygen pick-up.
- Reducing the proportion of oxygen in the bottle before filling by pre-evacuating the bottle or by treating it with flushing gas.
- Preventing air pick-up by using a long filling tube and thus providing a below-surface filling process.
- Using a high-pressure fobber before closing and thus displacing the air content in the bottle's head space by foaming beer.

These steps are common practice when planning a filling process for beer, with the use of a long filling tube mostly being rejected for output-related reasons (the time needed to raise the bottle underneath the filling valve reduces the filler's speed).

12.2.1.2 CO_2 Content

CO_2 is contained in numerous alcoholic and non-alcoholic beverages, and is water-soluble depending on temperature and pressure. Due to the natural fermentation processes involved, CO_2 is present in beer in a dissolved state (Figure 12.6). In a closed bottle, an equilibrium will occur after a certain time between the concentration in the liquid and the partial pressures in the gas phase. To calculate the equilibrium state of a gas, Henry's law must be used, which enables a gas's solubility at a defined pressure and a defined temperature to be calculated. The solubility of gases is expressed here in terms of the absorption coefficient. For beer, in accordance with these natural laws, a CO_2 equilibrium pressure of $2.26\,g/l \cdot bar$ can be calculated for a beer at a temperature of $10\,°C$ and a value of $1.65\,g/l \cdot bar$ at a temperature of $20\,°C$.

CO_2 must be prevented from coming out of solution during filling – by setting the filling pressure to approximately 1 bar higher than the saturation pressure at every point in the filling process. Nevertheless, CO_2 bubbles may be carried into the beverage during filling: CO_2 bubbles are formed, for example, by any gas bubbles present from beverage preparation, due to gas residues at the surfaces coming into contact with the liquid or at the beverages' surfaces [2].

Counter-measures here are gentle guidance of the product and also wetting the inner bottle wall prior to filling (rinsing). This results in avoidance of CO_2 losses

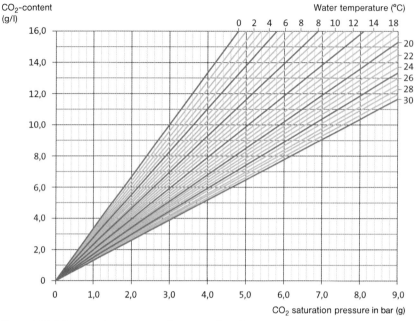

Figure 12.6 CO_2 saturation pressure of water as plotted against temperature.

in the beverage, and concomitant quality impairment, optimization of filling, settling and snifting times, and avoidance of product losses due to excessive foaming during the snifting function.

12.2.2
Filling Pressure

The filling pressure depends on the CO_2 equilibrium pressure. Setting a suitable filling pressure will prevent CO_2 from coming out of solution in the beverage during the filling process. If CO_2 comes out of solution after the bottles have been withdrawn from the filling valve, this will be manifested by the beverage's foaming, with concomitant filling losses. As already explained, the filling pressure is mostly set to 1 bar above the saturation pressure, which in practice, with no-filling-tube filling systems, leads to a preset value of 1–1.5 bar above the saturation pressure, while with long-tube filling systems, due to the below-surface filling process involved, values of 0.5–1 bar above the saturation pressure will suffice. In addition, the settling and snifting times must be brought into a correlation with the head space volume and the shape of the bottle, that is, bottles with a long, narrow neck (long-neck bottles) offer worse conditions during filling than bottles with a wide neck. This means that longer snifting times are required for bottles with a long neck and a small head space than those for bottles with a wide neck and a large head space.

12.2.3
Temperature

A low temperature goes hand in hand with improved CO_2 solubility. This creates a lower saturation pressure in the beverage, which in turn directly affects the filling pressure to be set. This means that shorter pressurization and snifting times can be selected, and there is less fobbing. In addition, a low filling temperature goes hand in hand with reduced gas consumption. Figure 12.6 provides information on the saturation pressure of the product depending on temperature and CO_2 content. So the temperature influences the treatment time and thus the size of the filler involved. From a filling viewpoint, the temperature selected should be as low as possible. Conversely, the energy costs are higher for cold filling, and water condensing on the outside of the bottle may cause difficulties with labeling and downstream packaging.

12.3
Process Steps When Filling Beer

When the above-mentioned framework conditions are taken into due account for designing the filling process concerned, appropriately matched process steps [3] will not only ensure accurate fill quantities but also maximally gentle filling of the product involved. 'Gentle' in this context signifies that the filling operation must not change or impair the product's characteristics at all (or only to a minimized extent). The salient requirements when filling beer are to:

- Preserve the saturation pressure in the case of carbonated beverages.
- Minimize oxygen pick-up in the product during filling.
- Maximize the filling accuracy.
- Meet the microbiological requirements for filling.
- Avoid drip formation after withdrawal of the bottles.

Depending on the type of filling system involved, the following filling steps (Figure 12.7) can be performed:

- Evacuation.
- Flushing with ring-bowl or pure gas.
- Pressurizing with ring-bowl or pure gas.
- Filling at one speed or at two speeds.
- Settling and snifting.

12.3.1
Evacuation

To reduce the amount of air and oxygen in the bottle before filling, in most cases double pre-evacuation of the bottle is performed when oxygen-sensitive products are involved, which of course includes beer (Figure 12.8). Since plastic containers

12 Filling

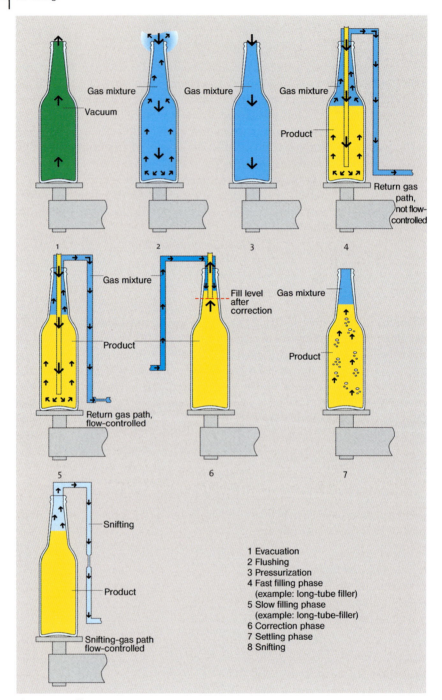

Figure 12.7 Filling steps.

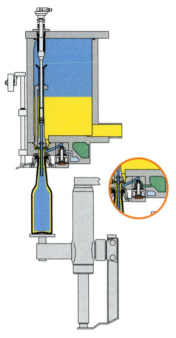

Figure 12.8 Evacuation.

do not possess sufficient stability to cope with an outside overpressure, this process can be used for glass bottles only.

By connecting the centering bell (and thus the pressed-on bottle) to a vacuum channel, the air present in the bottle will be extracted from it, depending on the level of the vacuum set. A brief intermediate flush with ring-bowl gas then re-establishes atmospheric pressure in the bottle. In the second evacuation step, the remaining proportion of air is removed from the bottle, except for a small percentage – the bottle is thus largely oxygen-free and offers optimum filling conditions for beer. In the field, a residual air content of 1–1.5% after the second evacuation, given a vacuum level of 100 mbar, is regarded as sufficient and practicable.

12.3.2
Flushing with Ring-Bowl or Pure Gas

In the case of plastic bottles, particularly, which are not suitable for a pre-evacuation step, the low-oxygen atmosphere is achieved by a flushing step for the bottle prior to pressurization and filling. Here, the bottle is lifted against the centering bell and the filling valve; depending on the filling system involved, it is flushed with ring-bowl or pure gas in either a pressed-on or non-pressed-on configuration.

Some systems offer a switch-over option between these two alternatives, with the user able to choose whether to flush from the ring bowl or from the separate pure gas channel. Flushing with ring-bowl gas is advisable here, since this will already produce very low oxygen values in the bottle. In particular, it is financial reasons (the high consumption of CO_2 when using pure gas) that counter-indicate flushing with pure gas.

12.3.2.1 Flushing with Ring-Bowl Gas

In the case of no-filling-tube filling systems, flushing is carried out using ring-bowl gas via the flushing and pressurization valve at the ring bowl. The gas is taken from the gas phase of the ring bowl and passed into the bottle. In the case of a long-tube filler, a separate channel and flushing valve establish the link to the ring bowl's gas phase.

In both cases, the ring-bowl gas flows through the opened pressurization and flushing valve into the bottle. The air present in the bottle is released into the surrounding atmosphere and replaced by the CO_2/air mixture from the ring bowl; the oxygen content in the bottle prior to filling is thus reduced.

12.3.2.2 Flushing with Pure Gas

For using pure gas to flush the containers, a filling system with a separate CO_2 supply or with a multi-chamber concept is required. Through the opened flushing valve and the vent tube (with the no-filling-tube variant) or via the filling tube and a special link to the CO_2 channel (with a long-tube filler), the pure CO_2 is passed into the bottle. The air present in the bottle is displaced to the outside atmosphere.

12.3.3
Pressurization

In order to create the same pressure conditions in the bottle as in the ring bowl, the pressed-on bottle is filled with ring-bowl gas until the pressure has been equalized between the bottle and the ring bowl. This step is required for carbonated beverages, which of course include beer.

In the case of a filling valve without a filling tube, the gas phase of the ring bowl is connected to the bottle via the gas needle in the vent tube; with electronic filling systems, this is done through what is called a pressurization valve. Owing to the pressure differential between the pressure in the ring bowl and in the bottle, the ring-bowl gas flows via the vent tube into the bottle until the pressures have been equalized. In the case of a long-tube filler, a filling valve is additionally opened at the filling tube, through which the pressure differential is equalized.

12.3.4
Filling

The filling valve opens and in a no-filling-tube filling system the product flows along the vent tube into the bottle (Figure 12.9). A small spreader and/or a deflec-

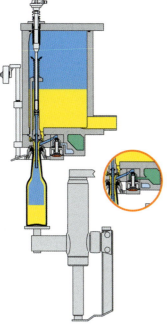

Figure 12.9 Filling phase in a tubeless system.

tor collar or a swirl insert guides the liquid onto the bottle's inside wall, with the liquid film flowing down along it and bubble formation thus being largely avoided. The return gas is displaced from the bottle via the vent tube into the ring bowl or into a separate channel. The end of filling in a no-filling-tube filling system is determined by the vent tube or a probe integrated into it. As soon as the liquid level reaches the vent tube, return gas can no longer flow into the ring bowl; in the probe-controlled system, the probe signals 'fill level reached' and the valve stem is closed.

In filling systems with a filling tube, the filling valve is opened and the product flows through the filling tube into the bottle. Here, the return gas is passed back from the bottle neck into a snifting channel via a snifting valve. In long-tube filling systems, the filling phase is controlled with regard to filling speed in such a way as to ensure that the product flows slowly into the bottle until the filling tube actually dips into the beverage, thus preventing beverage foaming. By switching yet another return gas valve into the circuit, return air exit from the bottle is then accelerated, thus creating a fast-filling phase. The slow-filling phase before reaching the pre-selected fill level ensures minimized foaming of the products and maximized fill-level accuracy. As soon as the fill level (which is measured by a probe) has been reached, the filling valve will close and product can no longer flow into the bottle.

In addition, a correction phase can be provided at no-filling-tube filling systems after filling has been completed, with CO_2 being injected into the bottle's neck so as to force the product that has risen into the vent tube back into the ring bowl.

12.3.5
Settling and Snifting

This is followed by the settling and snifting phase, which serves to lower the filling pressure so as to prevent bubbles rising in the beverage, and avoid the concomitant filling losses due to foaming and CO_2 coming out of solution. During the settling time, the bubbles carried in can rise in the beverage, so that they cannot lead to foaming during the snifting function. Depending on the requirements involved, snifting can also be implemented in several stages. After the bottle's head space has been snifted via the return-gas paths, there will again be atmospheric pressure in the bottle, which can be lowered and passed to the closer via the transfer starwheel.

12.4
Filling Systems for Beer

In the beverage industry, rotary fillers are the usual choice, able to handle up to 84,000 bottles an hour, for example, or up to 120,000 cans an hour. This kind of filler speed will of course always be directly dependent on the size and shape of the containers being handled, and on the nature and composition of the product being bottled/canned.

Assessment of a filler will be based on the features specific to the particular type of filler involved. These relate to:

- Actuation of the filling valves.
- Pressure situation in the ring bowl and in the container.
- Measurement of the fill quantity or level (i.e. definition of the end of filling).
- Number of control valves and the functions of the filling valve.
- Design of the filling valve.

The functions of the various types of filler are depicted with the aid of typical filling systems from the Krones company.

12.4.1
Mechanical Level-Controlled Filling Systems

Control of the individual filling functions by means of control cams or pins enables the bottles to be run through the filler carousel without electronic pulses. The bottle is fixed in place in the centering bell under lifting-cam control and pressed against the filling valve. Actuators located on the outside of the filler carousel cause the vacuum and the gas supply to open, and the bottle is pressurized up to the ring bowl's own pressure. As soon as the pressures in the ring bowl and the bottle are the same, the valve stem opens and the filling process begins. During filling, return gas flows back into the ring bowl through the vent tube. The end of the filling step is reached when the liquid level in the bottle has risen as far as the vent tube – thus stopping return gas from flowing out – and the filling valve is closed by another actuator. The subsequent steps in the filling system, too, like snifting and lowering of the bottle, are controlled by actuators and the lifting cam (Figure 12.10).

In view of the ever-growing number of different products and containers filling machines are nowadays required to handle, a system variant has been developed that enhances user-friendliness in performing the tasks needed for changing over to a different product. In this variant, the purely mechanical system has been expanded to include an electro-pneumatically controlled component, which enables some of the work steps involved in change-overs to be performed electronically. In this version, a pneumatically operating control cylinder operates the gas valve spring, so that the opening and closing steps of the valve stem can be actuated electronically. This electronic actuation by the electronic component cabinet causes the filling steps to be triggered under time control or angle control. End of filling is still defined by the end of the gas return flow into the ring bowl. This variant of a mechanical filling system manages without all further mechanical control elements on the ring bowl's exterior – except the cam roller for pressing the bottle against the filling valve – thus enhancing suitability for hygienic design and largely obviating the need for manual change-over of pins and mechanical elements, since

Figure 12.10 Filling with a mechanical filling system.

thanks to the electronic control system the pre-programmed time changes stored in memory for opening and closing the valve stem can be conveniently altered.

12.4.1.1 Mecafill VKPV – The Mechanical System for Bottling Beer

In a mechanical filling system with electro-pneumatic control (Figure 12.11), a separate channel has been integrated at the ring bowl which serves to build up the vacuum atmosphere prior to filling. A pneumatically operated control valve – the vacuum valve – links the channel located underneath the ring bowl to the pressed-on bottle. This valve version furthermore features a snifting valve, which re-establishes atmospheric pressure inside the bottle after filling. The snifting nozzle is provided with a throttling feature, enabling the snifting time to be specifically matched to the product concerned and preventing the product from foaming. The Mecafill VKPV is a no-filling-tube filling system in which a small spreader at the vent tube is used to guide the product onto the bottle's wall. The gas needle has been integrated in the vent tube, which ensures gas exchange between the ring bowl and the pressed-on bottle for pressurization. After the vacuum valve has opened, the air content is reduced as an initial step. After that, the bottle is flushed with CO_2 to produce almost atmospheric pressure inside it. In the second evacuation step, the vacuum valve opens again and the remaining oxygen content in the bottle is reduced still further. After the second evacuation, the bottle is pressurized up to the ring bowl's pressure. The opened gas needle allows ring-bowl gas to flow into the bottle, thus creating the same pressure inside the bottle and in the ring bowl itself. After this pressure equalization procedure, the valve stem now opens, due to the equal pressure levels in bottle and ring bowl, and filling begins. As soon as the liquid level has reached the vent tube, gas can no longer flow out of the bottle and into the ring bowl: the fill level has been reached. An electronic pulse closes the valve stem, thus ending the filling procedure. In turn controlled by an electro-pneumatic pulse, the snifting valve now opens and the pressure in the bottle's head space is slowly bled off until an atmospheric pressure has again been reached. After that, the bottle is lowered under lifting-cam control and passed to the discharge starwheel.

The salient features of VKPV at a glance are:

- Mechanical level-controlled filling system with electro-pneumatic control.
- Fill level is reached when the liquid level stops return gas flowing out.
- Counter-pressure process.
- No-filling-tube filling, with vent tube.
- Single-chamber system.
- Double evacuation reduces oxygen content in the bottle to approximately 1%.
- Filling system for glass bottles.

12.4.2
Electronic Level-Controlled Filling Systems

Although the mechanical filling systems are still of widespread interest worldwide, there is increasing demand for filler solutions featuring electronics. This

12.4 Filling Systems for Beer

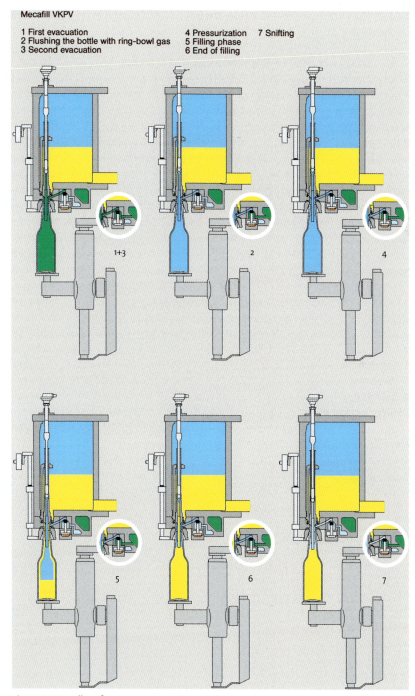

Mecafill VKPV

1 First evacuation
2 Flushing the bottle with ring-bowl gas
3 Second evacuation
4 Pressurization
5 Filling phase
6 End of filling
7 Snifting

Figure 12.11 Filling functions in a mechanical filling system with electro-pneumatic control.

also describes the concept of probe-controlled filling systems with fill-level measurement.

12.4.2.1 Sensometic VPVI – Filling With an Electronic Probe

The Sensometic VPVI is a single-chamber filling system (Figure 12.12) with an additional channel (fitted outside at the ring bowl) for creating a vacuum inside the bottle. Filling valve design is based on the principle of no-filling-tube filling. This means that the product flows from the ring bowl along the vent tube into the bottle. The small spreader at the vent tube guides the product onto the bottle's inside wall as a thin film, with the probe integrated here into the vent tube. In the basic position, the valve stem is held in its closed setting by spring force, thus closing the product channel at the bottom. In addition to the control cylinder triggering the open/close function for the valve stem, control valves have been fitted to the ring bowl, for returning to the ring bowl the gas escaping from the bottle during filling. Two more control valves ensure snifting of the return gas paths – upper gas paths and bottle head space – after the end of filling. A vacuum is created inside the bottle directly underneath the valve, using the vacuum control valve at the bottom of the ring bowl.

As soon as the pneumatic lifting unit has pressed the glass bottle onto the centering bell for a gastight fit, the first evacuation step is performed under time control. The bottle is then flushed with ring-bowl gas in order to establish an

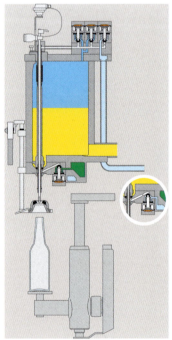

Figure 12.12 Probe-controlled filling system.

almost atmospheric pressure. The same pressure can be produced in the bottle as in the ring bowl by pressurizing the bottle with ring-bowl gas taken from the ring bowl via an upper outlet aperture. The filling process is then initiated by opening the filling valve. Note that both return gas valves are open, thus enabling return gas to flow into the ring bowl swiftly. After the pre-set initial filling time has elapsed, the return gas control valve closes and only the return gas valve with flow control remains open, thus enabling the product to rise slowly and without foaming as far as the probe's operating point. As soon as the probe's operating point has been reached, the filling valve will close. After a settling phase for the beverage, the snifting valves at the top of the ring bowl and at the bottom of the filling valve will open – the overpressure in the vent tube and the bottle's head space is bled off until the pressure is back at atmospheric level. The lifting unit will then lower the bottle and transfer it to the discharge starwheel (Figure 12.13).

The salient features of VPVI at a glance are:

- Probe- and level-controlled filling system.
- No-filling-tube filling, with vent tube.
- Double evacuation reduces the amount of oxygen in the bottle to approximately 1%.
- Filling system for glass bottles.

12.4.2.2 Sensometic VPL-PET – Probe-Controlled Long-Tube Filler for Single-Chamber and Multi-Chamber Operation

Sensometic's long-tube version offers good options for bottling beer with a low oxygen pick-up and minimized CO_2 consumption. The long filling tubes are used

Figure 12.13 Sensometic VPVI with crowner.

to fill the bottle with product from below the surface, thus significantly reducing oxygen pick-up during filling: the values achievable may be below 0.03 mg/l. In view of the long filling tube, the probe is fitted on the filling tube's exterior. At product or bottle change-over, the filling tube may have to be replaced, since the length of the filling tube and the probe's position have been matched to the bottle height involved. Moreover, the probe-controlled filling system can also be fine-tuned to suit the type of beverage being handled, as the client can use either single-chamber or multi-chamber mode. Prior to filling, the bottle (not pressed on, of course) is flushed with used gas from the ring bowl. In single-chamber operation, the PET bottle can then be pressurized with pure gas before filling, via an additional pure-gas channel; In multi-chamber operation, this is not necessary, in view of the purity of the ring-bowl gas, since the return gas from the bottle is discharged into a separate return gas channel, thus preserving an almost pure CO_2 atmosphere in the ring bowl, which in turn means ring-bowl gas is used here for pressurization. After the bottle has been pressurized, slow filling will start under electronic time control. In single-chamber operation, the return gas valve is used to control the amount of return gas flowing back into the ring bowl. In multi-chamber operation, this is done by means of a valve controlling the flow of return gas into the snifting channel. The slow-filling phase lasting until the filling tube has been immersed into the liquid will prevent the product from foaming at the bottle's base. This is followed by the fast-filling phase, using both return gas valves. After that, another slow-filling sequence will start under time control, so as to avoid foaming and achieve high fill-level accuracy; this will last until the probe's operating point at the filling tube is reached. After the end of filling, the flow-controlled snifting valve is used to re-establish atmospheric pressure in the bottle's neck space. While the bottle is being lowered at the lifting unit, the 'pipe air release' control valves and the vent tube at the filling tube are opened, and the contents of the filling tube are drained into the bottle (Figure 12.14).

The salient features of VPL-PET at a glance are:

- Probe- and level-controlled filling system.
- Filling with long filling tubes.
- Option for selecting either single-chamber or multi-chamber operation.
- Pure gas channel.
- Single-chamber operation: bottle is flushed and pressurized either with ring-bowl gas or pure gas from the pure gas channel.
- Multi-chamber operation: bottle is flushed and pressurized with pure gas from the ring bowl.

12.4.3
Electronic Volumetric Filling Systems

The Volumetic family of fillers covers two kinds of filling system: the Volumetic VOC can filler exemplifies a metering filler that pre-doses the fill quantity required in a metering chamber under electronic control, while the Volumetic VODM filler

Figure 12.14 Long-tube filler.

family provides models using flow meter technology, where an inductive flow meter defines the fill quantity prior to filling. Quantity metering in a flow-meter filler is assigned to the volumetric filling systems. A system of this kind can meter an accurate quantity of product which remains the same despite a possible tolerance in the containers waiting to be filled and is reproducible at will.

12.4.3.1 Volumetic VOC – The Can Filler

The Volumetic VOC metering filling system (Figure 12.15) features a metering chamber for each individual filling valve. In this chamber, product is pre-dosed and then passed into the container. This concept has been developed for canning, since product pre-dosing enables a high level of accuracy to be achieved – higher than is possible in cans when using a level-controlled filling system with vent tube. The reason for this is the large cross-sectional area of a can compared to the narrow bottle necks, where the liquid level mostly exhibits a small area, depending on the cross-sectional area of the bottle neck concerned. This is why in a conventional filling system for cans, a greater range of fluctuation in fill quantity may be encountered than is the case with a volumetric filling process. Can diameter

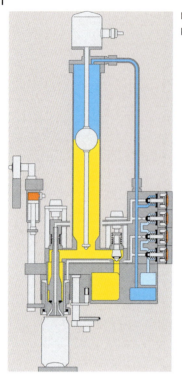

Figure 12.15 Concept of a filling system with pre-dosing chamber.

tolerances of ±0.15 mm will in the can already lead to an over/underfill of 1 ml. A metering chamber renders the filling process independent of foaming or bubbles in the beverage since the requisite product quantity is pre-dosed and not measured in terms of reaching a vent tube inside the container. This means that for changing over to different products and/or can dimensions, it will be sufficient to modify the appropriate settings in the control system, without – as is necessary in mechanical canning systems – actually having to replace the vent tube.

Integration of a differential pressure chamber enables the cans to be handled in pressed-on mode, despite their low axial pressure resistance. The overpressure in the can is bled off into a differential pressure chamber during canning and the only force acting on the can, which is being supported by its internal pressure, is the axial load. The product is kept available in a ring bowl and flows into the metering chamber. Note that filling of the metering chamber is determined by the static liquid pressure gradient in relation to the product storage tank. The fill level in the metering chamber is controlled by a trans-sonar probe. The filler's control unit is used to select the minimum and maximum values in each case for the individual can size being handled. The crucial point for gentle and foam-free filling is that the product flows into the metering chamber from the bottom, meaning that the metering chamber – controlled by the probe's lower operating point – is

Figure 12.16 Can filler.

emptied only as far down as a defined low level and the valve space proper is always filled with product. (Figure 12.16)

The centering unit now descends under cam control onto the can placed under the filling valve and the can is flushed with CO_2, first in non-pressed-on and then in pressed-on mode. After pressurization, the filling valve will open and the product will flow into the can without turbulence via the closed annular gap (360°). The valve stem has been designed so as to ensure that the liquid is passed onto the can's wall. The pressure situation in the empty, pressurized can and in the can with a rising liquid level is equalized by the differential pressure chamber, so that the axial stress on the can is matched to the fill level. As soon as the transsonar probe has reached the lower limit, the pre-dosed quantity of product will have flowed into the can and the valve stem will close. The differential pressure chamber and the return gas path are snifted via the open snifting valves and the snifting channel.

The widget technology used in some markets is implemented with the aid of an additional channel, through which the can is flushed with nitrogen in pressed-on mode before being pressurized, thus displacing the air content in the widget by nitrogen. After the can is opened by the consumer, the nitrogen will escape due to the pressure drop in the can, allowing the fine-pored foam to form, a characteristic for various beers (canned draught beer).

The salient features of VOC at a glance are:

- Electronic metering filling system.
- Fill quantity is metered by a metering chamber.
- End of filling electronically controlled after the probe's lower operating point has been reached.
- Counter-pressure process with differential pressure chamber.
- No-filling-tube filling system.
- Single-chamber system.
- Filling system for canning.

12.4.3.2 Volumetic VODM-PET – Filling With Flow-Metered Quantitative Measurement

In contrast to a level-controlled filler, the advantage of a volumetric filler with flow-meter technology is that the bottle is filled with a product quantity which is always accurately reproducible. A filling system featuring flow-meter technology (Figure 12.17), like the Volumetic VODM, always fills the container with the desired quantity of product. Correspondingly, this may – again depending on container tolerances – result in different fill levels. Another aspect of the flow-meter system is that the fill volume is not measured by components protruding into the bottle – vent tube or fill-level probe. Integration of a flow meter between the ring bowl and filling valve permits a construction for the ring bowl featuring no internals. It also means that any changes in the fill quantity at bottle change-over can be programmed using the control system, without any physical modifications being necessary.

To be able to use a flow meter, the product must exhibit a conductivity of more than $20\,\mu S/cm$, so as to ensure reliable volume flow measurements. The VODM for beer is available with long filling tubes and also in no-filling-tube design.

Following an electronic pulse to the control system, the pneumatic control cylinder opens the valve stem and the product flows into the bottle (Figure 12.18). The return gas is passed from the container directly into the atmosphere or back into the ring bowl. As soon as the flow meter detects that the pre-programmed product quantity has been reached, the valve is closed via a pulse sent by the electronic component cabinet to the pneumatic control system.

The salient features of VODM-PET at a glance are:

- Electronic flow meter system.
- End of filling is reached when the flow meter has detected the pre-programmed fill quantity.
- Counter-pressure process.
- No-filling-tube filling, with vent tube.
- Single-chamber/multi-chamber system.
- Filling system for PET bottles.

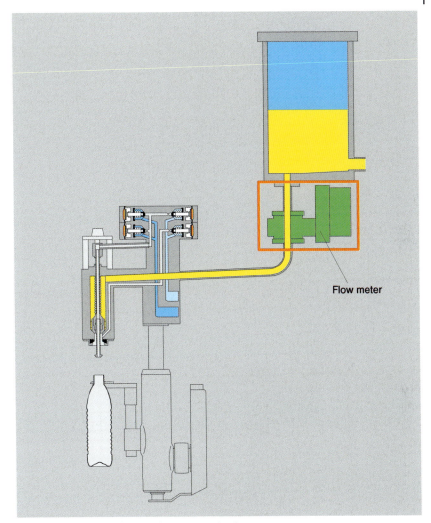

Figure 12.17 Filling system with flow-meter technology.

12.4.4
Associated System Modules

12.4.4.1 Fobbing

In order to reduce the oxygen content in the neck of the bottle after filling, the air remaining there is displaced by a fobber (Figure 12.19). This is done by a nozzle mounted in a fixed position immediately downstream of the bottle discharge, which sprays hot water (80–85 °C) at a pressure of approximately 10 bar into the bottle passing underneath it. This thin jet of water (0.09 ml/min) is designed so as

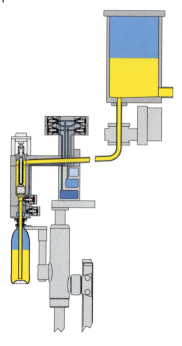

Figure 12.18 Filling phase in the electronic flow-meter system.

Figure 12.19 Fobber for reducing oxygen in the bottle neck.

12.4 Filling Systems for Beer

to penetrate into the liquid inside the bottle and thus cause foaming of the product involved (mostly beer). This foaming displaces the air from the bottle neck, thus enabling the oxygen content in the bottle neck to be reduced prior to closing.

12.4.4.2 Crowners

After the crown has been passed from the crown chute into the transfer segment, a magnet takes over for guiding the crown (Figure 12.20), which is then positioned at the crowning element's ejection plunger using a pushing notch. The crowning element descends until the crown comes up against the bottle in the crowning throat and is held by it. After this, only the crowning throat continues to move, in a downward direction (Figure 12.21). In this initial phase, it is solely the force of

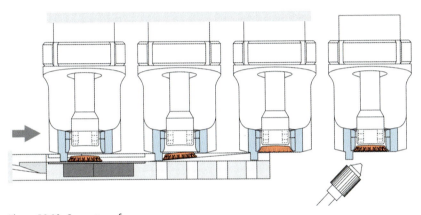

Figure 12.20 Crown transfer.

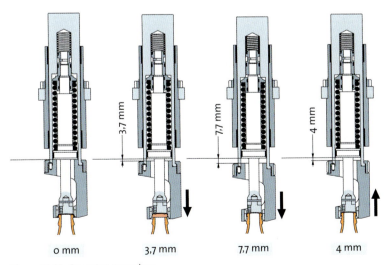

Figure 12.21 Crowning procedure.

the holding spring that acts on the crown. In the second phase of crowning, the ejector spring is compressed and the pressure on the bottle increases. The crowning operation is completed when the crown has moved 7.7 mm into the crowning throat, which means that it is 1 mm inside the cylindrical section of the crowning throat and the requisite crowning diameter of 28.6–28.7 mm is precisely reached. The crowning operation is thus concluded and the crowning force drops again.

12.4.4.3 Screw-Cappers for Plastic Screw Caps

The screw-caps are spaced in the closure chute and passed to the capper from above. The closure is accepted by the pick-up device and fixed in position inside the closing element. The glass bottle is secured against turning in the machine by a clamping belt and a central starwheel. The PET bottle is held in position by spikes glued to the neck starwheel disk. The capping element places the closure on the bottle, whereupon it is pressed against the bottle by spring pressure. As this happens, the PET bottle is pressed onto the spikes at its neck ring and secured against turning.

The screw-on operation begins. (Figure 12.22) As soon as the closure reaches its specified screw-on torque, the hysteresis clutch slips, and the closing operation is terminated. Using this clutch ensures an approximately identical torque at all closing element speeds.

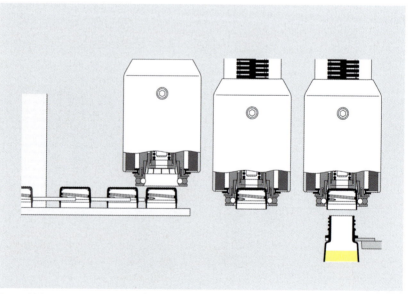

Figure 12.22 Screw-on operation for PET bottles.

12.5
Constituent Parts of a Bottling Line

In addition to the filler, a bottling line incorporates an empties inspector, a bottle washer, monitoring systems for the washed bottles in the case of a returnables line, and inspection systems for the filled, closed and labeled bottles. Once the bottles are passed to the packing and palletizing section, they have left what is called the wet end of the line. (Figure 12.23)

Appropriate conveyors link the various elements in the line together and ensure that the bottles are transported in a cluster or spaced out. Points or ejection systems enable bottles tagged as defective to be rejected downstream of the inspection unit concerned.

12.5.1
Bottle Washer

According to Sinner's Circle, there are four influencing factors for container cleaning. Temperature, chemicals, mechanisms and time are deployed in the appropriate combinations required to feed the returnable bottles back into the filling process after being cleaned. The aim of all treatment steps is to provide a bottle that is:

- Free from beverage-spoilage germs and bacteria.
- Free from bottle dress residues (labels, foils).
- Free from beverage and content residues.
- Free from odors.

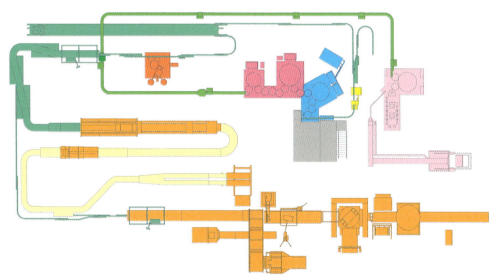

Figure 12.23 Concept of a complete bottling line.

- Free from chemical residues.
- Gleaming.
- Drip-free.
- At the appropriate temperature for further handling.

The primary cleaning agent and carrier medium used is water. Since the cleaning effect of conventional water by itself is not particularly good, chemical additives are employed in order to improve it. High temperatures of the cleaning liquids are particularly necessary for the microbiological cleaning effect. Combining a suitable contact/processing time with mechanical cleaning methods (e.g. internal and external jetting) ensures comprehensive interaction of all the factors involved.

Apart from the cleaning process proper, gentle treatment of the containers is also vital in the bottle washer, irrespective of whether glass or plastic returnable bottles are involved. It is particularly important to avoid mechanical stresses on the bottles when they are being fed in and discharged, and when selecting the treatment temperature care must be taken to ensure that the material exhibits the requisite thermal stability.

Depending on the concept of the bottle washer concerned, a distinction is drawn between single- and double-end machines. The salient feature of a single-end machine is that the bottle infeed (the point where the soiled bottles enter the machine) and the bottle discharge (where the bottles leave the machine) are at the same end of the washer, one above the other. In a double-end machine, by contrast, the bottle infeed and the bottle discharge are separated, one at each end of the machine. The decision to opt for a single- or double-end machine will depend on the space and the budget available to the brewery concerned. In terms of hygiene, the double-end machine basically scores better, since entrainment of soiling between the bottle infeed and the bottle discharge is inherently ruled out, by reason of the separated locations.

12.5.1.1 Treatment zones

Depending on the model concerned, three different treatment zones (Figure 12.24) are provided in a bottle washer:

- *Pre-cleaning section.* The principal task of the functions in the pre-cleaning section is to remove from the bottles any coarse soiling residues (cigarette ends, straws, product residues, etc.) and to warm the bottles up to the main cleaning temperature. This avoids major temperature jumps for the bottle material at the beginning of the main cleaning stage.
- *Main cleaning zone.* One or more caustic immersion baths are provided here, depending on the design of the bottle washer involved. The preferred temperature in the caustic immersion baths is 75–82 °C for glass bottles and 58–75 °C for PET bottles. The limiting factor for the temperature setting in the case of glass bottles is the surface coating; in the case of PET bottles, it is the shrinkage characteristics of the plastic involved. The task of the caustic immersion baths is to achieve chemical and microbiological purity of the bottles,

12.5 Constituent Parts of a Bottling Line | 307

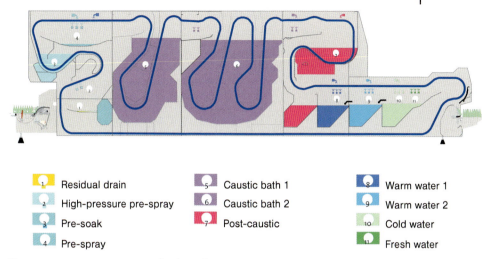

1	Residual drain	5	Caustic bath 1	8	Warm water 1
2	High-pressure pre-spray	6	Caustic bath 2	9	Warm water 2
3	Pre-soak	7	Post-caustic	10	Cold water
4	Pre-spray			11	Fresh water

Figure 12.24 Treatment zones in a bottle washer.

and to detach adhering label material, collect it and remove it from the machine.

- *Post-treatment zones.* Bottle treatment in the warm-water zones, featuring jetting devices, immersion bath plus subsequent cold-water and fresh-water jetting, removes the caustic residues still adhering to the bottles and reduces the bottle temperature slowly to the level required for further processing. Since while the machine is running there is continuous entrainment of liquids via the drive chain, the bottle carriers and the bottles themselves, the post-treatment section has to continually counteract and monitor this entrainment.

Depending on the machine design concerned, the various treatment zones are concatenated with horizontal or vertical loop guidance of the bottle carriers, thus maximizing the dwell time of the bottles in the baths and treatment zones involved.

12.5.1.2 Components of a Bottle Washer
Irrespective of whether the machine concerned is a single-end or double-end design, bottle washers feature the same constructional details.

- *Bottle infeed and discharge (Figure 12.25).* Apart from gentle, trouble-free insertion of the bottles into the machine, this component also has to be designed for optimum handling of different bottle types and minimized workload for changeovers. Furthermore, in order to meet the requirements of occupational safety, noise minimization measures are also mandatory at this point. Infeed conveyors ensure a continuous flow of bottles, which are then diverted into lanes before the bottle infeed by agitators on the jostler belt. The bottles are then gently guided, without pressure and without falling over if at all possible, along the infeed rails

Types of infeed

Example for discharge

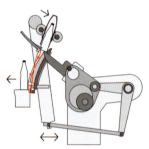

Figure 12.25 Types of infeed and discharge in the bottle washer.

before being pushed into the bottle pockets. The continuous movement of the individual conveyor belts at different speeds ensures the desired forward motion of the bottles. The bottles are discharged after the washing process by rotating fingers, which accept them from the bottle pockets and deposit them on the conveyor.

- *Bottle pockets and bottle transport through the machine.* The bottles are conveyed through the machine in the bottle pockets, which are made of plastic or a steel/plastic material combination. The bottle pockets fix the bottles in position, enabling them to be properly targeted by the jetting units. Appropriate openings in the bottle pockets mean the sprayed-in water can circulate for external cleaning of the bottles (Figure 12.26). Depending on the rating of the bottle washer involved, 16–40 bottles are arranged next to each other along the bottle pocket carriers, so that rows of bottles fixed in position next to each other are treated inside the machine. The steel bottle pocket carriers are fixed at their sides to the transport chain, which conveys them through the machine. Depending on the speed of the machine concerned, in a double-end machine with three caustic immersion baths up to 958 bottle pocket carriers are integrated into the washer.

12.5 Constituent Parts of a Bottling Line

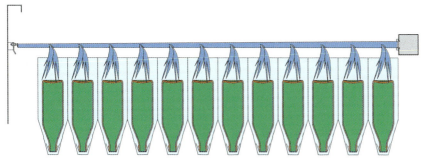

Figure 12.26 External cleaning of the bottles in the bottle washer.

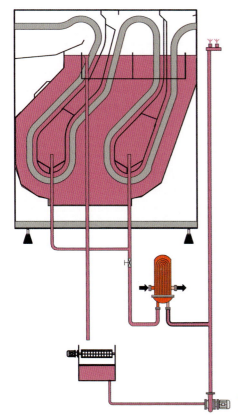

Figure 12.27 Label and dirt removal system.

- *Label and dirt removal (Figure 12.27).* After being immersed in the hot caustic bath, the labels and foils are detached. A pump system creates a defined circulation direction of the caustic, by which the label residues are carried away and transported into a strainer unit, where they are arrested and can be removed in a collection container.

- *Bottle jetting (Figure 12.28).* The jetting units installed in the post-treatment zone provide targeted treatment for the inside of the bottle, so as to flush out caustic residues and other cleaning constituents completely. A synchronized spraying nozzle on a nozzle shaft sprays first the wall of the bottle, then its base and then the opposite wall (Figure 12.29).

- *Heat distribution and preparation.* The water is passed through the bottle washer, and the temperatures required in the sections concerned are achieved by the liquid path configuration and the integration of plate heat exchanger systems. Internal piping circulates the water, so that low-temperature sections can be supplied with the cooled-down water from the warm-water sections.

- *Machine drive.* The disadvantages of the Cardan drives previously used, such as increased wear and tear on the chain, heavy workload for maintenance and adjustment, and difficulties in detecting unequal load distribution, have been eliminated by the use of synchronous drive technology for the bottle washers. The basic idea involved here is that the overall drive power is distributed among the step-down gear units involved. In total, however, the drive power is the same as would be provided by a single drive. The system is conditional upon the gear units operating in synchronization. Since a frequency converter is used to syn-

Figure 12.28 Bottle jetting system.

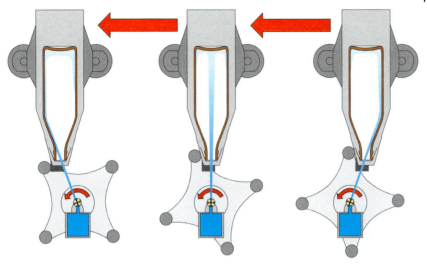

Figure 12.29 Treatment of the inside of the bottle through spraying nozzles.

chronize the individual motors concerned, they run at precisely the same speed, even if different models or rating classes are involved.

- *Extractor devices for gases.* The different temperature zones inside the bottle washer cause condensate and drip water to form, which may lead to soiling of the already washed bottles. In order to avoid this, the vapor is extracted from the machine in the vicinity of the bottle discharge. When aluminum foil is used on the bottles, hydrogen may be formed if the foil gets into the caustic soda solution. In order to avoid the formation of a potentially explosive gas mixture, the hydrogen is collected and extracted at individual positions in the machine.

- *Head section disinfection.* A high level of hygiene in the surroundings and interior of the bottle washer ensures the best possible treatment of the bottles and of course optimum functioning of the machine. The machine is divided into different cleaning zones, accessible through inspection openings. In addition, jetting systems for cleaning and disinfection can be integrated or automatic dosing of cleaning agents and disinfectants can be used.

- *Machine control system.* The programmable-logic control system of a bottle washer is handled from an operator terminal that communicates all the machine's status displays. The cleaning parameters are also displayed, as are malfunctions and (where appropriate) suggestions for correcting these faults.

12.5.1.3 Typical Bottle Treatment Sequence
See Figures 12.30 and 12.31.

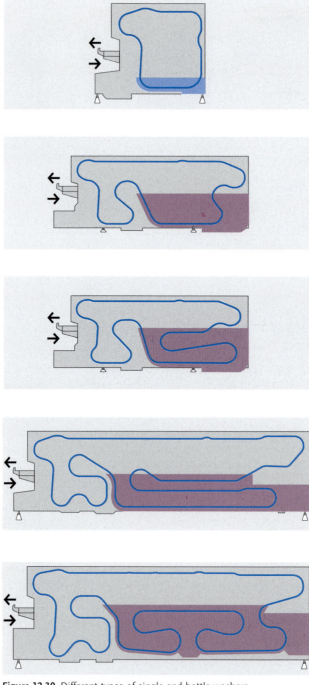

Figure 12.30 Different types of single-end bottle washers.

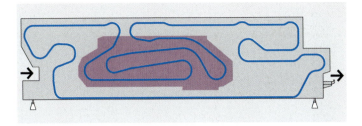

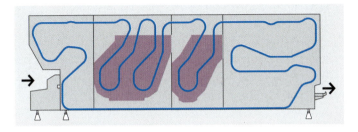

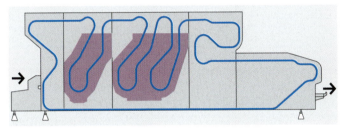

Figure 12.31 Double-end bottle washers with various different layouts for caustic immersion baths.

12.5.2
Inspection and Monitoring Units

The filling and packaging process for beverages is monitored by appropriate inspectors at several positions in a line. The aim here is to reject faulty bottles and cans, and to check on the filled containers. This is of crucial importance, not least for reasons of product liability, plus of course checks are also necessary for the optimum visual appearance of the products concerned on the supermarket shelves. The machines that can be used for this purpose can be directly integrated into a transport line as in-line models, installed at the requisite positions as flexible units in modularized construction, or provided as rotary machines.

12.5.2.1 Machine Types
The in-line model (Figure 12.32) is a compact machine that integrates several different inspection modules. The bottles are passed individually through the machine

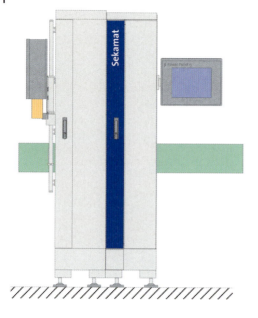
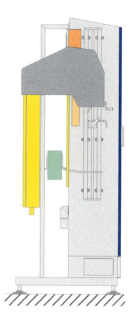
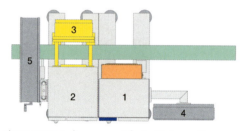

1 Basic Sekamat unit
2 Camera unit
3 LED illumination (highly-ajustable)
4 Operator panel with touch-screen
5 Optional inspection unit from the Checkmat 731 series

Figure 12.32 In-line concept for bottle inspection.

on a conveyor and checked for various defects (Figure 12.33). Rotary machines (Figure 12.34) are mounted on an infeed carousel, which spaces out the bottles as they enter and subjects them to the requisite checks in the inspection carousel.

More and more now, compact-size inspection units are also being used, which do not (like the above-mentioned in-line or rotary models) integrate a large number of inspection modules, but are designed for one specific application, handling one to three inspection tasks. This means these units are in most cases more affordable and space-saving than in-line or rotary machines.

Numerous positions in the line can be chosen to install the inspectors:

- Empties detection.
- Empty-bottle inspection.
- Full-bottle inspection after filling, including closure inspection.
- Inspection of the labeled bottles and of any codes printed on them.

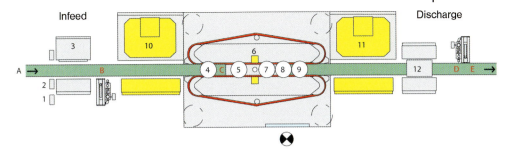

Inspection units:
1 Color detection sensor
2 Height detection photoelectrode sensor assembly
3 Contour, height, color and scuffing detection camera
4 Variostation (lateral neck finish/neck finish)
5 Infrared residal liquid
6 High-frequency residal caustic
7 Base, inner sidewall
8 Variostation (screw thread inspection)
9 Sealing surface inspection
10 Side wall module, infeed (incl. scuffing and scuffing ring)
11 Side wall module, discharge
12 Lateral neck finish mirror PET/glass

A Machine protection (machine stop with overheight and fallen-over containers)
B Rejection at machine infeed
C Base blow-off unit
D Rejection of broken containers
E Rejection of dirty containers

Figure 12.33 Inspection units in an in-line machine.

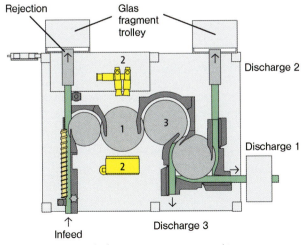

Figure 12.34 Example for a rotary inspection machine.

Appropriate rejection units are used to eject the bottles tagged as defective straight away, thus removing them from any further processing.

12.5.2.2 Inspection Tasks

Depending on the positioning of the machine involved, what are called inspection modules are used, operating with a variety of different technologies. They are able to inspect the following positions:

- *Base.* The base inspection function detects beverage residues, covered bases, foils, glass splinters or glass damage at the base zone.

- *Sidewall.* Sidewall inspection detects foil and label residues at the external sidewall, soiling or scratches – and also scuffing, depending on the performance and sensitivity of the camera concerned. The inner sidewall inspection function enables damage to the inside of the bottles to be detected, such as damage behind inscriptions or applied ceramic labels.

- *Neck finish – sealing surface/thread/lateral neck finish.* Damage to bottles is frequently encountered in the neck finish area, since this is one of the most frequently stressed sections of the bottle. First of all, fissures or cracks at the threads constitute a problem for renewed closure of returnable bottles, since the desired sealing effect is questionable and, moreover, consumers are put at risk by any glass splinters that might get into the beverage when the bottle is opened. The inspection routines accordingly cover the thread, the sealing surface of the mouth and the side areas of the neck finish that may be damaged by any chipping.

- *Bottle shape, size, contour and color.* This inspection routine is particularly important when a large number of different bottle types are involved, which in a returnables operation have to be sorted before being fed into the bottling process concerned. The system is required to distinguish between foreign bottles and the firm's own bottles, which are then grouped appropriately in the sorting system. After this, the firm's own bottles are fed into the line, while the others are made available to the other bottlers involved through the empties pool. Sorting criteria in this context can be the bottle's shape, its contour, its size and its color.

- *Residual liquid, foreign substances.* After the bottle has been washed, in exceptional cases it may happen that residues of caustic or other foreign substances nonetheless remain in the bottle. These deleterious substances must be detected before any further handling of the bottle and the bottles concerned must be reliably rejected so as to prevent any harm coming to consumers.

- *Foreign bodies.* Glass splinters in the bottles (Figure 12.35), particularly, may cause substantial injuries to consumers of the product involved. Detection of these and other foreign bodies in the filled bottle provides reassurance for the producer.

- *Fill-level inspection.* Bottles that have not been filled (Figure 12.36) with the preprogrammed fill volume have to be rejected: (i) over/underfilled bottles are a

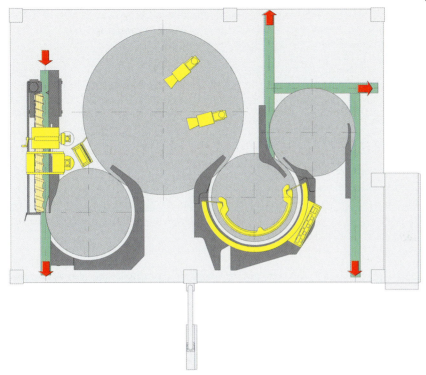

Figure 12.35 Detection of glass splinters in a rotary system.

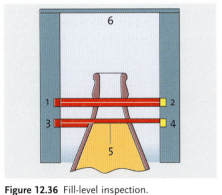

1 Overfill transmitter
2 Overfill receiver
3 Underfill transmitter
4 Underfill transmitter
5 Fill level
6 Measuring head

Figure 12.36 Fill-level inspection.

contravention of the German Packaging Ordinance, and (ii) overfills entail competitive distortions and underfills result in image impairment. In the case of visibly underfilled bottles, detection of this error also enables any malfunctioning filling valves to be identified.

- *Closure inspection.* Bottles with slanted closures and missing closures have to be rejected for safety reasons (in the case of bottles containing a carbonated product) and quality assurance considerations. Otherwise, there will be a risk that the overpressure inside the bottle may cause accidents in connection with an incorrectly fitted closure or the quality of the beverage will be impaired.

- *Label type, label positioning*. Optimum visual appearance and the right information on the bottle are crucial considerations for bottlers dealing with high levels of product diversity and concomitantly frequent container and product change-overs.

- *Dating/barcode*.

12.5.2.3 Inspection Technology

The technologies used in the various inspection systems (Figure 12.37) concerned are always linked to a high-accuracy image-processing program. Thus, the data transmitted are evaluated in accordance with the algorithms provided and the resultant good/bad decision arrived at.

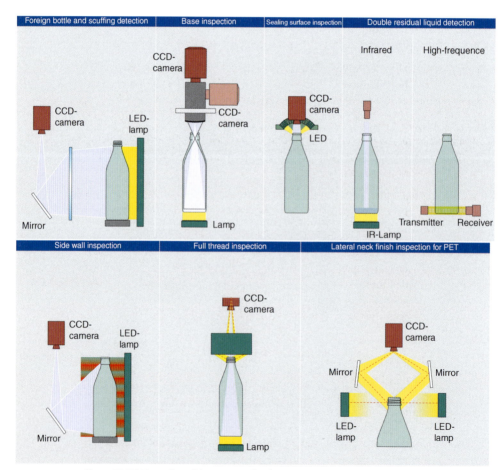

Figure 12.37 Examples of detection technologies.

- *Camera technology.* For inspecting sidewall, base, neck finish and thread, high-resolution charge-coupled display (CCD) cameras are employed. These camera signals are imaged with high grey-scale value resolution, so that deviations from the predefined standard image are detected and evaluated. Using several cameras from different sides and mirror systems enables an all-round picture of the bottles under inspection to be generated. Alternatively, the bottles are turned on the container plates in a rotary machine, so that the camera makes a comprehensive image of the entire bottle in a fast series of pictures. The area to be inspected is illuminated by an appropriate choice of light sources, the requisite polarization filters and the light guidance systems, so that the camera can supply the digitized information needed on the bottle. For label detection, too, the CCD camera system offers numerous options for image processing and also for code detection (e.g. European Article Numbering system, Universal Product Code). The labels are detected by a wide-area pattern comparison function, and checked for position, placement and quality; error tolerance is user-settable. Label inspection using optical character recognition enables plaintext and letters (e.g. for the best-before date) and a concomitant correctness check to be run. Quality monitoring using optical character verification checks the presence of the date, monitors the quality of inscriptions and compares the characters involved.

- *High-frequency technology.* High-frequency technology is used for caustic detection, based on the different dielectric constants of liquids in comparison to glass. For detecting caustic, more energy is transmitted to the sensor, thus achieving an indication for whether the bottle is to be rejected.

- *Infrared technology.* By illuminating the base of the bottle with IR radiation using an IR transmitter, liquid residues at the bottom of the bottle are detected and the bottle passed to the gate-type rejection unit.

- *Radiation technology featuring γ-rays and X-rays.* In this process, the bottles are passed under a measuring bridge, conceived as a transmitter/receiver system for the type of radiation involved.

- *Mass and color spectrometers.* Mass spectrometers are used for detecting foreign substances in returnable plastic containers after washing. Gas samples from the bottles are divided into molecular masses, detected by quantity, and evaluated. Foreign substances that cannot be discovered by means of a gas sample are detected with the aid of a color spectrometer. If defined limit values are exceeded, the containers concerned will be rejected.

12.5.2.4 Reliability of the Inspectors

Proper detection of defects at the individual inspection routines involved is initially assured by the parameters preset at the detection and evaluation software. Nonetheless, it is vital to subject precisely these reliability devices as well to regular self-check routines. This task is performed by the test bottle program, which is passed through the systems concerned at regular intervals, with specially prepared bottles being fed in, and it is only reliable detection of these deliberately defective bottles that is an indicator for dependable function of the inspection units concerned.

References

1 Hertlein, F. (1997) Characteristic properties of plastic packaging materials and packaging. *Brauwelt International*, **III**, 204–216.
2 Rammert, M. (1993) Zur Optimierung von Hochleistungsfüllanlagen für CO_2-haltige Getränke, Paderborn, p. 45.
3 Blüml, S. and Fischer, S. (2004) Process steps, in *Manual of Filling Technology* (V. Kronseder, ed.), Behr's, Hamburg, pp. 69–417.

13
Labeling

Jörg Bückle

The options for product dress today are many and varied: body, shoulder, neck-around, back and wrap-around labels made of numerous different materials are available to the marketing specialists (Figure 13.1). This is supplemented by pointed or round neck foils, application of pressure-sensitive labels (PSLs) and sleeves. The range of labelers on offer is thus comprehensive and varied.

13.1
Some Basic Remarks on Machine Construction

The majority of present-day labelers operate on the rotary principle (Figure 13.2). This type of labeler's advantages over an in-line configuration are its smaller footprint in relation to output, its higher labeling accuracy and the option it provides for applying all dress variants commonly in use. The basic construction of a rotary model comprises a machine table, container guidance system, machine head, container table, labeling station, drive and control system. The container table, also referred to as a 'carousel' due to its shape, the bottle plates 'sitting' on it and the rotational movement it performs transports the containers through the following treatment sections: labeling proper, brush-on and label inspection, with special control elements here imparting a defined rotational movement to the bottle plates, which is generated in the traditional way by a fixed table cam and gear sections. These mechanical control elements are specifically designed and manufactured to suit the labeling job involved. The use of servomotors, by contrast, driving every single bottle plate separately, enables rotational movements preprogrammed for the labeling job on hand to be run under computerized control.

While retaining this basic construction, each labeler is individually configured for the labeling job specified (Figure 13.3). For example, rotary machines can likewise be fitted with stations for hotmelt or PSLs, in addition to the wet-glue labeling equipment. In most cases, labeling is also linked to a container coding or dating function, performed either in the labeling station or at the bottle table. The

Figure 13.1 The options for dressing products in labels and foils are many and varied.

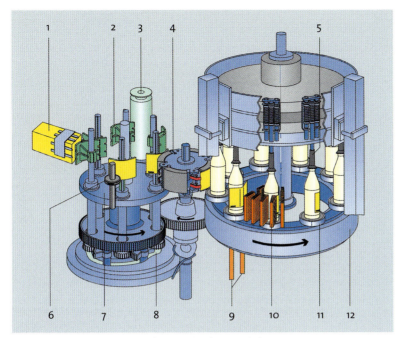

Figure 13.2 Working principle of a rotary labeler. 1, Label magazine; 2, Gluing pallet; 3, Glue roller; 4, Gripper cylincler; 5, Centring bell; 6, Date-cooling device; 7, Pallet carousel in oil bath; 8, Control cam; 9, Oil-circulating lubrication system of bottle table; 10, Brush on unit; 11, Container plate; 12, Container table.

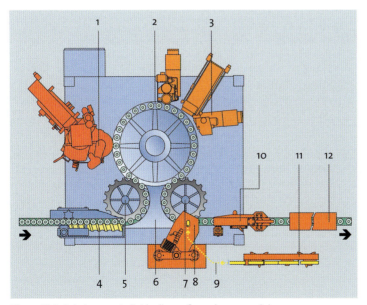

Figure 13.3 Labelers are individually configured on a modular system. 1, Cold glue labelling station for body and shoulder labelling; 2, Bottle table in spoked wheel design; 3, Hot glue labelling station for wrap-around labelling; 4, Infeed worm; 5, Infeed star wheel; 6, Centre guide; 7, Discharge wheel; 8, Station for shrink sleeve application; 9, Sleeve; 10, Hot air nozzles; 11, Sleeve supply unit; 12, Shrink tunnel.

label inspector integrated at the machine's discharge will prevent any incorrectly labeled containers, or those without any label at all, from reaching the shops.

13.2
Wet-Glue Labeling

Wet-glue labeling technology continues to be the most inexpensive option for dressing containers. This is due to the materials used for label and adhesive, and to the machine technology involved, which provides top efficiency levels right up into the ultra-high performance range. Machines with two stations, for example, achieve an output of 120 000 labels an hour in practical operation. Demonstration machines, whose stations simply pick up the labels and then deposit them in a container, even manage outputs of up to 160 000 labels an hour.

The main category of use for wet-glue labeling is the application of bottle dress to damp containers. There are different label types (body, shoulder, neck-around, seal, band, back, etc.), label formats (length, height) and label papers to choose from.

Casein-based glue, whose basic constituent is obtained from milk protein, is the most widespread choice for processing paper labels on glass. With returnable containers, the labels can easily be removed again in the bottle washer, thanks to the water-soluble wet glue. However, if the label is not supposed to come off when exposed to water, as is the case with sparkling wine bottles in an ice-bucket, this can also be ensured by selecting a special wet glue.

Using optimized types of wet glue, moreover, also enables the containers to be dressed in plastic labels made of oriented polypropylene (OPP) material. Above and beyond this, research work is currently ongoing, tackling the option of applying transparent films with wet glue as well.

Nowadays, all rotary machines work with labeling stations with a stationary label magazine and oscillating pallets (Figure 13.4). The system featuring a swiveling label magazine and a label guide is installed only very rarely in in-line labelers.

Normally, one of these stations is sufficient for up to three labels if these are positioned on a vertically symmetrical axis. This applies, for example, to beer bottles with body and shoulder labels, plus foiling. For dressing the bottles in back labels as well, a second station must always be integrated.

When a multiplicity of dress variants is involved, dedicated-station design is well worth adopting. Here, each label type is handled by its own station. Dedicated-station design renders the machines user-friendlier, more easily accessible, with concomitantly simpler change-overs, thus dramatically reducing non-productive make-ready times.

One fundamental precondition for a good labeling result is a flawless glue pattern. A distinction is made here between strips (Figure 13.5), honeycombs or full-surface gluing. The optimum glue film thickness depends on the bottle and label material, and on the operating conditions involved. If the glue film is too thin, the labels will be only partially glued, they may not adhere properly or even

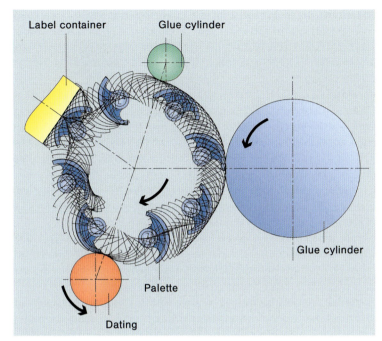

Figure 13.4 Movement sequence of the oscillating gluing pallets.

Figure 13.5 The process of strip gluing has become widely accepted.

fall off completely. If the film is too thick, the labels will float on the bottles and be susceptible to slipping. In this case, glue drops may be spun off the gluing pallet and soil the machine. Furthermore, a glue film set to an excessive thickness causes unnecessary costs. The glue film thickness should therefore be easy to re-adjust during operation.

Thanks to the glue, the label adheres to the pick-up pallet. The pallets' movement has been precisely matched to suit the movement track of the rotating gripper cylinder, which possesses one or several gripper fingers, depending on the label type involved. The gripper finger(s) reach(es) into the recesses on the circumference of the pick-up pallets and gently pull off the glued label. To minimize stress on the labels, the pull-off function must be performed at a very low speed, in spite of a high operating speed. If the labels are pulled off too fast, they will tear. Furthermore, this will put unnecessary stress on cam rollers, bearings and other components.

What are called synchronous systems provide major benefits here, because these systems' pull-off speed (with identical output ratings) is substantially lower than in counter-flow systems. Moreover, the pull-off angle between the labels and the pallet surfaces is also in a range that is gentle on the material.

Once transferred, the label is clamped in place between the gripper finger and the associated anvil bar, and then relocated smoothly onto the passing container. During transfer to the container, the label is pressed on by the gripper sponge while simultaneously being blown upon by the gripper cylinder's blower air jets, so as to make sure it is transferred with proper alignment and adheres quickly.

Following a precisely defined rotational movement of the container around its own vertical axis, the label is smoothed by brushes and/or sponge rollers installed on the bottle table at right angles to the container, and brushed on.

13.2.1
Foiling

Bottle-neck foils as a decorative supplement to the other dress, or for tamper-evident sealing, are handled with wet glue like paper labels. The foil is transferred with the body or body/shoulder label. Alternatively, the pre-cut foil can also be applied by a dedicated station.

The gluing pallets for picking up the foil are designed to be 'glue-free' in the bottle mouth area, for reasons of hygiene (Figure 13.6). In addition, if there is a combined transfer of the foil together with the body label, for example, a divided glue scraper must be used, since the foil requires a somewhat thicker film of glue. With dedicated-station designs, this divided glue scraper can be dispensed with, because the station will have been specifically designed for foiling. A special container table control system ensures defined rotational movements of the bottle, which are required for the brushing-on function, and differ from the rotational movements of the machines used for labeling pure and simple.

The foil's snug-fit propensity can also be utilized for tamper-evident sealing; for this purpose, a station installed vertically above the container places a special foil

Figure 13.6 The gluing pallets for picking up the foil are designed to be 'glue-free' in the bottle mouth area.

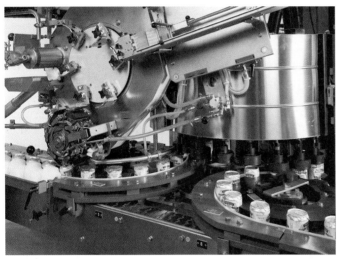

Figure 13.7 For tamper-evident sealing, a station located directly above the container places a foil blank on it.

blank on the closure. Another option with the station positioned vertically above the container is dressing beverage cans in a hygienic protection foil (Figure 13.7). The primary focus here is on guarding the can closure against soiling.

13.3
Hotmelt Labeling

The demand for hotmelt labeling has of course been rising to match the steady increase of polyethylene terephthalate (PET) usage in the beverage industry, as hotmelts are indeed optimally suited for this container material. As the name implies, adhesives whose processing temperature lies between 120 and 170 °C are used for hotmelt labeling.

The special strengths of hotmelts include their very good 'carrying function', their high initial bond, and their extremely fast and permanent gluing properties. They are used for labeling both cylindrical and polygonal containers made of glass, plastic and metal. The machine output achievable with hotmelt wrap-around labeling is up to 85 000 containers an hour. The labels used are both paper ones and an ultra-wide variety of foil/film types. With bulgy bottles, moreover, special film labels can also be shrunk onto the container's contour for a shape-hugging fit.

The container materials and their condition have a very marked influence on the labeling result. If the labels must be applied to glass bottles, these must not enter the labeler in too cold a state. A glass temperature of below 28 °C will cool down the adhesive much too fast, resulting in a 'peeling effect', where the adhesive 'flakes off' from the container. With plastic, in contrast, the container temperature is not of such crucial importance. For example, labeling PET bottles at 16 °C with hotmelts poses no problems.

Above and beyond this, difficulties may possibly arise with heavily surface-finished glass bottles. If the bottles have been given an excessively high finish, only inadequate gluing will be possible, due to the changes in surface characteristics, unless the adhesive concerned has been matched to suit this particular application. The same is valid for moist containers. To obtain an optimum gluing result, therefore, glass bottles coming out of the filler slightly misted up, for example, should be dried with a blower prior to gluing. For plastic bottles, drying is likewise recommended.

13.3.1
Hotmelt Labeling with Pre-cut Labels

For hotmelt labeling with pre-cut wrap-around labels, the finished labels (made of paper or plastic film) are provided in a magazine. In contrast to wet-glue labeling, the labels are not picked up by a glued pallet, it is the glued bottle that withdraws the label from the magazine.

The labeling station proper consists of two gluing units located before and behind the label magazine respectively (Figure 13.8). The first hotmelt unit uses a silicone glue roller to apply glue strips approximately 15×30 mm in size at the circumference of the container; the number of glue strips applied depends on the label height involved. The container passes the label magazine, withdraws a label with its glued strip and, turning, wraps the label around itself.

At the same time, the second hotmelt unit uses a steel glue roller to glue a strip approximately 15 mm wide at the label's trailing end, which reaches across the entire label height. This hotmelt strip serves to glue the overlap.

Once the label is actually on the container, it is smoothed and brushed on. For each type of container, the appropriate brush-on kit must be installed in the machine. This concludes the labeling process. The centering bell releases the

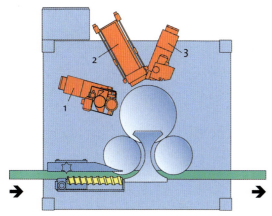

Figure 13.8 The first gluing unit is for the container, the second for the label. 1, Hotmelt unit 1; 2, Label magazine; 3, Hotmelt unit 2.

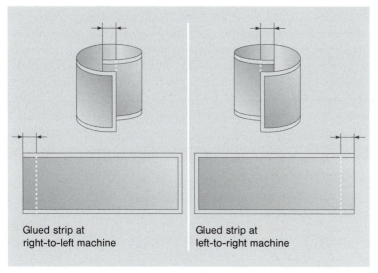

Figure 13.9 The overlap required will depend on the container to be dressed.

container, which is accepted by the discharge starwheel and placed on the discharge conveyor.

When processing plastic film labels, care must be taken to ensure that the film labels used have been matched to suit the hotmelt's melting temperature and to prevent them becoming statically charged, so as to avoid malfunctions caused by labels clinging to each other during sheet-by-sheet separation out of the label magazine. As a rule, labels of this type must be given an antistatic coating, which is applied during label production.

The label length will depend on the container to be dressed (Figure 13.9). The correct length for cans or glass containers is their circumference plus an overlap of 8–10 mm. The correct label length for plastic bottles holding carbonated beverages, by contrast, is their circumference plus an overlap of 12–15 mm. For reel-fed labeling, moreover, the recesses required for cutting registration control must be factored in.

13.3.2
Reel-Fed Hotmelt Labeling

The operating staff's workload is considerably eased by reel-fed labeling (Figure 13.10) where the paper or film label is fed to the machine on a web, with the machine cutting it precisely to the desired length only immediately before labeling begins. Here, the label supply is significantly greater than that provided in a label magazine. When the reel is changed, all the operator has to do is splice the end of the first reel to the beginning of the second, thus obviating the need to rethread the label web. The reel can also be replaced automatically. In this case, the first reel's end is spliced to the beginning of the second, without this interrupting production.

As far as PET bottles are concerned, it is reel-fed wrap-around labeling with OPP labels, in particular, that is gaining steadily in importance. However, this process is just as suitable for handling other plastics and paper labels as well.

For reel-fed labeling, the containers arriving on the conveyor belt are accepted by the infeed worm, spaced, taken over by the infeed starwheel and placed on the

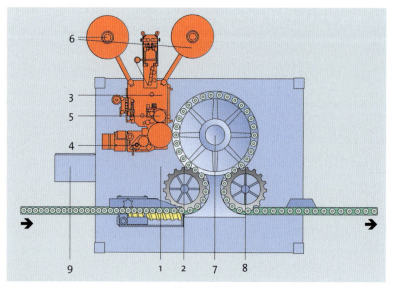

Figure 13.10 Rotary machines are used for reel-fed hotmelt labeling. 1, Infeed worm; 2, Infeed starwheel; 3, Labelling station; 4, Hotmelt unit; 5, Cutter drum; 6, Label reels; 7, Container table; 8, Discharge starwheel; 9, Coutrol cabinet with operating panel.

container table's centering plates. They are firmly clamped in position between centering bell and bottle plate, whereupon a rotational movement is imparted to them. Guidance by neck-handling is available as an optional extra, something that provides major advantages for pre-labeling empty PET bottles.

The labeling station consists of the label transport with cutting registration control, cutting device, gluing station and a vacuum-supported gripper cylinder for label transfer to the containers waiting to be labeled.

First of all, the gluing station applies a strip of glue to the leading end of the pre-cut label, with which it adheres to the container (Figure 13.11). Since the container is turned during label transfer, the label is pulled on tightly, with trailing-edge gluing ensuring a reliable bond. Finally, the labels are then precisely snug-fitted to the containers, with the overlapping label ends, in particular, being firmly pressed on.

13.3.3
Roll on/Shrink on

Wrap-around labels made of shrinkable film are an attractive alternative to the sleeving process. The labeler applies the reel-fed labels to the containers using UV cross-linkable hotmelt for leading and trailing-edge gluing. The glue is activated by UV irradiation in a tunnel. The UV light ensures a stable bond for the UV

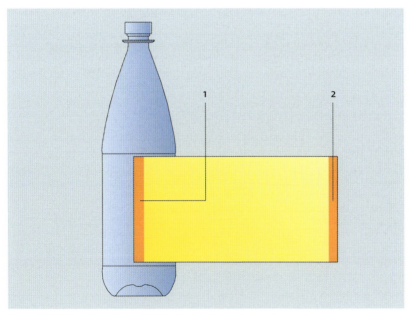

Figure 13.11 The glue strip applied to the label's leading end serves to transfer the label onto the container. 1, Leading-edge gluing; 2, Trailing-edge gluing.

cross-linkable hotmelt – one that will no longer break even at high temperatures. This is crucial for the subsequent heat treatment in the shrink tunnel, in which the plastic film is closely snug-fitted to the container concerned.

The options for using the more affordable label material enables the unit costs involved to be reduced by up to a third compared to sleeve labels. Shrinking is possible up to a maximum film shrinkage of 14%, so it is not yet suitable for very markedly contoured containers.

13.3.4
Tamper-Evident Seals

For tamper-evident seal application, the labeler is fitted with a hotmelt unit as an additional station, which can be removed without problems when not in use. Its job is to apply hotmelt to part of the tamper-evident strip so the latter is securely affixed to the closure. For this purpose, the metal closures should have a special varnish, but not a wax-like coating. If this is not the case, suitable methods like flaming, arc or plasma treatment can be used to pre-treat the surfaces appropriately.

13.4
Pressure-Sensitive Labeling

PSLs are not only visually attractive, their technical properties commend them as well. These include their wet strength or the absence of label-dependent handling parts at the labeling station. Apart from multi-trip labels made of paper or plastic film, PSL systems can also cope with transparent dress in the no-label-look and multi-layered labels like recipe booklets. Labels are also available on the market for returnable bottles, which can be washed off in the bottle washer.

In the case of a PSL, the adhesive is already there on the label and does not have to be applied separately. The label is removed directly from a carrier web and then affixed. In turn, PSLs are used for all commonly encountered container or pack materials, like glass, plastic, cardboard or metal. All container shapes can be handled (round, oval, rectangular, polygonal, etc.). It is essential that the containers arrive in the labeler absolutely dry and free of dust, at a temperature of not less than 16 °C, and also exhibit a good surface quality. This applies most particularly to handling transparent labels, since any unevenness in the container's surface will here result in visible air inclusions. These visual defects cause higher costs compared to wet-glue labeling, unless special wash-off labels are used. After all, if multi-trip labeling is not performed correctly, the container concerned often also has to be rejected, since the label can no longer be removed.

In the pre-labeling process, by contrast, the labels are applied to the bottles before these are filled. The advantage of this is that the bottles need not be blow-dried so as to produce the dry bottle surface required for labeling. The PET bottles thus pre-labeled with PSLs then pass through an aseptic filling process without any problems. All that is left to do when a PSL is to be applied as a tamper-evident seal is to blow off any condensed water in the closure area in a very short blowing tunnel once the bottles have been filled and closed.

The heart of a labeler for PSLs is the label applicator (Figure 13.12), which transports the carrier web with the PSLs to the container with the aid of a drive-driven roller mechanism. The label is detached over the peeling plate using a photo-cell signal and transferred to the container. When the label has been fully applied, it is then accurately snug-fitted by special components at the container table, whereupon the container exits from the machine. In state-of-the-art labeling systems, it is at this stage of the process that the adhesive is activated by heat, pressure or UV light.

The number of label applicators may vary between one for front side only, two for both sides, or four and more for upmarket dress variants. The output will depend on the length of the label involved. In practice, the machines nowadays achieve hourly outputs of more than 60 000 containers an hour.

In addition to stationary applicators, there is also an option for using autonomous ones, which have been designed for easy removal from the machine, to be fitted again at need. The modularized principle applied for the labelers enables label applicators to be fitted in place of the cold-glue labeling stations, or as additional stations on cold-glue and hotmelt labelers.

The applicators for PSLs, too, can be equipped with an automatic reel-changing feature (Figure 13.13). At the end of the first label reel, the machine decelerates to minimum speed. The beginning of the second label reel is spliced automatically to the end of the first one, and the machine accelerates again. As a rule, automatic splicing will proceed so fast that no output losses are incurred but production continues to run uninterrupted. The empty label reel can then be changed, and the adhesive web inserted without any time pressure.

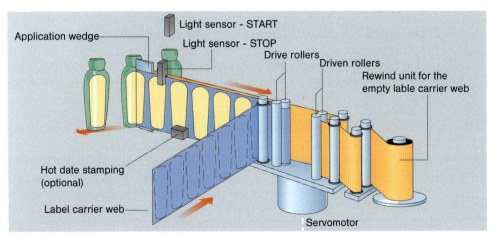

Figure 13.12 In the label applicator, the label is detached by the application wedge (S. 255).

Figure 13.13 The PSL applicators can be equipped with an automatic reel change feature.

13.5
Sleeving

Sleeves enclose the container all the way round, thus offering plenty of room for advertising messages and information of all kinds. Sleeves can also reduce the risk of breakage or improve barrier properties. Sleeving machines can be installed either upstream or downstream of the filler, or also directly on the container manufacturer's premises. They are suitable for the speed range between 3000 and 52 000 containers an hour.

Plastic film sleeves are always applied to the container without any adhesive, with two different processes being used for this. In the stretch-sleeve process, the labels made of stretchable film are pulled over the container, with the film hugging the container's contour for a snug fit. In the shrink-sleeve process, by contrast, the film is initially wrapped loosely around the container and is then shrink-fitted precisely to the container's contour by exposure to heat.

In both processes, the containers arriving on the conveyor are first grasped by the infeed worm, accepted by the infeed starwheel and passed to the container table, where they are fixed precisely in position on the container plates with a centering ring and a holding clamp. The sleeves are now applied using the process involved. The containers are then passed through the discharge starwheel onto the downstream conveyor.

To ensure that there are no malfunctions when the end of the reel is reached, the machine is fitted with a sleeve supply unit. When the sleeve reel is replaced, the end of the first reel is spliced to the beginning of the second one. This obviates the need for threading in the new sleeve. A cutting table located at the sleeve magazine makes this task a whole lot easier. The number of labels can be increased many times over by using an additional reel magazine, which accommodates up to eight reels.

13.5.1
Stretch-Sleeve Process

In the stretch-sleeve process (Figure 13.14), feed rollers unwind the stretchable sleeve film from the reel, before it is transported via the sleeve supply unit to the cutting station, where the sleeve is pulled over a mandrel to open it.

In the cutting station, a blade cuts the sleeve precisely to the selected length. While cutting is still ongoing, the sleeve is passed to the applicator fork, which pulls the film in its stretched state over the container to its final position. After this, a clamp holds the film section in place while the applicator fork is lowered out of the way back to its original position. The holding clamp will not open until the container has been passed to the discharge starwheel, thus ensuring that the sleeve can neither slip nor shift.

Thanks to what is called a memory effect, the film will return to its original shape even after being stretched by up to 10%. This 'contraction' causes the sleeves to fit snugly against the contours of the containers involved.

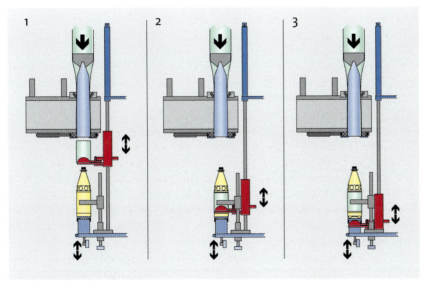

Figure 13.14 Stretch-sleeve process.

13.5.2
Shrink-Sleeve Process

In the shrink process, too, container feed and sleeve cutting follow the principle already described (Figure 13.15). When the sleeve blank is transferred to the container, however, there are two variants to choose from.

In the first variant, the already-opened sleeve is placed on the container by a forward feed unit immediately after cutting. A cylindrical transfer element pushes the sleeve over the container to its final position, where it is securely held by a clamp until the container is passed to the discharge starwheel. This type of transfer is advisable for round containers.

In the second variant, an applicator fork pulls the sleeve over the container, in a similar way to the stretch process. This means the sleeve can be accurately applied even to special-shaped containers.

In a steam or hot-air tunnel, the sleeve is then shrunk onto the container's contour. By the nature of the process, steam shrinking can be performed in significantly shorter heating tunnels than a hot-air process and is also more accurately controllable.

However, steam treatment is not unrestrictedly possible with all materials. PET bottles, for instance, can be exposed to steam only in their filled condition, since with empty bottles the steam has an adverse effect on the material's properties. Thus, if empty PET bottles are to be dressed in a shrink-sleeve, a hot-air tunnel is the only possible option.

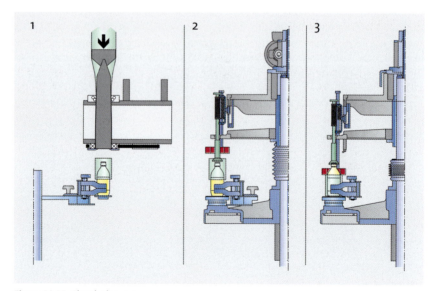

Figure 13.15 Shrink-sleeve process.

Table 13.1 Overview of date coding systems.

Type of coding/dating	Positioning	Label side	Hourly output
Front-of-label coding/dating	gluing pallet	front of label	60000–66000
Back-of-label coding/dating	gripper cylinder	back of label	60000–66000
Perforation date coding	gripper cylinder	back of label	60000–66000
Laser date coding	gluing pallet/bottle carousel	front of label	72000
Inkjet date coding	gluing pallet	front of label	single-line 35000/ two-line 25000
	bottle carousel	front of label	single-line 60000/ two-line 40000
Hot date stamping	applicator rail of PSL station	front of label	28000
Heat transfer, intermittent	applicator rail of PSL station	front of label	20000

13.6
Date Coding and Identification

Statutory regulations demand that the sell-by date and/or a batch identifier be stated for particular products. Bottlers/canners additionally require intra-company markings like code numbers and similar. However, this must not be permitted to impair the visual appearance of the dress. These are tasks that modern-day date coding technology can be relied upon to master, even at continually increasing line speeds. Although in recent years there has been a perceptible shift towards non-contact processes, such as inkjet or laser, in the date coding systems employed, mechanical date coding with stamping or perforation continues to be a viable option for certain applications, so that it has also been covered in Table 13.1.

14
Beer Dispensing
Reinhold Mertens

14.1
Beer Quality in the Draft Beer System

Many different factors influence the quality of draft beer. The breweries put all their efforts into producing and delivering a high-quality product, which should not be altered by the service staff. A well-poured beer is still the highlight of a good pub. Some of factors that have a great influence on beer quality as covered in the following sections.

14.1.1
Temperature

The keg storage temperature should always stay the same to prevent an early ageing process within the keg. The temperature of the beer at the tap should be close to 5 °C, but not higher! The drinking temperature is in the range of 5–8 °C, which varies depending on regional habits. However, the cooling technology has to be adjusted to maintain the preferred drinking temperature throughout seasonal variations in temperature (Figure 14.1).

14.1.2
Time on Tap

Time on tap should be as short as possible. The optimum time is not longer than 3 days. A longer time will always have a negative influence on quality.

14.1.3
CO_2 Content

The CO_2 content of the beer is natural to the brewing process, and is determined within the brewery according to beer type and style. Modern beers have a relatively high CO_2 content of 4.0–7.0 g/l (i.e. wheat beer). Ideally, this content should

Handbook of Brewing: Processes, Technology, Markets. Edited by H. M. Eßlinger
Copyright © 2009 WILEY-VCH Verlag GmbH & Co. KGaA, Weinheim
ISBN: 978-3-527-31674-8

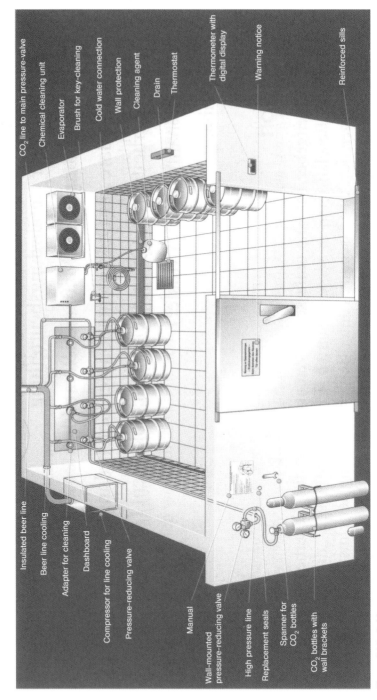

Figure 14.1 Cold room.

not be changed while the keg is on tap. The main influence of the natural CO_2 pressure is the applied pressure to propel the beer to the tap (Section 14.2.5). The ideal gauge pressure is the same as the natural saturation pressure of the beer in the keg.

14.1.4
Foamhead

The head on a beer brings out the flavor and the aroma. The foam should consist of very small bubbles of the same size and cling close together, and should be long-lasting. Important factors are the correct pressure, technical and hygienic state of the dispensing system, and the condition of the beer glass.

14.1.5
Pouring the Beer

This is the last point of the beer journey with a important impact on quality. The service staff have to observe the parameters of a high-quality draft beer at all times, and answer over and over again the following questions:

- All parameters observed? Correct pressure? Temperature? Clean beer lines?
- Are the glasses clean?
- Are the glasses flushed before pouring?
- Proper pouring technique?
- Pouring at the correct flow rate?

14.2
Design of Draft Beer Systems

The basics to preserve the quality of the beer are determined by the design of the dispensing system. The physical parameters (CO_2 pressure) and hygienic design and conditions are the key points relating to draft beer quality.

14.2.1
Requirements for Rooms

The location of the keg storage, cabinet or walk-in closet and the distance to the bar are decisive for the installation of beer lines. The location where the beer kegs are stored should preferably be refrigerated to 5 °C (dispensing temperature). Hygienic condition of the insulation as well as of the stored and tapped kegs is very important. All surfaces and the floor should be easy to clean, with a water supply, sink and sewer installed or nearby. The illumination should be at a minimum of 100 Lux for safe and correct working conditions. For safety reasons, the doors of refrigerated rooms must be designed to be always opened from the inside.

Beer line installation through walls or ceilings should always be done by using protection pipes. Large temperature changes between rooms require the installation of steam blocks.

14.2.2
Requirements for Refrigeration

All refrigeration units should be installed in suitable rooms. The rooms must be equipped with large enough openings for fresh and exhaust air. If the room is heated up by exhaust air, the output of the refrigeration unit is drastically reduced. The temperature control must be installed near the warmest spot in the room. There are three types of cooling in a dispensing system.

14.2.2.1 Storage/Cabinet Cooling
This cooling has to be designed strong enough to keep the beer at a temperature of 5 °C for dispensing. It is very important to understand how many kegs have to be cooled in what time – a freshly delivered keg at the regular outside temperature needs about 24 h to be cooled down to 5 °C!

14.2.2.2 Ancillary Cooling
This cooling keeps the beer cold over a wider distance from the keg storage to the tap. It is *not* designed to cool down the beer! The Python water cooling system should be designed large enough and specified to the amount of beer dispensed over 1 h. Installation of dispensing tap chilling prevents CO_2 from being dissolved and causing foam pockets and pressure instability. However, installing this tap chilling can cause the tap and font to swet, producing condensed water outside and inside of the font.

14.2.2.3 Bar Cooling
Bar cooling is mainly used for bottled beverages. If beer lines and Pythons run inside and through the cooled bar, the temperature should be kept the same as inside the beer lines (5 °C as an optimum).

14.2.3
Requirements for Beer Lines

The requirement for beer lines are: hygienic design, pressure-resistant material and good insulation, especially for Pythons. A Python is designed as follows: beer lines, ancillary cooling coils, CO_2 tube, vapor stop, insulation tape and minimum of 13 mm of insulation tube. Both ends of the Python have to be vapor-insulated after installation, otherwise condensed water will be produced inside of the Python and all cooling effects will be lost. Beer lines have to be installed elevated for self-draining to avoid hygiene problems during periods of non-use of the draft system (e.g. closed periods, etc.).

14.2.4
Requirements for CO_2 Lines

Again, pressure resistance and hygienic design are the main points. Very often CO_2 lines are installed too close to the warm exhaust air of refrigeration units. Temperatures of above 25 °C are very common. Not all lines are resistant to high temperatures.

The inside diameter of CO_2 lines should be a minimum of 7 mm, preferably 10 mm to ensure enough flow volume and avoid dispensing problems. The gas lines need to have a large enough diameter to maintain correct pressure inside the keg during times of high demand. The quality of pouring a perfect draft beer depends largely on the quality of the gas supply.

14.2.5
Requirements for Beer Bars/Bar Counter

The work surface should be stainless steel for easy cleaning and hygienic handling. There should be no blind spots for hygienic reasons. All piping, water supply, sinks, drains, electricity and refrigeration should be thoroughly considered in the planning stages. A condensed water drain inside the bar with an odor stop, drip-tray with drain and odor stop, and inside bar illumination have to be installed.

A water supply with warm and cold water and a hand cleaning station for the service personal at the bar or nearby are important hygienic requirements.

14.2.6
Requirements for Glass-Washing Equipment

The hygienic condition of the beer glasses should be of the same quality as the drinking water used for flushing the glasses prior filling. All equipment used must be in proper condition (e.g. sinks, brushes, glass-cleaners, etc.). Always follow local health regulations. The following three methods are widely used.

14.2.6.1 Two-Sink Installation
Sink 1 contains warm water, cleaning brushes and a special beer-glass cleaning detergent. The second sink is filled with fresh, clear drinking water for flushing. Change the contents of sink 1 frequently according to use. Allow water to flow slowly through the underwater tap continuously into sink 2. This skims dirty parts to the water surface into the overflow tube.

14.2.6.2 Glass Cleaning System
The system consists of a cleaning and a flushing part. In the cleaning pot there is a brush installed and a special beer glass cleaning detergent (mostly cleaning tab) used with tab water. The second part is the fresh water flushing of the glass: inside and outside! During the flushing cycle fresh water is always filled into the cleaning

pot and the dirty water is pushed overboard through an opening on the upper edge of the cleaning-pot.

14.2.6.3 Glass Washing Machines

Always refer to the manufactures recommended procedures for handling and maintenance. Washing machines are always used with two detergents: cleaning and flushing. It is important that both detergents are used. The flushing detergent (mild acid) is responsible for neutralizing the remains of the cleaning detergent (strong alkaline) and providing a sparkling glass. The temperature of the hot cleaning water must be restricted to a maximum of 60 °C. The quality of the beverage glass will very quick deteriorate with higher temperatures. A very smart device for all glass washing types is called a glass shower. This shower is built in the working surface on top of the bar and used for every glass before filling. In this way all dust and other particles are removed, and the glass is cooled down for a sparkling beer.

14.2.7
Calculation of Applied Gauge Pressure

A very important ingredient of a carefully brewed beer is CO_2. It has special flavor characteristics, and is responsible for the fresh taste and a good foam. The CO_2 content is defined in the brewery according to the type and style of the beer. It ranges generally from 4.0 to 7.0 g/l. During the time of drafting this defined content should be maintained until the keg is empty. There are three different pressures to consider: saturation pressure, working pressure and applied gauge pressure.

- The saturation-pressure depends on the CO_2 content intended by the brewery and the temperature. Within the keg there is an equilibrium between the keg's headspace and the beer.

- The working pressure is the gas pressure necessary to propel the beer from the keg to the font and tap. Factors that influence the pressure are beer tubing (length and diameter), gravity (height difference from keg to tap) and system hardware (unions, couplers, shanks, etc.).

- The applied gauge pressure is the actual pressure (in bar) applied to the keg and set at the pressure regulator. If two or more different beers are on tap, the individual pressure for each beer line is set on the secondary pressure gauge.

If the applied gauge pressure is equal to saturation pressure, we have the ideal conditions for drafting a perfect beer. There is no change in taste or character – the beer reaches the customer as intended by the brewmaster. If the applied pressure is higher than the saturation pressure, CO_2 is imparted in the top surface layer where the beer and the headspace meet, over-saturating the beer all the way to the bottom of the keg within 2 days. This will finally result in taste and dispensing problems. If the applied gauge pressure is lower than the saturation pressure, CO_2 is allowed to leave the beer and enter the keg's headspace, resulting in a flat beer. Again, there is a negative impact on character and taste.

During regular drafting of beer the desired pressure at the tap should be between 0.1–0.2 bar with a volume flow of about 3 l/min. If a compensator tap is used, the pressure at the tap must be equal to the saturation pressure to avoid dispensing problems (heavy foaming with low pressure). The calculations below show the different pressures needed (the parameters listed in Table 14.1 are used). Here, we use a beer with a CO_2 content of 4.8 g/l at a temperature of 6 °C = saturation pressure of 0.88 bar.

- Each 1 m of vertical distance has a restriction pressure of 0.1 bar.
- Each 1 m of beer line with a diameter of 4 mm has a restriction pressure of 0.72 bar.
- Each 1 m of beer line with a diameter of 7 mm has a restriction pressure of 0.05 bar.
- Each 1 m of beer line with a diameter of 10 mm has a restriction pressure of 0.01 bar.

Restriction is defined as the pressure in bar of resist that the beer encounters as it flows through the system. We can calculate the applied gauge pressure for the given parameters as:

Vertical distance = 0.40 bar

Length 9 m with 7 mm diameter = 0.45 bar

Working pressure = 0.85 bar

The working pressure is very close to the saturation pressure (i.e. good parameters!).

Table 14.2 shows the change in gauge pressure by simply changing the hardware. In this dispensing system the applied pressure of 0.95 bar would be the optimum. Using this setting, the volume flow of about 3 l/min is set. To test the correct setting, the tap should be opened for about 15 s. The measuring cup should then contain 750–850 ml. This shows a flow of 3–3.5 l/min. Depending on the size of the beer outlet, volumes of 5 and 15 l/min are used.

Table 14.1 Calculation parameters for gauge pressure.

CO_2 content of the beer (from the brewery)	4.8 g/l
Temperature of beer storage	6 °C
Calculated saturation pressure	0.88 bar
Beer line diameter	7.0 mm
Beer line length	9.0 m
Gravity (vertical distance keg–tap)	4.0 m
Faucet: compensator tap	no
Volume flow	3 l/min
Special restricting hardware (pumps, volume control, etc.)	no

Table 14.2 Change in gauge pressure (bar) by changing the hardware.

		Beer line diameter			
		7 mm		10 mm	
Vertical distance (m)	4.0	0.40	0.40	0.40	0.40
Length of beer line (m)	9.0	0.45	0.45		
Length of beer line (m)	9.0			0.10	0.10
Compensator tap	yes	0.88		0.88	
Compensator tap	no		0.10		0.10
Applied gauge pressure (bar)		1.73	0.95	1.38	0.60

14.3
Dispensing

14.3.1
Types of Dispensing

There are thousands of different draft beer systems utilizing three basic types of dispensing.

14.3.1.1 Dispensing from Underneath the Beer Bar
The keg is located underneath the beer bar in a cooled storage room. Here, we have to overcome gravity, which is referred to as the vertical lift from the bottom of the keg to the tap. A typical direct draw system from the basement has a vertical distance of 3 m. This requires a calculation of 0.3 bar (0.1 bar/m) (Figure 14.2).

14.3.1.2 Dispensing from above the Beer Bar
If the keg location is above the beer bar, the same principles are applied. The vertical drop and the beer line resistance cancel each other, so this distance is not taken into consideration (Figure 14.3).

14.3.1.3 Dispensing Direct from the Beer Bar
The beer is dispensed direct from the beer bar, if possible out of a cooled compartment. The beer tubing can be used to balance the system (length and diameter). Other means of reducing the volume are restricting coils or a compensator tap (Figure 14.4).

14.3.2
Dispensing with Beer Pumps

Beer pumps are being used more and more, and are well known from the soft drinks business, especially from post-mix dispensing.

14.3 Dispensing | 347

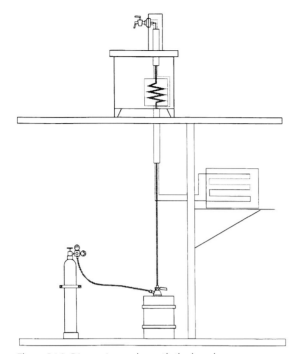

Figure 14.2 Dispensing underneath the beer bar.

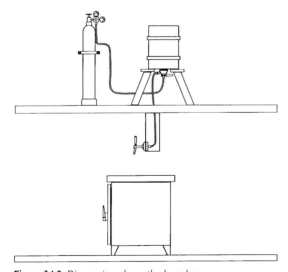

Figure 14.3 Dispensing above the beer bar.

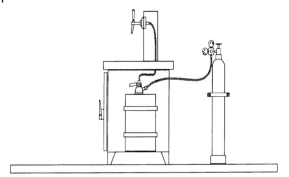

Figure 14.4 Dispensing direct from the beer bar.

All beer pumps use a pressure sensor and a pressure switch. When opening the tap, a sudden drop of pressure in the beer line starts the beer pump. The desired flow volume is set at the pump. This installation is widely used with long beer lines or large vertical distances, or both. The kegs or beer tanks are pressurized only with the original ideal saturation pressure as determined by the brewery. The pump does the rest of the work to overcome the resistant pressures (height and lines). No over-carbonization or decarbonization take place and beer is enjoyed as the brewmaster intended.

Beer pumps are driven by electricity or gas (air, CO_2, N_2). When using CO_2, a ventilation duct for fresh air has to be installed to avoid high concentration of CO_2 in the storage room. When using air compressors, oil-less air compressors have to be installed in a separate dry room. An air filter with a water separator must be incorporated into the line after the compressor and before the first pump.

During the cleaning cycles the beer pumps are cleaned chemically with a special line cleaner, while the beer lines after the pump can be cleaned chemically and mechanically.

14.3.3
Dispensing Beer with Pre-mixed Gas (CO_2/N_2)

Mixed gas from cylinders and on-site gas blending are the sources of this medium for dispensing beer. Mixed gas in a cylinder with a ratio of up to 30% CO_2/70% N_2 is appropriate for stout beer, but when applied to ales, lagers, Pilsner or wheat beers this allows the beer to go flat because the partial pressure of CO_2 is too low. It is difficult to raise the ratio of CO_2 in the mix as this gas eventually liquefies under high pressure within the cylinder. This mix is rather expensive and the very high pressure within the cylinder increases the accident risk. Unfortunately, many outlets improperly apply mixed gas in a cylinder to their beers on tap. This is an attempt to offset system inefficiencies (temperature, length, height) with high applied gas pressures and the wrong gas blend ratios, thus affecting the beer's flavor, quality and sale.

The last thing a bar owner and the brewery wants is to cause the customer to switch to a bottle or a can, or go somewhere else to enjoy a different beer!

14.3.4
Use of Gas Blenders

Pre-mixed gas in the cylinder does not have the necessary ratio to propel the beer to the tap and at the same time to support the required CO_2 content. To achieve all these goals, on-site blending is used more and more. On-site blending entails a source of CO_2 such as a gas cylinder and a nitrogen source such as a nitrogen cylinder or nitrogen generator. Both sources are connected to a blending device. The ratio is set accurately to the necessary blend. The blend is customized to deliver the correct CO_2 amount to keep the beer fresh and the correct nitrogen amount to propel the beer from the keg to the tap (Figure 14.5).

14.3.5
Computerized Beer Dispensing

Computerized beer dispensing is mostly used in large restaurants such as discotheques, cinemas, and so on. There are different types of dispensing units on the market. Basically they are used to measure the correct amount of beer dispensed:

- How much beer was sold by the individual service person? Who served how much beer?
- Dispensing with a push of a button the correct amount.

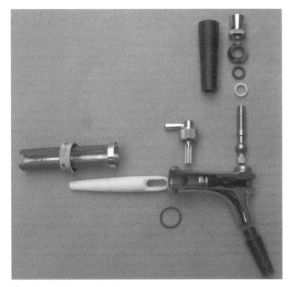

Figure 14.5 Dispensing tap.

The amount is measured with flow turbines or with induction control.

Maintenance of the unit and cleaning of the beer lines and measuring devices have to be done at least every week to ensure correct measuring and beer quality.

14.3.6
Beer-Dispensing Tanks

Large beer outlets do not use the regular 50-l kegs, instead they use beer-dispensing tanks. The beer is delivered in trucks and pumped into the storage tanks. Again, we have two types of tanks:

- *Tanks without an inliner.* The beer in these tanks has direct contact with the inside wall of the tank. This requires a thorough cleaning of the tank before each filling. The empty tank has to be pressurized with the correct CO_2 pressure of the beer before pumping the beer into the tank. The beer is transported with CO_2 pressure or beer pumps to the dispensing tap.

- *Tanks with an inliner.* The beer is pumped into an inliner which lies between the inside wall of the tank and the beer. With each filling there comes a new, sterile inliner made of an airtight plastic. The beer is pumped with air pressure to the dispensing points. This air pressure is applied to the outside of the inliner and does not affect the stored beer. The advantage of this type of dispensing is that the CO_2 content of the beer is not affected.

14.4
Parts of Draft Beer Systems

14.4.1
Requirements for Gas-Pressurized Parts

Owing to the relatively high pressures and the CO_2 content within the gas lines of the dispensing system, they need the same attention as the beer lines.

We differentiate between high pressure and regulated pressure. The high-pressure side is between the CO_2 cylinder and the primary CO_2 beer regulator. If the pressure regulator is not installed directly to the gas cylinder, a high-pressure line has to be installed. This line must withstand a pressure of up to 200 bar. All gas cylinders in use must stand in an upright position and be secured against falling. All the parts on the gas side of a draft beer system must withstand a pressure of a maximum of 3 bar. The safety valve of the pressure regulator is set to a maximum of 3 bar. A higher pressure than 3 bar is released via the safety valve. The safety valve has to be checked regularly for proper function. The pressure gauge (0–6 bar) at the primary regulator has a red safety mark at 3 bar. The high-pressure cylinder gauge indicates the inside pressure for CO_2 (0–250) and nitrogen/blended gas (0–315 bar).

The primary regulator divides the high pressure side from the regulated or low-pressure side. Each individual beer line has to be regulated with the secondary beer regulator. To ensure enough pressure to the individual kegs, the primary beer regulator has to be set as high as possible (2.0–2.9 bar). The lower pressure for each keg is set at the secondary regulator. The diameter of the gas lines should be no smaller than 7 mm to ensure sufficient gas volume during times of high demands. For safety reasons the gas lines should withstand a pressure of 7 bar. All parts of the gas side bust be secured against pressure deformation.

14.4.2
Requirements for Beverage Parts

The main focus on this side of the dispensing system is the hygienic design of the parts and the overall hygienic condition. All accessories such as tube nipple, tube connecting parts or taps should be manufactured from stainless steel with face-sealed connections.

For the installation of the beer lines (Python), a multi-layer plastic tube (polyamide, polyethylene, ethylene vinyl acetate, Nylon) should be used and polyvinylchloride avoided. The inside surface of the tube must be smooth. Again, beverage tubes with a working pressure up to 7 bar should be installed for safety reasons. All the parts like beer taps and keg couplers are easily connected with tube connecting parts. The placement of the clamps is very important. CNS clamps with 'one ear' and an inside ring should be installed. When push-in fittings are used, the tube must be cut square, and burrs and sharp edges must be removed. The tube is then pushed into the fitting up to the tube stop. An O-ring inside provides a permanent leak-free seal.

It is important for all connecting methods to have a face-sealed connection to minimize hygienic hazards. The risk of a contamination has to be kept to an absolute minimum.

All parts of the beverage side have to withstand deformation due to higher pressure.

14.4.3
Keg-Tapping Equipment

There are many different keg systems, but the following three systems are most widely used:

- *The European Sankey ('S' system)*. The keg coupler handle is in the upright position and the two coupler lugs are aligned with the corresponding openings in the keg valve. The keg coupler is turned 90° clockwise and locked into position. Pressing the handle downward will allow gas to enter the keg and beer to flow out of the keg to the tap. Pulling the handle in the upright position will close the gas and allow the extractor tube to close the beer supply.

- *The Germans slider ('A' System)*. The keg coupler handle operates the same in all systems. The keg coupler is aligned with its base to the side of the keg valve and simply slid over the keg valve. The handle opens and closes the gas and beer supply.

- *The slider 'M' system.* This system works like the 'A' system with a unique probe design and a security combination extractor tube. It allows one to connect several kegs 'in series' and dispense at the same time without changing kegs.

14.5
Hygiene Requirements in Draft Beer Systems

Keg beer, delivered from the brewery to the outlet, should be stored at the correct temperature without large temperature changes. It tastes great and has no hygienic deficiencies. Once connected to the draft system the keg is opened and negative factors can start affecting the beer. In particular, bacteria can enter the beer tubes and enter the keg. This will affect beer quality, taste and stability. The whole draft beer system must be cleaned and disinfected (Figure 14.6).

14.5.1
Hygiene Target

The hygiene target is to minimize the input of any negative materials into the system. Draft beer can be affected by the build-up of bacteria, yeast, mold and beer stone within the beer lines and the accessories. This should be kept on a minimum level. To help maintaining a perfect sanitary level, special cleaning and disinfecting procedures have to be performed.

Figure 14.6 Dispensing head on top of a keg.

14.5.2
Cleaning and Disinfecting Procedures

Unclean beer lines and dispensing equipment have a negative effect on the beer. This contamination has to be removed. A cleaning interval of 7–14 days is necessary to keep the system in a perfect hygienic condition. An appropriate cleaning procedure has to be implemented to reach this hygiene target and eliminate any risk to quality and the customer.

The type, size, location and use of the system determines the correct procedure. Whenever the use of a cleaning sponge is possible, the combination of chemical and mechanical cleaning procedures should be applied. The diameter of the cleaning sponge should be a minimum of 2 mm larger than the diameter of the beer line. Cleaning sponges are intended for single use.

The basics of cleaning are specified by Professor Sinner (the Sinner Circle). The four parameters of cleaning chemical, temperature, mechanics and time are closely related. If one parameter is reduced, other parameters have to be adjusted accordingly. For example, if the temperature of the draft system has to stay low (below 10 °C) other parameters have to be increased (more chemicals, more time or more mechanic work).

- Cleaning chemical: according to the manufacturer's instructions (1–5%).
- Temperature: cold or warm (preferably warm!).
- Time: contact minimum 10–15; better 20–30 min.
- Mechanical: use sponges whenever possible or at least recirculation.

The combined chemical/mechanical cleaning procedure provides the best results – so whenever possible use cleaning sponges for beer lines.

The recommended procedure for chemical/mechanical cleaning is:

1. Connect all beer lines to a circuit.
2. Connect the beer lines to a cleaning device.
3. Flush beer lines with water until no beer is left in the lines.
4. Use alkaline or acid line cleaner as required.
5. Fill the line circuit with cleaner.
6. Add the sponge into the circuit.
7. Recirculate chemical and sponge (15–20 min/according to manufacturer).
8. Flush the chemicals from the line with water.
9. If cleaner without sanitation is used, repeat steps 3–8 with disinfectant.
10. Clean and sanitize tap with a brush and flush with water.
11. Clean and sanitize keg coupler with a brush, and flush with water.
12. Re-connect taps and couplers.
13. Re-connect the keg.
14. Flush the water from the lines with beer until ready for pouring beer.
15. Check the system for leaks.

The recommended procedure for recirculating cleaning is:

1. Connect all beer lines to a circuit.
2. Connect the beer lines to a circulating pump.
3. Flush the beer from the lines with water.
4. Use alkaline or acid line cleaner as required.
5. Fill the line circuit with cleaner.
6. Recirculate cleaner as recommended.
7. Flush the cleaner from the line with water.
8. Repeat steps 3–8 with disinfectant if necessary.
9. Clean and sanitize the tap and keg coupler, and flush with water.
10. Re-connect the tap and coupler.
11. Re-connect the keg.
12. Flush water from the lines with beer until ready for pouring
13. Check the system for leaks.

If cleaning with sponges or recirculating cannot be done use the following recommended procedure for soaking:

1. Connect all the beer lines to the cleaning container.
2. Flush the beer from the lines with water.
3. Use alkaline or acid line cleaner as required.
4. Fill the beer lines with cleaner.
5. Allow chemicals to soak in the line as recommended.
6. Flush chemicals from the lines with water.
7. Repeat steps 3–6 with disinfectant if necessary.
8. Clean and sanitize the tap and keg coupler, and flush with water.
9. Re-connect the tap and couplers.
10. Re-connect the keg.
11. Flush water from lines with beer until ready for pouring.
12. Check system for leaks.

The hygiene goal can only be achieved if parts in daily use like the tap and coupler are cleaned regularly. The tap should be cleaned daily with a tap cleaning ball which is available in suppliers. The coupler must be cleaned with every keg change to avoid possible cross-contamination.

These procedures can be executed by the bar owner or by a professional cleaning agent. Some breweries provide this service to their customers.

14.5.3
Hygiene Problem Areas

The requirements for the parts of a dispensing systems are shown in Section 14.5.2. Each part or accessory constitutes a weak point in the system. The weakest points of a dispensing system are at both ends: the tap and the keg coupler. These two parts have contact with air and beer at all times. Other parts like turbines in beer lines have to be uninstalled and cleaned manually at regular intervals. Parts like pumps have a limited lifetime. and must be maintained and checked as recommended by the manufacturer. However, most problems arise from tube connecting parts. Only face-sealed connections provide proper hygienic conditions within the lines.

14.6 Testing

There are frequent tests necessary for continuously used equipment to certify the correct technical and hygienic condition of the dispensing system. The system has to be checked and the results documented to reflect the present status to the owner and the authorities.

14.6.1 Hygiene Testing

The biological condition of the system is certified through hygiene testing. A small amount of beer is poured into a sterile glass, inoculated on a culture medium and incubated at different temperatures. We differentiate between typical and non-typical organisms. Each country allows a specific germ count. *Escherichia coli* bacteria are definitely not allowed. Apart from the biological test, there are also non-biological tests available. They have to be validated and compared to the biological results.

14.6.2 Leak Tests

As shown in Section 14.5.2, the gas-pressurized parts need the same attention as the hygienic portion. The whole system has to be gas-tight. Leaking CO_2 can be life-threatening for people working with dispensing systems, especially with a large gas supply. Leaks are also the reason for deteriorating beer quality and malfunctioning systems. A simple means for checking for leaks is to close the gas cylinder. When observing the pressure gauges, there should be no decline in indicated pressure. No beer should be dispensed during this time. If there is a decline in indicated pressure, the search for a leak must be started immediately. A certified manometer has to be used for thorough testing. Every new installation should have a pressure check. A pressure gauge is installed at the end of the line and a defined gas pressure is applied to the system. Again there should be no decline in indicated pressure.

14.6.3 Temperature Tests

There are many different temperatures within the whole dispensing system:
- Beer temperature within the keg.
- Room temperature within the cooling cabinet or storage room.

- Beer temperature at the tap.
- Temperature for ancillary cooling.

Initially, minimum and maximum temperatures are given and checked every 6 month to keep the beer dispensing free of quality problems. Maintenance of the room refrigeration units and the ancillary cooling unit and change of water supplies is necessary.

14.7
Safety Precautions

There are certain safety precautions necessary whenever CO_2 is used within a dispensing system in order to prevent accidents or loss of life to people working in this environment. CO_2 is odorless, tasteless and invisible, but it can kill. Danger starts at a air concentration higher than 3% CO_2 in breathing air. Sudden death will occur at a concentration of about 10% CO_2.

14.7.1
CO_2 Gas Alert Units

In addition to technical air ventilation to prevent high CO_2 concentrations, a gas alert unit is a very practical and useful device. Alert units are only to be installed by authorized personal, and require periodic maintenance and service. The placement of the sensor unit has to be close to the source of contamination. Mechanical damage and damage due to rinse/spray water must be avoided. The alert unit has to be installed outside of the danger zone. An optical as well as an acoustic signal should alert the staff. Malfunctions of any kind must be indicated to avoid unintentional entry of the danger area. The staff have to be briefed on all aspect of the detection and alert features of the unit.

The CO_2 concentration in a room is calculated as follows:

- CO_2 concentration (%) = CO_2 available $m^3 \times 100$/room volume m^3
- CO_2 available = 1 cylinder of $10\,kg = 5\,m^3$
- Room volume = L $10\,m \times$ W $5\,m \times$ H $2\,m = 100\,m^3$
- Calculation: $5 \times 100\,m^3/100 = 5\%$ CO_2 content in room = danger!

The installation of a warning unit is mandatory in many countries (e.g. under the German 'Workman's Protection Regulation').

14.8
Final Remarks

The last meters of the long journey of the beer to the customer are very decisive in relation to beer quality. Maintaining the quality as it comes from the brewery should be the aim of every dispensing facility. Thorough planning, using the correct parts and accessories, and servicing and cleaning will assure a high-quality product and allow the provider to maintain a brewery-fresh beer day after day. Serving a great-tasting beer will keep the brewery, the customers and last but not least the outlet-owner happy!

Reference

1 German Brewing Association (••) Standards for dispensing (Inbev) DIN Standards IMI. Cornelius Micro-Matic Selbach John Guest Python Fachverband Getränkeschankanlagen eV.

15
Properties and Quality
August Gresser

15.1
Composition of Finished, Bottom-Fermented Beer

Beer with an original extract of 12 wt% (12 °P) consists of:

- 4.0–4.5 wt% extract consisting of:
  - 80–85% carbohydrates
  - 4.5–5.2% proteins
  - 3.5% glycerin
  - 3–4% minerals
  - 2–3% tannins, bitter substances and coloring matters
  - 0.7–1.0% organic acids
  - a relatively small amount of vitamins
- 3.8–4.2 wt% alcohol or 4.7–5.2 vol% alcohol
- 0.42–40.55 wt% CO_2
- 90–92% water.

In addition to water, beer contains other volatile compounds:

- Alcohols and their derivates:
  - higher alcohols: 60–120 mg/l
  - volatile organic acids:
    - acetic acid: 120–200 mg/l
    - formic acid: 20 mg/l
  - esters: 20–50 mg/l
  - aldehydes: 5–10 mg/l
  - total diacetyl (diacetyl and 2-acetolactate): <0.10 mg/l
  - acetoin (3-hydroxybutanone): <3.0 mg/l.
- Gases:
  - CO_2: 4–6 g/l
  - air:

Handbook of Brewing: Processes, Technology, Markets. Edited by H. M. Eßlinger
Copyright © 2009 WILEY-VCH Verlag GmbH & Co. KGaA, Weinheim
ISBN: 978-3-527-31674-8

- oxygen
- nitrogen.

15.2
Overall Qualities of Bottom-Fermented Beer

The specifics of beer can be described as follows:

- Specific weight: 1.01–1.02 g/cm^3
- Viscosity: 1.5–2.2 mPas
- Surface tension: 42–48 dyn/cm
- pH: 4.25–4.60.

15.3
Redox Potential

The redox potential is defined by the contents of reductones, melanoidins, as well as the beer's oxygen content. Polyphenols, sulfhydryls, nitrogen substances and bitter substances of hops possess reducing capacity. A low rH value (real or apparent hydrogen pressure) is important for flavor stability as well as for the physicochemical and biological stability of beer. To a certain elegree the beer is protected from oxidation by its reducing substances [1].

15.4
Beer Color

The beer color influences the beer type substantially. The mashing technology must be performed in accordance with the beer type depending on the quality of the raw materials (water, malt) and the existing brewhouse equipment. The malt used has a substantial influence on the aroma and the color [2].

The color for Pilsner-type beers ranges between 5.3 and 7.5 European Brewery Convention (EBC) units. Dortmunder and strong export beers show colors between 9.5 and 12 EBC units, Wiener and Märzen beers between 18 and 30 EBC units, and dark beers between 45 and 95 EBC units.

In order to obtain dark beer colors, dark malt, dark caramelized malt between 3 and 5%, 5% of color malts or less and roast malt beer are used. To guarantee a sufficient activity of enzymes it is advantageous to add 10–20% of pale malt. When harder water with a high residual alkalinity is used, darker beer colors will result. The use of soft water promotes the malt aromatic character [2, 3].

The tone of color should be brilliant and correspond to the beer type. Changes in color may result from the use of unsuitable raw materials or water quality, mistakes during wort production, insufficient trub separation or sluggish fermentation. With a high content of oxygen in bottled beer the color will increase by 0.5–1 EBC units, also giving a slightly red nuance [1].

15.5
Taste of Beer

The beer taste should correspond to the beer type and remain constant in the racked beer as long as possible. The beer taste is characterized by different perceptions that occur to the taste organs, shortly following one after the other they can melt together and then decline more or less quickly. Also, smell is directly involved in the taste sensation. One distinguishes between first taste, carbonation and final taste. Thus, the impression as a whole should be well balanced. The intensity of the taste impression depends on the temperature, CO_2 content and personal disposition of the taster.

Smelling indicates, first, the aroma of the beer. The first taste also conveys the impression of a more or less present body (palate) which correlates with the original gravity and beer color. The liveliness (carbonation) occurs in the middle of the taste impression. It depends on the pH value, amount of dissolved buffer substances and on different phosphates, which are defined by malt quality and brewing liquor. The final taste is mostly induced by the bitterness. The final taste can also fade away slightly sour or sweet, depending on the ratio of the bittering substances to the remaining colloids. The total taste impression of a beer is smooth when its different perceptions are in harmony and therefore blend well with each other. The bitterness should fade away more or less quickly, depending on its original intensity and without leaving an astringent aftertaste [1].

Taste and aroma are not stable [4]. Therefore, flavor stability is an important property for beer quality. Flavor stability can be improved by technological means and by addition of antioxidative substances (see also Chapter 16).

15.6
Beer Foam

The foam head on a beer glass is not only an objective qualitative characteristic, evidently and noticeably, but principally indirectly an important indicator for the quality of beer. Furthermore, the foam layer has a particular importance for chemical and physical aspects. First of all, the foam decreases the temperature rise of freshly tapped beer. Moreover, the foam decelerates the release of CO_2 and avoids the oxygen uptake of the beer by a factor of 10^3. This function of protection by the foam already becomes apparent by the overflow of the foam when bottles are filled [1, 2].

15.6.1
Basis of Beer Foam

Many surface-active ionic substances are primarily located in the shield of the gas bubbles. In the main part there are polypeptides with a high molecular weight and with pronounced hydrophobic properties that ensure a large viscosity and surface elasticity of the film of gas bubbles. Nitrogen compounds with a low molecular weight like amino acids and oligopeptides are able, when in higher concentrations, to banish polypeptides from the film, thus decreasing foam stability. Certain bitter hop substances and heavy metal ions can react with polypeptides with a stabilizing effect on the film. In terms of good foam stability, substances present in the fluid spaces between the gas bubbles, particularly non-ionic hydrophilic substances like α- and β-glucans, should have a high molecular weight. That way they increase the liquid viscosity and decelerate the migration from the foam lamella, increasing the foam stability and head retention [5, 6].

15.6.2
Influence of Gas

Nitrogen and oxygen have a positive effect on beer foam; CO_2 influences beer foam in a negative way. A high CO_2 concentration in combination with a fast gas release will produce large gas bubbles in the film in which less foam-active substances can exist.

15.6.3
Influence of Foam Stability

Apart from the foam-positive substances already mentioned, there are other foam-positive compounds like melanoidins, glycoproteins, pentosans and oligomer polyphenols with a low molecular weight. Amino acids with pronounced hydrophilic properties, as well as peptides, glycerides, stearine, medium-chain fatty acids and polymer polyphenols with a high molecular weight, lead to a deterioration of foam stability. This is valid also for compounds like fats, esters, sulfites, and detergents used for cleaning and disinfection [1, 7–14].

Specific non-ionic and cationic detergents may react negatively on the beer foam. Ethyl alcohol, which diminish the surface tension, can positively affect foam stability, but only in low concentrations, whereas it has a negative effect at higher concentrations [15].

15.6.4
Influence of Brewing Liquor

The brewing liquor plays a secondary and indirect role with regard to its influence on foam. Foam stability in beer is improved by reducing the residual alkalinity (and thus the pH) of the mash [16].

15.6.5
Influence of Hop Products

By means of the polyphenol fractions, the type of hop products has a great influence on foam stability. Practical experiments have shown that foam stability in beer increases using hop powders, standard extracts or extracts with bitter substances [1].

15.6.6
Influence of Malt

It is the malt that brings in the real carriers of the foam. It influences the beer foam, even if indirectly, by its influence on several steps during the production process of beer (mash filtration, fermentation, maturation and beer filtration). Beers brewed with malts from winter barley varieties like Escourgeon and Sonja show slightly reduced foam properties than beers from the varieties Nymphe and Hydra and from spring barley varieties. From the analytical characteristics of malt, like soluble nitrogen compounds, viscosity of wort and of beer, fine–coarse difference, content of β-glucans, and filterability of beer, it is not possible to make a clear prognosis on foam stability. On the contrary, 'foam-positive' malts often bring disadvantages with respect to extract yields, filterability of mash and beer, and also the tendency to haze formation [1, 16].

15.6.7
Influence of Mash Filtration

Experiments have shown that foam stability is not clearly affected by wort turbidity and/or lautering time. Even beers brewed from extreme cloudy worts showed only a tendency to a lower foam stability; however, in this case the beer filterability was poor. Higher filtration temperatures (78 °C instead of 70–72 °C) result in better foam values, but also in an inferior filterability and in a higher tendency to haze formation. This has been realized during experiments using mixtures of malt from spring and winter barleys.

Beers from winter barley malts showed inferior foam values. The reduced foam properties of beers from winter barley may be induced by a higher concentration of foam-negative substances. As small-scale experiments have shown, the foam-positive substances dissolve in the liquid in fair amounts at a mash filtration temperature of 70 °C. Higher temperatures did not cause any sensitive variances [1, 16].

15.6.8
Wort Boiling

During wort boiling, both foam-positive and foam-negative substances are reduced, especially in the case of intensive boiling at higher temperatures and when hop-derived polyphenols are added. Experiments have shown that subdivision of the hop addition and reduction of the total evaporation rate from 5 to 3% do not have any influence on foam stability. In another brewery, the change from atmospheric boiling to vapor compression and a reduction of the evaporation rate from 10 to 5% have not shown any statistically significant decrease of foam stability. In a third brewery, the use of CO_2 extract and the reduction of the total evaporation from 12.5 to 6.5% slightly improved the foam stability. Boiling at a counter-pressure up to 0.5 bar and reducing the total evaporation from 12 to 6%, respectively from 6 to 3%, did not result in any significant variation in the stability of the beer foam [1, 16].

15.6.9
Cold Break Removal

Several examinations performed over long periods in many breweries with and without cold break removal have not shown any statistically reproducible effect on foam stability. Also, elimination of oxygen by pressurizing the whirlpool with CO_2 has not resulted in any improvement of the foam [1, 16].

15.6.10
Main Fermentation

Using normal flocculating yeast lead to a better foam stability contrary to using powdery yeast. In general, all measurements during a cold main fermentation that lead to a reduction of the pH also have a positive effect on foam stability. These factors lead to the following conclusions:

- Intensive wort aeration.
- Limited pitching rate.
- Low enrichment of CO_2.
- Intensive yeast propagation.

High fermentation temperatures in general affect foam stability, because more fermentation byproducts with foam-negative effects are formed. A too strong convection in the tank and setting a CO_2 counter-pressure too early, as well as a too high original gravity (high-gravity brewing), have a negative effect on foam stability [1, 16].

15.6.11
Storage Conditions

Beer storage conditions can greatly influence the foam properties of the finished beer. Beer storage temperatures of 2 °C or even more and high yeast cell concentrations (too early green beer transfer) will lead to a reduced foam stability. The reason is an excretion of foam-negative substances by the yeast as medium-chain fatty acids (C6–C12) and glycerides. The worst case scenario is yeast autolysis [1, 7, 10, 12, 16].

15.6.12
Beer Filtration

A diminution of the foam values can also be caused by a filter whose sheets were replaced or if the absorbing products are varied. Several practical tests have shown that the foam stability changes relatively little, but this 'little' is statistically very significant, from a kieselguhr filter – coupled at polyvinylpolypyrrolidone (PVPP) – and a disk filter. Presumably, this can be blamed on the treatment with PVPP, that detains not only turbid active anthocyanogens, but also foam-positive polypeptides [1, 16].

15.6.13
Precocious Indicators for the Foam Appearance

Statistic correlations between foam stability and other parameters of wort and beer have been calculated from the results of numerous examinations and observations. Thereby, a statistically positive relation between foam values and total coagulable nitrogen of the wort has been noted and also a confirmed correlation with the content of anthocyanogens in beer. On the contrary, no relation between foam and the viscosity of wort and beer has been confirmed.

There is a significant negative correlation between the original gravity of a beer and its alcohol content and the foam stability. This indicates that in normal beers the ethanol concentration and trace amounts of higher alcohols that correlate positively with the original gravity of a beer may also influence foam stability.

The correlations between bitter units and the α-acid concentration, on the one hand, and foam stability, on the other hand, are also statistically well defined. While iso-α-acids may contribute to an improvement of foam stability and to its adherent properties, by an increase of the surface viscosity, a high concentration of non-isomerized α-acids in beer probably is a sign of insufficient elimination of foam-reducing substances [1, 16].

15.6.14
Conclusion

Analyses of the multiplicity of the available data permit the formulation of the following theses:

- A precocious indicator for the foam appearance of a beer does not exist. The earlier during beer process the prognosis concerning the foam is made, the less precise it is, because the foam sustains the most important influences only at the end of the production process.

- We have no analytical indicator to predict the foam properties of a beer in the early stage of production.

- Often the foam-negative factors have more influence than the foam-positive factors.

- Foam-positive substances like polypeptides, glycoproteins, and α- and β-glucans often have a negative effect on the beer production and on the final quality.

- Foam-negative substances like fatty acids, glycerides, higher alcohols and esters also negatively influence the overall beer quality.

The following points should always be considered in relation to good foam stability of a beer:

- The characteristics of a malt require a sufficient aliquot of at least 0.65%/SS of soluble nitrogen. Pay particular attention to the limiting values for the cytolytic solubility: differences in extract below 1.8% SS, viscosity below 1.58 m Pas, and iodine value of the laboratory threshing machine (date 1986) for finely ground malt below 2.5 ΔE_{578nm} and for coarsely ground malt below 7.5 ΔE_{578nm}.

- Malt should contain at least 0.65% soluble nitrogen under consideration, for example, of the cell wall solution (fine–coarse difference below 1.8%; wort viscosity below 1.58 mPas) and further analytical parameters.

- When choosing the hop products, prefer those with little tannins, if permitted by the colloidal beer stability. The latter has to be confirmed at any rate.

- When selecting hop products, prefer those that are free from polyphenols; take into account the colloidal stability of the beer.

- Normal breakdown of starch and proteins, a clear wort after mash filtration, iodine value of the wort below 0.3 ΔE_{578nm}; sufficient wort boiling (15–25 mg/l of coagulable N in cast wort); sufficient wort aeration (above 6 mg/l O_2); good yeast propagation; low main fermentation temperatures (below 10 °C); low storage temperatures (below 1 °C); avoid an increase of the pH value (too much yeast, temperatures that are too high, time duration too long); controlled beer stabilization (avoid treatment with bentonite or proteolytic enzymes); avoid contamination with fats, oils, detergents and supersaturation with CO_2.

15.7
Bitter Substances in Hops

Hops are analytically well characterized compared to other plants. However, only little is known about the influence of the bitter acids (approximately 100), aroma substances (approximately 400) and hop polyphenols (approximately 100) on beer quality. In fact, there is a notable lack of knowledge about the different hop compounds and their reactions in the final beer. In this context, high-performance liquid chromatography (HPLC) applications have imparted considerable knowledge and the technique is already established for many routine measurements. It offers the possibility to characterize not only different hop varieties, growing areas and staling characteristics of hops, but also the different hop products processed from raw hops (see Figures 15.1–15.3) [1, 17]. Some important questions still need to be answered, however:

- Within the hop-derived compounds in beers, which affect and which do not affect the sensory characteristics of beer?

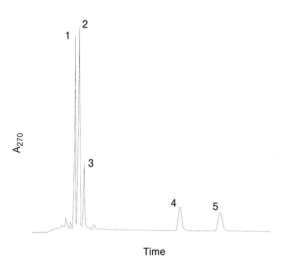

Figure 15.1 HPLC analysis (A_{270}) of isomerized hop pellets: 1, iso-cohumulones; 2, iso-humulones; 3, iso-adhumulones; 4, colupulone; 5, lupulone + adlupulone.

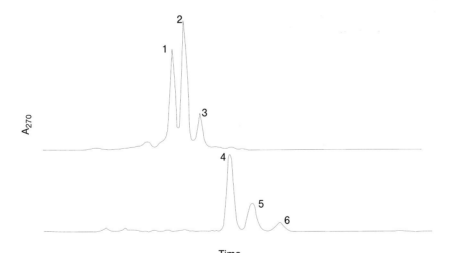

Figure 15.2 HPLC analysis (A_{270}) of tetrahydroiso-α-acids (derived from α-acid) with iso-α-acids for comparison: 1, iso-cohumulones; 2, iso-humulones; 3, iso-adhumulones; 4, tetrahydroiso-cohumulones; 5, tetrahydroiso-humulones; 6, tetrahydroiso-adhumulones.

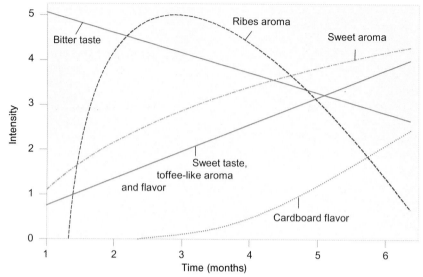

Figure 15.3 Flavor stability of beer (from [4]).

- Are those compounds stable in beer and to what extent do they have an impact on flavor stability?

In order to answer these questions a logical approach would be to separate all hop-derived components by means of HPLC and to spike beers with those isolated components for a sensory evaluation. However, there are already very interesting results reported concerning the cohumulone fraction of the α-acids [18–34].

15.7.1
Influence of Cohumulone on the Bittering Quality

Only until a few years ago various opinions on the contribution of cohumulone on the bittering quality existed. Very simple trials were carried out to approach this subject. From hops with different contents of cohumulones (Hallertau Hersbruck hops with 20% and Brewers Gold hops with 45%), two isomerized extracts were processed having a content of iso-cohumulones of approximately 25% (Hallertau Hersbruck) and approximately 50% (Brewers Gold). Those extracts were used as the sole bittering source for brewing trials in different European breweries. The results of the sensory evaluation revealed a preference for the beers produced with the extract low in iso-cohumulone due to a more pleasant bitterness. These trial brews confirmed that iso-humulones and iso-adhumulones generate a softer taste. Consequently, for a balanced bitterness in beer, the varieties with a low content of cohumulones are to be preferred, for example, Hallertauer Magnum (around 28%), Taurus (24%), Nugget (30%) and the new variety Merkur (21%).

15.7.2
Influence of Cohumulone on Foam Stability

Iso-humulones accumulate in the foam of the beer, exhibiting a stabilizing effect. Therefore, beers brewed with a higher amount of hops will have a more voluminous and stable foam. Due to the better solubility of iso-cohumulones in beer their stabilizing effect in foam is minor compared to the other homologs (Table 15.1).

Hence the content of iso-humulones and iso-adhumulones is higher and improves the foam stability. Hop varieties with a low content of cohumulone will impart a higher portion of iso-humulones and therefore influence the foam stability positively.

15.7.3
The Influence of Cohumulones on Beer Aging

The different homologues of the iso-humulones are prone to degradation during beer aging. To investigate the differences, beers in clear bottles hopped with two hop varieties in the form of extract and pellets were stored at 25 °C for 10 and 20 days. Table 15.2 shows the degradation of the iso-α-acid contents (the given values are the mean values). Those trials confirmed that the *cis*-isomers of the iso-humulones are more stable than the *trans*-isomers, although the ratio of *cis*-isomers to

Table 15.1 The content of iso-cohumulones in beer and foam.

	Beer		Foam		Foam : beer ratio
	mg/l	% rel.	mg/l	% rel.	
Iso-co-α	8.6	36	11.2	22	1.30
Iso-n-α	12.0	50	29.8	57	2.48
Iso-ad-α	3.3	14	11.0	21	3.33
Total-iso-α	23.9	100	52.0	100	2.18

Table 15.2 Iso-α-acid homologues (decrease in percent relative to fresh beer).

Trans-co-iso-α	−38
Cis-co-iso-α	−20
Trans-n-iso-α	−24
Cis-n-iso-α	−12
Trans-ad-iso-α	−7
Cis-ad-iso-α	−4

trans-isomers depends on technological means. Hence, it is of practical interest to know that the iso-adhumulones are more stable than the iso-humulones and both more stable than the iso-cohumulones. It is generally acknowledged that degradation reactions during beer storage have a negative impact on the beer taste. In this regard one should take advantage of any technological means that succeed in reducing the degradation reactions [25–34].

Considering that beer before it is actually consumed might be stored up to several months, hop varieties of low cohumulone content should be used for the beer production. Table 15.3 shows the typical cohumulone contents of 16 different hop varieties. Tables 15.4 and 15.5 show the problems of sensory evaluation with regard to beer aging caused by bittering substances as well as conserving substances.

Table 15.3 Content of cohumulones in different hop varieties (relative percentage of cohumulones of total humulones).

Bitter hops (Hersbrucker, Hallertauer mfr. Tettnanger, Spalter, Saazer)	20–25
Succeeding breeding varieties (Hallertauer Tradition, Spalter Select)	~25
Cultivated aroma varieties (Perle, Golding)	27–30
High-quality bitter hops (Northern Brewer, Hallertauer Magnum, Nugget, Taurus)	23–30
Bitter hops (Target, Brewers Gold, Columbus)	37–45

Table 15.4 How are the bitter substances appraised qualitatively corresponding to aging?

'Good/soft'
    often a quantitative problem, e.g. comparison with apples and oranges:
        beer 1 (fresh hops): 25 IBU: 27 mg iso-α/l
        beer 2 (old hops): 25 IBU: 17 mg iso-α/l
'Neutral'
'Bad'
    crude bitterness: aged polyphenols?
    old, musty: aged aroma components?

Table 15.5 Possible causes for different judgments.

Conditions of ageing is unclear (with/absence of O_2)
Degree of aging not defined (dynamic process)
Simultaneous decomposition of other components (tannins, oils, etc.)
Influence of other staling components:
    Polyphenols responsible for a harsh bitterness?
    Aroma substances responsible for a old/musty note?

15.8
Aroma Substances in Hops

Within the numerous effects of hop aroma substances in beer, we focus on one specific aspect. A lot of brewers aim for the production of a beer with a distinct hop aroma, although the limited stability of the hop aroma has to be taken into account (Figure 15.4). A typical German possibility is to add hops very late during boiling in the form of hop pellets (1–10 min before the end of boiling or addition in the whirlpool). The dosage is based on the α-acid content of the hops. The following suggestions are given in order to obtain a distinct hop aroma [19, 24, 35]:

- In order to achieve a distinct hop aroma the hop addition should not be based on the α-acid content, but on the hop oil content (given for all varieties in ml hop oil/100 g hops). Normal quantities depending on the desired intensity are 0.1–1.0 ml hop oil/hl beer.

- Also, products other than pellets (e.g. oil-rich hop extracts) can give a distinct hop aroma and may even provide a better product stability. The oil dosage may as well be split in two additions – one addition with pellets and the other with an oil rich extract. To learn how oil-rich hop extracts are produced, see the relevant references. An overview is given in the Tables 15.6–15.8 together with an indication of the points that still need to be clarified.

- It is also possible to add the hop oil after the wort boiling, in between fermentation and filtration. To avoid an unwanted perfume character, the contents of di-, mono- and sesquiterpenes may not be excessive.

15.9
Polyphenols in Beer Production

Polyphenols influence, directly or indirectly, beer production because of their chemical properties and, in particular, because of their multiplicative reactivity.

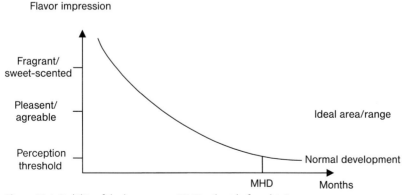

Figure 15.4 Stability of the hop aroma. MHD = 'best before date'.

Table 15.6 Ester fractions in different hop varieties (relative percentage from three esters in the oxygen fraction).

Classic aroma hop varieties (Hersbrucker, Hallertauer mfr. Tettnanger, Spalter, Saazer)	1–5
Breeding with existing varieties (Hallertauer Tradition, Spalter Select)	5–8
Cultivated aroma varieties (Perle, Golding)	10–20
High-quality bitter hops (Northern Brewer, Hallertauer Magnum, Nugget, Taurus)	15–25
Bitter hops (Target, Brewers Gold, Columbus)	25–35

Table 15.7 Advantages of fractionating CO_2 extraction.

Defined, stable product (product equal to whole hops without thermal influence)
Dosing of resins (α) and oil independently
Possibility of a flexible technology, e.g. bitter, soft, distinct aroma
The hop aroma in the beer is more stable
Better yield of bitter substances of aroma hops
Reduced costs

Table 15.8 Changing parameters.

Raw materials

Variety

Treatment at harvest

Degree of aging

Necessary characterizations, e.g. [epoxide of (caryophylene + humulene)]/(caryophylene + humulene)

Product typology: combination of products e.g. pellets (0.3 ml oil/hl) and extract richer in oil (0.2–0.5 ml oil/hl)

Time of addition

Wort kettle

Whirlpool

Before/after fermentation

Before filtration

Various possibilities; however, a lot of research is still to be carried out.
No sufficient knowledge for possible applications.

We mean by this especially the filterability of beer, colloidal stability, foam stability, color, reduction potentiality, aroma and stability. The following observations, which where made during practical experiments shall help to better evaluate the role of polyphenols in the beer process [1, 16].

15.9.1
Definition of Polyphenols

Polyphenols have been known for a long time for their tannin potential (thus, also the name 'tannins'). They react with iron salts forming colored complexes. This property is used in their analytical determination.

During the brewing process the monomer polyphenols (molecules of flavan) as basic elements play the most important role. One can distinguish between catechines (one OH group in the medial ring) and anthocyanogens (two OH groups in the medial ring).

Anthocyanogens are analyzed by adsorption to PVPP and subsequent heating with hydrochloric acid solution. Thereby, red complexes are formed which can be measured by a spectrophotometer. These haze-active monomer polyphenols can react to form oligomer and polymer polyphenols, which are known as tannins, tannoids and flobaphenes. With increasing molecular size, the oligomer and polymer polyphenols become more and more haze-active, due to the increasing number of hydroxyl groups. They react more and more easily with proteins, forming insoluble complexes. This can be seen by a distinct increase of wort and beer color, and also has a certain influence on beer taste. These polyphenols are considered to have an unpleasant bitterness. It seems that they are involved in beer aging and participate in the formation of undesired carbonyls. Polyphenols also have reducing properties. They influence the filterability and the foam stability of beer by the reaction with proteins, carbohydrates, melanoidins and hop bitter substances.

15.9.2
Origin of Polyphenols

15.9.2.1 Malt Polyphenols

Approximately 70–90% of the polyphenols of beer are derived from malt. They are located in the aleurone layer, endosperm and husk. Malt from protein-rich six-row winter barley contains more polyphenols than malts from spring barley. With increasing malt modification and an intensive kilning process, the content of soluble polyphenols is being increased.

Beers from malts with higher contents of anthocyanogens have a lower tendency to haze formation. However, beers with a reduced anthocyanogen level also showed a lower haze risk. This contradiction can be explained by the fact that the breakdown of high molecular carbohydrates and proteins dominates the negative effect of anthocyanogens. The concentration of soluble polyphenols in malt varies between 0.3 and 0.5% of dry matter. Anthocyanogens are the most important fraction.

15.9.2.2 Hop Polyphenols

Up to 30% of polyphenols present in beer are derived from hops. The highest content of polyphenols in the hop plant is within the cone bract. The content of polyphenols in the hop cones can vary between 4 and 14% of dry matter. Within the polyphenols, the anthocyanogens are the most important group due their distinct activity compared with the corresponding substances present in malt.

Some polyphenols may be haze-active in filtered beer and may cause an irreversible turbidity. Therefore, polyphenols are regarded as a disturbance and are at least partly removed by means of PVPP. At the same time, the food industry takes stock in the production of natural polyphenolic antioxidants as, for example, present in tea or seeds (Table 15.9). The polyphenol content in those plants may exhibit two different effects: antioxidative effects, protecting sensory characteristics in food (e.g. in fats) or anticarcinogenic effects as scavengers of free radicals (e.g. in rosemary, tea and read wine). Hops contain up to 4–6% polyphenols and an application as indicated above appears quite realistic; however, extensive research is still needed.

Polyphenols and, in this regard flavonoids, are present in all plants. There are 10 groups of flavonoids with approximately 5000 different molecular structures. One hundred of these compounds have been identified in hops. The flavonoids were originally used as natural dyeing agents, and are gaining more and more importance with regard to health-related benefits (e.g. blood circulation, diuretic effects, cardiovascular diseases etc.).

Until recently, polyphenols were measured using unspecific methods (e.g. colorimetric methods), whereas nowadays HPLC applications with diode array detec-

Table 15.9 Comparative antioxidative effects in comparison (*J. Agric. Food Chem.* 1995).

Substance	Concentration needed for an inhibition of 50% (micro;M)
Butylhydroxytoloule	0.27
Ascorbic acid	1.45
α-Tocopherol	2.40
Quercitine	0.22
Camphor oil	1.82
Catechine	0.19
Gallic acid	1.25
p-Cumaracid	>16

Test = inhibition of oxidation of lipoproteins.(*J. Agric. Food Chem.* 1995)

tion offer very precise measurements of single polyphenolic components. Table 15.10 shows the contents of different groups of polyphenols in hops.

The influence of polyphenols on beer is still an ongoing argument since many brewers believe that polyphenols have a negative impact on colloidal stability, foam and flavor of beer. To clarify the issue trial brews were carried out in the research brewery in St Johann, Germany.

Wort was hopped with pure resin extract and, additionally, polyphenols in form of the bractole fraction (derived as spent hops from the pellet 45 production) were added. No polyphenols were added to the control boil. The added amounts of the polyphenol fraction in the first trials were 0, 300 and 600 g/hl derived from Saaz hops, and these amounts were added both at the beginning and at the end of the boil.

For the second trials, between 50 and 300 g/hl of the polyphenol fraction was added to the wort. The hop varieties and the duration of the wort boil was varied. Afterwards, the following parameters were measured: foam stability, color, colloidal stability, reduction capacity, content of polyphenols and sensory evaluation.

The results of both series of trials can be summarized as follows:

- Color and foam stability are independent of the addition of polyphenols

- The polyphenol fraction exhibited a negative effect on the colloidal stability, increasing with prolonged boiling.

- The reduction capacity (assessed with MEBAK 2.20.1) proved to be higher due to the addition of the polyphenol fraction. The values increased from 45 (control) to 50 (addition of 100 g/hl polyphenol fraction) and 55 (addition of 150–300 g/hl).

The contents of polyphenols are shown in Table 15.11. The results of the sensory evaluation of the fresh beers are given in Table 15.12 and lead to the following conclusions:

- The results were significant with regard to differentiation and preference. The beers with the polyphenol fraction contained a surprising balance in taste, and a very fine hoppy and fruity aroma.

Table 15.10 Polyphenol groups of hops.

Group of substances	Content (mg/100 g)
Hydroxybenzoic acid	1–10
Hydroxycinnamic acid	100–450
Proanthocyanidins	100–600
Flavonols	30–200
Flavonols	1–10
Quercetin flavonoids	50–250
Camphorol flavonoids	50–300
Prenylflavonoids	100–1000

Table 15.11 Polyphenol contents in the beers of the second trials.

	Dosage of polyphenol fraction (g/hl)					
	0	50	100	150	300	600
Polyphenols HPLC (mg/l)	35	40	44	47	55	90
Polyphenols EBC (mg/l)	170	183	191	205	230	300

Table 15.12 Triangle tests with a trained tasting panel.

	0/600 g/hl: start of boiling	0/300 g/hl: end of boiling	0/600 g/hl: end of boiling
Number of tasters	12	11	14
Correct responses	12	10	14
Preferences	4 × 0 I 8 × 600 IB	2 × 0 8 × 300 FB	2 × 0 12 × 600 FB

- It is not clear if those sensory effects can also be ascribed to glycosidically linked components derived from hops.

- The sensory results proved to be reproducible. A second tasting confirmed the preference for the beer with the polyphenol fraction. Hence, the polyphenol fraction has a noticeable impact on sensory effects, which was rated positively by the major part of the tasters.

- Those sensory results were confirmed even after a storage period of 4 weeks at 27 °C.

The beers (with an oxygen content of 0.8–1.0 mg/l) underwent an extra long stability test. The control beer was judged to be not enjoyable, the beers with an addition of 600 g/hl at the beginning of the boil were evaluated better, although the beers with the addition of 300 and 600 g/hl at the end of the boil obtained the best sensory results.

In contrast to the sensory-positive results for the addition of the polyphenol fraction are the negative findings with regard to the physical stability, which are shown in Table 15.13.

Trials carried out at Joh. Barth & Sohn lead to some interesting conclusion which are that polyphenols [36–39]:

- Do not have any negative impact on beer foam or beer color.
- Do not impart a harsh bitterness if the duration of boiling is not excessive, but rather tend to improve bitterness.

15.9 Polyphenols in Beer Production | 377

Table 15.13 Negative results.

Turbidity (after 15 months)

    Sample at 5 °C showed haze at cold temperature; at 22 °C, almost clear:
    all other beers were hazy: 600 g/hl more haze than 600 g/hl: 600 g/hl slightly more haze than 300 g/hl

Nitrate: comparison: 4 mg/l

300 g/hl: 40 mg/l

600 g/hl: 70 mg/l

Table 15.14 Sum of polyphenols in four different hop varieties (in mg/100 g hops air-dried).

Group of substances	Tettnang Tettnanger	Hallertau Hallertauer	Hallertau Tradition	Hallertau Magnum
Hydroxybenzoic acids	323	335	349	124
Hydroxycinnamic acids	9	9	7	5
Flavanole	137	99	113	32
Proanthocyanidins	448	330	372	108
Quercetin flavonoids	185	206	132	68
Campherol flavonoids	235	264	197	71
Other flavonoids	4	10	8	8
Sum[a]	1609	1507	1473	532

a Including unidentified components.

- Allow for a well-defined flavor.
- Increase the reduction capacity of beers and the flavor stability (at least of the trial beers, which had a relative high oxygen content).

However, they also:

- Increase the susceptibility of haze in the beers (at a higher rate if boiling time increases)
- Increase the nitrate content in beers.

Those findings lead to the following recommendations:

- It is advisable to choose a hop variety rich in low-molecular-weight polyphenols or flavonoids (e.g. aroma varieties grown in a moderate climate; Table 15.14).
- To maintain the characteristics of the low molecular polyphenols, the better choice is for hops stored for a shorter period.
- The duration of the boil should be limited.

- If the addition of pellets is limited due to the nitrate content it is recommended to add pure resin extract at the beginning of the boil and the pellets only after half of the boiling time.

- In order to assure constancy throughout the years it is possible, despite the crop-dependent variations of the α-acid content, to use mechanical enrichment (pellets with a changing resin content). In this way the ratio of α-acids to polyphenols stays constant.

- The enrichment of lupuline offers various benefits, since a higher portion of the bractole fraction is eliminated, which goes along with a better polyphenols: nitrate ratio.

Another aspect to consider is the solubility of the polyphenols, which varies between the different hydrophilic components such as hydroxybenzoic acids and hydroxycinammic acids. Flavanols and proanthocyanidins dissolve easier than prenylflavonoids (due to their lipophilic character), whereas the flavonoids ability to solubilize is rather average. The low molecular weights, determined by HPLC, are also the reason for a better extraction during boiling compared to the polyphenol measurements with the EBC method (Table 15.15). Losses during beer production can be ascribed to thermal transformations and to precipitation within the hot or cold trub, yeast, tank sediment or losses during filtration. The group of prenylflavonoids reacts differently, represented by xanthohumol. Found within all polyphenols, xanthohumol exhibits a very poor solubility and has to undergo a isomerization process, similar to that of α-acids. The high losses of xanthohumol during the brewing process can be explained by the poor solubility but also with adsorption reactions (e.g. in the trub). Even if the boiling time is increased, and the losses in trub and during filtration are decreased, only a yield of 30% can be obtained, as indicated in Table 15.16.

Experiments carried out at Oregon State University have shown that xanthohumol exhibits anti-canceroginic effects *in vitro* at a dose of 30 mg/person a day [40]. Although the normal contents of xanthohumol and iso-xanthohumol in beer have no physiological relevance, it could be of interest for the pharmaceutical industry to increase the possibilities of an improved extraction [41–45].

Table 15.15 Dosage of polyphenol fraction and content of polyphenols.

Dosage of polyphenol fraction	Units	300 g/hl: end of boiling	600 g/hl: end of boiling
Polyphenols EBC in wort	mg/l	156	312
Polyphenols EBC in beer	mg/l	60	130
Yield	% rel.	38	42
Polyphenols HPLC wort	mg/l	36	72
Polyphenols EBC beer	mg/l	20	55
Yield	% rel.	56	76

Table 15.16 Xanthohumol and iso-xanthohumol during the brewing process (dosage in wort: 5 mg/l).

	Wort	Green beer	Final beer
Xanthohumol (mg/l)	0.4	0.2	0.04
Iso-xanthohumol (mg/l)	1.5	1.3	0.73
Sum (mg/l)	1.9	1.5	0.77
Yield (% relative)	38	30	15

15.9.4
Polphenolic-Related Reactions during Brewing

Polyphenols can form many different compounds. As mentioned earlier, monomer and oligomer polyphenols can interact both in the aerobic and in anaerobic phase.

15.9.4.1 Reaction with Proteins

Reactions with proteins are particularly significant. These compounds have weak hydrogen bridges. Furthermore, the polyphenols can bind on proteins via weak hydrophobic bonds. The proline content in a polypeptide chain is responsible for this. The ionic linkage is stronger than others, for example, between ε-amino groups in a lysine sequence of a polypeptide chain and a polyphenolic molecule. All these interactions are reversible initially and result in irreversible covalent bonds. Furthermore, it seems that polyphenols are able to associate also with α- and β-glucans, like melanoidins and even with hop resins via hydrogen bridges. However, it appears that for at least one part of polymer polyphenols, the analysis does not result in valuable data.

15.9.4.2 Influence of Hops

The content of anthocyanogens in hops varies according to the hop product used. Lack of anthocyanogens lead to beers with a high attenuation, but with poor filterability due to a higher content of α- and β-glucans, low free nitrogen content, a lower Ludin B fraction, with less tendency to turbidity and lower reducing capacity. In relation to a good colloidal stability, the elimination of polyphenols of the hop is particularly important along with a simultaneous elimination of the malt tannins. The differences in the content of anthocyanogens in CO_2 extracts or pellets 45 did not prove to have a clear influence on the foam stability nor the flavor or color of the resulting beers.

15.9.4.3 Advantages and Disadvantages

The use of Galant malt free of anthocyanogens in combination with CO_2 extracts 'free of tannins' provides the following benefits:

- Lower tendency of turbidity in beer.
- No additional costs for stabilization.

- No issues with any food regulations.
- Independence from PVPP suppliers.

However, there are disadvantages to take into account in terms of production and in terms of beer quality. This disadvantages are not caused by the absence of anthocyanogens in malt, but rather caused by the general shortcoming of Galant malt. Future cultivation trials will focus on possible solutions.

15.9.5
Value of Anthocyanogens and Other Beer Characteristics

Within the statistical evaluation of 55 beers produced in different breweries, correlations between the content of anthocyanogens and other quality parameters have also been assessed. Summarizing the different correlations we can affirm that a lower content of anthocyanogens in beer will result in a beer with less tendency to turbidity, a relatively better taste and a better stability, a slightly lower foam stability, and a slightly clearer color.

15.9.6
Reaction Path of Polyphenolic Components during the Brewing Process

Today a brewer has a wide range of possibilities to influence the content of polyphenols during the brewing process. In malt, the content of soluble anthocyanogens increases with higher amount of proteins, but also with a higher modification of the malt and a more intense kilning. The influence of hop tannins is primarily determined by the choice of the hop products considered suitable. With regard to the process water, it is known that a lower residual alkalinity goes along with a higher content of soluble anthocyanogens.

During milling, the solubility of anthocyanogens decreases according to the following order: wet milling > dry milling of fine parts > dry milling of coarse parts > conditioned > with separation of the grist. Furthermore, lower temperatures in the milling process and during lautering, short process times, and the oxygen content will result in lower values of anthocyanogens in wort and in beer. A more intense extraction of the spent grain and the recovery of the last runnings will increase the polyphenol content.

If possible, the wort should be boiled without hop tannins in the beginning to achieve a better flocculation of the malt polyphenols.

In addition to oxygen, an intensive boiling and a good separation of the trub may also reduce the content of anthocyanogens. Furthermore, a vigorous fermentation with a fast decrease of the pH value, a long and cold storage period without any addition of process beers, and a sharp filtration, all lead to a reduction of the content of anthocyanogens in beer. However, all these possible and theoretical interventions are secondary to other aspects regarding the brewing process and the quality of the beer.

15.9.7
Conclusions

Various trials that have investigated the role of polyphenols in the brewing process lead to the general confirmation that anthocyanogens are not necessary to maintain the most important beer characteristics. Hop anthocyanogens can easily be avoided by the appropriate choice of the hop product. The anthocyanogens of the malt also can be avoided with the use of the new malt Galant; however, this goes along with major disadvantages. Therefore, the most effective way to remove anthocyanogens is by PVPP filtration of the finished beer.

15.10
N-Heterocycles

Beer flavor is determined by a manifold of substances, as for example, higher alcohols, acids, esters, diacetyl and dimethylsulfide (DMS) but also by a series of heterocyclic compounds. These are formed mainly during thermal processes and to a lower extent during fermentation or by raw materials.

By heterocyclic compounds we mean organic molecules in which one or more carbon atoms are substituted by so-called hetero atoms (e.g. nitrogen, oxygen or sulfur). *N*-Heterocycles are of particular interest because they have partly very low flavor thresholds (parts per billion level or even below) and have typical flavor notes like bready, caramel or burnt. Among others routes, they arise from thermal reactions (see Figure 15.5) between reducing sugars and amino acids, and they are formed during withering and kilning of malt, wort boiling, and pasteurization. This is the reason why the concentration of these compounds is elevated in dark worts.

More than 80 different *N*-heterocycles are known; the structures of the most important are shown in Figure 15.6(a and b) and the formation of the *N*-heterocycles during wort boiling is shown in Figure 15.7 [46].

15.10.1
Presence of Heterocycles

N-, *O*- and *S*-Heterocycles are ubiquitous and are also present in numerous foodstuffs (e.g. coffee, boiled meat, potato products, vegetables, popcorn, roasted nuts, rum, whiskey, roasted barley, bread, wort and beer). Common to all these foodstuffs is that they (or even the raw materials) have undergone thermal processes during production. Thermal treatment promotes the formation of numerous heterocycles (e.g. furfurylalcohol, furfural derivatives, hydroxymethylfurfural and maltol). It is well known that 2-furfural and furfurylalcohol are indicator substances for a thermal charge of wort and beer. Other heterocycles (e.g. oxazole, thiazoles, pyrrol, maltoxazine pyrroline, indole, etc.) are formed under heat from amino acids and sugars (see Figure 15.8). Figure 15.9 shows the chemical structure of some *N*-heterocycles.

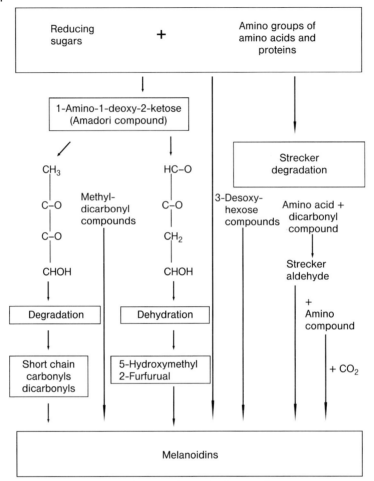

Figure 15.5 The formation of melanoidins and N-heterocycles.

N-Heterocycles can be analyzed by gas chromatography using a nitrogen-selective detector. A gas chromatography chart of a dark malt shows approximately 200 different peaks of which around 80 are identified and around 40 could be quantified up to now. As this complexity is confusing, one tries to evaluate the flavor impression and the flavor value of these individual components using the so-called 'sniffing technique'. Hereby, the effluent of the gas chromatography column is split at the end of the column; one part goes as usual to the detector and the other part enters a 'sniffing port' for the evaluation of the flavor properties of the individual peaks by the human nose.

Most of these compounds are described as bready, coffee, caramel, cracker, boiled wort or burnt. N-Heterocycles are desirable in dark malts and dark beers as they increase the flavor sensation. In pale beers, an excessive concentration may cause a serious off-flavor.

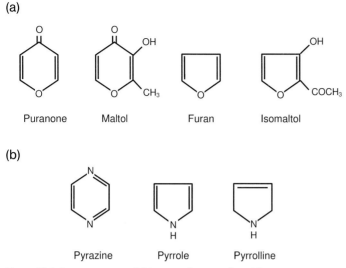

Figure 15.6 Base structures of (a) oxygen heterocycles with the taste of caramel and malt, and (b) nitrogen heterocycles with the taste of cereals, bread and nuts.

15.10.2
N-Heterocycles in the Malting Process

Germination time influences the formation of N-heterocycles in a simple way, since during germination a lot of low molecular compounds are formed that may act as reaction partners in the following processing steps (e.g. kilning, wort boiling). Low germination temperatures (12–15 °C) showed the same effect, whereas the influence of moisture during germination is indifferent. The most important influence on the formation of N-heterocycles are the kilning temperature and kilning time.

15.10.3
Mashing Conditions

The influence of the pH during mashing on the formation of N-heterocycles is not uniform; an influence of oxygen could not be observed. The most important influence comes from the share of dark malt in the grist load. At a dark malt share up to 25% during wort boiling, more N-heterocycles are formed than evaporated. Evaporation dominates at higher levels of dark malt. 2-Methyl-pyrazine may serve as indicator for heat input. During wort boiling it shows a strong increase, but is also dependent on the amount of dark malt. It is also formed during fermentation from a still unknown precursor. A similar behavior is seen for 2-acetyl-pyrrole. Pyrazole, however, is mainly formed during wort boiling and therefore it may serve as an indicator for the evaluation of wort boiling systems.

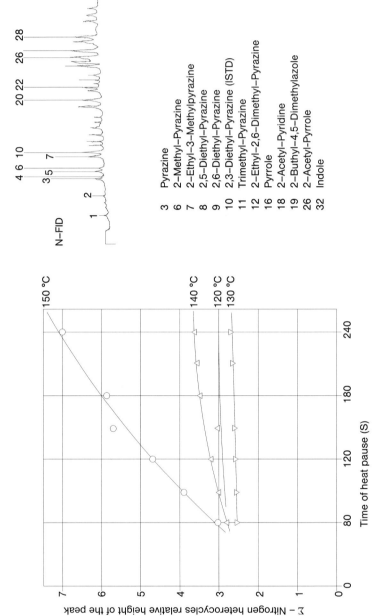

Figure 15.7 N-Heterocycles of wort (according to Schwill-Miedaner).

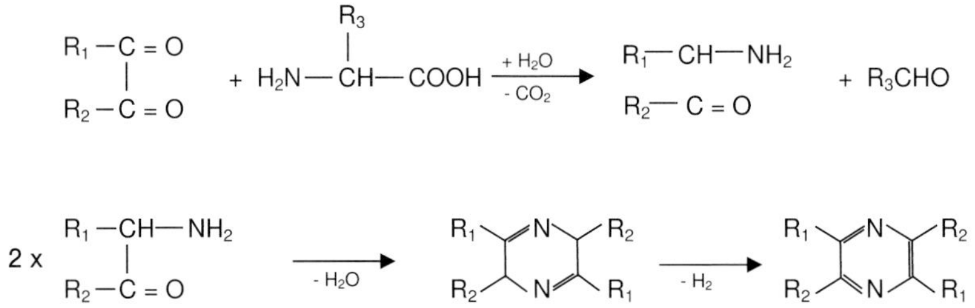

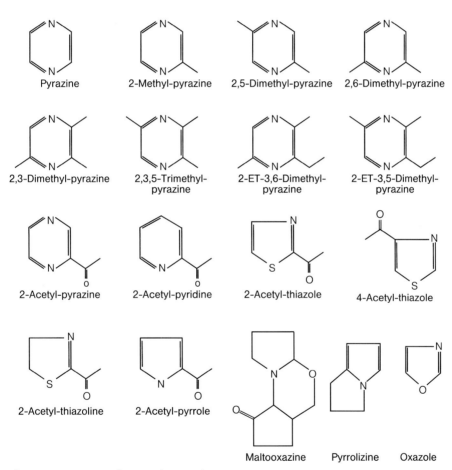

Figure 15.8 Formation of N-heterocycles from amino acids.

Figure 15.9 Structures of some N-heterocycles.

15.10.4
Wort Boiling

Wort boiling is, besides kilning, the most important process concerning the formation of N-heterocycles. The pH of wort and its oxygen charge have no influence. During boiling we have an increase of N-heterocycles at the beginning, but evaporation dominates the formation in the later course of boiling. The most important factor is the evaporation rate. Investigation of the vapor condensate showed that for, for example, 2-acetyl-thiazole (responsible for a pasteurization off-flavor; flavor threshold 10 ppb) 90% of amount must be removed to obtain the desired level in the wort. If during boiling with a low counter-pressure (NDK) too high temperatures are applied, one gets increased concentrations of several N-heterocycles (2-methyl-pyrazine, 2-acetyl-pyrrole, 2-acetyl-thiazole). Temperatures should not exceed 104 °C. High-temperature wort boiling systems showed serious problems with the formation of N-heterocycles ('boiled flavor').The thermal stress of boiled wort can easily be evaluated by the determination of the thiobarbituric acid index (TBI) or the determination hydroxymethylfurfural.

15.11
DMS

DMS – a highly volatile sulfur compound – can effect beer seriously with respect to its flavor properties; thus, the brewer aims at low levels in beer. The flavor impression is described as cabbagy, like cooked vegetables or sweet-grainy and is strongly dependent on the existing concentration [47].

The flavor threshold is not quite clear; it seems to be in the range of 40–60 ppb and even higher for all-malt beers. A DMS off-flavor occurs often in combination with other volatile sulfur compounds like mercaptans. Japanese investigations on Japanese and European beers have shown concentration ranges between 8 and 118 ppb for lager beers and 10–16 ppb for stouts. German all-malt beers showed a higher concentration of 60–100 to 200 ppb; in this latter case the beer had a significant flavor defect. Adjunct beers have low DMS levels of 10–20 ppb.

15.11.1
Formation of DMS

In unmalted barley neither free DMS nor its precursor can be detected. During germination an inactive precursor of DMS is formed depending on the germination parameters such as moisture content, germination temperature and germination time. Barley variety also plays a distinct role. This inactive precursor [DMS precursor (DMS-P)] is generally seen to be the amino acid S-methylmethionine. This DMS-P is unstable at higher temperatures and degrades under the impact of heat stress (in this case during kilning), forming free DMS and the active precursor dimethylsulfoxide (DMSO). DMSO can be metabolized by yeast (strongly strain dependent) forming DMS. See Figure 15.10.

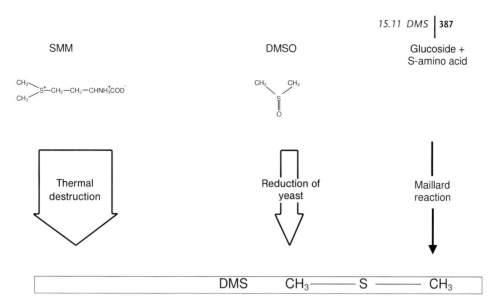

Figure 15.10 The formation of DMS.

15.11.2
Barley and Malt

The 'active' precursors can also split in free DMS by the influence on other temperatures. Barley does not contain DMS. Only by its modification (i.e. via germination) and at the end of the drying process the amount of DMS can be influenced. Some investigations have confirmed that the humidity of the germing mass exercises an influence on the content of DMS precursors. The higher the moisture of the germinating barley, the higher the content of DMS-P will be. A Kolbach index of 40 implies a content of precursors of about 3000 µg/kg, 44% implies 4000 µg/kg and 48% implies 6000 µg/kg. See Figure 15.11.

15.11.3
Temperature

Furthermore, the germination temperature plays an important role: the higher the temperature, the higher the content of the DMS-P.

15.11.4
Withering and Kilning

As already mentioned, the DMS-P is unstable at higher temperatures. Therefore, the withering and kilning temperatures have the strongest impact on the DMS-P level of the finished malt. At a final kilning temperature of 70 °C, the DMS-P level is reduced to approximately 40% compared to the end of withering; at 80 °C the reduction is already 68%. The strongest degradation of DMS-P can be achieved during kilning of dark malts at 100–105 °C. The applicable final temperature level

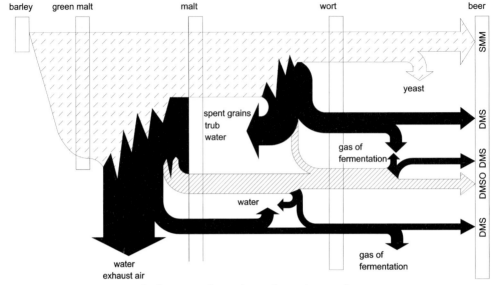

Figure 15.11 The formation of DMS during the production of malt and wort. SMM = S-methylmethionine.

and time must be selected carefully as there are also serious influences on the color and TBI of the malt.

15.11.5
Malt Cleaner

The majority of DMS-P in malt grains is situated in the acrospire and the rootlets. Therefore, a careful malt cleaning process results in a distinct reduction of DMS-P.

15.11.6
Brewhouse

The mashing procedure and the intensity and duration of wort boiling play an important role in the reduction of free DMS and DMS-P. Worts produced by infusion mashing contain more DMS than worts derived using a decoction mashing method. It is remarkable that the duration of wort boiling and the pH of the wort are responsible for the degradation of the DMS-P, whereas the intensity of boiling promotes the evaporation of free DMS. After the end of boiling, during the rest in the whirlpool tank, the degradation of the DMS-P continues and free DMS is increased. Using classical boiling systems with external or internal boilers and long boiling times (90–120 min), no problems with DMS will arise. With classical low-pressure boiling (LPB) systems, worts with low DMS concentrations can be obtained by applying temperatures up to 110 °C; high-temperature boiling systems result in worts nearly free of DMS. Modern boiling systems with very short boiling times, however, often show problems achieving the desired level of free DMS in wort and beer. Beer quality limits our possibilities with respect to high kilning temperatures of the malt and energy saving in the brewhouse.

15.12
Gushing (Uncontrolled Overflow or Overfoaming of Beer)

15.12.1
General

The first article on the phenomenon of gushing of beer was published in 1924. The evaluation of numerous publications and contributions concerning the reasons for gushing can be summarized in four groups:

- Metal ions in the bottled beer.
- Precipitations of calcium oxalate crystals.
- Filter media.
- Malt-induced gushing (contaminations with *Fusarium*).

The classification into 'primary gushing' (malt induced) and 'secondary gushing' (caused technologically) is meanwhile extended by the term 'technical processes' [1, 16, 48–52].

15.12.2
Determination of the Gushing Behavior of Beer Induced by Raw Materials

Years ago at Weihenstephan a method was developed that allows a prediction of cereal-induced gushing by extracting the grains of malt or adjunct by water. The influence of cereals and its contamination by molds and the effect of calcium and magnesium on gushing is shown in Tables 15.17 and 15.18 [53].

15.12.3
Metal Ions in Bottled Beer

Investigations conducted 40 years ago have shown a relationship between iron ions and gushing [54]. In small-scale trials it could be detected that $[Fe^{3+}]$ could cause gushing [52], whereas $[Fe^{2+}]$ did not have any effect. In fermentation trials it could be established that in the course of fermentation the available iron is taken up nearly completely by the yeast (Figures 15.12 and 15.13). As can be seen in Table 15.19, the iron concentrations of all the beers are below the limit of potential gushing. Increased iron concentrations in finished beer are exclusively caused by an extraction from filter media. The following checks are recommended:

Table 15.17 Infection of green malt during germination with mold spores.

	Volume of overflowed beer (ml)	
	From bottle with cereal solved in water	From bottle with beer produced with these cereals in a pilot plant
Non-infected cereal	0	0
Cereal infected with:		
Epicoccum nigrum	0	0
Alternaria alternata	11	8
Rhizopus stolonifer	13	6
Mucor mucedo	4	0
Fusarium culmorum	32	98
Fusarium graminearum	25	108

Table 15.18 Influence of calcium and magnesium on gushing.

	Volume of overflowed beer after (ml)					
Addition of protein extract (ml)	2		6		10	
+addition of calcium (mmol)	0	5	0	5	0	5
overflowed beer (ml)	5	7	36	160	166	176
+addition of magnesium (mmol)	0	5	0	5	0	5
overflowed beer (ml)	5	13	36	72	166	170
+addition of EDTA (mmol)	0	5	0	5	0	5
overflowed beer (ml)	5	0	36	5	166	100

- Comparison of the iron level before and after filtration.
- Checking the iron concentration several times during filtration.
- Large bright beer tanks for blending.
- Change kieselguhr if necessary.

15.12.4
Precipitation of Calcium Oxalate Crystals

Precipitated calcium oxalate may have different crystal structures and induce gushing in bottled beer [55]. The oxalic acid concentration is determined by the selection of raw materials such as water, cereals and malt as well as hops. It must be the aim to eliminate as much of calcium oxalate as possible by precipitation before filtration. This can be achieved by the addition of $CaCl_2$ or $CaSO_4$ to the brewing liquor or the mash, respectively.

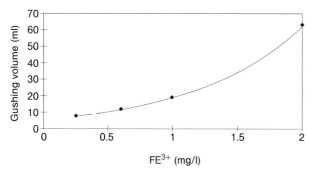

Figure 15.12 Correlation of [Fe^{3+}] and gushing.

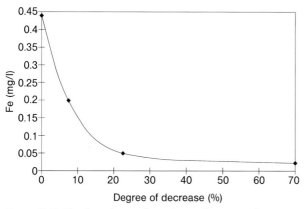

Figure 15.13 The degradation of iron during the primary fermentation.

Table 15.19 Concentration (mg/l) of total iron during the brewing process.

Cast wort	0.10–1.00
Unfiltered beer after maturation	0.03–0.07
Confectioned beer	0.09–0.28
Limiting value	0.12–0.14

Immediately after filtration there is an equilibrium between calcium and oxalate ions, and thereby oxalate is dissolved. The deciding difference is the 'stability' or 'instability' of this equilibrium. At a 'stable' equilibrium, a certain pick-up of calcium during filtration by the extraction of filter media can be compensated for; at an 'unstable' equilibrium, a minor pick-up of calcium already leads to a precipitation of calcium oxalate.

A decrease of the solubility limit of calcium oxalate can be observed during the brewing process. While 60 mg/l of calcium oxalate can be solved in wort, in beer it is only 15–30 mg/l.

For a better judgment of a beer's tendency to form calcium oxalate precipitations and thus be able to take the correct measures for their prevention, the determination of the 'free calcium ratio' Q ($CaSO_4/CaC_2O_4$) according to Schur has proved useful [55]. It is the quotient of Ca^{2+} ions (calculated as calcium sulfate) and the oxalate value (expressed as calcium oxalate).

Stable conditions are given if the calcium oxalate concentration is below the solubility limit L_p ($[Ca^{2+}] \cdot [C_2O_4]$) in beer. There will be calcium oxalate precipitations in the beer if the initial quotient Q is already unfavorable or the balance is displaced by the subsequent access of calcium ions. Calcium reacts with oxalic acid and precipitates as calcium oxalate.

Jacob, on the other hand, suggests a content of oxalic acid below 12 mg/l with a calcium content between 50 and 70 mg/l in the beer to achieve stable beers [56].

In any case, the goal must be to eliminate as much oxalic acid as possible before the filtration in the form of calcium oxalate through a surplus of calcium ions in order to prevent a later reaction in the filtered beer [56].

The following measures can be recommended:

- Elimination of calcium oxalate before filtration by the addition of calcium to the brewing liquor [57] and/or long and cold beer storage [58].
- Avoid calcium uptake during and after filtration.
- Determination of the 'free calcium ratio' Q and calcium oxalate in unfiltered beer [59].

15.12.5
Filter Media

Practical observations and defined investigations confirm an influence of particles of kieselguhr, perlite and charcoal on gushing behavior. There is a strong dependence of the particle size on the overflowing beer volume. With increasing amounts of filter media, the overflowing beer volume increases, but only to a distinct limit that depends on bottle shape, bottle volume (beer content) and CO_2 concentration. As Figure 15.14(a–c) shows, using active carbon the limit will be reached already at low concentrations [60].

Very small CO_2 bubbles with a so-called 'critical diameter' may also contribute to the gushing risk [60]. Hairline cracks at the inner bottle surface may act as a source for a sudden CO_2 release (Figure 15.15).

15.12.6
Malt-Induced Gushing

The only technological possibility to avoid a malt-induced gushing problem is to examine the malt for its gushing potential before using it in the brewhouse. Defined contamination tests and the results are listed in Table 15.20. There are clear differences between the individual mold strains with respect to gushing

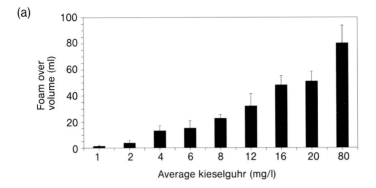

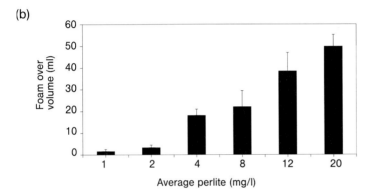

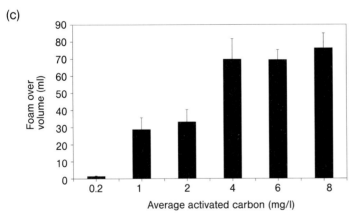

Figure 15.14 The foam over-volume: a bottom-fermented beer in relation to an average content of (a) kieselguhr, (b) perlite and (c) activated carbon.

Figure 15.15 Micro-split at the inner bottle surface as a source for a sudden CO_2 release.

Table 15.20 Stimulation of the gushing effect caused by different species of fungi.

Fungus	Gushing tendency
Non-infected spent grains	0
Fusarium graminearum	3
Fusarium culmorum	3
Fusarium sporotrichoides	2
Fusarium avenaceum	2
Fusarium croockwellense	1
Fusarium tricinctum	0
Fusarium equiseti	3
Fusarium oxysporum	0
Fusarium poae	0
Fusarium sambucinum	3
Drechslera	0
Alternaria alternata	0
Penicillium	2

0 = no gushing tendency, 3 = marked gushing tendency.

induction. In addition to the well-known *Fusarium* species *F. graminearum* and *F. culmorum*, a number of other strains can also cause gushing. Improper storage conditions can force the growth of these microorganisms after harvest in the silo; in this case no *Fusarium* species could be detected on the grains directly taken from the field.

15.12.7
Chemical Components Causing Gushing

It has not been possible to isolate individual compounds and to identify their chemical structure that will cause gushing from contaminated cereals. Trials with

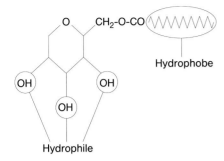

Figure 15.16 Chemical structure of a substance that can cause gushing.

reference substances gave some information about this problem. Investigations with surface-active substances showed that only a few of these components that have a special chemical structure are able to cause gushing in carbonated liquids. Figure 15.16 shows an example of a chemical structure of a component that may induce gushing. The deciding structural characteristics for this property are:

- No electrical charge.
- Long chain (more than C12) and saturated fatty acid (hydrophobic part).
- Two or more OH groups (hydrophilic part).

References

1 Narziss, L. (2004) *Abriss der Bierbrauerei*, 7th edn, Wiley-VCH Verlag GmbH, Weinheim.
2 Back, W. (2005) *Ausgewählte Kapitel der Brauereitechnologie*, Fachverlag Hans Carl, Nürnberg.
3 Narziss, L. (1992) *Die Bierbrauerei: Die Technologie der Würzebereitung*, 7th edn, Ferdinand Enke Verlag, Stuttgart.
4 Dalgliesh, C.E. (1977) Flavor stability, in *Proceedings of the 16th European Brewery Convention Congress, Amsterdam*, pp. 623–59.
5 Ronteltap, A.D., Hollemans, M., Biaperink, G.J. and Prins, A. (1991) Beer foam physics. *Master Brewers Association of the Americas Technical Quarterly*, **28**, 25–32.
6 Aronson, M.P. (1986) Influence of hydrophobic particles on the foaming of aqueous surfactant solutions. *Langmuir*, **2**, 653–9.
7 Narziss, L., Reicheneder, E. and Barth, D. (1982) Über den Beitrag von hochmolekularen Proteinfraktionen und Glycoproteiden zum Bierschaum. *Monatsschrift für Brauerei*, **35**, 213–21.
8 Lusk, L.T., Goldstein, H. and Ryder, D. (1995) Independent role of beer proteins, melanoidins and polysaccharides in foam formation. *Journal of American Society of Brewing Chemists*, **53**, 93–103.
9 Roberts, R., Keeney, P.J. and Wainwright, T. (1978) The effects of lipids and related materials on beer foam. *Journal of the Institute of Brewing*, **84**, 9–12.
10 Wilde, P. (2003) The physical mechanisms responsible for the destabilization of the beer foam by fatty acids, in *Proceedings of the European Brewery Convention Congress*, pp. 875–82.
11 Reicheneder, E. and Nariss, L. (1987) Untersuchung über die Schaumschädigung des Bieres durch Hefeproteinasen. *Brauwelt*, **127**, 956–63.
12 Narziss, L., Reicheneder, E., Jogasuria, P., Eichhorn, P., Mayer, H. and Gommelt, T.

(1988) Neueste Erkenntnisse zum Thema Bierschaum. *Brauwelt*, **128**, 1630–42.
13 Nariss, L. and Reicheneder, E. (1989) Untersuchungen über die Auswirkungen von Hefeproteinasen auf den Bierschaum, in *Proceedings of the European Brewery Convention Congress*, pp. 577–84.
14 Röttger, W. (1973) Schaum und Molekülfraktionen, Dissertation, TU München, Freising.
15 Glas, K. and Scheuing, R. (2000) Tenside und Bierschaum. *Brauwelt*, **46/47**, 2024–5.
16 Heyse, K.-U. (1995) *Handbuch der Brauerei-Praxis 3*, Fachverlag Hans Carl, Nürnberg.
17 *Hops and Hop Products, Vol. 3, Manual of Good Practice EBC*.
18 Forster, A. (1993) Varietà luppolo – Birra & Malto 53/4.
19 Forster, A. (2000) Le nostre attuali conoscenze sul luppolo – Birra & Malto 75/17.
20 Forster, A. and Schmidt, R. (1994) Zur Charakterisierung und Gruppierung von Hopfensorten. *Brauwelt*, **133**, 2036–8, 2049–57.
21 Forster, A. (1976) *Brauwelt*, **116**, 1141–5.
22 Forster, A. (1978) *Master Brewers Association of the Americas Technical Quarterly*, **15**, 163–7.
23 Forster, A. (1982) *Vortrag der 69, VLB-Oktober-Tagung, Berlin*.
24 Forster, A. (1998) Hopfen ein natürlicher Rohstoff oder Basis für maßgeschneiderte Moleküle. *Mitt-Österreichisches Getränkeinstitut*, **34**, 28–34.
25 De Cooman, L., Aerts, G. and Overmeire, H. (2000) Alterations of the profiles of iso-alpha-acids during beer ageing, marked instability of trans-iso-alpha-acids and implications for beer bitterness consistency in relation to tetrahydroiso-alpha-acids, *Journal of the Institute of Brewing*, **106** (3), 169–78.
26 De Cooman, L., Aerts, G., Witters, A., De Ridder, M., Boeykens, A., Goiris, K. and De Keukelaire, D. (2001) Comparative study of the stability of iso-alpha-acids, dihydroiso-alpha-acids and tetrahydroiso-alpha-acids during beer ageing. *Cerevisia and Biotechnologie*, **26** (3), 155–60.
27 Diffor, D.W., Likens, S.T., Rehberger, A.J. and Burkhardt, R.J. (1978) The effect of isohumulone/isocohumulone ratio on beer head retention. *Journal of the American Society of Brewing Chemists*, **36** (2), 63–5.
28 Hughes, P. (2000) The significance of iso-alpha-acids for beer quality. *Journal of the Institute of Brewing*, **106** (5), 271–6.
29 Hughes, P.S. and Simpson, W.J. (1996) Bitterness of congeners and stereoisomers of hop-derived bitter acids found in beer, *Journal of the American Society of Brewing Chemists*, **54** (4), 234–7.
30 Laws, D.R.J., Shannon, P.V.R. and John, G.D. (1976) Correlation of congener distribution and brewing performance of some new varieties of hops. *Journal of the American Society of Brewing Chemists*, **34** (4), 166–70.
31 Meilgaard, M. (1960) Hop analysis, cohumulone factor and the bitterness of beer: review and critical evaluation. *Journal of the Institute of Brewing*, **66** (1), 35–50.
32 Ono, M., Hashimoto, S., Kakudo, Y., Nagami, K. and Kumada, J. (1983) Foaming and beer flavor. *Journal of the American Society of Brewing Chemists*, **41** (1), 19–23.
33 Wackerbauer, K. and Balzer, U. (1992) Hopfenbitterstoffe in Bier Teil 1: Veränderungen in der Zusammensetzung der Bitterstoffe während des Brauprozesses. *Brauwelt*, **122** (5), 152–5.
34 Wackerbauer, K. and Balzer, U. (1992) Hopfenbitterstoffe in Bier Teil 2: Der Einfluß des Cohumulons auf die Bierqualität. *Brauwelt*, **122** (10/11), 396–8.
35 Gresser, A. (1985) Beiträge zur Kenntnis der Hopfenaromastoffe unter besonderer Berücksichtigung technologischer Verfahren, Dissertation, TU München, Freising.
36 Forster, A., Beck, B. and Schmidt, R. (1995) Untersuchungen zu Hopfenpolyphenolen, in *Proceedings of the 25th European Brewery Convention Congress*, Oxford University Press, New York.
37 Forster, A., Beck, B. and Schmidt, R. (1999) Hopfenpolyphenole – mehr als nur Trübungsbildner in Bier [Hop polyphenols

do more that just cause turbidity in beer], *Hopfenrundschau International*, ••, pp. 68–74.
38 Forster, A. et al. (2002) Xanthohumol in Bier – Möglichkeiten und Grenzen einer Anreicherung. *Monatsschrift für Brauwissenschaft*, **9/10**, 184–94.
39 Forster, A. and Köberlein, A. (1998) Der Verbleib von Xanthohumol aus Hopfen während der Bierbereitung. *Brauwelt*, **138**, 1677–9.
40 (1998) Hopfen-Rundschau n. 6, pp. 152–3.
41 Buckwold, V.E., Wilson, R.J.H., Nalca, A., Beer, B.B., Voss, T.G., Turpin, J.A., Buckheit, R.W., Wei, J.Y., Wenzel-Mathers, M., Walton, E.M., Smith, R.J., Pallansch, M., Ward, P., Wells, J., Chuvala, L., Sloane, S., Paulman, R., Russell, J., Hartman, T. and Ptak, R. (2004) Antiviral activity of hop constituents against a series of DNA and RNA viruses. *Antiviral Research*, **61** (1), 57–62.
42 (a) Forster, A., Gahr, A., Ketterer, M., Beck, B. and Massinger, S. (2002) Xanthohumol in beer-possibilities and limitations of enrichment. *Monatsschrift Fur Brauwissenschaft*, **55** (9–10), 184. (b) Gerhauser, C. (2005) Beer constituents as potential cancer chemopreventive agents. *European Journal of Cancer*, **41** (13), 1941–54.
43 Stevens, J.F. and Page, J.E. (2004) Xanthohumol and related prenylflavonoids from hops and beer: to your good health! *Phytochemistry*, **65** (10), 1317–30.
44 Walker, C.J., Lence, C.F. and Biendl, M. (2003) Studies on xanthohumol levels in stout/porter beer. *Brauwelt*, **143** (50), 1709–12.
45 Wunderlich, S., Zurcher, A. and Back, W. (2005) Enrichment of xanthohumol in the brewing process. *Molecular Nutrition and Food Research*, **49** (9), 874–81.
46 Schwill, A. (1983) Versuche zur Optimierung der Hochtemperatur-Würzekochung, Dissertatio, TU München, Freising.
47 Anness, B.J. and Bamforth, C.W. (1982) Dimethyl sulfide – a review. *Journal of the Institute of Brewing*, **88**, 244–52.
48 Hartmann, K., Kreisz, S., Zarnkow, M. and Back, W. (2004) Identifizierung von Filterhilfsmitteln nach den einzelnen Filtrationsschritten. *Der Weihenstephaner*, (4), 141–3.
49 Winkelmann, L. (2004) Das Gushing-Puzzle eine Erfolgsgeschichte. *Brauwelt*, (25), 749–51.
50 Zarnkow, M. and Back, W. (2001) Neue Erkenntnisse über gushingauslösende Substanzen. *Brauwelt*, (9/10), 363–70.
51 Fischer, S. (2000) Bestimmung gushingverursachender Partikel in Versuchslösungen und in Bier mittels Photonenkorrelationsspektroskopie. *Schlussbericht für das Forschungsvorhaben R*, **314**, 9–11.
52 Zepf, M. (1998) Neuere Erkenntnisse bei der Untersuchung von Gushingursachen, in 31st Technologisches Seminar, Weihenstephan.
53 Donhauser, S., Weideneder, A., Winnewisser, W. and Geiger, E. (1990) Test zur Ermittlung der Gushingneigung von Rohfrucht, Malz, Würze und Bier. *Brauwelt*, (32), 1317–20.
54 Guggenberger, J. and Kleber, W. (1963) Über den Mechanismus des Wildwerdens von Bier, in *Proceedings of the European Brewery Convention Congress, Brüssel*, pp. 299–319.
55 Schur, F., Anderegg, P., Senften, H. and Pfenninger, H. (1980) Brautechnologische Bedeutung von Oxalat. *Brauerei Rundschau*, **91**, 201–7.
56 Jacob, F. (1998) Calcium-Oxalsäure – Technologische Relevanz, in 31st Technologisches Seminar, Weihenstephan.
57 Kieninger, H. (1983) Calcium and gushing. *Brauwelt*, 14–25.
58 Burger, M. and Becker, K. (1949) Oxalate studies on beer. *Proceedings of the Society of Brewing Chemists*, **7**, 102–15.
59 (1973) *Brewers Digest*, **48**, 67.
60 Back, W. and Zarnkow, M. (2001) Auswertung des Gushing-Fragebogens, in 34th Technologisches Seminar, Weihenstephan.

16
Stability of Beer
August Gresser

16.1
Flavor Stability

16.1.1
Introduction

In addition to taste, foam, and biological (see Chapter 15) and colloidal stability, the most important characteristic of the beer quality is the flavor stability. For beer flavor, stability means its ability to keep its characteristics unaltered from the time of filling to the time of consumption.

The 'staling' process has been defined several times in different languages. However, nearly all countries that produce beer use the English term as a conventional vocabulary, and they refer to a sequence of degradations, difficult to summarize in one word. De Clerck used the term 'cooked bread', whereas most Italian authors used the term 'taste of paper'.

The staling process is characterized by oxidation reactions of natural beer components (higher alcohols, melanoidins, amino acids, fatty acids, hop resins); several components are formed, of which the carbonyl compounds, particularly aldehydes with six to 12 carbon atoms, are the most important. Table 16.1 shows some of the carbonyl compounds best known for influencing the degradation of beer aroma [1, 2].

The changes in beer flavor during its conservation are divided into two groups:

- Changes in freshness and in bitterness, less harmony in taste compared with the initial taste.
- Changes in aroma, appearance of lightstruck flavor.

These phenomena do not appear simultaneously. Some changes appear quite early (during beer transport or during improper storage) and other changes appear after weeks or even after months of storage.

Handbook of Brewing: Processes, Technology, Markets. Edited by H. M. Eßlinger
Copyright © 2009 WILEY-VCH Verlag GmbH & Co. KGaA, Weinheim
ISBN: 978-3-527-31674-8

Table 16.1 Flavor threshold of same basic 'stale-flavor-carbonyl' compounds in increasing order.

Components	Concentration (ppb)	Threshold value (ppb)	Aroma
Trans-2-cis-6-nonadienal: $CH_3CH_2CH=CH(CH_2)_3CH=CHCHO$		0.05	
Trans-2-nonadienal: $CH_3(CH_2)_5CH=CHCHO$	in fresh beer: n.; in 'stale beer': 3.6	0.1	cardboard, oxidized
Trans-2-trans-4-decadienal: $CH_3CH_2CH=CH(CH_2)_3CH=CHCHO$		0.3	aldehyde, oily
Trans-2-nonadienal: $CH_3(CH_2)_3CH=CHCH=CHCHO$		0.5	aldehyde, oily, rancid
Trans-2-decenal: $CH_3(CH_2)_6CH=CHCHO$		1	bitter, oxidized, rancid
9-Undecenal: $CH_3CH=CH(CH_2)_9CHO$		1.5	aldehyde, bitter, orange peel
n-Undecanal: $CH_3(CH_2)_9CHO$		3.5	aldehyde, bitter, orange
n-Dodecanal: $CH_3(CH_2)_{10}CHO$	in non-pasteurized beer: 2–9; in pasteurized beer: 4–16; in 'stale beer': 3–8	4	aldehyde, acidic, fatty
n-Decanal: $CH_3(CH_2)_8CHO$	in non-pasteurized beer: 8–14; in pasteurized beer: 6–16; in 'stale beer': 7–12	6	aldehyde, bitter, orange peel
n-Nonal: $CH_3(CH_2)_7CHO$	in non-pasteurized beer: 4–11; in pasteurized beer: 6–16; in 'stale beer': 7–12	18	aldehyde, bitter, astringent
Cis-3-exanal: $CH_3CH_2CH=CHCH_2CHO$		20	grassy, green grass, burnt, green oats, almond
Methional: $CH_3-S-(CH_2)_2CHO$		250	burnt, almond
n-Hexanal: $CH_3(CH_2)_4CHO$		350	aldehyde, bitter
Trans-2-hexanal: $CH_3(CH_2)CH=CHCHO$	in non-pasteurized beer: 4.5; in 'stale beer': 8.6	600	bitter, astringent, green oats
5-Methylfurfural		20 000–150 000	burnt, almond, cardboard

16.1.2
Reasons for Beer Aging

Even if there is no gustatory perception of changes in beer flavor, the chemical beer composition is changing (e.g. in the case of polyphenols or melanoidins); however, no aromatic nuances different from the primary flavor are produced, yet. Brownian motion plays an important role, and is supported by the oxygen dissolved in beer and also by the agitation of the beer during the transport; brief temperature changes intensify these phenomena.

Reterences regarding reasons for the staling process indicate that it is impossible to refer to only one mechanism or to a limited series of mechanisms identifying the processes inducing the degradation of the beer aroma. Several mechanisms lead to the formation of carbonyl compounds. There are suggestions for different schemes of the degradation processes: the simpler scheme is shown in Figure 16.1 and the more complicated process developed by Drost et al. is shown in Figure 16.2. The purpose of these schemes is to show how complicated the staling process is.

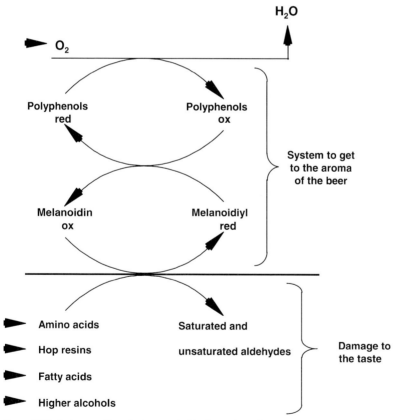

Figure 16.1 Redox reactions that promote the formation of staling flavor (source: Blockmanns et al.).

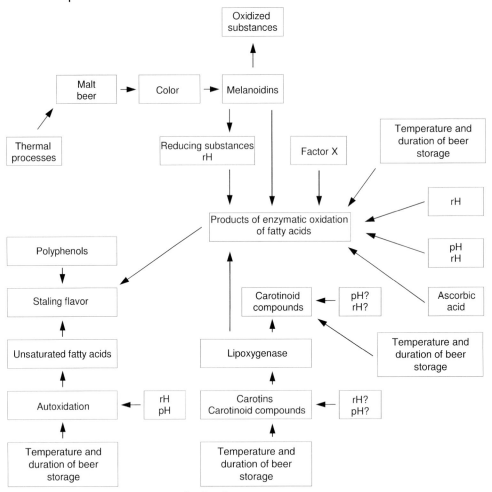

Figure 16.2 Formation of staling flavor [22].

16.1.3
Changes of Aromatic Compounds

Numerous reactions are considered to be important. Carbonyl formation is enhanced by the following factors [3, 4]:

- Strecker's degradation of amino acids (temperature/time).
- Oxidative degradation of iso-humulone (oxygen).
- Oxidations of alcohols (melanoidins, oxygen, Maillard products).
- Autoxidation of fatty acids.
- Enzymatic oxidation of lipids (oxygen).
- Aldol condensation of aldehydes (proline, CH_3, CHO).
- Oxidative degradation of carbonyl groups (oxygen, time).

- Reduction of mercaptans over the 3-methyl-2-butenile group in bitter acids from hops.

One can see that oxygen plays an important role; on one hand directly for oxidation of compounds; on the other hand indirectly for the reduction of reducing compounds contained in wort and beer. Oxygen causes reactions that influence beer quality negatively. Oxygen is indispensable for yeast only during the pitching process. As the influence of oxygen on flavor stability is also important for small and medium sized breweries, possibilities to avoid an excess oxygen intake during wort production, fermentation, filtration and filling are dealt with.

A summary of some factors influencing beer quality in comparison with the initial beer flavor is shown in Figure 16.3.

16.1.4
Definition of Indicator Substances

Carbonyl compounds known for their influence to beer aging are summarized in Figure 16.4. They interact by sensorial synergies in beer and, among others, they are formed by oxidation of higher alcohols. Reducing agents, hop components and polyphenols act as inhibitors in these reactions. Their flavor impression is slightly sweet, malty or like bread.

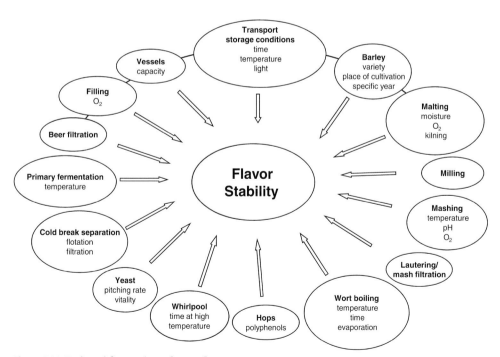

Figure 16.3 Technical factors that influence flavor.

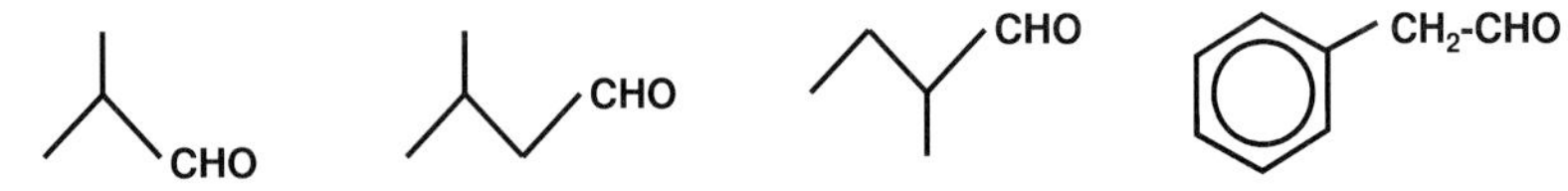

Figure 16.4 Stale flavor carbonyls.

Methylpropanal, Methylbutanal, Methylbutanal, Phenylacetaldehyde

Figure 16.5 (E)-2-Nonenal.

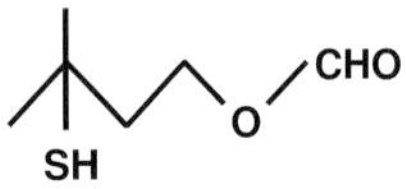

Figure 16.6 3-Methyl-3-mercaptobutylformiate.

Figure 16.5 shows (E)-2-nonenal, whose negative influence on taste is not generally accepted, because it is not known if aldehydes themselves give an unpleasant taste to beer or if they appear only temporarily during reactions. However, aldehydes are of analytical use for tracing reactions that are important for beer aging. The threshold value of (E)-2-nonenal is 0.1 µg/l and its flavor is called a 'cardboard flavor'.

Figure 16.6 shows 3-methyl-3-mercaptobutylformiate. Its exact mechanism of formation is still unknown, but it is already apparent within a few days, if a great deal of oxygen is present in the beer. Its threshold value is 5 ng/l and its flavor is called 'ribes flavor' (or 'catty flavor'), which is transformed into a distinct odor and taste of bread. Simultaneously, the color appears darker.

Table 16.2 shows an overview of indicator substances for the evaluation of organoleptic beer stability and Table 16.3 shows the normal values in pale beer [3–8].

16.1.5
Technological Measurements to Preserve Organoleptic Stability

16.1.5.1 Barley Variety
It is known that both provenience and barley variety have very important influences on flavor stability of bottled beer. Figure 16.7 shows the catechin fraction of a malt extract of the varieties Alexis and Caminant. An extreme difference between the composition of Alexis and that of Caminant, a species free of proanthocyanidin, can be observed.

Figure 16.8 shows the influence of heat and the formation of staling components after a forced aging process of beer [9, 10].

Table 16.2 Definition of the indicators for evaluation of flavor stability of beer.

Staling components	Heat indicators	Oxygen indicators
3-Methylbutanal	2-Furfural	3-Methylbutanal
2-Methylbutanal	γ-Nonalactone	2-Methylbutanal
2-Furfural		Benzaldehyde
5-Methylfurfural		2-Phenylethanal
Benzaldehyde		
2-Phenylethanal		
Estere dietilico		
Succinyl acid diethylester		
2-Phenylacetyl acid		
Ethylester		
2-Acetylfuran		
2-Propionyl		
γ-Nonalactone		

Table 16.3 Concentration range of staling components in bottom-fermented lager beer (µg/l).

Indicator	Fresh	Forced aged
Staling components	50–100	150–250
Heat indicators	10–50	70–150
Oxygen indicators	10–50	50–80

16.1.5.2 Germination

16.1.5.2.1 Introduction Fragmentation caused by the activity of lipoxygenases and peroxidases occurs in the presence of oxygen during the germination process. Thermal degradation of fatty acids takes place simultaneously [5, 11–13].

A reduction of oxygen after the third day of germination has the following advantages:

- Reduction of germination losses.
- Reduced protein modification.
- Increased lipoxygenase activities.
- Higher lipoxidation potential.

The improvement of flavor stability is shown in Figure 16.9.

16.1.5.2.2 Withering Lower drying temperatures and longer drying times enhance the concentration of degradation products from fats in malt and the flavor stability of beer improves. Withering temperatures higher than 60 °C increase the

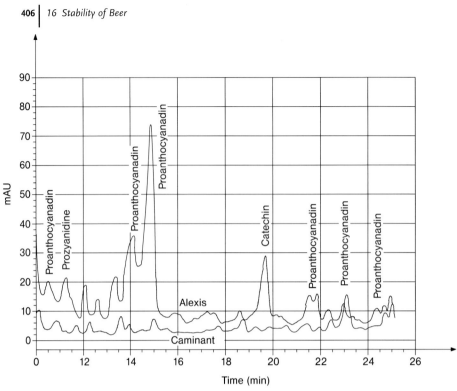

Figure 16.7 The catechin fractions in malt extract of the varieties Alexis and Caminant.

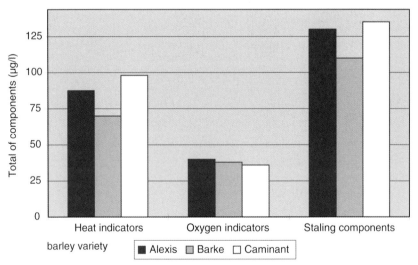

Figure 16.8 Aroma components of beer after a forced aging test. Barley variety: black bars, Alexis; grey bars, Barke; white bars, Caminant.

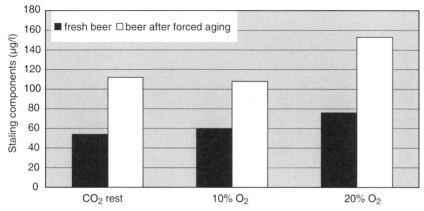

Figure 16.9 Influence of oxygen content during germination on flavor stability [11].

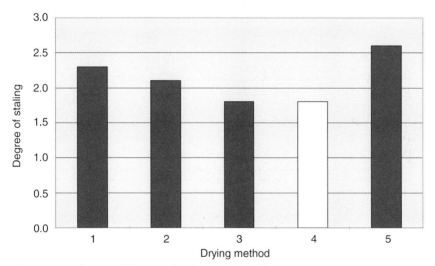

Figure 16.10 Influence of kilning on the flavor stability of beer [14]. Drying method: 1, comparison (2 hrs at 50 °C, 2 hrs at 55 °C, 8 hrs at 60 °C); 2, kiln (20 hrs at 50 °C, 50% of maximum ventilator speed); 3, kiln with two floors (kilning for 30 hrs); 4, 2 °C/h; 5, 12 hrs at 65 °C.

formation of Maillard products in malt, which leads to a deterioration of flavor stability (Figure 16.10) [14, 15].

16.1.5.2.3 **Kilning** An important factor that increases stale flavor formation is the non-enzymatic browning reaction during kilning, and also during mashing and wort boiling (Figure 16.11 and Table 16.4).

The final kilning temperature is very important regarding the formation of precursors of stale flavor components. An increase of the final kilning temperature

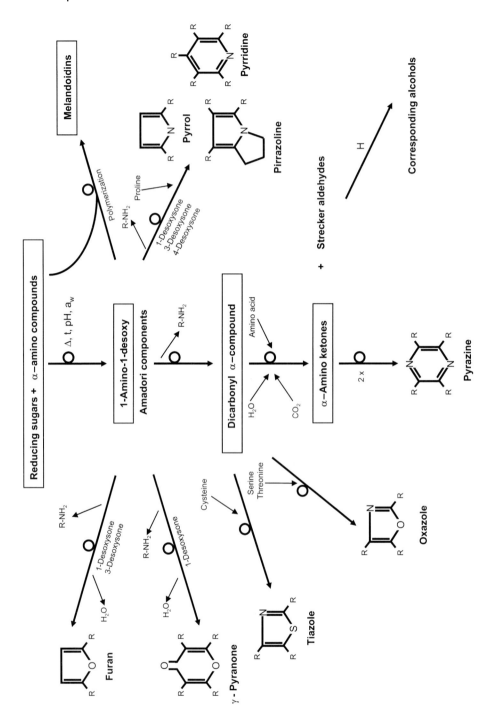

Figure 16.11 Flavor formation during the kilning-process in the course of Maillard reactions.

Table 16.4 How can the bitter substances be qualitatively evaluated according to aging?

'Good/soft'
  often a quantitative problem, e.g. comparison with apples and oranges:
    beer 1 (fresh hops): 25 IBU: 27 mg iso-α/l
    beer 2 (old hops): 25 IBU: 17 mg iso-α/l
'Neutral'
'Bad'
  crude bitterness: aged polyphenols?
  old, musty: aged aroma components?

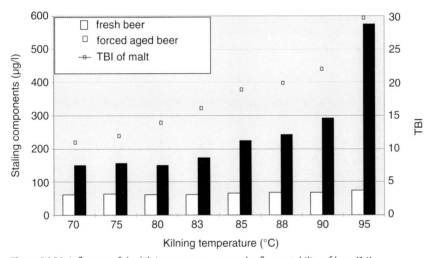

Figure 16.12 Influence of the kilning temperature on the flavor stability of beer [14].

causes an exponential elevation of the thiobarbituric acid index (TBI) in malt and of staling components in a forced aged sample of beer (Figure 16.12) [14–18].

16.1.5.2.4 **Malt Quality** Depending on the variety of barley and on the technologies of germination, the malt quality influences the flavor stability of beer directly. The malt has to be of good quality, but it should not have a too advanced proteolytic modification. The drying temperatures should be moderate and during kilning a large thermal stress should be avoided. At the end of kilning, a suitable compromise to obtain an optimal degradation of dimethylsulfide (DMS) precursor must be found.

16.1.5.3 **Wort Preparation**

16.1.5.3.1 **Malt Milling** A suitable malt milling with conditioning is important to produce an optimal starting point for the flavor stability of beer. The advantages of good conditioning are the following:

- Less crushing of husks.
- Better grist composition.
- Less extraction of husk components.
- Better composition of polyphenols.
- Softer beer taste.
- Brighter beer color.
- Quicker filtration of mash.
- Better yield of raw material.

16.1.5.3.2 **Mashing Process** It is known that oxygen at higher temperatures reacts with polyphenols, anthocyanes and tannoids. Brighter beers that are richer in polyphenols are more sensitive to oxygen in comparison with medium-colored beer and, especially, darker beers. For this reason, the brewer should carefully monitor the presence of oxygen with instruments during wort filtration and also during mashing, wort boiling or in general in the field of higher temperatures. Oxygen cannot be determined during the mashing process—one can take values from empirical procedures [13, 15, 19]. It is of general importance to avoid under more thermal stress, shear stress, and cavitation.

16.1.5.3.3 **Temperature During the Mashing Process** At higher mashing-in temperatures the activity of proteolytic enzymes is reduced. This results in higher amounts of high molecular proteins and better foam properties of the beers. Furthermore, there are lower quantities of amino acids, which cause a minor formation of Maillard products. Thus, a better flavor stability is achieved (Figure 16.13).

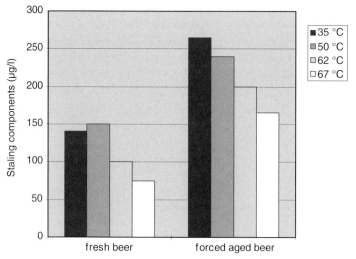

Figure 16.13 Influence of the mashing-in temperature on the flavor stability of beer [20].

The oxygen influence during the mashing-in process can be controlled by the following:

- In wet mills the mash level must be controlled carefully to avoid air suction.
- Dry grind: use large amounts of water; maybe degas water; mash inlet from the bottom of the mash tun.

During mashing one should generally avoid an excessive oxygen uptake. Therefore, infusion mashing is better than decoction. The velocity of the agitator is also of importance. During temperature rests the velocity should be low or the agitator switched off.

During the mash transfer to the lauter tun the following points are of importance:

- Mash inlet from the bottom of the lauter tun or with several pipes from the side.
- Avoid whirls (air suction) in the mash tun by means of large tubes or reduction of the flow velocity.
- The mash should not spring on the main arm of the raking machine of the lauter tun.

Cloudy wort with more than 200 European Brewery Convention (EBC) turbidity units should be recirculated below the liquid level in the lauter tun. The level of wort and sparging water should not sink below the spent grain surface. Collecting the first wort and the spargings in a closed system is now state of the art, otherwise a serious oxygen uptake results.

16.1.5.3.4 Biological Acidification A reduction of the pH value in the mash by biological acidification reduces oxidative reactions during the mashing process. In general, most of the enzymes, especially α-amylases, are activated. There is an inhibition of lipoxygenase activity at a pH of 5.2. This means that at a pH of 5.2 in the mash, less carbonyl compounds (degradation products of free fatty acids) are formed and this increases the concentration of reducing substances in the finished beer. The beer character is well balanced and its flavor stability improves (Figure 16.14) [15, 20].

16.1.5.3.5 Oxygen During Mashing Oxygen should be excluded from the mash as far as possible. Using nitrogen, for example, as protection gas will decrease the oxidation of fatty acids and of polyphenols. Gas-flashing with nitrogen improves the index of polymerization and enhances the concentration of reducing substances, and in that way generally improves flavor stability (Figure 16.15). Practical measures for the brewer include:

- Mashing-in and mash transfer from the bottom of the mashing vessel.
- Well-adapted speed of the agitator.
- Agitators should be frequency controlled to avoid oxygen uptake through whirls.
- Low oxygen content of the brewing liquor.

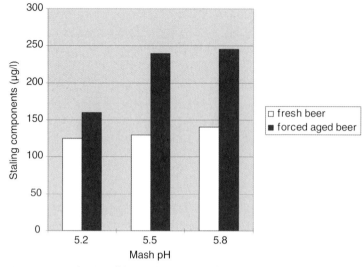

Figure 16.14 Influence of the mash pH on the flavor stability of beer [20].

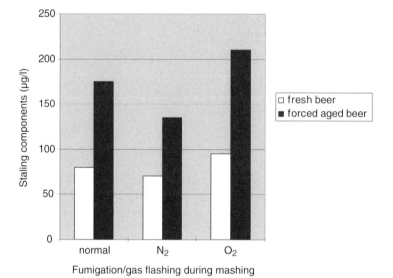

Figure 16.15 Influence of oxygen during mashing on the flavor stability of beer [20].

- Mix as homogenously as possible.
- Avoid the formation of dough.
- Short mashing-in times.
- Heating-up rate as high as possible.
- During temperature rests decrease agitator speed by half.

- The mash pump should be frequency controlled:
  - to avoid an input of air
  - to avoid shear stress.
- Pipe bends should also be wide enough to avoid shear stress (see below):
  - proteins are protected
  - obtain better foam
  - husks remain complete
  - achieve a better iodine value.

16.1.5.4 Mash Filtration

The lauter tun can have the following weak points:

- Leaks.
- Air pipe for pressure balance (continuous aeration!).
- Oxygen uptake in the wort collecting vessel.

The wort collection tank should be filled from below. The drain should be dimensioned very wide to limit wort flow velocity below 1 m/s during discharging.

Decanting the wort quickly from below in short contact times enhances flavor stability [5, 21–23]:

- Suitable raking machine.
- Clear worts by controlling the raking machine.
- Frequency controlled raking machine (speed 0.5–2.5 m/min).
- Continuous sparging
  - clear wort: <100 EBC turbidity units
  - less lixiviation of husks
  - less oxygen uptake: 0.1–0.3 mg O_2/l
  - saves time
  - better beer quality.
- Control filtration velocity by means of a continuous turbidity measurement (turbidimeter) of the wort.

Brewing liquor may contain changing amounts of dissolved oxygen. All systems for water treatment to eliminate oxygen suggests the saturation of water with CO_2. The advantages of eliminating oxygen as far as possible are:

- Less formation of disulfide links.
- Better structure of the spent grain layer.
- Better wort and beer quality.
- Less oxidation of polyphenols due to the activity of phenoloxidases.
- Less oxidation of fatty acids to hydroxyl acids and hydroperoxides (precursors of staling components).

16.1.5.5 Wort Boiling

Figure 16.16 shows the formation of reducing substances from the first wort to pitching wort (sample taken halfway through wort cooling). We can deduce that

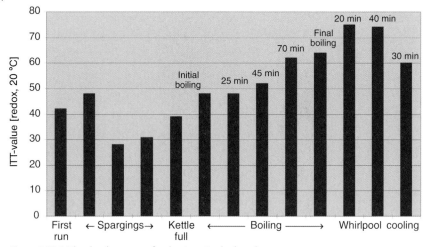

Figure 16.16 The development of reductions in the brewhouse.

there are thermal influences already before and also after wort boiling. We also have to consider the elimination of DMS.

Concerning the wort kettle, we have to consider:

- Wort inlet at the bottom.
- Keep the kettle closed (no suction of air). In general, the formation of foam in the wort kettle is always an indicator of aeration.
- External and internal boilers are advantageous.
- Reduce the pumping velocity at the end of wort transfer.

16.1.5.6 **Hot Break Removal**
When using a whirlpool for hot break removal, the whirlpool rest should be kept as short as possible to avoid thermal stress of the wort. The result is a lighter color and a lower TBI of wort and beer, and finally an improved flavor stability of the beer (Figure 16.17) [3, 15].

16.1.5.7 **Flotation (Removal of Cold Break)**
Table 16.5 shows the influence of the flotation process on important beer flavor components involved in flavor stability.

16.1.5.8 **Yeast Handling**
Normally no problems will arise when aerating pitched wort. However, the wort may become oxidized in the case where the yeast is added after flotation and aeration is at higher pressures. The over-pressure applied to mixing systems ranges from 0.5 to 6.0 bar. The dissolved oxygen in wort depends on the corresponding pressure. At higher pressures the yeast is added instantaneously since it immediately consumes excessive oxygen. Seen from the biological point of view this

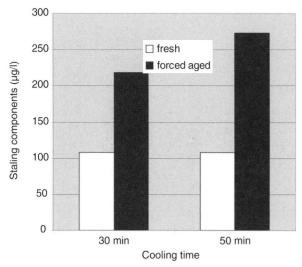

Figure 16.17 Influence of hot wort rest in the whirlpool on flavor stability [3].

operation has to be preferred. Inadequate yeast handling will result in the following disadvantages:

- Sulfury/yeasty flavor of the beer.
- Reduced head retention (bad foam).
- Unpleasant bitter taste.
- Increased pH value of the beer.
- Fermentation is slowed down.
- Increased levels of 2-acetolactate and slow maturation.
- Risk of contamination with beer-spoiling microorganisms.
- Filtration problems.
- Higher turbidity in filtered beer.

Poor flavor stability will result due to the decreased presence of reducing substances. The follow measures are advisable to obtain a yeast activity of 100%:

- Handling of the pure culture:
  - Application of the assimilation technology in the brewery by means of regular production cycles (Figure 16.18)
  - suitable assimilation technology in small breweries with a variety of different beer types.
- Cropped yeast:
  - pressure release and degassing of the yeast
  - cooling down to 1–4 °C
  - addition of water
  - mixing with efficient agitators.
- Handling of the fermentation:

Table 16.5 Influence of flotation on flavor stability.

Aroma substance (µg/l)	With flotation	Without flotation
1-Penten-3-ol	20.9	31.4
1-Octen-3-ol	1.2	1.2
2-Furfurylalcohol	1163	2408
2-Acetylfuran	3.9	10.5
2-Acetylpyrrol	88.5	98.4
3-Methylbutan-2-one (I)		
2-Methylbutanal (I, S)	37.2	43.2
3-Methybutanal (I, S)	4.6	15.7
2-Furfural (I, C)	42.6	59
Benzaldehyde (I, S)	0.5	4.0
Phenylethanal (I, S)	12.0	34.1
Acetylfurfuryl	1.7	3.9
Nicotinic acid (I, C)	0	13.3
n-Hexalactone	11.5	12.1
n-Octalactone	2.5	3.9
n-Nonalactone (I, C)	56.5	53.0
n-Decalactone	1.2	1.6
n-Dodecalactone	4.6	2.5
Sum of C	99	125
Sum of S	54	97
Sum of I	153	222
Sum of C without 2-furfural	111	163
Total sum	1452	2796
Total sum without 2-furfural	1410	2737
Total sum without 2-acetylpyrrol	1364	2698
Degustation based on staling scheme		
Odor (O)	2	3
Taste (G)	2	3.4
Bitterness (A)	1.8	2.6
Degustation value $(2 \times O + 2 \times G + A)/5$	2.0	3.1
Acceptance (%)	100	20[a]

Staling scheme: 1, fresh; 2, slightly stale; 3, stale; 4, extremely stale.
I, staling compounds; A, Strecker aldehydes; C, heat indicators.
Beer type: birra normale italiana (% di Mais: 25).
Storage of both beers: 6 months.
Storage conditions: 20 °C, without light.
a   Insufficient acceptance after 6 months.

- blending of collected yeast with assimilation yeast (e.g. 60–40%)
- reduction of the yeast cell concentration necessary for pitching
- reliable and reproducible pitching process
- adequate removal of cold break by means of flotation and a following secondary aeration
- early collection of the settled yeast (from the fourth fermentation day).

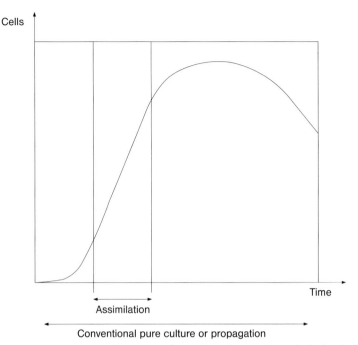

Figure 16.18 Physiological condition of yeast depending on the method of multiplication.

Optimal aeration of wort and yeast at the right moment with optimal yeast technology are prerequisites for a good flavor stability of beer. Insufficient yeast handling and beer storage conditions (too much yeast, too long, too high temperatures) will cause the following problems:

- Increased proteinase activity (Figure 16.19).
- Yeast autolysis during fermentation and during beer maturation.
- Foam problems in the bottled beer.
- Unpleasant taste properties and astringent bitterness.

16.1.5.9 Fermentation and Maturation

Direct contact with air during open fermentation may be detrimental. A second aeration of the young beer realized too late can cause production of 2-acetolactate (diacetyl). It is important to remove the sedimented yeast as soon as possible during beer fermentation and beer maturation. The principle 'as little oxygen as possible has to be in contact with beer' has to be considered [4, 24, 25].

The following points are important when beer is transferred:

- Pipes must be pre-pressurized with CO_2.
- Appropriate flow velocity (not too high) at the tank inlet.
- Avoid whirl formation during waste discharge of open fermenters or tanks.

Low maturation temperatures and maturation period, which are not too long, also have positive effects on the flavor stability of beer.

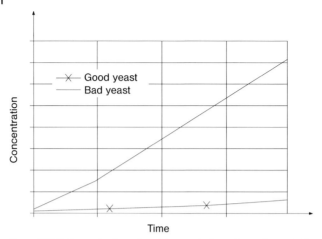

Figure 16.19 Concentration of proteinases during primary fermentation and maturation.

Possibilities to avoid more protease activity in bottled beer are:

- Optimal yeast handling.
- Flash pasteurization of the beer.

16.1.5.10 Filtration

16.1.5.10.1 Kieselguhr Filter If the process water is not degassed, beer from the initial filtration has oxygen values that will not guarantee sufficient biological stability. Often the oxygen content in these beers is 2–3 mg/l. In fact, oxygen content in water is 10 mg/l and more. At the moment of transferring the beer into bottling tanks values of 1.5–0.5 mg/l are measured. If the filter heads and tails are separated from the rest of the beer, the oxygen values are reduced to 0.02–0.05 mg/l. Naturally it has to be considered that a suspension of dosed kieselguhr can increase the oxygen content by 0.1 mg/l in beer if the kieselguhr suspension has an oxygen content of 10 mg/l. For this reason, dosing units are now treated with CO_2. Beer is also used for the preparation of the kieselguhr suspension [26, 27].

16.1.5.10.2 From the Storage Tank to the Bright Beer Tank During all these steps the oxygen content of the beer must be controlled carefully and if possible continuously independent of the brewery size. Normally beer ready for bottling contains 0.03 mg/l of oxygen. If air is used as counter-pressure gas, the following difficulties can occur in this phase:

- Oxygen absorbance on surfaces that are directly in contact with air (dimension, pressure, time, temperature).
- Formation of swirls while emptying the tank (flow velocity, mixture of oxygen and CO_2, pressure, time, temperature).

- Air bubbles in the filter, pipes, hoses, valves, fittings and so on (volume, turbulence, time, pressure, temperature),
- Air intake in pumps, valves and separators by formation of vacuums (e.g. pressure differentials).
- Changes from beer to water and vice versa (oxygen content in water, intermixture).
- Blending with waste beer (e.g. overdosing units).

Air bubbles can be formed with the following consequence:

- At a pressure of 1 bar an air bubble of 1 ml contains 0.28 mg of oxygen.
- 1 l of air at 1 bar contains 280 mg of oxygen, at 2 bar it contains 580 mg of oxygen.
- If this last air bubble dissolves slowly in beer, 580 l of beer are contaminated with 1.0 mg/l of oxygen.

16.1.5.10.3 **Measurements when Beer is Transported with Air** The applied pressures have to be slightly above the saturated vapor pressure of CO_2, also taking the temperature into consideration. Where beer is directly in contact with air, surfaces have to be as small as possible. The flow velocity at the inlet and at the outlet of the different tanks and instruments should be below 1 m/s. Flow velocities of 1.5–2 m/s are advisable in tubes to eliminate air bubbles as quickly as possible. Practical measures for the practical brewer include:

- Pre-pressurize all pipes and tanks with CO_2 during beer transfer.
- Use oxygen-free water for any contact with beer.
- Together with CO_2, the use of degassed water is saturated with CO_2 is advantageous, in particular to pre-coat filters (pre-coating with beer from the filtration of the day before is also possible).
- Iron, manganese and copper are oxidation catalysts, and have to be kept away from beer.
- Avoid leakages and air bubbles in pipes, bends, swing bend panels and so on.

When CO_2 is used it has to be considered that there can be a supersaturation because of the swirls while emptying tanks. The reaction of CO_2 with alkaline solutions can cause an implosion, a neutralization of active substances and thus a lower cleaning efficiency.

16.1.5.10.4 **Bottling Hall** The following sources cause higher oxygen contents in beer if the bright beer tanks are pre-pressurized with air:

- Formation of layers with different oxygen concentrations.
- Beer surfaces.
- Outlet of the bottling tank.
- Change from beer to water in the pipes.
- Beer blenders, panel pumps.

- Filling the bottle.
- Air in the bottle neck.

Therefore, the bright beer tank and the following pipes should exclusively be handled with CO_2. With new filling systems it is possible to realize dissolved oxygen contents below 0.05 mg/l and to reach 0.3 ml of air in the bottle. Neck beer losses by over-foaming are limited to 1 ml/l.

16.1.5.11 Filling

Serious weak points during bottling:

- Air bubbles in the filler inlet pipes.
- Leakages in reducing valves and valves for pressure balance.
- Intervals during filling.
- Pressure fluctuations during pre-vacuuming.
- Distance between filling element and ground (turbulences).
- Pre-pressure
- Sloping beer level in the bottle (velocity too high).
- Size of the foam bubbles.
- Air under the crown.
- Surface water in the bottle from the bottle washer.
- Beer temperature during bottling.

With respect to beer quality and stability, the following figures should be aimed at during bottling:

- Dissolved oxygen at filler inlet: <0.05 mg/l.
- Oxygen uptake during filling: <0.05 mg/l.
- Air in the bottle neck: <0.4 ml/500 ml (<0.23 mg O_2/l).
- Good values for total oxygen: <0.33 mg O_2/l.
- Very good values: <0.20 mg O_2/l.

Measures to reduce oxygen intake include:

- Use modern filling stations and modern filling techniques (double pre-vacuuming, high-pressure injection).
- High purity of CO_2.
- Target value for total oxygen content in bottled beer: <0.2 mg/l.

Oxygen elimination is expensive, but it is very important to order to obtain a high-quality final product. Oxidation indicators and decomposition products also increase with increasing oxygen content during bottling (Figure 16.20):

- 3-Methylbutanal.
- Phenylethanal.
- Iso-butanal.

Figures 16.21 and 16.22 show the result obtained by technological measurement on the formation of aging compounds and taste tests concerning the flavor stability of bottled beer [3, 28, 29].

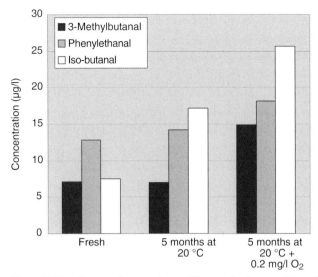

Figure 16.20 Influence of oxygen during filling on the flavor stability of beer (source: Thum).

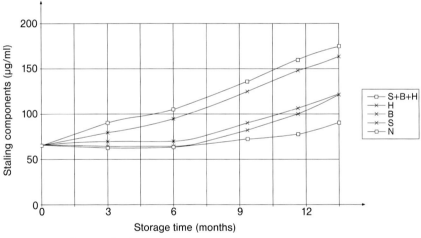

Figure 16.21 Influence of technological means on staling components of beer depending on the storage time. Normal beer (N) in comparison to optimization in the brewhouse (S). Means of optimization combined (S + B + H). Biological acidification (B). Assimilation technology (H).

Since the mid 1980s, plastic has claimed a steadily increasing share among the available package alternatives. Beverages packaging has become a marketing instrument more and more. Thus, it stands between advertising (marketing), packaging and logistics costs (controlling), and product quality. The trend towards bottling in polyethylene terephthalate (PET) bottles can no longer be stopped. Less-sensitive beverages, such as cola beverages or mineral water, can be bottled in standard PET bottles using returnable bottles as well as one-way bottles.

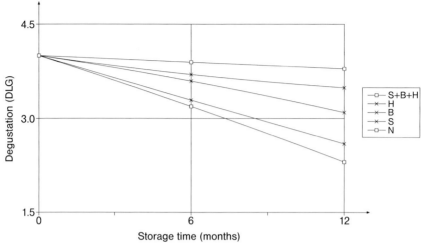

Figure 16.22 Influence of technological means on the flavor stability of beer depending on the storage time. Normal beer (N) in comparison to optimization in the brewhouse (S). Means of optimization combined (S + B + H). Biological acidification (B). Assimilation technology (H).

However, custom-made PET bottles, which are adjusted to the demands of product stability, must be used for sensitive products. These custom-made plastic bottles are not suitable for use as returnable bottles, except for polyethylene naphthalate (PEN) bottles. Oxygen permeability, among other things, is an important factor in beer packaging. There seem to be two alternative types of plastic bottles for bottling beer:

- PET bottles with an inside coating.
- Multilayer bottles with integrated oxygen scavengers.

Table 16.6 gives an overview of the protection against gas permeation of different packing materials. In addition, protection against light plays an important role regarding the function of beverage packing (see Section 16.2).

16.1.5.12 Analytical Control of Flavor Stability and Stale Flavor Compounds

Thermal treatment of beer (flash pasteurization, bottle pasteurization) can be proved by the absence of any enzymatic activity (sucrose test). The progress of flavor deterioration can easily be proved by sensory analysis by a carefully trained taste panel. The TBI is an excellent indicator of thermal treatment of malt and wort, but will not give very reliable results for beer. For beer, the aniline index (AI) is a better indicator to prove a thermal treatment of beer or improper storage conditions of the bottled beer in the market. A further application of this method is for the determination of the conditions during storage of bottled beer. In cases of 'returns', an increased AI is an indication of improper storage of bottled beer (see Table 16.7).

Table 16.6 Oxygen conditions during beer filling (mg O_2/l beer).

	Keg	Can	Glass bottle	PEN bottle	PET bottle (multilayer)
Beer after filtration in bright beer tank	0.04	0.04	0.04	0.04	0.04
Intake during filling process	0.01	0.02	0.02	0.02	0.02
Air in head space	0.0	0.01	0.01	0.01	0.01
Oxygen permeation through compound mass of closure during 6 months	0.0	0.0	0.40	0.40	0.20
Oxygen migration from material into the product finished after ...	0.0	0.0	0.0 (2–3 months)	0.35 (2–3 months)	0.90 (3–4 days)
			0.47	0.82	1.17

Table 16.7 AI and storage conditions.

Storage time (months)	AI		DLG	
	20 °C	6 °C	20 °C	6 °C
0	2	2	4.8	4.8
1	15	10	4.7	4.7
3	48	25	4.0	4.4
6	72	28	3.7	4.3

Very detailed information can be obtained by gas chromatography analysis. Individual staling components such as Strecker aldehydes, carbonyls from long-chain fatty acid breakdown, 2-furfural, esters and so on can be determined. These components can serve individually as indicators of thermal treatment (e.g. 2-furfural), excessive oxygen uptake (Strecker aldehydes) or a storage time that is too long (several esters) [3].

16.2
Lightstruck Flavor

Packaged beer has a finite shelf-life, as beer constituents undergo slow and progressive deterioration, manifested by deterioration in the beer flavor, color and colloidal stability. Since beer flavor is of paramount importance to its acceptability to the consumer, understanding and controlling flavor changes of packaged beer during storage is of clear commercial relevance. As packaging technology advances and the packaging options for beer become more diverse, so a better understanding of the role of the package and packaging operations in flavor stability becomes ever more important. It is known that beer is very sensitive to light.

A mercaptan is formed by the reduction of iso-α-acid under the influence of light, the threshold value of which is less than 1 ppb. Figure 16.23 shows the mechanism for the formation of the lightstruck flavor in beer.

The single most important characteristic of the bottle that sets it apart from other forms of packaging is that it is usually manufactured from light-translucent material glass. This translucency offers benefits over alternative forms of packaging in that it allows the product to be seen directly by the consumer. However, it is unfortunate that when beer is exposed to light it may developed an objectionable 'skunky' aroma. This is mainly attributable to 3-methyl-2-butene-1-thiol (MBT) formation from iso-α-acid (see Figure 16.23) and, thus, measures need to be taken to protect the packaged beer from the flavor-damaging effects of prolonged light exposure in order to use bottles as a viable packaging option. The effective range of wavelengths required to form MBT in beer is 350–500 nm. More recently, it has been demonstrated that the formation of MBT in bottled beer can be slowed down according to the color of the glass used. Colorless bottles were shown to inhibit the production of MBT the least, whilst brown bottles markedly improved the light stability of the packaged beer. The effect was explained in terms of the differing abilities of the glass used to filter out the damaging light wavelengths.

Differences in transmittance of light in the 350–500 nm range were also observed with green bottles. Figure 16.24 shows that the ability of the glass used to filter out the relevant wavelengths increases from beer bottle A to E. Figure 16.25 shows

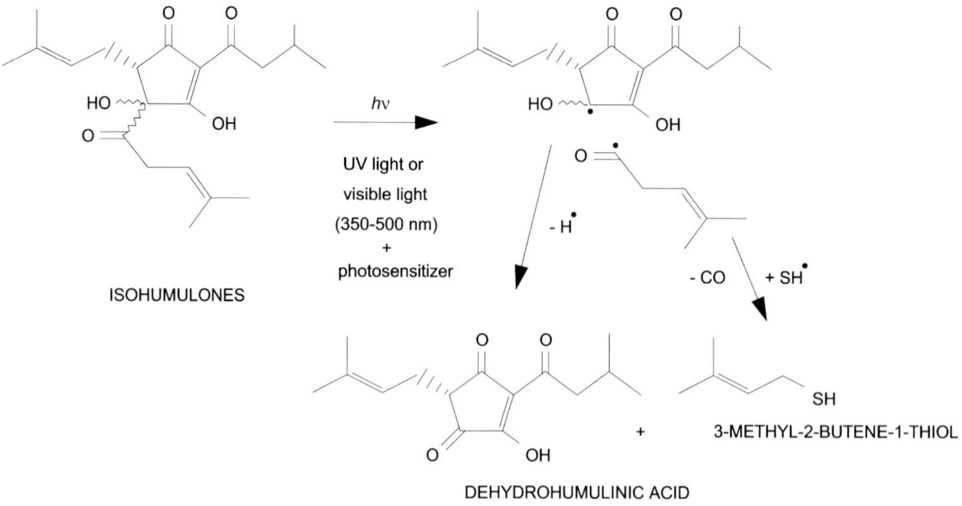

Figure 16.23 Formation of lightstruck flavor in beer [30].

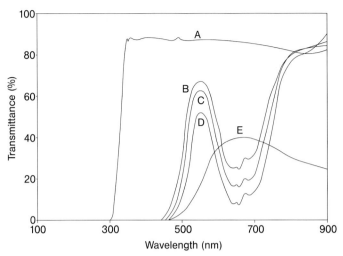

Figure 16.24 Light transmittance patterns of various beer bottles: A, colorless bottle; B, green bottle 1; C, green bottle 2; D, green bottle 3; E, brown bottle.

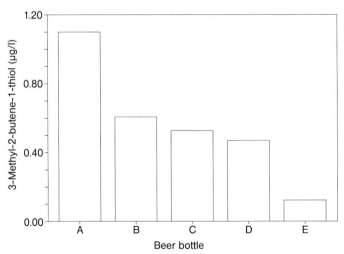

Figure 16.25 MBT formation in lager beer. A, colorless bottle; B, green bottle 1; C, green bottle 2; D, green bottle 3; E, green bottle 4.

that this trend corresponds to a decreased likelihood of MBT formation in the bottled beer.

The 'skunky' aroma and flavor can develop very quickly. Depending on the package transmittance and intensity of light, a noticeable aroma and flavor could develop in minutes [30].

Studies have also shown that the residual α-acids in beer can undergo isomerization to iso-α-acids upon exposure to light. It has been reported that the oxidative degradation products of α-acids, bittering in their own right, show some interesting photochemical properties. For example, the hupulones are very sensitive to light, degrading completely after sufficient exposure. In contrast, dehydrohumulinic acids increase in concentration upon sufficient exposure to light.

Clearly, packaging beer in bottles introduces a number of photochemical stability issues that need to be understood and addressed to ensure that the product reaches the consumer in the best condition possible. The development of light-struck off-flavors is of key importance and suitable protection of the product is of obvious necessity. However, when using bottles with limited or no UV-shielding properties (for whatever reason), and still wanting to avoid the lightstruck flavor as long as possible, the only serious solution is to reduce the iso-α-acids. Reduced hop products that are not stimulated by light have been used to bitter beer where light stability is required.

Figure 16.26 shows the chemical structure of reduced iso-humulones and of hydrohumulones, and Figure 16.27 shows the formation of tetrahydro-iso-α-acid from α-acids and β-acids. To guarantee the insensitivity to light, beers have to be bittered exclusively with reduced iso-extracts, as only traces of normal iso-α-acids are able to cause the lightstruck flavor (see Figure 16.28). Another necessary supposition is the separate conduction of yeast determined for beer produced like this and hence avoiding the adsorption of iso-α-acids from yeast of 'normal beers'.

Since the addition of reduced iso-extracts generally happens during filtration, until this moment in the brewing process the bactericidal effect of bitter substances is not in effect. Raw iso-α-acids have a less accentuated bitterness than the iso-acids, such that they also differ in the bitter odor sensation. Tetra- and hexa-hydro-iso-α-acids possess a clearly inferior capacity to dissolve than iso-α-acids, which can be schematized as follows:

- It is necessary to use a pressing metering unit, which often implies the application of a solvent (e.g. ethanol) or of an emulsifier.
- The permanence of the beer foam is improved.

The potential of the bitter taste is substantially higher with a clearly modified quality of bitterness. With an exclusive application of tetrahydro-iso-α-acids, the bitter taste and the beer foam are so unusual that in the most cases the addition of a combination of both is preferred. The use of reduced extracts is more expensive than the use of conventional products. In recent years the tetrahydro-iso-α-acid extract has been more used, as it is considered as an alternative to other foam stabilizers, which may have to be declared. A dosage of 2–4 mg/l is sufficient to obtain an improvement [30–36].

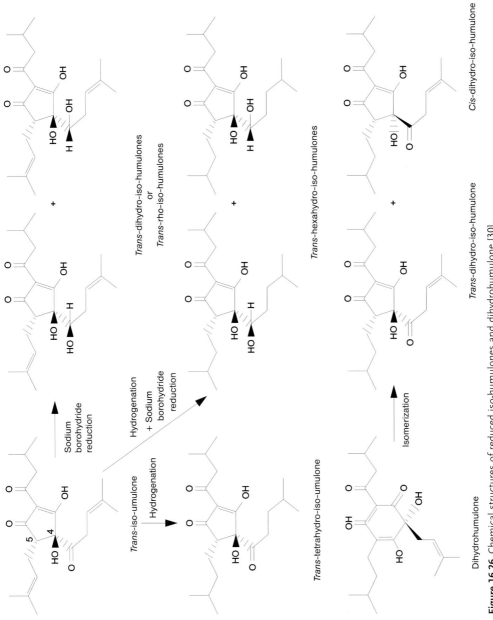

Figure 16.26 Chemical structures of reduced iso-humulones and dihydrohumulone [30].

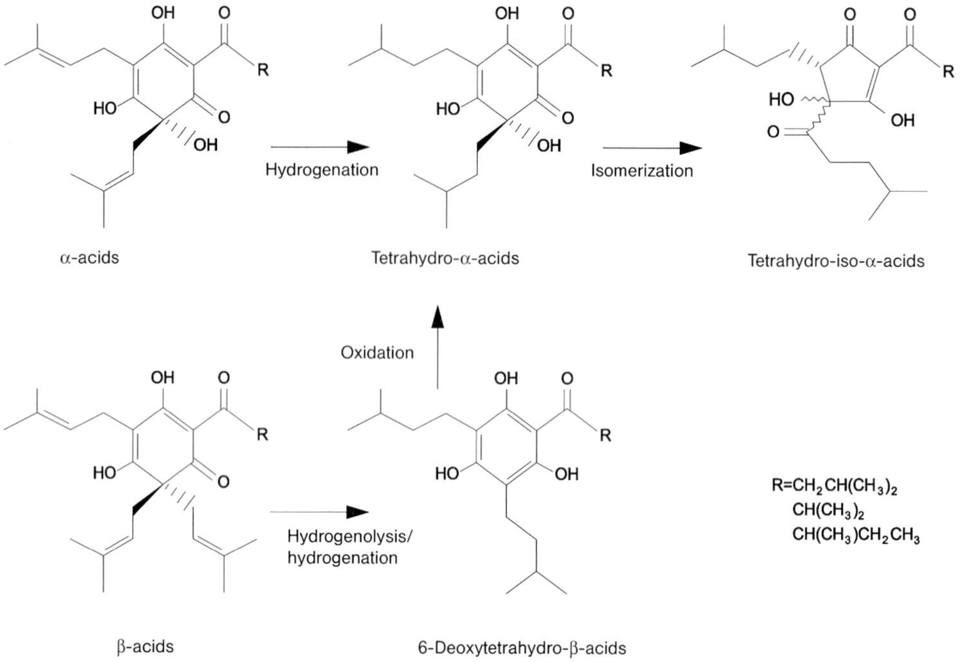

Figure 16.27 Preparation of tetrahydro-α-acids from α-acids and β-acids [30].

16.3
Colloidal Stability of Beer

16.3.1
Introduction

If bottled beer is exposed to extreme conditions during its shelf-life (e.g. serious temperature changes) and if it is stored for a too long time, dissolved proteins are precipitated by polyphenols (e.g. proantocyanidines) and cause turbidities (Figure 16.29). Protoanthocyanidins are constituents of barley and hops (but not in raw grains like maize and rice), and are extracted from them during wort production, simultaneously with some protein fractions of barley grains, which are detectable in the finished product.

With regard to beer, we have to distinguish between reversible turbidity (chill haze) and permanent irreversible turbidity (permanent haze). The first appears if beer is cooled down to 0 °C or even below, but it disappears if warmed up to 20 °C. It is a precursor of the permanent haze, which remains if beer is stored at room temperature for a longer time. The particles that form the permanent haze are bigger than those that form the reversible turbidity.

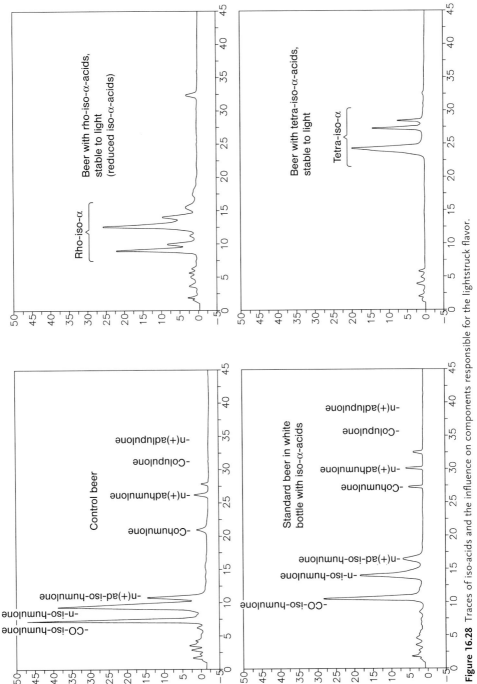

Figure 16.28 Traces of iso-acids and the influence on components responsible for the lightstruck flavor.

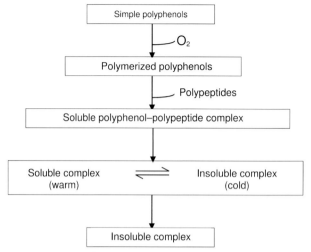

Figure 16.29 Haze formation in beer.

16.3.1.1 Composition of Turbidity in Beer

The qualitative composition, but especially the quantitative composition, of the turbidity in beer can vary over a wide range. There are three important groups of polymers:

- Polypeptides.
- Polyphenols.
- Polysaccharides.

Each of these three groups may have a share up to 75% of the turbidity. Mainly, one can find complexes between proteins and tannins. Each of these substances can form the main part of up to three quaters of the precipitate. Most common is the chelate formation between proteins and tanning agent. However, there are associations between single or primarily both components and carbohydrates (α- and β-glucane, pentosans) and/or minerals. Mineral substances play an important role in the composition of those turbidities. Mainly iron and copper are known as catalysts for oxidation processes. Hop resins, melanoidins, lignin, foam stabilizers and preservatives or detergents can also appear as components of the haze.

16.3.2
Mechanism of Turbidity Formation

Details of the mechanism of turbidity formation are not very clear. Polypeptides, polyphenols and polysaccharides cohere themselves by hydrogen bonds. Metals bind on hydroxy, amine and thiol groups of the different polymers.

These two types of coherence appear in chill haze. The bonds are quite weak and can be broken easily. It is commonly assumed that the formation of irreversible turbidity is often caused by oxidation, especially the formation of covalent bonds.

Chemical turbidities normally consist of carbohydrates or micro-crystals of calcium oxalate. Turbidity of carbohydrates for the most part can be avoided using homogeneous and well-modified malt, and by precise work in the brewhouse. Addition of calcium sulfate or calcium chloride to brewing water and to the boiling wort may avoid calcium oxalate problems. The contact of filtered beer with calcium should be strictly minimized.

16.3.3
Influence of Raw Materials and Auxiliary Materials

Colloidal stability for the most part depends on the raw and auxiliary materials. However, the brewing plants, the systems of production and of bottling, the special methods of stabilization, and the transport and conservation conditions of finished beer also play a very important role. Brewing water with less than 5°dH German hardness of residual alkalinity and well-modified malt low in proteins (below 11%) are known as most important prerequisites for good colloidal stability. The disadvantages of low quality malt can be compensated to a limited extent only by a suitable brewhouse procedure and eventually by the use of industrial enzymes.

16.3.4
Brewhouse

16.3.4.1 Malt Milling
If the mill is operated under optimal conditions its influence on colloidal stability is negligible. Malt quality dominates.

16.3.4.2 Mashing Procedure
Mashing at a temperature of 40–50 °C improves the proteolysis and decreases the presence of high-molecular-weight proteins in wort. Well-adapted infusion systems are advantageous concerning the filterability of beer and energy consumption.

16.3.4.3 Mash Filtration
Very turbid mashes, especially with long filtration times, lead to less colloidal stability. Last runnings of 1% extract or below should be avoided because they contain more polyphenols with a high polymerization index.

16.3.4.4 Wort Boiling
Through experiments it was found that based on the same boiling duration only worts with a lower total evaporation were shown to result in beers with less stability, while a reduction of boiling time from 70 to 35 min was shown to be harmful in this respect. Oxygen has no important effect during wort boiling.

16.3.4.5 Hot Break
The hot break has to be removed quantitatively to obtain stable beers.

16.3.5
Fermentation and Storage

Alcohol formation, pH drop and final cooling of green beer cause the precipitation of potential turbidity-forming components during fermentation, maturation and cold storage. The cold storage phase at −1 or −2 °C should last more than 7 days.

In conventional old cellars small compartments with only a few tanks allow a well-adapted temperature profile for maturation and cold storage. An exceeding storage time mostly combined with a slight increase of pH value has a negative effect on the colloidal stability of beer.

16.3.6
Filtration

Filtration difficulties caused by insufficient malt quality can be neutralized by blending with malt of excellent quality and by a more intensive mashing process. Shear forces in pumps, valves and plate separators can decrease the natural clearing ability of the beer and thus its colloidal stability. Shock cooling of the beer immediately before the filtration does not offer advantages for the colloidal stability. Contact with oxygen during filtration and when beer is racked has to be minimized.

16.4
Stabilization Systems

The measures discussed above to increase colloidal stability are only sufficient for smaller breweries with short distribution pathways and times. To assure colloidal stability in beers destined for national or even international markets with long distribution times, beer must be stabilized using different techniques. They act on tannins, proteins, carbohydrates, oxygen and heavy metals.

The stabilization of beer by polyvinylpolypyrrolidone (PVPP) has found a large area of applications, particularly since it is possible to recycle PVPP. Today, xerogels and hydrogels of silicic acid are widely used. Hydrosols of silicic acid are added to wort and permit a more intense fermentation and a better pre-clarification, and provide a better filterability and better colloidal stability. A similar stabilizing effect could be obtained by using directly a variety of barley without proantocianidine (Figure 16.30); Figure 16.31 shows the foam stability in samples stabilized with PVPP, but which were not pasteurized at different conservation temperatures.

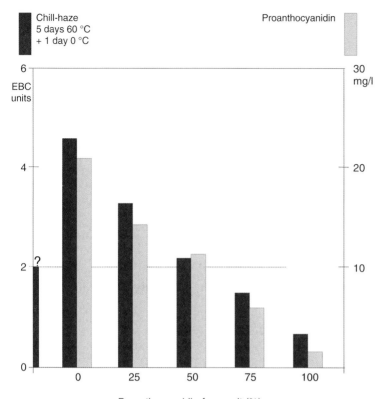

Figure 16.30 Turbidity in EBC units of beers brewed with different percentages of proanthocyanidine-free malt after 5 days at 60 °C plus 1 day at 0 °C.

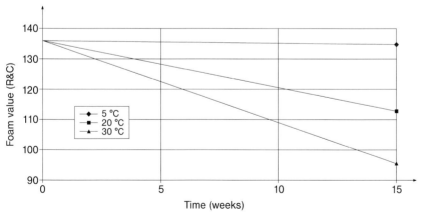

Figure 16.31 Foam stability in non-pasteurized samples at different storage temperatures.

References

1 Dalgliesh, C.E. (1977) Flavor stability, *Proceedings of the 16th European Brewery Convention Congress, Amsterdam*, pp. 623–59.
2 Miedaner, H., Narziss, L. and Eichhorn, P. (1991) Einige Faktoren der Geschmacksstabilität – sensorische und analytische Bewertung, *Proceedings of the 23th European Brewery Convention Congress, Lisbon*, pp. 401–8.
3 Lustig, S. (1994) Das Verhalten flüchtiger Aromastoffe bei der Lagerung von Flaschenbier und deren technologische Beeinflussung beim Brauprozeß, Dissertation, TU München.
4 Narziss, L. (2004) *Abriss der Bierbrauerei*, 7th edn, Wiley-VCH Verlag GmbH, Weinheim.
5 Tressl, R., Bahri, D. and Silwar, R. (1979) Bildung von Aldehyden durch Lipidoxidation und deren Bedeutung als 'Off-Flavor-Komponenten' in Bier, *Proceedings of the 17th European Brewery Convention, Berlin*, pp. 27–41.
6 Belitz, H.-D., Grosch, W. and Schieberle, P. (2001) *Lehrbuch der Lebensmittelchemie*, 5th edn, Springer-Verlag, Berlin.
7 Komarek, D. (2001) Key odorants in beer – influence of storage on the flavour stability, Dissertation, TU München.
8 Hill, P.G., Lustig, S. and Barklage, H.J. (2001) Seeing the light – using innovative instrumental analysis to assess new packaging, *Proceedings of the 28th European Brewery Convention Congress, Budapest*, pp. 910–15.
9 Boivin, P. (2001) A review of pro- and anti-oxidant of malt, *European Brewery Convention Symposium Flavour and Flavour Stability, Monograph 31*, No. 1.
10 van Waesberghe, J. (1996) Anti-oxidants and pro-oxidants in preprocessed brewing ingredients. *Master Brewers Association of the Americas Technical Quarterly* **33** (2), 96–101.
11 Kretschmer, H. (1996) Der Einfluß physiologischer Vorgänge beim Mälzen auf das Verhalten der Lipide und deren Einfluß auf Bieraromastoffe und Bieralterung, Dissertation, TU München.
12 Ketterer, M. (1994) Untersuchungen zum Fettstoffwechsel bei der Keimung und der hieraus resultierenden Metabolismen, Dissertation, TU München.
13 Zürcher, A. (2003) Der Einfluß des Blattkeims von Gerstenmalz auf die Geschmacksstabilität und weitere Qualitätsmerkmale von Bier, Dissertation, TU München.
14 Forster, C. (1996) Der Einfluß der Darrtechnologie auf die Malz- und Bierqualität, Dissertation, TU München.
15 Back, W., Forster, C., Krottenthaler, M., Lehmann, J., Sacher, B. and Thum, B. (1997) Neue Erkenntnisse zur Verbesserung der Geschmacksstabilität. *Brauwelt*, **38**, 1677–92.
16 Forster, C., Narziss, L. and Back, W. (1997) Untersuchungen zum Bieraroma und der Geschmacksstabilität mit verschiedenen Malzschüttungen hergestellter dunkler Biere, *Proceedings of the 26th European Brewery Convention Congress, Maastrich*, pp. 561–8.
17 Preuss, T. (2001) Technologische Maßnahmen zur Erzielen malzaromatischer dunkler Biere hoher Geschmacksstabilität nach Charakterisierung der Schlüsselaromastoffe in ausgewählten dunklen Bieren, Dissertation, TU München.
18 Cantrell, I.C. and Griggs, D.L. (1996) Malt: its role in oxidation. *Master Brewers Association of the Americas Technical Quarterly* **33** (2), 82–6.
19 Muts, G.C.J. and Pesman, L. (1986) *European Brewery Convention, Monograph XI. Wort Production*, Fachverlag Hans Carl, Nürnberg.
20 Takahashi, Y. (1996) Pilotversuche über den Einfluß der verschiedenen Maischparameter auf die Eigenschaften der Würze und des Bieres unter besonderer Berücksichtigung der Geschmacksstabilität, Dissertation, TU München.
21 Deutsche Norm Sudhausanlagen in Brauereien DIN 8777 (1996) *Normenausschuß Maschinenbau (NAM) im*

DIN Deutsches Institut für Normierung e. V. Berlin, Beuth Verlag.

22. Drost, B.W., van Eerde, P., Hoekstra, P. and Strating, J. (1971) Fatty acids and Staling of Beer, *Proceedings of the 13th European Brewery Convention Congress, Estoril*, pp. 451–8.

23. Nordlöv, H. (1985) Formation of sulphur dioxide during beer fermentation, *Proceedings of the 20th European Brewery Convention Congress, Helsinki*, pp. 291–8.

24. Back, W. (2005) *Ausgewählte Kapitel der Brauereitechnologie*, Fachverlag Hans Carl, Nürnberg.

25. Heyse, K.-U. (1995) *Handbuch der Brauerei-Praxis, 3, verbesserte und erweiterte Auflage*, Getränke-Fachverlag Hans Carl, Nürnberg.

26. Manual of Good Practice: European Brewery Convention (1999) *Technology an Engineering Forum – Beer Filtration, Stabilisation and Sterilisation*, Hans Carl Verlag, Nürnberg.

27. Manual of Good Practice: European Brewery Convention (2000) *Technology an Engineering Forum – Fermentation and Maturation*, Hans Carl Verlag, Nürnberg.

28. Franz, O. (2004) Systematische Untersuchungen zur endogenen antioxidativen Aktivität von hellem, untergärigem Bier unter besonderer Berücksichtigung technologischer Maßnahmen beim Brauprozeß, Dissertation, TU München.

29. Eichhorn, P. (1991) Untersuchungen zur Geschmacksstabilität des Bieres, Dissertation, TU München.

30. Manual of Good Practice: European Brewery Convention (1997) Hops and Hop Products Vol. 3 Manual of Good Practice EBC.

31. Forster, A. (2000) Le nostre attuali conoscenze sul luppolo – Birra & Malto 75/17.

32. Forster, A. (1998) Hopfen ein natürlicher Rohstoff oder Basis für maßgeschneiderte Moleküle. *Mitteilungen Österreichisches Getränkeinstitut*, **34**, 28–34.

33. Forster, A., Beck, B. and Schmidt, R. (1999) Hopfenpolyphenole – mehr als nur Trübungsbildner in Bier [Hop polyphenols do more that just cause turbidity in beer], *Hopfenrundschau International*, **00**, pp. 68–74.

34. Guzinski, J.A. (1994) Considerations of reduced hop extracts, *EBC Symposium on Hops, Zoeterwoude, Monograph XXII*.

35. Seldeslachts, D. et al. (1999) The use of high tech hopping in practice, *Proceedings of the 27th European Brewery Convention Congress, Cannes*.

36. Weiss, A. et al. (2002) Sensory and analytical characterisation of reduced, isomerised hop extract and their influence and use in beer. *Journal of the Institute of Brewing*, **108** (2), 236–42.

17
Analysis and Quality Control

Heinz-Michael Anger, Stefan Schildbach, Diedrich Harms, and Katrin Pankoke

17.1
Introduction

This chapter deals with the area of analytics and its role in quality control in breweries. Of course it is not possible to cover all aspects of analytics in breweries within one chapter, and it is not the authors' aim to substitute the precise description of the analyses in the referring compilations issued by the European Brewery Convention (EBC), Mitteleuropäische Brautechnische Analysenkommision (MEBAK), American Society of Brewing Chemists (ASBC), Institute of Brewing and Distilling, and Brewery Convention of Japan (see below). However, the authors hope to give some additional background and hints amending these compilations, and therefore to provide some exciting information for the interested reader and a reference for first information as required.

Analytical methods have improved significantly within recent years, opening further possibilities in process control during beer preparation and in beer quality control. Also, public opinion has become more aware of contaminants in food in general, which are nowadays detectable more precisely. Examples of contaminants in beer are hydrophobins, being related to the occurrence of gushing, *N*-nitroso dimethyl amine mainly as a result of non-proper kilning procedures, mycotoxins, and allergens like sulfur dioxide and gluten. However, not only contaminants, but also ingredients of the beer with further benefits come into analytical focus. Examples of additional benefits of ingredients in beer are, with respect to possible health improvement, xanthohumol from the hops, polyphenols, vitamins, fibers and minerals. Many of these ingredients and their impact have to be examined in more detail in the future. All these points broaden the number of analyses carried out in beer.

As analytical methods become more and more sophisticated, it is difficult to keep control of the number of analyses performed under the paradigm of sufficient process control. As analytics is a tool, it is the authors' aim to provide hints from their professional background as members of a service laboratory and former head of a laboratory in a large German brewery, respectively, on how to use analytics reasonably within the process and in day-to-day quality control of beer production.

There are some obvious reasons for doing quality control:

- Legislation for the consumer's safety, within the home country, but also for the countries the beer is going to be exported to.
- Evaluation and improvement of one's own product quality.
- Plant control.
- Control of the incoming raw materials and aids.

The legislation is manifold. In more detail it may be grouped in the following areas:

- Food safety.
- Hygienic food ('Hazard Analysis of Critical Control Points' concept).
- Mycotoxin thresholds (e.g. ochratoxin A, deoxynivalenol, zearalenone and frimonisins).
- Contaminants in food (e.g. arsenic, lead, cadmium and mercury).
- Allergens and their declaration on the label.
- Threshold values for pesticides.
- Radioactivity.

However, also more or less optional commitments like the ones listed below have gained importance in recent years:

- DIN [ISO 9000 (emphasis on quality control)].
- International Food Standard (emphasis on hygiene).
- Quality and Security (yeast and spent grain as feed).

Last, but not least, the marketability of the product has to be ensured. In this context, various information must be displayed on the label and this has to be controlled analytically:

- Information about the product (kind of product, producer).
- Declaration of the alcohol content.
- Declaration of the expiry date.
- Declaration of the production lot.
- Declaration of the nominal filling volume.
- Calorific value of the product.
- Declaration of allergenic ingredients.

Analytical methods related to wort, beer and raw materials used for beer production are described in detail in a number of compilations. The most important ones are:

- *Analytica-EBC* [1], issued by the European Brewery Convention (in English).
- *Brautechnische Analysenmethoden von MEBAK* [2–6], issued by MEBAK (in German).

- *Methods of Analysis* [7], issued by the ASBC (in English).
- *Methods of Analysis* [8], issued by the Institute of Brewing and Distilling (in English).
- *Methods of Analysis* [9], issued by the Brewery Convention of Japan (in Japanese).

These compilations are brought up to date at more or less regular intervals. In order to achieve comparability and as quite a number of the analytical methods described are traditional ones based on conventions, it is important to specify and cite the source of the analysis precisely. Within the compilations of analytical methods, in most cases additional information about the analyses are given, for example, the basic principle of the analysis, some hints about the purpose of the analysis, frequent problems arising while performing the analysis, interpretation of the results in technological respect, standard values and – very importantly – figures about the precision of the analysis (repeatability, r, and reproducibility, R).

Repeatability r and reproducibility R describe the range covering 95% of the results obtained analyzing the same sample several times under repeatability and reproducibility conditions. Definitions for repeatability and reproducibility conditions are given in the literature, such as the following [10]:

- *Repeatability conditions*: conditions where independent test results are obtained with the same method on identical test items in the same laboratory by the same operator using the same equipment within short intervals of time.
- *Reproducibility conditions*: conditions where test results are obtained with the same method on identical test items in different laboratories with different operators using different equipment.

In general, repeatability and reproducibility are determined in collaborative tests carried out before the analysis becomes part of the referring compilation. In many cases they are calculated simply by multiplication of the standard deviation within a laboratory (r) and the standard deviation between laboratories (R) by a factor 2.8, respectively. The information about the repeatability and reproducibility of an analysis is of great importance for a proper judgment, and therefore for a serious interpretation of analytical results. Although this is the case, no values for r and R are given in the following sections, as these values are subject to changes. Much more background information about the mode of their determination (proficiency tests, round-robin tests) and more up-to-date values are published within the compilations cited.

17.2 Analyses

In the following the most relevant analyses for beer and its preliminary products and byproducts are described. They are summarized according to their analytical principle.

17.2.1
Density, Extract, Alcohol Content, Original Gravity and Degree of Fermentation

One of the most frequent parameters analyzed in beer and wort is the extract. The extract by nature is not homogenous, made up of mainly sugars and dextrin, but also other components like nitrogenous compounds and minerals. As it is very laborious to analyze component by component and therefore not practical in day-to-day routine analysis, the extract is referred to a solution containing only sucrose. The physical property analyzed is the density; the relationship between density of a solution and the amount of sucrose in a solution with the same density is described in the sugar table according to Goldiner *et al.* [11]. Sucrose is chosen as the reference, as the main ingredient of the extract is maltose, a disaccharide like sucrose, and therefore very similar in its physical and chemical properties. Thus, the knowledge of a certain extract in wort or beer inherits only the information that the wort or beer has the same density as a solution containing only sucrose in the concentration, which we call extract. Although at first glance this seems to be a little bit puzzling, it does make sense, as this procedure makes routine analysis much easier.

Considering this procedure, the importance of proper density measurement in brewing analytics has to be emphasized. In this context some terms related to density have to be described and carefully distinguished. The definitions cited are taken mainly from *Analytica-EBC* [1].

- True density ($d_{20,9}$). The mass of a substance divided by its volume [g/cm³ or kg/m³]. The true density of water at 20 °C is 0.998 207 1 g/cm³ or 998.2071 kg/m³ and the true density of dry air at 1 atmosphere (1013 mbar) is 1.2046 kg/m³.

- Apparent density. The weight of a substance, when compared to standard weights, divided by its volume [g/ml]. The apparent density of water at 20 °C is 0.997 15 g/ml. As the measurement of the weight of the substance is carried out in air, the weight compared with the mass of the substance is decreased by the buoyancy caused by the surrounding air.

- Specific gravity ($d_{20/20\,°C}$). The density of a substance at 20 °C divided by density of an equal amount of water at 20 °C. Air has a specific gravity of 0.001 208 and water has a specific gravity of 1.0000 at 20 °C. Specific gravity has no units.

- Apparent specific gravity ($s_{L20/20\,°C}$). The apparent density of the sample at 20 °C divided by the apparent density of water at 20 °C. Air has an apparent specific gravity of 0.0000 and water has an apparent specific gravity of 1.0000 at 20 °C. Apparent specific gravity has no units.

- 'Gravity figure', specific gravity at 20/4 °C ($d_{20/4\,°C}$). The density of a substance at 20 °C divided by density of an equal amount of water at 4 °C. As the density of water at 4 °C is 1.0000 g/ml, the figure is identical to the true density. 'Gravity figure' has no units.

Different ways of determining the density are now common in the brewing industry. The most traditional way is carried out using a pycnometer, where a defined volume is first filled with pure water and afterwards with the sample. After each filling, the mass is determined. The result by dividing the two masses is the apparent specific gravity ($s_{L20/20°C}$).

Another way of determining the density is with a hydrometer. Calibration of hydrometers is based either on the density or on the extract.

The most common and most precise way of determining the density is the use of an electronic density meter. The principle of the electronic density meter is based on the resonance oscillation of a glass tube in a U-shape (the same principle as a tuning fork). As the resonance frequency of the U-tube is dependent of its mass and the mass is dependent of the density of the liquid or gas within the U-tube (sample), the resonance frequency is directly related to the density of the sample. The result is given as the specific gravity ($d_{20/4°C}$), which has the same value as the density.

Depending on the principle used, some conversions are of interest. First of all the relationship between density and extract may be subtracted from the sugar table according to Goldiner and Kleemann, or, more conveniently, determined using a polynomial like the one given in MEBAK [4]:

$$E_r = -460.234 + 662.649 * s_L - 202.414 * s_L^2,$$

where E_r is the real extract [% m/m].

As the polynomial according to MEBAK is based on the apparent specific gravity ($s_{L20/20°C}$), a formula converting the specific gravity ($d_{20/4°C}$) into the apparent specific gravity ($s_{L20/20°C}$) is essential when using an electronic density meter [4]:

$$s_L = \frac{d_{20} - 0.0012}{(0.99823 - 0.0012)} = \frac{d_{20} - 0.0012}{0.99703}.$$

The extract in beer increases the density, whereas the alcohol in beer decreases the density. Therefore, the extract calculated out of the density is not the real extract, but an extract called the apparent extract (which has nothing to do with the apparent density or the apparent specific gravity described above). For the calculation of the apparent extract out of the density, the same formula or table applies as for the real extract.

In order to obtain the real extract as well as the alcohol content, extract and alcohol may be separated by distillation. Amounts of 100 g of beer are distilled and the distillate and the residue are filled up to 100 g respectively by distilled water (distillation method, used for the determination of the original gravity). The distillate contains the alcohol, whereas the extract stays in the residue. By measuring the density of the residue filled up to 100 g, the real extract may be determined. Corresponding to the sugar table, a table describing the relationship of the density of a solution containing only ethanol with its ethanol content is available (according to the MEBAK polynomial for the extract cited above). This table is issued by the International Organization of Legal Metrology (www.oiml.org) and is the basis for the ethanol taxes in a number of countries. Again, MEBAK provides an alternative to the table in the form of a polynomial [4]:

$$A = 517.4*(1-s_L) + 5084*(1-s_L)^2 + 33503*(1-s_L)^3,$$

where A is the alcohol (ethanol) content [% m/m].

By knowing the alcohol content as well as the real extract, it is now possible to determine the original gravity (OG) of the beer, which is the theoretical extract the beer had before fermentation. For this, the famous Balling formula is used:

$$OG = \frac{(A*2.0665 + E_r)*100}{100 + A*1.0665}.$$

The Balling formula is based on the empirical knowledge that during fermentation 2.0665 g extract are converted into:

- 1 g alcohol.
- 0.9565 g CO_2.
- 0.11 g yeast.

Thus, the extract before fermentation is:

- Fermentable extract = $A * 2.0665$ [%-m/m]
- Non-fermentable extract = E_r [%-m/m].

Calculation of the concentration of extract before fermentation:

- Fermentable extract = $A * 2.0665$
- Non-fermentable extract = E_r
- Total = $A * 2.0665 + E_r$.

That concentration refers to beer. As we are interested in the theoretical concentration of the wort prior to fermentation, it is necessary to consider the losses during fermentation due to yeast and CO_2 formation. Therefore, the following mass balance is used:

- 100 g beer refers to:
- (100 g + yeast + CO_2) wort.
- (100 g + $A * 1.0665$) wort.

That is the reason for the denominator and completes the Balling formula.

Another approach is the measurement of the alcohol content by a patented method using near-IR absorption, which has become commercially available and very popular in recent years. Knowing the alcohol (via near-IR absorption) and the density of the apparent extract (via an electronic density meter), the density of the real extract may be calculated according to the Tabarié relationship (e.g. quoted in MEBAK [4]):

$$s_{L20/20°C}(E_r) = s_{L20/20°C}(E_a) + (1 - s_{L20/20°C}(A)).$$

where $s_{L20/20°C}(E_r)$ is the apparant specific gravity of the real extract [], $s_{L20/20°C}(E_a)$ is the apparant specific gravity of the apparent extract [] and $s_{L20/20°C}(A)$ is the apparant specific gravity of the alcohol [].

One thing is important to mention in respect of the Tabarié relationship – the specific gravity ($s_{L20/20°C}$) of the alcohol refers to a mixture made out of ethanol and water with the same concentration of ethanol in % m/m as the beer (as described in the distillation method by MEBAK II, 2.10.4). The specific gravity ($s_{L20/20°C}$) of a mixture made out of ethanol and water with the same concentration of ethanol in % v/v as the beer is different. The conversion of the alcohol content in beer from % m/m to % v/v can be achieved by [4]:

$$A[\%v/v] = \frac{A[\%m/m] * d_{20}(\text{beer})}{0.789},$$

where 0.789 is the density of pure alcohol.

As the distillation method is labor intensive and near-IR spectroscopy is quite a new technique, a number of further ways for the determination of the original gravity, alcohol and real extract of the beer have been developed. All of these methods have in common the fact that they consist of two different measures. This is necessary, as two parameters are not known: the real extract and the alcohol content. If the results of both measurements are influenced by the real extract and by the alcohol content as well, but in different a manner, it is possible – by the application of empirical formula – to calculate both. The best known example of this analysis is the combination of the determination of the apparent extract by density measurement and the determination of the refraction index, which again is influenced by the extract as well as by the alcohol content. By using empirical formulas (e.g. cited in MEBAK) it is possible to calculate all the necessary values (real extract, alcohol content, original gravity and degree of fermentation). For different types of beer – related to their original extract – different empirical formulas apply.

Other combinations common in the brewing industry are:

- Density plus ultrasonic measurement.
- Density measurement plus catalytic combustion.

Further approaches are possible (e.g. thermo-analytical measurement).

The result of the ultrasound, measurement is influenced approximately twice as much by the alcohol content than by the extract; a ratio which is reflected in the Balling formula as well. This allows to determine the original gravity by ultrasonic measurement only. Calibration curves for different types of beer have to be generated individually, so this is not an option for a laboratory analyzing many different beers. However, it is used for the in-line determination of the original gravity after filtration for adjustment of the original gravity, as in this case the type of beer to be analyzed is known and therefore this knowledge can be used for selecting the right calibration curve.

A further figure calculated out of the original extract and the present extract (apparent E_a or real E_r) is the degree of fermentation. This is the amount of apparent or real extract fermented referred to the original gravity. Some different formulas are given in literature. The one given for the apparent degree of fermentation (ADF) is taken from MEBAK [2], the one for the real degree of fermentation (RDF) from the EBC [1]:

$$ADF\,[\%] = \frac{(OG - E_a)*100}{OG}$$

$$RDF\,[\%] = \frac{(A*2.0665)*100}{A*2.0665 + E_r}.$$

Apart from the fact that the real degree of fermentation is calculated using the alcohol content, it has to be emphasized that in the formulas given the apparent degree of fermentation refers to the original gravity in % m/v (g/100 ml) and the real degree of fermentation refers to the original gravity in % m/m (g/100 g).

The final degree of attenuation is determined by adding fresh yeast to the beer in the laboratory in order to ferment all fermentable extract still present in the beer. Through day-to-day control, the minimum extract is determined. After reaching the minimum, the extract starts to rise again due to autolysis of the yeast.

17.2.2
Photometric Measurements

Photometric measurements are based on the diminution of light passing a sample in a cuvette. The diminution is the result of absorption as well as of scattering of the light. The sum of absorption and scattering is called extinction. Transmission (T) is the intensity of light passing through the sample (I) in relation to the intensity of light entering the cuvette (I_0) and is given in percent:

$$T = \frac{I*100}{I_0}.$$

The following relationship exists between transmission and extinction (E, no units):

$$E = \log_{10}\frac{I_0}{I} = \log_{10}\frac{100}{T}.$$

As long as the scattering of light is negligible – which is true unless the samples are turbid – the extinction is made up only from absorption. The absorption itself is dependent on a number of influences described by the Beer–Lambert law:

$$A = \varepsilon * c * d$$

where A is the absorption (no units), ε is the absorption coefficient, c is the concentration and d is the thickness of the cuvette. By application of the Beer–Lambert law, it is now possible to determine the concentration of dissolved organic matter (c) out of the absorption (A), the specific absorption coefficient of that matter (ε) and the thickness of the cuvette. Several points are important:

- No other matter may interfere with the measurement of the absorption at the wavelength chosen.
- As already stated, the sample has to be free from turbidity.
- The measurement has to take place within the linear range of the photometer.

The linear range is dependent on the photometer, with modern photometers allowing the measurement of absorption up to a value of 3, whereas measurements with older photometers should be limited to an absorption below 1.5. In that circumstance it has always to be kept in mind that at an absorption of 1, 10% of the light reaches the detector, at an absorption of 2 only 1% and at an absorption of 3 only 0.1% is detected, leading to high demands on the quality of the photometer.

17.2.2.1 Color

The photometric measurement of the color of wort and beer is carried out at 430 nm, as the absorption at that wavelength corresponds best with the impression obtained by the human eye. The absorption spectrum of the beer does not show a peak at this wavelength, but in contrast has quite a decline in that range (see example in Figure 17.1). Therefore, to obtain reproducible and repeatable results it is important to match the wavelength accurately. In the example shown in Figure

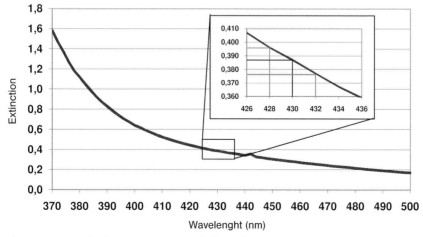

Figure 17.1 Example absorption spectrum of Pilsener beer.

17.1, a deviation of the wavelength of ±2 nm leads to a difference in extinction of 0.02, which is equal to 0.5 EBC units.

As the wavelength accuracy of the photometer is of major importance it has to be monitored at regular intervals. Furthermore, the turbidity of the sample should not exceed 1 EBC unit. In case of higher values the sample has to be filtered. Dark worts and beers have to be diluted to an absorption value below 1.0.

17.2.2.2 Free Amino Nitrogen (FAN)

Another photometric analysis often carried out in breweries is the determination of FAN in wort. The principle of that analysis is based on the addition of ninhydrin, a coloring agent, to the sample. By boiling, parts of the ninhydrin are reduced by amino acids from the sample. In a subsequent reaction, non-reduced ninhydrin together with reduced ninhydrin and ammonia set free again from amino acids form a colored substance, which is measured photometrically at 570 nm, providing information about the amount of nitrogen derived from amino acids in the sample.

17.2.2.3 Bitter Units

Bitter units are extracted from the sample (wort or beer) by a liquid–liquid separation with isooctane. The bitter substances (iso-α-acids) are transferred into the non-polar phase of iso-octane, which is improved by the addition of hydrochloric acid. As only very few substances apart from the bitter substances are dissolved in the non-polar iso-octane phase, the absorption measured at 275 nm is proportional to the concentration of bitter substances. The result is expressed in 'bitter units', which approximately match the concentration of iso-α-acids expressed in parts per million.

Clarification of the sample prior to analysis has to be carried out by centrifugation instead of filtration, as filters tend to absorb bitter substances. Another likely cause of poor results is insufficient shaking during the liquid–liquid separation.

17.2.2.4 Photometric Iodine Reaction

Iodine reacts with dextrin and starch to form a reddish-blue color. This reaction gives valuable information about the completeness of saccharification in the brewhouse. Residual starch and dextrin are precipitated by ethanol and the precipitate redissolved in a buffer. After addition of iodine solution, the color is measured photometrically.

The same background forms the basis of the simple iodine test that is carried out in the brewhouse. No precipitation is done and the color after addition of iodine solution is evaluated visually.

17.2.2.5 Thiobarbituric Acid Index (TBI)

The determination of the TBI provides information about the intensity of heat the wort or beer was exposed to. Apart from hydroxymethylfurfural, which itself is a heat indicator as well, Maillard products and other organics are covered. After reaction of the sample with acetic thiobarbituric acid a yellow color is formed that may be quantified by photometry [12].

17.2.2.6 Total Polyphenols and Anthocyanogens

After reaction with iron ions under caustic conditions, polyphenols form brown complexes that are measured photometrically at 600 nm. The name of the anthocyanogens comes from of their ability to form red anthocyanidins by cooking with hydrochloric acid. This reaction is used for the determination of the anthocyanogens. After adsorption on polyamide, hydrochloric acid and butanol are added and heated. The red color is measured at 550 nm.

17.2.2.7 Ions

A number of ions may be determined photometrically. These determinations are based on different color-forming reactions. They are especially applicable to water, as it has a simple matrix with only minor interferences in most cases.

In more complex matrices like beer, the color of the sample has to be compensated for. Additionally, competing reactions may occur, so the applicability of the analysis has to be validated for each matrix, which quite often outweighs the advantage of an easy-to-perform analysis like a photometric determination, especially with low sample frequency.

Examples of the photometric determination of ions are listed in Table 17.1.

Special cases are the determination of sulphate and potassium. No color is measured, but turbidity is formed by precipitation (barium sulphate and potassium tetraphenylborate, respectively). Photometers only measure the absorption, but not the scattering of light. In the case that no interference occurs with the absorption by colored substances at the wavelength used for determination, the absorption may be used for the quantification of the turbidity. However, this results in higher detection limits of 40 (sulphate) and 8 ppm (potassium), which is not sufficient for all water types. The limits of detection of the other ions are much lower and, therefore, generally sufficient for the major ions found in water.

Thus, the photometric determination of these ions is a good alternative to more sophisticated methods like atomic absorption spectrometry (AAS) or ion chromatography, which makes them suitable for on-site-analysis. Ready-to-use test kits are available commercially.

17.2.2.8 α-Amylases According to EBC/ASBC: Dextrinizing Units

In malt analyses, the activity of the α-amylases is determined optically. An extract from the malt to be investigated is used to digest a limit dextrin substrate solution down to a defined grade, characterized by a certain iodine color. The time necessary is measured.

17.2.2.9 Other Photometric Measurements

Oxidizing agents like ozone, chlorine and chlorine dioxide may be defermined photometrically. The principle behind the determination is the reaction of the

Table 17.1 Examples for photometric ion determinations.

Analyte	Reaction	Wavelength (nm)
Chloride	release of thiocyanate ions out of mercury silver thiocyanate and reaction to iron(III) thiocyanate	468
Sulphate	reaction with barium chloride to barium sulphate (turbidity!)	430
Phosphate	reaction with molybdate and antimony ions to a complex that is reduced by ascorbic acid to phosphoric molybdenum blue	850
Nitrate	reaction with 2,6-dimethylphenol to 4-nitro-2,6-dimethylphenol	324
Nitrite	reaction with primary aromatic amines to diazonium salts, which form intensively colored azo dyes with aromatic reagents with amino or hydroxyl groups	535
Potassium	potassium is precipitated by tetraphenylborate (turbidity!)	690
Ammonia	reaction with salicylate ions and hypochlorite ions to indophenole	694
Iron	formation of an orange to red complex with 1,10-phenanthroline	510
Manganese	formation of a red complex with formaldoxime	450

oxidizing agent with *N,N*-diethyl-1,4-phenylenediamine, forming a red dye, which is measured at 510 nm.

The chemical oxygen demand in waste water can also be determined by photometry. Organic compounds are oxidized by potassium dichromate. The resulting color is measured photometrically. Although this method has its origin in the waste water area, it may also be applied in the fresh water or process water sector, giving valuable information on the organic content of the referring water [mg O_2/l]. Ready-to-use kits are available from different suppliers. In the original method, the quantification takes place by titration.

17.2.3
pH Measurement

Measurement of the pH in water, wort and beer plays an important role in quality assurance, as the pH influences the efficiency of enzymes and is an excellent and easy to obtain indicator of various problems (e.g. fermentation difficulties or microbiological spoilage of the beer).

The analysis is quick and easy to carry out, as the potential difference (electrical current) between an outer and a reference electrode is measured. The outer electrode includes a glass membrane, which causes a potential depending on the pH of the surrounding solution.

A number of factors have to be taken into consideration for proper measurement:

- The pH value of the sample should be between the pH of the buffer solutions used for calibration; in general, for wort and beer, buffer solutions of pH 7.00 and 4.00 are used.
- If it takes a long time for the result to be displayed unchanged, the electrode may be dirty and has to be cleaned (special cleaning solutions for pH electrodes are available) or even substituted.
- In water samples with little buffer capacity it is difficult to obtain a constant value (pH drift is likely).

Further remarks and more detailed information about pH measurement in different matrices can be found in a number of specialized books or brochures (e.g. [29]).

17.2.4
Conductivity

Conductivity is an important measurement in water analyses, as it is related to the overall mineral content of the water. It is carried out by measuring the electrical current between two electrodes of a defined area over a defined distance, submerged into the sample. The analysis is easy to be carry out, as only the conductivity electrode has to be held in the sample, with immediate display of the result. As the conductivity is strongly influenced by the temperature of the sample, the result has to be referred to a standard temperature, generally 25 °C.

17.2.5
Titration Methods

Titration methods are easy to perform and therefore quite common, wherever they are applicable. Typical titration methods are acid–base titration, complexometric titration and manganometric titration.

17.2.5.1 Acid–Base Titration
Titrations with acid or caustic are typically carried out for the determination of the buffer capacity of different liquids like beer or water or the strength of cleaning solutions (acidic or caustic).

In water chemistry, titration to pH of 4.3 and 8.2 plays an important role. This is due to the fact that these are the inflexion points of the titration curve of CO_2, as shown in Figure 17.2.

At pH 4.3, the total inorganic carbon (sum of CO_2, hydrogen carbonate HCO_3^- and carbonate CO_3^{2-}) exists as CO_2, whereas at pH 8.2, the total inorganic carbon

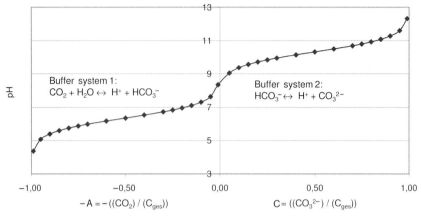

Figure 17.2 Titration curve of CO_2.

exists as hydrogen carbonate. Most common waters will show an initial pH value between 4.3 and 8.3. Therefore, by titration with acid to pH 4.3, all inorganic carbone present in the water will be transformed into CO_2 and the amount of acid used in the titration is equivalent to the amount of hydrogen carbonate originally present in the water. Titration with caustic to pH 8.2 will transform all CO_2 into hydrogen carbonate and the amount of caustic used is equivalent to the amount of CO_2 originally present. Now by performing these two simple analyses, the water may be characterized with respect to the content of hydrogen carbonate and CO_2, which is essential for the characterization of a water's properties.

Of course these titrations are only applicable when no other buffering substances like phosphates are present in the water.

17.2.5.2 Complexometric Titration
Complexometric calcium and magnesium with a solution of ethylene diamine tetraacetic acid, titration forms the bains for the determination of removing the analytes from a weak complex by forming a stronger complex and leading to a change in the color of the original complex.

17.2.5.3 Manganometric Titration
The determination of the permanganate index is widespread in water analytics. Organic substances in the water are oxidized by $KMnO_4$. The amount of $KMnO_4$ is equivalent to the oxygen demand for the oxidation of the organic substances oxidizable under the analytical conditions. It has to be taken care whether the result is given in the unit mg $KMnO_4$/l or, more commonly, mg O_2/l.

17.2.5.4 Diastatic Power: Iodometric Titration
The determination of the diastatic power in malt is also a titration method. A malt extract is prepared and the amount of maltose formed out of a standardized starch solution under defined conditions is measured by iodometry.

A huge number of further titration methods are available with many of them having their origin in drinking water analysis. The reason for that is that they are quite vulnerable to interference with other matrix compounds, which by nature are restricts them to the drinking water sector.

17.2.6
Determination of Nitrogenous Compounds

Nitrogenous compounds play an important role, as the degradation of proteins during malting and mashing has an influence, for example, on yeast behavior during fermentation, taste of beer, head retention and non-biological stability. Several methods for f determining nitrogenous compounds exist, and are common in malting and brewing analytics:

- Determination according to Kjeldahl.
- Determination according to Dumas (combustion method).
- Near-infrared (IR) transmission spectroscopy.

17.2.6.1 Determination According to Kjeldahl

The method according to Kjeldahl is based on a digestion of the sample (oxidation of the nitrogenous organic compounds to H_2O, CO_2 and NH_3), the distillation of NH_3 and the subsequent determination of NH_3 in the distillate by titration. Only nitrogen bound in organic substances and ammonia is detected by this method; other inorganic nitrogens like nitrite and nitrate are not determined.

The result of the analysis is given in mg N/l or mg N/kg. Assuming an average content of nitrogen in the nitrogenous compounds (proteins, peptides and amino acids), this result may be converted into protein by using a constant factor. The factor used for barley and, therefore, for wort and beer is 6.25 (e.g. given in MEBAK [2]).

17.2.6.2 Determination According to Dumas (Combustion Method)

During the determination according to Dumas the sample is burnt at 1000 °C in a pure oxygen atmosphere. By doing so, nitrogen as a compound of organics is transformed into nitrogen oxide (NO_x), which is reduced to N_2 in a subsequent catalytic step. N_2 is measured by using a thermal conductivity detector. In addition to organic nitrogenous compounds, this analysis detects inorganic nitrogens like nitrite and nitrate as well, which has to be kept in mind when interpreting the results. As in the Kjeldahl analysis, the factor 6.25 is applied for conversion of nitrogen into protein.

17.2.6.3 Near-IR Transmission Spectroscopy

This method is mainly used for the determination of total nitrogen in cereals. Based on the principle of measuring the transmission of near-IR light (wavelength 850–1050 nm), a broad range of data for calibration is necessary. However, there is no special sample preparation, allowing fast and precise analysis, due to the large sample size of approximately 500 g.

17.2.6.4 Fractions of Nitrogenous Compounds

The total protein content in beer varies over a wide range and is between around 300 to 900 mg N/l (2000–5500 mg/l as protein). As already stated, the process of malting and mashing is a degradation procedure of proteins. Therefore, it is of interest to obtain information about the degree of degradation and the size of the proteins present. This is achieved by different precipitation methods, precipitating proteins of different sizes. The most common precipitation methods are:

- Coagulation.
- $MgSO_4$ precipitation.
- Phosphorous molybdic acid precipitation.

As precipitation is not only dependent on the size of the protein, it has to be kept in mind that these methods may only give rough information about the protein size. More precise information may be obtained by other methods like protein electrophoresis, although these methods do not play an important role in day-to-day analytics. They are mainly used for research and development purposes.

By boiling under defined conditions, high molecular proteins are precipitated by coagulation and subsequently removed by filtration. The filter with the precipitated protein is analyzed according to Kjeldahl. As a result of coagulation, proteins with a molecular mass of more than 60 000 g/mol are detected. The usual concentration in beer is approximately 10–20 mg N/l (60–130 mg/l referred to protein), resulting in approximately 2% of the total protein content (see Table 17.2).

The addition of a surplus $MgSO_4$ salt to a sample of wort or beer results in the precipitation of proteins as well. As in the determination of the coagulable nitrogen, the precipitate is removed by filtration, washed carefully and the residue analyzed according to Kjeldahl. The size of the proteins precipitated by $MgSO_4$ is roughly speaking above 2600 g/mol. Therefore, the $MgSO_4$-precipitable nitrogen covers a much bigger share of the total protein content than the coagulable nitrogen. The $MgSO_4$-precipitable nitrogen content is also called high molecular nitrogen. Average values range from 120 to 240 mg N/l (750–1500 mg/l referring to protein), which is approximately 33% of the total protein content.

The third method of precipitating proteins from a sample is by the addition of phosphorous molybdic acid. The precipitate is again removed by filtration, but in contrast to both precipitation methods described above, the concentration of the nitrogenous compounds in the filtrate is analyzed. The precipitation with phosphorous molybdic acid, although very hard to define precisely, starts at a molecular mass of even below 2600 g/mol. The filtrate nitrogen is also called low molecular nitrogen and contributes 450–520 mg N/l (2800–3250 mg/l referred to protein), which is approximately 53% of the total nitrogen content.

As precipitation starts below 2600 g/mol, the sum of the filtrate nitrogen from phosphorous molybdic acid precipitation (low molecular nitrogen) and $MgSO_4$-precipitable nitrogen (high molecular nitrogen) is less than 100% of the total nitrogen, leaving a difference called the middle molecular weight nitrogen with a molecular mass of around 2600 g/ml. The middle molecular nitrogen makes up

Table 17.2 Distribution of the nitrogenous compounds in beer according to different criteria and analyses.

Size		Lundin fractions		Coaguable N/FAN		Total	
(g/mol)	(%)		(%)		(%)		(%)
>60000	2.2	MgSO4-precipitable N = high molecular N	33	coaguable N	2.2	total N	100
30000–60000	0.6						
12000–30000	0.7						
4600–12000	12						
2600–4600	18						
150–2600	50	middle molecular N	14				
		not phosphorous molybdic acid-precipitable N = low molecular N	53				
<150	17			FAN	17		

Shaded areas describe analyses covered in this chapter.

approximately 120–150 mg N/l (750–940 mg/l referred to protein), which is approximately 14% of the total protein content.

FAN (the analysis is described in Section 17.2.2.2) is part of the low molecular nitrogen and mainly covers the amino acids. Amino acids have a molecular weight of around 150 g/mol. The concentration in beer is approximately 100 mg N/l (600 mg/l referred to protein) and therefore adds approximately 17% to the total protein content in beer.

An overview provides Table 17.2. The numbers given represent a rough estimation and are just given as an example. They may vary over a wide range from beer to beer. More precise information depending on the type of beer is available in books like that by Krüger and Anger [14] or in more recent statistics published by institutes [15].

17.2.7
Carbon Dioxide

The measurement of CO_2 in most cases is carried out by using manometric methods, based on the Henry–Dalton formula:

$$c_S = K_S * dp$$

where c_S is the concentration of dissolved substances, K_S is the coefficient of solubility and dp is the partial pressure. This equation describes mathematically the relationship between the concentration of a dissolved substance in a liquid phase and its partial pressure in the corresponding gas phase, as long as the liquid and gas phase stay within their equilibrium. The coefficient of solubility is dependent on the temperature. If the temperature is known and the total pressure is made up of only one major component, which is the case for CO_2 in carbonized beverages, the concentration of the dissolved substance in the liquid may be determined. In the case of CO_2, the equilibrium is achieved, for example, by intensive shaking of the sample. The overpressure is measured manometrically. Calculation is carried out according to the so-called Haffmans formula:

$$CO_2 = (p+1.013) * e^{\left[-10.73797 + \frac{2617.25}{\delta + 273.15}\right]}.$$

This describes the relationship between CO_2 content [% m/m], overpressure p [bar] and temperature δ [°C]. The analytical devices have to undergo regular checks and calibrations, at least once per year.

Other formulas apply to other matrices like water and soft drinks. These formulas use only slightly different factors. The main pattern of the formula stays the same.

In rare cases, other dissolved gases like nitrogen may occur in beverages in a non-negligible amount. This is the case, for example, when nitrogen as a pressure gas is used for improving the head retention of beer. As these gases contribute to the overpressure as well, the determination of CO_2 based on manometric methods leads to wrong (too high) results.

In this case, CO_2 measurement based on wet chemistry is one approach. It is carried out by adding 12 N NaOH to the beer in order to transform CO_2 into carbonate. An aliquot of the beer sample is then transferred into a closed apparatus according to Blom and Lund (e.g. described in MEBAK [4]). By adding sulfuric acid, CO_2 is set free from the carbonate and transferred within the apparatus by a stream of air into a solution of barium hydroxide, where it is bound as barium carbonate. Titration of the surplus barium hydroxide with hydrochloric acid results in the calculation of the CO_2 concentration.

Another approach developed recently is multiple volume expansion, based on the fact that the solubility of gases other than CO_2 is much lower. By expanding the volume of the measuring chamber, the pressure drop is higher in the case of the presence of other gases. Therefore, the difference in pressure measured at two different volumes may be used to compensate for the CO_2 content and to obtain an estimated value for the content of other gases.

17.2.8
Measurement of Oxygen

Although being present in beer in much lower concentrations than CO_2, the measurement of oxygen is of great importance. This is due to the fact that oxygen uptake deteriorated flavor stability as well as shelf-life of beer. A number of different sensors are available for the determination of oxygen:

- Membrane-covered polarographic sensors (Clark electrodes).
- Electrochemical measurement with a potentiometer.
- Optodes ('luminescence of dissolved oxygen' sensor).

In the case of membrane-covered polarographic sensors, oxygen diffuses through a membrane and is reduced at an electrode. A metal is oxidized on a corresponding electrode. The electrical current necessary for the oxidation is measured. Electrochemical measurement with a potentiometer, although based on the same reduction of oxygen, does not need a membrane pervious to oxygen. The electrochemical potential is kept constant and the electric current resulting from the reduction of oxygen is measured. Optodes are the most recent development, using membranes that change their luminescent behavior in relation to the oxygen content of the solution they are in. Their applicability in brewing analytics has yet to be proven.

17.2.9
Measurement of Chlorine Dioxide by a Sensor

Similar to the measurement of oxygen, chlorine dioxide may be determined by a covered polarographic sensor.

17.2.10
Head Retention (Foam)

Beer foam is a colloidal system, being built from a gas (CO_2) and a liquid (beer). It is formed by CO_2 release, when bubbles rising in the beer are loaded with surface-active substances from the beer, which form a more or less stable foam when reaching the surface of the liquid. Immediately after formation the foam starts to collapse, building liquid beer again. The head retention as an analytical value now describes the ratio of foam formation to foam destruction, trying to transform the consumer's impression of the stability of the foam into measurable figures.

Although most of the factors influencing head retention are known, the measurement itself has its problems. This is mainly due to poor repeatability r and especially poor reproducibility R, making the comparison of the values obtained from different laboratories somewhat difficult.

Apart from a large number of historical approaches to measuring head retention, two main principles are common nowadays. One describes the time it takes until a defined height of the foam collapses (method based on Klopper [15]); the other measures the time it takes for a defined volume of beer to be formed out of the foam again (based on Ross and Clark [16]). The first principle is realized, for example, in the NIBEM apparatus, whereas the second one is used in the Lg and Steinfurth foam tester. In both cases, the main challenge is the formation and subsequent destruction of foam under defined and reproducible conditions.

Improvements may be achieved by transforming the whole process of foam formation and destruction into a closed apparatus, minimizing the influence of the environment. This was realized in the Lg foam tester, based on an invention from Carlsberg laboratories, and further improved by the Steinfurth foam stability tester. By doing so, it was possible to improve repeatability and reproducibility of the method from Ross and Clark significantly [17]).

17.2.11
Turbidity and Non-Biological Stability

Turbidity is caused by the scattering of light. The occurrence of turbidity limits the shelf-life of beer; hence, prediction of the time until a visible turbidity is detected is of major interest in the brewing industry.

The measurement of turbidity is carried out using turbidity photometers, which detect the light scattered by the sample. As the angle at which most light is deviated from the axis is dependent on the size of the particles (small particles deviating the light at a bigger angle than big particles), it is important to determine at which angle the detection should be carried out. However, it is possible to gain some clues about the size of the particles causing the turbidity. Thus, during filtration turbidity is measured at two angles: 25° (forward scattering) and 90°. High values at a small deviation angle (25°) indicate the presence of bigger particles like yeast cells or kieselguhr particles, whereas high values at larger angles (90° or even 135°) hint towards small particles like colloidal protein polyphenol complexes.

Another parameter influencing the result is the wavelength of the light used. In order to maintain comparability of the results, the wavelength has to be chosen according to the referring analytical instructions. The usual wavelength for the determination of turbidity in beer is 650 nm, whereas for drinking water 850 nm is common. Furthermore, the matrix plays a role, making it difficult to compare results, for example, from dark beers with bright beers. Last, but not least, the apparatus itself is of certain influence.

A dispersion of the substance formazin is used in most cases as the standard for turbidity (e.g. in beer, but in water as well). The EBC unit is common in the brewing industry, whereas in water turbidity is mainly measured in nephelometric turbidity units (NTU). As the basis of both units is the same, it is possible to transform them mathematically according to the simple relationship: 1 EBC unit = 4 NTU. Other units occur, but are less common in the brewing and beverage industries.

Several methods exist for the prediction of the non-biological stability of beer. The most common one is the forcing test, in which the beer is stored at high (40 or 60 °C) and low (0 °C) temperatures alternately, accelerating the speed of beer aging. Turbidity is measured at regular intervals after the cold storage and the result is given in most cases in days of warm storage until turbidity exceeds 2 EBC units. The main disadvantage of the forcing test is the fact that the relationship between the result of the test and the non-biological stability has to be determined by each brewery individually, making it again difficult to compare results between different beers. However, within one brewery the forcing test is applicable and is in most cases a reliable method for the prediction of the shelf-life.

Other tests are based upon the addition of some sort of agent, leading to a turbidity, which is evaluated visually or by a turbidity photometer. Examples are the addition of ethanol (alcohol chilling test according to Chapon), formalin (formalin test), ammonia sulphate (precipitation limit by ammonia sulfate) or picrinic acid plus citric acid (test according to Esbach). Although giving some information about the non-biological stability to be expected, their main disadvantages are their sensitivity on only one side of the partners forming turbidity (either polyphenols or proteins) and that an agent has been added to the beer, disturbing the reactions in the beer usually leading to the turbidity. Therefore, it is not surprising that these tests only supply limited information about the shelf-life of beer.

17.2.12
Viscosity

Viscosity is measured in the wort and especially in laboratory worts (congress wort and 65 °C wort) as part of the malt analysis, and allows us to draw conclusions from the cytolysis of the malt used. Mainly insufficiently degraded, high molecular β-glucans originating from the cell walls of the endosperm contribute to high viscosity. Viscosity is either given in the unit mPas (milliPascal × seconds) – so-called dynamic viscosity η – or in the unit mm^2/s – so-called kinematic viscosity v. It is possible to transform one unit into the other, which is done by multiplication with the density ρ [g/cm^3], as is shown by:

$$\eta = v * \rho.$$

Viscosity is dependent on the extract of the wort. In order to achieve better comparability of the results measured, it is possible to mathematically refer the viscosity to standard values for extract, which is done by mathematical transformation. Unfortunately this is not a linear relationship. The formulas necessary for the transformation were developed by Zürcher [18, 19]:

$$\eta_{norm.(x)} = \frac{e^{a*x^{1.20}} - e^{-a*x^{1.20}}}{2} + 1.002,$$

with:

$$a = \frac{\ln\left(\eta_{sol.} + \sqrt{(\eta_{sol.})^2 + 1}\right)}{C^{1.20}},$$

and:

$$\eta_{sol.} = \eta_{meas.} - 1.002,$$

where $\eta\text{norm}(x)$ is the dynamic viscosity normed at extract x [$m Pas$], $\eta_{sol}(x)$ is the dynamic viscosity soluble [$m Pas$], η_{meas} is the dynamic viscosity measured [$m Pas$], x is the extract reference the viscosity is referred to [% m/m] and C is the extract of the sample [% m/m].

17.2.13
Congress Mash

Congress mash is an important starting point for a number of chemical analyses carried out to describe the quality of malt. In order to get the malt ingredients into a dissolved form so they are ready for analysis, the malt undergoes a standardized mashing scheme. That mashing scheme starts at 45 °C; after a rest of 30 min, the temperature is increased to 70 °C within 25 min, where a second rest at 70 °C for 60 min takes place. After that rest, the mash program is finished. The congress mash program is described in detail in the analyses compilations. As this mash program is a very intensive one, differences between the malts may be quite small. Therefore, some authors suggest changing the mashing scheme to a less intensive one with the aim of getting bigger differences.

A number of analyses may be carried out from the congress mash:

- Extract and extract difference between coarse and fine grist.
- Odor.
- Iodine test.
- Filtration time.
- Turbidity.
- pH.
- Viscosity.
- Color.
- Color of boiled wort.
- Soluble nitrogen.
- FAN.
- β-Glucans.
- Final degree of attenuation.
- Dimethylsulfide (DMS).
- Further soluble ingredients.

These analyses are described in other sections within this chapter in more detail. The color of the boiled wort is measured after boiling the congress wort. This value correlates with the expected beer color much better than the color of the congress mash.

Other mashes prepared out of the malts are the mashes obtained by using isothermal mashing procedures. Their extract is referred to the extract retrieved by the congress mash procedure. The most common is the isothermal mash at 45 °C (VZ45), which is related to the proteolysis. Recently, some authors have challenged the meaning of this figure, relating it more to amylolysis than to proteolysis [30].

17.2.14
Spent Grain Analysis

Spent grains from the lauter tun or the mash filter are analyzed in order to find out whether the extract was completely retrieved. Two sorts of extract in the spent grains are distinguished from each other: one is the soluble extract, which may be extracted by washing the spent grains (soluble extract); the other is the extract that can still be solubilized by enzymes (digestable extract).

Within the analysis, the spent grains are mashed with and without enzymes, similar to the preparation of a congress mash, and the extract measured in the usual way. The extract retrieved by the use of enzymes is the total extract still present in the spent grains, the extract from the mash without enzymes is the soluble extract and the difference between these two is the digestable extract.

17.2.15
Friabilimeter

Analysis with the so-called friabilimeter produces two main results: mealiness and glassy kernels.

The friabilimeter is a standardized apparatus, in which the malt kernels are crushed between a rubber roll and a rotating sieve cage. The small particles of the malt kernels will pass through the sieve. The residue consisting of bigger particles from the kernels and even whole kernels is weighed. These are the undermodified parts of the sample. The difference between the original and the undermodified sample is the better modified part of the sample. Related to the original sample this figure represents the so-called mealiness.

From the residue, the kernels still intact down to a size of three-quarters of the original kernel size are manually separated and weighed. These represent the unmodified kernels. Their mass referred to the weighted sample is called the glassy kernels.

The results are dependent on the water content, as a higher water content makes the kernels more flexible and therefore the results will be worse (lower mealiness, higher amount of glassy kernels).

The analysis is based on the design of the analytical apparatus. Therefore, it is necessary to secure the comparability of the results from friabilimeter to friabilimeter, which is done in regular collaborative tests with a collective of devices of identical design (Friabilimeter Calibration Network) and published [20]. Intermediate calibration may be performed by using standardized malt samples (e.g. EBC standard malts).

The friabilimeter analysis is a quick and simple method that gives first information about the modification, and is therefore suitable for immediate intake control. Normal specification for Pilsener malt is at least 80% for mealiness and maximum 3% glassy kernels.

17.2.16
Grading

Grading of the malt is carried out by simply using a standardized set of sieves.

17.2.17
Hand Assessment

Hand assessment is carried out with barley, malt and hops. It is quick to carry out. Although it needs some practice, and therefore some training of personnel, it should also be part of the intake procedure for auxiliary materials like kieselguhr. Although human senses are not very precise and subject to change under daily working conditions, they are much more open to any kind of unexpected failures in terms of smell, taste and appearance than any specific analyses, and therefore suitable for an overall evaluation of the quality of the raw and auxiliary materials. Further information on how to perform hand assessments properly is given in the compilations already cited [1–9].

17.2.18
Homogeneity and Modification

Homogeneity and modification of the malt kernels are determined by dying the residual cell walls with the dye calcofluor. Before dying, the malt kernels are fixed and abraded down to half. After the dying procedure the samples are transferred to a chamber illuminated with UV light, letting the cell walls fluoresce. In most cases evaluation takes place by computer programs. The ratio of the fluorescing area to the total endosperm area within every kernel leads to the modification figure. The distribution of the ratio provides information on the homogenity of the sample.

A good overview about the judgment of the results of different malt analyses related to cytolysis, including information about the precision of the analysis, is given by [21].

17.2.19
Protein Electrophoreses

Protein electrophoresis is used for the separation of proteins. The separation is achieved by applying the sample after extraction on a gel, which is put into an electrical field that allows the proteins to separate from each other. After a defined span of time the electrical current is stopped and the gels are dyed, making proteins visible as colored spots. By comparison with patterns produced by standards, it is possible to identify protein fractions. Protein electrophoresis is not used for routine analysis of nitrogenous compounds in beer (see section 17.2.6), but it is a magnificent tool for the maltster as well as for the brewer, as it allows the identification of barley and malt varieties by the typical patterns of their proteins within a defined fraction [22].

17.2.20
Gushing

A reliable quantification of gushing tendencies in the raw materials is still problematic from the analytical point of view. Gushing tests of raw material for determining primary gushing are generally based on extracting the raw material in an appropriate manner. In the so-called modified Carlsberg test, the extract is added to mineral water, which is then shaken for several days and the fobbing after opening the bottles is measured. In another approach, more or less complete beers are prepared out of the raw material.

17.2.21
Hop Bitter Substances

The hop bitter substances are extracted from the hops or the hop product by diethyl ether. They are further separated by using their different solubility.

The total resins are determined gravimetrically, after purifying the resins in diethyl ether solution by drying and redissolving it in methanol, and then drying an aliquot of that solution. The soft resins are extracted from the methanol fraction by the use of hexane and then also determined gravimetrically.

The α-acids are determined by conductometric titration with lead acetate, taking the diethyl ether phase of the hops as a basis. The hard resins are determined by subtracting the soft resins from the total resins unsafe mathematically. The so called β-fraction is the difference between the soft resins and the conductometric value. More specific analysis of the α- and β-acids may be carried out by using high-performance liquid chromatography (HPLC) (see section 17.2.23.1).

17.2.22
Continuous Flow Analysis (CFA)

CFA is a method of performing the analyses in a largely automated way. The principles of the referring analyses offered by the different manufacturers of these systems are predominantly based on the compilations like *Analytica-EBC*. Table 17.3 shows an excerpt of the analyses available and relevant for the brewing industry.

17.2.23
Chromatographic Analyses

All chromatographic analyses have in common a separation of the analytes by the interaction between a mobile and a solid phase (column), and the subsequent detection using a suitable detector. The identification of the peaks is based mainly on their retention time. The combination of the pre-treatment of the sample, the separation by the column and the specificity of the detector lead to high accuracy and specifity of the measurement.

Table 17.3 Selected CFA analyses available and relevant for the brewing industry.

FAN
Anthocyanogens
β-Glucan
Sulfur dioxide
Polyphenols
Bitterness
Acetaldehyde
Nitrate
Density
Amylases
Ascorbic acid
Benzoic acid
Chlorine
Color
Maltose
Reducing sugars
Acetic acid

17.2.23.1 High Performance Liquid Chromatography (HPLC)

In HPLC analyses, the mobile phase is a liquid. It is suitable mainly for the determination of non-volatile compounds. The possible applications are manifold. In the brewing and beverage industry it is used, for example, for the determination of specific bitter substances, fermentable sugars and polyphenols, as well as for preservatives, anions and mycotoxins. There are many more examples and Table 17.4 just shows a short excerpt, listing the detector generally used for the referring analyte. Further details of the sample preparation, column type (solid phase) and eluent (mobile phase) may be obtained from analytical compilations like MEBAK or by information from the supplier of HPLC devices and columns.

The determination of mycotoxins by HPLC with previous sample preparation by using immunoaffinity columns proved to be a very powerful analytical tool. In this analysis, a highly specific sample preparation by attachment of the analyte onto an immunoaffinity column with subsequent elution is combined with HPLC, allowing a highly sensitive and specific analysis for contaminants occurring in natural products like grains. This method is suitable as a reference method, for example, for the determination of mycotoxins with enzyme-linked immunosorbent assay (ELISA) (see Section 17.2.28).

One possible method of analyzing β-glucans in (congress) wort and beer is by using a fluorescence detector. A fluorochrome named calcofluor is added into a stream of the sample. Calcofluor forms short-lived complexes with β-glucans with a molecular mass above approximately 10 000 Da, which give a signal in the fluorescence detector. Strictly speaking, this analysis does not belong to the HPLC methods, as no column is used, but to the type of analyses called flow injection analysis. However, the devices used for the analysis are similar or even the same as in HPLC.

Table 17.4 Analytes and detectors in HPLC analyses.

Analyte(s)	Principle of detection
α- and iso-α-Acids	absorption (UV)
Xanthohumol	UV
Fermentable sugars	refraction index
Amino acids	fluorescence
Amines	UV, fluorescence
Ferulic and cumaric acid	absorption (UV)
4-Vinylguaiakol and 4-vinlyphenol	absorption (UV)
Organic acids	UV
Vitamins (B1, B2, C)	UV
Benzoic acid	UV
Hydroxymethylfurfural	UV
Formaldehyde	UV
Sulfite	conductivity
Anions	conductivity
Mycotoxins: deoxynivalenol	absorption (UV)
Mycotoxins: ochratoxin A	fluorescence
Mycotoxins: zearalenon	absorption (UV)
β-Glucan	fluorescence

17.2.23.2 Gas Chromatography (GC)

Most of the factors listed in connection with HPLC analyses also apply to GC analyses. The main difference is that the mobile phase is a gas. Therefore, it is mainly used for the determination of volatile components. Different principles of detection are generally used in GC analyses:

- Flame ionization detector (FID). Substances are burnt in a hydrogen-air-flame, by which free electrons occur. These free electrons are measured by an electrode. The FID is suitable for all kind of carbon-hydrogen substances, like higher alcohols and esters.

- Flame photometric detector (FPD). Substances are burnt in a hydrogen flame. Parts of the substances emit light, which is detected. Mainly substances containing sulfur or phosphor are detected (e.g. DMS).

- Electron capture detector (ECD). The detector uses ^{63}Ni as a radioactive source, emitting electrons that are measured by two electrodes. Substances passing the detector may absorb these free electrons and therefore change the signal of the detector. The detector is very sensitive towards substances containing halogens or substances containing conjugated C=O double bonds, like diacetyl and pentandione.

Depending on the analytical task there are many more detectors available, like the thermal conductivity detector, the thermal energy analyzer (TEA-detector), the mass selective detector or the chemiluminescence detector.

Table 17.5 Analytes and detectors in GC analyses.

Analyte(s)	Principle of detector
Fermentation byproducts (e.g. alcohols, esters, aldehydes)	FID
Vicinal diketones (diacetyl, pentanedione), acetoin	FID or ECD
DMS	FPD
Fatty acids	FID
Hop oil	FID
Nitrosamines (N-nitrosodimethylamine)	TEA or chemiluminescence
Sulfur dioxide	FPD or MSD
Acrylamide	ECD
Trihalomethanes (THMs)	ECD
Benzene	ECD plus FID
Vinyl chloride	ECD plus FID

An overview of the most common analyses performed with GC in the brewing and beverage industry is given in Table 17.5, showing the principle of detection used for the corresponding analyte as well. More detailed information again may be found in the compilations like MEBAK or obtained by the suppliers of GC equipment.

17.2.24
Enzymatic Analyses

Many enzymatic reaction are coupled with the transfer of energy to or from the NAD(P)/NAD(P)H complex. In contrast to NAD(P), NAD(P)H has a distinct maximum in its spectrum at a wavelength of 340 nm (Figure 17.3), so the difference between these two can be easily determined photometrically.

This may be used for enzymatic analysis. During the determination, the analyte is transformed enzymatically in a coupled reaction with the formation of NAD(P)H out of NAD(P) or vice versa. The enzyme used for the conversion of the analyte is responsible for the specificity of the reaction and therefore of the analysis, allowing the determination of an analyte within a complex matrix with only minor sample preparation effort and still achieving very low detection limits.

As up-to-date spectral photometers possess a continuous spectrum, they allow adjustment of the optimal wavelength of 340 nm. Spectrum line photometers, however, only apply distinct wavelengths. The ones closest ones to 340 nm with a mercury spectrum photometer are 334 or 365 nm. Although this is not optimal regarding the spectrum of NAD(P)H, they can be used for the enzymatic determination, as their wavelength accuracy is excellent. Their light emission is caused by electron transfer within one atom of a specific element like mercury, emitting light with only the specific wavelength corresponding to the referring electron transfer.

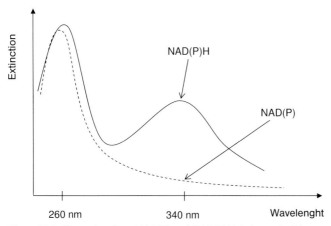

Figure 17.3 Absorption from NAD(P) and NAD(P)H (schematically).

Table 17.6 Enzymatic analyses.

Ethanol (in low concentrations)
Sulfite (sulfur dioxide)
Organic acids, e.g. oxalic acid
Glucose, fructose, sucrose, maltose, lactose, raffinose
β-Glucans
Acetaldehyde
Acetic acid
Nitrate
Ascorbic acid
Citric acid
Lactic acid

Examples of enzymatic analyses used in the brewing industry are given in Table 17.6. The specificity of the enzymatic determination of ethanol is a good example of the important role enzymatic analyses may play. As the usual ways of determining the alcohol content described above (Section 17.2.1) are less specific – being influenced by higher alcohols present in the beer as well – and less accurate at low concentrations, the enzymatic determination of ethanol is a reliable alternative, especially when legal aspects like the maximum concentration of ethanol allowed in alcohol-free beverages play a role. Another example is the determination of sulfur dioxide in beer. The enzymatic determination of sulfur dioxide in beer is the method legally advised, for example, in Germany.

β-Glucan determination using the enzymatic method is more complex, as of all β-glucans are first precipitated by ammonia sulfate and then washed with ethanol. Lichenase is an enzyme that splits endo-(1–3) and -(1–4) β-linkages of polysaccharides. After further digestion of the resulting tri- and tetrasaccharides, the resulting glucose is determined using the enzymatic test as usual.

Ready-to-use kits with all of the chemicals and solutions necessary are commercially available, allowing a wide range of analytes to be determined all in a similar matter, needing only a photometer as an analytical apparatus. The correct dosage of the sample and the agents using high-precision pipettes plays an important role.

17.2.25
Determination of the Calorific Value of Beer

The calorific value of beer may be calculated using the analyses already described. This method was presented by Drawert and Hagen in 1971 [23] and is only described in brief here. Analyses to be carried out before the calculation are:

- Original gravity, alcohol, apparent extract.
- Final attenuation.
- Total protein (calculated out of total nitrogen).
- Total glucose.

Total glucose after hydrolyses is determined enzymatically. The fermentable carbohydrates are calculated out of the difference of the apparent extract to the apparent extract after final attenuation, multiplied by the attenuation coefficient of 0.81, considering the apparent decline of the extract by the alcohol content. The next step is the calculation of the dextrins (i.e. the difference between total glucose and the fermentable carbohydrates). This difference is again multiplied by a factor of 0.915, the dextrination factor, which considers that the dextrins are formed out of four to 20 glucose units on average. Finally, the usable carbohydrates (important for people with diabetes) are calculated by adding the fermentable carbohydrates to the dextrins, plus another 0.05% as a general value for pentosans. The amount of total protein is analyzed according to Kjeldahl.

The calorific value now is calculated by assuming 38 kJ for each % m/m of alcohol, 17 kJ for each % m/m of carbohydrates, 17 kJ for each % m/m of protein and another 4 kJ for other material contributing to the calorific value.

In the case of other beverages, such as some malt beverages or beer mix drinks, containing sugars other than glucose (e.g. saccharose or fructose), these other sugars have to be determined as well and considered in the calculation of the dextrins, as they are included in the fermentable carbohydrates, but not in the dextrins.

17.2.26
Atomic Absorption Spectroscopy (AAS) and Inductively Coupled Plasma Optical Emission Spectrometry (ICP-OES)

AAS is suitable for the determination of metals and semi-metals, achieving low detection limits. This is of interest, for example, for the analyses of the cations in water and beer, but also for the determination of heavy metals.

AAS consists mainly of a light source, an atomization unit for the sample and a detector. A sample is vaporized and the analyte atomized, for example, by injection of the sample into a flame (flame AAS) or by electrically heating the sample in a small graphite tube (graphite tube AAS). The atomized analyte now absorbs light of a number of wavelengths specific for the referring analyte, which is measured by the detector. Detection limits down into the parts per billion and parts per trillion range are possible; graphite AAS is more sensitive than flame AAS.

AAS is limited to the detection of only one element at a time, which makes it quite time consuming. A related principle, ICP-OES, avoids this disadvantage. In ICP-OES, the sample is injected into an argon plasma, so that different analytes start to emit light of different wavelengths specific for every element, which can be detected in parallel. This allows the determination of different analytes at the same time. Detection limits are higher than in graphite tube AAS.

As the experimental effort is quite high and the number of routine analyses limited, AAS or even ICP is not very common for in-house analyses in brewery laboratories. Typically these analyses are done by service laboratories.

17.2.27
Sulfur Dioxide: Distillation Method

Beer contains sulfur dioxide in both a free and a bound form. In the bound form, sulfur dioxide is attached to acetaldehyde. Free sulfur dioxide is present in the total content (free plus bound sulfur dioxide) in only small concentrations and therefore only plays a minor role.

The content of sulfur dioxide in beer is of interest in two respects. On the one hand, sulfur dioxide improves by its anti-oxidative properties the flavor stability of the beer. On the other hand, some people tend to react poorly towards the uptake of sulfur dioxide in food. This is why in the European Union the content of sulfur dioxide in food has to be declared if it exceeds 10 ppm [24]. Thus, the brewer is interested in an increase of the sulfur dioxide in beer, but has to take care about legal requirements and limits making precise analysis of sulfur dioxide essential.

The distillation method originates from the wine sector, where it is used as the reference method in the European Union [24]. It is based on setting free bound sulfur dioxide by the addition of acid, and subsequent heating and distillation of the sulfur dioxide. The sulfur dioxide is retained in a recipient containing H_2O_2 that oxidizes sulfur dioxide to SO_3, which dissociates in water into sulfuric acid:

$$SO_2 + H_2O_2 + 2 H_2O \rightarrow 2 H_3O^+ + SO_4^{2-}.$$

The amount of sulfuric acid is proportional to the total amount of sulfur dioxide in the sample and determined by titration with sodium hydroxide.

As the concentrations in beer are much lower than, for example, in wine, the method has to be carried out carefully in order to avoid even small losses, as this considerably alters the result [25]. Examples of point to notes are the precise adjustment of the carrier gas nitrogen and the proper purging of the apparatus after the end of the distillation. The method is described, for example, in MEBAK volume II.

Further methods for the determination of sulfur dioxide in beer are GC/FPD or GC/MSD, CFA, HPLC after enzymatic sample preparation and ion chromatography (free sulfur dioxide) [Harms 2006].

17.2.28
Emzyme Linked Immuno Sorbent Assay (ELISA)

ELISA belongs to the group of immunoassays. Immunoassays are based on the antigen–antibody reaction. In an immunoassay, the analyte – being in general the antigen – is attached to a corresponding antibody and afterwards detected after further reaction. In an ELISA, the further reaction is an enzymatic color reaction, allowing the detection to be carried out by simple photometry (see Section 17.2.2).

Mainly two types of ELISAs are common:

- Sandwich ELISA.
- Competitive ELISA.

In a sandwich ELISA, the analyte (antigen) is attached on an immobilized antibody. After washing, a second antibody attaches to the antigen (analyte), carrying an enzyme. The antigen is trapped between the two antibodies, like in a sandwich. A chromogen, which is added after repeated washing, undergoes a reaction catalyzed by the enzyme attached to the second antibody, forming a distinct color that is determined photometrically. The competitive ELISA works in a slightly different way. A second antigen, on which an enzyme is attached and which reacts with the immobilized antibody as well, is added to the sample, competing with the analyte for the antibody. The higher the concentration of the analyte, the less competing antigen is bound onto the antibody and the smaller is the corresponding photometric signal. Further variations of ELISAs are possible and in use.

The huge advantage of an ELISA is the fact that the analyte – due to the specificity of the antigen–antibody reaction – is easily selected out of a complex matrix like beer. This effect is used in the sample preparation by immunoaffinity columns as well (see Section 17.2.23.1 on HPLC). A disadvantage is the sensitivity of an antibody towards other analytes as well, which may lead to false results, and this has to be investigated carefully beforehand.

ELISAs are used, for example, for the determination of mycotoxins in routine control. Commercial test kits are available, with immobilized antibodies in so-called microtiter plates, allowing the determination of a large number of samples simultaneously.

Due to their origin in the antigen–antibody reaction, ELISAs may be used for the determination of specific proteins or protein fractions. This approach is used in research from time to time. One example is the attempt to determine protein fractions responsible for foam and non-biological turbidity [26]. Another example is the determination of gluten – a protein fraction derived from cereals like barley. Some people tend to react poor upon the uptake of gluten in food (celiac disease). This analysis is not very common and its applicability towards beer still has to be proven. However, some interesting approaches using new antibodies have been developed recently, so this analysis may play a more important role in future [31].

17.2.29
Electron Spin Resonance Spectroscopy (ESR)

Free radicals are detected by ESR. It is used for measuring free radicals in beer under the conditions of forced aging (forcing test). By the increase of free radicals, conclusions may be drawn about the flavor stability of the beer. This method is still improving and is mainly used in research and development, but not in routine analysis. Further information is available in the recent literature [27, 28].

17.3
Analyses in Daily Quality Control

Table 17.7 shows the most frequent chemical analyses carried out at the different stages of the process. This is only an example and has to be adapted to individual circumstances. Not all analyses described in the first part of the chapter occur in the table. This is due to the fact that some analyses like the determination of oxalic acid, hop oil or some analyses based on ELISAs may play a role in research and development or in finding the cause of problems like filtration difficulties or gushing, but have not found their way into day-to-day quality control.

In order to find the optimum between reasonable costs for quality control without having to sacrifice important information, it is worthwhile understanding the individual situation, and determining which analyses should be carried out by one's own personnel and which analyses should be sent to an external laboratory. This may be done systematically, working along the following steps:

- Fixation of an analytical control scheme, including elements such as:
  - Enumeration.
  - Name of the sample.
  - Quantity of the sample.
  - Where to take the sample.
  - How often to take the sample.
  - What analysis to be carried out.
  - Who does the analysis.
  - Fixation of the normal range of the values.
  - Limit values.
  - Who to inform (generally and in the case of any exceptional data).
  - Storage of data.
- Fixing the cost frame (e.g. 1 Euro/hl).
- Determination of all necessary analyses.
- Decision from an economic point of view of each analysis to be done by one's own lab or an external laboratory.

Table 17.7 Example of a control scheme for breweries (day-to-day control).

No.	Sample	Analyses	Frequency
1.	Raw materials		
1.1	Malt	friabilimeter	every delivery
		sieving test (grading)	every delivery
		hand assessment	every delivery
		water content	each supplier every 3 months
		congress mash:	each supplier every 3 months
		extract	
		wort color	
		color of boiled wort	
		viscosity	
		pH	
		final degree of fermentation	
		β-glucan	
		DMS precursor	
		homogeneity and modification	every supplier every 3 months
		contaminants	every supplier at least once per year/every 3000 tons of malt consumption
		mycotoxins	every supplier at least once every 6 month
		N-nitrosodimethylamine	
		heavy metals, e.g. As, Cd, Hg, Pb	
		variety identified by electrophoresis	
1.2	Hops	α-acid content	each delivery
1.3	Water		
1.3.1	Water intake	smell and taste	once per week
		conductivity	once per week
		turbidity	once per week
		pH	once per week
		Complete analyses:	once per year or according to referring legislation
		all ions	
		heavy metals	
		contaminants:	
		pesticides, etc.	
		trihalomethanes (THMs)	
		further organochlorides	
		polycyclic aromatic hydrocarbons (PAHs)	
		benzene	
		acrylamide	
		epichlorhydrin	
		vinyl chloride	
		color	
		total organic carbon	

Table 17.7 Continued

No.	Sample	Analyses	Frequency
1.3.2	Brewing water	smell and taste	once per week
		conductivity	once per week
		pH	once per week
		total hardness	once per week
		alkalinity (m value)	once per week
		residual alkalinity	once per week
1.3.3	Service water	disinfectant, e.g. CO_2	once per week
		pH	once per week
		total hardness	once per week
1.3.4	Boiler feed water	total hardness	daily
		conductivity	daily
1.4	Auxiliary material/aids		
1.4.1	Kieselguhr	odor	every delivery
1.4.2	Polyvinylpoly-pyrrolidone (PVPP)	odor	every delivery
1.4.3	Acids	concentration	once every 3 months
		purity	once every 3 months
1.4.4	Caustics	concentration	once every 3 months
		purity	once every 3 months
1.4.5	Disinfectants	concentration	once every 3 months
		purity	once every 3 months
2.	Process control		
2.1	Grist	visual evaluation	every charge
2.2	Mash	saccharification (iodine test)	every brew
2.3	Lauter wort	extract	every brew
		turbidity	every brew
		odor and taste	every brew
2.4	Spent grains	soluble and digestable extract	once every 3 months
2.5	Pfannevollwürze	extract	every brew

Table 17.7 Continued

No.	Sample	Analyses	Frequency
2.6	Cast wort	extract	every brew
		final degree of fermentation	every month
		pH	every month
		color	every month
		bitter units	every month
		TBI	every month
		viscosity	every month
		β-glucan	every month

		nitrogenous compounds
		  FAN
		  MgSO4-precipitatle N
		  coaguable N
		  total N
			every month

		DMS and DMS precursor	every month
		photometric iodine reaction	every month
2.7	Green beer	original gravity, extract, alcohol, degree of fermentation	every fermentation
		vicinal diketones	every fermentation
2.8	Unfiltered beer	original gravity, extract, alcohol, degree of fermentation	every tank, 1 day before filtration
		pH	every tank, 1 day before filtration
		CO_2	every tank, 1 day before filtration
		turbidity	every tank, 1 day before filtration
		sensory analysis	every tank, 1 day before filtration
2.9	Filtered beer	oxygen (in-line)	every filtration charge
		CO_2 (in-line)	every filtration charge
		turbidity (in-line)	every filtration charge
		sensory analysis	every filtration charge
2.10	Beer in Bright Beer Tank (BBT)	oxygen	every tank
		CO_2	every tank
		sensory analysis	every tank
		original gravity, extract, alcohol, degree of fermentation	every tank
		pH	every tank
		color	every tank
		bitter units	every tank
2.11	Filling	original gravity, extract, alcohol, degree of fermentation	every charge
		color	every charge

Table 17.7 Continued

No.	Sample	Analyses	Frequency
2.12	Filled beer	original gravity, extract, alcohol, degree of fermentation	every charge
		pH	every charge
		color	every charge
		bitter units	every charge
		CO_2	every charge
		foam	every charge
		turbidity	every charge
		forcing test	every charge
		sulfur dioxide	every product every 3 months
		polyphenols and anthocyanogens	every product every 3 months
		higher alcohols	every product every 3 months
		steam evaporable fatty acids	every product once per year

Criteria for the decision to carry out the analysis in-house or by an external lab are:

- The necessity of the analysis (legislation, demands from the retailers, quality).
- The time span in which the results have to be available.
- Availability of in-line measurements.
- One's own laboratory capacity and qualification.

When using in-line measurements, it has to be kept in mind that the availability of qualified personnel and of a reference measuring system in the lab is absolutely essential. Therefore, in-line measurements may help the laboratory in its daily work, but will never be able to substitute it as a whole.

References

1 European Brewery Convention (2008) *Analytica-EBC*, Verlag Hans Carl, Nürnberg.
2 Anger, H.-M. (ed.) (2006) *Brautechnische Analysenmethoden, Rohstoffe Gerste Rohfrucht Malz Hopfen- und Hopfenprodukte, Methodensammlung der Mitteleuropäischen Brautechnischen Analysenkommission (MEBAK)*, Selbstverlag der MEBAK, Freising-Weihenstephan.
3 Anger, H.-M. (ed.) (2006) *Brautechnische Analysenmethoden, Wasser, Methodensammlung der Mitteleuropäischen Brautechnischen Analysenkommission (MEBAK)*, Selbstverlag der MEBAK, Freising-Weihenstephan.
4 Miedaner, H. (ed.) (2002) *Brautechnische Analysenmethoden, Band II, Methodensammlung der Mitteleuropäischen Brautechnischen Analysenkommission (MEBAK*, 4th edn, Selbstverlag der MEBAK, Freising-Weihenstephan.
5 Pfenninger, H. (ed.) (1996) *Brautechnische Analysenmethoden, Band III, Methodensammlung der Mitteleuropäischen Brautechnischen Analysenkommission*

(MEBAK), 2nd edn, Selbstverlag der MEBAK, Freising-Weihenstephan.
6 Pfenninger, H. (ed.) (1998) *Brautechnische Analysenmethoden, Band IV, Methodensammlung der Mitteleuropäischen Brautechnischen Analysenkommission (MEBAK)*, 2nd edn, Selbstverlag der MEBAK, Freising-Weihenstephan.
7 Technical Committee and the Editorial Committee of the ASBC (2008) *Methods of Analysis of the American Society of Brewing Chemists*, American Society of Brewing Chemists, St Paul, MN.
8 Institute of Brewing and Distilling (2002) *Recommended Methods of Analysis*, Vols **1 and 2**, Institute of Brewing and Distilling, London.
9 Brewery Convention of Japan (1998) *Methods of Analysis*, revised edn, Brewery Convention of Japan, Tokyo.
10 Magnusson, B., Näykki, T., Hovind, H. and Krysell, M. (2003) *Handbook for Calculation of Measurement Uncertainty in Environmental Laboratories*, Nordtest, Espoo.
11 Goldiner, F., Klemann, H. and Kämpf, W. (1966) *Sugar Table*, Institut für Gärungsgewerbe, Berlin.
12 Thalacker, R. and Birkenstock, B. (1982) Eine neue Kennzahl in der brautechnischen Analyze – die Thiobarbitursäurezahl (TBZ). *Brauwissenschaft*, **35** (6), 133–7.
13 Piendl, A. (1999) *Physiologische Bedeutung der Eigenschaften des Bieres*, Fachverlag Hans Carl, Nürnberg.
14 Krüger, E. and Anger, H.-M. (1990) *Kennzahlen zur Betriebskontrolle und Qualitätsbeschreibung in der Brauwirtschaft*, Behr's Verlag, Hamburg.
15 (a) Anger, H.-M. (2007) Durchschnittswerte bei Bieranalysen, Brauerei Forum, VLB, Berlin.
(b) Klopper, W.J. (1973) Foam stability and foam cling, *Proceedings of the European Brewery Convention Congress*, pp. 363–71.
16 Ross, S. and Clark, G.L. (1939) On the measurement of foam stability with spezial reference to beer. *Wallerstein Lab Communications*, **46**, 46–54.
17 Potreck, M. (2004) Optimierte Messung der Bierschaumstabilität in Abhängigkeit von Milieubedingungen und fluiddynamischen Kennwerten, Dissertation, Technische Universität Berlin.
18 Zürcher, C. (1973) Tabellen zur Umrechnung der Viskosität von Würze und Bier auf einen einheitlichen Gehalt bzw. Stammwürzegehalt. *Monatsschrift für Brauerei*, **26**, 242–251.
19 Zürcher, C. (1974) Ein neuer Spezialrechenschieber für den Brauereitechniker. *Monatsschrift für Brauerei*, **27**, 127–132.
20 Anger, H.-M. (2002) Netzwerk Friabilimeter, 1. Durchgang einer Kalibriergemeinschaft. *Monatsschrift für Brauwissenschaft*, **3/4**, 60.
21 Wackerbauer, K., Hardt, R. and Hirse, U. (1996) Bewertung von Lösungseigenschaften von Malzen mittels Friabilimeter, Calcofluor- und FIA-Methode. *Monatsschrift für Brauwissenschaft*, **7/8**, 220–5.
22 Schildbach, R. and Burbidge, M. (1979) Identifizierung von Gerstensorten an Einzelkörnern durch Flachgel-Elektrophorese der Proteine und Aleuronfärbung. *Monatsschrift für Brauerei*, **32** (11), 470–80.
23 Drawert, F. and Hagen, W. (1971) Beurteilung von Diätbieren, *Brauwelt*, **47**, 991–997.
24 Verordnung (EWG) Nr. 2676190 der Kommission vom 17. September 1990 zur Festlegung gemeinsamer Analysemethoden für den Weinsektor.
25 Schiwek, V. and Kunz, T. (2006) Untersuchungen zur Analytik von Schwefeldioxid in Bier, Praktikumsbericht, TFH und VLB, Berlin.
26 Ishibashi, Y., Terano, Y., Fukui, N., Honbou, N., Kakui, T., Kawasaki, S. and Nakatani, K. (1996) Development of a new method for determining beer foam and haze proteins by using the immunochemical method ELISA. *Journal of the American Society of Brewing Chemists*, **54** (3), 177–82.
27 Kunz, T., Stephan, A., Methner, F.J., Kappl, R. and Hüttermann, J. (2002) Grundlegendes zur

Elektronenspinresonanz-Spektroskopie (ESR) und Untersuchungen zum Zusammenhang zwischen oxidativer Bierstabilität und dem SO_2-Gehalt. *Monatsschrift für Brauwissenschaft*, **7/8**, 140–153.

28 Franz, O. and Back, W. (2002) Erfahrungen zur Messung von freien Radikalen mittels Elektronenspinresonanz-Spektrometer in der Brauerei. *Monatsschrift für Brauwissenschaft*, **7/8**, 156–162.

29 pH-Fibel, Einführung in die pH- und Redox-Messtechnik, Selbstverlag wissenschaftlich-Technische Werkstätten GmbH (1989), Weilheim.

30 Keßler, M., Kreisz, S., Zarnkow, M., Back, W. (2006) Braucht der Brauer einen Stärkelösungsgrad? *Brauwelt*, **10**, 266–269.

31 Thalacker, R., Bößendörfer, G., Birkenstock, B. (2006) Über den Glutengehalt des Bieres, *Brauwelt*, **41/42**, 1230–1235.

18
Microbiology

Werner Back

18.1
Microflora in the Brewery

Due to the following selective characteristics of beer, only a few microorganisms can proliferate [1–4].

- Low pH around 4.5.
- CO_2 concentration (anaerobic atmosphere).
- Alcohol concentration around 5 vol%.
- Bitter principles of hops.
- Deficiency of readily utilizable sugars and amino acids.
- Low temperatures in the production area.

For manufacturing control, mainly manufacturing cultures (bottom-fermenting and top-fermenting yeasts as well as 'Sauergut' cultures) and pest organisms are of importance. As the latter need to be analyzed in the trace region, special indicator germs are relevant for biological manufacturing control. In particular, capsular slime-forming acidic acid bacteria *Gluconobacter frateurii* and *Acetobacter pasteurianus* play an important role here as, like beer pests, they grow preferably in a beer milieu. Since they need oxygen, they have no chance of growing in the filled, almost oxygen-free beer.

These slime formers generate in miscellaneous niches in contact with beer biofilms, in which (in addition to other germs) typical beer pests can settle down, propagate and eventually adapt to the beer milieu (Figure 18.1).

Pathogens and heat-resistant germs cannot develop in beer. Therefore, moderate temperatures (62–72 °C) can be used during pasteurization to achieve microbiological safety [2, 3].

In general, mold fungi [5] in the brewery are seen as harmless contaminants; merely the unattractive growths of the so-called cellar mold (*Aureobasidium*, *Cladosporium*, *Moniliella*, etc.) on damp walls and casks occasionally necessitates particular cleaning measures. Fusaries (especially *Fusarium graminearum*, *F. culmorum*) found on barely, wheat and malt can be classified as indirect beer pests. These

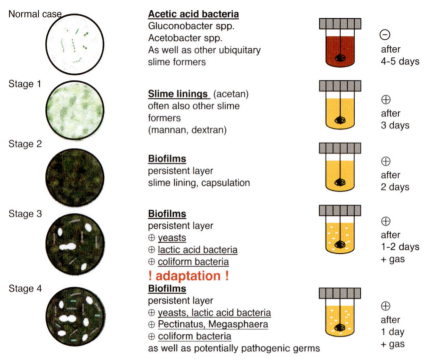

Figure 18.1 Development of beverage pests.

form toxins and other undesirable substances, which are carried on to the production processes and in parts to the beer.

In order to avoid interference during fermentation and problems inside the beer, special attention must be focused constantly on beer-spoilage contaminants in the factory. These beer pests are, besides different foreign yeasts, predominantly some lactic acid bacteria as well as certain Gram-negative species, mainly *Pectinatus* spp. and *Megasphaera cerevisiae*.

Beer pests are differentiated into the categories of harm: obligate, potential and indirect beer pests. Obligate beer pests are the most dangerous. They tolerate the beer-specific selectivity characteristics, and can often grow without adaptation in beer and cause quality defects. Potential beer pests can cause problems with reduced selectivity in the beer. After prolonged contact with beer in the brewery, they can also adapt with time and adopt an obligate beer-spoilage character. Indirect beer pests cannot proliferate in the filled beer. However, they can grow in the production area and cause pre-spoilage in wort, yeast or green beer. Such flaws can even be carried on to the filled beer and also cause problems there.

Apart from the differentiation into categories of harm, a classification as primary and secondary contaminants is helpful for biological manufacturing control. The former can be found in the production area (*Pediococcus damnosus, Lactobacillus lindneri*, etc.) and possibly can reach the filled beer via filtration. A high diminishing occurs due to the filters, hence the test results are often detected with no abnormalities and yet turbidity can occur in the individual casks after an increased incubation time (e.g. 4 weeks). With primary contaminations, location of the contamination sources is very difficult as the germs have already spread throughout the whole production area when the problems occur.

Secondary contaminations occur in the filling area [6], especially at the filler or capper. Due to the permanent beer contact, biofilms with beer pests cultivate preferentially there. Through rotation of machinery and the moist environment, germs or even whole biofilms can be sprayed and get into open bottles. Consequently, more or less pronounced spread contaminations develop in several bottles at a time. The most common secondary contaminants are *Lactobacillus brevis, L. casei* as well as *Pectinatus* spp. and *Megasphaera cerevisiae*.

18.2
Manufacturing Cultures

Of special relevance are bottom-fermenting and top-fermenting manufacturing yeasts. The ultimate ambition is the maintenance of purity and vitality of these strains, as the condition of the yeast has a crucial influence on the beer quality [7]. Vital pitching yeast guarantees speedy fermentation, rapid and extensive diacetyl decomposition, a pleasant taste profile as well as a good taste and foam stability. These kinds of strong fermentation manufacturing yeast are also able to compete with beer pests.

Brewery culture yeasts are certain manufacturing strains or cultivated forms of *Saccharomyces cerevisiae* or *S. carlsbergensis*, which due to their fermentation properties are especially well suited for beer production. Mainly bottom-fermenting (*S. carlsbergensis*) and top-fermenting yeasts (*S. cerevisiae*) are utilized. The former ferments in the original wort at low temperatures of 5–15 °C and, at the end of the fermentation, precipitates more or less intensely to the bottom. Depending on the sedimentation behavior, they are distinguished as flocculent and non-flocculent yeasts. The flocculent yeasts sediment during fermentation in a relatively compact form. Consequently, yeast harvest and filtration after storage are facilitated. Non-flocculent yeasts which remain suspended usually produce a better fermentation result; however, they are harder to harvest and can cause problems during filtration.

Top-fermenting yeasts generally ferment at a higher temperature of 15–25 °C and rise to the top during intensive fermentation, resulting in a strong top layer. In the case of top-fermenting yeasts, different strains are also used depending on the type of beer. Hence, one can differentiate between, for example, wheat beer and alt beer yeasts. In contrast to bottom-fermenting yeasts, a pronounced development of clusters is characteristic for many top-fermenting strains.

Bottom-fermenting and top-fermenting brewery yeasts generally form round to oval cells with a size of 5–10 × 5–12 µm, to some extent elliptic and cylindrical cells of 3.5–9.5 × 5.0–20.0 µm, and seldom also elongated or tube-shaped cells. The average cell size often varies considerably in the different strains. Inside the cells, lipid drops and, particularly in older cells, vacuoles are often found. Other apparent or taxonomic important cell structures are lacking. The clustering is multi-lateral.

The bottom-fermenting yeasts (*S. carlsbergensis*) exist as single cells and cell pairs. Larger cell formations occur only rarely. In contrast, agglomerates (random cell attachments) can be seen frequently.

Bottom-fermenting and top-fermenting yeast do not grow on crystal violet, lysine or copper sulfate agar and cannot assimilate nitrate. *S. carlsbergensis* differs from *S. cerevisiae* in terms of melibiose fermentation and complete raffinose fermentation (*S. cerevisiae* ferments only one-third), missing growth at 37 °C, and missing or rare ascosporulation [4].

Sauergut cultures (mainly *Lactobacillus amylolyticus*) [8] have gained great importance in recent years, as they not only allow the adjustment of a desirable pH value of the mash or wort, but also offer numerous additional technological and qualitative advantages [3].

18.3
Foreign Yeasts

Certain foreign yeasts can occasionally proliferate in filled beer and cause flaws in taste and odor; consequently, they need to be regarded as beer pests. Mostly, these are different wild strains of *S. cerevisiae* and other species of the genus *Saccharomyces* as well as *Brettanomyces* spp. [2, 3]. Of utmost importance are fermentation strong species as, not uncommonly, they produce an unpleasant appearance as obligate beer pests.

These foreign yeasts often differ morphologically from culture yeasts. Most commonly, small, longish cells develop which form more or less large clusters. For example, *S. cerevisiae* var. *diastaticus* predominantly forms small, oval, elliptical to cylindrical cells, which can also be elongated. The clusters are more linear and are made up of only a few cells. The fermentation spectrum is similar to top-fermenting yeasts; however, with the aid of amylases and amyloglucosidases, dextrin and starch are additionally fermented. Consequently, the contaminations in the filled beer cause over-fermentation, which causes post-turbidity, development of sediments and often also flaws in taste and odor. Furthermore, the simultaneous reduction of the extract leads to a slightly flat taste with a harsh bitterness note. These yeasts are present mostly in the unfiltrate and, as they often possess small cells, can pass through the filter easily. Occasionally, they appear as secondary contaminations during filling.

Other important wild yeasts are *S. cerevisiae* var. *bayanus* and the closely related *S. cerevisiae* var. *pastorianus*. The cells are mostly cylindrical or irregularly club- or sausage-shaped. Clusters are rarely formed. The cells occur predominantly in pairs, whereby characteristic 'signpost formations' often exist. Generally, the

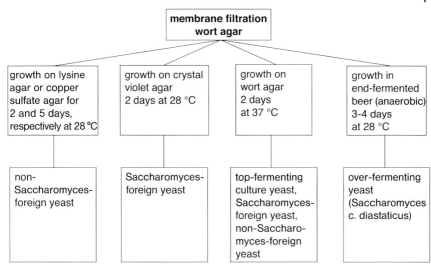

Figure 18.2 Rough differentiation of yeasts on membrane filter cultures.

yeasts have a bottom-fermenting character, but ferment no melibiose and only one-third of raffinose, similar to top-fermenting yeasts. However, maltotriose and maltotetraose can often be utilized. Beers with a low degree of fermentation are at particularly at risk, where in the case of contaminations post-turbidity and raspy-bitter taste discrepancies can be generated. These yeasts occur predominantly in the unfiltrate area and as impurities in culture yeast.

Non-*Saccharomyces* yeasts appear more infrequently than beer pests. Problems can also be triggered by *Brettanomyces* species (e.g. *B. anomalus*). These slow fermenters occur frequently in Belgian geuze and lambic beers, in ales and porters, as well as in Berliner Weisse, where they produce the desirable characteristic ester aroma. In common beers they cause massive flaws in odor and taste through strong acidic acid and ester formation. The cells are oval, elliptical, cylindrical or elongated with ogival clustering (shoulder clusters on the wide-side of the mother cell). Furthermore, pseudomycels with in parts single or wispy arranged blastospores are also frequently formed.

The rough differentiation of yeasts occurring in the brewery is shown in Figure 18.2. Although only five test solutions are employed, the brewery biologist is in an optimal position to determine the relevance of the grown yeasts. However, numerous physiological–biochemical tests are necessary for an accurate identification [2, 3, 9, 10].

18.4
Beer-Spoilage Bacteria

Beer-spoilage bacteria are mostly lactobacilli and pediococci; however, dangerous Gram-negative species such as *Pectinatus* and *Megashaera* can also be found (see Figure 18.3) [1, 3].

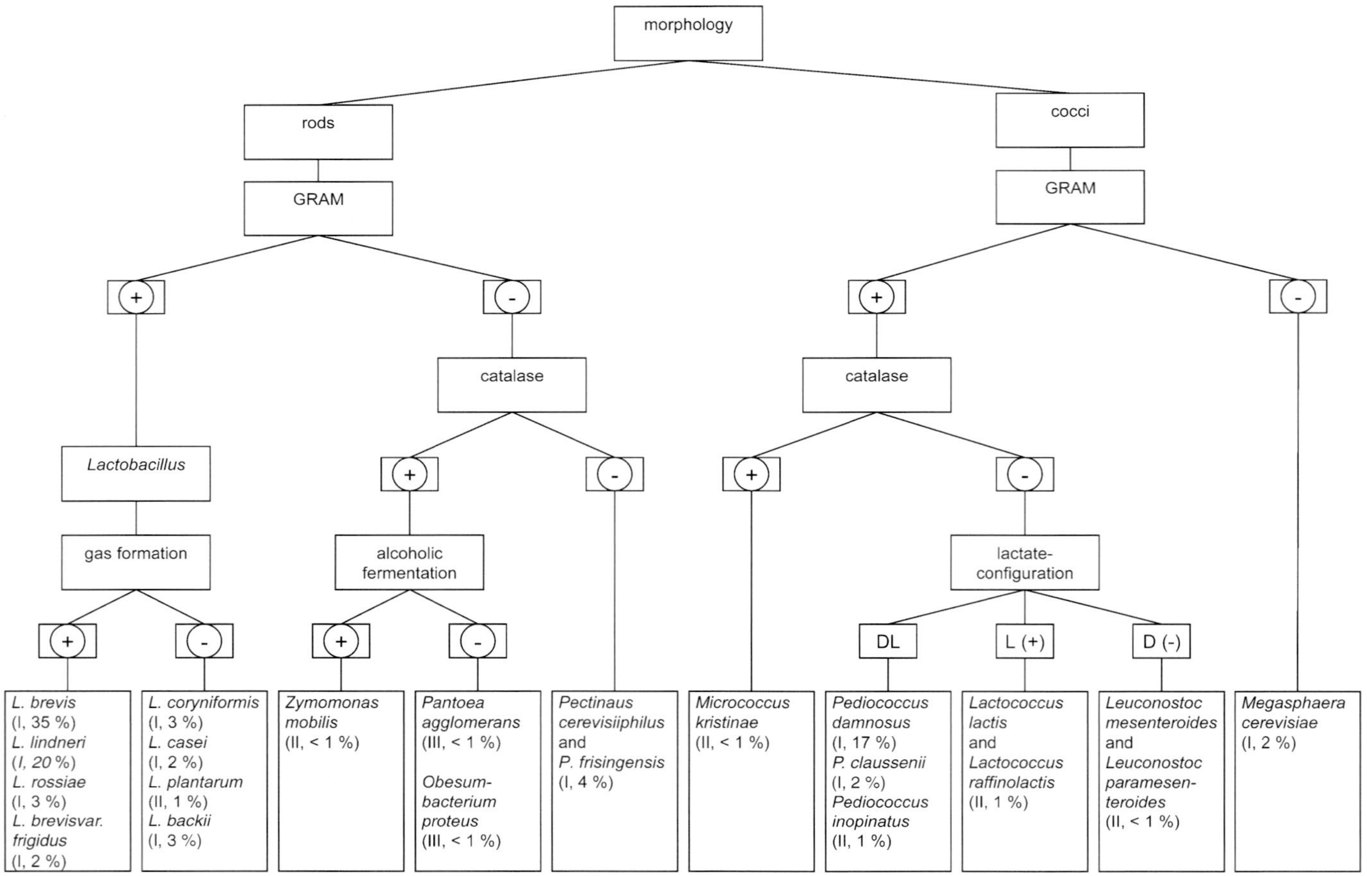

Figure 18.3 Classification of beer-spoilage bacteria [1, 2]: I = obligate beer spoilage; II = potentially beer spoilage; III = indirect beer spoilage. The percentage represents the frequency of the species in the beer by problems (period 1980–2007).

The most important *Lactobacillus* species are *L. brevis* and *L. lindneri*; less common are *L. rossiae, L. buchneri, L. coryniformis, L. casei* and *L. backii* [3, 11, 12].

L. brevis frequently develops longer, parallel-walled, single or pair-wise rods with round ends (0.7 × 4 µm). The double-rods are often bent ('signpost form'). Cell chains are not formed. Extremely long rods (up to 50 µm) can occasionally be found in beer. Common characteristics for *L. brevis* are the development of gas (hetero-fermentative), the fermentation of pentoses and melibioses, as well as arginine cleavage. This most common beer pest causes turbidity of the beer and sediments as well as a noticeable reduction of the pH, giving the beer an acidic taste. However, diacetyl is not formed. Frequently, these are secondary contaminations. *L. brevis var. frigidus* forms slime in the beer. *L. bucherni* differs mostly from *L. brevis* by additional melicitose fermentation.

L. lindneri forms short, slightly irregular or coccoid cells arranged in longer chains. Long rods are occasionally formed, especially in beer samples. The hetero-fermentative species ferments mainly glucose and maltose, and does not cleave arginine. Slight sediments and turbidity are formed easily in the beer, yet generally no flaws in taste occur. This typical primary contaminant is frequently found in yeast cellars, fermentation cellars and storage cellars, and can be carried on through the filters as a result of its very small cells. *L. rossiae* also has similar properties and produce slime.

The facultative hetero-fermentative species *L. casei, L. coryniformis* and *L. plantarum* form shorter rods that are often arranged in chains. The species occur more frequently in weaker hopped beer (wheat beer) and cause an apparent flaw in taste as a consequence of diacetyl formation. These are often secondary contaminants. The homofermentative obligate beer-spoilage species *L. backii* ferments mannose, mannitol and sorbitol. It differs from the other species in, among other things, the missing fermentation of maltose and gluconate [12].

Tetrads formation is characteristic in *P. damnosus*. These are typical primary contaminants, which appear often in culture yeast and in the unfiltrate as well as in wheat beers. The cells can also be carried on via the filter to the filled beers. A strong diacetyl formation (butter flavor) and a drop in the pH value occur in the beer. Furthermore, the beers often show slight turbidity and noticeable sedimentations. Two other beer-spoilage *Pediococcus* species, *P. inopinatus* and *P. claussenii*, behave similarly. The latter forms slime in the beer. Both species are considerably less common [3, 13].

The Gram-negative, catalase-negative, strict anaerobic species *Megasphaera cerevisiae* forms large oval or round cells (1.2–1.6 µm), which exists as diplococci and short chains. In particular, fructose, pyruvic acid and lactic acid are fermented. Primary metabolites are butyric acid, carpronic acid, acetic acid, propionic acid, valeric acid as well as CO_2 and hydrogen gas. Only a slight turbidity in the beer is formed; however, massive flaws in odor and taste (cloaca-like smell) are produced as a consequence of the aforementioned metabolites. The species is sensitive to alcohol (below 5 vol%) and also prefers higher pH (above 4.4). These are typical secondary contaminants which develop preferentially in the environment of the filler/capper.

Pectinatus cerevisiiphilus and *P. frisingensis* are also strict anaerobic, Gram-negative, catalase-negative and possess similar negative effects as the previous species. The cells are slender (0.8 × 4 µm), parallel-walled with pointy ends, slightly bent or snake-like and corkscrew-like, respectively, and mono-laterally flagellated. Similar to *M. cerevisiae*, they grow in a range of 15–40 °C (optimum 28–32 °C). Different sugars, sugar alcohols and organic acids (especially pyruvate and lactate) are fermented. The primary metabolites are propionic acid, acetic acid, pyruvic acid, acetoine and CO_2. The contaminated beers (pH above 4.3, alcohol content below 5 vol%) show not only severe sedimentation and turbidity or clotting, but also unpleasant flaws in odors and taste (cloaca-like smell). As in *M. cerevisiae*, these are typical secondary contaminations, which occur preferentially in the bottle cellar.

18.5
Detection of Beer Pests

A manufacturing-specific control plan is a prerequisite for high microbiological safety (Figure 18.4).

The detection of beer-spoilage bacteria is carried on with NBB (nutrient media for beer spoiling bacteria) medium [1–3], which is evaluated internationally and recommended by experts due to the high accuracy of the measurements, speedy detection and beer-specific selectivity [11]. Simple handling and unambiguous analysis of the samples are also important. On the one hand, this is possible due to the strong growth of the beer pests; on the other hand, manufacturing yeasts and harmless concomitant organisms are widely inhibited.

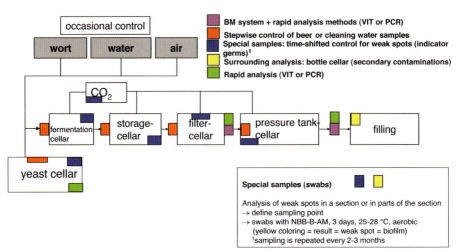

Figure 18.4 Analysis of the surroundings in the production–unifiltrate. Primary contaminations. PCR = polymerase chain reaction. The BM system = continuous sampling after filtration or before the filler [14]. VIT = molecular biological rapid analysis [15].

The beer-specific selectivity of NBB is adjusted such that all grown germs cultivated according to the instructions are meaningful for biological manufacturing control.

Trace analysis (preferably using large sample volumes) is made possible as the different sample types are examined with sample specific detection methods. Thus, clear beer samples and wash-water samples are filtered through a membrane and incubated on NBB agar in an anaerobiosis box at 27 °C for 5 days.

Yeast samples are mostly analyzed in NBB broth. Generally, 0.5 ml yeast is inoculated to 10 ml bouillon and incubated as above, although no anaerobic atmosphere is necessary. After 2–3 days, the presence of beer pest is shown by a yellow color change (see Figure 18.5a). For larger yeast samples (around 10 ml) it is also possible to work with NBB concentrate (Figure 18.5b).

Generally NBB concentrate is used for unfiltrate and wheat beers; 5% NBB-C is given to the beer sample (e.g. brim-full 180 ml swing stopper sample bottles) and incubated at 27 °C for around 7 days. The detection is very selective for obligate beer pests. The samples should be microscopically controlled during analysis since occasionally turbidity caused by proteins can occur and an indicator such as NBB-A or NBB-B is lacking. Potential and indirect beer pests can also be determined by the addition of water (10–50%) (see Figure 18.6).

Weak spots in the production and filling area are checked with a sterile swab.

In addition to beer pests, primarily indicator germs (in particular, acetic acid bacteria and foreign yeast) need to be determined and an aerobic incubation with less-selective NBB-AM bouillon is recommended. The incubation is carried out at 27 °C for a maximum of 3 days. Only test tubes with an indicator change from red to yellow are considered for the analysis. Such a result suggests a biofilm and thus a weak spot at which beer pests can develop sooner or later (see Figure 18.1). Particular weak spots are cover buttons, blind tubes, tube sacks, CO_2 tubes, pumps, bypasses, carbonation and measuring facilities as well as seals.

The described swabbing method with NBB-B-AM has proved of value in the filling area of the beverage industry (and also in the mineral well and soft drink industry) worldwide, since it is easy to handle and the analysis has a high significance (Figure 18.7). Swabs are particularly taken on direct and indirect contact points with the beer (e.g. stars, valves, lifting elements, bolted connections, stay bolts, hollow spaces). On statistical grounds it is reasonable to cover about 20–30 of such weak spots. Deposites on the filler and capper are particularly dangerous since these biofilms offer ideal conditions for beer pest owing to the permanent beer contact. Persistent slime films can also develop at the head of washer, at the bottle inspector and on bottle transportation belts, in which beer pests can also settle down.

The advantage of this method is also that with regular sampling, weak spots can be found through the determination of indicator germs before beer pest appear and adapt. Therefore, swabs should be taken (preferably) weekly during ordinary fillings (e.g. Tuesdays, analysis Friday) to capture the actual biological circumstances. The results (yellow tubes) should be on average under 30% or even under 20% (threshold), extending over several weeks. In addition, the effect of cleaning

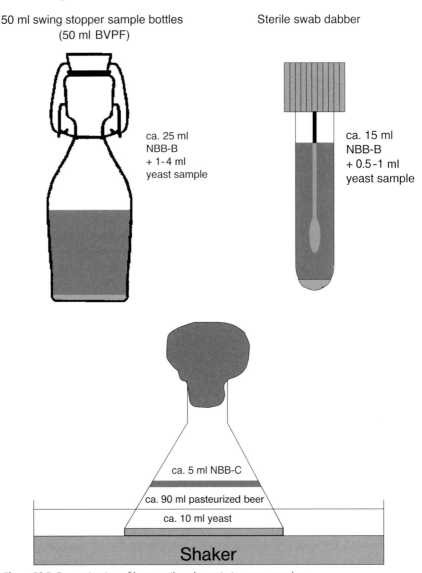

Figure 18.5 Determination of beer-spoilage bacteria in yeast samples.

on the plant can be checked occasionally. With swabs after cleaning, the yellow tubes should be under 10%.

A microscopic control of the swabs is not really necessary since the use of NBB-B-AM gives a color change to yellow within 3 days (traffic light principle) that generally can be assessed as a relevant biofilm (Figure 18.8). On direct contact points, only indicator germs and no beer or beverage pests should be present.

obligate
NBB
Beer sample
ca. 170 ml
ph 4.5–4.8
obligate ⊕ potential
Sterile H_2O
NBB
Beer sample
ca. 150 ml
ph <5.0
obligate ⊕ potential ⊕ indirect
Sterile H_2O
NBB
Beer sample
ca. 100 ml
ph >5.0
obligate ⊕ potential ⊕ indirect ⊕ indicator germs
Air
Sterile H_2O
NBB
Beer sample
ca. 100 ml
ph >5.0

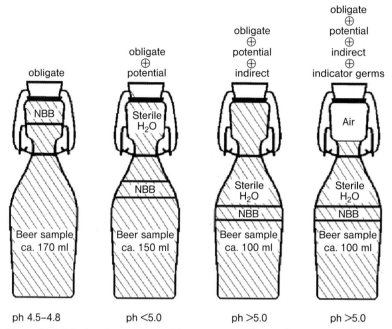

Figure 18.6 Selective determination of beer pests according to their category of harm.

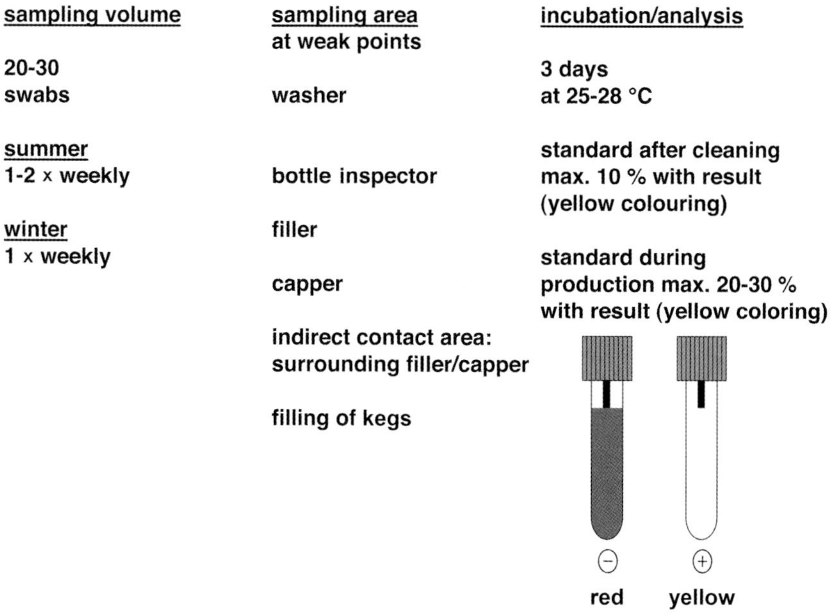

sampling volume
20-30
swabs
summer
1-2 x weekly
winter
1 x weekly
sampling area
at weak points
washer
bottle inspector
filler
capper
indirect contact area:
surrounding filler/capper
filling of kegs
incubation/analysis
3 days
at 25-28 °C
standard after cleaning
max. 10 % with result
(yellow colouring)
standard during
production max. 20-30 %
with result (yellow coloring)
⊖
red
⊕
yellow

Figure 18.7 Evaluation of wiped samples.

Figure 18.8 Biological situation in the bottle cellar as presented on swab samples. Three days aerobic incubation at 25–28 °C (six of 24 samples with result = 25%).

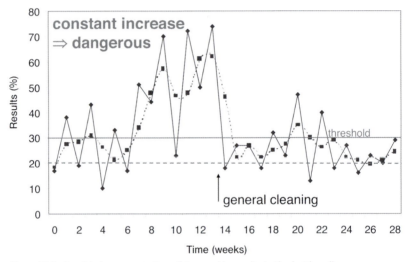

Figure 18.9 Graphical representation of the weekly results in the bottle cellar.

The weekly swab results can be presented clearly and impressively in a chart. It can be immediately recognized if the threshold of 30% is exceeded with increasing tendency. This suggests the presence of persistent films and the risk of secondary contaminations. In contrast, one-time outliers are unproblematic if the following controls again measure less than 30% (Figure 18.9). In addition, the graphical presentation has the advantage that not only negative influences (e.g. construction measures), but also the efficiency of the cleaning measures can be followed directly. Thus, the state of hygiene in the filling area can be judged at any time. Furthermore, corresponding counter-measures with air-germe collectors can be initiated in time.

The determination and cultivation of yeast is generally carried out with wort agar at 25–28 °C and aerobic incubation. To differentiate between strongly fermenting yeasts and respiratory yeasts, parallel tests are carried out with aerobic and anaerobic incubation.

The detection of culture yeasts after filtering is of importance to judge the filtration efficiency. Furthermore, the detection of foreign yeasts in the culture yeasts (pure culture, cropped yeast, tank-bottom yeast) is occasionally necessary. There, the difference in temperature maxima of bottom yeast (around 32 °C) and most foreign yeast (37–45 °C) is taken advantage of. Consequently, foreign yeast can be selectively accumulated in bottom yeast samples at an incubation temperature of 37 °C. A multi-stage enrichment is recommended to carry out reliable trace analysis. To differentiate between *Saccharomyces* foreign yeasts and non-*Saccharomyces* foreign yeasts, they are subsequently incubated on crystal violet and lysine agar, respectively (Figure 18.10)

Another possibility to determine foreign yeasts in bottom-fermenting yeast is presented in Figure 18.11. Pantothenate agar, which does not contain pantothenic acid, is additionally used for the determination of foreign yeast in top-fermenting yeast. Since pantothenic acid is essential for top-fermenting yeasts, in a contamination with bottom-fermenting yeast the latter can proliferate without competition. The recommended test media are described in [2].

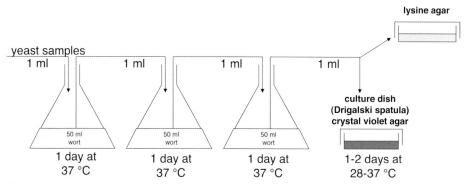

Figure 18.10 Trace determination of foreign yeasts in bottom-fermenting culture yeasts.

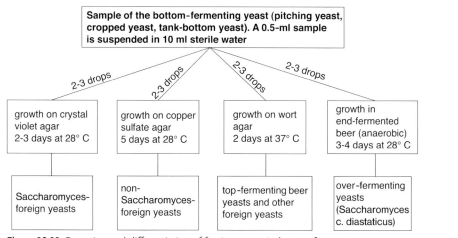

Figure 18.11 Detection and differentiation of foreign yeasts in bottom-fermenting yeasts.

References

1. Back, W. (1980) Bierschädliche Bakterien. Nachweis und Kultivierung bierschädlicher Bakterien im Betriebslabor. *Brauwelt*, **120**, 1562–9.
2. Back, W. (2000) *Farbatlas und Handbuch der Getränkebiologie, Teil II*, Fachverlag Hans Carl, Nürnberg.
3. Back, W. (2008) *Color Atlas and Handbook of Beverage Biology*. Fachverlag Hans Carl, Nürnberg.
4. Back, W. (2005) *Ausgewählte Kapitel der Brauereitechnologie*, Fachverlag Hans Carl, Nürnberg.
5. Samson, R.A., Hoekstra, E.S., Frisvad, J.C. and Filtenborg, O. (1995) *Introduction to Food-Borne Fungi*, 4th edn, Central Bureau voor Schimmelkultures, Baarn.
6. (a) Back, W. (1994) Sekundärkontaminationen im Abfüllbereich. *Brauwelt*, **134**, 686–95. (b) Back, W. (1994) Secondary contaminations in the filling area. *Brauwelt International*, **12**, 326–33.
7. Back, W. (2008) *Mikrobiologie der Lebensmittel, Getränke*, Behr's, Hamburg.
8. Bohak, I., Back, W., Richter, L., Ehrmann, M., Ludwig, W. and Schleifer, K.H. (1998) Lactobacillus amylolyticus sp. nov., isolated from beer malt and beer wort. *Systematic and Applied Microbiology*, **21**, 360–4.
9. Kreger–van Rij, N.J.W. (1984) *The Yeasts, A Taxonomic Study*, 3nd edn, Elsevier, Amsterdam.
10. Barnett, J.A., Pankhurst, R.W. and Yarrow, D. (1990) *Yeasts: Characteristics and Identification*, 2nd edn, Cambridge University Press, Cambridge.
11. Sneath, P.H.A., Mair, N.S., Sharpe, M.E. and Holt, J.G. (eds) (1986) *Bergey's Manual of Systematic Bacteriology*, Vol. 2, Williams & Wilkins, Baltimore, MD.
12. Bohak, I., Thelen, K. and Beimfohr, C. (2006) Description of Lactobacillus backi sp. nov., an obligate beer-spoiling bacterium. *Monatsschrift für Brauwissenschaft*, **60**, 78–82.
13. Dobson, C.M., Denner, H., Lee, S., Hemmingsen, S., Glaze, S. and Ziola, B. (2002) Phylogenetic analysis of the genus Pediococcus, including Pediococcus claussenii sp. nov., a novel lactic acid bacterium isolated from beer. *International Journal of Systematic and Evolutionary Microbiology*, **52**, 2003–10.
14. (a) Back, W. and Pöschl, P. (1998) Bypass-Membranfiltration (BM-System) – Verbesserung des Spurennachweises nach der Filtration. *Brauwelt*, **138**, 2312–15. (b) Back, W. and Pöschl, P. (1999) Bypass-Membranfiltration (BM-System). *Brauwelt International*, **17**, 202–4.
15. Thelen, K., Beimfohr, C., Bohak, I. and Back, W. (2002) Specific rapid detection test for beer spoilage bacteria using fluorescently labelled gene probes. *Brauwelt International*, **20**, 156–9.

19
Certification

Bernd Lindemann

Any successful company will have a management system in place. The management system documents the operational and organizational structure. Management system is the term applied to all the organizational rules and instructions. They should be built up in an orderly workflow. References to procedural methods provide the management standards. It is seen as prestigious and state-of-the-art if a company is able to state its position with reference to international standards such as DIN EN ISO 9001.

19.1
Management Systems and Business Management

Systematically produced management systems describe the build up of the company alongside the workflows. The aim of such specifications is the allocation of accountability and responsibilities. In certain cases, these responsibilities identify specific activities. Without these responsibility definitions, leadership in companies would be impossible. In ideal cases, an integrated management system concerning all aspects of a company, from financial quality to the environment and job security, is supported by the management.

19.2
Management Systems Standards

There are a number of standards in place as support for companies that wish to increase their quality system. These standards should clarify the rules in place to ensure universal validity. They are internationally accepted, and should help to improve the effectiveness and performance of the company.

Handbook of Brewing: Processes, Technology, Markets. Edited by H. M. Eßlinger
Copyright © 2009 WILEY-VCH Verlag GmbH & Co. KGaA, Weinheim
ISBN: 978-3-527-31674-8

19.2.1
DIN EN ISO 9000, 9001 and 9004

The ISO 9000 family is classified into three ISO Standards: ISO 9000, 9001 and 9004. ISO 9000 illustrates the guidelines for the selection and implementation for ISO 9001 and 9004. These models aim to generate confidence in the quality of the goods produced by the manufacturer. Requirements for the real quality of products do not exist. ISO 9004 advises that using these standards with other relevant industry-specific recommendations will, among other things, facilitate the implementation of the quality management systems.

ISO 9000 describes the fundamentals of quality management systems for the application of the standard series (ISO 9000-1) and, in turn, essentially describes the ground rules for quality management systems (ISO 9000-2). Furthermore, it also defines the terminology for these systems (ISO 9000-3).

DIN EN ISO 9001 basically contains five requirements that incorporate everything from the planning and realization of a product to customer service. This standard is the only one in the ISO 9000 series that contains demands on the quality management requirements that have to be fulfilled. In order for this system to be adaptable for all companies, there is the possibility under an explanatory statement to exclude the demand in the product realization chapter.

ISO 9004, in contrast to ISO 9001, contains no demands, but instead merely provides instructions for the use of a management system in the form of recommendations. Its contents are in line with the organization of ISO 9001. Moreover, it focuses on a further area of aims, with which the overall performance such as efficiency and effectiveness of a company should constantly improve. In order for ISO 9004 to best serve the companies, ISO 9001 is intended to be continually and selectively improved. The fact that ISO 9004 can be independently applied to ISO 9001 means it is cannot be audited for certification purposes as it contains no obligatory content.

19.2.2
DIN EN ISO 14001

DIN EN ISO 14001 is built similarly to DIN EN ISO 9001. It contains demands for the implementation of an environment management system. Companies must attempt to improve their environmental activities. Therefore, environmental targets need to be formulated in order for the established management system to provide the necessary resources so that environmental targets can be attained. A major emphasis lies on continuous improvement.

19.2.3
DIN EN ISO 22000

ISO 22000 is a standard designed with a specific aim in mind – the guarantee of food safety. The standard has been in force since 30 September 2005 and is used to unify the numerous standards within the food industry. This field was previously regulated by various standards such as Hazard Analysis of Critical Control Points (HACCP), the British Retail Consortium (BRC), the International Food Standard (IFS), EurepGAP (Good Agricultural Practices) and Good Manufacturing Practice. Henceforth, ISO 22000 now combines all of these single standards into one overall standard. The main objective of this standard is the creation of a worldwide, uniform standard, particularly for the food industry. ISO 22000 corresponds with the quality management standard ISO 9001:2000 and thus provides a special, factual widening of this standard for the food industry.

19.2.4
Global Food Safety Initiative

The Comité International d'Enterprises à Succursales (CIES; International Committee of Food Retail Chains) was founded in 1953, with the aim of representing the interests of food retail chains. The CIES called for the Global Food Safety Initiative (GFSI) to be brought to life in 2000. The mission of the GFSI is the continuation of the management system improvement processes, to ensure the safety of food and to maintain the trust of the consumer by guaranteeing safe food. The GFSI has formulated specifications and standards for the guarantee of safe food. Up to now, four standards have fulfilled the GFSI's guidelines: the BRC, IFS, Dutch HACCP and Safe Quality Food.

19.2.5
IFS

In order to establish a uniform food safety standard, German Retailers developed the IFS in 2002. In 2003, the French Retailers of the IFS Working Group joined and collaborated with the development of Version 4 of the IFS. The aims of the IFS are: the generation of an assessment foundation for all producers of own brands, standard formulations, the implementation of audits and the authentication of reciprocal audits, and a high level of transparency within the supply chain. The IFS is applicable to all stages of production tied to agricultural commodities, to food products that are manufactured and to the transportation that supplies them. The IFS defines the content of audits, including the requirements, procedures and validation thereof, as well as specifications for the certifying body and auditors.

The structural organization of the IFS (catalogue of requirements) is:

- Specifications for the quality management system.
- Accountability of the management.
- Resource management.
- Production processes.
- Measurement, analysis and improvements.

19.2.6
BRC

The BRC has, in parallel with German and French Retailers, developed a catalogue of requirements that serves as a basis for the auditing and validation of own brand suppliers. This catalogue encompasses HACCP, quality management, the state of repair of buildings, process sequence, product testing, process knowhow and personal hygiene.

19.2.7
DIN EN ISO 17025

This standard details the general requirements for the quality management system, and the function of the testing and calibration laboratories. It is internationally acknowledged and is used by respective national accreditation bodies as the basis for the expertise of the laboratory. The goal is the safe, secure analysis of results within known and established error boundaries. The direction and leadership requirements are based on the philosophy of the ISO 9000 series, such as the depiction of the mode of operation for the performance of competent analysis from calibration of equipment and validation of measurement processes to the point of the compilation of significant test reports, including the interpretation of the findings.

19.3
Principles and Similarities

All the aforementioned standards follow one general philosophy: a systematic, working management system works in accordance with the following principles:

- Installation, implementation and care of a management system.
- Liability of the leadership team.
- Management of resources.
- Realization of processes and products (services).
- Measurement, analysis and improvement.

Basically, all of the standards make the consumer/customer the focus of activities. The focus of requirements differ in the subjects quality, environment and food safety. DIN EN ISO 9001, 14001, 22000, IFS and BRC all serve to demonstrate the management system to external parties (customers, consumers, businesses, etc.) and detail the requirements for certification audits.

19.4
Legal Requirements

The protection of the consumer is the essential concern of legal policies. On an European level, a number of series of regulations and guidelines for the food sector have been passed. Two essential policies are Regulation 178/2002, the so-called 'basis regulation', and Regulation 852/2004, the so-called 'food hygiene regulation'.

19.4.1
Regulation (EU) 178

Three important matters were determined at this stage:

- The term 'food' was defined
- The establishment of the European Food Safety Authority was fixed
- The tracking of food products became statutory.

The regulation holds for all production, processing and assembly levels of food and animal feed. It should produce a high level of protection for consumers.

19.4.2
Regulation (EU) 852

Two important matters are defined under the food hygiene regulation:

- Food producers and processing plants need to guarantee 'Good Hygiene Practices'.
- Food producers and processing plants need to recognize and adhere to the HACCP work tool for the identification of risks for food products.

19.5
Certification According to ISO Standards

Obtaining a certification indicates the acknowledgment of conformity to the underlying standards through an independent accredited certification body. The certification process normally follows the following scheme:

- Choice of area to be certified.
- Registration by the certification area.
- Questionnaire and project discussion.
- Checking of the documentation.
- Report regarding the documentation check.
- Pre-audit (only if demanded and by appointment).
- Certification audit in the company.
- Certification.

19.6
Certification through IFS and BRC

A certificate issued through the private standards IFS and BRC can proceed in the same way as for ISO certification. The fundamental difference lies in the fact that the private standard issuer conducts the accreditation of the certifying body and the accreditation of the auditors. The auditors will see the business from the point of view of the standard issuer and also from the point of view of the retailer.

19.7
Certification through HACCP

HACCP is the obligatory tool to identify and handle risks to food products. Therefore, it cannot be used as the basis for certification. However, some companies still 'buy into' the external authentication to show externally that they utilize this tool.

20
World Beer Market

Jens Christoph Riese and Hans Michael Eßlinger

20.1
Introduction

Beer has always been a very special product and one can assume that it will carry on to be greatly desired. This leads to some special features that shape the world market of beer and make it so unique. During the last decade beer consumption has increased by more than 20% worldwide. Despite stagnation in several large beer-consuming nations, the total market size of 1 698 938 000 hl reached a new peak in 2006 [1]. Consequently, the global annual per capita consumption also increased to reach 26.4 l in 2006. However, although local traditions always have to be taken into account, even amongst the 25 largest markets the gap between the highest and lowest consumption amounts to several hundred percent. Furthermore, the raw materials used for brewing depend on local availability. Nearly all cereals can be used for brewing, which is one of the reasons for the wide variety of different beer flavors.

Apart from the beer drinkers, governments have discovered beer as a source of taxation. All over the world beer tax is highly appreciated and the beers are classified into various tax groups on the basis of their extract of original wort or alcohol content. The groups differ from country to country. There are different means of tax collection: raw material taxation (based on weight or volume of the brewing materials used), intermediate product taxation (assessed according to the volume of the wort) or finished product taxation (levied on sales beer).

Handbook of Brewing: Processes, Technology, Markets. Edited by H. M. Eßlinger
Copyright © 2009 WILEY-VCH Verlag GmbH & Co. KGaA, Weinheim
ISBN: 978-3-527-31674-8

20.2
Statistics

20.2.1
Raw Materials for Brewing

20.2.1.1 Barley and Malt Market

After wheat, rice and maize, barley takes rank fourth in the worldwide production of cereals [2]. The annual production of barley is about 140 million tons (Table 20.1) of which about 25 million tons is required for malting. This portion consists of varieties that differ considerably in quality and thus also in price from the remaining barley (R. Schildbach *et al.*, unpublished).

The main production areas of barley are the mild climate zones in the northern hemisphere with centers in Europe and Near East. Primary gene centers of barley are central Asia and Ethiopia, secondary gene centers are situated in the Near East and North Africa; there are thousands of different barley types.

For the brewing industry only the following four types are of special interest:

- Two-row barley:
  - Spring barley: most important barley in the world, preferred as malting barley.
  - Winter barley: grown mainly in Europe partly as winter malting barley (lower quality for brewing).
- Six-row barley:
  - Spring barley grown as malting barley mainly in the United States and Canada.
  - Winter barley mainly grown in Europe as feed barley (except malting barley Esterel in France).

Table 20.1 Barley production (average 2001–2003).

	Area (10^6 ha)	Yield (tons/ha)	Production (10^6 ton)
World	54.1	2.5	140.3
Europe	29.1	3.1	89.1
France	1.7	6.0	10.2
Germany	2.1	5.7	11.7
Spain	3.1	2.5	7.8
Ukraine	4.2	2.2	9.2
Russia (Europe + Asia)	9.5	2.0	18.9
Asia	12.0	1.9	22.2
North and Central America	6.0	2.7	16.5
Canada	4.0	2.5	10.0
United States	1.8	3.0	5.4
South America	0.8	1.8	1.4
Oceania	3.8	1.9	7.2
Australia	3.7	1.8	6.7

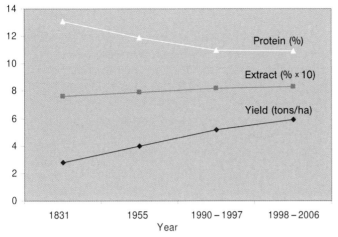

Figure 20.1 Development of spring malting barley in central Europe.

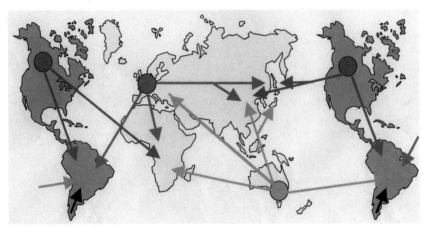

Figure 20.2 Barley and malt trade worldwide.

Barley has first bred in Europe in 1860. Since then the yield has more than doubled, the raw protein content has lowered and the extract increased by 7% in absolute term (Figure 20.1). As a result of the breeding progress, every year new varieties with better yield and quality enrich the market, and older ones disappear. A comparison of barley varieties is shown in Table 20.2 [3, 4].

Beer consumption, malting capacity and barley production run parallel in all countries. In addition, prices influence the worldwide barley and malt trade (Figure 20.2). The main barley and malt production centers of Europe, Canada/United States and Australia are competing for the breweries in Eastern Asia, South America and Africa. New malting facilities in Russia and China will change this

Table 20.2 Some important malting barley varieties (2003–2005).

Region	Country of origin	Mainly grown in
Europe	Prestige 2 RS United Kingdom	France; The Netherlands; Belgium; Denmark; Sweden; Portugal
	Scarlett 2 RS Germany	Germany; France; Belgium; Czech Republic; Poland; Spain; Hungary
	Barke 2 RS Germany	Germany; Sweden; Denmark; Belgium; Austria
	Optic 2 RS United Kingdom	United Kingdom
	Power 2 RS Denmark	Denmark
	Braemer 2 RS United Kingdom	Germany; United Kingdom
	Cellar 2 RS United Kingdom	United Kingdom; Belgium
	Auriga 2 RS Germany	Germany
	Jersey 2 RS Czech Republic	Slovakia
	Esterel 6 RW F	France; Spain; Hungary
	Vanessa 2 RW Germany	France; Spain; Germany
	Regina 2 RW Germany	United Kingdom; Italy
	Pearl 2 RW United Kingdom	United Kingdom
North America	AC Metcalfe 2 RS Canada	Canada; United States
	Harrington 2 RS Canada	Canada, United States
	CDC Kendall 2 RS Canada	Canada
	CDC Stratus 2 RS Canada	Canada
	Robust 6 RS Canada	Canada; United States
	Excel 6 RS Canada	Canada
	B 1602 6 RS Canada	Canada
	Lacey 6 RS United States	United States
South America (Argentina, Uruguay and Brazil)	Scarlett 2 RS Germany	
	Q Alfa 2 RS Argentina	
	Q Palomar 2 RS Argentina	
	Perun 2 RS Czech Republic	
	Musa 016 2 RS Uruguay	
	Musa 936 2 RS Uruguay	
	MN 698 2 RS Brazil	
	BRS 195 2 RS Brazil	
	Embrapa 127/128 2 RS Brazil	
	BR 2 2 RS Brazil	
Australia	Schooner 2 RS Australia	WA; VI; NSW
	Sloop 2 RS Australia	SA; VI
	Gairdner 2 RS Australia	SA; VI; WA
	Stirling 2 RS Australia	WA
	Arapiles 2 RS Australia	VI; NSW
	Grimmet 2 RS Australia	NSW; Q

2 RS = two-row Spring Barley; 6 RS, = six-row Spring Barley; 2 RW = two-row Winter Barley; 6 RW = six-row Winter Barley.
WA = West Australia; VI = Victoria; SA = South Australia; NSW = New South Wales; Q = Queensland.

Table 20.3 Barley and malt trade in 2005.

	Import (1000 tons/year)		Export (1000 tons/year)	
	Barley	Malt	Barley	Malt
China	2141			
Japan	1429	514		
Australia			4731	486
United States	117		605	
Canada			2226	547
Brazil		650		
Europe				4210

situation in the coming years. Imports and exports are described in Table 20.3 (S. Franzi, 2007, personal information).

20.2.1.2 Global Malt Production

Malt, a processed form of top-quality barley, is the basic ingredient used in the production of beer. It provides most of the complex carbohydrates and sugars necessary to give beer its distinctive flavor and alcohol content. For the same reasons, malt is used in making whiskey and other distilled spirits. It is also gaining considerable popularity in the food industry as a flavoring component and a source of nutrition.

Global malt production amounted to approximately 17 million tons in 2003 [5]. As malt is quite sensitive to transport costs, it is not surprising that the top three malt-producing countries are also the three major beer markets in the world. Roughly 26% of the total malt volume is sold in foreign markets. As seen in Table 20.4, the most important exporting country is France, which is responsible for 20% of total exports. Due to their significant beer industries and lack of barley growing areas and malting facilities, Japan and Brazil are the biggest importers of malt. Due to a lot of new Greenfield maltings, Russia will cover its demand domestically and will export additional quantities in the very near future.

After having seen the huge French production volume it is not surprising that the two largest malting companies are located in France. Whereas the Soufflet Group has focused its plants in the European Community and Russia, the Malteurop Group is furthermore represented in the Chinese market with its own production facilities. In comparison to the top 20 brewers, the global malting industry is less concentrated. The cumulative market share amounts to 63.4% and is therefore 9% smaller than that the brewing industry (Table 20.5).

20.2.1.2.1 Hop Market The hop plant (*Humulus lupulus*) is native to northern temperate zones, including Northern Europe, North America and West Central Asia. Hops only grow between 35° and 55° latitude in either the northern or

Table 20.4 Global production, export and import of malt in 2003.

Region	Country	Production		Exports		Imports	
		10⁶ tons	%	10⁶ tons	%	10⁶ tons	%
Europe	Germany	2,040,000	12	495,000	11	270,000	6
	United Kingdom	1,530,000	9	315,000	7		
	France	1,360,000	8	900,000	20		
	Belgium	850,000	5	585,000	13	180,000	4
	Spain	340,000	2				
	Denmark	170,000	1	135,000	3		
	The Netherlands	170,000	1				
	Czech Republic	340,000	2	135,000	3		
	Russia					495,000	11
	Poland					135,000	3
	Rest of European Union	850,000	5				
	Total	7,650,000	45	2,565,000	57	1,080,000	24
Asia/Pacific	China	3,060,000	18				
	Australia	680,000	4	405,000	9		
	Japan					540,000	12
	Thailand					135,000	3
	Vietnam					135,000	3
	Total	3,740,000	22	405,000	9	810,000	18
America	United States	2,380,000	14			180,000	4
	Canada	1,020,000	6	405,000	9		
	Argentina			225,000	5		
	Uruguay			135,000	3		
	Brazil					540,000	12
	Venezuela					180,000	4
	Mexico					90,000	2
	Total	3,400,000	20	765,000	17	990,000	22
World	Rest of the world	2,210,000	13	765,000	17	1,620,000	36
Total		17,000,000	100	4,500,000	100	4,500,000	100

southern hemisphere since they require a relatively lengthy period of daylight during the growing season. In South Africa, Zimbabwe and certain other areas closer to the Equator than 35°, lights are used to extend the daylight period. Hops also require a cold dormant period each year. See Table 20.6.

One interesting note: in Germany the average hop farm is 16.5 acres, in the entire European Union the average farm size is only 12.9 acres, whereas in the US the average amounts to over 400 acres per grower.

20.2.1.2.2 **Use of Hop and Hop Products** Over the years brewers and farmers have found ways to manipulate the hop through selection and cross-breeding, pelletization, extraction, and isomerization to impart just the desired character to the beer. Due to high utilization rates as well as to technological advantages like

Table 20.5 Top malting companies 2006 [6](S. Menu, personal information).

Rank	Name	Foundation	Country of origin	Production capacity (tons)	Market share (%)
1	Groupe Soufflet	1952	France	1,500,000	8.2
2	Groupe Malteurop	1984	France	1,330,000	7.3
3	Cargill Inc.	1976	United States	1,290,000	7.1
4	United Malt Holdings	2006	United States/Canada	1,125,000	6.2
5	IMC	1996	United States	925,000	5.1
6	Boortmalt Group	1924	Belgium/France	550,000	3.0
7	Russky Solod	2003	Russia	540,000	3.0
8	Rahr Malting	1847	United States	510,000	2.8
9	Greencore Group	1991	United Kingdom	500,000	2.7
10	Ausmalt/Joe White	2002	Australia	500,000	2.7
11	Guangdong Enterprises	1988	China	440,000	2.4
12	Global Malt	2000	Germany	350,000	1.9
13	Viking Malt	1883	Sweden	340,000	1.9
14	Cofco-Dalian	1996	China	300,000	1.6
15	J. P. Simpson	1866	United Kingdom	270,000	1.5
16	Bavaria Holland Malt	1719	Netherlands	245,000	1.3
17	Ireks	1856	Germany	220,000	1.2
18	Crisp Malting Group	1890	United Kingdom	210,000	1.2
19	Durst Malz	1824	Germany	200,000	1.1
20	Muntons	1921	United Kingdom	195,000	1.1
Total				11,540,000	63.4

light stability and foam enhancement, the market shares for isomerized hop products and downstream products like the tetrahydro-iso extract are increasing constantly (Table 20.7). When looking at the key players in the global hop market it is noteworthy that 85% of the total volume is controlled by only four companies, which are moreover privately owned (Table 20.8).

20.2.2
Beer Consumption

The world population is still growing steadily and the annual per capita consumption of beer has risen slowly to the current level of 26.4 l. It is evident that Europe and America have an average consumption of around 60 l, whereas the value for the rest of the world ranges between 10 and 25 l [8]. See Table 20.9.

20.2.3
Beer Styles

When dividing beers into different categories it turns out that lager is the most favored beer in the world. About 94% of the beers marketed are lagers. Lagers are

Table 20.6 Main hop-growing areas and their production in 2005 [1, 7].

Country	Region	Acreage (ha)	Production (10^6 tons)	Production (10^6 tons α)	Average (% α)
Germany	Hallertau	14,221	29,640.6	2,599	8.8
	Elbe-Saale	1,332	2,486.2	305	12.3
	Tettnang	1,193	1,702.8	71	4.2
	others	415	637.2	30	4.7
	total	17,161	34,466.8	3,005	8.7
Czech Republic	Saaz	4,225	5,462.2	211	3.9
	others	1,445	2,369.0	91	3.8
	total	5,670	7,831.2	302	3.9
Poland		2,289	3,413.7	237	6.9
Slovenia		1,511	2,539.0	189	7.4
England		1,071	1,593.3	125	7.8
Ukraine		1,464	1,473.0	66	4.5
Rest of Europe		3,537	5,209.0	369	7.1
Europe		32,703	56,526.0	4,293	7.6
United States	Washington	8,537	17,903.1	2,022	11.3
	Oregon	2,089	3,653.2	287	7.9
	Idaho	1,330	2,445.3	224	9.2
	total	11,956	24,001.6	2,533	10.6
Argentina		184	257.4	17	6.6
The Americas		12,140	24,259.0	2,550	10.5
China	Xinjiang	1,830	5,500.0	387	7.0
	Gansu	1,656	4,272.5	270	6.3
	total	3,486	9,772.5	657	6.7
Rest of Asia		317	538.1	35	6.5
Asia		3,803	10,310.6	692	6.7
Africa		506	937.0	121	12.9
Australia/Oceania		852	2,082.5	221	10.6
World		50,004	94,115.1	7,877	8.4

bottom-fermented pale beers with an alcohol content between 4.5 and 5.5% by volume and a low bitterness between 15 and 25 European Brewery Convention units.

Stouts and ales are described in detail in Chapter 10 – ales amount to 2% and stout to about 1% of the beers worldwide. Specialties (1.5%) include mainly wheat/white beers, fruit beers, higher strength beers, seasonal beers and also flavored beers. Non-alcohol beers contain less than 0.5% alcohol by volume and low alcohol beers contain up to 2.8% alcohol by volume. They are included in one group

Table 20.7 Hops and hop products (H. Schwarz, 2007, personal information).

Type	Utilization rate (%)	Market share (%)
Hop cones	25–30	5
Hop pellets	30–35	45
Pre-isomerized hop pellets	50–55	10
Hop extracts	30–35	27
Pre-isomerized hop extract	50–55	3
Downstream products	70–90	10

Table 20.8 Top hop companies and their market share (%) by turnover.

Barth-Haas Group	35
Hopsteiner Group	30
Yakima Chief	10
HVG	10
Others	15

together with malt beverages, for example Malta. Their market share is about 2%. See Table 20.10.

20.2.4
Packaging

Beer packaging can be produced from metal, glass or plastic. Kegs, cans, glass bottles or plastic bottles [polyethylene terephthalate (PET)/polyethylene naphthalate PEN] are used to package beer (the latter just in recent years). Draught beer takes 12% of the beer volume, bottles 64%, cans 20% and PET bottles about 4%. See Table 20.11.

20.3
Beer Markets and Their Key Players in 2004

The global beer production increased by 24.3% between 1995 and 2004. When looking at these figures more closely, it is apparent out that in countries where beer played a minor role in the past, it is now becoming getting more and more popular. In contrast to these changes, the traditional beer markets are facing stagnation or decline [9, 10].

Table 20.9 Beer production, exports, imports, consumption and per capita consumption in 2004 of selected countries (ranking in the order of decreasing overall consumption).

	Production (10^6 hl/year)	Exports (10^6 hl/year)	Imports (10^6 hl/year)	Consumption (10^6 hl/year)	Per capita consumption (l/year)
China	289.9 [1]	1.4	2.8	288.8 [1]	22.1
United States	214.4 [2]	2.3	27.8 [1]	241.3 [2]	84.3
Germany	108.8	13.8	2.9	97.8	118.5
Brazil	89.6	0.3	0	89.3	50.3
Russia	81.4	1.5	2.3	82.3	56.3
Japan	65.3	0.2	0.5	65.6	51.6
United Kingdom	56.2	3.3	7.0 [2]	59.9	100.1 [4]
Mexico	68.5	14.5 [1]	1	55	52.6
Spain	31.3	0.7	3.6	34.2	82.8
Poland	29.5	0.4	0.2	29.2	75.3
France	16.8	2	5.4	20.2	33.8
Czech Republic	18.9	2.8	0.2	16.3	158.2 [1]
The Netherlands	24.8	13.9 [2]	2	12.8	78.5
Austria	8.7	0.6	0.6	8.8	108.6 [3]
Ireland	7.7	3.3	0.8	5.2	134.2 [2]
Europe	533	61.4	32.8	512.8	58.3
Americas	480.8	22.1	41	493	56.9
Asia/Pacific	460.1	5.2	6.2	461	12.9
Africa	72	1.4	5.4	76	7.1
World total		90	85.4	1,543.0	24.1

Table 20.10 Global beer styles 2004.

Style	Volume (THL)	Growth rate 1999–2004 (%)	Market share (%)
Specialties	22,497	2.90	1.46
Stout	14,181	2.10	0.92
Ale	34,389	−3.20	2.23
Premium lager	138,929	3.80	9.00
Mainstream lager	1,303,042	2.80	84.46
Non/low alcohol beer (including Malta)	29,791	3.20	1.93
Total market	1,542,829	2.70	
All premium	191,735	3.30	12.43

Table 20.11 Global beer packaging 2004.

Type of packaging	Volume (THL)	Growth rate 1999–2004 (%)	Market share (%)
Draught	184,410	0.00	11.95
Bottles	992,240	2.70	64.31
Cans	314,205	1.80	20.37
PET	51,974	46.00	3.37
Total market	1,542,829	2.70	

The global beer market expanded in 2005 by about 2.4% to 1.55 billion hectoliters. The main growth areas in the world in 2005 were Asia (+6.2%) and Central/South America (+2.9%). A decline was seen in North America (−1.2%), due to the growing popularity of spirits and wine, especially in the on-trade. In Western Europe (−0.5%) the ageing population is drinking less beer and looking for alternatives like wine and mineral water. Australia and New Zealand had a decline of 0.4%.

20.3.1
Beer Markets

The following sections give the key facts about the most important beer markets per region.

20.3.1.1 Europe

With a total beer production of 528.9 million hectoliters, Europe is still the largest beer-producing region. Between 1995 and 2004, the growth rate was 22.8%. It should be noted that this growth is driven by countries from Central and Eastern

Europe like Poland and Russia. Furthermore, it is noticeable that former wine-orientated countries like Spain and Italy are discovering beer, whereas the production in traditional beer countries is stagnating or declining.

Western European beer markets 2004

Country	Population (10^6)	Production volume (10^6 hl)	Change 1995–2004 (%)	Per capita consumption (l)	Market share 1 + 2 (%)	Number of breweries	Global rank
Germany	82.4	106	−9.2	117	27	1279	3
United Kingdom	59.2	59	+0.2	99	45	60	8
Spain	41.1	31	+21.3	88	63	21	9

Central and Eastern European beer markets 2004

Country	Population (10^6)	Production volume (10^6 hl)	Change 1995–2004 (%)	Per capita consumption (l)	Market share 1 + 2 (%)	Number of breweries	Global rank
Russia	143.2	85	+378	58	47	300	5
Poland	38.5	28	+82.2	77	65	65	10

20.3.1.2 America

Having a production volume of 492 million hectoliters, America is the second largest beer-producing region. Between 1995 and 2004 the total market increased by 7%.

North American beer markets 2004

Country	Population (10^6)	Production volume (10^6 hl)	Change 1995–2004 (%)	Per capita consumption (l)	Market share 1 + 2 (%)	Number of breweries	Global rank
United States	294	232.4	−0.6	79	68	1800	2
Canada	31.5	23.1	+1.3	89	85	75	13
Mexico	103.4	68.5	+53.9	52.5	100	22	6

Central and South American beer markets 2004

Country	Population (10^6)	Production volume (10^6 hl)	Change 1995–2004 (%)	Per capita consumption (l)	Market share 1 + 2 (%)	Number of breweries	Global rank
Brazil	178.4	85.6	+1.9	50	85	43	4
Venezuela	25.6	20.8	+30.8	77	93	12	16
Columbia	44.2	16.0	−10.1	34.4	99		20

20.3.1.3 Asia

With an increase of 158.5 million hectoliters to a total production volume of 439.7 million hectoliters, Asia is the fastest-growing region. The growth rate amounts to 56.4%.

Asian beer markets 2004

Country	Population (10^6)	Production volume (10^6 hl)	Change 1995–2004 (%)	Per capita consumption (l)	Market share 1 + 2 (%)	Number of breweries	Global rank
China	1304	291.0	+88.2	17	23	500	1
Japan	127.5	65.5	−2.5	56	74	33	7

20.3.1.4 Africa

Until now South Africa and Nigeria have dominated the beer market in this region. The total production volume in 2004 amounted to 70.7 million hectoliters, corresponding to a growth rate of 29.5%.

African beer markets 2004

Country	Population (10^6)	Production volume (10^6 hl)	Change 1995–2004 (%)	Per capita consumption (l)	Market share 1 + 2 (%)	Number of breweries	Global rank
South Africa	45	25.0	+2.0	55	99		11
Nigeria	124	9.4	+108.9	7.6	94		23

20.3.1.5 Australia and Pacific Beer Markets

Between 1995 and 2004, the beer production in this region declined by 6.3% to a total of 20.9 million hectoliters.

Australian beer market 2004

Country	Population (10⁶)	Production volume (10⁶ hl)	Change 1995–2004 (%)	Per capita consumption (l)	Market share 1 + 2 (%)	Number of breweries	Global rank
Australia	19.3	17.2	−5.6	90	96	114	18

20.3.1.6 Profitability

After having seen the structure of the major beer markets is also very interesting to look at their profitability. Table 20.12 shows the most lucrative markets ranked by their earnings before interest, tax and amortization (EBITA). The market in the United States is by far the most profitable in terms of sales as well as EBITA; 24% of the sales and 27% of the global EBITA are earned in this region. Altogether, these 10 countries are responsible for 52% of the global volume and at the same time for 69% of the revenues and 69% of the global profits.

Table 20.12 Most profitable beer markets [11].

Country	Volume (10⁶ hl)	Sales (10⁶ US$)	Sales (US$/hl)	EBITA (10⁶ US$)	EBITA (US$/hl)	EBITA (margin %)
United States	241	25,347	105	3,194	13	12.6
United Kingdom	59	9,757	165	937	16	9.6
Brazil	81	2,583	32	594	7	23.0
Germany	98	9,218	94	593	6	6.4
Canada	21	3,808	178	586	27	15.4
Japan	66	8,750	132	542	8	6.2
Mexico	52	3,998	76	524	10	13.1
Australia	17	3,144	183	503	29	16.0
Spain	32	3,796	120	444	14	11.7
Russia	75	2,980	40	432	6	14.5
Top 10	742	73,381	99	8,349	11	11.4
Share	51.4%	68.5%		68.9%		
World	1,444	107,158	average 74	12,124	average 8	average 11.3

20.3.2
World's Top Brewers

The consolidation in the global beer industry only began in 1990 and has not yet finished. Apart from a few brewers that left their home markets very early, the industry was affected by national and not by international groups. Global players are now gaining more and more weight after the recent mergers

Table 20.13 World top 20 brewers in 2004.

Rank	Company	Country	Annual output (10^6 hl)	Global market share (%)	Cumulative market share (%)
1	InBev	Belgium/Brazil	183.7	11.9	11.9
2	SAB Miller	South Africa/United States	178.3	11.6	23.5
3	Anheuser-Busch	United States	144.2	9.3	32.8
4	Heineken	The Netherlands	117.5	7.6	40.4
5	Carlsberg	Denmark	67.6	4.4	44.8
6	Coors and Molson	United States/Canada	57.6	3.7	48.5
7	Scottish & Newcastle	United Kingdom	51.4	3.3	51.9
8	Modelo	Mexico	42.8	2.8	54.6
9	Tsingtao (group)	China	37.1	2.4	57.1
10	Kirin	Japan	36	2.3	59.4
11	Asahi	Japan	31.8	2.1	61.4
12	Beiing Yanjing	China	28.9	1.9	63.3
13	FEMSA	Mexico	25.7	1.7	65.0
14	San Miguel Corp.	Philippines	20.7	1.3	66.3
15	Diageo (Guinness)	United Kingdom/Ireland	20	1.3	67.6
16	Fosters	Australia	18.5	1.2	68.8
17	Radeberger	Germany	17	1.1	69.9
18	Polar	Venezuela	14.4	0.9	70.9
19	Anadolu Group (Efes)	Turkey	13.7	0.9	71.7
20	Schincariol	Brazil	13.4	0.9	72.6

between Interbrew and AmBev, SAB and Miller, and Coors and Molson. See Table 20.13.

A better understanding for the globalization of the beer industry is evident when the volumes are broken down into international and domestic sales. Third ranked Anheuser-Busch – whose growth has always been internal – sells more than 86% in its home market. In comparison, Carlsberg and Heineken sell approximately 5% and SAB Miller around 18% in their respective domestic markets. Table 20.14 shows the leading brewers by regions.

20.3.3
Branding in the Global Brewing Industry

When comparing the market share of single brands with those of their owners, a 'global brand' is hardly evident. Budweiser, the world's top brand, is only responsible for 3% of global beer sales. The cumulative market share of the top 20 single brands only amounts to 25% (Table 20.15).

In the portfolio of global market leader InBev, more than 200 brands are competing for limited research and development, marketing and distribution budgets.

Table 20.14 World top brewers by region 2004.

Rank	Europe		America		Asia/Pacific		Africa/Middle East	
	company	output (10^6 hl)	company	output (10^6 hl)	company	output (10^6 hl)	company	output (10^6 hl)
1	Heineken	80.0	Anheuser-Busch	124.9	SAB Miller	37.3	SAB Miller	33.1
2	InBev	69.1	InBev	91.2	Tsingtao (Group)	36.9	Heineken	13.5
3	Carlsberg	58.2	SAB Miller	73.1	Kirin	34.6	Castel/BGI	12.2
4	Scottish & Newcastle	50.0	Coors and Molson	45.1	Asahi	31.6	Diageo (Guinness)	7.7
5	SAB Miller	34.8	Modelo	41.8	Beijing Yanjing	28.9		
6	Radeberger	16.8	FEMSA	25.6	InBev	22.9		
7	Anadolu Group (Efes)	13.6	Heineken	14.5	San Miguel Corps	20.7		
8	Coors and Molson	12.4	Polar	14.4	Anheuser-Busch	16.2		
9	Mahou	10.6	Schincariol	13.4	Henan Gold Star	12.3		
10	Diageo (Guinness)	8.4	Pabst/S & P	8.8	Suntory	11.4		
Total		353.9		452.8		252.8		

Table 20.15 World top 20 beer brands (2004).

Rank	Brand name	Brand-Owner	Output (10^6 hl)	Market share (%)	Cumulative market share (%)
1	Bud (range)	Anheuser-Busch	48.20	3.12	3.12
2	Budweiser	Anheuser-Busch	47.40	3.07	6.20
3	Skol (Brazil volume only)	InBev	31.20	2.02	8.22
4	Corona	Modelo	28.20	1.83	10.05
5	Heineken	Heineken	22.80	1.48	11.52
6	Coors Light	Coors and Molson	21.20	1.37	12.90
7	Miller LITE	SAB Miller	20.70	1.34	14.24
8	Brahma Chopp	InBev	19.70	1.28	15.52
9	Super Dry	Asahi	19.60	1.27	16.79
10	Busch (range)	Anheuser-Busch	15.80	1.02	17.81
11	Beijing Beer	Beijing Yanjing	12.70	0.82	18.63
12	Carlsberg	Carlsberg	11.80	0.76	19.40
13	Tsingtao	Tsingtao	11.70	0.76	20.16
14	Schincariol/Nova Schin	Schincariol	11.40	0.74	20.90
15	Carling Black Label (Africa)	SAB Miller	11.30	0.73	21.63
16	Michelob (range)	Anheuser-Busch	11.30	0.73	22.36
17	San Miguel Pale Pilsen	San Miguel Corp.	11.10	0.72	23.08
18	Amstel	Heineken	11.10	0.72	23.80
19	Polar (range)	Polar	11.10	0.72	24.52
20	Guinness Stout	Guinness/Diageo	10.50	0.68	25.20

References

1 Barth-Haas Group (2006/2007) *The Barth Report – Hops*, 7. Barth-Haas Group, Nuremburg.
2 Schildbach, R. (1999) Braugerste weltweit, *Proceedings 27th European Brewery Convention Congress*, Cannes, pp. 299–312
3 Schildbach, R. (2002) Braugerstensorten in Europa – Verbreitung und Qualität. *Brauwelt*, **142**, 15–18 and 27–32.
4 Schildbach, R. (2006) Aktuelle Braugerstensorten in Europa 2005/2006. *Brauwelt*, **146**, 221–5.
5 Rabobank (2004) *World Beer and Malt Map*, Rabobank, Utrecht.
6 www.rahr.com.
7 Hopsteiner Guidelines for Hop Buying in www.hopsteiner.com.
8 Plato Logic Ltd (2005) *World Beer Report*, October edn, Plato Logic Ltd, Reading.
9 (a) Hansmaennel, G. (2002) Beer world monopoly, *Brauindustrie*, **87** (6), 26–8. (b) Hansmaennel, G. (2002) Beer world monopoly, *Brauindustrie*, **87** (7), 26–9.
10 Rodwan, J.G. (2003) Recent developments in the global beer market. *Brauwelt International*, **••**, 405–7.
11 Ebneth, O.J. and Theuvsen, L. (2005) Globalization of the Brewing Industry: 5th International PENSA Conference, Ribeirao Preto.

21
Physiology and Toxicology
Manfred Walzl

Until recently, medical research has frowned upon alcohol in any form. However, alcohol-based tinctures and drops have always been traditional resources in popular medicine and were important assets of 'grandmother's medicine chest'. Moreover, our ancestors have always believed in the positive health effects of their daily ration of wine or beer and this conviction has found its way into numerous proverbs. There was only one catch to it – these supposed virtues of beer had not been backed up by actual scientific findings.

Now something has changed because in the last few years hardly any food has experienced such a reputation boost like beer. This natural product, known for several millennia, has become a highly interesting focus of medical research and the first results are becoming available. Even though research is still in its initial phase (up to now about 3500 scientific studies on the issue of beer have been published), quite remarkable findings have already been brought forward [1].

21.1
Astounding Health Benefits of Beer

It has, for instance, been found that beer dilates the coronary arteries and improves the blood levels of 'good cholesterol', inhibits fatty deposits in the internal artery walls (atherosclerosis), and thereby helps to ward off myocardial infarction [2–13]. For urologists beer is an ideal prophylactic against the formation of kidney stones [1, 14]. Multiple studies have shown that beer has cancer-preventing properties [15–21].

It should be emphasized that all of those positive effects are independent of the beer's alcohol content. This is shown by the observation that beer with a low alcohol content and even non-alcoholic beer also show health benefits [1].

However, could it be that a moderate consumption of alcohol has positive effects on our health? The answer to that question has been emerging slowly but consistently and is 'yes'. Data compiled by researchers in, amongst others, Germany, the United States and Japan prove without doubt that drinking moderate amounts of

Handbook of Brewing: Processes, Technology, Markets. Edited by H. M. Eßlinger
Copyright © 2009 WILEY-VCH Verlag GmbH & Co. KGaA, Weinheim
ISBN: 978-3-527-31674-8

beer reduces your risk of heart attack and stroke, and increases your life expectancy [1–6, 8–10, 12, 13, 22–25] as it has, for example, been confirmed by a German study in which more than 1000 men have been closely monitored over a period of several years [1]. The aim of this extensive study was to confirm that moderate alcohol consumption can have a positive effect on the development of atherosclerosis and its associated health risks (e.g. heart attack or stroke).

The results of the study leave no doubt: moderate beer drinking reduces the risk of a heart attack. These data have been confirmed by a study undertaken by a research team at the University of New South Wales in Australia that obtained nearly identical results [1]. Compared to subjects who drank no alcohol, the rate of heart attacks in the group of beer drinkers was lower by about 50%. What is more, beer delivered its heart-protective effect irrespective of age, blood pressure, fitness level and body weight of the study subjects.

However, it seems that it is not only the heart that profits from beer drinking. When one considers all death causes together, one finds something really amazing: beer drinkers live longer. With an alcohol consumption of 20–40 g/day in men (and of less than 20 g in women), the overall mortality was lowest [1, 10, 24].

Hence, recent years have seen a rush of new research being published on beer and health – most referring to the elderly, dementia, cardiovascular disease and all-cause mortality. This is of greatest interest, since we are facing an enormous health economic problem. With the aging of the population, the prevalence of dementia is expected to increase significantly. Unfortunately, there is no established treatment or preventive measure for dementia, but dealing with its onset could significantly decrease its prevalence, with important health implications.

Hence, this news is guaranteed to raise a cheer among those who enjoy a glass or two: drinking 1 l of beer or half a bottle of wine can make the brain work better, especially in women. Research published by several groups has found that even those who drink only small amounts of beer and wine have significantly sharper thought processes than teetotalers [1, 13, 26, 27].

Many researchers believe that the cardiovascular benefits associated with moderate drinking may translate into cognitive benefits, since cognitive impairment and cardiovascular disease share many common risk factors. Indeed, moderate alcohol intake improves lipoprotein metabolism, for instance, thereby reducing the cardiovascular mortality risk [1, 28, 29].

Thus, a study evaluated the association between alcohol intake and cognitive function in older woman, and found that drinking up to approximately one to two drinks per day was associated with better cognitive function and less cognitive decline over 2 years compared with non-drinkers. This study involved 12 480 women from the Nurses Health Study in the United States, which included over 120 000 registered nurses who were aged 30–55 when the study began in 1976. The scientists used reports of alcohol intake, which were part of biannual questionnaires, to classify the women as non-drinkers, those who drank 1.0–14.9 g of alcohol (up to one drink per day) and those who drank 15.0–30.0 g (up to two drinks a day). Women with unstable drinking patterns and those who drank more than

30 g of alcohol per day were excluded. As a good practice in science, factors that might influence cognitive function and alcohol intake, including education, cardiovascular risk factors, physical activity, body mass index, smoking, mental health scores, energy fatigue scores, social integration, and the use of certain medications and supplements, had been adjusted by the scientists.

The findings showed that 51% of the participants were non-drinkers, 44% were moderate drinkers with up to one drink per day and 5% had higher levels of drinking. The researchers found that the moderate drinkers had better cognitive scores than non-drinkers. The relative risk of impairment was reduced by 23% in the test of general cognition and by 19% in the global cognitive score. The results for cognitive decline were similar. For example, on the test of general cognition, the relative risk of a substantial decline in performance over a 2-year period showed a 15% better improvement, as compared with non-drinkers. In contrast, the scores of women with higher levels of drinking were not significantly different from those of the non-drinkers. The authors therefore concluded: 'The data suggest that up to one drink per day does not impair cognitive function and may actually decrease the risk of cognitive decline' [1].

Overall, beer seems to have an interesting impact on different neurological structures. A relatively new study investigated even positive influences of a regular but moderate beer drinking pattern on the development of Parkinson's disease [30].

All this might be a better insight into the dosage, but which alcoholic beverage should be used? While the inverse association between red wine consumption and cardiovascular risk is globally recognized as the 'French paradox', many epidemiological studies have now concluded that beer and red wine are equally beneficial [1]. This was also confirmed by the study cited above.

However, papers dealing with moderate alcohol consumption and the risk of dementia are increasing. The results obtained are similar in many studies performed all over the world – from the United States to Japan, from Germany to Australia. They all show a decrease in the risk of cognitive dysfunction and dementia for moderate drinking. This was also seen in the Rotterdam Study, where the research group examined the relation between alcohol and dementia in 7983 individuals aged 55 years and older taking part in a prospective population-based trial. They studied all participants who did not have dementia at baseline between 1990 and 1993, and who had complete data on alcohol consumption [1, 5].

The results of the trial now do not seem very surprising. During an average observation period of 6 years, 197 individuals developed dementia. The median alcohol consumption was 0.29 drinks per day. Light-to-moderate drinking (one to three drinks a day) was significantly associated with a lower risk of any dementia, including Alzheimer's disease, which decreased by 58%, and vascular dementia, which decreased by 29%. Again, the effect was unchanged by the source of alcohol. The effect of light-to-moderate drinking seemed more prominent among men in this trial.

Dementia mainly depends on two developments: either Alzheimer's disease or atherosclerosis, which minimizes the blood flow to (and within) the brain.

One very important risk factor for atherosclerosis is the formation of free radicals (i.e. atoms that have lost an electron). These atoms intend to grab the missing electron from another atom, leading to a chain reaction, and thereby producing free radicals, which are one of the reasons for the development of either atherosclerosis, dementia or even cancer [1, 26].

Given the fact that the cited diseases are associated with oxidation processes, substances in beer and wine – known as anti-oxidants – are assumed by numerous studies to have a positive and preventive effect. This is not just a hypothesis as there is indeed evidence for it, confirmed by the following example. Nine measures of cognitive function in drinkers compared to abstainers were analyzed in 883 men and women aged over 65 years. Results showed that participants who consumed alcohol had significantly better mean scores on seven of the nine cognitive function tests and less frequently had scores below selected 'cut points' compared to those who abstained from all alcohol intake [1].

This leads us to a fact we have known over the centuries – beer increases intelligence. In Japan it was noted that within a group of 2000 persons, those who drank moderate amounts of beer showed an increase in the intelligence scale: men by average 3.3 points, women by 2.5 points [1].

Moderate alcohol consumption seems to have not only positive effects on psychological and psychiatric structures, but also on physiological conditions. The National Institute on Alcohol Abuse and Alcoholism has released an important position paper in the United States on drinking, which will serve as the National Institute of Health's formal position on the health risks and potential benefits of moderate alcohol consumption. The paper was commissioned in support of the 2005 update of the United States Department of Agriculture and the Department of Health and Human Services Dietary Guidelines. The report makes several important statements. Namely, that there is no raised risk of ischemic stroke from moderate drinking, and drinking provides protective effects in terms of vascular disease among older individuals and those otherwise at risk for stroke or heart disease [1].

This in accordance with a trial in which scientists examined 3176 subjects with an average age of 69 years between 1993 and 2001. Over a follow-up period of around 6 years, 190 persons taking part in the study experienced a first stroke; in 172 of them the strokes were caused by obstructed blood supply. After adjusting for other risk factors, compared with those who did not drink in the past year, moderate drinkers had a reduced risk of ischemic stroke by 33% and vascular diseases (including ischemic stroke, myocardial infarction or vascular death) by 26% [1].

Although epidemiologic studies consistently link moderate alcohol consumption with a lower risk for myocardial infarction, the relationship between alcohol consumption and ischemic stroke is less clear. However, it is clear that the cardiovascular risk factors are the same as those for ischemic stroke. Thus, it is worth reviewing some of the risk factors known to be affected by moderate drinking [1–13, 22, 24, 28, 29].

(i) Blood fat deposition into the walls of arteries is a prime step in the development of atherosclerosis. Alcohol stimulates the liver to increase the production of high-density lipoproteins ('good cholesterol'), which purges arteries of low-density lipoproteins ('bad cholesterol'). Other potentially noxious factions, such as triglyceride and lipoprotein(a), may also be favorably impacted by moderate drinking. Researchers gave evidence for these considerations. One of the predictors for the development of stroke had been measured in 460 men by using ultrasound. The results showed a clear relation between alcohol consumption and carotid artery disease. Compared to abstainers, light drinkers faced a 44% lower risk. This was in contrast to heavy drinkers who showed an increase of risk by 278%.

(ii) A blood clot is a coup de grace of ischemic stroke, obstructing the flow of blood with its oxygen and nourishment to the unfortunate tissue beyond. As I have mentioned already, both alcohol and polyphenols in beer and wine beneficially modulate several facets of an overactive coagulation cascade. However, it should be emphasized that the risk of hemorrhagic stroke is increased by too much drinking.

(iii) Homocysteine, an amino acid, when overabundant due to genetic factors or lifestyle, encourages harmful blood clotting and may directly injure arterial walls. Supplemental folic acid can protect against homocysteine excess. As seen in a relatively new trial, stroke and mortality dropped rapidly in the United States and Canada after folic acid fortification of enriched grain products was fully implemented in 1998. With respect to foregoing studies, stroke mortality was already falling in United States as well as in Canada from 1990 to 1997, but in 1998 a precipitous drop began, the report indicates. In the United States, the annual decrease in mortality during the early period was 0.3%, whereas starting in 1998 the reduction was 2.9%. Similar results were seen in Canada. Sensitivity analysis suggested that these trends were probably not due to changes in known stroke risk factors, but may contribute to supplemental folic acid which can be found (besides being present in tomatoes or broccoli), primarily in beer – as the research work of the *Caroline Walker* show indicates.

(iv) Elevated blood pressure (i.e. hypertension), a stealthy, potentially disabling and lethal disorder, may be ameliorated by moderate drinking and, in keeping with the J-shaped curve, exacerbating by excessive drinking.

Each one of these factors, and there are many more, can lead to a decline in cognitive function or to a decrease in blood flow, too. Subclinical magnetic resonance imaging findings on the brain, such as white matter changes, small infarcts and enlarged ventricles or sulci, are of great prognostic importance to evaluate the potential risk of stroke and/or dementia in older adults. To determine the relationship of alcohol to magnetic resonance imaging findings, the association of alcohol consumption with different brain lesions has been studied. The researchers found – less surprisingly – a J- or U-shaped relationship between alcohol and the lesions. Compared to abstainers, individuals consuming between one and up to seven drinks weekly had a 32% less risk of white matter abnormalities.

In additional, from a relatively new point of view, alcohol in moderation can become a new challenge for fighting social problems regarding the elderly. In work mainly performed by Professor Dr Jan Snel from Amsterdam University, it was found that moderate users are happier, are less prone to commit suicide, have less health complaints, sick leave and hospitalizations, recover better from illness, are better educated, and enjoy higher incomes and a richer social life [1, 27].

We can clearly see that most people drink for enjoyment. Surveys show that 60% drink alcohol for conviviality – something that is mostly missed within groups of aged people – and 40% for relaxation. In other words, pleasure (i.e. the stimulation of the senses) is the common element that make people consume certain products or do some activities. Thus, the question is whether the pleasure of having a drink rather than the alcohol might be the cause of the effects on physical, mental and self-reported health. Unfortunately, evidence on this aspect of alcoholic drinks is scarce; there is much more understanding on the effects of pleasure from music, leisure or humor.

According Jan Snel, the findings show that humor and music boost the immune system, lower stress and make people relax. Leisure activities contribute to good mental health, mood and work satisfaction. Daily uplifts strengthen the immune system, keep people mentally in balance and are an antidote to everyday stress. However, enjoying a drink is a combination of all these elements – after all it is a pleasant activity, mostly done in company, in leisure time, and is associated with laughter and humor. One could say that a life that contains these elements is a happy lifestyle – with pleasure at its core. Pleasure strengthens good health. Whether it comes from drinking, humor, physical activity, chatting with friends, going out or shopping is irrelevant. The pleasure itself is relevant.

The definition of this health-related lifestyle is quite different from the one used by epidemiologists and that is used in everyday life. This latter definition of health is disease-based. A pleasant example? A lifestyle that is associated with stroke and cardiovascular disease consists of risk factors such as obesity, physical inactivity, smoking, fatty diet or an irregular or hectic life. To lower the risk of cardiovascular disease, changing this behavior pattern and lifestyle is inevitably recommended.

With regard to Professor Snel, why not define health from the factors that define good health and help people to strengthen these health-promoting factors by enjoying life by taking a drink, relaxing, visiting friends, having hobbies, eating out and traveling? Following this reasoning, it is logical to assume that people who drink for pleasure should differ from those who use a drink to drown their sorrows. Indeed, a large-scale study showed this difference: 50–60 year olds were followed for 13 years; 2.5% of moderate drinkers died from coronary heart disease against 3.4% of the abstainers and 5.6% of the problem drinkers. Similar differences were found for stroke, cancer, accidents, aggression and crime.

The beneficial effects of moderate alcohol use will doubtless be much stronger when studies focus on those who drink for pleasure. Hopefully, the future will show us more of these trials [27].

However, when taking all this in account an important problem still remains. Despite promising aspects and despite numerous confirmations that moderate alcohol consumption is beneficial for our health, this matter is mostly neglected from political and also psychological points of view.

21.2
Beer and Alcohol

It has been proved that elderly drinkers reach higher blood alcohol concentrations at lower levels of consumption than younger drinkers. Hence, are the doses indicated an overdose? This brings us to a point that is discussed in medicine and the public in equal shares. Since the introduction of the 0.5 per mille blood alcohol limit for drivers of motor vehicles in Austria, the impact of beer consumption on the ability to drive has been widely debated.

As a matter of fact, people are not aware of how much beer they may consume over a certain period of time without violating the law on impaired driving. The objective of one of our studies was to compare beer with an alcohol content of 5.5% with low-alcohol beer of 3%.

The trial was comprised of 216 individuals (138 men and 78 women). The body mass index and body fat index were taken into consideration. They had to consume 0.5 l of beer each within a period of 20 min after 24 h of alcohol abstinence and 6 h of food abstinence. After a 15-min, break, their breath alcohol concentration was measured by means of breathalyzers used by the Federal Police. The test consisted of a total of three consecutive cycles, which meant that 1.5 l of beer were consumed within 90 min. The same test was repeated 2 weeks later with low-alcohol beer. Six months after that, a similar trial took place – again after 24 h of alcohol abstinence, but following consumption of a standardized meal, consisting of 180 g 'Wiener Schnitzel' and 150 g of potato salad, immediately prior to beer consumption.

It was found that after consumption of 0.5 l of beer with a 5.5% alcohol content, the average breath alcohol concentration was 0.31 per mille; after 1 l of beer, the concentration was 0.50 per mille and after consumption of 1.5 l of beer, the concentration was 0.71 per mille. The corresponding values for low-alcohol beer were significantly lower at 0.16, 0.27 and 0.40, respectively. Of course, all of these differences were highly statistically significance [1]. The most interesting results could be obtained after consuming food. Following food intake, all values were reduced on average by 14–35% in the group consuming 5.5% alcohol beer, but by 42–50% in the group consuming low-alcohol reduced beer.

21.3
Beer and Cancer

A few years ago you would find in the yellow press regular warnings of the following kind: 'Beer drinking can provoke cancer'. Today's scientific studies confirm without any doubt that the exact opposite is true. Ingredients of beer are considered as a sort of hope in cancer prevention. This realization, of course, is not new. Already in the 1930s it was shown that Irish brewery workers seemed less likely to suffer from cancer than other groups of individuals and this observation has been also been confirmed in other countries (e.g. Germany and Denmark).

The latest studies have produced considerable results. Certain beer ingredients coming from its hops content, in particular xanthohumol, probably play a very major role in the inhibition of cancer growth. It has been shown in animal studies that a hops-rich diet resulted in the regression of a tumor within an extremely short timeframe [15–17, 19, 21].

International studies have proven that individuals who regularly drink moderate amounts of beer are much less likely to suffer from carcinomas of the bladder, prostate, stomach and lungs. Beer even seems to provide substantial protection against the dangerous substances ('mutagens') that are generated in the course of certain food preparation techniques like grilling and frying.

Research work on this topic is being stepped up right now and first results are extremely promising. That is one of the reasons why recently the first 'anti-cancer beer' was been approved by a United States regulatory body. Even if some skeptics still remain to be convinces, one thing is clear: in the coming years, research work on beer will continue to get maximum attention [1].

21.4
Beer Helps to Protect the Stomach and the Arteries

There are also other health benefits attributed to beer. A person who consumes beer in a moderate but regular way has a lower risk of suffering from *Helicobacter pylori* infection. This bacterium is at the origin of inflammations and other disorders of the gastrointestinal tract, and probably also contributes to the development of atherosclerosis. A study comprising 425 subjects came to the conclusion that *Helicobacter* infections more often affect persons abstaining from alcohol. A weekly consumption of 75 g alcohol (corresponding to about 2 l of beer) resulted in a decrease of the infection rate by one-third; with a higher consumption of alcohol, the rate improved even by two-thirds [1, 31].

21.5
Lower Risk of Developing Kidney Stones

Over the centuries it has already been suggested that beer may have a protective effect against kidney stones, but now scientific evidence has come in, for example, from Finland: beer considerably reduces the risk of formation of renal calculi. Over a period of 5 years, a group of scientists at the National Health Institute in Helsinki followed up no 27 000 Finns aged between 50 and 69. Although none of the subjects had been suffering from urinary calculi at the time of the inclusion in the study, some time later more than 300 of the participants were affected by this extremely painful condition. The subjects were smokers and participated in a prevention programme against pulmonary diseases.

A thorough analysis of the results of this study led to a conclusion that surely must be a pleasant one for beer drinkers: 'Each daily bottle of beer reduces the risk of developing kidney stones by 40%', was the message that emerged from this recently published study. A group of researchers in Italy made similar observations. Other researchers go even further and recommend 'beer as the most reasonably priced prophylaxis for renal calculi' [1, 14].

21.6
Ideal Sports Drink

Beer and sports could be a remarkable combination. In one survey, 92% of 360 top athletes interviewed indicated that beer was their favorite drink before and after sports activities; 63% drink beer in the evening before a competition to be able to fall asleep, 41% because of its relaxing effect. At least every second athlete regards beer as the ideal thirst quenching drink. However, above all, physically active persons prefer beer because of its refreshing qualities and its purity: beer has a high water content (but is relatively low in alcohol) and is rich in minerals (potassium, magnesium, etc.) as well as in vitamins. It is a practically isotonic beverage that replenishes the water reserves depleted by exercise and its readily available carbohydrates act as a valuable energy source [1, 10, 32].

21.7
Improved Concentration, Better Performance and Quicker Reactions

Italian sports physicians discovered that 1 l of beer per day boosts the performance, concentration and reaction of athletes, and strengthens their muscles. Heart specialists found that beer consumed after endurance sports (marathon, jogging, cross-country skiing) replenishes the depleted fluid and energy reserves in an ideal way. Already some decades ago it was found that beer increases the activity of the lungs and thus accelerates oxygen uptake [1].

In addition to water, beer contains valuable minerals and vitamins, and its readily available carbohydrates can provide us with an energy lift. That beer is free of fat and cholesterol is an additional asset.

Carbohydrates are an important part of our diet. Our muscle cells, in particular, depend on these energy providers. When one's intellectual capacity decreases, when the reactions slow down, when the concentration falters and the muscle coordination is affected – then it is time for a beer. There is less doubt: beer is a wholesome and easily digestible beverage, and most of all a fast-acting energy provider.

Beer is also a source of soluble fiber which is derived from the cell walls of barley. Two glasses of beer contain an average of 10% of the recommended daily intake of soluble fiber and some beers can provide up to 30%. Other than keeping you regular, fiber has a further benefit by slowing down the digestion and absorption of food, and reducing cholesterol levels, which may help to reduce the risk of heart disease [1].

21.8
Against Bacteria

Some research has additionally shown that people who drink beer moderately have a degree of protection from *Helicobacter pylori*, which is known to cause the majority of stomach ulcers and may be a risk factor for stomach cancer. Beer (and wine, too) is thought to facilitate eradication of the organism possibly due to an antibacterial effect [1, 31]. While beer obviously has numerous positive effects on the human body, there it is a big contrast for bacteria in general. It was found out that pathogenic germs could not survive in beer over a certain period of time. Hence, beer does not bear pathologic compounds, as also seen in the following example.

21.9
Beer Removes Metals from the Organism

It is well known that our life is influenced by complex factors, among which an important role play environmental factors (30–40%). Such risk factors are air pollution, food and water contamination with ions of heavy metals, radionuclides and other toxic compounds. Many of them can cause the formation of malignant neoplasms and other pathologies. Owing to such a tendency, scientists started to explore the processes of organism ageing and began to look for means to avoid this. Numerous studies have shown that products rich in vitamins, and biologically active substances, are able to increase activity of the secretory system, and have antitoxic characteristics. As a means of speeding up taking toxins out of the organism, scientists have paid attention to beer, because it contains a large amount of protein and amino acids, a lot of mineral substances (1–2 g/l) and vitamins of group B [1].

Organic acids and extractive substances of hops play another important role in the vital functions of the organism. It is well known that biologically active substances in beer widen microvascular vessels of the mucous membrane of the digestive apparatus and in that way speed up metabolism in the organism.

Beer has antimicrobial and anodyne characteristics. Scientists chose beer for research after studying the publications of the American researcher Lesli Klevi, who announced that laboratory animals (rats) that drank beer according to a certain scheme lived longer than usual. It is important to pay attention to such a fact that beer not only helped take heavy metals out of the organism of the animals, but also normalized the function of other organism systems [1].

Thus, accumulation of the heavy metals in the organism causes a breach of protein exchange in the organism, which in turn increases the risk of cardiovascular system diseases. When drinking beer regularly during 5–7 days protein exchange in the organism normalized and over the same time the risk to the cardiovascular system decreases. The results prove that drinking beer in order to take toxins out of the organism according to the scheme 250–500 ml every day before meal over 5–7 days (then pause for the same period) drinking favors taking out heavy metals from the organism.

Research on the influence of beer on human organism made scientists begin a complex study of beer characteristics as a prophylactic means of the organism clearance from toxic substances that influence the life [33].

21.10
Beer is 'Clean'

As already mentioned, beer is a very 'clean' food and extremely low in toxins. Apart from a very low concentration of metals (0.001–0.006 mg/l), no residues of plant-protective agents can be detected. Pathogenic germs do not have any chance to grow in beer, hence it is free of toxic elements.

At the most there are just a couple of ingredients in a very low concentration to be considered. For example, nitrate (below 25 mg/l), N-nitrosodimethylamine (below 0.5 µg/l), sulfur dioxide that can cause headaches or reddening of the skin (below 10 mg/l; a lower concentration as compared to wine) or biogenic amines, sometimes leading to migraine (8–140 mg/l; also a lower concentration as compared to wine).

On the other hand, beer is completely free of cholesterol. The concentration of fatty acids is extremely low, thus a caloric overload cannot be observed.

Additionally, there are compounds considered as contributing to health, such as amino acids (around 1000 mg/l), vitamins (around 2 mg/l), minerals (1100–2100 mg/l, but very low in sodium), CO_2, organic acids (around 600 mg/l), tanning agents (around 200–250 mg/l) or bitter acids of hops (20–30 mg/l) which improve digestion. Finally, beer is a rich source of fiber [1, 10].

21.11
Beer Makes Beautiful

There is an extra bonus: due to the concentration of vitamins and minerals, beer makes you beautiful – at least according to the ladies in ancient Egypt. The Roman historian Pliny wrote over 2000 years ago that they 'use beer foam to freshen up their complexion'. How can this effect be explained? The answer lies in the vitamins contained in the beer [1, 24].

The natural raw materials hops, malt, yeast and water supply the body all with those vitamins of the B complex that are essential for the health of our skin and hair. These are not only considered as 'beauty vitamins', but are also important for metabolism in general. In addition to vitamin B6 (pyridoxine) and considerable quantities of vitamin B2 (riboflavin), beer also comprises pantothenic acid (provitamin B5) and niacin. Pantothenic acid has been called the 'queen of the skin vitamins' since it facilitates the metabolism of skin cells and ensures their optimal nutrition. Vitamin B5 soothes and regenerates the skin after sunburn. However, niacin is equally vital for our largest organ, the skin. It protects us from the damaging effects of UV rays, and has an impact on pigment and collagen production. Vitamin B2 is required for human growth, in general, and for hair and nails growth, in particular, and promotes the healing process in the case of skin lesions.

The vitamins of the B complex can be stored in the body only in very small amounts and that is why we have to supplement them from our daily diet. Here again beer proves to be advantageous: the consumption of 1 l of beer is already sufficient to cover 20% of the daily requirement of vitamin B2, more than a third of the B6 we need daily and around 25% of the necessary daily intake of pantothenic acid. Likewise, half an adult's daily requirement of niacin is supplied in that way [1, 10].

Due to these beneficial effect, s modern beauty care has revived an almost forgotten traditional cure: the beer bath. Spas are increasingly offering the 'Beer Schaffelbath'. This is an ideal method for relaxing (mostly due to the ingredients provided by the beer's hops content) and is also good for the skin.

21.12
Beneficial Minerals

Beer is so rich in minerals that it must be considered a 'nutriceutical' – a health food with specific clinical effects. In particular, it contains the phosphoric acid molecules that our body uses as building blocks for the cell system. We need potassium (1 l of beer covers about 20% of our daily requirement) and magnesium (1 l of beer covers about 45% of our daily requirement) in order to maintain normal muscle and nerve function. Magnesium is indeed essential for muscle function and nerve conduction, and potassium helps our muscles (especially the heart muscle) to contract, as well as playing an important role in fluid excretion and blood pressure control. On the other hand, beer is very low in sodium. This is a further point in beer's favor. An excessive intake of sodium can lead to high blood pressure with all its well-known dangerous complications [1, 10].

Regarding minerals, there are new developments in the field of osteoporosis (i.e. the decrease of bone density after middle age). The facts are alarming – in Austria alone about 800 000 persons suffer from this condition that predominantly affects women (although there are 20% male patients). The incidence is rising and costs likewise. Each year about 240 000 million Euros have to be spent on osteoporosis therapy, since the disease causes 30 000 fractures annually.

Thus the need for an efficient prophylaxis becomes more and more urgent. Jonathan Powell has a simple recommendation for beer, as a moderate but regular consumption of beer has protective effects against osteoporosis. The explanation can be found in beer's significant silicon content, which acts as 'bone cement' and can thus inhibit the gradual loss of bone mass. We need about 50 mg of silicon a day. One liter of beer already provides up to 40 mg. This is considerably more than what you would find, for instance, in bananas, which have always been considered a particularly rich source of silicon. We know now for sure that a daily liter of beer brings about at least two different benefits: it prevents osteoporosis and at the same time ensures an increased silicon intake [1, 10, 34, 35].

21.13
Legend of the Beer Belly

Over the centuries it was believed that regular consumption of beer would inevitably lead to an overweight body and an unsightly beer belly. This is why the beer drinking citizens of the early 20th century were always depicted with a pot belly. This tale of the unavoidable beer belly has been transmitted from generation to generation until scientific research succeeded in unmasking it as a completely unfounded legend.

What causes a pot belly? Certainly not the beer's ingredients. On the other hand, it is known that alcohol stimulates the appetite. That is why aperitifs are served before an elegant meal and not milk shakes. It is considered that a dash of alcohol promotes salivation and the production of gastric juices, which is the body's routine for preparing itself for the work of digestion (hence the proverbial saying: 'Looking at the food made my mouth water'). This routine is by no means a bad thing. On the contrary, the meal is thus much better assimilated and digested.

Researchers were able to establish that drinking beer is not necessarily associated with a pot belly. The reverse is true: on average, beer drinkers have a lower body mass index than typical wine drinkers. However, what is most fascinating is that female beer drinkers have a lower body mass than, for instance, teetotalers. Meanwhile, scientific studies performed in Italy, the Czech Republic and England have revealed that a belly is correlated with a special gene (the so-called DD gene) that can be found in the blood pressure control center of men. This gene increases the probability of a belly up to 200% [1, 4, 24, 36].

21.14
'Beer Prescription'

Finally, below is a brief general prescription that is largely common sense and mostly heart healthful, but which cannot be taken as individual advice [1]:

- If no contraindication and you are so inclined, enjoy your beer or wine moderately: say up to 1 l of beer and 0.5 l of wine per day for men; half for women.
- Drinking regularly with meals is probably best. Prefer beverages with a higher content of polyphenols, like beer or red wine.
- Do not binge.
- Do not forget: drinking excessively risks damage to your brain, liver and heart, and even your neighbor's automobile!
- Beer should be used to enrich life, not medicate it.

Thus, the challenge of new strategies regarding health and alcohol remains. Hence, the 'Race for Quality' has no finishing line – a sentence devoted equally to brewers and scientists.

References

1 Walzl, M. and Hlatky, M. (2004) *Jungbrunnen Bier [Beer–the Fountain of Youth]*, Verlagshaus der Ärzte, 3rd ed. Vienna.
2 Brenner, H., Rothenbacher, D., Bode, G. et al. (2001) Coronary heart disease risk reduction in a predominantly beer-drinking population. *Epidemiology*, **12**, 390–5.
3 Dimmitt, S.B., Rakic, V., Puddey, I.B. et al. (1998) The effect of alcohol on coagulation and fibrinolytic factors: a controlled trial. *Blood Coagulation and Fibrinolysis*, **9**, 39–45.
4 Hoffmeister, H., Schelp, F.P., Mensink, G.B. et al. (1999) The relationship between alcohol consumption, health indicators and mortality in the German population. *International Journal of Epidemiology*, **28**, 1066–72.
5 Grabauskas, V. (2003) From classical epidemiological research to health policy formulation: contribution of Kaunas-Rotterdam Intervention Study. *Medicina*, 1184–92.
6 Kiechl, S., Willeit, J., Rungger, G. et al. (1994) Quantitative assessment of carotid atherosclerosis in a healthy population. *Neuroepidemiology*, **13**, 314–17.
7 Mennen, L.I., Balkau, B., Vol, S. et al. (1999) Fibrinogen–a possible link between alcohol consumption and cardiovascular disease? *Atherosclerosis, Thrombosis and Vascular Biology*, 887–92.
8 Mukamal, K.J., Conigrave, K.M., Mittleman, M.A. et al. (2003) Roles of drinking pattern and type of alcohol consumed in coronary heart disease in men. *New England Journal of Medicine*, **348**, 109–18.
9 Mukamal, K.J., Kronmal, R.A., Mittleman, M.A. et al. (2003) Alcohol consumption and carotid atherosclerosis in older adults. *Arteriosclerosis, Thrombosis and Vascular Biology*, **23**, 2252–9.
10 Piendl, A. (1999) *Physiologische Bedeutung der Eigenschaften des Bieres*, Fachverlag Hans Carl, Nürnberg.
11 Tavani, A., Bertuzzi, M., Negri, E. et al. (2001) Alcohol, smoking, coffee and risk of non-fatal acute myocardial infarction in Italy. *European Journal of Epidemiology*, **17**, 1131–7.
12 Walker, C. (2001) Folate in beer and the prevention of cardiovascular disease. *Proceedings of the International Health seminar, BRI.*
13 Walker, C. (2001) Evaluation of vitamins, including folate, in beer. *Proceedings of the 2nd CBMB Beer and Health Symposium, Brussels.*
14 Borghi, L., Meschi, T., Scianchi, T. et al. (1999) Urine volume: stone risk factor and preventive measure. *Nephron*, **81**, 31–7.
15 Castellsague, X., Quintana, M.J., Martinez, M.C. et al. (2004) The role of type of tobacco and type of alcoholic beverage in oral carcinogenesis. *International Journal of Cancer*, **108**, 741–9.

16 Gerhäuser, C., Alt, A., Heiss, E. et al. (2002) Cancer chemopreventive activity Xanthohumol, a natural product derived from hop. *Molecular Cancer Therapeutics*, **1**, 959–69.

17 Gerhäuser, C. and Frank, N. (2005) Xanthohumol, a new all-rounder? *Molecular Nutrition and Food Research*, **49**, 821–3.

18 Monobe, M., Koike, S., Uzawa, A. et al. (2003) Effects of beer administration in mice on acute toxicities induced by X rays and carbon ions. *Radiation Research*, **44**, 75–80.

19 Nozawa, H., Yoshida, A., Tajima, O. et al. (2004) Intake of beer inhibits azoxymethane-induced colonic carcinogenesis in male Fischer 344 rats. *International Journal of Cancer*, **108**, 404–11.

20 Shimamura, M., Hazato, T., Ashino, H. et al. (2001) Inhibition of angiogenesis by humolone, a bitter acid from beer hop. *Biochemical and Biophysical Research Communications*, **289**, 220–4

21 Woodson, K., Albanes, D., Tangrea, J.A. et al. (1999) Association between alcohol and lung cancer in the alpha-tocopherol, beta-carotene cancer prevention study in Finland. *Cancer Causes Control*, **10**, 219–26.

22 vd Gaag, M.S., Ubbink, J.B., Sillanaukee, P. et al. (2000) Effect of consumption of red wine, spirits, and beer on serum homocysteine. *Lancet*, **355**, 1522.

23 de Luis, D.A., Fernandez, N., Aller, R. et al. (2003) Relation between total homocysteine levels and beer intake in patients with diabetes mellitus Type 2. *Annals of Nutrition and Metabolism*, **47**, 119–23.

24 Schwarz, A. and Schweppe, R. (2000) *Gesund und schön mit Bier*, VGS Verlagsgesellschaft, Köln.

25 Starck, P. and Voigt, A. (2000) *Bier trinken*, **7**, Tomus-Verlag, München.

26 Seshadri, S., Beiser, A., Selhub, J. et al. (2002) Plasma homocysteine as a risk factor for dementia and Alzheimer's disease. *New England Journal of Medicine*, **346**, 476–83.

27 Snel, J. and Lorist, M.M. (1998) *Nicotine, Caffeine And Social Drinking – Behavior and Brain Function*, Harwood, New York.

28 Andersen, M.L. and Skribsted, L.H. (2001) Modification of the levels of polyphenols in wort and beer by addition of hexamethylenetetramine or sulfite during mashing. *Journal of Agricultural and Food Chemistry*, **49**, 5232–7.

29 Sillanaukee, P., vd Gaag, M.S., Sierksma, A. et al. (2003) Effect of type of alcoholic beverages on carbohydrate-deficient transferrin, sialic acid, and liver enzymes. *Alcoholism, Clinical and Experimental Research*, 57–60.

30 Hernàn, M.A., Chen, H., Schwarzschild, M.A. et al. (2003) Alcohol consumption and the incidence of Parkinson's disease. *Annals of Neurology*, **54**, 170–5.

31 Brenner, H., Rothenbacher, D., Bode, G. et al. (1999) Inverse graded relation between alcohol consumption and active infection with Helicobacter pylori. *American Journal of Epidemiology*, **149**, 571–6.

32 Watten, R.G. (1995) Sports, physical exercise and use of alcohol. *Scandinavian Journal of Medicine and Science in Sports*, **5**, 364–8.

33 Bozhkov, A.I. (2006) Beer takes out metals from the organism. http://www.suninterbrew.ua/eng/nt-2006-july-23.php.

34 Jugdaosingh, R., O'Connell, M.A., Sripanyakorn, S. and Powell, J.J. (2006) Moderate alcohol consumption and increased bone mineral density: potential ethanol and non-ethanol mechanisms [review]. *Proceedings of the Nutrition Society*, **65**, 291–310.

35 Powell, J.J., Mc Naughton, S.A., Jugdaosingh, R. et al. (2005) A provisional database for the silicon content of foods in the United Kingdom. *British Journal of Nutrition*, **94**, 804–12.

36 Dixon, J.B., Dixon, M.E. and O'Brien, PE (2002) Alcohol consumption in the severely obese: relationship with the metabolic syndrome. *Obesity Research*, **10**, 245–52.

22
Automation

Georg Bretthauer, Jens Uwe Müller, and Markus Ruchter

22.1
Introduction

Automation technology is understood to be the use of methods, strategies, processes and installations (hardware and software) as well as of aids that are capable of fulfilling defined objectives in a largely independent manner (i.e. automatically) without the constant interference of man [1]. Such objectives include:

- Improvement of product quality.
- Increase of productivity.
- Enhancement of reliability and safety.
- Improvement of the use of existing resources.
- Conservation of the environment.

As well as to:

- Make working conditions more humane.
- Increase the quality of life.

Since its beginnings in ancient times, automation technology has served man in order to better satisfy his needs. Examples are the control of the fluid level in vessels by Heron in ancient times [2], centrifugal force control of a steam engine by Watt in the eighteenth century [3], temperature control of a refrigerator, electronic stability programs in modern private vehicles as well as control systems in power plants or breweries [4].

This chapter provides a detailed review of the automation technology applied in breweries. Beer is the final product of breweries. It is a beverage containing alcohol, extract and CO_2, which is prepared from barley malt, raw hops or other hop products in addition to brewing water and top- or bottom-fermenting yeast. Figure 22.1 gives a schematic representation of the production process of beer.

This process consists of different subsystems (units): malt silo, malt mill, mash tun, lauter tun, wort kettle, whirlpool, wort cooler, fermenting tank, storage tank, filter and filling station. The mash tun, lauter tun, wort kettle and the whirl pool

Handbook of Brewing: Processes, Technology, Markets. Edited by H. M. Eßlinger
Copyright © 2009 WILEY-VCH Verlag GmbH & Co. KGaA, Weinheim
ISBN: 978-3-527-31674-8

532 | *22 Automation*

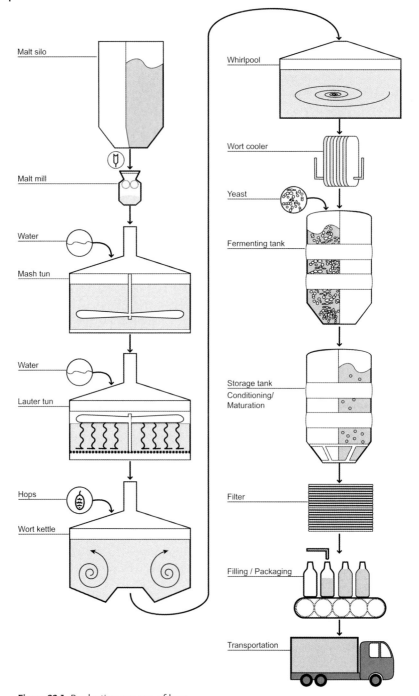

Figure 22.1 Production process of beer.

are commonly located in the brewhouse, whereas the wort cooler, fermenting tank and storage tank are situated in the fermentation and storage cellar. As displayed in Figure 22.1, a brewery represents a technical system consisting of several subsystems. Apart from the production process, supply and disposal facilities are needed as well, constituting further subsystems. Hence, a brewery is a very complex system, the primary objective of which is the production of beer of continuously high quality. In addition, production and energy costs as well as pollution of the environment should be minimized. In order to guarantee that these objectives will be met, a powerful automation technology is required that consists of suitable hardware and software components. The most important components will be discussed below.

In order to achieve the above objectives it is necessary to measure the essential parameters of the beer production process at definite times. These parameters encompass the levels of the containers, in/outflow of the units, pressure in tanks, temperature at particular points of interest in the units, pH measurement and conductivity for analyzing the flow. Different measurement equipment is required for the assessment of these parameters. The measurement technology will be addressed in Section 22.2. Following the measurement, the attained quantities have to be compared with the desired values for these parameters. If a difference is detected, open- and closed-loop controllers are necessary to reduce or completely compensate for this deviation. Here, classic controllers like PID controllers or advanced controllers such as knowledge-based controllers or fuzzy controllers are used, and will be further described in Section 22.3. Due to the high complexity of the brewery process (see Figure 22.1), many open- and closed-loop controllers are necessary and have to be coordinated. A process control system (PCS) can serve this purpose and should guarantee that the objectives defined above will be fulfilled. The description of such a PCS is the subject of Section 22.4. A lot of information is available in a brewery for controlling, monitoring, diagnosis and asset management. As a consequence, information and communication systems are installed that are discussed further in Section 22.5. This chapter will concludes with a discussion of future trends of automation technology in the brewing industry.

22.2
Measurement Technology

In the brewery, measurement technologies of the production process and of the supply and disposal facilities can be distinguished. Throughout the production process the measurement technology should assure the hygienically perfect production of beer in a wet environment. The crucial parameters to be measured are discussed in the following sections. Of these parameters, which are summarized in Table 22.1, the measurement of oxygen and turbidity are of particular importance to assure product quality. Therefore, both parameters are subsequently addressed in more detail.

Table 22.1 Parameters of interest.

	Level	Temperature	pH	Pressure	Flow	Conductivity	Oxygen	Limit monitor	Dosing	Turbidity
Malt silo	×	×						×		
Mash tun	×	×	×		×			×		
Lauter tun		×		×	×			×	×	×
Wort kettle	×	×	×	×	×			×		
Whirlpool		×						×		
Wort cooler	×	×		×	×					
Fermentation	×	×		×	×			×	×	
Filtration	×	×		×	×			×		×
Filling	×	×		×	×			×		
Packaging									×	

22.2.1
Level

The level is the basis for the calculation of the content (volume, mass) of a silo or other subsystems. Several measuring principles are known. Usually, radar technology (ultrasound, microwaves) is preferred for continuous level measurement. In silos that are not higher than 30 m, a guided radar with a rope antenna is used. In higher silos, radar freely emitted with a special antenna is most commonly applied. However, these methods are confronted with a challenge in the form of the pouring cone. As a consequence, the calculation of the content remains an approximation. A more precise determination of the amount of malt is achieved by using automatic scales at the input and the output of the respective subsystem. A further approach is to weigh the empty container as well as the container filled with malt. See Figure 22.2.

22.2.2
Temperature

Temperature has an influence on all of the processes (biological, chemical, physical) in the production process of beer. In order to produce high-quality beer, the process has to be maintained within defined temperature ranges. Consequently, temperature is controlled in nearly all units. PT 100 and, rarely, PT 1000 sensors are used as a standard for temperature measurements. Measurements at points critical to quality of the product may be carried out with redundant sensors. See Figure 22.3.

Figure 22.2 Capacitive level sensor Liquicap FMI 51.

Figure 22.3 Temperature sensor Omnigrad TR 45.

22.2.3
pH Value

One essential parameter is the pH value, ranging from 0 (acidic) via 7 (neutral) up to 14 (alkaline). Enzymatic and chemical or physical processes often have optimal pH ranges within which they proceed more rapidly or completely. Brewing lime or lactic acid can be added for the adjustment of such ranges. Generally, glass or ISFET pH sensors (Figure 22.4) are applied for pH measurements. As the mash with its solids tends to block the diaphragms of the pH transducers or their reference electrodes, exchangeable fittings should be used. This makes it possible to clean the sensor regularly without having to interrupt the process. Special systems are available for automatic cleaning and, if necessary, calibration.

It should be noted that some breweries principally seek to avoid the use of glass equipment in the production process to prevent safety hazards. However, since the beer is passed over at least two filters before filling, a hazard for the consumer caused by glass fragments can be excluded.

22.2.4
Pressure

In some units, (atmospheric) pressure also has an influence on the optimal conditions for producing a good beer. Different pressure sensors are available that can measure absolute pressure as well as differential pressure. According to physical principles, resistance strain gauges and inductive/capacitive measuring methods are used. There are sensors with ceramic cells and metal membrane. The former are used in abrasive media or under rough circumstances such as steam, and need more time than a metal diaphragm sensors to recover following changes in temperature. Sensors with metal diaphragms are made from stainless steel and can be flush-mounted. See Figure 22.5.

22.2.5
Flow

Volume and mass flow are two important parameters that have to be measured at the input and the output of each unit. In most cases hydrostatic systems are

Figure 22.4 ISFET pH measurement sensor Tophit CPS 471 D.

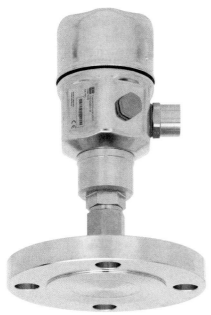

Figure 22.5 Pressure sensor Cerabar M flansch.

applied for the volume measurement, since they can be easily integrated in the container wall. However, the exact volume can only be determined by taking the actual density of the mash and its temperature into account. If permitted by rotating parts and the container cleaning system, antenna-supported systems or capacitive probes can be used. In this case the measurements are not affected by vapor, turbulent surfaces or foam.

Standard volume measurements of the water reservoir and wort are conducted in-line using magnetically inductive flow meters. They allow sufficiently precise measurements also in the case of larger line diameters at reasonable costs. The measurement is further improved by sensors compensating for, for instance, pulsed flow and detecting partly filled lines.

Coriolis-type mass flow meters are used for an exact determination of quantity and quality. Such sensors measure the flow rate much more precisely because the temperature of the product is taken into consideration (Figure 22.6).

22.2.6
Conductivity

Conductivity is an important parameter for in-line quality control and the separation of different products. Two types of measurement technologies are employed – conductive and inductive sensors. In practice, inductive measurement is mostly used. It should be noted that conductivity is affected by temperature and concentration.

Figure 22.6 Coriolis flow sensor Promass 83 F.

22.2.7
Oxygen

Oxygen plays an important role in different stages of the beer production process. Whilst during the initial phase of fermentation oxygen is essential for the propagation of the yeast culture, in most other stages of the production process oxygen pick-up is not desirable. After the fermentation too much dissolved oxygen (DO) can result in an oxidation of sensitive ingredients, producing bready and cardboard off-flavor notes that reduce the quality of the product. Further, it is important that as little DO as possible enters the beer, in particular during filling, since oxidation processes will reduce the shelf-life of the final product. Therefore, DO monitoring throughout the production process is of crucial importance to quality control and increases the shelf-life of the final product [5].

The measurement of oxygen can be performed based on physical, chemical, electrochemical or optical principles [5]. Oxygen meters commonly consist of two components – the sensor and the meter. The sensor is the part that reacts with oxygen and produces an electrical signal that is proportional to the oxygen concentration. This signal is amplified and converted to concentration units that are then displayed by the meter.

Among the oxygen meters most widely used in breweries are different types of electrochemical instruments, which encompass polarographic–amperometric sensors (Clark sensor), non-membrane potentiostatic sensors as well as potentiometric and amperometric sensors using solid electrolytes. These sensors usually operate as electrochemical cells with a positive electrode (cathode) and a negative electrode (anode), which are connected by a saturated electrolyte.

In a polarographic–amperometric sensor, the electrodes are separated by a permeable membrane. When a suitable polarization voltage is applied, oxygen passes through the membrane and is chemically reduced within the sensor, which generates an electrical current that is proportional to the partial pressure of oxygen.

In galvanic–amperometric sensors, no polarization voltage is necessary, since the galvanic voltage between the cathode (usually silver or platinum) and anode (lead, iron or zinc) is already sufficient. The sample medium itself acts as the electrolyte and the electrodes are in direct contact with the sample medium. The electrical current between the electrodes depends on the concentration of the DO in the sample. While galvanic sensors have faster response times, they have the disadvantage that the surface of the electrodes needs to be cleaned continually, which can be achieved by an abrasive block.

In potentiostatic sensors, a reference electrode is included that keeps the potential of the oxygen electrode at a constant level. This results in the advantage that the measured electrical current is directly proportional to the oxygen concentration.

Solid electrolyte sensors make use of solid-state electrical conductors such as zirconium dioxide doped with calcium oxide and yttrium oxide, which at high temperatures become conductive for oxygen ions. The two electrodes are placed in separate gas chambers and are isolated by the solid electrolyte. An electric voltage develops with different partial pressures of oxygen between the outer and inner electrodes.

However, most electrochemical sensors have the disadvantages that oxygen is consumed during measurement and the sample needs to be stirred. Optical oxygen sensors offer an alternative that can help to avoid these issues. These optical sensors consist essentially of an oxygen-sensitive membrane and an optical system. An oxygen-sensitive dye is embedded in this membrane within a supporting matrix. The optical system includes an excitation light source like a light-emitting diode or laser and a photon detector such as a photodiode or photomultiplier as well as an optical fiber serving as excitation and emission light waveguides [6]. The molecular oxygen changes the fluorescence of the dye proportional to the oxygen concentration [5].

DO measurement takes place prior to and after filtering as well as before filling.

When measuring oxygen it has to be taken into consideration that the amount of oxygen in the sample is constantly reduced due to reactions with the ingredients of the sample or consumption by microorganisms [5].

22.2.8
Turbidity

Turbidity is the cloudiness or haziness of a fluid caused by individual particles (i.e. suspended solids) that are generally invisible to the naked eye, thus being much like smoke in the air. The most common particles making up this beer haze include yeast, colloids of proteins/tannins, carbohydrates and filter aids such as diatomaceous earth. In addition to solids, fluids and gases like CO_2 can also contribute to the haze measurement [7]. Since product clarity is one of the characteristics most apparent to the consumer of the beer it is of great importance to the brewer and therefore monitored at different points of the production process.

The measurement of turbidity is commonly based on an optical principle and is an expression of the optical property that causes light to be scattered and absorbed rather than transmitted in straight lines through the sample [8]. Consequently, turbidity can be determined based on measuring scatter and/or absorption. Both options are based on the same measuring method. A light beam is emitted from a light source to pass through a sample container or cell and a photodetector then senses the light that results from the interaction with the sample [8]. For the absorption measurement, a detector is placed on the optical axes of the incident beam and measures the transmitted light. When measuring scattered light, the detectors are placed at different angles from the optical axes [7].

Photodetectors produce an electronic signal that is then converted to a turbidity value. A variety of light sources and different types of detectors are further discussed in [8]. The unit for turbidity commonly used in breweries is the European Brewery Convention (EBC) unit. Usually absorption is measured to detect high particle concentrations (e.g. during the yeast removal procedure), while scatter measurement is applied at lower concentrations (e.g. following the filtration procedure).

The distribution of scattered light is influenced by the particle concentration and size, configuration, color, and refractive index, resulting in specific patterns. Consequently, the turbidity detected by light detectors can strongly depend on their angle to the axis of incident light. Scattered light detectors are commonly based on the nephelometric measurement principle measuring 'side scatter' with the detector being placed at a 90° angle. Next to the 90° detector, 'forward-scatter' detectors are also commonly employed that are placed at a smaller angle (10°–25°) from the axis. Forward-scatter measurement is highly sensitive to larger particles like diatomaceous earth, while side-scatter measurement is less influenced by particle size and thus also suitable to detect colloidal haze [7]. Thus, forward-scatter detectors allow for an optimal control of the filtration process and help to efficiently identify filter leakages, which is essential to protect the subsequent units from contamination with yeast or filter aid, and to guarantee product clarity and stability. At the same time, the 90° detector is important for monitoring the colloidal characteristics of the product, which is of importance to beer stability and can detect leakages after sterile filtration.

Since side-scatter and forward-scatter measurements yield different types of information on the turbidity of the product, it is recommended to combine both methods. This can be achieved either by utilizing a combined multiple-angle turbidimeter or two separate side-scatter and forward-scatter sensors. Based on a thorough comparison of the two alternatives, Philipp [7] recommends the deployment of separate units that can be deployed more flexibly at the most suitable locations in the production line. For example, a forward-scatter sensor is best deployed following powder filtration based on diatomaceous earth, while a side-scatter sensor should be used subsequent to the sterile filtration.

22.2.9
Dosing

Dosing plays an important role in breweries. For instance, the addition of hops takes place fully automatically in modern breweries. Either hop pellets or, to an increasing extent, hop extracts are used. Dosage of the extremely sticky product must be very exact. For this purpose the flow measurement equipment described above is also well suited. Dosing is of further relevance during the filling procedure.

22.2.10
Limit Monitors

In order to meet high security standards for the production process, upper and lower threshold values need to be defined for the described parameters. Hence, for displaying these limiting values, indicators (also referred to as limit monitors) are employed in almost all units of the beer production process. With regard to these indicators, only a 'Yes' or 'No' answer is given. For example, in a malt silo, the maximal limiting value is used to avoid overfilling. In this case the filling process must be stopped. The minimal limiting value is used to detect the empty silo and subsequently to turn off the pumps. Examples of such indicators are vibronic systems or mechanical sensors like rotary announcers. Appropriate limit indicators are used, for instance, for flow and pressure in the other units.

22.2.11
'In-Line' Measurement

When transferring quality-relevant measurements from the laboratory to a field setting, at an actual production site, the realization of 'in-line' measurement is of paramount importance. Real-time measurement allows for an automatic response and the detection of the tendencies towards errors at an early stage, before they may have a serious effect. The fact that field measurement technology does not achieve laboratory accuracy is compensated for by the possibility of *in situ* deployment, the exclusion of manual errors and the high reproduction accuracy. 'In-line' measurement technology has to operate in field settings, which implies in particular that it has to cope with real environmental influences, such as moisture, vibrations and temperature fluctuations. Therefore, sensors developed as a result of the process automation procedure are much better suited for utilization in breweries than sensors commonly used in laboratory technology.

The efficiency of the different units of the brewery can be evaluated and the losses can be defined by installing the same technology in all sections and comparing the measured values.

It should be pointed out that all the sensors have to meet hygiene requirements and need to be designed for continually high temperatures. Thus, very high requirements have to be met especially by the elastomers of the sealings. Additionally, in the silo area, the equipment has to be designed to meet the dust explosion protection regulations. A detailed description of the different sensor concepts used in breweries is given in [5]. Special equipment needed for particular sensors in breweries is, for instance, presented in [9–11].

22.3
Control Strategies

The primary aim of each brewery is the production of high-quality beer at minimal costs, including a reduction of energy consumption and avoiding negative effects on the environment. This calls for control algorithms that can ensure this goal is achieved and at the same time help to meet the above requirements. Generally, these control algorithms can be classified into (i) classic or standard algorithms and (ii) advanced algorithms.

22.3.1
Classic Algorithms

In order to control a system there are two different types of control:

- Open-loop control.
- Closed-loop control.

An open-loop control is given if the system to be considered is influenced by technical means according to a predefined schedule. In this case, there is no disturbance on the system. Examples are elevators, washing machines or dosing. An open-loop control is illustrated in Figure 22.7.

Typically, however, realistic circumstances imply that a system is influenced by disturbances. To compensate these disturbances, a closed-loop system is used. In this case, the output of the system is compared with a set point, and the controller is used to minimize the difference between the output and set point or to bring this difference to zero. This is illustrated in Figure 22.8.

In classic control the most widely used controller is the PID controller (PID = Proportional plus Integral plus Derivative). It is described by the following equation between the control deviation $e(t)$ and the control variable $u(t)$:

$$u(t) = K_P e(t) + K_I \int e(t) dt + K_D \frac{d}{dt} e(t)$$

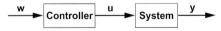

Figure 22.7 Open-loop control.

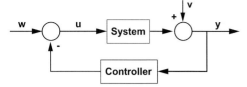

Figure 22.8 Closed-loop control.

with K_P, K_I and K_D as the parameters. The PID controller is very flexible and can be used for almost all subsystems to be controlled in breweries. It should be noted that the D part amplifies high-frequency disturbances. Thus, in some cases, only P, PI or PD controllers are used, depending on the characteristics of the system.

Apart from the PID controller, static linear feedbacks like state feedback and output feedback are also used.

22.3.2
Advanced Algorithms

In order to achieve a high quality of control, the use of advanced control algorithms is recommended. The most important ones are adaptive control algorithms, fuzzy control algorithms and knowledge-based control algorithms.

Adaptive control algorithms are used if the system to be controlled includes parameters that slowly vary over time or are non-linear. These algorithms encompass a basic closed loop and a second loop – referred to as the adapting loop – that compensates for the additional deviations. The two most important schemes are:

- Model reference adaptive control (MRAC).
- Model identification adaptive control (MIAC).

They are illustrated in Figures 22.9 and 22.10, respectively. Both the schemes have their advantages and disadvantages, which are address in more detail in [12, 13].

Fuzzy control algorithms are used if the knowledge of the operators of a complex system is given in form of fuzzy rules. A fuzzy rule is described by:

IF 'conditions' THEN 'action(s)

The following rule offers a very simple example:

IF 'maximum value of level is exceeded' THEN 'turn off the pumps'

Here, in contrast to classic control and also to adaptive control, qualitative knowledge is used. Fuzzy set theory was developed for the evaluation of this knowledge [14]. Its application to control problems is called fuzzy control. It allows the development of fuzzy controllers utilizing qualitative knowledge for

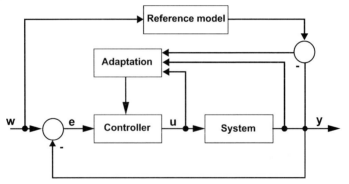

Figure 22.9 Adaptive control system with reference model (MRAC).

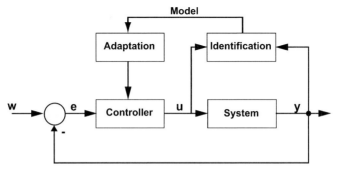

Figure 22.10 Adaptive control system with identification model (MIAC).

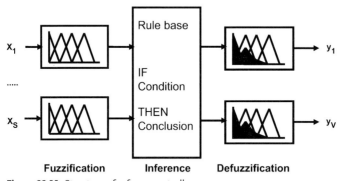

Figure 22.11 Structure of a fuzzy controller.

optimizing the system to be considered. The structure of a fuzzy controller is shown in Figure 22.11.

Accordingly, a fuzzy controller consists of three parts: (i) the fuzzification component, (ii) the inference component for the evaluation of the rule base with aggregation, activation and accumulation, and (iii) the defuzzification component. Different mathematical methods are used for each of these parts.

A fuzzy controller has two main properties: on the one hand, it is a static controller and, on the other hand, it is a non-linear controller. A detailed description of fuzzy control and some powerful applications can be found in [15, 16]. In addition to these advantages, each fuzzy controller has three disadvantages: (i) static behavior, (ii) great number of degrees of freedom, and (iii) problems with the rule base, such as completeness and consistency.

A combination of classic and advanced control algorithms is favorable in complex systems like breweries. A suitable control scheme is given in Figure 22.12. Such a structure is normally realized as a fuzzy PID control loop [15].

Another version of this controller is discussed in [17] and is referred to as a knowledge-based analytical controller (KAC). The basic idea is to eliminate the disadvantages of a fuzzy controller given above, and to take an analytical relationship between input and output variables into account. The principle structure of the KAC is shown in Figure 22.13.

The represented structure comprises basic modules for the input–output relationship that are combined with the rule base. Furthermore, a search algorithm determines the best solution within definite control ranges. Hence, analytical and qualitative knowledge is integrated. The procedure is similar to the behavior of an experienced operator. This approach is illustrated in more detail in the following section based on a case study for the control of the lauter tun as a crucial subsystem in the brewery process.

22.3.3
Advanced Control of the Lauter Tun

22.3.3.1 Description of the Lauter Tun
The lautering follows after the mashing in the brewery process. Its purpose is the separation of the extract (i.e. the sugar-enriched wort) from the mash solids (such as husks, starch grit). This wort separation procedure is frequently performed employing a lauter tun (see Figure 22.14).

Once the mash has been pumped to the lauter tun, the mash solids or 'spent grain' gradually deposit on a wedge wire surface forming the false bottom of the tun. The solids form a dense filter bed, also referred to as 'spent grain cake', on

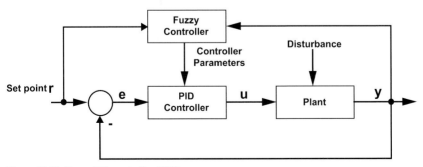

Figure 22.12 Fuzzy PID controller (PID).

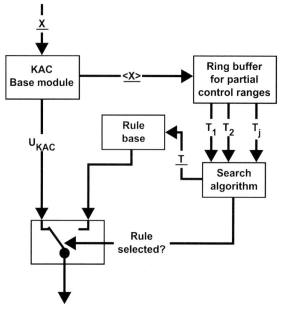

Figure 22.13 Structure of a KAC.

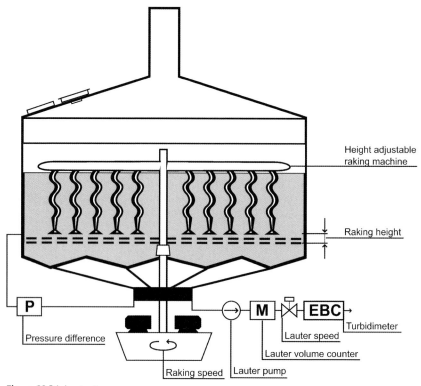

Figure 22.14 Lauter tun.

the top of this false bottom. A suction device is installed below the false bottom that extracts the wort from the filter cake and pumps it to the wort kettle.

The lautering process encompasses two stages. Initially, the lauter pump creates suction to pump off the liquid extract from the mash until only that liquid remains that has soaked into the filter cake. This procedure is followed by the sparging stage during which the filter bed is repeatedly sprinkled with sparge water for optimal extract recovery. Sparging primarily helps to extract sugars from the grain and drains the wort completely from the mash.

During this filtering process the spent grain cake is compacted. At a constant volume flow rate this results in negative pressure, since the lauter pump has to increase suction. This effect can in part be counter-balanced by breaking up the spent grain cake with a set up of multiple rake arms. An array of waved cutting plates is attached to each rake arm that helps to agitate the grain bed. The entire rake system can be raised or lowered depending on the state of the spent grain cake.

22.3.3.2 Process Characteristics
The control of the lauter tun should aim at achieving:

- Low wort turbidity (1–5 EBC).
- Minimal lauter duration.
- Maximum extract recovery (remaining extract below 1%).

The manipulating variables for this control system are:

- Raking height.
- Raking speed.
- Volume flow rate of the extract.

This implies that the control variables in the form of lauter pump performance, raking speed and height are coordinated with the target variables, including lauter duration, wort turbidity and extract recovery.

The relationships between the filter effect, texture of the spent grain cake and performance of the lauter pump (also taking the negative pressure right below the spent grain bed into account) are of particular interest with regard to the separation procedure. As mentioned above, the suction results in a compaction of the spent grain cake and consequently changes its characteristics, which can, to some degree, be compensated by grain cake agitation via raking. The effects are, however, non-linear and time-variant. Therefore, advanced control algorithms are needed. In addition, these effects and characteristics of the grain cake are influenced by the raw materials used in the brewery process. For instance, the filtering qualities of the grain cake can vary considerable depending on the amount of dust contained in the malt mixture, which again is linked to the level of extraction from the malt silo.

It needs to be taken into consideration, though, that the stated objectives partly call for contradicting control steps. For instance, in order to optimize wort recovery and turbidity, a slow lautering procedure (i.e. long lauter duration) is recommended. This is due to the fact that the longer the wort is exposed to the spent grain, the more complete the extract recovery. Still, according to the objectives, the lauter duration should be minimized by the control process as well.

22.3.3.3 Structure of the Controller

A KAC (see Figure 22.15) was developed based on the standard structure for knowledge-based analytical control methods. It was built on the assumption that the regulation of the control variables is based on the given volume of mash to be lautered.

Each of the control variables is composed of a PI controller depending on the set point and a balance variable. During balancing the control variables are further influenced by the coordination modules of the respective target variables (duration, turbidity, recovery). The main effect of the controller is induced via the balancing modules. The mash volume-dependent profiles (i.e. character lines) that are contained in these modules are varied corresponding to the stated objectives. The controller action depends on the resistance of the spent grain cake and thus adapts the control variables to the characteristics of the malt and the filter qualities of the spent grain cake, respectively.

22.3.3.4 Results

The lauter control method presented here has been installed in several breweries. The main objective for this deployment in operating breweries was the reduction

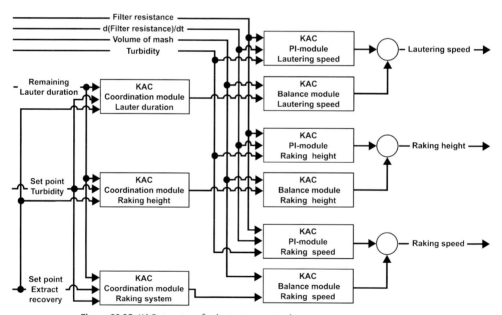

Figure 22.15 KAC structure for lauter tun control.

of the lauter duration, which should allow for an improved handling of peak loads as well as a shortening of production periods.

Figure 22.16 shows the results obtained for a conventional controller (PID), while Figure 22.17 shows the corresponding results for the KAC. A comparison shows that a reduction in lauter duration of approximately 30% can be obtained by applying the new controller. The different modes of operation between the conventional and the new approach can best be distinguished when observing the pressure difference. The KAC aims at maintaining the system under the maximum feasible negative pressure in order to avoid deep cuts and irreversible compaction of the spent grain cake, and as a result achieves the desired reduction in lauter duration.

It is further possible to adapt the technological objectives such that, next to a reduction in lautering time, the procedures in the brewery process can be automated in a form that allows for a flexible adjustment to the requirements of the market, such as summer or winter production. From the results obtained for different breweries it can be concluded that by using this new type of controller, time savings of the order of 10–30% could be achieved.

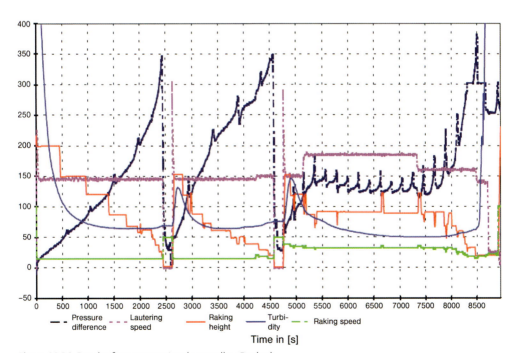

Figure 22.16 Results for a conventional controller. Dashed blue line, pressure difference; purple line, lautering speed; red line, raking height; solid blue line, turbidity; green line, raking speed.

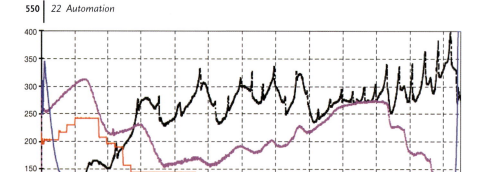

Figure 22.17 Results for a KAC. Dashed blue line, pressure difference; purple line, lautering speed; red line, raking height; solid blue line, turbidity; green line, raking speed.

22.4
Process Control System (PCS)

According to Figure 22.1, a considerable number of open and closed loop controllers are necessary in a brewery. These controllers also have to be coordinated which is accomplished by means of a PCS. Since in a brewery the controllers are not all installed in one central location, but are distributed throughout the system with each subsystem controlled by one or more controllers, this also calls for a distributed control system (DCS). The DCS guarantees that the pre-defined objectives will be fulfilled. The basic structure of such a system is shown in Figure 22.18.

A PCS consists of the following components:

- Process stations.
- Operator stations.
- Engineering stations.
- Process bus.
- Open bus system.

The process stations take on the input and output signals of the process. They are directly connected with the sensors and actuators. This function is usually fulfilled by a programmable logic controller (PLC). In contrast to other computers, a PLC is armored against severe conditions (dust, heat, cold, moisture) and is

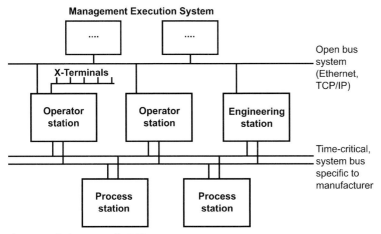

Figure 22.18 Structure of a PCS.

equipped for extensive input/output arrangements. The operator stations are the interface between the DCS and the operator. As the human–machine interface, they serve for displaying, monitoring, analysis, optimization and archival storage. The engineering stations are ideal for fieldwork, including off-line engineering, on-line programming, maintenance and fault tracing.

The process bus connects the process stations with the engineering stations. It operates in real-time and is a redundant bus. The second system bus is an open bus system (Ethernet, TCP/IP), and constitutes the interface between the engineering stations and other electronic devices used in the brewery (e.g. a laboratory control system or a management execution system).

A detailed review of PCSs can be found in [18, 19]. There are several renowned PCSs that have been developed for breweries. These include, for instance, Botec [20], Braumat [21] or brewmaxx [22]. However, other PCSs like Advant OCS [23] or WinCoS [24] can also be used.

22.5
Information Technologies

A lot of information is available in a brewery for controlling, monitoring, diagnosis, asset management and statistical analysis. As a consequence, information and communication systems need to be installed. They include modules for recipe-controlled processes, flexible production planning, diagnostics and maintenance, asset management, archiving and visualization as well as horizontal and vertical integration.

Recipe-controlled processes are used universally in the brewing industries. Advanced modules are not only capable of controlling individual operations, they also offer solutions for the entire production process. Hence, they allow for the convenient creation of recipes for new brands. The brewing expert can modify and adapt the individual steps of the respective process, and can adjust the essential parameters such as volumes, temperatures and time intervals.

Flexible production planning is of great relevance as it grants an optimal production flow, and enables immediate and competent reactions to changes in production at any time.

A high plant availability is necessary to protect the production process from unnecessary downtimes. Therefore, modules built for easy maintenance and with powerful diagnostics capabilities are employed that, at the same time, grant user-friendly operations (clear, simple, reliable). Here, an important aspect is that authorized personnel can intervene manually whenever it is necessary.

Asset management systems are computer software and/or hardware systems that aid in the process of digital asset management (i.e. the storage and management of the data). They permit a reduction of time and cost of content production, and also a cost-effective optimization of the asset management across an entire organization.

Batch and time-related archiving guarantees a complete documentation of all data available. This is the basis for a high-quality management system that is of crucial importance to any brewery.

Modules for horizontal integration enable optimal data acquisition and consequently warrant a process control of the production from start to finish (i.e. from the malt intake to the filling stations). The individual units of the production process (see Figure 22.1) are, as a consequence, no longer isolated, but become inherent components of the overall production process.

Equally, a vertical integration is of great relevance since it ensures that the data collected in the PCS are connected with the manufacturing execution system and also the enterprise resource planning system. Consequently, the manufacturing execution and enterprise resource planning systems can automatically access these data for statistical analysis and efficiency statements.

The modules described above belong in part to the components of the PCS, and to the components of specific information and communication systems like manufacturing execution or enterprise resource planning systems (see Figure 22.18).

22.6
Conclusions

Automation technology plays an important role in the improvement of the production process of beer. For this purpose, better sensor technology, more advanced control algorithms, more powerful PCSs, and more sophisticated information technologies are required [25].

In the future, sensors will become increasingly integral constituents of the subsystems addressed above. They will no longer be constrained to supplying information, but will become a part of the process information system or complete company information systems. Sensor intelligence will allow the interpretation of the values measured at respective locations, which will relieve the system from routine tasks and, as a result, simplify system architecture. In this way, 'intelligence' will progressively move from the upper level to lower levels of the automation hierarchy. The (partly wireless) communication infrastructure will permit remote diagnosis, remote and self-monitoring, as well as the self-calibration of sensors. Networks of sensors instead of coupled point-to-point sensor applications will simplify planning, installation and maintenance of sensors. Traditional measurement transducers will increasingly be replaced by sensor arrays, spectrometric principles and contact-free solutions.

Due to significant progress in the fields of computer technology (e.g. storage capacity, computing time) and software technology, it is now possible to implement advanced control algorithms like adaptive control schemes or knowledge-based control schemes that allow the automation of entire production chains. It is no longer the partial process with an individual solution, but the complete production process of beer that is in the focus of analysis. While the fraction of hardware in the individual automation solution was predominant in the past, the proportion of software is increasing today.

Information technologies must meet the requirements of man. This calls for a user-friendly design of the software architecture and modules. By tailoring to several human senses (multi-modality), operation will become safer and more intuitive. Wireless networks will allow a mobile and permanent access to facilities, while at the same time making use of methods and processes of 'ambient intelligence'. The corresponding standards for the connection between process and operation system will have to be device independent, context related and task oriented. Self-explanatory systems will replace conventional operation instructions. To a growing extent, language input rather than haptic input modalities will characterize these novel information technologies.

Future work will focus on the integration of automation technology solutions in the complete production and business process in breweries as well as on the development of a system- and level-independent open automation on the basis of inter-linked components with clear reliabilities and access rights along with the use of the internet. Furthermore, the increasing interaction of mechanics, electronics and software (mechatronics, adaptronics, bionics) will be addressed. The trend towards an enhanced user orientation, more standards, openness and modularity will also be part of future endeavors. Industrial and business processes that were considered separately in the past will continue to merge with automation technology.

In addition, distributed automation systems will grow beyond 'closed', company specific, LAN-based structures and will utilize various types of 'open', public networks of various types as an integral component. Aspects of availability, temporal behavior, transmission safety and information protection of such networks as well as the reduction of energy comsumption will be in the center of attention. To solve these tasks in breweries, close collaboration of the brewery experts with specialists from the disciplines of automation technology, computer science and information technology will be required.

References

1 Bretthauer, G., Richter, W. and Töpfer, H. (1996) *Automatisierung und Messtechnik. Mittag Praxislexikon*, Mittag-Verlag, Maria Rain.
2 Mayr, O. (1969) *Zur Frühgeschichte der technischen Regelungen*, Oldenbourg-Verlag, München.
3 Kriesel, W., Rohr, H. and Koch, A. (1995) *Geschichte und Zukunft der Mess- und Automatisierungstechnik*, VDI-Verlag, Düsseldorf.
4 Esslinger, H.M. and Narziss, L. (2002) *Beer*, Wiley-VCH Verlag GmbH, Weinheim.
5 Manger, H.-J. (2005) *Kompendium Messtechnik. Online-Messgrößen in Brauerei, Mälzerei und Getränkeindustrie.* VLB, Berlin.
6 Xiao, D., Mo, Y. and Choi, M.M.F. (2003) A hand-held optical sensor for dissolved oxygen measurement. *Measurement Science and Technology*, **14**, 862–7.
7 Philipp, R. (2008) Trübungsmessung bei der Filtration/Vergleich 90° Streulicht und 11° Vorwärtsstreulicht, http://www.braulotse.de/Texte/Optek/Truebung.html (accessed 07. 12. 2008)
8 EPA (1999) Guidance Manual for Compliance with the Interim Enhanced Surface Water Treatment Rule: Turbidity Provisions. EPA 815-R-99-010. US Environmental Protection Agency, Office of Water, Washington, DC.
9 www.de.endress.com (accessed 07. 12. 2008).
10 www.prozessanalytik.de (accessed 07. 12. 2008).
11 www.mtshop.com (accessed 07. 12. 2008).
12 Aström, K.J. (1982) Theory and applications of adaptive control a survey. *Automatica*, **18**, 471–86.
13 Krstic, M., Kanellakopoulos, I. and Kokotovic, P. (1995) *Nonlinear and Adaptive Control Design*, John Wiley & Sons, Inc., New York.
14 Zadeh, L.A. (1965) Fuzzy sets. *Information and Control*, **8**, 338–53.
15 Jäkel, J., Mikut, R. and Bretthauer, G. (2004) Fuzzy control systems, in *Control Systems, Robotics and Automation* (Unbehauen, H., ed.), EOLSS, Oxford, paper 6.43.23, pp. 1–28.
16 Jäkel, J. and Bretthauer, G. (2004) Fuzzy system applications, in *Control Systems, Robotics and Automation* (Unbehauen, H., ed.), EOLSS, Oxford, paper 6.43.23.4, pp. 1–24.
17 Müller, J.-U. (2001) Eine neue Methodik zur Integration von qualitativem und quantitativem Prozesswissen für den Entwurf von komplexen Steuerungs- und Regelungsalgorithmen, Dissertation, Universität Karlsruhe.
18 Polke, M. (2001) *Prozessleittechnik*, Oldenbourg-Verlag, München.
19 Früh, K.F., Maier, U. and Schaudel, D. (eds) (2008) *Handbuch der Prozessautomatisierung*, Oldenbourg Industrieverlag, München.
20 www.krones.de (accessed 07. 12. 2008).
21 www.automation.siemens.com (accessed 07. 12. 2008).
22 www.brewmaxx.de (accessed 07. 12. 2008).
23 www.abb.de/ProductGuide (accessed 07. 12. 2008).
24 www.buhlergroup.com (accessed 07. 12. 2008).
25 Bretthauer, G. and Westerkamp, D. (2006) Automation technology 2010: challenges and changes. *at* **54** (9), 459–61.

23
Malthouse and Brewery Planning
Walter Flad

23.1
Malthouse Planning

23.1.1
Introduction

The main units of malting plants are the barley and malt storage, and the steeping, germination and kilning facilities. In the following sections we give short overviews about the essential technological demands of the installations and refer to the leading solutions in malting plants, followed by dimensioning calculations.

The connections between these units (e.g. raw material acceptance, conveyors, dust removal technique, barley and malt purification machines, shipping devices, etc.) are not considered here. It is also beyond the scope of this chapter to dwell on the wide field of process details such as enzyme activities and all the rich variations of malting technology. For all those who want further information about this we recommend the standard reference books of Professor Dr L. Narziss [1, 2].

23.1.2
Storage of Barley and Malt

Barley intake must be monitored by heavy-duty scales and pre-cleaning should be done during storage. For proper handling of barley and malt and for good automation preconditions (batch tracing), it is necessary to use round or rectangular silos made from concrete or steel. The outlet funnel should have an interior angle of 90°. The barley quality will be stable if the water content is below 15% (ideally below 12%). To confirm these conditions, silos can be aerated and cooled by external installations, if necessary. The silo volume demand depends on storage possibilities around the malthouse and can reach up to 80% of the annual production capacity. It is used mostly for barley at the beginning of a harvest campaign and for malt at the end.

23.1.3
Steeping

The stored barley is induced to germinate by addition of water. This procedure takes place in the steeping room. There the barley picks up water from an original amount of 12% to a water content of 38% or even more. The corns simultaneously begin to respire, producing CO_2 and small amounts of alcohol. Technological parameters must be set properly in this step to assist the start of germination and to avoid unwanted CO_2 enrichment.

Traditional steeping tanks have been cylindroconical, but it has proved more effective to use so-called flat steep vessels, which are round tanks for barley with a maximum height of 3.0 m above a slotted floor bottom. Under the floor there are injectors for compressed air and a CO_2 exhaust system. Thus, the barley can be alternately flooded with water and set dry, aerated and cleared of CO_2.

A typical steeping cycle is shown in Table 23.2.

23.1.4
Germination

The germination is connected to the steeping procedure. During germination, the barley begins to grow and to activate the enzymes that are so important for the brewing process. However, germination must be controlled and accurately moderated to avoid extract losses and undesired assimilation products that would cause difficulties in the brewing process.

The ancient practices were based on floor malting houses. Today these installations are insignificant on the industrial scale. Instead there are many variations of pneumatic malting systems, which are characterized by slotted floors and well-directed air flow systems. Due to this, process parameters are controlled exactly by the air stream, which can be varied in terms of speed, temperature and humidity.

The germination takes place at temperatures of 12–18 °C and water contents of 30–48%. To avoid corn dehydration, the air must be humidified during nearly the whole of the germination time. In summer, fresh air must be chilled to prevent high temperatures in the barley.

The technical solutions for the design of germination boxes are innumerable. Just to mention few of them, we will point out:

- Classical germination boxes (Saladin Box).
- Germination drums.
- Moving pile systems (Seeger).
- Conversion box process (Lausmann).
- Germination-kilning boxes.
- Germination towers.
- Germination-kilning towers.

Section 23.1.6 gives an example of a process cycle in a combined germination-kilning tower.

One significant advantage of these combined systems must be highlighted: The combined systems – in contrast to all traditional solutions with separate germination and kilning device – have the big advantage of better hygienic conditions, because the whole system gets quasi–sterile during each kilning cycle.

23.1.5
Kilning

At the end of the malting process it is necessary to bring the biochemical and chemical reactions to an end and to preserve the malt. This is achieved by drying the corns to a humidity of 4%. At the same time the maltster adjusts the typical desired attributes of their product, such as color, smell, taste and other technological values.

Decades ago kilns were directly fired by wood or other combustibles. However, since the discovery of the cancer-causing potential of nitrosamines, which are generated under conditions as mentioned above, the heating of kilns has converted to indirect methods (air heating by steam or hot water).

The kilning process can be divided into three sections:

- Withering (50–65 °C).
- Heating up (65–80 °C).
- Curing (80–85 °C).

The characteristic temperatures of these stages and the need to save money in this relatively energy-intensive procedure have effected many concurrent kilning techniques. The best evaluated techniques are:

- Single-floor high-capacity kilns.
- Double-floor kilns.
- Germination-kilning boxes or towers.

See Figure 23.2 for a more detailed view of the parameters of a typical kilning process.

In general, all systems with short processing times need high ventilation power and have higher consumption values than slower cycles. Double-floor kilns have the benefit of using the warm outgoing air of the curing kiln as delivery air for the withering kiln. For energy-saving reasons, all kilns should have an integrated cross-flow heat exchanger, which can recover about 60–80% of the heat content of the outgoing air for pre-heating the delivery air.

23.1.6
Show Case Malting Plant

23.1.6.1 Design of the Steeping, Germination and Kilning Tower
Among many other possible options, for this showcase malting plant we chose a combined round malting tower, consisting of one flat steep vessel in the top floor

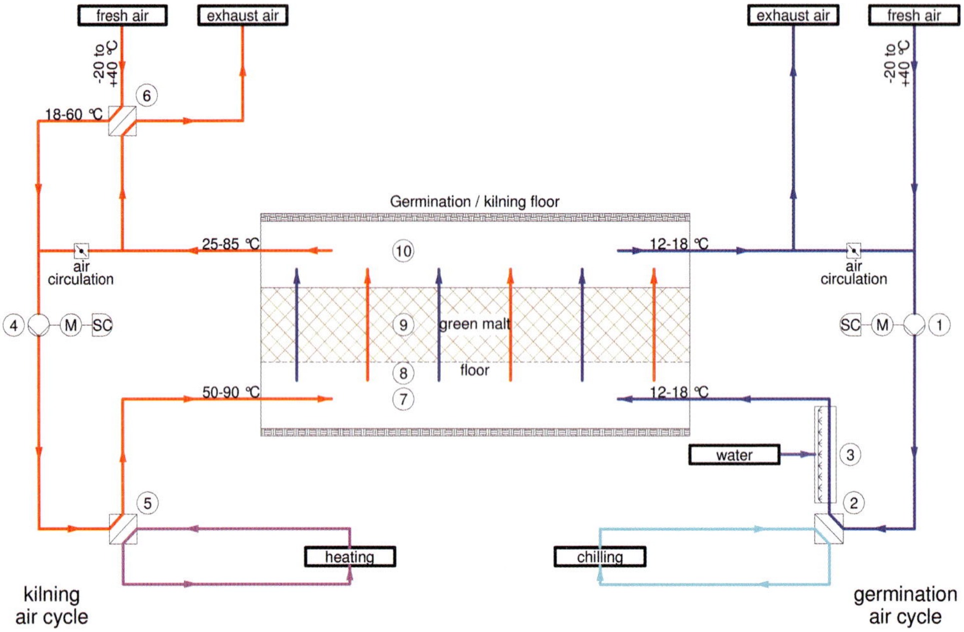

Figure 23.1 Functional flow sheet for one germination/kilning floor. *Germination air cycle*: 1, fan, frequency controlled (450 m³/ton · h); 2, chilling exchanger (chilling power: 1.8 kW/ton; electrical power: 0.45 kW/ton); 3, air humidification. *Kilning air cycle*: 4, fan, frequency controlled (2000–3100 m³/ton · h); 5, heat exchanger (30 kW/ton · h); 6, cross-flow heat exchanger (60–80% efficiency). *Germination/kilning floor*: 7, space under the floor (2.5 m); 8, slotted floor (specific loading: 450 kg barley/m²); 9, green malt (height of green malt: about 1 m); 10, space above the floor (headroom: 3.5 m).

of the tower and four combined germination/kilning floors below. Other technical equipment is installed in the basement, such as water reserves and a ventilation system.

In the horizontal projection the floors are designed as rings around a kernel building (6–12 m diameter), which includes the germinating air ducts such as warm air, cold air, circulation the control system and the adjustment of air and an installation channel. The temperatures in each floor are set up by air shutters. The air ducts for the kilning cycle can be arranged as extensions of the building. Preferably, the cross-flow heat exchanger is installed on top of the tower. For a better understanding of the functions, see Figure 23.1.

In combined germinating-kilning systems it is important to adjust the cycle times of the different process stages such that all steps of the process match each other. The steeping process scheme is shown in Table 23.1. The steeping process time of 34 h fits in the cycle time of 36 h. So in 1 week 4 batches of barley to 100 tons can be steeped-in, and 1 day (Sunday) is free of work. For germination, we choose the process scheme shown in Table 23.2. After these 106 hours the kilning process begins in the same floor by switching the ventilation ducts. Figure 23.2 shows the functional interrelations in more detail. The kilning period is designed for 36 h and can be explained by Figure 23.2 in which all important data

Table 23.1 Steeping process scheme in a flat steep vessel.

Action	Time (h)	Approximate value (m³/ton·h)
Rest	2	
CO_2 exhaust	10	50
Immersion steeping	2	
Pressure aeration	2	50
Rest	2	
CO_2 exhaust	14	50–150
Cast (steeped barley)	2	
Total	34	

Table 23.2 Germination process scheme.

Action	Time (h)	Approximate value (m³/ton·h)
Interval aeration	6	250
Aeration 100%	50	500
Aeration 80%	25	400
Aeration 70%	15	350
Aeration 60%	10	300
Total	106	

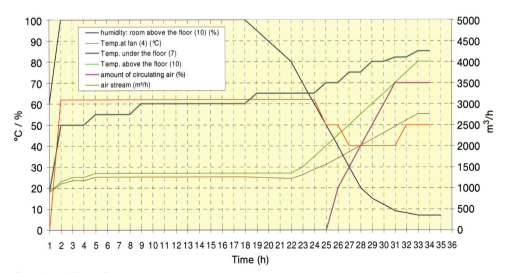

Figure 23.2 Kilning diagram.

are monitored. It can be seen that the moisture content of the outlet air persists for a long time at about 100% (withering section). If the moisture falls below 65%, heating-up section starts, followed by the curing section. At the beginning of heating up ventilation power can be lowered and the aeration is gradually switched from pure fresh air feed to air circulation.

23.1.6.2 Calculations

Annual Capacity

For the following calculations we consider a showcase malting plant, for which we make use of some suggestions:

- Batch size: 100 tons
- Batch cycle: 4 times/week.

The annual capacity is:

$$4 \times 100 \text{ tons/week} \times 52 \text{ weeks/year} = 20\,000 \text{ tons barley/year}.$$

Furthermore, we have to introduce some basic data for barley and malt as shown in Table 23.3. The hectoliter weight is usually declared as kg/hl. Due to better handling and to the correct use of SI units, these values are converted into tons/m³.

Table 23.3 Hectoliter weights of barley and malt.

	Abbreviation	Hectoliter weight (tons/m³)
Barley (air dry, 15%)	$W_{hl,\,B15}$	0.65–0.70
Barley (moisture expanded, 40%)	$W_{hl,\,B40}$	0.5
Malt (air dry, 4%)	$W_{hl,\,M4}$	0.55

Malting Losses

Malting losses occur in connection with the process, which are composed of steeping losses, respiration losses and germination losses. These losses accumulate to a value of 8.5–9.0% (water free), from which the aspiration losses are eminent (about 5.2%, water free). For calculations with barley and malt, the water content of each needs to be considered.

Conversion Example (Barley to Malt)

- Humidity of barley: $H_B = 15\%$.
- Humidity of malt: $H_M = 4\%$.
- Losses (water free): $L_{WF} = 8.5\%$.

23.1 Malthouse Planning

Question: How much malt will be produced from 20,000 tons of barley?

$$\text{Annual malting capacity} = \frac{m_B \times (100 - H_B) \times (100 - L_{WF})}{(100 - H_M)}$$

$$= \frac{20\,000 \text{ tons barley}_{15\%} \times (100 - 15.0) \times (100 - 8.5)}{100 - 4.0}$$

$$= 16\,200 \text{ tons malt}_{4\%}.$$

Silo Capacity

Characteristics:

- Silo demand factor: f_s = 80% of annual production (AP).
- Capacity of one silo: C_{silo} = 400 tons (quadratic layout, side length a = 6 m).
- Interior funnel angle: $\alpha = 90°$.

Dimensions of one silo cell:

$$\text{Edge height } h_{edge} = \frac{m_{silo}}{W_{hl.B15} \times a^2} = \frac{400 \text{ ton}}{0.7 \text{ tons/m}^3 \times 6^2 \text{ m}^2} = 15.8 \text{ m}.$$

$$\text{Funnel height } h_{funnel} = \frac{a/2}{\tan \alpha/2} = \frac{6/2}{\tan 90/2} = 3.0 \text{ m}.$$

The total height is 15.8 + 3.0 = 18.8 m (plus discharge height).

$$\text{Number of silos } n = \frac{AP \times f_s}{C_{silo}} = \frac{20\,000 \text{ tons/year} \times 0.8}{400 \text{ tons/silo}} = 40 \text{ silos}.$$

Calculation of the Floor Area

Characteristics:

- Specific barley loading: SL_B = 450–500 kg/m².
- Batch size: BS_t = 100 tons.

$$\text{Floor area } A_{fl} = \frac{BS_t \times 1000 \text{ kg/ton}}{SL_b} = \frac{100 \text{ ton} \times 1000 \text{ kg/ton}}{450 \text{ kg/ton}} = 222 \text{ m}^2.$$

23.1.7 Consumption Data

For malting systems like this, the consumption data can be assumed as shown in Table 23.4.

In some cases it makes sense to reflect on combined heat and power systems CHP. For the investigation of that diagram shown in Figure 23.3 was designed.

Table 23.4 Consumption data.

	Units	Value
Electric power (winter)	kWh/ton	50–60
Electric power (summer, with chilling power)	kWh/ton	80–100
Heating energy (kiln)	kWh/ton	600–700
Water	m³/ton	4
Waste water	m³/ton	3

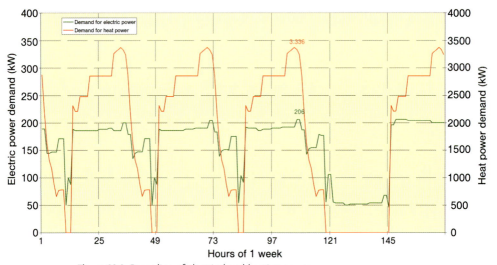

Figure 23.3 Recording of electrical and heating power.

It is a recording of the electrical and heating power values in a 1-week period for the showcase malting plant. With the knowledge of these energy demand progressions it is possible to dimension such units. Under efficiency aspects it is mostly suitable to design block heat and power plants using gas or fuel engines. The economic data depend strictly on the investment costs of the plant, and the actual costs of the combustibles and electric power.

23.2
Brewery Planning

The following sections contain the reflections and calculations necessary to dimension the machinery in breweries. As an example the authors have chosen a showcase brewery with an annual output of 1 000 000 hl sales beer (SB) for their calculations.

Additional abbreviations used are: gross tank space (GTS), net tank space (NTS) and unfiltered beer (UB).

Characteristics of the Showcase Brewery

In order to have a consistent and simple-to-handle calculation model, it is necessary to define some conventions *a priori*:

- 100% malt, no malt substitutes like corn/maize, rice or raw grain.
- Only bottom-fermented beer types.
- 800 000 hl/year bottled beer (80%), average bottle size: 0.45 l.
- 200 000 hl/year draft/keg beer (20%), average keg size: 35 l.
- Up to 100 000 hl SB in 1 month: 0.10 years/month (annual peak factor).
- 8% total wort/beer loss [from hot wort (HW) to SB].
- Up to 6 (5.5) working days in a week, including 0.5 cleaning days.

Wort/Beer Losses

The volume (not extract) losses in the particular production departments must be considered in the calculations. The following assumptions given in Table 23.5 are based on empirical values of comparable brewery sizes. It is self-evident that different technologies may produce different loss values (e.g. high-gravity brewing).

The required dimensions of machinery in the showcase brewery are calculated in the following sections.

Table 23.5 Wort/beer losses.

Production step		Wort/beer loss (%)	Factor to SB
From	To		
Brewhouse (HW)	fermentation (CW)	4.0	0.920
Fermentation (CW)	maturation/storage	1.0	0.960
Maturation/storage	filtration	1.0	0.970
Filtration	bright beer tanks	1.0	0.980
Bright beer tanks	filling plants	0.5	0.990
Filling plants	SB	0.5	0.995
Brewhouse (HW)	SB	8.0	0.920

23.2.1
Brewhouse

A modern brewhouse is able to produce up to 12 brews/day. A brew cycle of 12 brews/day needs at least the following brewing vessels: one mash tun kettle, one lauter tun, one wort prerun tank, one wort copper/kettle and one whirlpool tank.

Other breweries may need other constellations depending on annual output, brew cycle (3, 6, 8, 10 or 12 brews/day), beer types, raw materials, working time, automation and energy technology.

The brew size of the showcase brewery with consideration of all conventions above is calculated as follows:

$$\text{Brew size} = \frac{1\,000\,000 \text{ hl SB/year} \times 0.10 \text{ years/month}}{12 \text{ brews/day} \times 5.5 \text{ days/week} \times 4.2 \text{ weeks/month} \times 0.920 \text{ hl SB/hl HW}}$$

$$= \sim 400 \text{ hl HW/brew}.$$

The grist load for 12°P beers is about 16.5 kg/hl HW, that means in this case:

$$\text{Grist load} = 400 \text{ hl HW} \times 16.5 \text{ kg/hl HW} = 6600 \text{ kg}.$$

23.2.2
Wort Cooling

The wort cooler should be dimensioned to cool the brew in not more than 50 min. Hence, the flow rate amounts to 480 hl/h.

23.2.3
Malt Silos and Malt Treatment

Characteristics:

- 2 weeks stock keeping (depending on delivery conditions).
- 16.5 kg/hl HW specific grist load.
- 75% average filling level.

$$\text{Malt silo capacity/space} = \frac{400 \text{ hl HW/brew} \times 16.5 \text{ kg/hl HW} \times 66 \text{ brews/week} \times 2 \text{ weeks}}{1000 \text{ kg/ton} \times 0.75} = \sim 1200 \text{ tons}.$$

Of course, the total silo volume should be split to separate different varieties and qualities from each other (e.g. 8 × 120 tons and 4 × 60 tons).

Within the silo the malt will assume a resting angle of 30° on top, which must be accounted for in volume calculation. The specific weight of the malt can be estimated at 0.55 kg/m³.

The output of the malt treatment units (grading screens, plansifter, magnet, malt balance) should follow the cycle time of the brewhouse. At 12 brews/day the grist load has to be treated in a maximum of 1.5 h. Thus, the output needs to exceed 6.6 tons/brew/1.5 h = 4.4 tons/h. If a wet mill is to be used, a 15–20 min mash-in duration requires an output of about 20 tons/h.

23.2.4
Fermenting, Maturation and Storage Tanks

It is state of the art to start fermenting without flotation and removing the cold break. This is possible when using vital yeast in connection with cylindroconical tanks.

Fermenting Characteristics

- 384 hl cold wort (CW)/brew.
- 7 days fermentation time (includes 1 day for manipulations: emptying, cleaning and filling).
- 25% extension space (top-fermenting beer types: 35%).
- Two-stage process: after fermentation the green beer will be transferred into the maturation and storage tanks (many other technologies are possible, including unitank methods).
- Half-day tanks (6 brews/tank).

$$\text{Fermenting capacity/space} = \frac{1\,000\,000 \text{ hl SB/year} \times 0.10 \text{ years/month} \times 1.25 \text{ hl GTS/hl NTS} \times 7 \text{ days}}{0.96 \text{ hl SB/hl NTS} \times 4.2 \text{ weeks/month} \times 7 \text{ days/week}}$$

$$= \sim 31\,000 \text{ hl GTS}.$$

Tank size = 6 brews × 384 hl CW/brew × 1.25 hl GTS/hl CW = ~2880 hl GTS.

It is necessary to install 11 cylindroconical tanks at least. If the brewery produces different beer types with low annual output and seasonal beers it is recommended to install additional tanks with a smaller size (e.g. 3 brews/tank).

The dimensions of tanks can be calculated by using an interactive website sheet, programmed by the authors (www.tbw-freising.de).

Maturation and Storage Characteristics

- 14 days maturation and storage time (includes 1 day for manipulations).
- 10% extension space.

$$\text{Maturation/storage capacity/space} = \frac{1\,000\,000 \text{ hl SB/year} \times 0.10 \text{ years/month} \times 1.10 \text{ hl GTS/hl NTS} \times 14 \text{ days}}{0.97 \text{ hl SB/hl NTS} \times 4.2 \text{ weeks/month} \times 7 \text{ days/week}}$$

$$= \sim 54\,000 \text{ hl GTS}.$$

It is convenient to use the same type of tank to ferment, mature and store beer. This ensures a good flexibility. The brewery has to install at least 19 cylindroconical tanks for maturation and storage. Seasonal beers and beer types with low annual output need additional tanks with a smaller size.

23.2.5
Yeast Management

It is desirable to have good yeast vitality and growth. Yeast management consists of the propagation plant, the yeast crop and the waste yeast treatment. Yeast management is very complex. Several very different methods are in use in the brewing industry. Due to this, a general calculation basis cannot be defined. In principle, all tanks should be cylindroconical and fitted out with cooling jackets. The necessary headspace can exceed 50% of the net tank space.

23.2.6
Filtration

The most common filtration method is still kieselguhr (powder) filtration. Recently, however, kieselguhr-free filtration methods have become more relevant (e.g. continuous cross-flow membrane methods). These techniques run round the clock with several modules, which are individually cleaned clockwise, while the others continue the filtration process.

Filtration Characteristics

- Batch filtration with kieselguhr during 5 days/week.
- Gross filtration time: 16 h/day with 12 h pure filtration time.

$$\text{Necessary filtration output} = \frac{1\,000\,000 \ \text{hl SB/year} \times 0.10 \ \text{years/month}}{0.98 \ \text{hl SB/hl UB} \times 4.2 \ \text{weeks/month} \times 5 \ \text{filtrations/week} \times 12\,\text{h/filtration}}$$

$$= \sim 400 \ \text{hl UB/h}.$$

23.2.7
Bright Beer Tanks

Characteristics:

- 1.5 filtration days stock keeping.

$$\text{Necessary tank capacity/space} = 400 \ \text{hl bright beer/h} \times 12\,\text{h/filtration} \times 1.5 \ \text{filtrations} \times 1.1 \ \text{hl GTS/hl bright beer} = \sim 7920 \ \text{hl GTS}.$$

The tanks can be split into $3 \times 1500\,\text{hl GTS}$ and $4 \times 750\,\text{hl GTS}$. The bright beer tank size should match with the maturation and storage tank sizes and the filling batches. Seasonal beers and beer types with low annual output need additional tanks with a smaller size.

23.2.8
Bottling Plant

Characteristics:

- 24 h/day bottling time, 5 days/week (3 shifts/day).
- 70% overall line efficiency (varies from 50 to 75%).

$$\text{Necessary bottling plant output} = \frac{800\,000 \text{ hl/year} \times 0.10 \text{ years/month} \times 100\,\text{l/hl}}{0.995 \times 0.70 \times 4.2 \text{ weeks/month} \times 5 \text{ days/week} \times 24 \text{ h/day} \times 0.45\,\text{l/bottle}}$$

$$= \sim 50\,000 \text{ bottles/h}.$$

It is reasonable to choose a filling line with 55–60 000 bottles/h output to have a reserve. The overall line efficiency will reduce if the brewery fills many different beer types with various bottle types, bottle sizes, crates and packages. In such cases the efficiency can decrease to 50% or even lower.

23.2.9
Kegging Plant

Characteristics:

- 16 h/day kegging time, 5 days a week (2 shifts/day).
- 85% overall line efficiency (efficiency factor varies from 75 to 85%)

$$\text{Kegging plant output} = \frac{200\,000 \text{ hl/year} \times 0.10 \text{ years/month} \times 100\,\text{l/hl}}{0.995 \times 0.85 \times 4.2 \text{ weeks/month} \times 5 \text{ days/week} \times 16 \text{ h/day} \times 35\,\text{l/keg}}$$

$$= \sim 200 \text{ kegs/h}.$$

The over all line efficiency will reduce if the brewery fills many different beer types with various different keg types and sizes.

23.2.10
Space Requirement of Full Packs and Returned Empties

The necessary areas depend on:

- Product range.
- Layer patterns.
- Annual output.
- Filling batches.

- Pallet dimensions.
- Stack height.

It is necessary to have a detailed look at every individual case because of so many possible variants. An interactive calculation sheet can be used (www.tbw-freising.de).

23.2.11
Utilities and Power Supply

In this section we introduce a rough calculation method, based on a simple estimation with specific benchmark values for all departments of energy and utilities plant. This method can be used to get a first indication of the required capacities. (In detail, the calculations are more complicated and different greatly from case to case, such that specific explanations would require a separate volume.) Attention should be paid to the fact that this method is only suitable for pre-planning purposes and calculations need to be more detailed in further progress. Furthermore, it usually makes sense to have additional reserve capacities installed.

23.2.11.1 **Supply with Heat**
The heating plant generally consists of one or more boilers to generate steam or hot water with a system pressure of about 3–5 bar. All heat exchangers must match to this pressure level. Main heat consumers are the brewhouse (mash and wort heating and cooking), filtration (sterilization), bottle and keg filling lines, room heating, and possibly special supplies (e.g. tunnel or flash pasteurizers). If heat recovery systems are consistently used, the connection values can diverge from the following standard calculations. Most important is the energetic brewhouse design, which determines ultimately which benchmark value is applicable.

Range of specific installed heat capacity	8–20 kW/1000 hl SB/year
Estimated specific heat capacity	10 kW/1000 hl SB/year
Heat capacity to install	*~10 MW*

23.2.11.2 **Supply with Coldness**
In most cases, refrigeration plants consist of a centralized chilling compressor station with ammonia as refrigerant. Main consumers are wort cooling, tank cooling during fermentation, green beer cooling and possibly special supplies (e.g. flash beer coolers, etc.).

Range of specific installed refrigeration capacity	2.0–4.0 kW/1000 hl SB/year
Estimated specific refrigeration capacity	2.5 kW/1000 hl SB/year
Coldness capacity to install	*~2.5 MW*

23.2.11.3 CO_2 Recovery

Range of specific CO_2 recovery capacity   0.5–0.8 kg/h/1000 hl SB/year
Estimated specific CO_2 recovery capacity   0.75 kg CO_2/h/1000 hl SB/year
CO_2 recovery capacity to install   *750 kg/h*

23.2.11.4 Supply with Compressed Air

Range of specific installed compressed air capacity   1.5–4.5 m^3/h/1000 hl SB/year
Estimated specific compressed air capacity   2.0 m^3/h/1000 hl SB/year
Compressed air capacity to install   *~2000 m^3/h*

23.2.11.5 Supply with Electrical Power

Range of specific installed electrical power   2–6 kVA/1000 hl SB/year
Estimated specific installed electrical power   3.0 kVA/1000 hl SB/year
Electrical power to install   *~3000 kVA*

23.2.11.6 Supply with Fresh Water

Range of specific installed water supply   0.1–0.15 m^3/1000 hl SB/year
Estimated specific installed fresh water supply   0.1 m^3/1000 hl SB/year
Fresh water supply to install   *100 m^3/h*

23.2.12 Key Figures for New Breweries

In an early planning status it is helpful to have a rough vision of the required land area and the required total costs for the construction of a new brewing plant. For that purpose the authors have summarized experiences from many comparable projects worldwide. From that data they have extracted certain key figures, which have been related to the plant size. As a matter of course each plant needs to satisfy the criteria of its special market, the applicable laws and many other necessities. Thus, these key data are merely suitable for approximate calculations at the launch stage of a project. In particular, the costs can vary over a wide range depending on the country of realization, automation standards and technological specifications. The figures cover breweries with an annual output between 25 000 and 4 000 000 hl SB/year. These breweries are equipped to produce beer as characterized in the previous sections.

23.2.12.1 Required Land Area

We assume that the new production departments are built as single-storey buildings. Only a few buildings with special functions (lavatories/toilets, ventilation plant, laboratory, office building) are built with two or three floors.

The space requirement of the bottling plant is calculated on a returnable bottle base without a separate bottle sorter plant. Sorting machines are integrated in the bottling plant. Full packs and returned empties are stored in block storage. Additional to the annual output, up to 10% of other trading products can be stored. The calculations contain site roadways, site services, approach roads, loading docks and undeveloped areas for extensions up to about 20%.

Table 23.6 shows the space requirements for newly built breweries in relation to their annual output. When converted into a diagram (Figure 23.4), the values in the table show a declining curve shape. The specific space requirements are given in Figure 23.5. Figure 23.6 shows the space requirement for some production departments, such as filtration, yeast management, cleaning in place (CIP) system and the associated stores. The curve behaves like the overall space requirement. Figure 23.7 shows the space requirement for the fermenting, maturation/storage and bright beer tanks. The curve is mainly linear. Figure 23.8 shows the space requirement for the filling plants with the necessary stores. The curve contains jumps where the plants reach their capacity limits. Figure 23.9 shows the space requirement for storage and distribution. Due to the fact that big breweries are producing round the clock and the trick loading takes place in two shifts there is an inflection at a certain point of the curve.

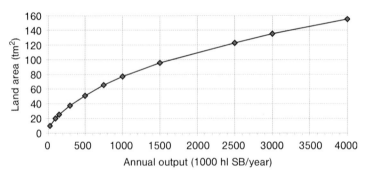

Figure 23.4 Total space requirement for breweries.

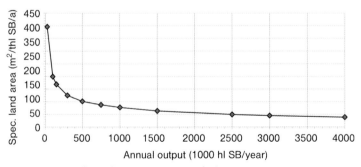

Figure 23.5 Specific total space requirement for breweries.

Table 23.6 Space requirement for a new brewery (m²).

	Output (1000 hl SB/year)											
	25	100	150	300	500	750	1000	1500	2500	3000	4000
Malt, brewhouse, water, store	250	500	600	1000	1350	1700	1900	2200	2550	2700	2850
Yeast, filtration, CIP, store	200	600	750	1000	1300	1500	1700	2000	2450	2600	3000
Cylindroconical and bright beer tanks	350	600	750	1000	1250	1650	1900	2500	3500	4300	5500
Filling plants, side rooms	600	1400	1700	3000	4700	7000	8500	11000	13000	13000	15000
Stack shed, loading docks, store	1300	2300	3000	5000	7500	9000	10000	12000	16000	18000	22000
Utilities and power supply	150	300	400	600	800	1000	1150	1400	1750	1900	2150
Office building, parking places	600	1400	2000	3600	5000	6300	7400	9000	12000	13000	15000
Site roadways, approach roads	600	2100	2700	4300	6400	8500	9500	12500	17000	19000	22000
Green space	1000	1500	2000	3000	4000	5000	6000	7500	9500	11000	12500
Distance areas	2400	4000	4500	6000	7000	8500	9000	11500	14000	15200	17000
Extension areas	2000	4000	5000	6000	7000	10000	12000	16000	19000	22000	25000
Asphalted areas	400	1000	1200	2000	3000	4000	5000	6300	8000	10000	11600
Waste water treatment	150	300	400	500	700	850	950	1100	1250	1300	1400
Sum	10000	20000	25000	37000	50000	65000	75000	95000	120000	134000	155000
Specific space requirement m²/1000 hl SB/a	400	200	167	123	100	87	75	63	48	45	39

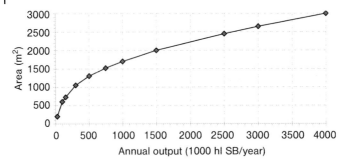

Figure 23.6 Space requirement: filtration, yeast, CIP, store.

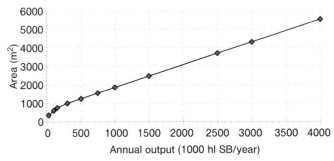

Figure 23.7 Space requirement: tanks.

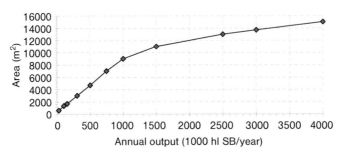

Figure 23.8 Space requirement: filling plants and side rooms.

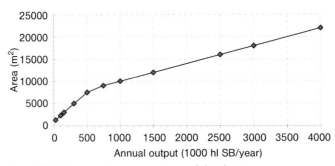

Figure 23.9 Space requirement: storage and distribution.

23.2.12.2 Required Investment Costs

The plant investment costs are based on fully automated breweries with reasonable technology standards. The buildings are 'turn-key' calculated under the postulation of adequate foundation soil. It is understood that the architecture is normal without excessive constructions. The costs do not contain furnishings, computer equipment and software in the offices. Also excluded is the initial acquisition of crates, bottles, kegs and all sorts of vehicles. Table 23.7 shows the investment costs in relation to different brewery sizes.

In diagrammatic form, the results look as shown in Figures 23.10 and 23.11. These show declining curve shapes. Figure 23.12 shows the costs in relation to each other. It is evident that the amount of technical installations (60% of total costs) exceeds the costs of building and building services (40% together). However, this ratio has been detected for European countries and depends on the region of realization.

It must be pointed out again that the key figures from this chapter are only for rough orientation. Exact and detailed planning needs the analysis and implementation of innumerable individual parameters in every single case.

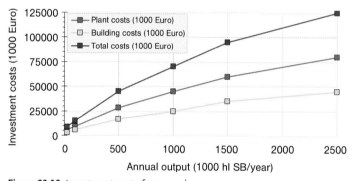

Figure 23.10 Investment costs for a new brewery.

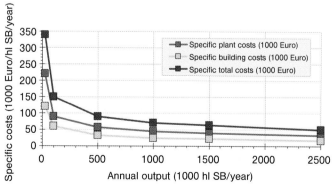

Figure 23.11 Specific investment costs for a new brewery.

Table 23.7 Investment costs for a new brewery.

	Output (1000 hl SB/year)					
	25	100	500	1000	1500	2500
Plant costs (1000 Euro)						
malt, brewhouse, wort cooling	1 200	1 800	5 250	7 500	9 000	10 500
Yeast, filtration, CIP	425	800	2 150	3 000	4 250	6 000
cylindroconical and bright beer tanks with pipe system	1 000	1 600	5 900	10 750	16 000	24 000
bottling plant	1 500	2 500	8 000	12 500	16 000	20 000
kegging plant	250	400	950	1 250	1 500	2 000
utilities and power supply with pipe system	375	900	4 000	7 000	9 000	12 500
on-site accomplishments	750	1 000	2 250	3 000	4 250	5 000
total plant costs (1000 Euro)	5 500	9 000	28 500	45 000	60 000	80 000
specific plant costs (Euro/hl)	220	90	57	45	40	32
Building costs (1000 Euro)						
malt silos, malt treatment, brewhouse, water	175	375	1 050	1 350	1 500	1 800
filtration, yeast management, CIP, store	125	350	1 000	1 200	1 500	1 750
cone room tanks	250	500	1 000	1 500	2 000	3 000
paneling of tank building	400	650	1 350	2 000	2 900	4 000
filling shed (bottle and keg), store	425	1 000	3 250	6 000	7 500	9 000
stack shed, loading docks, store	675	1 250	4 000	5 000	7 000	8 500
boiler plant and engine room	65	125	350	500	650	800
office building	75	150	700	850	1 250	1 400
site roadways, approach roads, parking places	75	200	700	1 000	1 400	1 750
layout of green space	60	100	200	350	450	500
asphalted areas	25	50	150	250	350	500
miscellaneous (additional building expenses, waste water, environmental protection, etc.)	650	1 250	2 750	5 000	8 500	12 000
total building costs (1000 Euro)	3 000	6 000	16 500	25 000	35 000	45 000
specific building costs (Euro/hl)	120	60	33	25	23	18
Total investment costs (1000 Euro)	8 500	15 000	45 000	70 000	95 000	125 000
Specific total investment costs (Euro/hl)	340	150	90	70	63	50

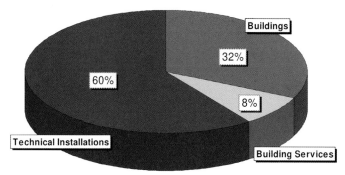

Figure 23.12 Split investment costs for a new brewery (100000 hl SB/year; total investment costs around 15 million Euro).

23.2.13
Documentation and Specifications

All documents worked out during the planning and realization/implementation of a project are very important for the brewery or malthouse. They should be collected and archived. Particularly important components are the execution documents, for example:

- Mimic diagrams.
- Flow diagrams.
- Process diagrams.
- Function flow sheets.
- Construction plans.
- Assembly plans/instructions.
- Operating/instruction manuals.
- Maintenance manuals.
- List(s) of replacement parts.
- Material certificates.
- Welding certificates.
- Examination and acceptance reports.
- Structural engineering.
- Building permits.
- Operating permits.

After implementation and commissioning, it is necessary to collect and archive the following additional documents:

- Declaration of conformity (CE marking).
- Construction drawings (building).
- Construction drawings (assembly).
- Piping conception/system (overall view plan).

- Electrical installation.
- MCR (measure–control–regulate) installation.
- Wiring diagrams.
- Current automation software.
- Waste water system plans.
- Warranty documents.

References

1 Narziss, L. (1999) *Die Bierbrauerei. Band 1: Die Technologie der Malzbereitung*, 7th edn, Ferdinand Enke Verlag, Stuttgart.

2 Narziss, L. (2004) *Abriss der Bierbrauerei*, 7th edn, Wiley-VCH Verlag GmbH, Weinheim.

24
Packaging

Jörg Bückle

Wrapping bottles, jars, cans and multipacks in films, or packing them in trays, cartons or plastic crates entails ultra-stringent requirements for the machines and lines involved. For example, in a line rated at 50 000 bottles/hour, 2500 crates holding 20 bottles each have to be packed and unpacked per hour. This corresponds to 42 crates/min or 0.7 crates/s. Precision, speed, reliability, flexibility and cost-efficiency are the watchwords here. Packing technology is accorded maximum priority in modern-day plants.

24.1
Selecting the Suitable Machine Configuration

The starting point for choosing the machinery configuration in the dry end is the primary package. This, in conjunction with the non-returnable or returnable secondary packaging selected, determines the necessary or possible line concepts involved. This means that the containers and packs concerned will entail quite different approaches to machinery selection.

Returnable glass and returnable polyethylene terephthalate (PET) are characterized, for example, by a high degree of standardization for containers and packs. The machinery configuration is for this reason largely pre-determined. One special feature is the system required for handling the returning crates of empties. Owing to increasing individualization for crates and bottles, sorting the empties is assuming ever-greater importance. One special case is cartons for returnables, which are filled only once.

For kegging lines, too, the containers are mostly standardized. This means that basic machines comparable to those for returnable glass are used. This line, however, comprises only a palletizer and a depalletizer; there is no packer for the secondary packaging. When dimensioning the machinery, the heavy weight of the units being handled has to be allowed for. This is particularly applicable for robots, which are being used with increasing frequency for palletizing kegs. See Table 24.1.

Table 24.1 Possible machine configurations for returnables.

Empties sorting	manual: rotary machines (for sorting containers) automatic: linear machines (for sorting containers) 'sorting block' (for sorting packs and containers)
Depalletizer for plastic crates	classical robot
Unpacker for empties	intermittent (classical or robot) continuous
Empty-crate storage magazine	vertical horizontal (classical or robot)
Packer for fulls	intermittent (classical or robot) continuous
Palletizer for fulls including load stabilization (strapping)	classical robot

In the case of non-returnable PET, by contrast, the containers involved are not standardized and the variety of different packs is enormous. This means the entire system can be optimized in terms of logistics. The aim in this approach is to achieve optimum pallet capacity utilization, so as to minimize the logistical costs incurred. The PET line is additionally differentiated, in terms of the type of bottle feed involved, into lines with blow-molders, lines without blow-molders and mixed lines. See Table 24.2.

24.2
Packing into Packs Open at the Top

Packs open at the top, like bottle crates or cartons, are packed and unpacked using what are called pick-and-place packers, which operate either discontinuously or continuously. Increasing use is also being made of robots for this kind of application.

24.2.1
Classical Machine Design

In discontinuously operating intermittent packers (Figure 24.1), the gripper head moves horizontally between the pack and the container infeed. This horizontal travel is either positively controlled by mechanical means or can be user-defined

Table 24.2 Possible machine configurations for non-returnables.

Bottle feed	sweep-off depalletizer for empty bottles when handling finished bottles
	classical
	robot
	in the case of blow-molders, the bottles are fed in directly
Packer for packs (secondary packaging)	cardboard multipacks
	plastic carriers
	shrink
	wrap-around
Packer for tertiary packaging where necessary	wrap-around
	trays
	plastic crates
Palletizer	classical
	robot
	variations possible with or without layer pads
Load stabilization	shrink shrouds
	stretch-wrapping
	stretch shrouds

by means of a linear axle with its associated drive. The second, state-of-the-art system provides the (un)packer with more flexible application options, rendering it suitable for specialized tasks as well, for example, as a sorting packer.

When reaching the end of its horizontal travel, the gripper head descends and uses its gripper cups (Figure 24.2) to grasp the containers. It then moves back both vertically and horizontally, is lowered onto the waiting crates or cartons, and places the containers into these. The cycle then starts anew.

For packing and unpacking the containers, the machine is fitted with one or more gripper heads. The number and configuration of the gripper cups correspond to the packing unit concerned. If there are 20 bottles (4 × 5) in the crate, the gripper head will likewise possess 20 gripper cups (4 × 5). Each gripper head is consequently suitable for only one particular type of packing unit. Other types of pack (but also significantly divergent bottle types) will require the packer to be converted for change-over. Apart from the gripper system, additional product-specific components must also be replaced in this change-over routine, grouped together under the collective term of 'handling parts'.

Intermittent packers are limited in terms of their maximum output, since, in line with their design principle, the pack has to be motionless when the containers are being withdrawn or inserted. In the high-speed range, it is therefore often impossible for the containers and the packs to keep pace in the time available

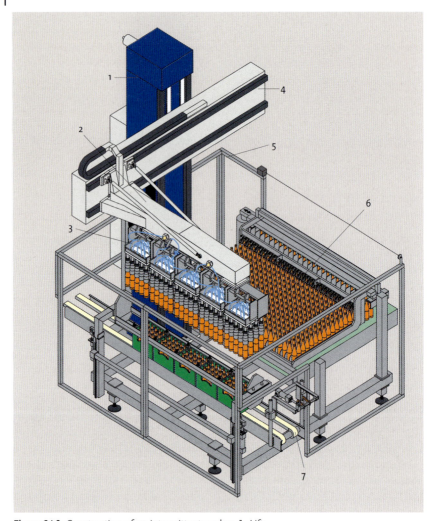

Figure 24.1 Construction of an intermittent packer. 1, Lift column; 2, Gripper head carrier; 3, Handling parts; 4, Horizontal axle; 5, Safety device; 6, Container table; 7, Pack conveyor.

('stop-and-go' operation). The output ranges between 300 and 5000 units/hour, depending on the type of pack and size of machine in question.

Rotary packers, by contrast, achieve hourly outputs of up to 6000 units, depending on the particular pack involved, because they perform the packing operation continuously in transit (Figure 24.3). For this purpose, container transport and pack conveyors are installed on opposite sides of the machine. The gripper heads are mounted on gripper-head carriages, which rotate horizontally around the machine tower on a precisely defined packing curve that has been exactly synchronized with the container and pack conveyors. The rotary packer is thus able to pick

24.2 Packing into Packs Open at the Top

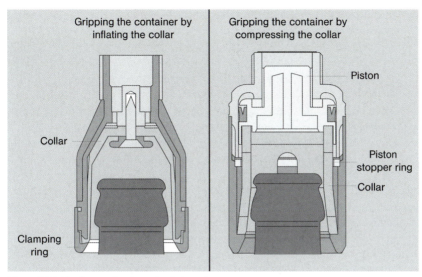

Figure 24.2 The heart of each packer is its gripper cups.

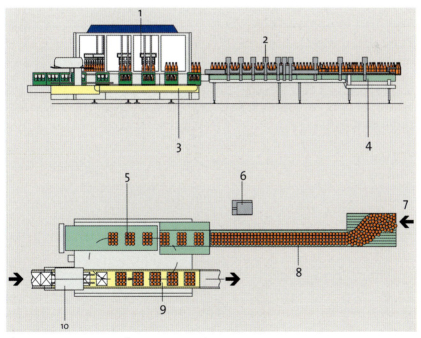

Figure 24.3 Construction of a continuous packer. 1, Gripper head; 2, Lane guide system; 3, Pack conveyor; 4, Container conveyor; 5, Container pick-up; 6, Coutrol panel; 7, Container infeed; 8, Container conveyor, 6-lane; 9, Pack conveyor; 10, Pack spacer.

up the containers continuously and deposit them gently in the packs waiting on the opposite conveyor track.

Rotary packers operate with the tried-and-tested standard gripper heads. In addition, gripper heads can be used that have been designed specifically for the particular product being handled. Apart from individual containers, multipacks can also be packed in crates, cartons or trays. Using functionally optimized head carriers and quick-release fastenings, the gripper heads can be swiftly replaced. Movable magazines are available for storing the gripper heads.

For small line ratings and for confined space situations, combined packers are the appropriate choice: they are able to unpack and pack crates in a single operating cycle. In these machines, the handling parts feature gripper heads of which half are used for unpacking only and the other half for packing. This ensures hygienically flawless handling of the fulls. The unpacker gripper heads pick up the bottles and the gripper head carrier moves to the container discharge table, where the bottles are set down. Simultaneously, the packer heads pick up the waiting bottles at the container infeed table. The gripper head carrier moves back to the pack conveyor and the bottles are packed in the waiting empty packs.

24.2.2
Robot Technology

Filling and packing operations in the food, beverage and consumer goods industries are characterized by a high degree of automation and optimum utilization of the integrated machinery involved. The trend towards small, flexible packaging units, however, encourages the use of multifunctional robots instead of specialized machinery designs. In this context, the three variants of gantry, articulated-arm and SCARA (Selective Compliance Assembly Robot Arm) robots are those mainly used (Figure 24.4).

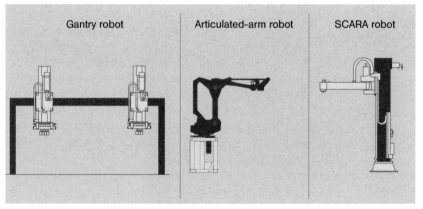

Figure 24.4 Robot designs for use in the beverage industry.

24.2 Packing into Packs Open at the Top

Gantry robots are available as two-, three- or four-axis systems with different axis lengths. Depending on the axis length involved, the maximum load capabilities will vary. A gantry robot usually operates with three linear axes, at right angles to each other, thus covering a cuboid-shaped working area. With modern-day drive systems, Cartesian robots are very fast and accurate. Moreover, they are smaller in construction than their conventionally driven predecessors.

The articulated-arm robot basically features four or six axes. One of these functions as the main rotary axis, on which the other movement axes are based. A kidney-shaped working area is derived from the lengths of the traversing axes and the maximum turning and articulation angles or arm lengths, with the robot able to reach any desired pick-up and set-down point inside this working area.

The SCARA robot is also designated as a horizontal articulated-arm robot. SCARA robots are particularly suitable for pick-and-place jobs. Being able to achieve extremely high speeds and accelerations, they make for very short cycle times, with a concomitantly high hourly output. SCARA robots are constructed as either three- or four-axis swivel-arm or two-axis linear types.

As in the classical pick-and-place packers, the gripping principle is matched to the container and the layer pattern concerned, so that mechanical grippers, suction cups or a combination of these can grasp the products securely and gently (Figure 24.5). The gripper head is mounted on a gripper head carrier with a central lock. As an optional extra, both articulated-arm and swivel-arm robots can be fitted with a system for fully automatic gripper head change-over.

Four programmable axes are usually sufficient for the handling tasks encountered in the beverage industry, enabling the requisite points/movements to be

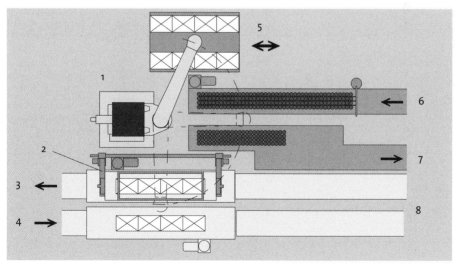

Figure 24.5 Functional principle of a combined unpacker/packer. 1, Robot; 2, Insertion frame for fulls; 3, Fulls; 4, Empties; 5, Gripper head change-over; 6, Container conveyor for filled bottles; 7, Container conveyor for empty bottles; 8, Pack conveyor.

implemented quite easily. A six-axis robot is only really necessary for handling kegs: the swiveling movements for keg turning at low to medium speeds can be executed directly by the robot.

When dimensioning and planning gripper heads, the robot's maximum permissible carrying capacity also has to be factored in. The figure of relevance here is always the total weight of the gripper and load together.

The third important parameter to be considered is the output. Practical experience has shown that in the beverage industry, by reason of the gripper technology, a maximum of six to seven cycles per minute is possible. Higher accelerations are feasible only when simple travel curves and non-sensitive loads are involved. The minimum cycle times specified in the data sheets of the robot vendors can for this reason not be adopted without qualification for use in the beverage industry either. In the high-speed range, conventional machine technology with complete-layer handling is accordingly still a viable alternative to robots.

24.3
Wrap-around Packaging

The classical process of wrapping a product in cardboard boxes comprises three phases: the pre-glued carton blank is erected and its base glued, then the containers are inserted, and finally the flaps are sealed, again using hotmelt. Thus, for this type of package, no fewer than three machines are required in the medium- to higher-output range. The wrap-around method, though, performs all the three packing steps mentioned above in a single continuously operating machine.

The wrap-around packer folds a flat carton blank (usually B- or E-flute and not pre-glued) directly around the containers and does not glue the package until after this has been done. This means the secondary packaging makes for firmer wrapping of the contents than a folding carton, thus providing better protection. These fully automatic, continuously operating packers handle 30–60 packs/min.

The basic machine comprises the following main assemblies: container feed with lane divider, machine frame, blank magazine, container inserter, pack conveyor and folding station. Thanks to the modularized machine concept and the servodrive technology utilized throughout, the packer can be accurately fine-tuned to the packaging job involved in each case (Figure 24.6). When these machines are supplemented to include a film-wrapping station with shrink tunnel, they are suitable for producing shrink-wrapped packs standing on a tray or a pad. Soft packages and multipacks can also be handled as an optional extra. Thanks to the packer's modular design, it is likewise possible to install a divider inserter above the container infeed.

State-of-the-art wrap-around packers operate predominantly on the in-line principle where container infeed and blank infeed are located one above the other (Figure 24.7), which makes for a 'slimmer' machine footprint than is the case with side loaders. In each cycle, the in-line packer's suction gripper heads remove the topmost blank from the lift magazine and place it horizontally on the blank

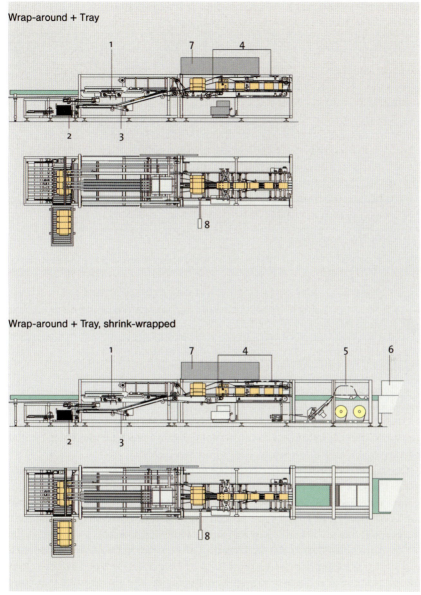

Figure 24.6 Variants of wrap-around packers. 1, Container spacer; 2, Blank magazine; 3, Blank infeed; 4, Pack formation unit; 5, Film wrapping station; 6, Shrink tunnel; 7, Control cabinet; 8, Control panel.

guide unit. The pusher bars then push the container groups onto the blanks fed in from below.

Guide bars located on both sides are used to passively erect the blanks as they continuously pass through. After that, a rotating flap closer tucks in the front and

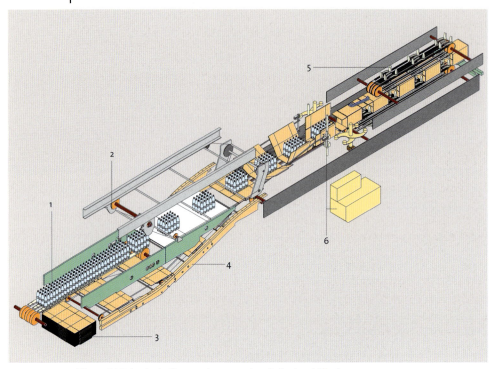

Figure 24.7 In the in-line version, container infeed and blank infeed are located one above the other. 1, Container infeed; 2, Container spacer; 3, Blank magazine; 4, Blank infeed; 5, Pack formation unit; 6, Gluing unit.

rear side flaps. Nozzles spray hotmelt onto the flaps, which are then glued to the bottom flaps in the downstream pack formation unit.

Circulating sealing chains press the bottom flaps against the side flaps, and also guide the pack thus produced through the machine. The top flaps are folded and glued to the manufacturer's joint. To complete the process, closing chains (engaging from above) seal the top flaps. Special pressing devices stabilize the pack until the hotmelt has set. When it has, the pack is ready for dispatch.

The blanks handled in the wrap-around process are delivered lying flat on a pallet in stacks. For product-friendly transport, there must not be an overhang and the blanks must be stacked with alternate sides facing upwards. In this case, approximately 50 blanks each are piled on top of each other in a cambered configuration. Ideally, a top sheet will protect the stack, with the base and the top sheet of a pallet carrying blanks additionally being strapped. The pallet's strapping should be cut open only shortly before the blanks are actually used at the packer and pallets whose load has not been completely used up should be covered with top sheets again, so as to preserve the blanks' as-delivered condition.

The storage conditions to be aimed for are 20 °C at 50–60 relative humidity. The blank material itself must not exhibit a moisture content of more than 11%, since otherwise the cardboard will lose part of its flexural strength.

It is advisable to take the blanks into the packing hall and keep them there for at least 1 day before they are processed, so they get used to the temperature inside the hall. Nonetheless, not more than 1 day's supply should be kept in the production hall, so as to prevent the ambient conditions prevailing there from causing changes in the material which might impair the packer's performance capabilities.

24.4 Shrink-Wrap Packaging

Shrink-wrappers form cans, glass or plastic containers into numerous multipack variants. In addition, several multipacks can be formed into a larger multipack. In the single-lane version, the machines can be run at between 18 and 120 packs/min, depending on the version involved. Thus, on two lanes the outputs can range up to 240 packs/min and on three lanes up to 360 packs/min.

The basic machine is modular in construction (Figure 24.8), with retrofit options to meet freshly perceived requirements as the years go by. If, for example, a dedicated shrink-wrapper is required to provide an option for subsequent tray handling, the basic machine can already be delivered with an extended machine frame. Thus when the plant subsequently upgrades to tray production, by installing the pack formation unit and the blank magazine, no more modifications will be needed at the container infeed and discharge sides.

The tray or pad blanks are removed from a magazine located underneath the container infeed and fed to a position below the container groups (Figure 24.9). The machine folds the tray blanks around the container groups and glues them at the flaps with hotmelt. Like all other packs, the trays pass through the machine long-side leading. When film-wrapping is performed without trays or pads, the containers are passed over plastic-coated slide plates through the pack formation unit.

The film magazine accommodates two film reels in its standard version. When changing over from reel 1 to reel 2, only an extremely short time is required for splicing together the film ends. An additional unit for automatic film reel splicing enables change-overs to be carried out without any interruption to the production process.

In the servo-controlled film-cutting module, a rotating blade separates the film into parts of precisely equal size. An electronically controlled film supply unit ensures that the film tension remains constant throughout. When the film concerned features cutting marks, meaning it contains product advertising, optical sensors guarantee precise positioning of the cut. After being cut the film is transported upwards by a vacuum-assisted conveyor to the film wrapper.

In the film wrapping station, a wrapping bar takes the film along with it and wraps each pack so as to ensure that the two ends of the cut section overlap underneath the pack. For output-related reasons, several wrapping bars are used, driven by a servomotor, which enables different traversing curves to be generated. Moreover, in the event of a malfunction, the film wrapping station can be moved to its

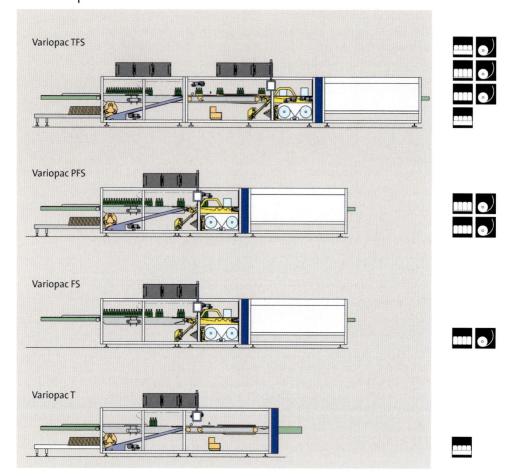

Figure 24.8 Variants of shrink-wrappers.

reference position independently of the rest of the machine, which means it is unnecessary to empty the entire machine first.

To ensure that the film reaches its optimum shrinkage temperature (which lies between 105 and 120 °C at the film's surface) in the downstream shrink tunnel, it has to be heated up with hot air from ambient temperature by approximately 80 °C. This process can be influenced, given a defined infeed speed, using the hot-air temperature and tunnel length parameters. If, for example, the tunnel is to be a short one, this will inevitably reduce the amount of processing time available. Heat-up is possible in this case only with a large temperature differential between the hot air and the pack. This entails a risk that the film will overheat, and begin to flow. The desired pack stability is then no longer assured.

By comparison, a longer tunnel (again with the same total output) offers not only an increased processing time, but also space for two heating zones, in which

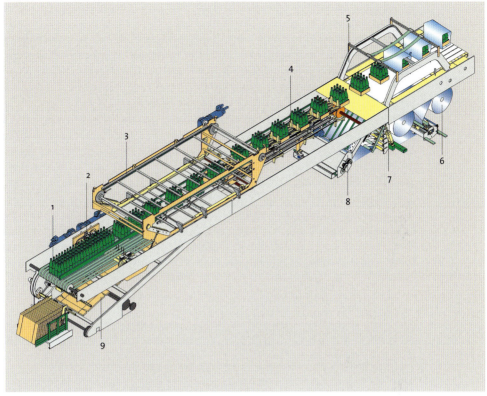

Figure 24.9 Functional principle of a tray shrink-wrapper.
1, Container infeed; 2, Container spacer; 3, Pusher chain unit; 4, Pack formation unit; 5, Film wrapping; 6, Film feed; 7, Film dancer; 8, Film cutting; 9, Carton feed.

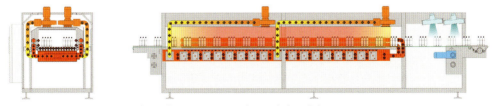

Figure 24.10 Shrink tunnel with two heating zones and extended cool-down section.

the air flows can be regulated independently of each other (Figure 24.10). This precise air-flow (and thus also temperature) control in the entire tunnel assumes particularly high importance with a multilane design, since here the middle track is protected by packs advancing in parallel.

Film-wrapped packs can subsequently be finished in a tray packer. Alternatively, the finished multipack is passed directly to the palletizer. In addition, carrier systems can be fitted to the pack downstream of the shrink-wrapper using a separate machine.

This state-of-the-art machine design has enabled significant cost reductions to be achieved in the packaging aids, thanks to increased non-susceptibility of the machines to quality fluctuations in the raw material. For example, the machines used to operate only if the carton pads involved were of optimum quality – nowadays the use of recycled cardboard is a viable option. The requisite wall height of the trays has also decreased: 40 mm is the current standard; less than 20 mm is possible.

Thanks to improved technology for film feed and film cutting, the film thickness has likewise shown a substantial decrease over the years. The films in common use today are 30–80 μm thick, with a maximum permissible tolerance for film thickness of ±5%. Films of 30–50 μm can be recommended only for small packs and/or when very high-quality material is being used. With a thinner film, however, there are more running meters on the reel, since the client buys the film by weight. The additional price for the thinner film may thus in fact pay off, because more packs can be wrapped in it (Figure 24.11).

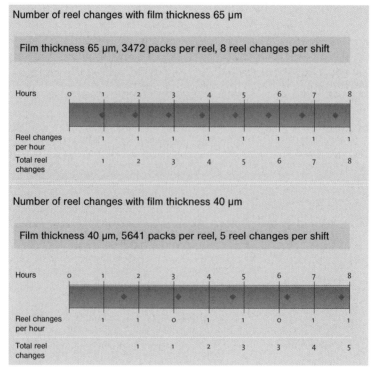

Figure 24.11 Reel changes required with various film thicknesses.

The film's shrinkage characteristics are stated in longitudinal and transverse directions. The values targeted will depend on the size and nature of the pack concerned. As the speed of the packaging machine increases, the film's stiffness assumes ever-greater importance for the handling reliability involved. Moreover, the films should slide easily against each other. The film-to-film friction coefficient must not exceed a value of 0.4. This figure applies solely for the film wrapping station and does not relate to downstream equipment. Stacking and transfer characteristics must also be given due attention in this context.

The shrink film is processed in either transparent or printed form. The finalized specification of the film's dimensions and its printed area will depend, for example, on the container shape and film properties involved. It is advisable to specify this only after testing and a trial run with filled original containers and sample films on the actual machine.

The film width is specified in accordance with the pack width plus the amount of film required to be protruding at each side. The film length is determined from the length and height of the pack concerned, plus the overlap. This overlap should be approximately 80 mm in order to achieve optimum machine running characteristics. The length determined must be verified by a trial run using original containers and corrected if necessary.

For a film with cutting marks, it is the printing roller that determines the effective or real film length. For reasons of production engineering (the film has to be tensed for printing and will contract slightly afterwards), the distance from printed image to printed image does not correspond exactly to the cylinder's circumference, but can be as much as 2% smaller. The film length parameter in the machine's control system must accordingly be calibrated against the actual film length and appropriately corrected if any deviation is found.

To ensure that the register mark sensor always scans only one signal, only one register mark is permitted on each section of the film. The precise position and the minimum size of the register mark will be entailed by the design characteristics of the system involved. In addition, it has to exhibit an edge at right angles to the direction of travel, which can be either printed or unprinted. If the printed image is serving as a register mark, the area being scanned must be rendered non-transparent throughout. In the overlap area, neither a printed image nor print pre-treatment is permitted, so as to assure optimum sealing of the film ends.

The area for the printed message or the barcodes ideally extends over an approximately level surface, meaning it does not continue into the contoured area of the containers. This prevents the printed image from becoming distorted or illegible because of the shrinking process or creasing.

Moreover, film reels should – if at all possible – be stored next to the shrink-wrapper for about 1 day before being processed, so as to even out any temperature differentials between the reel of film and the ambient air. This means that no water can condense onto the film's surface and cause production malfunctions. Care should be taken to avoid storing the film reels in locations exposed to sunlight or high temperatures.

24.5
Multipacks Made of Paperboard

The material used for cardboard multipacks is a sheet of heavy thick paper. It is a homogeneous material and thus significantly differs from the corrugated board used for wrap-around packaging. What users have come to appreciate about paperboard multipacks is their adaptability (Figure 24.12), plus the fact that the cardboard material can be imprinted with almost photographic quality, lending the multipacks an extremely attractive visual appearance.

For each type of multipack, the filling company is supplied with a fully printed blank, which is then handled on special packers. In preparation for the actual packing process, such a packer's container infeed first separates the containers into lanes. In the case of six-packs, for example, two conveyors feed the containers into the packer, with the container flows then being divided into units of 2×3.

Once the blank and container speeds have been synchronized, the folding function can begin: for simple blanks, this can be handled mechanically using guide plates. Complex folding operations necessitate the interaction of several mechanical and pneumatic components. The package is then glued at the bottom or closed using interleaving joins.

Since it is inherent in the system that the maximum output will decrease with increasing length of the packaging unit concerned, it is always the longest packaging unit that is the determinant parameter in the planning stage. If, for example, a four-pack and an eight-pack are to be handled on the same machine, this means that the four-pack, too, has to be run at the lower speed entailed by the eight-pack.

Conveyor lubrication may also affect the performance of the multipack line. Insufficient lubrication will cause problems with container feed, such as transport gaps in a canning line. If too much lubrication is provided, the multipacks will no longer be glued together properly, because the properties of the adhesion surface will alter. Too much lubrication will also in certain cases soften the cardboard, which may result in losing a container or in ugly patterns on the pack. Moist multipacks, too, are more easily damaged while being conveyed to the palletizer. Cardboard should therefore already possess certain wet-strength properties, which is achieved by admixing additives to the raw cardboard material involved.

Figure 24.12 Various kinds of multipacks.

Something that also has to be allowed for when designing the paperboard blank is the fact that PET bottles holding highly carbonated beverages expand after being filled. A second salient aspect of handling PET bottles is their elasticity, which plays a role in the context of pack transport. The backed-up packages may exert such severe pressure that the first multipacks in the row will fall apart. Milk cartons are even too flexible for being firmly and safely enclosed with normal material. That is why they are handled using blanks, provided with a contact adhesive, for example.

When it comes to storing the blanks, it should be borne in mind that paperboard loses or adsorbs moisture more quickly through its unprinted than its printed side. Opened pallets should accordingly be covered, so that the blanks do not 'corrugate'.

The storage conditions, moreover, should be kept as constant as possible. For this reason, the blanks are not suitable for lengthy storage in the infeed to the packer either.

Above and beyond this, empirical feedback has shown it is best to always begin production with paperboard blanks from a new pallet. If the machine is running perfectly, some blanks from already-opened pallets can be fed in as a test: if they do not cause any problems, they can continue to be handled. If malfunctions do occur, these blanks must be abandoned at the end of the packing operation.

In the event of suddenly occurring malfunctions, it is therefore also advisable to use a new pallet of blanks. This renders it easy to detect whether the paperboard or the machine is causing the malfunction involved. Only when a machinery malfunction has definitely been detected should the operator make any adjustments to the system.

24.6
Multipacks with a Plastic Carrier

Automated grouping of beverage containers using a lightweight plastic carrier to form a conveniently transportable multipack is another option for end-of-the-line packaging. The most widely used type for this packaging system is what is referred to as a rim-applied carrier fitted to the top edge of a can. However, there are a host of other types available, such as those fitted to the middle of the container (Figure 24.13), which renders them suitable for PET bottles as well. This type of carrier simultaneously increases the stability of the finished multipacks to such a degree that they can be palletized directly, meaning an additional tray can be dispensed with. In this case, a pallet's entire packaging material consists of one and the same substance, which makes disposal much easier. In addition, side-applied carriers protect the containers against scratching, because the containers in different multipacks do not come into direct contact with each other.

The disparate carrier types are handled in a continuous process. The types of machine offered differ, for example, in terms of the container feed, which directly affects the system's output. In the most widely used version, two cans are always

Figure 24.13 Plastic carriers group together beverage containers to form a conveniently transportable multipack.

fed in simultaneously. On a two-abreast machine, two-, four-, six- and eight-packs are formed. Alternatively, three cans also be fed in simultaneously. This three-abreast machine forms six- and 12-packs. However, it is also able to create other packages with a multiple of 3: nine- and 15-packs are likewise possible, for example. On high-speed machines, a 2 × 3 container feed is used: this system is able to pack six or 12 containers in parallel, which significantly increases the output of the packaging machine concerned.

In the packaging machine, the containers are passed under a rotating drum, via which the carriers are supplied: they are fed in continuously from a reel. Special grippers stretch the plastic carrier so that it can be fitted over the can or bottle concerned.

After the carrier has been transferred from the drum to the container, the plastic material contracts, and grips the container snugly. The carrier is then cut to size for the packaging unit desired and fed to the tray packer. As an option, the packages can also be cut to size only in the tray pack section itself.

The machines can also be combined with a container alignment feature, which turns the labeled or printed side of the container to face outwards. Each container, in addition, can be aligned individually, so as to create large-area advertising images if the container design is suitable for this purpose.

25
Cleaning and Disinfecting
Udo Praeckel

In the food industry, cleaning and disinfecting are of overriding importance when it comes to the quality and shelf-life of the product. Comprehensive and regular cleaning and disinfecting measures, which must always be considered to go hand-in-hand, are therefore an important requirement for proper process management. Necessary measures are always to be carried out based on a fixed schedule that takes the type of product and operational anomalies into account. Cleaning and disinfecting have two different goals: (i) product-contaminating substances are to be fully removed and (ii) microorganisms are to be rendered inactive.

Basically, contamination can be divided into two types: inorganic and organic. Inorganic contamination includes:

- Rust.
- Metal grit.
- Glass grit.
- Dust.
- Water scale.
- Beer scale.
- Soil (inorganic part).

Contamination of an organic nature includes:

- Beer residues.
- Hop resins.
- Residue (organic parts) of cleaning and disinfecting agents.
- Fibers.
- Body greases.
- Soil (organic part).
- Bacteria.
- Yeast.
- Molds.

Handbook of Brewing: Processes, Technology, Markets. Edited by H. M. Eßlinger
Copyright © 2009 WILEY-VCH Verlag GmbH & Co. KGaA, Weinheim
ISBN: 978-3-527-31674-8

25.1
Cleaning

Cleaning is the removal of contamination or undesired residues from hard surfaces with the aid of chemical and/or physical cleaning methods and agents. Factors for successful cleaning include:

- Temperature (hot cleaning, cold cleaning).
- Cleaning time (the longer the cleaning time, the greater the cleaning success).
- Mechanics (pressure, volume flow, flow speed).
- Chemical (type and concentration of the cleaning agent).

The most important factor is the selection of a suitable cleaning agent. The following minimum requirements must be met by a suitable cleaning agent:

- Quick and completely solubility in water.
- Quick swelling and detachment of the specific main components of the contaminant (e.g. protein, beer scale, hop resins and fur).
- High soiling-carrying capacity.
- Easy rinsing, so as to reduce the final rinsing times.
- Non-foaming or foam-reduction property for other foams.
- Compatibility with materials used in production systems.

An important requirement for effective cleaning is the ability to determine the concentration of the cleaning agent solution. This occurs either via simple titration or, in the case of automatic cleaning in place (CIP) systems, the concentration is determined via conductivity.

Mechanics are the physical conditions required for cleaning. These include the pressure (minimum 3–5 bar), volume flow and flow speed (3–4 m/s). These technical requirements must be taken into account during the planning and installation of the systems. The temperature of the cleaning agent depends on the technical options, stubbornness of the contamination to be removed and, of course, chemical composition of the cleaning agent solution itself.

The following temperature ranges are common:

- Fermentation tanks, storage tanks and bright beer tanks: cold to 40 °C
- Brewhouses, lauter tuns, mash tuns, wort coolers, all pipelines: 70–90 °C

The chemical processes of breaking up soiling (e.g. dissolving stone via acids), swelling of soil (e.g. with dried-on starch and protein residues), saponification (e.g. with greases), dispersion (e.g. hop resins) and rinsing clean are subject to rate laws. Strictly speaking, the soaking time means the contact time of the cleaning agent with the soiling at the right concentration and temperature.

Within certain limits, the temperature, time, mechanics and chemical factors can be exchanged with one another. It is thus possible, for example, to compensate for a low cleaning temperature by increasing the cleaning agent concentration and/or the flow speed.

25.1.1
Cleaning Agents

Basically, cleaning agents can be divided into three large groups: neutral, alkaline and acidic. Cleaners within each group have different compositions (Table 25.1). Neutral cleaning agents are used when easily water-soluble, dispersible or emulsifying contamination is to be removed from surfaces. Alkaline cleaning agents are then used when organic contamination that is not easily water-soluble is to be removed. In this case, a chemical transformation occurs – water-insoluble residues are changed to water-soluble fragments. Poorly water-soluble mineral contamination is removed when acidic cleaning agents chemically transform it into water-soluble salts.

25.1.1.1 Alkaline Cleaning Agents

Sodium hydroxide exhibits an excellent emulsifying capacity for protein. Accordingly, it is used in a wide variety of applications in breweries. Caustic potash has an even greater capacity for breaking up soiling than sodium hydroxide. It is only used in limited applications, however, due to the fact that is several times more expensive.

The disadvantages of sodium hydroxide include:

- No dispersion properties.
- No surface-active effect.
- No sequestering power.
- Not easily rinsed out.
- Attacks aluminum.
- Foams at high pressure.

To reduce these disadvantages to a minimum, various additives are added to the caustic soda solutions depending on the requirements of the situation. Since alkaline solutions strongly tend to foam up, defoamers must be added to them. A differentiation between hot and cold defoamers is made here. While hot defoamers are used for bottle cleaning, cold defoamers are used as foam inhibitors in brewhouse and pipe cleaning.

In many cases, more than just pure sodium hydroxide and an additive are used. In these cases, so-called prepared cleaners are used (i.e. the user receives a prepared concentrate, which he must then dilute to a specific usage concentration).

Alkaline products are not suitable for use with aluminum, however. Special cleaning agents containing silicates must be used for aluminum system parts. Silicates are also contained in alkaline cleaning agents. They have very good cleaning, dispersion and emulsifying properties. Sodium metasilicate is a very good corrosion inhibitor for aluminum. When using high temperatures, the silicate content in the cleaning agent may not be too high, otherwise there is a risk that calcium silicate may be precipitate. This is very hard to remove.

Table 25.1 Cleaning agents and additive substances.

Component	Active ingredient	Remarks	Use
Alkalis	NaOH/KOH	removal of organic soil; carbonate formation via CO_2	CIP cleaning; bottle cleaning; keg cleaning; foam cleaning; floor cleaning
Acids	phosphoric acid	inorganic acid; removal of inorganic and organic soiling; release of phosphates in waste water	CIP cleaning; keg cleaning; foam cleaning
	nitric acid	inorganic acid; removal of inorganic contamination (beer scale); passivation of stainless steel; release of nitrates in waste water	CIP cleaning; keg cleaning
	sulfuric acid	inorganic acid; removal of inorganic contamination (beer scale); pH-lowering agent; concentrate attacks stainless steel	CIP cleaning; bottle cleaning
	hydrochloric acid	inorganic acid; removal of inorganic contamination (hardness-mineral deposits); high corrosion potential	stone removal of systems; neutralization of waste water
	citric acid	organic acid; removal of inorganic soiling	CIP cleaning; foam cleaning
Additive substances			
oxidizing cleaning booster	hydrogen peroxide; chlorine	for the detachment of stubborn organic incrustations; products decompose during cleaning process	CIP cleaning; special cleaning in circulation process
dispergators; sequestering agents	polycarboxylates; phosphonates	increase the soil-release capacity of the cleaning solution; improve cleaning success	bottle cleaning; CIP cleaning
surfactants; defoamers	non-ionic surfactants	decrease the surface tension of the cleaning solution; prevent disturbing foam forming	bottle cleaning; CIP cleaning
complexing agents	NTA; EDTA; gluconate; phosphate	complexing of the water-hardness and metal ions in the solution; prevent precipitation and deposits	bottle cleaning; CIP cleaning; alkaline foam cleaning
solubilizers	alcohols; glycols	stabilize individual ingredients in the concentrate of prepared cleaning agents	prepared cleaning agents

Another ingredient of alkaline cleaning agents is sodium carbonate. It does not exhibit great cleaning, dispersion and emulsifying properties, but can be used as a component of the cleaning agent formulation for aluminum.

The primary purpose of the inclusion of complexing agents in alkaline cleaners is to prevent the formation of deposits on surfaces. They complex the builders of water hardness so that they cannot be deposited on heat-exchanger surfaces, for example. Previously, sodium polyphosphate was the most important of these, especially due to the fact that it also exhibits active cleaning properties. These types of complexing agents do not only sequester the minerals causing hardness dissolved by water, but can also dissolve mineral precipitations when used for a longer period of time.

Dispersion agents enhance the soil-release ability of the cleaning solution and prevent the growth of hardness-mineral crystals in water. They are advantageous in that they do not react with metal ions 1:1 as do complexing agents. Rather, they can be used in the sub-stoichiometric range (threshold effect).

Surfactants are usually water-soluble, surface-active chemicals added to cleaning agents as wetting and dispersion agents. Differentiation is made between anionic, cationic, amphoteric and non-ionic surfactants. Anionic and non-ionic surfactants primary function as cleaners.

The advantages of using surfactants include:

- Surfactants allow the cleaning solution to penetrate into narrow gaps by lowering its surface tension.
- Surfactants facilitate the penetration of the cleaning solution and thus accelerate the swelling of residue.
- Surfactants emulsify grease in the aqueous phase.
- Surfactants also transform water-insoluble soiling into an apparently soluble form.

25.1.1.2 Acidic Cleaning Agents

The most suitable acidic cleaners are products with phosphoric acid. Phosphoric acid far exceeds nitric acid and sulfuric acid in its cleaning power. Almost all cleaning agents found on the market (except for pure beer scale removal agent) are based on phosphoric acid. Mixtures with other acids are also available, which both occurs for reasons of greater specific conductivity and allows for better dissolving of inorganic coatings. In conjunction with surfactants, products based on phosphoric acid are not harmful to stainless steel, even at greater temperatures (over 80 °C).

Mineral deposits (e.g. beer scale and water scale) are best removed with nitric acid cleaning agents. They transform water-insoluble salts into a water-soluble, easily rinsing form. Nitric acid is not harmful to stainless steel. These products can also be used on aluminum. They are harmful to non-ferrous metals, however. The use of nitric acid products is not without risks, since the reaction with organic substances can release poisonous nitrous gasses.

25.2
Disinfecting

Food and beer residues are the ideal nutrient media for every type of microorganism; whilst they are necessary for the production of foods, but they can also cause spoilage and food poisoning. To prevent biologically induced spoilage, food must be protected from contamination with germs. Sufficient protection is only achieved through hygienic production and packaging.

Disinfecting agents used in the food industry are tasked with making production equipment free of microorganisms after use and subsequent cleaning. Microorganisms can be killed physically and chemically. Physical elimination involves heat treatment, UV and X-rays and other methods. Disinfecting with chemicals is possible via a host of disinfecting agents.

Efficient disinfecting with any agent can only be ensured when a clean, physically intact smooth surface is being treated. All pores, cracks, deposits and other surface damages hinder meaningful disinfecting.

The requirements of disinfecting agents are as follows:

- Range of microbicidal effectiveness.
- Effectiveness at low temperatures.
- Toxicity.
- Effectiveness under organic load.
- Easily rinsed out.
- Material compatibility (danger of corrosion).
- Stability when stored.
- Environmental friendliness.
- Economy.

The following active disinfecting substances are important in the beverages industry (Table 25.2):

- Substances containing active chlorine.
- Oxidizing agents.
- Aldehydes.
- Biguanides.
- Quaternary ammonium compounds (QACs).
- Chlorine dioxide.
- Halogenated carboxylic acids.

In the past, disinfecting products containing phenol derivatives were used often. Today, however, they are of very little importance in this field.

Products containing active chlorine have long been used to disinfect systems in the food industry. Generally, products based on sodium hypochlorite are used. In the weak acidic range, chlorine disinfecting occurs much quicker than in the alkaline range. For one thing, the decreasing stability of the chlorine carrier on the acidic level greatly reduces its usefulness. There is also a risk of pitting of

Table 25.2 Disinfecting substances.

Active ingredient	Remarks	Use
Peroxyacetic acid	acidic disinfecting agent with oxidizing effect (destroys cell membrane); conditionally stackable due to loss of effectiveness; automatic dosing via inorganic conductive acids only; sealing materials may be harmed with extended contact; very broad range of effectiveness	bottle cleaning; CIP cleaning
Hydrogen peroxide	neutral disinfecting agent with oxidizing effect; very environmentally and waste water friendly, since it decomposes with organic material in water and oxygen; high usage concentration; very broad range of effectiveness	CIP cleaning; spray disinfecting
Active chlorine (sodium hypochlorite)	alkaline disinfecting agent with oxidizing effect; danger of chlorophenol formation (negatively effects taste of the product); very broad range of effectiveness; ATTENTION: when mixing with acidic solutions, chlorine gas is released!	bottle cleaning; CIP cleaning; drinking-water disinfecting
Chlorine dioxide	disinfecting agent with oxidizing effect; two-component system that is mixed on-site when used; economical operating costs, but high investment costs; very broad range of effectiveness	bottle cleaning; CIP cleaning; drinking-water disinfecting
Quaternary ammonium compounds	neutral disinfecting agent (surfactants); destroys the cell membrane; heavily foaming (not suitable for CIP); surface-active; relatively difficult to rinse out due to the surface activity (adheres well to the surface)	static disinfecting; spray disinfecting
Biguanides	neutral disinfecting agent; destroys the cell membrane; forms deposits in alkaline medium; well suited to manual applications	head space disinfecting; static disinfecting; spray disinfecting
Aldehydes	neutral disinfecting agent; broad range of effectiveness and thus highly effective; oxidizes with air and then forms brown deposits	static disinfecting; head space disinfecting
Halogenated carboxylic acids	acidic disinfecting agent; very broad range of effectiveness; cannot be processed manually due to high toxicity; release of AOX in waste water; offered as a combined cleaning and disinfecting agent together with an inorganic acid	CIP cleaning

AOX indicates the sum of all 'absorbable organic halogen substances'.

chrome-nickel-steel and aluminum surfaces on the acidic level. It is absolutely imperative that the pH range be between 10 and 12 here.

If organic contamination is present, the effectiveness of the chlorine is reduced considerably (chlorine degradation). Chlorine products are also unstable at high temperatures (over 40 °C).

Products containing peroxyacetic acid are increasingly being used for disinfecting in the food industry. Such products are advantageous in that they are practically undetectable in the case of insufficient final rinsing. Peroxyacetic acid is a strong oxidizing agent comprised of a mixture of peroxyacetic acid, acetic acid and hydrogen peroxide. Practically all microorganisms are killed with these products. Room temperature and a low concentration kills not only vegetative bacteria of all types, yeasts and molds, but also endospores of the otherwise difficult-to-combat *Bacillus* spp. and *Clostridium* spp. To a certain extent, a disinfecting agent solution containing peroxyacetic acid can even be stacked. It is imperative, however, that it be absolutely free of organic contamination here.

QACs are cationic, surface-active substances that can actively penetrate gaps and pores thanks to their low surface tension. QACs are characterized by their effectiveness across a very broad pH range. They are effective in both the acidic and alkaline ranges. They are indifferent to metals and plastics and do not cause corrosion. At the concentrations used, QACs are not dangerous to handle, have no odor and are not harmful to skin. Gram-positive bacteria are easily killed, whereas greater concentrations or longer contact times are required for Gram-negative bacteria in general. QACs foam very heavily and can usually not be used in CIP systems.

Due to their substantivity, these products adhere strongly to the surface, which makes the products very difficult to rinse off.

Halogenated carboxylic acids (bromoacetic acid, chloroacetic acid) have long been favored as disinfecting agents, especially in CIP systems in the beverages industry. As with peroxyacetic acid, these products have a very wide range of effectiveness. In contrast, however, they do not lose their effectiveness when organic soiling is present. Halogenated carboxylic acids are also used in acidic media and thus formulated with phosphoric acid and/or sulfuric acid.

The law requires that cleaning and disinfecting agents must be removed from production plants in the food industry through sufficient final rinsing to the point that only technically unavoidable contamination is present. Final rinsing must occur with water, which fulfills the requirements for drinking water.

Using beer as an example illustrates the negative effects of insufficient final rinsing on the beer. QACs cause beer to cloud (protein precipitation) and destroy beer foam. They are very difficult to rinse out due to their substantivity and minimal surface tension. Aldehydes also precipitate proteins and cause beer to cloud. Products containing active chlorine can form chlorophenols through reactions with organic substances and thus considerably damage the sensory experience of beer and alcohol-free beverages. Peroxyacetic acid causes oxygen absorption and thus promotes oxidizing processes in beer. Disinfecting agent residue that does not visibly affect the beer can otherwise affect the senses unpleasantly. For this reason, the law requires that final rinsing be carried out for each agent. This is also absolutely necessary for quality reasons.

25.3
Cleaning Methods

CIP is the automatic internal cleaning of production plants, such as tanks, containers, heaters, hoses and pipelines. Production plants covered with product residues are cleaned and disinfected without the need for disassembly. Cleaning and disinfecting agent solutions are sprayed onto the surfaces to be cleaned using pumps via suitable sprayers. A differentiation between 'loss' cleaning and disinfecting and 'stack' cleaning and disinfecting is made here. In both cases, the cleaning and disinfecting process is automatic.

Automation means to save time and money, facilitated working, and greater working and product safety. CIP cleaning and disinfecting methods must ensure that all parts contacted by the foodstuff are clean. The condition here is that the production plant and CIP system must be constructed in such a way that this requirement is fulfilled.

Isolated CIP systems are required for the following production areas of a brewery:

- Brewhouse
- Pre-filtrate (fermentation tank, storage tank).
- Filtration and special systems.
- Filtrate (bright beer tank, filling).

To clean and disinfect a production plant with the CIP method, cleaning and disinfecting agent solutions that are pumped through these systems in the circuit are required. These cleaning solutions must be provided by a supply system according to the consecutively occurring rinsing, cleaning and disinfecting steps.

Two method variants are available here:

- 'Non-recovery' CIP cleaning with single use of a freshly prepared cleaning solution.
- 'Recovery tank' CIP cleaning with multiple reuse of prepared cleaning solutions.

25.3.1
Non-Recovery CIP Cleaning Method

The 'non-recovery' CIP cleaning system is comprised of a buffer container, pressure pump, valves, dosing equipment, heating system and a controller, and operates as follows. For pre-rinsing, water is taken from a buffer container, pumped through the system or tank to be cleaned via the pressure pump and discharged via the return line for waste water treatment. In the following cleaning step, water is also taken from the buffer container and then circulated through the system in the circuit with added cleaning agents. Once the cleaning step is complete, the used cleaning agent is discharged. The non-recovery CIP cleaning system (loss cleaning) is more economical for small circuit volumes (less than 750 l) than recovery tank CIP cleaning.

Advantages of non-recovery (new preparation) CIP cleaning include:

- A cleaning solution with a defined concentration is added for each rinsing step. This also allows different concentrations in different cleaning paths or cleaning programmes and thus optimum adaptation to the type of soiling.

- New preparation means that a cleaning solution that is not loaded with soiling particles from previous cleanings is used.

- By discharging the cleaning solution after every rinsing step, cross-contamination is prevented.

- The plant expenditure is minimal and thus the fixed costs of cleaning are considerably reduced.

25.3.2
Recovery Tank CIP Cleaning Method

Recovery tank cleaning systems are generally used as centralized systems – different production areas are cleaned with various cleaning circuits from a centralized supply system. The CIP cleaning plant contains stacking containers for alkaline and acidic solutions and disinfecting agent solutions that are prepared ready for use. If possible, they are split up into application and concentration areas (Figure 25.1).

The pre-selected concentration is maintained in each container via dosing systems comprised of a pump, mixer and conductometer. After use, these cleaning and disinfecting agent solutions are fed back into the stacking container to be used again during the next cleaning cycle. The concentration is maintained by adding the corresponding cleaning or disinfecting agent concentrate before being reused.

The CIP plant also includes tanks for fresh-water pre-run, rinse water stacking, and pipe systems with valves for pre-run and return and pumps for the circuit. If necessary, heating registers are installed in the stack tanks so that they can be heated to the desired temperature directly when each rinsing step begins. If the cleaning media are used at an increased temperature, the stack tanks should be provided with insulation.

If the cleaning medium is fed to the pre-run at temperatures above 80 °C, heat exchangers should be installed in the cleaning agent pre-run. Before the CIP process is carried out, the contents of the stacking containers can be heated one-by-one to the desired application temperature in the circuit via these heat exchangers.

The advantages of the recovery tank CIP cleaning system include:

- Lower consumption of water and heating energy for cleaning solutions thanks to reusability.
- More economical utilization of cleaning solutions that must be used in higher concentrations.

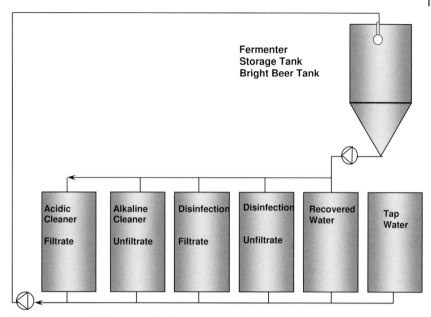

Figure 25.1 Example for a centralized CIP station.

- Shorter cleaning times thanks to the preparation of a ready-to-use cleaning solution at the required temperature.
- Shorter cleaning times thanks to the use of greater cleaning agent concentrations.

25.3.3
Combined CIP Cleaning Method

Combined CIP cleaning systems couple the advantages of recovery tank CIP cleaning plants and non-recovery tank CIP cleaning systems, and represent state-of-the-art technology.

Combined cleaning plants are different from recovery tank cleaning systems due to the following options:

- Bypassing of the stack tanks in/during the cleaning circuit.
- Cleaning-specific heating during the cleaning circuit.
- Option of direct dosing of cleaning agent for cleaning-specific adaptation of the required cleaning agent concentration.

These technical system conditions achieve the following:

- Stack or loss cleaning can occur.
- The cleaning solution needs only be heated to the required temperature relevant for the soiling.

25 Cleaning and Disinfecting

- Different cleaning agent concentrations can be used in the different cleaning circuits, since the stack tanks are only used with the concentration required for the 'simplest' soiling.

Permanently installed low-pressure spray balls or spray jet cleaners (Figures 25.2 and 25.3) are currently the optimum technical solution for the spraying of cleaning

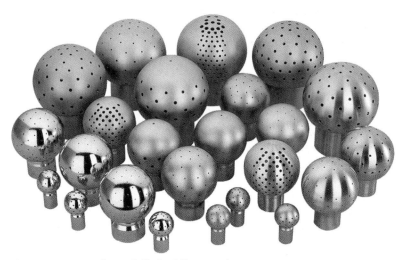

Figure 25.2 Variety of spray balls for different applications (Photo: Tuchenhagen GmbH, member of the GEA Group).

Figure 25.3 Jet spray head (photo: Tuchenhagen GmbH, member of the GEA Group).

solutions in CIP plants. The dimensioning, type of installation and selection of spray heads must be carried out very carefully. When doing so, the following must be taken into consideration:

- Tank shape (upright, horizontal, rectangular, cylindrical, cylindro-conical).
- Tank dimensions.
- Position and size of manholes.
- Position and shape fixtures for mixers, heating coils, fluid baffles and connection sockets.
- Tank materials, including seals.
- Size of the outlet connection.
- Type of soiling and assessment of the difficulty of its chemical detachment.

The installation depth and attachment depend on sufficient flow volumes at both the top and the side tank walls. Spray balls that are too low do not achieve sufficient spray width of the spray jet in the upward direction, and spray balls that are too high deflect the spray jets so heavily due to the unfavorable angle of incidence that the quickly descending cleaning solution could upset the entire spray jet distribution of the spray head. In the case of horizontal tanks, the installation depth is determined by the tank filling plan of the production, or the spray balls are to be installed so low that the spray jets do not influence one another. The goal with both upright and horizontal tanks is optimum overwhelming in the heavily soiled area. The lower area is cleaned easily by the downward-flying cleaning solution. For a horizontal fermentation tank, a spray angle greater 180° is to be selected, and for a horizontal storage tank and bright beer tank, below 180° (Figure 25.4).

Selection of the type, shape, power, positioning and number of spray balls must occur based on the methods of process engineering, since the hydraulic system comprised of the sprayers and cleaning pump and the proper selection of the cleaning and disinfecting agent are the most important factors for successful cleaning and disinfecting.

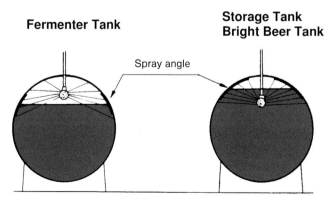

Figure 25.4 Spray angle for different applications.

The cleaning circuits are of basically two types: 'open cleaning circuit' and 'closed cleaning circuit'. An open cleaning circuit is a process in which not all units (tanks, containers) to be cleaned are completely filled with cleaning fluid. This process is usually required for the pre-run and return. In the closed cleaning circuit, all pipelines and units are completely filled with water and the cleaning fluid can be lead through the system depending on the pressure. A typical cleaning program for an 'open' and 'closed' cleaning circuit is specified in Tables 25.3 and 25.4.

Depending on the type and quantity of soiling in the individual areas of a brewery, there are a variety of options for cleaning and disinfecting (Table 25.5). Heavily soiled vessels in the brewhouse and pipelines are usually cleaned with an alkaline solution after pre-rinsing, and acidic cleaning occurs after the corresponding intermediate rinsing.

Fermentation tanks, storage tanks and pipelines in this area can be cleaned with an alkaline cleaner and then an acid with subsequent disinfecting or, if the con-

Table 25.3 Typical CIP program for open cleaning circuit (tank program).

Rinsing with fresh water/return water
Pumping down, drain
Rinsing with recovered solution
Pumping down, drain
Filling circuit
Creating circuit
Heating
Dosing (alkali)
CIRCUIT
Pumping down, recovering
Rinsing with fresh water (tap water)
Pumping down, recovering
Rinsing with fresh water (tap water)
Pumping down return water/drain
Filling circuit
Creating circuit
Dosing (acid)
CIRCUIT
Pumping down recovering/drain
Rinsing with fresh water (tap water)
Pumping down, recovering/drain
Filling circuit
Creating circuit
Dosing (disinfecting)
CIRCUIT
Pumping down, recovering
Rinsing with fresh water (tap water)
Pumping down, recovering/drain
Rinsing with fresh water (tap water)

Table 25.4 Typical CIP program for closed cleaning circuit (pipeline program).

[Rinsing with fresh water (tap water)/return water]
Rinsing stacked solution
Creating circuit
Heating
Dosing (alkali)
CIRCUIT
Rinsing with fresh water (tap water), recovering/drain
Creating circuit
Dosing (acid)
CIRCUIT
Rinsing with fresh water (tap water), recovering/drain
Creating circuit
Dosing (disinfecting)
CIRCUIT
Rinsing with fresh water (tap water), recovering/drain
Rinsing with fresh water (tap water)

tamination is less, cleaning with an alkaline cleaning step followed by a subsequent combined acid cleaning and disinfecting product. The use of alkaline cleaning agents requires a CO_2-free environment to prevent the sodium hydroxide reacting with the CO_2 and forming soda and sodium hydrogen carbonate. Soda and sodium hydrogen carbonate exhibit only minimal cleaning power. The conductivities only differ minimally, however, and are like that of sodium hydroxide. For this reason, it is necessary to monitor the concentration of sodium hydroxide via titration. The reaction between sodium hydroxide and CO_2, if it was not previously blown out with air, can lead to negative pressure (vacuum) in closed containers (cylindro-conical tanks) and thus to container damage.

Previously, bright beer tanks were cleaned with an acid and then disinfected after intermediate rinsing. The development of so-called 'one-step cleaners' (i.e. acidic cleaning agents containing one or more disinfecting substances) enables the cleaning and disinfecting process to be carried out in a single step. In comparison to the process described above, this process allows the elimination of two process steps. This means that, in addition to corresponding time savings, water is saved due to the elimination of intermediate rinsing.

Mobile CIP systems (Figure 25.5) were developed for smaller, horizontal tanks or tank systems for which upgrading with CIP spray heads was not worthwhile. A nozzle head that sprays the bundled water jets against the interior of the tank is installed on a four-wheeled base. Both the nozzles themselves (horizontal rotational axis) and the entire head (vertical rotational axis) rotate. This results in an even spray pattern over the entire interior surface of the tank. Deposits are dissolved in the cleaning fluid, and the spray jet supports this process mechanically. A pump outside the tank sucks up the provided liquid in the tank or in a separate container and feeds it to the nozzle head inside the tank through a hose lead

Table 25.5 Overview of CIP program variants.

Applications	Programs				
	Alkaline and acidic cleaning	Alkaline and acidic cleaning and disinfecting	Alkaline cleaning and combined acidic cleaning/disinfecting	Acidic cleaning and disinfecting	Combined acidic cleaning/disinfecting
	brewhouse vessels; pipelines	fermentation tanks; storage tanks; pipelines	fermentation tanks; storage tanks; pipelines	storage tanks; bright beer tanks	bright beer tanks
Program steps					
Pre-rinsing	×	×	×	×	×
alkaline cleaning (1.5–2.0% NaOH; 0.3% additive)	×	×	×		
intermediate rinsing	×	×	×		
acidic cleaning (1.0–1.5% acid)	×	×		×	
combined acidic cleaning and disinfecting (1% acidic cleaner/disinfecting)			×		×
intermediate rinsing		×		×	
disinfecting (0.3% peroxyacetic acid)		×		×	
final rinsing	×	×	×	×	×

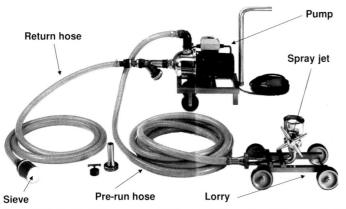

Figure 25.5 Mobile CIP equipment 'TC-Tank Jet' (Photo: Tensid-Chemie GmbH).

through the manhole. The position of the carriage in the large container can be changed with the progression of the cleaning process from the outside.

From an economic and, just as important, ecological standpoint, CIP systems should be operated optimally. Cost savings are possible through the reduction of energy costs, and time can be saved in rinsing procedures, reduction of the quantities of fresh water used and thus the waste water quantity, and through the proper usage concentration of the cleaning and disinfecting products. Analysis tools that can measure the temperature, pressure, conductivity value, volume flow and valve setting parameters during a cleaning and disinfecting step are available today. They are then displayed graphically after the analysis (Figure 25.6(a)).

After the analysis and interpretation of the measurement results (actual state), it can be determined whether the cleaning and disinfecting programmes are set optimally or whether there is potential for savings. In the example shown in Figure 25.6(b), rinsing with fresh water in a brewery was carried out for far too long after a cleaning step. In this example, you can see that a reduction in the fresh-water rinsing time from 10 to 5 min is sufficient. This corresponds to a savings of $1.6\,m^3$ of water per cleaning step. Accordingly, the amount of fresh water saved during 1 year's worth of cleaning would be $2,880\,m^3$. Considering the costs of tap water and waste water, approximately 10,000 Euro could be saved per year with this plant alone.

25.4
Material Compatibility

The most common material used in the manufacture of tanks, equipment and pipelines is chrome-nickel-steel. Aluminum, copper and standard steel with linings (for tanks) are also still used, albeit to a lesser degree. Chrome-nickel-steel is characterized by its very good chemical resistance to all alkaline and acidic

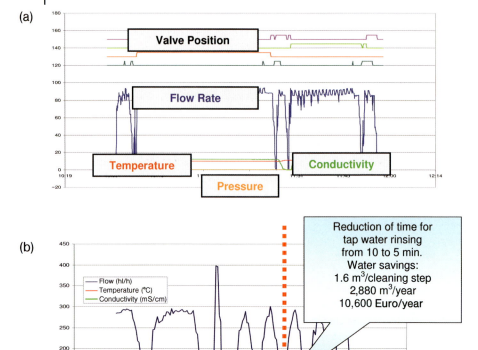

Figure 25.6 (a) Parameters that are important for a successful cleaning. (b) Measurement record of a CIP cleaning process (Tensid-Chemie GmbH).

cleaning agents in common concentrations, with the exception of cleaning agents containing sulfuric acid. Disinfecting agents containing hypochlorite (active chlorine) and water with a heavy concentration of chloride ions can lead to pitting, depending on the concentration, temperature and contact duration. When using cleaning agents containing active chlorine, it is absolutely imperative that they not be allowed to mix with acidic cleaning solutions (due to the formation of chlorine gas).

Aluminum is generally attacked by all acidic and alkaline cleaning agents. Cleaning agents containing sulfuric acid and nitric acid create chemically-safe passive coatings under certain circumstances, however. In the case of cleaning agents containing phosphoric acid, the amount of weight lost can be categorized as consistent. Alkaline cleaning agents are especially harmful to aluminum unless special corrosion inhibitors are contained in these cleaning agents.

System parts made of copper are to be viewed critically from a corrosion standpoint if oxidizing cleaning solutions are to be used. Oxygen released in the cleaning solutions also accelerates the harm done to the material. In CIP circuits that contain system parts made of copper or copper alloys, only cleaning agents with special corrosion inhibitors should be used for this reason; oxidizing cleaning agents, such as cleaning agents containing nitric acid, chlorine or active oxygen lead to increased surface attack or discoloration.

Coated tanks and equipment in particular are designated as being highly susceptible to corrosion from a cleaning standpoint if damaged linings (hair fractures) come into contact with acidic cleaning solutions. The resistance of lining materials to cleaning agents must be obtained from the respective manufacturer due to their differing chemical compositions (epoxy resins, phenolic resins, epoxy–phenolic resins, glass enamels). In general, glass enamels and epoxy resin coatings, phenolic resin coatings and epoxy–phenolic resin coatings exhibit good resistance to acidic cleaning agents at room temperature. Alkaline cleaners, on the other hand, should not be used above a specified upper temperature limit. Essentially, they cannot be used with phenolic resin and epoxy–phenol resin linings. Oxidizing cleaning agents (active chlorine, active oxygen) may only be used after preliminary testing or agreement with the manufacturer of the tank lining.

If sealing materials come into contact with food, they must exhibit certain levels of cleanliness. Since cleaning and disinfecting can almost never be carried out with only a single chemical type, the optimum material must be selected from the available range. In the case of seals, consideration of the chemical resistance and the swelling behavior are often decisive for the proper selection of a suitable material. In addition to the multitude of factors affecting the seal, other influential values must be taken into account (e.g. the temperature, resistance to grease, pressure, speed, surface composition of metallic parts and the material type of the machine parts to be sealed). Compatibility tests should also be carried out in this case before making a decision on the sealing materials to be used.

25.5
Cleaning Glass Bottles

Industrial bottle cleaning is a complex, engineered process that is subject to many dependencies. Cleaned bottles in the beverages industry are subject to the following requirements:

- Freedom from typical empty bottle soiling, such as mold, dust and dried-on beverage residues.
- Freedom from any bottle covering (labels).
- Freedom from beverage-spoiling germs.
- Clear, shiny appearance.
- No odors present.
- No chemicals present.
- A bottle temperature suitable for filling.

Bottle cleaning in the brewery and beverages industries has always been important. Several types of machines with different structures and functions have been developed over time. Modern bottle-cleaning machines operate with the combined soak-rinse process. Here, the soaking zones are used for loosening the soiling and spraying removes the loosened soiling so that the cleaning solution can be optimally effective on the remaining soiling. Essentially, a differentiation is made between single-end and double-end machines. Both machine types have specific advantages and disadvantages, but these have no bearing on the effectiveness of cleaning. While German breweries predominantly use single-end machines, double-end machines are often used for bottle cleaning in breweries in other countries.

25.5.1
Bottle-Cleaning Machine

A modern bottle-cleaning machine is constructed based on the following treatment zones:

- Residue emptying.
- Pre-soak.
- Caustic soak zone (several possible).
- Caustic spray (several possible).
- Intermediate spray zone.
- Hot-water zone (several possible).
- Cold-water zone.
- Fresh-water zone.

The individual zones in the bottle-cleaning machine (Figure 25.7) carry out the following tasks:

25.5.1.1 Residual Draining
By rotating the bottles, any liquid found within can run out and be removed from the machine, and thus does not affect downstream zones.

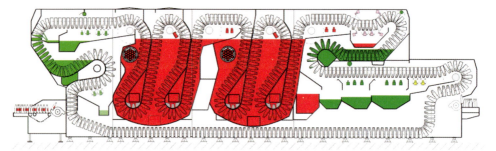

Figure 25.7 Double-end bottle washing machine Innoclean DM 52/87 24 (KHS AG).

25.5.1.2 Pre-soak with Pre-spray

In the pre-soaking zone, the bottles are filled with water (drained, heated water from the intermediate spray zone) and then emptied again. This removes most loose, clinging contamination and beverage residues from the bottles so that as little soiling as possible ends up in the main caustic soaker bath (which results to longer useful lye life and therefore less alkali consumption). In addition, the CO_2 gas found in the bottles is removed from the bottles and partially neutralized via the pre-heating water. Pre-spraying and vapor haul-off ensure that no CO_2 or only minimal CO_2 ends up in the alkaline area, which would neutralize the cleaning lye (sodium carbonate formation).

In addition, the bottles are slowly heated by the draining intermediate spray water. The pre-caustic is moved toward the post-caustic via a heat exchanger, whereby a large amount of the heat from the post-caustic can be recuperated. This heat exchanger saves lots of energy in heating up bottles.

25.5.1.3 Main Caustic Soaker Bath and Caustic Spray

Caustic soaker baths are the most important zones of the bottle-cleaning machine. These baths are responsible for the main cleaning due to their temperature being the highest (i.e. beer residue is broken up, greasy contamination is saponified and can be emulsified, and protein substances are denatured and detached from the bottle surfaces). Labels are detached from bottle surfaces through the combination of heat, sodium hydroxide, additive (concentrate) and contact duration.

The number of main caustic soaker baths is determined based on the degree of bottle soiling. In the case of slightly soiled bottles, an alkaline immersion bath with an approximately 6 minutes immersion time is sufficient to properly clean the bottles.

In the case of bottles with heavy contamination (e.g. mold deposits and dried-on soiling), at least two caustic baths with a spraying session inbetween them must be used to properly clean the bottles for an immersion time of 12–14 minutes. Caustic-spray zones and the label-discharge systems are located at the end of the soaker baths. Loose clinging contamination is rinsed off here, and labels located between the bottle and bottle cell are rinsed out of the cells and carried away via the label discharge system.

A very good cleaning result can be achieved within a short amount of time through the combination of a caustic soaker zone and caustic spray.

25.5.1.4 Intermediate Spray, Hot- and Cold-Water Zones

These zones are usually designated water zones of the bottle-cleaning machine. These water zones are used to rinse off alkalis and other residues and to cool the bottles at the same time. Fresh water is usually relatively cold. This ensures that the bottles exit the cleaning machine at low temperatures. The cascaded connection of the water zones causes the drained and slightly heated fresh water to end up in the cold-water zone, then in the hot-water zones and subsequently in the intermediate spraying zone. At this point, the water has a considerably high alkali content and temperature and runs into the pre-soak (Figure 25.8).

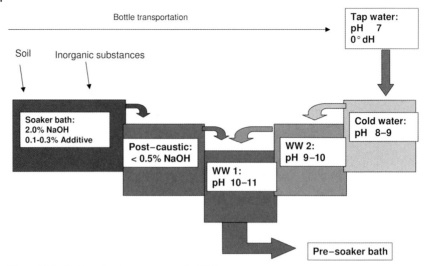

Figure 25.8 Outline plan lye-water cascade. WW = warm water.

25.5.2
Use of Chemicals in the Bottle-Washing Machine

Depending on the type of contamination and the consistency of the water used (hardness), different chemicals are used in the different zones of the bottle-washing machine. Sodium hydroxide is usually used as the base chemical for cleaning in the main soaker bath. In addition to the known properties of sodium hydroxide (Section 25.1), other negative properties are relevant in this area:

- The high surface tension means that penetration into labels is reduced (label removal takes too long).
- The labels can become frayed.
- The bottle material (glass and PET) can be damaged.
- Disturbing foam via saponified contaminants.

The alkaline cleaning solutions are comprised of a fully-prepared cleaner or from sodium hydroxide and a caustic soda-free additive (concentrate). As there are different types of contamination to be removed (beverage residues, label glue, label paper, printing ink, etc.), there can be no 'all-round cleaner'. The additive for cleaning refillable bottles must be selected based on the plant and cleaner types. There are additives that are geared toward, for example, water-hardness stabilization, mold removal, rust-ring removal, high complexing capacity and reduced scuffing to glass by sodium hydroxide. The contents of the additives are listed in Table 25.1.

Labels are to be discharged with as little decomposition as possible. If the labels are frayed due to poor paper quality or an excessive concentration of the alkali, undesired foaming may occur. In this case, a separate defoamer must be added in doses to prevent over-foaming of the lye and to ensure trouble-free discharge of labels.

In the water zones, the carried caustic soda is removed from the hot surface of the bottle. Increasing the temperature or pH value can cause hard-water deposits. This can be prevented by the use of acidic products containing sequestering and dispersing agents in addition to the acid. These products operate in the sub-stoichiometric range and prevent the formation of a hardness-mineral crystal grid structure. This means that they do not precipitate as solid, hard deposits, but rather as fine amorphous particles that suspend well and can be rinsed out.

Complexing agents are not used for economic reasons. They would prevent precipitation completely, but would have to be added in doses of a higher concentration, since they form complexes with the metal ions of the minerals in the water causing hardness 1:1.

25.6
Cleaning PET Bottles

Bottles made of PET material (poly-ethylene terephthalate) are cleaned using the same system as glass bottles. Chemically, PET is a polyester (plastic) with temperature- and chemical-resistance properties that must be taken into consideration. One particular problem in comparison to glass is the temperature resistance of the material. At temperatures over 60 °C, the material shrinks. Therefore, the cleaning process may not exceed a maximum temperature of 60 °C.

If PET material (polyester) is exposed to pure sodium hydroxide at greater concentrations, stress cracking and hazing occurs. As with the cleaning of glass surfaces, additives for lyes that support the cleaning process must also be used for PET bottle cleaning. Most additives for caustic solutions used with glass cannot be used, however, as they contain substances that harm the bottle material. Appropriate additives that support cleaning and effectively prevent the occurrence of stress cracking and hazing were developed for use in the cleaning of PET bottles.

Should the defoaming action of the lye additive be insufficient, a suitable defoamer must be used separately, as with glass-bottle cleaning. It must also protect the PET material from hazing and stress cracking.

25.7
Cleaning Barrels

Beer barrels and kegs are cleaned externally and internally in specially designed machines. They are usually connected with barrel lifter and automatic palletizing equipment and bung-hole finders and turning stations that move the containers into position for interior spraying. The barrels/kegs are brushed and/or sprayed with high-pressure jets on the outside. For internal cleaning, the barrels are placed one after another onto the rotating spray heads (low-pressure cleaning) for the various cleaning media and steam (disinfecting). In the case of kegs, the keg fittings at the opening are permanently screwed in so that the insertion of spray balls is not possible. For this reason, cleaning valves are coupled to the fitting, through which the cleaning agents flow into the keg. They exit from there as well.

There are different processes used for cleaning kegs. For soaking, the kegs are filled with cleaning solution so that they are wetted with the cleaning medium. With this method, mechanical forces do not act on the soiling. Intensive soaking is especially necessary to swell up existing organic precipitates with alkaline cleaners in the case of non-filtered wheat beer.

When the cleaning agent is pumped over the inner wall of the keg, it enters the container via the extractor tube. At the end of the extractor tube, the cleaning medium contacts the bottom of the barrel, where cleaning agent solution flows down the inner surface and collects at the neck of the keg. The rising internal pressure causes the fluid to exit through the ring conduit.

With counter-flow cleaning, the cleaning solution is introduced into the keg through the ring conduit. The cleaning action is limited to the flowed-through fitting valve body and the spray jets located on the inner keg wall and the extractor tube. Since the cleaning medium collects in the keg neck, spraying can only occur briefly. Normally, spraying through the ring conduit is combined with pumping over the inner keg wall and the extractor tube.

With turbulent cleaning, the keg is partially filled with the cleaning solution. By blowing compressed air through the ring conduit of the keg, the cleaning medium is spun. The moving fluid (mechanically) detaches the soiling from the surface. The keg is emptied at the end of the cleaning step. Beer residues and any remaining CO_2 are blown out before cleaning with an alkaline cleaner occurs. Cleaning solutions and water are to be removed with sterile air during cleaning.

Barrels are usually cleaned with an alkaline product, which is followed by acid cleaning after intermediate rinsing. The alkaline cleaner for light metal and stainless-steel containers contains sodium hydroxide, complexing agents, dispergators and silicates as corrosion inhibitors, and sometimes surfactants as well. Suitable products containing active chlorine can also be used for combined cleaning and disinfecting. An alkaline cleaning step with sodium hydroxide as the basic chemical and cleaning-support additive is also possible. Acidic cleaners normally contain phosphoric acid and/or nitric acid and surfactants and defoamers. They should not be used for anodized barrels on a regular base, however.

Due to their closed construction, kegs are returned with less soiling than open barrels. For this reason, alkaline cleaning can often be dispensed with depending on the previously contained product, whereby cleaning time, water and cleaning agent can be saved. The same alkaline and acidic cleaning products are used as with barrel cleaning.

The last step in the process is the sterilization with steam. This can also be replaced by chemical disinfecting based on active chlorine or peroxyacetic acid. Spraying with fresh water which fulfils the requirements for drinking water must be carried out, however.

25.8
Foam Cleaning

The cleaning of non-closed systems, such as the external surfaces of tanks, pipelines, machines, open containers, walls and ceilings, usually occurs via low-pressure foam cleaning. In the case of the listed objects to be cleaned, they are usually large-surfaced areas that may be only poorly or completely inaccessible. High tanks, areas under tanks, areas between pipes or very irregular, angular surfaces of machines (fillers) often pose problems when it comes to cleaning. With foam cleaning, relatively long contact times between contaminants and cleaning agents are achieved with low liquid consumption. This process is used especially for bulky objects and on vertical surfaces. The long soaking times are especially advantageous when dried-on and burned-in contaminants must be soaked and swelled before they can be sprayed off with water.

Foam cleaning has the following advantages:

- The detection of all surfaces to be cleaned is unambiguous, since untreated areas of foamed areas are easy to distinguish visually.

- In comparison to previously executed manual cleaning with brushes, scrubbers and cloths, the risk of infection with foam cleaning is reduced considerably.

- In comparison to manual processes, the capacity for treating large areas is greatly increased.

- The cleaning personnel do not come into direct contact with the cleaning solution.

This type of surface cleaning of large-surfaced objects is carried out with systems working with pressures up to 40 bar. Here, the water inlet pressure is increased by approximately 20 bar in the units. To generate the necessary foam, a partial current of the high-pressure water is deflected via an injector system and the previously set quantity of cleaning agent is drawn into the water stream via under-pressure there. The air required for the foam generation reaches the water/cleaner mixture via a dosing unit and creates the desired foam in the unit, which can then be applied via a hose no longer than 25 m, a gun or a special nozzle. Foam cleaning is usually combined with spraying procedures for pre- and post-cleaning.

Pre-cleaning often occurs at a slightly increased temperature (40°–50°C), mainly to increase the effectiveness of the subsequent foam by heating the surfaces. The temperatures specified for foaming the cleaning solution vary between room temperature and 85°–90°C. The soaking time of the foam on the contamination is usually between 10 and 20 minutes, whereby stubborn contamination can extend the contact time to 40 minutes if the foam is sufficiently stable.

The subsequent high-pressure cleaning usually occurs with hot fresh water at approximately 50°C. In many cases, it is recommended to use subsequent spray disinfecting (filler).

25.9
Work Safety and Environmental Protection

Cleaning and disinfecting agents are more or less dangerous chemicals. To sufficiently protect the health of the personnel handling these substances, the information on the material safety data sheet must be observed. In general, it should be mandatory to wear safety goggles and the recommended protective clothing when handling chemicals.

The product information sheet on cleaning and disinfecting agents should also be read thoroughly and observed. Recommended usage concentrations and temperature specifications are to be complied with, which also protects the environment.

Used cleaning agent solutions are often heavily alkaline (pH above 11) and must be neutralized to the legally prescribed pH value before being released into the public sewerage system.

26
Waste Water
Karl Glas

26.1
Introduction

Water has two functions in brewing industry: on the one hand, it is a raw material for beer and, on the other hand, it is needed as a cleanser. However, as water does not have unlimited availability, it is important for the future to save the raw material 'water'. Therefore, a functioning protection of water is necessary. This can only be achieved with systematic waste water disposal. The requirements of fresh water fluctuate, depending on technical equipment, product diversity and the kinds of filled casks. With today's state of the art, a specific need of between 0.35 and 0.60 m^3 fresh water/hl sellable beer (SB) can be assumed. Table 26.1 compares a practice example with literature values.

26.2
Characterization of Brewery Waste Water

The consequences of the tightened requirements on waste water quality for factories are the increasing cleaning costs and waste water fees, including miscellaneous high pollution surcharges. Consequently, reasonable waste water disposal or waste water treatment implies a critical ascertainment of the pollution load and waste water volume as well as an exact definition of the outlet's quality. Each measure for an internal solution should start with a systematic exhaustion of all possibilities to decrease waste water volume and pollutant loading.

26.2.1
Types of Waste Water

Different types of waste water can be differentiated in the brewing trade as follows:

Handbook of Brewing: Processes, Technology, Markets. Edited by H. M. Eßlinger
Copyright © 2009 WILEY-VCH Verlag GmbH & Co. KGaA, Weinheim
ISBN: 978-3-527-31674-8

Table 26.1 Fresh water requirements in a brewery.

Area	Tested brewery fresh water (m³/hl SB)	Literature values fresh water (m³/hl SB)
Water treatment	0.0186	NA
Brewhouse/wort cooling	0.177	0.130–0.236
Fermentation	0.029	0.032–0.080
Storage cellar	0.015	0.024–0.067
Filtration cellar/pressure tank cellar	0.011	0.010–0.109
Steam generation	0.0015	0.001–0.030
Air compressor/cooling water	0.052	0.012–0.050
Miscellaneous	0.02	0.100–0.300
Total	0.37–0.44	0.407–1.149

- Production waste water.
- Technical industrial water or cooling water.
- Sanitary waste water.
- Domestic sewage or canteen waste water.
- Storm water

The production waste water is composed as follows:

- Product residue.
- Side-products (yeast residues, residues of spent grains).
- Additives (e.g. kieselguhr, glue, etc.).
- Rinsing water with the corresponding residues of cleaning agents and disinfectants.
- Cooling water.

The yields of waste water in a brewery consist mainly of production residues as such, as well as cleansing waters from rinsing and disinfection of the containers and casks. Due to batch processing, the waste water situation is marked by an intermittent yield, containing highly fluctuating soil concentrations and different pH and temperature values.

In addition, backing of solid materials such as spent grains, kieselguhr or excess yeast is very important, not least due to reasons of high pollution loads. These substances are not to enter the waste water system.

26.2.2
Waste water Constituents

Generally, pollutants in waste water are present in soluble and insoluble forms as well as organic compounds (fats, proteins, carbohydrates). Consequently, waste water constituents are differentiated as:

- Depletion materials (e.g. uric acid and glucose are biologically degradable and lead through anaerobic degradations to odor nuisance; furthermore, these materials can trigger oxygen depletion, which in turn reduces the oxygen concentration in water and can lead to the extinction of fish).
- Nutrients (e.g. nitrogen and phosphorus compounds, that can lead to eutrophication particularly in stagnant waters).
- Polluting agents (e.g. heavy metals, synthetic organic compounds, rinsing agents and disinfectants, lubricants for chains and belts).
- Interfering substances (e.g. salts, fats, oils, clay, sand).

The properties of the waste water constituents determine the decisive principle of action of the employed processes or process combinations (Figure 26.1).

26.2.3
Analysis of Waste Water

As brewery waste water contains relatively high amounts of carbohydrates (carbon compounds) and proteins (nitrogen compounds), the sum of the biological oxygen demand (BOD_5), chemical oxygen demand (COD), organic nitrogen as well as total phosphorus plays an important role for the description of brewery waste water. Waste water constituents and analytical parameters are compared in Table 26.2. For additional evaluation, the dissolved organic carbon (DOC) can also be consulted.

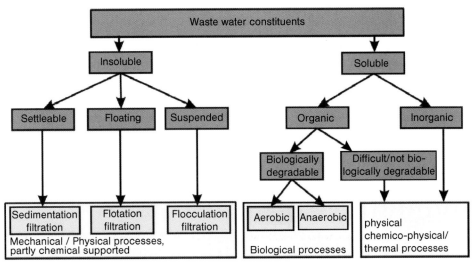

Figure 26.1 Classification of waste water constituents.

Table 26.2 Waste water constituents and the corresponding analysis.

Waste water constituents	Parameters/analysis
General	temperature pH value
Sensory analysis	coloring turbidity odor (taste)

Insoluble constituents
  crude solids
  settleable solids
  suspended solids
  floating solids
	settleable materials
	filterable materials, dry matter
	turbidity
	composition (N_{total}, P_{total}, BOD_5, COD, TOC, etc.)

Soluble constituents
  biological degradable
    C compounds
    N compounds
    P compounds
  not biological degradable
    salt or ions
    heavy metals
    problematic organic substances
    (halogenated hydrocarbons,
    pesticides)
	
	BOD_5, COD, TOC, DOC
	NH_4-N, NO_2-N, NO_3-N, $N_{inorganic}$, N_{total}, TBN
	PO_4-P, P_{total}
	
	conductivity
	ionic analysis, single analysis
	AOX, single analysis

Specific (fish) toxic constituents
  ammonia
  cyanide
  heavy metals
	
	NH_4-N
	Single analysis

AOX indicates the sum of all 'adsorbable organic halogen substances'.

On the basis of the rate of the sum parameters, a conclusion can be drawn about the biological degradability of the waste water. The COD:BOD_5 ratio gives information about the degradability of waste water under aerobic conditions. In general, a good degradability is given under aerobic conditions if the rate is 2 or under.

The following should be noted with regard to the term 'biologically degradable':

- Primary degradation: first degradation step, in which the chemical compounds lose their typical characteristics (e.g. surfactants).

- End degradation: complete mineralization of the matter.

- Examples of degradation tests to determine the biological degradability: screening tests (e.g. OECD screening test for surfactants) and simulation tests (e.g. small-scale waste water treatment plant).

Table 26.3 Rates of sum parameters for different brewery sections by the example of brewery waste water.

Section	Number of samples	COD:DOC	DOC:BOD$_5$	BOD$_5$:DOC
Malt house	14	4.12 ± 0–37 r = 0.85	1.53 ± 0.19 r = 0.69	2.70 ± 0.42 r = 0.64
Brewhouse	20	3.11 ± 0.60 r = 0.76	1.52 ± 0.17 r = 0.90	2.02 ± 0.24 r = 0.92
Fermentation and storage cellar	26	3.59 ± 0.50 r = 0.96	1.64 ± 0.23 r = 0.98	2.22 ± 0.24 r = 0.97
Bottle cellar	40	3.82 ± 0.55 r = 0.97	1.44 ± 0.23 r = 0.98	2.71 ± 0.51 r = 0.96
Total waste water	50	3.87 ± 0.69 r = 0.97	1.46 ± 0.16 r = 0.99	2.67 ± 0.50 r = 0.97

In contrast to domestic waste water, a COD:BOD$_5$ ratio of 1.5 can be given for brewery waste water. The rates for the sum parameters are compiled in Table 26.3 for the total waste water and individual split streams.

Examples for communal waste water:

- BOD$_5$ concentration: 300 mg/l.
- COD concentration: 600 mg/l.
- COD:BOD$_5$ ratio: 600 mg/l : 300 mg/l = 2.

Examples for brewery waste water.

- BOD$_5$ concentration: 1600 mg/l.
- COD concentration: 2400 mg/l.
- COD:BOD$_5$ ratio: 2400 mg/l : 1600 mg/l = 1.5 (range 1.4–1.6).

26.3
Preliminary Investigations to Determine Waste Water Pollutant Load and to Plan Waste Water Plants

Before planning in-house measures or waste water treatment or pre-treatment, respectively, a detailed survey of the water and waste water situation of the company should be carried out. Estimated product-specific literature values for the yield of waste water and pollutant loading are not adequate for the exact planning or comparison of economic viability, as the specific general conditions of the company are not considered.

For the planning of in-house measures or waste water plants, the volumes of product-specific water, waste water and pollutant loading need to be determined as well as the company's hazard potential of specific split streams. In this case a separation in different production areas is sensible, as generally not all split streams in the company can be covered simultaneously with a maintainable complexity of measurements.

To carry out a survey, a company-specific flowchart has to be worked out. This should show the exact production route, place of fresh water onset and waste water yield, respectively. In addition, the produced wort and beer quantities as well as average and peak values need be known in a measured time frame. During the analysis, the respective processing times and water consumption also need to be identified at the investigated places. The installation of additional water meters to determine water consumption in the different sections is also sensible for later control.

It is not enough to carry out a calculated estimation of the accumulated volumes for the waste water. In this case it is recommended to register waste water streams by means of suitable measuring devices (e.g. inductive volume counters). Consequently, with an adequate sampling device, it is possible to take simultaneous proportional waste water samples from the waste water streams. At high waste water flow it is necessary to take, in a given time frame, a higher sample volume than at lower flow.

Hence, an exact foundation is given to calculate the pollution load or the load of other harmful substances. If the survey for the specific split streams as well as the total production waste water of the company is carried out as described, exact readings are delivered of the product-specific water consumption as well as yields of waste water volumes, pollution load and harmful substances.

26.4
Practical Example of a Waste Water Measurement

26.4.1
Determination of Concentration and Volume

The volume proportional sampling is carried out over 24 h (24-h mixed sample) and the specific parameters are analyzed from this sample (Table 26.4).

26.4.2
Load Values

The load is calculated from the daily waste water volume and its corresponding parameters. The sum of the daily pollutant loading per parameter gives the weekly pollutant loading. Consequently, the ratio of the weekly pollutant loading and the weekly waste water volume yields the average mass concentration of the corresponding parameters (Table 26.5).

Table 26.4 Determination of waste water volume and corresponding concentrations.

Day	N (mg/l)	P (mg/l)	COD (mg/l)	BOD$_5$ (mg/l)	Waste water volume (m^3)	COD:BOD$_5$
Monday	35.0	9.5	1498	1095	843	1.37
Tuesday	42.0	9.3	1810	1172	803.0	1.54
Wednesday	42.0	11.2	1510	925	747.0	1.63
Thursday	56.0	10.3	2278	1236	791	1.84
Friday	24.5	6.1	1458	932	355.0	1.56
Average	41.7	9.7	1742	1092	707.8	1.59

Table 26.5 Determination of waste water volumes and corresponding daily load.

Day	N (kg/day)	P (kg/day)	COD (kg/day)	BOD$_5$ (kg/day)	PE$_{60}$
Monday	29.5	8.0	1,262.8	923.1	15,385
Tuesday	33.7	7.5	1,453.4	941.1	15,685
Wednesday	31.4	8.4	1,128.0	691.0	11,516
Thursday	44.3	8.1	1,801.9	997.7	16,295
Friday	8.7	2.2	517.6	330.9	5,514
Average	29.5	6.8	1,232.7	772.7	12,879

26.5
Specific Characteristic Parameters of Waste Water

26.5.1
Total Waste Water

According to literature values, the concentrations in brewery waste water fluctuate considerably. Totally different rates are obtained for different plants depending on the work operations and equipment (Table 26.6). These must be considered, for example, during the planning and construction of an in-house waste water treatment plant.

Table 26.7 presents the specific load of several parameters in reference to the amount of beer produced. Naturally, the fluctuation of the values from brewery to brewery is high as well. The organic loads in 1 hl beer are 892 and 561.2 g for COD and BOD$_5$, respectively, and the DOC load is 230.7 g/hl produced beer. The settleable materials account on average for 6969.1 ml/hl beer, which results in 137 g/hl of dry material and 63.7 g/hl of ignition residue. The total inorganic nitrogen (TIN) is on average 4.37 g/hl and the Kjeldahl nitrogen (TKN) 19.34 g/hl. The total bound nitrogen (TBN) is on average 23.12 g/hl. The phosphate load has a mean of 3.66 g/hl for ortho-phosphate and 4.57 g/hl for total phosphate.

Table 26.6 Composition of brewery waste water.

Parameter	Units	Mean	Minimum	Maximum
KMnO$_4$	mgO$_2$/l	2231	869	4840
COD	mg/l	2628	933	5515
BOD$_5$	mgO$_2$/l	1668	600	3671
DOC	mg/l	651	190	1620
Settleable material	mg/l	15.8	2.5	61.6
TRS	mg/l	426.1	106.6	1953.7
Ignition residue	mg/l	208.6	45.3	1236.5
TIN	mg/l	12.5	4.2	29.6
TKN	mg/l	58.2	26.6	126.8
TBN	mg/l	70.2	37.1	137.2
NH$_4$-N	mg/l	2.8	0.3	21.1
NO$_2$-N	mg/l	1.3	0.1	20.7
NO$_3$-N	mg/l	12.3	0.9	26.2
Ortho-P	mg/l	10.6	0.8	41
Total P	mg/l	11.8	1.0	33.6

Table 26.7 Specific pollutant load (per hl SB).

Parameter	Units	Mean/average	SD	Minimum	Maximum
Waste water quantity	m^3/hl	0.39	0.13	0.22	0.87
KMnO$_4$	g/hl	804.8	402.2	267	1793
COD	g/hl	892.6	379.1	286	1886
BOD$_5$	g/hl	561.2	251.1	184	1253
DOC	g/hl	230.7	140.9	58	554
PE$_{60}$	hl^{-1}	9.3	4.2	3.1	20.9
Settleable material	ml/hl	6969.1	9068.8	536	4289.2
TRS	g/hl	137	102	28.2	511.4
Ignition residue	g/hl	63.7	68.6	13	324
TIN	g/hl	4.27	2.72	0.9	10.4
TKN	g/hl	19.34	7.7	8.5	35.4
TBN	g/hl	23.12	7.52	11.5	37.6
NH$_4$-N	g/hl	1.01	1.74	0.1	7.2
NO$_2$-N	g/hl	0.42	1.53	0	7.1
NO$_3$-N	g/hl	3.91	2.77	0.4	9.9
Ortho-P	g/hl	3.66	3.05	0.2	12
Total P	g/hl	4.57	3.04	0.3	11.5

The specific waste water volume is on average about $0.35\,\mathrm{m^3/hl}$ beer, whereas realistically a fluctuation span of 0.25–$0.45\,\mathrm{m^3/hl}$ can be assumed. This means that of the required fresh water volumes of 0.15–$0.2\,\mathrm{m^3/hl}$ do not appear as waste water (deduction rate!).

26.5 Specific Characteristic Parameters of Waste Water

Table 26.8 Determination of the population equivalent.

	Fouling per 1 m³ household waste water in 24 h (mean in g/m³)				Per inhabitant (IH), with 200 l waste water/IH (g/IH)				
	mineralized	organic	total	BOD$_5$	mineralized	organic	total	BOD$_5$	COD
Settleable suspended solids	50	150	200	100	10	30	40	20	40
Not settleable suspended solids	25	25	75	50	5	10	15	10	
Solved matter	325	250	625	150 200	75	50	125	30 40	80
Sum	450	450	900	300	90	90	180	60	120

A common measure for the daily discharged waste volume per inhabitant is the BOD$_5$ and is expressed as the specific parameter population equivalent (PE). In the case of non-settleable homogenized waste water, 60 g BOD$_5$ corresponds to 1 PE (1 PE$_{60}$). For settled waste water (after 2 h in the Imhoff tank), a population equivalent of PE$_{40}$ (40 g BOD$_5$ per inhabitant) is specified (Table 26.8).

This becomes particularly noticeable in a comparison of the accumulated PE/hl produced beer. The values of PE$_{60}$ lie between 3.1 and 20.9 with a weighted mean value of PE$_{60}$ = 9.7/hl produced beer. The high fluctuation span is strongly dependent on the operation mode of the site, particularly efforts made to prevent highly loaded waste water split streams (e.g. waste water-independent disposal of hot trub or beer yeast). It is also dependent on the equipment and construction conditions such as the size of the container or technological preferences (e.g. re-mashing of last wort or use of a single-tank procedure).

For technical waste water, conditioned brewery PE$_{60}$ values between 6 and 9/hl SB apply nowadays.

26.5.2
Split Streams

In order to prepare for internal measures, the internal split streams need to be better understood. The average COB concentrations and the corresponding waste water volumes in m³/hl per section are presented in Table 26.9. In addition, the amount of waste water is based on production operations and the type of waste water. As already mentioned, the retention of brewing side-products as well as

Table 26.9 Amount of waste water with the corresponding COD concentrations.

Section	Average COD (mg/l)	Quantity (m^3/hl SB)	Waste water source	Type of waste water/ comment
Water treatment	<15	0.019	backwash multi-layer filter; regeneration ion exchange	contains particles, hardly any dissolved constituents; high salt content, free of particles

Brewhouse	3000	0.029	cleaning in place (CIP) purification:
			    mashing vessel
			    lauter tun
			    wort pan
			    separator, whirlpool
			    empty pipes
			    spent grain squeeze juice
			    last wort
			    hot trub
			    cold trub
				waste water containing particles from CIP processes; waste water containing particles with product or side-product residues; in parts very high COD loads

Fermentation cellar/storage cellar	800–2000	0.047	CIP purification:
			    pure culturing of yeast
			    yeast tank
			    fermenter, stock tank
			    separator
			    pipes
			    yeast squeeze juice
			    yeast–beer mixture (separating)
			    dreg
				waste water containing particles from CIP process or particle containing waste water with product or side-product residues. In parts very high COD loads

Filtration cellar/ pressure cellar tank	3000	0.033	CIP purification:
			    tanks
			    filter
			    kieselguhr press
			    filter forerun and backlash
			    kieselguhr squeeze juice
				partial very high product residues in the filtration fore shot and last cut; very high loads in the kieselguhr squeeze juice

Bottling	800–1200	0.097	Cleaning
			    bottles
			    cases
			    filler
			rinse and empty filler
			vacuum pump
			complete goods shower
			pasteurization
			labeling
			transport
				with the exception of bottle washer, generally particle-free waste waters with low product residues and low COD loads

Table 26.10 Pollutant load of brewery side-products.

	COD (mg/l)	BOD$_5$ (mg O$_2$/l)	COD$_{120}$ (120 g COD = 1 PE)	PE$_{40}$ (40 g BOD$_5$ = 1 PE)
Beer	120,000	80,000	1	2
Fermentation yeast	210,000	140,000	1.75	3.5
Storage cellar yeast	180,000	120,000	1.5	3
Hot trub	16,500	11,000	0.14	0.28
Cold trub	19,500	13,000	0.16	0.33
Kieselguhr	16,500	11,000	0.14	0.28

minimization of product loss play an important role in a brewery for the prevention of pollutant load. Every little product residue pollutes the brewery waste water, as shown in Table 26.10. Therefore, the ultimate ambition in producing beer and beverages should be the minimization of product losses, and a consistent discharge or retention of the listed side-products.

26.6
In-house Measures

26.6.1
Classification of In-house Measures

The in-house measures can now be classified as:

- Predominant reduction of water and waste water volumes.
  - Prevention of water loss (water saving fittings, equipment and processes, organization and control).
  - Change of water flow (multiple usages, circulation process, separate acquisition of split streams).
- Predominant reduction of pollutant load.
  - Prevention of product residues.
  - Recovery and utilization of liquid and solid byproducts and production wastes.
  - Recovery of resources.
  - Pre-treatment of waste water or in-house treatment (without recycling).
- Reduction of water and waste water volumes and pollutant load.
  - General measurements (training courses for staff, organization, automation, control).
  - Substitution of water (pneumatic or mechanical processes, replacement of water with the product).
  - Improvement of the manufacturing process in relation to water usage and the amount of waste water.

26.6.2
Goal of In-house Measures

- Smaller design of communal as well as in-house waste water treatment plant (low investment costs).
- Reduction of communal waste water charges and overheads.
- Decrease of high pollution surcharges.
- Reduction of waste water charges through compliance with the minimum requirements for discharge into the aquatic environment.
- Improved guarantee to comply with the discharge requirements of waste water in the public sewage system (indirect discharge regulation).

26.6.3
Practical Check-List of In-house Measures for the Production Steps

Practical check-lists for the respective brewery sections, taken from the literature and our own experiences, are listed in Table 26.11; however, this does not claim to be complete.

26.7
Waste Water Treatment

26.7.1
Neutralization

The discharge of waste water outside the pH range of 6.5–9.5 is prohibited by legislation. Consequently, it is often necessary to carry out a neutralization of some waste waters from the beverage industry. This happens with the addition of chemicals: at acidic pH values (0–7) by the addition of bases and at a basic pH value (7–14) by the addition of acids. Stirring devices are employed for mixing.

26.7.1.1 Neutralization of Carbonic Acid (CO_2)

Carbonic acid is the cheapest and most environmentally friendly neutralizing agent for basic waste waters. Often it develops in breweries as access fermentation carbonic acid. Carbonic acid is safe to handle, as there is no danger of corrosion and a super-acidification of the waste water can be excluded. Thus, the use of hydrochloric acid, which leads to an increased pollution of chloride, is unnecessary.

Organically contaminated waste water easily starts to rot and consequently develops unpleasant odors. The loading with oxygen avoids the fouling process. The cheapest source of oxygen is air.

Suspended solid materials deposited in the waste water retention reservoir necessitate an increased draining of the reservoir in order to clean it. Such types of sedimentation can be avoided with the aid of a good circulation system.

Table 26.11 Practical check-list for in-house measures.

Parameter	Avoidance potential
Brewhouse	
last wort	dry lautering, re-mashing, water meter
malt spent grain	preferably complete removal of spent grain, no particles in rinsing water
spent grain squeeze juice	transport of spent grain with compressed air, dry removal of spent grain, construction of spent grain silo
cleansing of false bottom	recycling of rinsing water for mashing
sedimentation cake	recycling of rinsing water for mashing
Wort production	
hot trub	compact separation, addition to lautering/spent grain
cold trub, flotation	compact separation, addition to lautering/spent grain, used yeast
Fermentation cellar	
yeast	complete absorption of the used yeast and disposal independent of canalization network, minimization of waste water, yeast press
rinsing waste waters	single-tank process, decreased cleaning
Storage cellar	
sludge	complete collection of sludge and disposal independent from sewage system, yeast press
residual beer	recovery of residual beer
Filtration	
filtration mode	technical equipment, filter combination
kieselguhr	discharge from filter (centrifuge, scoop off, squirting off)
forerun/backlash	addition to filtration, kräusening, re-mashing
Bottle filling	
start pour off	possibly processing
end pour off	recovery of beer–water mixtures, processing
labels discharge	label press, feedback of leach
chain belt greaser	adjustment of proportioning station, mean cod value used
purification leach	re-use of leach, sedimentation, other processes for leach treatment
leach sediment	filtration of sediment, land filling
Filtration	
beer loss, beer residues	collection, re-use
Problem areas	**Onsets**
Filtration	
water consumption	staff training, fitting of water meters/volume control, maintenance of machinery, preparation of pipeline plans; avoidance of 'hidden leakage' of fresh water, use of CO_2 to squeeze empty, re-use of slightly contaminated water (e.g. for car cleaning), re-use of uncontaminated cooling water, change of the cooling process
production planning	reasonable planning of production sequences to optimize cleaning processes, avoidance of product residues

Table 26.11 Continued

Parameter	Avoidance potential
facility planning	configuration of tanks and pipelines, avoidance of product residues, level indicator and anti-overflow device
unavoidable product residues	recovery from pipeline discharges, start and end pour off
waste water system	installation of buffering and batch tanks for the recovery of high contaminated split streams and/or pH fluctuations; avoidance of pH and COD waves
settleable material	installation of devices into canalization for reduction of solid materials
labels	saving of glue by point or streak gluing, leach-resistant labels, heavy metals!

All three problems (i.e. loading with carbonic acid, supply with air oxygen and good reservoir circulation) can be solved simultaneously with an adequate process. A submerged diffuser is simply inserted in the retention reservoir and the suction end connected with a carbonic acid feed (injection). If the pH value exceeds the appointed limits, a pH meter and control system will actuate a magnetic valve and carbonic acid will be fed in. After sufficient neutralization of the waste water, the carbonic acid feed is discontinued, but the submerged diffuser continues loading with oxygen and mixing of the reservoir.

26.7.1.2 Neutralization with Flue Gas

Flue gas also contains a considerable amount of carbonic acid that can be used to neutralize basic waste water. It is available free of charge from heating systems and power plants! Consequently, the neutralization with flue gases is interesting for industries that possess their own heating systems or power plants.

A submerged diffuser is inserted into a waste water retention tank. As a larger quantity of flue gas is needed for neutralization, it is not injected via the suction end, but is fed via the whole pipeline diameter via a T-fitting. In that way only flue gas or air can be added. The addition is done by an electrically driven selector valve. If the pH exceeds the appointed limits, a pH meter and control system will actuate the selector valve in such a way that only flue gas will be fed in. After sufficient neutralization of the waste water, the selector valve is set again on 'air supply', and the submerged diffuser ascertains the oxygen feeding and the mixing of the tank.

26.7.2
Mixing and Equalizing Tanks

Mixing and equalizing tanks are generally employed to reliably maintain the waste water peak quantities, BOD_5 and COD concentration, temperature, as well as pH. The contamination of the volume load is $1\,kg\;BOD_5/m^3/day$ or greater.

The aeration of the waste water avoids fouling and the development of unpleasant odors. Furthermore, sedimentation in the tanks is avoided. As a side-effect of the aeration, a part-biological degradation of the contaminated waste water is also achieved. The degradation rate is in that case up to 40%. Consequently, the burden on the following communal sewage disposal plant is reduced and the payment of high pollution surcharges with waste water taxes is mostly avoided.

An oxygen concentration of 0.5 mg/l is sufficient for the operation. Specific kind of bacteria is formed that uses a great part of the organic materials for building cellular matter. The flocculation of activated-sludge particles is inhibited. Hence, no excess sludge accumulates.

26.7.3
Aerobic Waste Water Treatment

Organic compounds found in water are metabolized by a mixed culture of bacteria in biological waste water treatment. The necessary bacterial mixed cultures are either found free-floating in the water or attached as a biofilm or microbial film on supporting materials. Therefore, aerobic waste water treatment is differentiated into activated-sludge processes and processes with fixed-bed reactors. Active-sludge processes can be operated as continuous or batch-wise sequencing batch reactor (SBR) processes. Trickling filters, submerged contact aerators, biological sludge filters as well as fixed- and fluidized-bed reactors serve as fixed-bed reactors.

In an aerobic operating sewage disposal plant the organic constituents of the waste water are worked up with oxygen in an aerated reactor. Here, the bacteria use up oxygen, and produce mainly CO_2 and water from the organic matter. Oxygen is the limiting factor in a pure BOD_5 degradation with sufficient biomass and substrate. To sustain the aerobic conversion an oxygen concentration of 2 mg/l must be available. This is achieved artificially. The energy converted in this process is used by the bacteria for growth and building up biomass. Thereby, about half of the offered substrate is converted into bacterial matter. In addition to a reduction of BOD, an important goal in aerobic waste water treatment is the removal of nitrogen and phosphate sources.

26.7.3.1 Activated-Sludge Plant
An activated-sludge plant always consists of an activated-sludge tank and a final sedimentation tank. The sedimented sludge of the final tank is for the most part returned as return sludge to the activated-sludge tank. Therefore, a continuous operating biological purification process with a high decomposition rate (99%) is possible. Only a small part of the sludge is removed from the system as excess sludge.

In larger plants, primary settling tanks are connected before the activated-sludge plant. Here, the waste water is freed from settleable, partly non-organic materials. In small to medium plants (until about 10 000 PE), primary settling tanks are mostly abandoned. In this case at least one good functioning rake screen with a slit width of 10–20 mm is used. If the waste water contains sand, a sand filter is also used.

To judge the contamination of the waste water, the biochemical oxygen demand BOD_5 or the oxygen demand COD is consulted. The BSB_5 or COD load is calculated by multiplying the BOD_5 or COD contamination by the waste water volume and thus the size of the submerged diffuser required can be determined.

Generally, an activated-sludge system is slightly contaminated if the volume load (ratio of BOD_5 load divided by the usable tank volume) is lower then $1\,kg$ $BOD_5/m^3/day$ and highly contaminated if the BOD_5 volume load is greater than $1\,kg\,BOD_5/m^3/day$. The combination of slightly contaminated and highly contaminated stages is often found in industrial waste waters. This is then called a two-stage activated-sludge plant. Depending on the waste water constituents, chemico-physical operating pre- and post-stages are also used.

26.7.3.2 High-Performance Reactors

High-performance reactors are used as pre-treatments for waste waters with very high BOD_5 or COD loads (e.g. yeast factories, distilleries). These can be equipped with a final sedimentation tank and returned-sludge pumping system (as in activated-sludge plants). This is always followed by a highly contaminated and often also by a slightly contaminated waste water treatment plant.

High-performance reactors are a closed systems and operate with a constant water level. The volume load in high-performance reactors can be up to $100\,kg$ $BOD_5/m^3/day$. A considerably high process heat is formed during the biological degradation. This can be led away with suitable fixtures and used for heating or power generation. This is an alternative to anaerobic reactors that generate biogas!

Aeration with high air input can lead to the formation of foam. Therefore, it may be necessary to employ a mechanical defrother in high-performance reactors.

26.7.3.3 SBR

SBRs are used for advanced semi-biological waste water treatments; for example, if the following communal waste water treatment plant is overloaded or a considerable amount of the waste water taxes should be saved (indirect discharger).

The two functions of a biological waste water treatment plant (activated sludge and final sedimentation) are carried out in only one tank. With this, two-thirds of the tank volumes are utilized as waste water collection space and one-third as settled sludge. The volume load is $1\,kg\,BOD_5/m^3/day$ or greater.

In the factory, an oxygen concentration of $0.5-1\,mg/l$ is sufficient. A full biology with activated-sludge floc is developed. The decomposition is 80% or higher. Excess sludge has to be removed from time to time. The sludge is stabilized and is often utilized in farming. It is also employed in agriculture for the production of biogas in existing slurry concentration plants.

SBRs can be applied in nearly all industrial areas. These are measured in Germany according to work sheet M210 of the German Association of Water, Wastewater and Waste. SBRs achieve identical decomposition rates as activated-sludge plants – they can be used for a total biological treatment of the waste water (direct discharger). The measurement is relatively complicated and should therefore be carried out by a competent engineering bureau.

26.7.3.4 Aerated Waste Water Ponds

Aerated waste water ponds are employed for waste water treatment in rural areas, such as in sugar factories (aerated storage ponds). These plants are very cheap to construct and enable a total biological waste water treatment. The plants often consist of two aeration tanks connected in series, followed by a final sedimentation tank and a polishing pond. The volume load in aerated waste water ponds is only 20–30 g BOD_5/m^3/day. Therefore, waste water ponds are noticeably bigger than activated-sludge plants. Very large areas are needed for their construction, which are only found in rural regions.

26.7.4
Anaerobic Waste Water Treatment

The treatment of highly organically contaminated industrial waste water has attained increasing significance. In recent years, the application of anaerobic plants has often developed to become the favorable mode of operation in the brewing industry, in particular for pre-treatment, and also due to the prospect of gaining biogas. Biogas fabrication is one of many possibilities to use regenerative energy to provide process heat and electrical energy.

26.7.4.1 Biochemical Basics

Reaction conditions that proceed in the absence of oxygen are called anaerobic. Organisms that are not dependent on oxygen for their metabolism are accordingly called anaerobes. Among other things, the terminology is important for the cultivation of microorganisms. The sensitivity of microorganisms against oxygen necessitates cultivation under oxygen-free conditions. Here, the so-called anaerobic technique is employed. An anaerobic chamber with a mixture of $H_2:CO_2:N_2$ of 10:10:80 stimulates the anaerobic conditions and allows the cultivation of anaerobic microorganisms.

Anaerobic reactions are redox reactions where no molecular oxygen is involved. Instead of oxygen, compounds with similar high redox potential serve as electron acceptors. With external electron acceptors the procedure is called anaerobic respiration. Here, nitrate, sulfate, Fe^{3+}, sulfur or fumigate serve primarily as electron acceptors. If no external electron acceptors are present, internal electron acceptors are used. This case is called fermentation.

There are organisms that can grow both under aerobic and anaerobic conditions. These microorganism are called facultative anaerobic. If a microorganism can only grow under anaerobic condition it is called a strict anaerobe.

The anaerobic process can be divided into four different, but interconnected stages. Three groups of bacteria are involved in these four different processes.

- *Hydrolysis*
  - Transformation of insoluble organic compounds into water-soluble products.
  - Degradation of proteins, lipids and carbohydrates into the building blocks of glycerin, fatty acids, amino acids and sugars.
  - Substrates that are not decomposed in the hydrolysis step are not available for anaerobic degradation.
- *Acidogenesis*
  - Development of methanogenic and non-methanogenic substances.
  - Methanogenic substances can be utilized by methane bacteria.
  - Composition of the products is dependent on the H_2 partial pressure and pH value.
    - High H_2 partial pressure: predominantly non-methanogenic substrates (ethanol, propionic, butyric, lactic and succinic acid).
    - Low H_2 partial pressure: predominantly acetic acid.
    - Acetogenic bacteria pH range 4–7, optimum pH 4.5–5.5.
    - High yield in acetic acid: low H_2 partial pressure and low pH.
- *Acetogenesis*
  - Degradation of non-methanogenic substrates to methanogenic substances (acetate) under formation of hydrogen and CO_2.
  - Sensitive against oxygen, temperature variations and low pH values.
- *Methanogenesis*
  - Formation of methane:

    $$CH_3COOH \rightarrow CH_4 + CO_2 \text{ slow}$$

    $$CO_2 + 4H_2 \rightarrow CH_4 + 2H_2O \text{ fast}$$

  - Most interference-prone step: sensitive to environmental interference (e.g. oxygen intrusion).
  - Inhibition of methane-producing bacteria, decrease in pH, as acetogenesis continues.
  - Optimal pH at 7 (±0.2–0.5).

The following parameters are critical for smooth anaerobic degradation (Figure 26.2):

- Temperature.
- pH.
- Mixing.
- Trace elements.
- Substrate composition.

Chemical reactions and hence biochemical reactions are strongly temperature dependent. However, experience shows that acidifying bacteria are relatively insensitive and flexible regarding the surrounding temperature. Two distinct temperature optima can be identified: in the mesophilic range at 35 °C and in the thermophilic range at 48–55 °C. The unhindered development of microorganisms

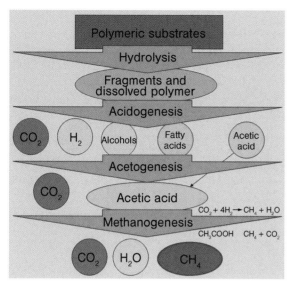

Figure 26.2 Anaerobic degradation steps.

is closely connected with an optimal pH. Primarily, anaerobic biocoenosis can be classified as very pH specific. A tolerance range for the pH from 6.8 to 7.5 is generally given in the literature. If industrial waste waters with extreme characteristics are to be handled, the preceding of pH regulation or a calamity tank is indispensable.

Important requirements for a high degradation activity of the bacteria are a sufficient supply of active biomass with degradable substrates and simultaneous removal of the metabolites. Thus, intensive mixing of the reactor is required. However, this is in opposition to the symbiotic living of acetogenic and methanogenic bacteria that can be disrupted by an overly intensive mixing.

Many factors play a role in the composition of the substrates. The most important factors are:

- Proportion of carbon, nitrogen and phosphorus.
- Influence of inhibiting and toxic compounds.
- Solid fraction.
- Sulfur compounds (especially sulfate).
- Organic acids.
- Nitrate and ammonium nitrogen.
- Heavy metals.

The presence of trace compounds is also indispensable for the normal activity of life. A complete lack of important trace compounds leads to severe deficiencies and can even lead to the death of the organisms.

More information on the anaerobic treatment of brewery waste water can be found in the relevant literature (e.g. Guidelines of the German Brewery Association).

26.7.5
Combination of Aerobic–Anaerobic Techniques

The combination of aerobic and anaerobic techniques is a two-step process. Here, the major part of the organic CSB is converted during the anaerobic step. The aerobic step is used in the final sedimentation.

Anaerobic steps:

- Conversion of the main part of organic material.
- energy recovery as biogas.
- Low production of excess sludge.

Aerobic steps:

- Oxidation of odor-intensive sulfur and nitrogen compounds.
- In part, further conversion of remaining organic compounds.
- In part, removal of nitrogen and phosphorus.

26.7.6
Comparison of Anaerobic and Aerobic Techniques

Anaerobic procedures have a lower degradation rate and bacteria growth than aerobic procedures. Only 5–10% of the offered carbon is converted into the bacteria mass. Hence, in the absence of oxygen, in an anaerobic reactor the organic material is converted into biogas, which can contain a methane proportion of up to 80% (Figure 26.3).

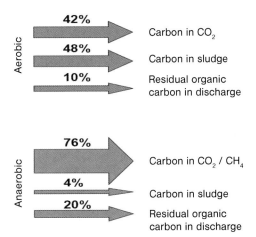

(Organic carbon in waste water 100%)

Figure 26.3 Carbon balance in aerobic and anaerobic degradation of organic compounds.

References

1 Böhnke, B., Bischofsberger, W. and Seyfried, C.F. (1993) *Anaerobtechnik, Handbuch der anaeroben Behandlung von Abwasser und Schlamm*, Springer-Verlag, Heidelberg.
2 Rosenwinkel, K.-H. (2005) Wasser- und Abwassermanagement in der Lebensmittelindustrie, Vortrag ZUK 'Anwendung von Membranverfahren in der Lebensmittelindustrie mit Schwerpunkt Wasserrecycling', Osnabrück.
3 Rosenwinkel, K.-H. (1997) Bestandsaufnahme von Abwasserteilströmen, Vortrag IUV 'Produktionsintegrierter Umweltschutz-Abwasserreinigung', Bremen.
4 Wissenschaftsförderung der Deutschen Brauwirtschaft, EV (1991) Basis-Forschungsvorhaben 15, Ermittlung und Abwasserteilströme in den einzelnen Abteilungen der Brauerei.
5 Glas, K. and Schmaus, B. (1999) Abwassersituation in der Brauwirtschaft Teil 1, in Anteil der Reinigungs- und Desinfektionsmittel an den Schmutzfrachten, *Brauindustrie*, **1**, 16–17.
6 Glas, K. and Schmaus, B. (1998) Abwassersituation in der Brauwirtschaft Teil 2, in Anfallende Abwassermengen und deren Zusammensetzung, *Brauindustrie*, **12**, 856–61.
7 Donhauser, S., Glas, K., Bromig, K.-H. and Schweiger, E. (1993) Summenparameter als Bewertungsgrößen für Brauereiabwasser, *Brauwelt*, **11**, 468–71.
8 Donhauser, S., Glas, K. and Schmaus, B. (1995) Reinigungs- und Desinfektionsmitteleinsatz und sein Einfluss auf die Abwasserzusammensetzung, *Brauwelt*, **1/2**, 8–16.
9 Rüttler, H. and Rosenwinkel, K.-H. (1985) Organische verschmutzte Abwässer der Lebensmittelindustrie, in *Lehr- und Handbuch der Abwassertechnik Band V*, Verlag für Architektur und Technische, Berlin.
10 Bank, M. (2000) *Basiswissen Umwelttechink*, Vogel Buchverlag, Würzburg.
11 Walter, S. (2005) Untersuchung verfahrenstechnischer Möglichkeiten zur Brauchwasserkreislaufführung in der Brauerei, Dissertation, TUM Weihenstephan.
12 Mientkewitz, U. (1997) Teilstromanalyse der Abwässer der Freiberger Brauhaus AG, Diplomarbeit, TU Dresden.
13 Geiger, E., Frank, T. and Koller, A. (2002) Reduzierung des Wasserverbrauches in der Brauerei, in *Weihenstephaner*, 3/2002.
14 Ahrens, A. and Schumann, G. (1996) Grundlagen der anaeroben Abwasserreinigung und ihre Umsetzung am Beispiel einer VLB-Modellanlage, *Brauwelt*, **3**, 84–6.
15 Wirtschaftliche Maßnahmen und Technologien zur Wassereinsparung in der Lebensmittelindustrie (2001) *Getränkeherstellung, Milch und Fleischverarbeitung*, Zentrum für Entsorgungstechnik und Kreislaufwirtschaft, Hattingen.

27
Energy
Georg Schu

27.1
Introduction

Brewing is considered to be an energy-intensive trade, where energy is required in a variety of forms:

- *Thermal energy* includes the heat produced by a process and the heat required to maintain that process.
- *Electrical energy* includes mechanical driving systems (motors), refrigeration and cooling, compressed air supplies, CO_2 recovery and illumination/ventilation.

Water is also considered to be of equal importance to heating and electricity as a basic component of the process.

27.2
Heat Requirements of the Brewery

Recent progress in both technology and techniques, motivated also by increasing fuel costs, have led to a decreasing trend in heat requirements for breweries that originated following the energy crises of the 1970s and 1980s. This downward trend is shown graphically for German breweries, between 1992 and 2007, in Figure 27.1.

There is a clear relationship between heat requirement and brewery size. For example, in breweries which produce up to 20 000 hl per year (class 1), the requirements range from 220 to 280 MJ/hl (equivalent to 61–78 kWh/hl). Class 2 breweries produce between 20 000 and 50 000 hl/year and class 3 between 50 000 and 100 000 hl/year, while those producing 100 000 to 500 000 hl/year and >500 000 hl/year represent classes 4 and 5, respectively. Among class 5 breweries, the heat requirements will range from 90 to 120 MJ/hl (equivalent to 25–33 kWh/hl).

Handbook of Brewing: Processes, Technology, Markets. Edited by H. M. Eßlinger
Copyright © 2009 WILEY-VCH Verlag GmbH & Co. KGaA, Weinheim
ISBN: 978-3-527-31674-8

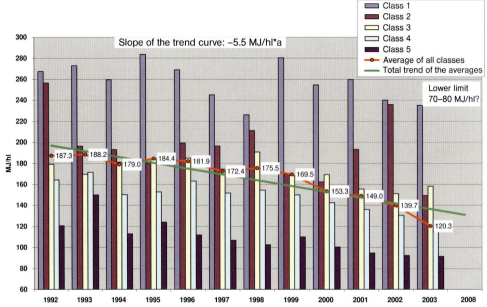

Figure 27.1 Trends in German brewery heat requirements, from 1992 to 2007.

While no valid data are available for those breweries producing several to many millions of hectoliters each year, the heat requirements for such processes would clearly be superior to those for a class 5 brewery, with an expected range of 19 to 25 kWh/hl.

27.2.1
Heat Consumption in the Brewery

Within the brewery, the heat provided is consumed by several major areas. On rare occasions, breweries will have their own malt house, or they may be connected to a district heating supply or other special consumer, but these will not be considered here. The typical distribution of heat consumption within the brewery is shown diagrammatically in Figure 27.2.

27.2.1.1 Brewhouse
In a conventional brewing process, the *brewhouse* is the major consumer of heat, accounting for an average of 40–45% of the total heat load during wort production. However, considerable differences in heat requirements are seen, depending on the brewing procedure and type of heating utilized (whether direct firing, high-pressure hot water or steam-heated systems).

The recent developments in modern brewing processes (i.e., reduced total evaporation with improved analyses) also reflect the energy requirements. As an example, in the modern-day brewery the specific energy requirements (based on

27.2 Heat Requirements of the Brewery

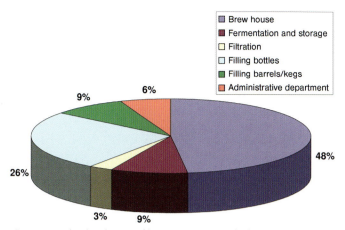

Figure 27.2 The distribution of heat consumers in the brewery.

fuel use) for brewing wort can be reduced to approximately 14 MJ/hl, assuming a total evaporation of only 4% and working with full wort kettles. The requirement for the total brewhouse will, therefore, be in the range of 30–35 MJ/hl. However, when this is compared to the typical heat requirement for a conventional process (55–56 MJ/hl), it becomes clear that the associated costs may be reduced by up to 50% [1].

The recent use of mechanical vapor compressors or vacuum boiling processes has, to some extent, transferred the energy requirements from the heating side to the electrical current side. Consequently, the close correlation between the brewhouse equipment and the 'warm water balance' must be taken into consideration when evaluating heating requirements. As a result, brewing processes should be compared only in relation to the total additional heating requirements for warm water production, whereas an energetically superior process will depend on both the demands of the brewery and its warm water requirements.

27.2.1.2 Service Warm Water

Service warm water, which is used for cleaning and, if necessary, also for sterilization, has a high heat requirement. Depending on the brewery's structure, very different specific requirement values may be found, and the extent of heat recovery measures may be equally wide.

According to an investigation conducted by Fohr and Meyer-Pittroff [2], considerable fluctuation occurs in specific service warm water requirements, with the majority of breweries requiring 0.6–0.93 hl/hl – much higher than the often-stated 'normal' value of 0.3–0.5 hl/hl. In fact, only 20% of breweries will have a low warm water requirement of <0.3 hl/hl.

Within the brewery, the major consumers for service warm water are the warm-fed cleaning in place (CIP) installations (e.g., brewhouse), the kegging plant and the filtration units (which often are cleaned to an excessive degree) (see Figure 27.3).

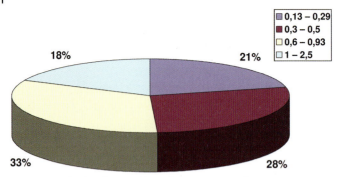

Figure 27.3 The distribution of specific requirements for service warm water in the brewery.

27.2.1.3 Keg Cleaning

The current downward trend in heating requirements is equally applicable to the keg cleaning stage, where modern systems achieve values of approximately 1 MJ per keg, compared to those of 4 MJ per keg that were considered to be efficient a few years ago.

27.2.1.4 Bottle-Rinsing Machinery

Although, in recent years bottle-rinsing systems have been optimized in terms of their energy consumption, such requirements will depend predominantly on the specific water consumption, with further savings achievable only by using fresh water. There is, however, a minimum requirement for water volume in relation to the maximum permitted bottle temperature and biological safety.

Typical modern bottle-washing machines have specific heat requirements of 38–50 kJ per bottle (0.5 l), but this may be partly compensated by heat recuperation [3].

On an annual basis, however, the heat requirements may often be higher and depend on the efficiency of the plant, the number of required preheating sequences and the filling frequency of the baths. Consequently, smaller breweries may be at a disadvantage in this respect due to their simple shift operation modes and short filling times.

27.2.1.5 Others (Pasteurization, Flash Pasteurization, CIP)

Although beer production will not always employ energy-intensive processes, systems such as pasteurizers, flash pasteurization, wheat beer maturation boxes, CIP programs, crate washers and other small heat-consuming components should be considered in detail.

The heat requirement of a tunnel pasteurizers, for example, may be considerably higher than that of a bottle cleaning machine if the operation is disturbed such that the process is discontinuous. From an energy-reequirement perspective, flash pasteurization plants have significantly better heat requirements than tunnel pasteurizers, due to their excellent ability to recover heat.

Although CIP plants may often be optimized to save heat, electrical current and water, any such changes will require the introduction of very close microbiological monitoring so as to ensure the effectiveness of the updated system.

27.2.1.6 Room Heating and System Losses

Depending on the geographical location of the brewery, and hence the local climate and temperature fluctuations, the brewery's requirements for thermal heat can vary widely. As an example, the nature and complexity of ventilator systems in the bottle filling area may be crucial, with the primary energy requirements being contained by limiting both transmission and ventilation heat losses, as well reducing any heat losses. Moreover, such measures are becoming increasingly important as the costs of energy continue to rise.

Maintaining the current pipework in good condition and deconstructing any unused pipework is essential in order to limit any fixed losses. The economy of condensates and the loss of condensate collecting points may have a huge effect on fixed heat consumption, with seasonal variations due to fluctuating ambient temperatures often being problematic.

27.2.2
Boiler House

In most breweries, the heat is generated within a central area – the boiler house – by the combustion of gas, liquid or solid fuels. The heat is then transported to the consuming area using appropriate transport media (heating water, high-pressure hot water, low-pressure vapor and high-pressure vapor).

27.2.2.1 Boiler Plant and Combustion

Within the brewery, both small room boilers (high-speed steam generators) and fire-tube boilers may be considered as boiler plants. Whilst it is unlikely that the large, water tube boilers used in power stations will be employed, steam-powered systems have been used in the past, though generally only in larger breweries.

The combustion system will vary, depending on the fuel used – solid, liquid or gas – as will the boiler construction characteristics and its specification. Recent developments have led to improved combustion processes, the main intention being to reduce the production of atmospheric pollutants, notably oxides of nitrogen [4].

27.2.2.2 Heat Carrier Systems

The most common heat carrier systems employed in the brewery are vapor/condensate and high-pressure hot water systems. As both have their advantages and disadvantages, the decision about which heat carrier to use must be considered on an individual basis.

When using a vapor/condensate, special attention should be paid to the condensate re-feed since, during the tension release of high-pressure condensates (as occurs post-evaporation), up to one-fifth of the energy input may be lost. However, good planning, or use of a suitable heat recovery plant or a sealed condensate re-feed, can reduce these losses to a minimum.

27.2.2.3 Fuels

Today, oil and/or gas are considered the predominant fuels, although in exceptional cases coal might still be used. The automated combustion of wood – and, more recently, of grain – continues to attract attention in light of the upward trends in fossil fuel costs and the aging of fossil-fuel combusting plants. Occasionally, biogas from the brewery's own wastewater treatment plant has been used in place of fossil fuel as an energy source.

In recent years, the proportion of heavy fuel oil used has decreased considerably, due mainly to the mandatory – and very expensive – emission threshold values that can only be achieved by using high-technology processes such as selective catalytic and noncatalytic reductions.

27.2.3
Optimization Possibilities: Exhaust Emission Heat Exchanger, Degassing, Oxygen Regulation, Water Treatment and Blow-Off

A number of techniques may be used in the brewhouse to improve efficiency, so as to obtain as much energy as possible from the available fuel [5]. Lambda controllers can be applied directly at the boiler to increase the efficiency, whilst by optimizing the water treatment the necessary blow-off (and hence accompanying energy losses) can be contained. The use of an exhaust emission heat exchanger reduces heat loss via the flue gases. Under certain conditions, the evaporation enthalpy of the resultant water ($2 H_2 + O_2 = 2 H_2O$) can be regained by condensing the water vapor present in the flue gas during firing. By using this approach the fuel's heat of combustion can be utilized in full.

Currently, systems are also available for the energetic optimization of degassing thermal drinking water; this may save up to 90% of the present losses, with vapors and/or expansion vapors being conveyed for use elsewhere in the brewery.

27.2.4
Possibilities for Heat Recovery

Within the brewery there are numerous possible waste heat sources that can be used for heat recovery, with various degrees of effort. However, the number and capacity of heat sinks is limited so that the reduction of waste heat must always be checked prior to any possible recovery. The details of heat recovery systems are presented in the following sections.

27.2.4.1 Wort Cooling

The recovery of warm brew water via wort cooling has long been practiced in breweries. In a well-constructed heat exchanger, the volume of warm water produced will provide the heat requirement for the warm brew water in the brewhouse. While the liquid ratio of brew water to wort is about 1.1 : 1, it may occasionally be sensible to couple this with the service warm water.

Insufficient attention is often paid to an adequate and individually adjusted dimensioning of the plate apparatus and, as a result, poor heat recovery with a simultaneously increased need for refrigeration may occur. To some extent, the liquid ratio must be drastically increased in order to attain the desired wort temperature. Consequently, an excess of warm water with a too-low temperature may occur that has been heated with primary energy and disposed of as overflow into the drain.

At this point, a distinction can be made between two-step and one-step wort cooling:

- For two-step wort cooling, the plate apparatus comprises two compartments, in the first of which the wort is cooled with cold brew water. This raises the temperature of the warm brew water, which is transferred to a reservoir. In the second compartment, the wort is cooled to the 'pitching temperature'. This second cooling is achieved with ice-water or glycol (and rarely by direct evaporation of the cooling agents) (see Figure 27.4).

- In one-step wort cooling, the cold brew water is first chilled such that the hot wort is cooled directly to the desired pitching temperature. The cooled brew water is produced between brews and held in an isolated tank until it is used in the process (see Figure 27.5).

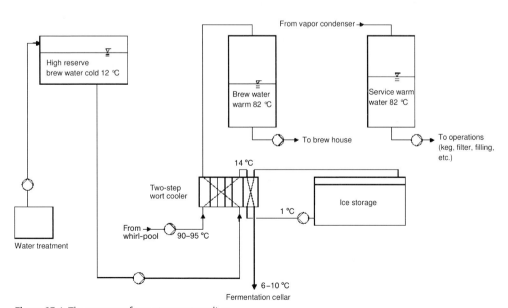

Figure 27.4 The process of two-step wort cooling.

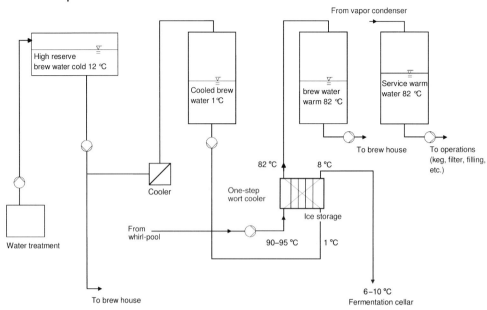

Figure 27.5 The process of one-step wort cooling.

27.2.4.2 Vapor Condenser

When passed off vapors are condensed during wort boiling, this causes the return water to be heated. It is possible to differentiate between *low-temperature vapor condensers*, which produce service water from cold water, and *HAT vapor condensers*, that load an energy storage with which, for example, external wort heating can occur. The potential for warm water production of a conventional vapor condenser with near-complete condensation is approximately 7.7 hl/hl evaporation. The total evaporation of 5% in respect of the full kettle and 100 hl AW will result in 40.8 hl per brew (see Figure 27.6).

In order to maintain these values and the high efficiency of the apparatus, air-free brewing, periodic cleaning of the heat exchanger surface and a sufficiently large dimensioning with area reserves are necessary.

27.2.4.3 Waste Heat from Vapor Condensate

The hot accumulated vapor condensate – especially that produced in high-temperature vapor condensers – is cooled with cold water in the vapor condensate cooler. This will lead to the production (again) of approximately 0.1 hl/hl AW of warm water at almost 80 °C.

27.2.4.4 Waste Heat from Compressed Air

The compression heat from compressed air production can also be transferred to the service water. The retrofitting of a heat recovery system is logical in water-

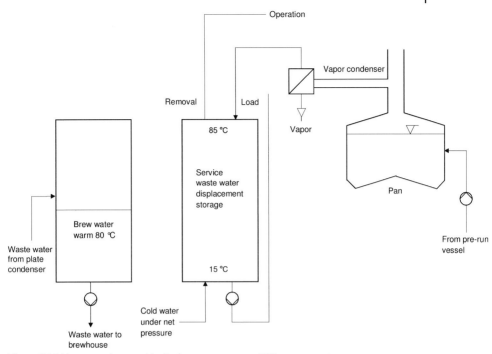

Figure 27.6 Vapor condenser with displacement storage. WW = warm water.

cooled compressors, since about 90% of the compression work can be recovered as heat. Oiled rotary screw compressors, although seldom used in brewing, can deliver warm water temperatures of up to 70 °C. In oil-free compressors, the attainable temperature level is often only 50–55 °C, and this requires an individual design of heat exchangers as no standard solutions are supplied. Although the potential (at 55 °C) is approximately 0.1–0.15 hl/hl AW, this is dependent on the compressed air or electrical power requirements, and also on the suitability of the existing compressors for the recovery.

27.2.4.5 Waste Heat from the Refrigeration System

Another possible means of energy recovery is possible by using (compressor) refrigerating systems. Here, the refrigerant hot gas is cooled with a so-called 'deheater' after being compressed and, depending on the refrigeration system and design of the heat exchanger, temperatures of up to 60 °C can be reached. The potential (at 55 °C) lies at approximately 0.1–0.3 hl/hl. The specific electrical power requirements of the system must also be considered here, but these must be calculated on an individual basis.

27.3
Power Supply

27.3.1
Requirement Figures

As for specific heat requirements, statistical data from intercompany comparisons on energy should also be consulted to determine the electrical current requirements for breweries (Figure 27.7). Although there has been a very slight tendency for small companies to increase their heat requirements, the tendency among breweries of all sizes during the past few years has been, on average, to decrease these requirements. This downward trend has been predominantly due to the ability of larger companies to introduce on-going improvements.

27.3.2
External Power Supply [6]

27.3.2.1 Supply and Measurement at the Release Point

Depending on the brewery's size, its power supply may be either low voltage (400/230 V) or medium voltage (5 kV, 20 KV). For a medium-voltage supply, the transformer may be owned either by the brewery or by the power company, and this in turn will affect 'standard rate' prices.

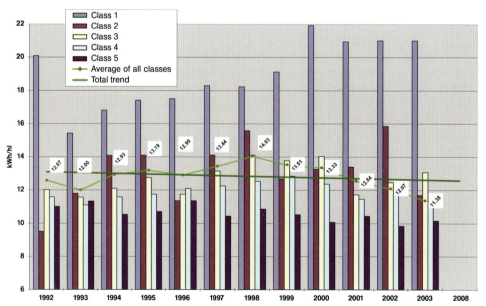

Figure 27.7 Development of specific electrical power requirements (Source: intercompany comparison, IGS 2004).

Transformers are optimally efficient at a load of approximately 70%, and a higher transformer rate can be divided into several modules, according to the company structure, providing both advantages and disadvantages that must be taken into account:

- Increase of the operational safety, silent power reserves.
- Adjustment to the actual demand.
- Operation with optimal load.
- Possibility of module extension.
- Reduction of no-load losses.
- Higher investment costs.

27.3.2.2 Power Factor Correction

In practice, a number of inductive resistances may cause a phase shift between voltage and electric power. A $\cos\phi = 0.9$, as is mostly demanded of a network operator, results in an apparent power of 1 kVA at an efficiency of 0.9 kW. The reactive power fraction accounts then already for 48.5%. A lower value of $\cos\phi$ will result in a higher cost for the reactive current, but this may be counteracted by the power factor correction.

27.3.2.3 Supply Contracts

The external power supply of breweries is mostly regulated by special customer contracts with the power companies. The prices comprise a customer-dependent (energy rate) and a customer-independent component (demand charge or base price).

27.3.3 Electric Power Consumption of the Brewery

Electric power is consumed by various areas within the brewery, as defined in the following sections (see Figure 27.8.)

27.3.3.1 The Brewhouse

Electric power consumption in the brewhouse depends on the dimensioning of the single components, the load, and the chosen process. Statistically, the brewhouse accounts for about 10% of the total requirement (i.e., 0.8–1.2 kWh/hl).

If possible, mechanical rather than pneumatic conveying equipment should be used for malt transport, as the power requirement of the latter system may be four- to 14-fold that of the former, depending on their realtive constructions.

With regards to the brewhouse equipment, every change of vessel requires places an additional effort on the operating power of the pumps. For example, if wort heating is achieved by an external heat exchanger, with the heating medium on one side and the wort on the other side, then both components will need to be pumped.

The choice of brewing process also has a major influence on power requirements and consumption. When using an external boiler, depending on the brew

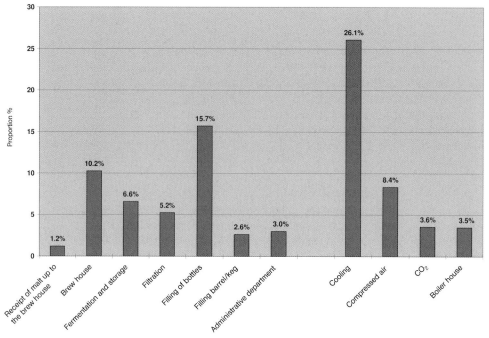

Figure 27.8 Allocation of the electric power requirements in the brewery.

temperature and evaporation, the wort must be circulated up to 12 times during the brewing process. Internal boilers, however, require no additional pumping energy.

A mechanical vapor condenser delivers considerable savings in terms of heat energy, but involves a high power consumption for compressing the vapors. If a thermal vapor condenser is used, then the electrical power requirement is omitted.

In a brewhouse with a mechanical vapor condenser with an internal boiler and an external wort boiler, the specific electrical power requirement is calculated as 0.55 kWh/hl VB. Depending on the equipment and methods used in the brewhouse, this value may fluctuate between 0.4 and 0.7 kWh/hl.

27.3.3.2 The Filling Area

Typically, the brewery's filling area will utilize 18–20% of the total power requirement (2–2.6 KWh/hl). The power ratings of filling columns, depending on the nominal output, are listed in Table 27.1. Here, the electrical performance was calculated from the installed performance, using a simultaneity factor of 0.7. The power requirements for both ventilation and lighting must be added in, although the plants which are used to produce and fill PET bottles – which require a considerable amount of additional power – are not taken into account.

Table 27.1 Power rating (kW) and electrical power requirements in the filling area.

	Nominal output (bottles/h)				
	20 000	40 000	60 000	80 000	100 000
Depalletizer		9.3	7.5	7.5	11.5
Pallet magazine		1.5	1.5	1.5	1.5
Palletizer	8.0	9.3	7.5	7.5	11.5
Pallet security	3.0	3.0	3.0	3.0	3.0
Uncasing machine	2.0	4.6	9.5	13.5	9.5
Case rinser	17.5	30.3	38.0	38.0	38.0
Casing machine	2.0	5.0	9.6	13.5	9.5
Cleaning machine	26.7	49.8	76.0	108.0	132.0
Inspection machine	2.0	2.0	4.0	4.0	6.0
Filler and seaming machine	6.4	17.0	23.0	34.0	46.5
Labeling machine	4.0	9.7	13.7	19.6	21.0
Pallet transport	2.1	3.1	3.1	4.3	4.3
Case transport	14.0	22.0	31.5	34.0	51.0
Bottle transport	28.0	48.0	55.0	68.0	122.0
Verifying device	2.0	2.0	3.0	3.0	3.0
Installed power	117.7	216.3	285.9	359.4	470.3
Electrical power	82.4	151.4	200.1	251.6	329.2

Source: Schreiner

27.3.3.3 Drive System and Components

The electric motors located throughout the brewery are used mainly to drive the machinery, conveying systems and/or pumps. If these motors require an alternating current (as is common with pump engines and conveying systems), then pole-changeable engines or, even better, infinitive-speed-governing engines should be used.

The use of speed-governing drives does not necessarily lead to a reduction in external peak power, but will lead to a considerable reduction in electricity costs if a wide adjustment range down to small loads can be covered. One sensible use of speed governing may be for a motor driving the combustion air blower at a boiler plant, as well as driving the engines for pumps.

27.3.3.4 Lighting

Good light is indispensable in the large majority of work processes, and a good light quality should be provided with the lowest energy input. The main criteria here include:

- Illuminance (the total luminous flux incident on a surface, per unit area).
- Utilance (the ratio of luminous flux received by the reference surface to the sum of the individual total fluxes of the luminaires of the installation)

- The type, numbers and arrangement of the lamps and their radiance/brightness.
- Switchability and controllability.
- Utilization of day light.
- Day light-dependent regulation.

With dark paintwork, up to 50% more light flux may be required to provide equal illuminance. The light efficiency, the color of the light and color rendition are very important. A high lifetime of the illuminants with a simultaneous high light efficiency ensures cost-effectiveness.

A ballast resistor should always be used with fluorescent lamps, because:

- there is no energy loss through heat loss;
- there is a 12% higher light efficiency;
- it is possible to dim the light;
- there is a flicker-free start;
- the light is flicker-free; and
- the fluorescent lamps' lifetime is increased by up to 50%.

Artificial lights should generally only be switched on if they are specifically being used in the work areas. Work places that are exposed to daylight can be automatically adjusted to the particular lighting conditions by using light sensors and electronic controls.

27.3.3.5 Heat Supply

Whilst high-pressure hot water has certain advantages as a heat carrier compared to high-pressure vapor, a major disadvantage relates to the considerably higher electric power requirement. Whereas, a high-pressure vapor will flow without any added current, due to the pressure drop between the vapor generator and consumer, high-pressure hot water must be conveyed with pumps.

The fraction of the electrical power requirement is, on average, 3.5%, but this is often significantly higher for systems using high-pressure hot water.

27.3.4
Optimization of the Electrical Power Supply: Load Management

First, the electricity bill for the entire year must be analyzed in order to identify any seasonal variations. From this, various characteristics can be differentiated that provide some initial information on cost-saving potentials:

- Low-tariff fraction of active energy.
- Wattless current. (i.e., current out of phase)
- Peak power.
- Efficiency factor.
- Load utilization period.
- Average electricity tariff.

In a second step, special costumer contracts should be checked.

The frequency and height of the peaks that occur can be determined in an operational load analysis. The level of disadvantage to be expected can be determined based on daily–or, even better, weekly–load profiles. In addition to organizational measures, technical measures also have an effect; these include the installation of a comprehensive surveillance system that can highlight very different technical standards.

Modern optimization computers, as well as energy or load management systems, will contain a trend calculation and can be programmed to specific operations. The latter enable the graphical presentation of data on a monitor, the acquisition of costs and the handling of several energy forms. An observation of the given disadvantage level is automatically accomplished by the load-shedding of consumer areas that are not absolutely necessary for the process. For each switching step, the minimum and maximum switching-off period–as well as break (recovery) periods–can be individually specified. Most importantly, all of the performance data and switching processes can be recorded in the computer.

27.3.5
Combined Heat and Power (CHP) [7]

Although, during the 1960s, self-power production with steam power plants contributed significantly to the energy supply of breweries, for economic reasons CHP has today become insignificant.

Currently, by using block-type thermal power station engines, high electrical efficiencies (for gas up to 38%, for diesel up to 44%) can be achieved such that a good adjustment of the heat and electrical power requirements is possible. Some (especially large) breweries operate gas turbine block-type thermal power stations, with CHP plants being built only if the correct basic conditions are provided. Thus, in Germany, it is now possible under certain prerequisites to operate steam power plants economically, as the Renewable Energy Law (Erneuerbare Energien Gesetz) offers an incentive with increased feed-in tariffs.

It should also be mentioned here that, due to the combined production of power and heat, there is a considerably higher fuel exploitation than occurs with separate production. There is also less of an environmental burden with currently available CHP techniques.

27.4
Cold Supply [8–11]

The cold supply constitutes the largest fraction of the electrical power requirement, at 30–50%. In the case of German breweries, the specific electrical power requirement will fluctuate between 1.7 and 5.0 kWh/hl, with an average value of approximately 3.6 kWh/hl. For this reason, optimization of the cold supply, as well as the cold production, is of major importance.

27.4.1
Cooling Requirements

Today, the cooling requirements of a brewery can be calculated as weekly loads, using specialized simulation tools, assuming that several basic requirements and the time flow of the brewery are accurately known.

27.4.1.1 Design of Fermentation and Storage Cellar
The older cellar sections of breweries generally involve a high cooling requirement, as these rooms are often poorly insulated against the surrounding and fairly warm ground. Any subsequent insulation is generally not possible, due either to a limited availability of space or to the instability of the ground. In this situation it is advantageous to use single isolated and cooled tanks rather than old cellars. There is also a noticeable disadvantage during cold production since, with a limited evaporation surface, lower evaporation and/or pre-run temperatures are required than with tank cooling.

27.4.1.2 Location/Climatic Zone
The higher the insulation standard of the tank, the lower is the influence of a higher outside temperature on tank cooling requirements in a warmer climate. In contrast, the influence of a higher fresh water temperature is increased if adjacent surface waters with a temperature of up to 30 °C are to be used. By using one-step wort cooling (which is prevalent in modern breweries) the cooling requirement for brew water may be almost double that of a German brewery using water at 12–14 °C. The cooling of degassed water for processes such as blending, pre-coating and non-alcoholic beverage production is also often necessary.

Another consequence of a high fresh water temperature is the increased need to provide cooling water for the machinery, especially in the case of pasteurizers, when a defined product temperature must be assured at the outlet. On occasion, it may even be necessary to pre-cool the cooling water.

The weekly cooling requirements for a modern, well-equipped small brewery (1500 hl/year) in a cold area of Northern Europe are shown graphically in Figure 27.9.

27.4.2
Cold Production

The highest possible evaporation temperature and the lowest possible liquidation temperature should be aimed for during cold production. The evaporation temperature is influenced by:

- Direct evaporation or the coolant system used.
- The design of the cooling area on vessels and other consumers.
- Alteration and regulation of the compressor power.

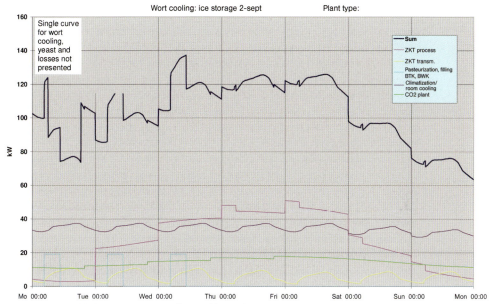

Figure 27.9 Weekly heat requirements for a cold, small brewery.

Today, in order to achieve a low refrigerant volume (especially with NH_3), coolant systems using glycol are invariably employed. In such cases, and in contrast to an optimally designed plant with direct evaporation ($t_0 = -4\,°C$), a 3–4 °C lower evaporation temperature (ca. 8 °C) is necessary. In areas using an oiled or contaminated heat exchanger, and where there is an inadequate adjustment of the compressor power requirements, the actual evaporation temperature will be noticeably lower, in the range of −10 to −12 °C. As a rule of thumb, a 3.5% increase in electrical power requirement can be expected for every 1 °C reduction in the evaporation temperature.

An additional point relates to the auxiliary power of the coolant recirculation, which is often not adjusted according to fluctuations in cold requirement, and will be operated all year round with a constant, high electrical power requirement. Pumping control here is especially effective and economical.

There are, of course, lower limits for the condensation temperature that are due not only to the climatic conditions of the location but also to the design of the condensers. Under German conditions, the unit should be designed to function at a condensation temperature of 30 °C, with a maximum 21 °C wet-bulb temperature. Air-cooled condensers are markedly inferior, especially during the summer, and should only be used for small systems, as a difference of 10–12.5 °C compared to the dry air temperature can be expected. Hence, although condensation temperatures of 45 °C can easily be achieved during the summer, this results in a high electrical power requirement and a high compressor abrasion.

With evaporation condensers, the condensation temperature is, even with a fractional load, often worse than the reference value. This is because coatings are easily formed on the coiled tubes of the condenser, such that the water flow rate through blocked spray nozzles is no longer sufficient. Strict attention should be paid here to maintaining the water quality of the circulation water.

As shown in Figure 27.10, the coefficient of performance (COP) of a system will not always be constant. In fact, under a fractional load operation, and taking into consideration the auxiliary equipment, values are often reached that are far below those specified by the producer. On average, COP values of between 2.5 and 6 are achieved for systems with a coolant supply, whereas in systems with a pure direct evaporation the COP should be ≥3.5.

The question of power division plays an important role with regard to investment costs, operational safety and energy requirements. In the example shown in Figure 27.11, four compressors are available, of which in the example week two are run constantly while a third is switched on sporadically. The fourth compressor needed to be switched on briefly on only one occasion. A long-lasting fractional load can be avoided by closing down a piston compressor, or by valve-regulating a screw compressors. In this way, the efficiency can be noticeably reduced under a fractional load with pistons and, in particular, with a screw compressor.

Consequently, as many compressors as possible should be operated at full load, and only one at fractional load, which implies the use of a centralized system with a higher ranking control. Recently, rotational speed has also been used sporadically for compressors, although this must be adjusted according to the compressor type in order to avoid damaging rotational speeds and vibrations.

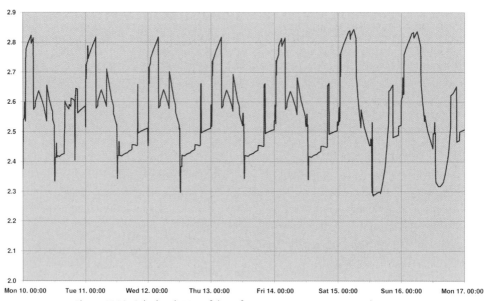

Figure 27.10 Calculated COP of the refrigerating system (average value: 2.56).

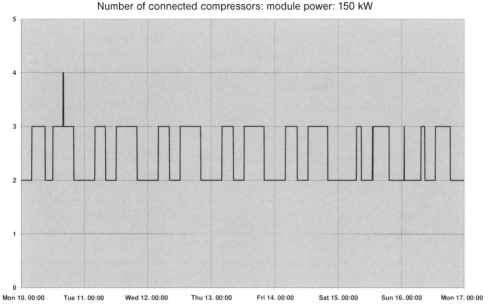

Figure 27.11 Calculated compressor numbers.

With poor power adjustments, fractional loads may often lead to undesirable reductions in the evaporation temperature, whereby the cooling power is reduced until equilibrium is regained.

Clearly, a continuous measurement and documentation of the operational pressures should be monitored in order to identify any extreme and unfavorable operational conditions. But, if this is not the case, then it must be left to chance whether a competent person will observe any pressure changes and recognize the conditions present.

27.4.3
Goals for an Optimal Cold Supply

In order to achieve an optimal energy requirement, the cold requirement must be minimized, and the refrigeration optimized. For this, the basic principles include:

- The optimization of wort cooling with regards to an as high as possible heat recovery and an as low as possible use of cooling for brew water pre-cooling or wort post-cooling.
- The use of single isolated and cooled fermentation/storage vessels with a high insulation strength and qualitative high-grade insulation foam (i.e., minimization of the surface area).
- The closure of poorly isolated cellar sections.

- An avoidance of room cooling, especially in ventilated rooms.
- An avoidance of lost heat and moisture sources in cooled areas (e.g., heated vessels, noninsulated pressure tubes, etc.).
- The optimization of flash pasteurization plants to high back-heating values and low cold requirements.
- To maintain favorable operation temperatures at the compressor.
- The spacious design of the heat exchanger areas.
- The use of a screw compressor with valves for smaller systems, and the avoidance of fewer than three compressors (fractional load COP unfavorable).
- The regular maintenance and control of compressors, evaporators and condensers.
- Regular degassing and de-oiling of the system.
- Controlling the electrical power requirements and operational parameters.
- Isolating the NH_3 network, and maintaining the glycol network complete and intact.
- Minimizing the pump power requirements, and avoiding throttle valve and bypass regulations.
- Complete refreezing (if possible) of the ice storage, avoiding block ice, and not operating the stirring device on a permanent basis.

For optimum production, both the concept and operation of the refrigeration system must be considered in detail, with the safety of the production unit and its power requirements playing important roles.

27.5
Compressed Air Supply [12, 13]

Whilst, during recent years, the requirements for operating air have shown a constant increase, those for compressed air have decreased, mainly as a result of technological advances, with compressed air having been largely replaced by CO_2. Yet, depending on the extent of such substitution, the requirements for compressed air will fluctuate widely.

The electrical power requirements to produce compressed air depend on the amount required, on the different compressed air qualities and, mainly, on the level of production pressure. Thus, a low production pressure is to be aimed at during planning, at which stage all pneumatic valves and fittings should be constructed to equate with a consumer pressure that is as low as possible.

Good planning of the compressed air network is essential, as a slight pressure reduction in the network will demand a respectively lower production pressure. If a pressure in excess of 1 bar is required during production, this will require a 10% increase in compressor power. However, by employing an optimum design for the compressed air distribution network, the pressure decrease between the producer and end consumer will be less than 1 bar. A low pressure reduction in the distribution network can be achieved by: (i) the use of smooth pipes; (ii) having fittings that are flow-technologically optimized; (iii) using flow rates of only 6–10 m/s; and (iv) dimensioning the pipes with a reserve capacity! In particular, the flow rate – and thus the chosen pipe diameter – will play important roles, since in turbulent flows (as are present in the compressed air network) the decrease in pressure is proportional to the square of the flow velocity.

Various investigations conducted in German breweries have demonstrated the potential for savings in compressed air supplies. The total network, including the attached consumers, should be regularly checked for leaks since, in some cases, simply by removing many leakages up to €0.05 per hl can be saved in terms of electricity costs. Capacity shortages can also be removed in compressed air production, such that expansion investments can be omitted.

It follows that employees must not treat permanent 'whistling and hissing' noises as normal, but rather recognize them as costly faults that should be fixed as quickly as possible. It is quite often the case that machine operators and other personnel are simply not aware that compressed air is one of the most expensive forms of energy in the brewery.

Today, compressed air production is dominated by oil-free screw or scroll compressors; however, due to their specifically reduced electrical power requirements and the simple power control with full or semi load switches, piston compressors also have a place and are used in the larger breweries.

With regards to energy requirements, compressed air production plays a significant role. For example, an over-dimensioned plant will often encounter idle time such that the specific energy requirements are worsened. Hence, a knowledge of the requirement profile will form the basis of the design for an effective and efficient compressor.

References

1 Schu, G.F. (1995) The balance of energy and warm water in the brewhouse. *Brauwelt International*, **IV**, 316–30.

2 Fohr, M. and Meyer-Pittroff, R. (1998) Neuentwicklungen auf dem Gebiet der Würzekochung – Thermische Brüdenverdichtung mit Innenkochung. *Brauwelt*, **12**, 460 ff.

3 Best, P. (2004) Energie- and Wasserbedarf der Flaschenreinigung, 7. Energietechnisches Seminar, Neufahrn.

4 Schu, G.F. (1988) Energieeinsparung and Luftreinhaltung mit modernen Öl- and Gasbrennern. *Brauwelt*, **128** (39/40), 1716–19.

5 Schu, G.F. (1987) Factors influencing the annual operating efficiency of factory boiler plants. *Brauwelt International*, **I**, 80–3.

6 (a) Schu, G.F. (1996) Duie Stromversorgung des Brauereibetriebes. *Brauwelt*, (38/39), 1831–4.
(b) Schu, G.F. (1996) *Brauwelt*, (44), 2044–6.
(c) Schu, G.F. (1996) *Brauwelt*, (48), 2340–2.

(d) Schu, G.F. (1997) *Brauwelt*, (1/2), 36–8.
(e) Schu, G.F. (1997) *Brauwelt*, (6), 203–4.

7 Schu, G.F. (1990) Die Kraft-Wärme-Kopplung mit Verbrennungsmotoren and Gasturbinen in der Brauerei. *Brauwelt*, **37**, 2048–58.

8 Cube, H.L. (1981) *Lehrbuch der Kältetechnik*, Verlag C.F. Müller, Karlsruhe.

9 Dölz, H. and Otto, D. (1992) *Ammoniak-Verdichter-Anlagen*, Verlag C.F. Müller, Karlsruhe.

10 Reisner, K. (1994) *Fachwissen Kältetechnik*, Verlag C.F. Müller, Karlsruhe.

11 Schu, G.F. Kälteversorgung, *in Praxishandbuch der Brauerei*, Behr's Verlag and Fachverlag Hans Carl, Nuremburg.

12 Wieczorek, J. (1995) *Die Wirtschaftliche Druckluftstation*. Resch-Verlag, Gräfelfing.

13 Feldmann, K.-H., Mohrig, W. and Stapel, A.G. (1985) *Druckluftverteilung in der Praxis*, Resch-Verlag, Gräfelfing.

28
Environmental Protection
Jochen Keilbach

28.1
Introduction

Prior to the 1990s, environmental consciousness was only poorly developed; however, the public discussion about climate change and environmental scandals has led to a rethink in society as a whole. In addition, beer is a product of high quality, naturalness and purity. It presents part of nature and the production should take place in agreement with environmental friendly demands. Ecologically negative reports nearly always result in damage to a brewery's image and, depending on the circumstances, can even result in its closure. With these requirements in mind, it should be an obligation for every brewery to be engaged with this subject.

28.2
Environmentally Relevant Subjects in Relation to Brewing

28.2.1
Waste Water

28.2.1.1 Avoidance of Waste Water by Reduction of Water Usage
Water, the main raw material in a brewery, is increasingly in focus due to cost pressures (i.e. increasing waste water charges). Today, depending on composition, the water usage in breweries is 3–4 hl/hl sales beer. This value has been reduced by different cost-cutting measures and new technologies in recent years. Examples of cost-cutting measures include:

- Specific phase separations in cleaning [cleaning in place (CIP)] by new measuring methods (conductance, pH) leading to minimization of rinsing waters.
- Recycling of different rinsing waters (e.g. post-rinsing waters are used in intermediate rinsings).
- Use of last wort and vapor condensate for mashing-in.

Handbook of Brewing: Processes, Technology, Markets. Edited by H. M. Eßlinger
Copyright © 2009 WILEY-VCH Verlag GmbH & Co. KGaA, Weinheim
ISBN: 978-3-527-31674-8

Examples of new technologies include:

- Vacuum pumps for bottle filling, which are not cooled by water.
- Cascade-like constructions of bottle cleaning machines.
- Use of air-cooled condensers instead of water-cooled condensers.

28.2.1.2 Composition of Brewery Waste Water

Brewery waste waters are classified by communal sewage treatment plants as non-problematic. Because of its high carbon contents, it is easily biological degradable. This waste water is often desirable as it ensures the maintenance of the vitality of the microorganisms present that are responsible for the biological purification of the waste water.

The pH value is often in the basic range due to the cleaning of bottles with caustic soda solution. This also has a positive effect on the communal sewage treatment plant since these contain mostly acidic waste waters from other industrial sectors.

Negative components can be introduced into the waste water through the use of cleansers and disinfectants. In accordance with the requirements of a brewery, cleansers and disinfectants need to be highly efficient, economical and environmental friendly. This is not always guaranteed and the main problem points are:

- Biological degradation.
- AOX problem.
- Phosphate load.

28.2.1.2.1 **Biological Degradation** Cleansers and disinfectants contribute very little to the total chemical oxygen demand (COD) load (about 5%). The COD-containing cleansers and disinfectants are, in particular, belt lubricants, surfactant-containing cleansers, disinfectants with high molecular additives such as quaternary ammonium compounds and organic fatty acids.

The German Detergent and Cleanser Law regulates the requirement of biological degradability of surfactants. Together with the European detergent regulation, it is regulated that all surfactant-containing cleansers need to show a total degradation (mineralization) of 60%. The primary degradation rate needs to be at least 80%.

28.2.1.2.2 **AOX Problem** AOX indicates the sum of all 'absorbable organic halogen substances'. These are halogenated hydrocarbons such as chloroform or methylene chloride. Such substances are not present in cleansers and disinfectants, but can form from other waste water constituents if active chlorine containing agents are used.

Such substances are not degradable and thus are undesirable in waste water. Consequently, in the German 'indirect discharger regulation', there is a very low threshold limit value of 0.2 mg/l for the 'free chlorine concentration' of the waste water. On the basis of halogenated acetic acids such as mono bromoacetic acid or chloroacetic acid, disinfectants are detected in waste water as AOX substances and are undesirable. There are many alternatives kinds of disinfectant; however, their very high effectiveness against yeast is undisputed. It is hence understandable that many breweries do not want to abandon halogenated disinfectants despite environmental concerns.

28.2.1.2.3 Phosphate Load Phosphoric acid forms the basis of acidic cleansers and disinfectants more often than nitric acid since its cleaning power is higher. By law, there is no constraint on the use of phosphoric acid in breweries. Phosphates are problematic for the environment, particularly if the communal sewage treatment plant has no phosphate elimination steps. Similar to other salts, phosphate serves as a nutrition source for algae and can contribute to a undesirable mass reproduction of algae – a phenomenon called eutrophication.

28.2.1.3 Waste Water Disposal

There are two ways for a brewery to dispose of waste water. The waste water can either be (i) discharged after suitable treatment directly in a river, stream and so on (direct discharger) or (ii) dispensed to the communal sewage treatment plant (indirect discharger). Most breweries operate on the basis of indirect discharge.

28.2.1.3.1 Minimum Requirements of a Brewery as a Direct Discharger [1]

1. COD: <110 mg/l.
2. BOD_5: <25 mg/l.
3. NH_4-N: <10 mg/l.
4. N total: <18 mg/l.
5. P total: <2 mg/l.

28.2.1.3.2 Requirements of a Brewery as an Indirect Discharger [1]
Here the corresponding communal regulations apply. Often the following threshold values can be found:

1. pH: 6.5–10.0.
2. Temperature: <35 °C.
3. Settleable substances: <10 ml.
4. NH_4-N + NH_3-N: <200 mg/l (with more than 5000 inhabitants, otherwise <100 mg/l).
5. P total: <50 mg/l.

The costs as an indirect discharger are regulated on a communal basis and can be vary dramatically. The range reaches from 1.20 to 2.50 Euro/m^3 waste water. Often the waste water rate is composed of a basic charge and a high pollution surcharge. This high pollution surcharge comprises mostly COD and biological oxygen demand (BOD_5), and sometimes also the content of nitrogen, phosphate or halogens.

28.2.1.3.3 Waste Water Treatment for Indirect Discharger

- pH neutralization:
  - Basic waste water: often CO_2 from fermentation is used, but acids (sulfuric acid, hydrochloric acid, etc.) can also be used.
  - Acidic waste water: caustic soda solution is nearly always added.

- Solid separation: appropriate filtering systems are employed.

- Biological treatment of waste water: A pre-treatment of waste water in the brewery is of importance for both the brewery and the sewage treatment plant. A homogenization as well as a first biological degradation of the pollution load is achieved. The result is a decrease of the COD and hence a reduction of the high pollution surcharge levied against the brewery. A differentiation can be made between aerobic and anaerobic pre-treatment methods:

- *Aerobic.* In general, aerated mixing and equalizing tanks are used. The size of these tanks should be at least one daily charge of waste water. The degradation rate of the BOD is up to 50%.

- *Anaerobic.* This kind of waste water pre-treatment is rare with indirect dischargers. Here, higher BOD degradation rates are obtained than with aerobic treatment by the selective degradation with microorganisms ('reactors'). The degradation rate can be up to 90%. The problem with the anaerobic method lies in the sensitivity of the system against pH, temperature and load fluctuations.

28.2.2
Energy

According to a prognosis from EWI/Prognose the developments in the international energy market until 2030 look as follows:

- The primary energy demand will be reduced by 15% in Germany; however, on the global market it will increase, especially in developing countries, by 60%.

- The energy demand of German industries will be reduced by 7%, whereas electric power consumption will grow.

- Due to abandoning nuclear power until 2020, natural gas takes on the main part with 33% of the gross electricity generation and hence lies ahead brown coal, wind energy and anthracite.

- Renewable energies will cover 26% of the gross electricity in Germany.

Due to high energy prices and cost pressures, the topic of 'energy' has come more and more to the fore in breweries. There, energy saving stands in the foreground since it is economically much cheaper than energy recover. The reduction of energy consumption, defined as a clear environmental goal, has reached the breweries' day-to-day business. A constant control of energy consumption and energy costs is carried out by installed meters. Here, comparative figure systems serve to evaluate the consumed units. This energy monitoring system needs constant maintenance in order to fully utilize the saving potential in this area by various optimizations.

A number of common weaknesses are discovered, predominantly energy losses (e.g. leakage in the compressed air network) and poorly constructed systems (e.g. pumps do not run at an optimum).

With investments into the energy sector, environmental considerations also need to be incorporated to protect the resources, in addition to the conventional 'return on investment' approach.

In recent years the following technologies and techniques have prevailed, and have succeeded in saving energy [2]:

- Thermal or mechanical vapor compression for heat recovery.
- Vapor condensers for warm water generation.
- Different forms of energy-saving wort boiling systems (thin film evaporation, forced blow internal boiler). With good quality malt the boiling time could also be considerably reduced.
- Optimization of the cooling device leads to an increase in the evaporation temperature which directly results in a reduction of the driving power.
- Biogas utilization from waste water.
- Usage of block-type thermal power stations in thermal/electrical energy coupling.
- Prevention of energy loss with optimal thermal insulation (in both hot and cold areas).
- Utilization of alternative fuels or solar energy.

28.2.2.1 Renewable Energy

Renewable energies in breweries can be found in the utilization of biogas taken from air discharged from the brewery's sewage disposal plant. This form of energy usage is not only interesting from an environmental point of view, but is also economical. Other renewable energy sources such as wind energy, solar heating, and photovoltaic and geothermal power are only used selectively in breweries. However, the breweries will need to face this discussion in the future. Apart from economic aspects, the use of renewable energies is seen very positively by the public and can also contribute to image improvement.

28.2.2.1.1 Renewable Energy: Production of Biogas from the Air Discharged from the Brewery's Sewage Disposal Plant
The usage of biogas developed during anaer-

obic treatment of waste water has gained much importance in the last years through different supporting programs (German Market Stimulation Programme, Renewable Energy Source Act, etc.).

The composition of biogas in an anaerobic reactor is:

- Methane: 60–90% volume.
- CO_2: 10–40% volume.
- Hydrogen sulfide: 0–2% volume.

Methane develops due to the basicity and high content of carbohydrates of the waste waters. The content of methane is responsible for the energy content of the biogas.

Here, chemically bound energy can only be transformed into electricity and heat thermally or electrically/thermally by block-type thermal power stations. In addition to an environmental factor, this kind of energy generation also has an economic advantage for a brewery.

More and more the network provider is obliged to take facilities for the generation of regenerative energy onto their network and to remunerate according to specific high rates.

28.2.3
Brewery Emissions

28.2.3.1 Gaseous Emissions

In the recent past, the reduction of greenhouse gas emission for climate protection has been pushed to the fore. International agreements such as the Kyoto Protocol have led to national (CO_2 emission reduction) and communal regulations (Agenda 21). As a result, renewable energies and combined heat and power plants are gaining more and more importance.

The 'Technische Anleitung zur Reinhaltung der Luft' (Technical Instruction for Air Pollution Prevention) of the German government is an administrative regulation for the Bundes-Immisionsschutzgesetz (German Federal Air Pollution Control Act), and defines immission and emission as follows:

- 'Immissions […] are air pollutants affecting humans, animals, plant, soil, water, atmosphere, or cultural and real assets'.
- 'Emissions […] are air pollutants that emanate from a plant'.

Gaseous emissions of a brewery can be divided into two groups [3]:

- Emissions through the combustion of fossil fuel (oil, coal, wood, etc.).
- Emissions specific to breweries (e.g. fermentation CO_2, air discharged from the factory's sewage disposal plant, brewhouse vapors and vapors from the bottle washer).

The threshold values of the first group are regulated by the Technical Instruction for Air Pollution Prevention. Odor expert reports on odor emissions of a brewery reduce the risk of problems.

An improvement of the emission can be reached in an existing plant with factors such as:

- Change in fuel.
- Optimization of the burner or change of burner.
- Smothering of flue gas.

With investing in new plant, the following should be considered:

- Choice of fuel.
- Type of burner.
- Type of boiler construction.

For the second group, in particular, vapors and fumes are typical, with high water concentrations and odor-intensive components contained therein. Boiling times have now been reduced to 60 min maximum with the currently available high-quality malts. With atmospheric boiling followed by vacuum evaporation, boiling times of 30–40 min are possible today and lead to a considerable reduction of emissions.

Furthermore, vapors and fumes can be reduced or avoided by:

- Vapor compressors.
- Decreased evaporation rate.
- Vapor condensers.
- Filters.

28.2.3.1.1 Emission of Fermentation CO2 Emissions of fermentation CO_2 can be reduced or avoided with a CO_2 recovery plant. The recovered CO_2 from the fermentation corresponds to about the CO_2 usage in a brewery. In exceptional cases a brewery can produce an excess of CO_2 and even sell it.

28.2.3.1.2 Emission of Air Discharged from the Factory's Sewage Disposal Plant
Odor developments from the factory's sewage disposal plant are a concern for all breweries. With an direct discharger with only one anaerobic step this is more problematic than with an indirect discharger with only an aerobic pre-treatment of the waste water. Different measures are available to avoid any odor nuisance in the environment. Here, the discharged air can be cleaned by biofilters, biowashers or chemical washers.

28.2.3.1.3 Emissions of Cooling Agents As a cooling agent, ammonia driven systems are nearly exclusively found in breweries. Ammonia belongs to the cooling agents, which are toxic, flammable or corrosive. This risk potential for humans is minimized by numerous requirements and regulations.

Ammonia cooling systems are driven in closed systems. Ammonia belongs to the natural cooling agents and is environmental friendly. Ammonia has the advantage that it can already be located at very low concentrations (from 5 ppm) by it sharp, pungent smell; hence, potential leaks can be repaired quickly. Escaping ammonia from a leak is always gaseous and, compared to fluorinated cooling agents (fluorinated hydrocarbons), has the advantage of not damaging the ozone layer and hence not contributing to the acceleration of the greenhouse effect.

Recently, the use of CO_2 as a cooling agent or refrigerant has again come to the fore. Compared to environmental friendly ammonia, CO_2 has the additional advantage that is neither toxic nor flammable.

28.2.3.2 Dust Emissions

Increased dust development can occur in breweries from malt sampling to grist mills or mashing tuns. Brewery-dependent different dedusters and filter systems are used for the reduction of dust emissions. These systems need to be designed effectively so that the approval value of the plant is maintained. This limit threshold value is often set at $20\,mg/m^3$.

28.2.3.3 Noise Emission

As noise are sounds marked, which by its sound volume and structure act harmful, disturbing or be a strain for the human and the environment.

In many countries the requirements on the work place are often like: 'In the work place the sound pressure level is kept as low as possible depending on the kind of workplace. The decision level at the work place in the working rooms is allowed to be at most 85 dB (A) taking into account the externally interfering noises. If this decision level cannot be kept even after the possible noise reduction it can be exceeded up to 5 dB (A)...'.

In February 2006, the new European Union noise control legislation was passed. This says that the value of 80 dB should not be exceeded; the value of 85 dB must not be exceeded.

Main sources of noise in breweries are:

- All areas of filling bottles and cans.
- Traffic (lorries, cars, forklifts, etc.).
- Production aggregates (e.g. mills).
- Compressors for the generation of air and cooling.

Causes of noise are:

- Running engines.
- Transporting bottles (especially caused by continuous collisions in the area of conveyor belts).
- Pumps (and the cavitation developed by these).
- Old plants with damaged bearings.
- Poor sound insulation (especially with open windows and doors).

Noise reduction is differentiated as active and passive noise abatement:

- *Active*. abatement of noise at the source (e.g. old plants) by new investment or technical repairs. Furthermore, improvement measures are used such as an increase of the quietness of bottles on belts by new concepts such as the principle of the inclined plane.

- *Passive.* Abatement of noise in the surroundings of the noise source by, for example, constructional measures such as sound-absorbing materials or greater distances between two plants to reduce the level overlap.

28.2.4
Waste

Everything that does not count as product is classified as waste. Consequently, brewery-specific feed such as spent grains and yeast are assessed as waste.

Every brewery should have a waste management concept to satisfy the demand of environmental friendly disposal. Apart from environmentally sound waste disposal, the economic benefit of a suitable waste management should also be kept in mind.

The nomination of a waste commissioner is not compulsory, but is advisable. They have to implement the minimum requirement of the legal regulations. Corresponding training courses as well as technical literature are offered in order to keep up to date.

The following points should be considered for the improvement of waste disposal:

- Extensive waste separation to improve recycling quality (e.g. recycling glass separated by color).
- Checking for the possibility to reduce the packaging for the distributors.
- Reducing one's own packaging material.
- Avoiding dangerous constituents in waste.

28.2.4.1 Special Brewery-Specific Production Waste

28.2.4.1.1 **Spent Grains and Yeast** Farming serves as the traditional method of disposal of spent grains and yeast, where they are used as high-quality animal feed. However, breweries have increasing sales problems due to the decreasing numbers of agricultural businesses. One consequence of this is a considerable decline in prices so that here hardly any noteworthy proceeds can be achieved. Thus, the question will arise as to how to dispose of the high-quality spent grains and yeast. Alternatives (combustor for the spent grains, biogas plant) are already available and in use, and will come more and more to the fore in the future; however, a real alternative to farming has not been found.

As a result of various agricultural scandals (e.g. bovine spongiform encephalopathy), certification is now also needed for animal feed spent grains and yeast. Hence, a brewery is not only a food maker, but at the same time an animal feed maker requiring a seal of approval. It is the goal to register and control the animal feed production chain from the field to the end consumer. In the register of European Union law, approval of single animal feed (positive list) spent grain is listed under 5.01.01 ('Side products of fermentation industry and distillation'). Yeast is listed under 14.03.01 ('Protein product from micro organism'). In Germany, predominantly 'Quality and Safety' is requested and applied as a certification standard; different certification standards are applied in neighboring countries (e.g. Good Manufacturing Practice in The Netherlands and Belgium).

28.2.4.1.2 **Kieselguhr** The use of kieselguhr for the filtration of beers still dominates the brewing sector. The reasons for this are obvious: it is simple, cheap and effective. Although kieselguhr-free filtration has been in research and development in industry for decades, no alternative has prevailed. However, it should be noted there are kieselguhr-free filtrations in breweries (on the basis of cross-flow technology). The costs of investment in such plants are, however, still very high.

The current problems for the use of kieselguhr as a filtering aid [4] are:

- Particulate matter emission during handling of kieselguhr.
- Limited resources of kieselguhr for beer filtration.
- Difficult to dispose of kieselguhr sludge.
- Disposal as hazardous waste increases the costs.
- Output in farming is still possible, but as supplement in spent grains and excess yeast it is problematic due to the animal feed regulations.

Methods of resolution:

- Kieselguhr-free filtration.
- Use of alternative filtering aids.
- Regeneration of kieselguhr and recycling.
- Composting of kieselguhr sludge.
- Additive for the construction industry.

References

1 Rosenwinkel, K.-H. and Jörg, B. (2004) Verfahrensentwicklung in der Brauereiabwasserbehandlung, 8. Hannoversche Industrieabwasser-tagung. Brauerei Seminar, pp. 47–68.
2 Russ, W. Energiemonitoring und -targeting, in *Praxishandbuch der Brauerei*, 7th edn, Behr's, Hamburg, *Akt.-Lfg.*, **10/02**, 7–9.
3 Russ, W. Gasförmige Emissionen, in *Praxishandbuch der Brauerei*, 7th edn, Behr's, Hamburg, *Akt.-Lfg.*, **10/02**, 9–11.
4 Methner, F. (2005) Kieselgursparende Technologie bei der Altenburger Brauerei. *Brauwelt*, (**31/32**), 931.

29
Sensory Evaluation

Bill Taylor and Gregory Organ

A professional sensory programme can contribute directly and indirectly to product quality, brand success, consumer loyalty and ultimately company profitability. This chapter will explore how a brewery can develop and implement a sensory programme designed to meet these aims.

When planning sensory evaluation in a brewery it is important to understand that expert sensory testing and consumer testing are not the same. The context for both types of testing is different. The consumer provides responses to variables, and provides information on preferences and purchase intent. Expert sensory evaluation provides information on beer flavor differences, the types and magnitudes of flavor attributes, and relationships between variables. To do this requires the use of all five senses. The senses referred to in this chapter will be the five classic senses of smell, taste, vision, audition and touch [1].

29.1
Introduction to the Five Senses

Sensory evaluation literally means an evaluation using the senses. The implication being that all the senses are involved. This is an important concept particularly as it relates to consumers. In some professional situations sensory evaluation is thought of as 'taste testing', which in effect reduces the scope of senses involved. There is no doubt that in certain situations the focus of a sensory evaluation may well need to be the aroma, bitterness or the mouthfeel of a product. However, in the normal course of drinking a beer the consumer, untrained in formal sensory evaluation, engages all the senses. This may not be a deliberate engagement. Many people do not realize that beer has an aroma worthy of careful appreciation. Drinkers do feel the coldness of a glass of beer as it is picked up, and can certainly see if the beer does not look as clear and bright as it usually does. A familiar taste may complete the involuntary sensory assessment.

Our sense of smell is believed to be 50 times more sensitive than our sense of taste. This is why it is relatively easy to smell the aromas of food coming from a restaurant as you walk by. Indeed, a lightstruck beer can often create a perceptible aroma in an entire room.

As the aromas from beer are often not as intense as some others, the beer panelist increases the effectiveness of olfactory stimulation by sniffing the beer. However, it has been shown that a few short sniffs is best for effective assessment. Repeated sniffing or long hard inhalation only dulls the response rather than concentrating the aroma or magnifying the response. A number of other factors can reduce the effectiveness of our sense of smell. For example, the impact of a head cold, spicy foods, medications and irritating odors.

Taste is a sense that often confuses people. Taste is often incorrectly thought to mean flavor. However, the sense of taste is but one component of flavor. Indeed it is the minor component and only contributes some 20%. The remainder is due to aroma. The primary tastes are the four sensations sweet, salty, sour and bitter. Although not often a factor in beer, the predominantly Asian *umami* (savory/spicy) taste is important to many foods and beverages. Taste buds are receptors found in high concentration on all areas of the tongue. The concept of a strict zonal mapping of taste receptors on the tongue has lost favor [2]. Since receptors are widely distributed, the whole surface of the mouth needs to be used in assessing taste. An additional factor is that while the sweet receptors respond quickly, the bitter receptors respond more slowly. Furthermore, when the bitterness is perceived, it will tend to linger. This lingering and the associated throat perception produce the sensation of after-bitterness.

Sight and hearing can play a part in assessing a beer. One can see that if the beer is darker than normal, has a sediment or has a lower fill than usual, then the panelist suspects a papery, aged beer. Likewise, if the beer does not release the expected hiss as the crown is lifted one can hear that the beer is low in carbonation.

The sense of touch is also important in the assessment of beer. The lips, mouth and tongue all play a part. The lips can respond to the feel of a dense creamy foam or liquid at first sip. The lips (as well as the fingers on a glass) tell us about temperature. The lips and tongue also respond to the carbonic sting of a lager or the soft mellowness of a nitrogenated ale. The tongue and the mouth respond to pressure and this sensation is translated into a description of body. A stout is sometimes described by drinkers as thick and chewy, while low-calorie beers may be described as thin. The effervescing carbonation in the mouth or the warming effect of higher alcohols create similar physical sensations. These sensory effects are referred to as the trigeminal sensation.

29.2
How to Assess the Flavor of Beer

A systematic and consistent approach to assessing the flavor of a beer sample is more likely to result in improved quality and validity of the data obtained.

29.2.1
Technique for Flavor Assessment

The following technique is recommended for the flavor evaluation of beer:

- Slowly swirl the sample around in the glass. This helps to increase the concentration of the more volatile aroma compounds into the 'headspace' of the glass by increasing the extent of beer/air interaction.

- Next slowly inhale a small quantity of the beer's headspace gases to assess the aroma characteristics. The panelist should then wait for 15–20 s before taking a second short sniff that can validate earlier perceptions and add any new dimensions of aroma. Do not take numerous small sniffs without pausing. This is not recommended as it rapidly leads to sensory fatigue and an incorrect assessment of the sample's aroma.

- Take a sample into the mouth and swirl it around. This ensures that all the taste buds are exposed to the sample. This is important as taste buds are located on the sides, front and back of the tongue. Some tastes (e.g. astringency) are also sensed on the cheeks and roof of the mouth. Panelists also need to assess the mouthfeel of the beer. Panelists should not rush back to the glass too soon. Wait another 15–20 s to avoid a desensitizing effect and then repeat the taste assessment.

When tasting the beer this sometimes leads to changes in the perceived aroma as a result of the volatile aroma components entering the nasal cavity. This happens because as the sample is warmed by being in the mouth aroma components diffuse from the mouth and into the back of the nasal cavity. If this happens the panelist is allowed to change their opinion of the sample's aroma attributes.

- While assessing the taste of the beer the panelist should also be aware of the trigeminal responses of the CO_2 level and temperature of the beer.

- The sample is then swallowed and the panelist concentrates on the after-taste of the sample.

29.2.2
Additional Points

As well as following a consistent technique for assessing the flavor of a beer, the panelist should be aware of the following points:

- The importance of concentrating on assessing the sample and not rushing the assessment.

- Each sample should be assessed on its own merit (i.e. without any expectation as to what the sample should smell or taste like). This does not apply to beer quality acceptance tests where the panelist is assessing 'trueness to type' of that beer sample.

- There is no preferred sequence of sample presentation. Indeed, samples should be presented in random order apart from strongly flavored or significantly different beers, which should be presented last. This does not apply to beer quality acceptance tests where the panelists generally assess samples from the mildest to strongest flavor.

- Panelists should not use the beer color (this is a reason why dark glasses are required) or the amount of foam as a guide to the nature of the sample. This does not apply to beer quality acceptance tests where clear glasses are used to allow the assessment of haze, clarity and foaming properties.

- The importance of not 'over-assessing' the sample (taking numerous smells and taste) as this quickly leads to fatigue and confusion.

- To help overcome sensory fatigue the smelling and sipping of pure water, or the use of a bland biscuit, before assessing each beer sample is recommended. This helps to 'clear' the aroma and taste receptors. Panelists can also be advised to smell the back of their wrists between samples as this helps to 'clear' the aroma receptors.

- The issue of sensory fatigue limits the number of samples that can be assessed in any one session.

29.2.3
Requirements for Attendance

The panelists should also be aware of the requirements for attendance at sensory evaluation sessions:

- Panelists must refrain from attending sensory evaluation sessions if they have a cold or other illness (e.g. allergy), or are taking medication, which might impair their sensory ability.

- Panelists must refrain from smoking, eating or drinking in the 30 min prior to a session so that their palates are clean and at their most sensitive. Wearing strong scents is not acceptable because this will interfere with the panel's assessment of the aroma of samples.

- It is preferable that lipstick not be worn because it can contaminate glasses and may affect the panelist's sense of taste.

- Panelists are not permitted to smoke in the sensory evaluation room or talk whilst any panelist is assessing a sample. Mobile phones should also be turned off to avoid distractions.

29.2.4
Overall Assessment

During formal sensory sessions the panelists focus on the individual aroma and taste characteristics of the sample. Nevertheless, the 'overall impression' is equally important. While the interaction of all sensory signals can build a positive or negative assessment, a mixture of signals will sometimes occur. A beer that looks and smells good is likely to taste good, but this is not certain. It could be that a beer that looks good and has a vibrant and appealing late hop character could however have a very high and harsh bitterness or a thin astringent finish that could rate the beer down. Equally, a good smelling and tasting beer could be let down by poor physical stability or foam stability that detracts from the appearance.

Furthermore, the balance of flavors is important. Individual characters that on their own may be thought of as good or bad can have a different place in the overall perception of flavor. A slight sulfur character in one beer can be negative, but may be acceptable in a complex full-bodied ale. A high bitterness may be in balance in a full-bodied beer, but be overpowering and harsh in a light-bodied beer. A phenolic character out of place in a lager may be a positive attribute in a specialty beer. Similarly, a diacetyl character may be out of place in a lager, but add to the complexity of a stout.

These examples illustrate that an extremely important part of sensory management in a brewery is to have a good understanding of what your beer should be like (the concept of 'true to type'). While a beer can be scored well during sensory evaluation, if it is not true to type for that brand the commercial acceptability of the beer could be brought into question. This trueness to type assessment should not be left to the vagaries of memory. Beer brand flavor has been known to drift and a sensory panel may be unable to identify the change if it occurs in small incremental steps. Thus, documenting a target sensory profile (preferably together with a range of analytical parameters) will help manage brand flavor integrity over the longer term.

29.3
Description of the Main Flavor Attributes

This section discusses a number of the flavors commonly found in beers [3]. As well as aroma and taste descriptions, some notes on the origin of the flavors are also given. Flavor thresholds are given where appropriate [4]. Procedures [3] that can be used to illustrate the flavor attributes are given in Table 29.1 and Figure 29.1.

29.3.1
Sulfur Dioxide

An older term for this attribute is sulfitic. Sulfur dioxide (SO_2) produces a pungent/choking aroma/sensation in the nose and towards the back of the mouth. The detectable presence of sulfur dioxide is considered to be a defect. Sulfur dioxide is produced naturally during fermentation. Higher sulfur dioxide levels can result when yeast is under stress during fermentation. Some beers receive sulfur dioxide addition after maturation, but breweries commonly aim for a total sulfur dioxide content in packaged beer below 10 mg/l.

Table 29.1 Procedures for illustrating the flavor attributes.

Attribute	Procedures[a]
Sulfur dioxide	add 25 mg/l sulfur dioxide[b]
Hydrogen sulfide/mercaptan	add 25 μg/l hydrogen sulfide; also FlavorActiv mercaptan pill[c] for mercaptan
DMS/cooked vegetable	add 100 μg/l DMS; also FlavorActiv onion pill for cooked vegetable
Solvent	add 40 mg/l ethyl acetate
Acetaldehyde	FlavorActiv acetaldehyde pill
Estery/fruity	separate additions of FlavorActiv isoamyl acetate and FlavorActiv ethyl hexanoate pills
Hoppy	separate additions of FlavorActiv hop oil pill and FlavorActiv kettle hop pill
Floral	FlavorActiv geraniol pill
Spicy	FlavorActiv spicy pill
Fresh grass	FlavorActiv freshly cut grass pill
Clove/4-VG	FlavorActiv phenolic pill
Grainy/straw	beer brewed using a mashing in temperature of 75 °C
Malty	add commercial malt extract
Caramelized	commercial beer such as Kilkenny Ale
Roasted	commercial stout beer
Fatty acid	FlavorActiv caprylic pill
Butyric	FlavorActiv butyric pill
Cheesy	FlavorActiv isovaleric pill
Diacetyl	FlavorActiv diacetyl pill
Yeasty	add diluted yeast extract
Oxidized	papery stage – FlavorActiv papery pill
Acidic/sour	FlavorActiv sour pill
Alcoholic	add ethanol to a total of 7% w/w
Body	low body – add 30% soda water
Sweetness	add 10 g/l sucrose
Bitterness	add 20 mg/l iso-α-acids
After-bitterness	add 40 mg/l iso-α-acids
Astringency	0.6 g/l tannic acid in water
Metallic	FlavorActiv metallic pill

a Amount of compound to be added to a relatively bland commercial beer.
b Before presenting the sulfur dioxide sample the panelists *must* be asked if they are asthmatic. This is vitally important as a very small percentage of the population is sensitive to sulfur dioxide.
c Add one FlavorActiv pill per liter of beer (see www.flavoractiv.com for further details).

29.3.2
Hydrogen Sulfide/Mercaptan

An older term for this attribute is sulfidic. The most common description of the aroma of hydrogen sulfide (H_2S) is rotten eggs. This compound may be present as a fermentation byproduct. It also occasionally occurs in freshly pasteurized beer due to the thermal breakdown of sulfur-containing amino acids and/or proteins. At its threshold (generally regarded as 5 μg/l) H_2S is hardly recognizable as such,

29.3 Description of the Main Flavor Attributes

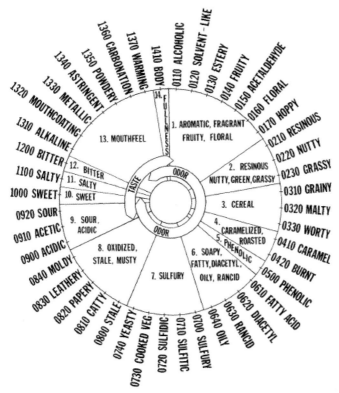

Figure 29.1 The Flavor Wheel. Reprinted with permission from [19].

but it does modify the general beer aroma, mainly by reducing the intensity of hoppy and estery aroma. Between 5 and 15 µg/l it gives rise to the well-known 'sulfury' aroma that is distasteful to many people, while above 20 µg/l it is decidedly offensive as the rotten egg aroma is clearly evident.

A wide range of other flavors are due to mercaptans (R-SH, also called thiols). These can be described as drain-like, meaty or burnt rubber. The threshold for thiols can be in the low µg/l range. Lightstruck flavor is also due to a mercaptan, 3-methylbut-2-ene-1-thiol.

29.3.3
Dimethylsulfide/Cooked Vegetable

This generally flavor-positive attribute is mainly due to the sulfur-containing compound dimethylsulfide (DMS). When DMS is present in high concentrations (e.g. greater than 60 µg/l) the main descriptor is cooked sweet corn. Other descriptors are cooked cabbage or processed tomatoes. DMS can occur in some European beers at up to 120 µg/l. DMS can also occur in beers produced from infected wort (e.g. *Citrobacter*).

DMS is mainly derived from malt. During the germination step in the production of malt the concentration of S-methylmethionine, $Me_2SCH_2CH_2CH(NH_2)COOH$, usually referred to as DMS precursor, increases greatly. During kettle boil the DMS precursor thermally breaks down into DMS. A large proportion of the DMS formed is subsequently lost during kettle boil. A small amount of DMS is formed during whirlpool stand (through additional thermal breakdown of DMS precursor) and/or fermentation (through the enzymatic reduction of DMSO). The final DMS concentration in the beer depends on the balance between how much DMS has been formed and how much DMS has been lost.

A cooked vegetable-type aroma could also be due to a range of other sulfur-containing compounds. These include polysulfides such as dimethyl disulfide (Me_2S_2) and dimethyl trisulfide (Me_2S_3), and thioesters such as S-methyl esters (R-COSMe). The threshold for these compounds can be in the low µg/l range. In some cases a garlic or onion like aroma might be due to high levels of dimethyl trisulfide.

29.3.4
Solvent

This attribute refers to a solvent-like flavor, which has also been described as airplane glue and nail varnish. It is generally regarded as a 'negative' estery note. It is usually due to excessive levels of ethyl acetate. Ethyl acetate levels generally range from 5 to 30 mg/l. The flavor threshold for ethyl acetate is usually regarded as 30 mg/l.

29.3.5
Acetaldehyde

This attribute refers to the characteristic flavor of acetaldehyde. Other descriptors include green apples, raw apple skin or bruised apples. Acetaldehyde can result from slow fermentations or use of low viability yeast leading to restricted yeast growth. The flavor threshold for acetaldehyde is usually regarded as 10 mg/l.

29.3.6
Estery/Fruity

This attribute refers to the fruity flavor of beer which is mainly due to the esters present in beer and covers a range of flavors including:

- A banana or pear-drops like flavor generally regarded as being due to iso-amyl acetate. Iso-amyl acetate levels typically range from 0.5 to 5 mg/l. The flavor threshold for iso-amyl acetate is usually regarded as 1.2 mg/l.

- An apple, boiled lollies or aniseed-like (at very high levels of ethyl *n*-hexanoate) flavor generally as being due to ethyl *n*-hexanoate. Ethyl *n*-hexanoate levels typically range from 0.05 to 0.45 mg/l. The flavor threshold for ethyl *n*-hexanoate is usually regarded as 0.21 mg/l.

Other esters that can contribute to an estery/fruity flavor include: iso-butyl acetate (described as banana/fruity with a threshold of 1.6 mg/l), ethyl *n*-octanoate (described as apple/fruity with a threshold of 0.9 mg/l) and 2-phenylethyl acetate (described as rose/honey/apple with a threshold of 3.8 mg/l). Ethyl *n*-butyrate (described as papaya/tropical fruits/mango/tinned pineapple with a threshold of 0.4 mg/l) can be indicative of *n*-butyric acid contamination.

Estery/fruity flavor is very dependent upon the nature and concentration of the esters present in the beer. The esters are formed during fermentation. The nature and concentration of the esters present (notably the ratio of iso-amyl acetate and ethyl *n*-hexanoate varies) depends on a large number of factors. An important factor is the yeast strain used. The condition of the yeast (i.e. yeast viability/vitality/health) is also extremely important as this influences the extent of yeast growth. It has also been found that hop-free fermentations generally give higher levels of esters compared to hopped fermentations (i.e. hopped prior to fermentation with addition of hop pellets or hop extracts). Other influences are the mashing temperature (higher temperatures usually give less esters), temperature of fermentation (lower temperature usually give less esters), amount of oxygen present at the beginning of fermentation (lower oxygenation usually gives more esters) and gravity of the pitched wort (higher gravity wort usually gives more esters). These factors are very inter-related and so the effect of any changes on the final ester concentration needs to be experimentally determined.

29.3.7
Hoppy

This general term refers only to the aroma resulting from hop pellet or hop oil addition rather than the bitterness coming from hop pellets, which is a purely taste attribute. Some hoppy aroma can also arise from the use of hop extracts.

The chemistry of hop aroma is extremely complex and the compounds responsible are still being investigated. In general terms, hoppy aroma is due to the volatile hop oil components or their transformation products. The original hop oil components can be transformed during kettle boil and/or fermentation. The sesquiterpene hydrocarbons humulene and caryophyllene (which are very common in most hop varieties) and their transformation products are believed to be responsible for the 'noble hop aroma'.

The general term hoppy is further described on the International Flavor Wheel [3] as 'fresh hop aroma' and 'kettle hop aroma'.

The term 'fresh hop aroma' applies to beers where the hoppy aroma is similar to that of fresh hops. These beers have considerable amounts of the volatile hop oil components, usually resulting from late or dry hopping. Late hopping refers to the addition of hop pellets to the kettle towards the end of kettle boil or directly into the whirlpool. Dry hopping refers to the addition of a portion of hop pellets to the beer after fermentation as is usually applied when brewing traditional English cask ales.

The hoppy flavor of most beers packaged in dark glass is best described by the term 'kettle hop' on the International Flavor Wheel. This type of hoppy aroma arises because most of these beers are brewed with wort that has received hop pellet addition at the beginning of kettle boil. Kettle boil then results in the loss of most of the more volatile hop oil components. The hop oil components remaining in the wort after kettle boil can be further transformed during fermentation.

The strength and type of the hoppy aroma also depends upon which hop variety is used. In general terms, hop varieties are used to add mainly aroma or bitterness (as these latter varieties have relatively high α-acid content). Some varieties are used for both aroma and bitterness addition.

Hop extracts are also widely used in brewing. Reduced hop extracts contain different aroma compounds from those typically present in hop oil and so beers using these extracts have very low 'hoppy' aroma. Some commercial hop extracts do have hop oil components added and so these extracts can give rise to a hoppy aroma in beer.

29.3.8
Floral

Floral can be described as the aroma of flowers, fragrant, rose-like and perfume-like. Floral flavor arises mainly as a result of hopping.

29.3.9
Spicy

This attribute refers to a general spicy flavor arising mainly as a result of hopping. It does not cover the clove-like flavor of 4-vinylguaiacol (4-VG).

29.3.10
Fresh Grass

The attribute fresh grass refers to the aroma of freshly cut grass and not to the straw or hay-like character of dried grass.

29.3.11
Clove/4-VG

This attribute refers to a clove-like flavor. The clove-like flavor is due to 4-VG, which can be produced by wild yeast infection. Thus, the presence of the clove/4-VG attribute may be considered a defect in some beers. However, clove/4-VG flavor is expected in some 'specialty' beers (e.g. wheat beers from Belgium or Germany). Wheat beers are traditionally brewed with a yeast strain that gives rise to high levels of 4-VG. The flavor threshold for 4-VG is usually regarded as 0.3 mg/l.

29.3.12
Grainy/Straw

Grainy/straw refers to the aroma of barley which can also be described as dried grass, straw or hay-like. A grainy/straw aroma may be present in the last runnings. A grainy/straw aroma present in beer may be indicative of an inefficient or too short kettle boil.

29.3.13
Malty

Malty refers to the characteristic flavor of malt or malt extract. The production of malt involves the steeping of barley grain in water to allow germination to begin. Once germination has proceeded for the desired time, the grain is gently dried to produce malt. Slight variations in the temperature regime used for drying the malt gives rise to the characteristic flavors of different malt types (e.g. compare the flavor of pale malt versus Munich malt). A blend of malts can be used in beer recipes to create differing malty flavors.

29.3.14
Caramelized

Caramelized attribute refers mainly to the characteristic flavor of crystal malt. It can also be described as caramel or toffee like. If the malt drying step is carried out at a high temperature while the grain moisture is high, partial caramelization of the sugars occurs. Since small 'crystals' of caramelized sugar are produced the product is called 'crystal malt'. Caramelized flavor could also arise from excessive kettle boil.

A caramelized flavor also appears as a specific stage during the oxidation of some beers.

29.3.15
Roasted

This attribute refers to the characteristic flavor of roasted barley or malt. Other descriptors for this attribute include burnt sugar, scorched toast, chocolate malt and smoky. An excessive amount of roasted malt can also give rise to a harsh/acrid aftertaste. Roasted malt is produced by a more intensive heat treatment of malt. Roasted barley can also be produced.

29.3.16
Fatty Acid

This attribute refers to the flavors arising from $C6$–$C12$ *n*-carboxylic acids (sometimes referred to as fatty acids, although this term more correctly applies to high-molecular-weight n-carboxylic acids, i.e. C16–C20). The flavor can also be described as soapy, waxy or sweaty. The presence of a fatty acid flavor is generally due to excessive levels of n-octanoic acid and n-decanoic acid. The flavor thresholds for n-octanoic acid and n-decanoic acid are usually regarded as 13 and 10 mg/l, respectively.

29.3.17
Butyric

The presence of a detectable level of n-butyric acid gives rise to a characteristic flavor described as vomit-like. Excessive levels of n-butyric acid can arise as a result of infected sugar syrup and infected wort or fermentations. The flavor threshold of n-butyric acid is usually regarded as 2.2 mg/l. The term rancid has also been used to describe this flavor, but the term rancid more correctly refers to the flavor of oxidized fats or oils.

29.3.18
Cheesy

A cheesy flavor is generally due to excessive levels of iso-valeric acid and arises from the use of aged hops. The flavor threshold of iso-valeric acid is usually regarded as 1.5 mg/l.

29.3.19
Diacetyl

The flavor of diacetyl can be described as butter, butterscotch or honeycomb. The presence of diacetyl in bright beer or packaged beer is considered a major fault. It can result from either faulty fermentation management (insufficient diacetyl scouring at the end of fermentation) or from early stages of lactic acid bacterial infections. Diacetyl can be a problem associated with the use of recovery beer. In the market diacetyl can result from poorly cleaned beer lines. The flavor threshold for diacetyl is usually regarded as 0.08 mg/l, which can be exceeded in some packaged beers. Note that diacetyl is also covered by the more general term vicinal diketones (R-COCO-R), which covers both 2,3-butanedione (better known as diacetyl) and 2,3-pentanedione. The flavor threshold of 2,3-pentanedione is usually regarded as 0.9 mg/l, which is much higher than the levels found in most beers.

29.3.20
Yeasty

This attribute covers the flavor of autolysed yeast or yeast extracts. Autolysed yeast can arise if beer is held too long in the fermenter or if yeast is held in storage too long at too high a temperature. It may also arise in packaged beer from the use of recovery beer of unsuitable quality.

29.3.21
Oxidized

Oxidized is a complex attribute as it covers a number of different flavors, which occur at various stages during the aging (another term is 'staling') of beer. Recently, the European Brewery Convention has published work directed to better describing the flavor changes that occur as beer oxidizes [5].

During the initial stage of beer staling the 'freshness/clean/crisp' of the beer slowly decreases. The overall flavor intensity also decreases and so the beer is dull or flat in flavor. During beer staling the decrease in 'positive' flavors is just as significant as the increase in 'negative' flavors.

A papery, cardboard-like flavor develops during the next stage of beer staling. The papery stage is believed to be due to the formation of unsaturated aldehydes, especially *trans*-2-nonenal, which are formed by the action of oxygen on the long-chain unsaturated fatty acids (which can be referred to as lipids) content of the beer. *Trans*-2-nonenal has a threshold level of only 0.1 µg/l.

The next stage of beer staling involves the slow disappearance of the papery flavor and the slow appearance of a characteristic flavor best described as caramelized/toffee.

The final stage of beer staling involves the slow appearance of a characteristic flavor best described as leathery or whisky-like.

The oxidation of beer is accelerated by the exposure of the beer or the wort to air (more correctly the oxygen in the air) and especially heat. Thus, important process concerns are dissolved oxygen content, pasteurizer temperatures and over-pasteurization. The sensitivity of beer to heat is very important with regard to storing beer during summer.

The formation of the oxidized attribute governs the flavor stability of beer and, hence, is one of the important determinants of the shelf-life of the beer.

Figure 29.2 illustrates in general terms the flavor changes that occur as beer ages. The times at which each of these flavors appears depends upon beer type and even the brewery producing the beer, oxygen content of beer, sulfur dioxide content of beer, metal ion content (in particular the iron and copper levels are these accelerate the oxidation of lipids through the Fenton reaction), temperature of storage and so on.

29.3.22
Acidic/Sour

This attribute is mainly a taste sensation, but some aroma can arise as a result of excessive levels of acetic acid. Other descriptors include tart, sharp, vinegar and lemon juice. Beer pHs are typically in the range of 3.8–4.3. Acidity levels can vary with the yeast strain used.

Acidic/sour can also be associated with infection and is, thus, generally considered a defect in most beers. However, acidic tartness is a desirable feature of some specialty beers. The most common causes of infection are *Lactobacillus* bacteria,

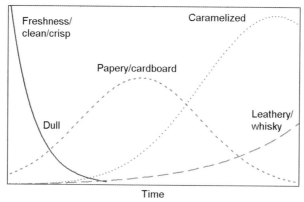

Figure 29.2 The main flavor changes, with regard to 'oxidized' flavor, which occur as beer ages. Arbitrary scale of intensity for each flavor on the y-axis. The time of the changes depends upon a variety of factors, see the text.

which produce lactic acid or a mixture of lactic acid and acetic acid. Acetic acid bacteria produce acetic acid, and may be found in cask beer and dirty dispensing systems as they require oxygen to grow. Beer containing acetic acid will have the acidic/sour taste as well as the characteristic vinegar aroma. Lactic acid has little aroma and so only affects the taste.

29.3.23
Alcoholic

This attribute describes the alcoholic/warming effect due to the ethanol content of beer. This attribute begins to become apparent in beers with greater than 5.0% alcohol content (e.g. Bock beer and barley wine).

Other higher-molecular-weight alcohols also occur in beer. The most important are 1-propanol, iso-butanol, iso-amyl alcohol and 2-phenylethanol. As a class, these alcohols are usually referred to as fusel alcohols. They usually do not occur in levels sufficiently high to have an impact on the overall flavor. The only fusel alcohol of some importance is iso-amyl alcohol, which has a slight fruity flavor. The flavor threshold for iso-amyl alcohol is usually regarded as 70 mg/l.

29.3.24
Body

In assessing this attribute panelists are instructed to *ignore* the overall flavor intensity and to concentrate on the mouthfeel of the beer – the effect of the beer on the inside of the mouth, including the after-palate effect. Beers that are thin in body could also be described as low body or watery. Beers that are full in body could also be described as high body, viscous, thick, satiating, creamy or chewy.

Body is partly related to the viscosity of the beer. One analytical parameter associated with body is Real Extract. Real Extract is an indication of the amount of sugars, dextrins and proteins remaining in the beer after fermentation. Most beers high in Real Extract would have a high body. Alcohol content is also important.

29.3.25
Sweetness

This attribute is purely a taste sensation. The sweet taste could be described as the taste of sucrose or honey. A beer high in sweetness might mask the bitterness/astringency of the beer. In these cases the beer might be described as smooth. The perceived level of sweetness will depend not only on the level of sugars in the sample, but also on the degree of bitterness, acidity and astringency. The latter tend to diminish the perceived sweetness.

The main sugars present in beer after fermentation, and their relative sweetness compared to sucrose (100), are glucose (70), maltose (50) and maltotriose (not sweet at all). After fermentation glucose is present in trace amounts, and maltose and maltotriose are the most abundant. Thus, the residual sugars do not contribute greatly to the perceived sweetness of beer.

Various other factors are also believed to influence the perceived sweetness of a beer. The ratio of chloride to sulfate ions is believed to be important. Chloride ions are believed to add a fullness and sweetness to beer flavor, while sulfate ions are believed to add a dryness and harshness. Thus, a beer with a higher chloride:sulfate ratio might have a higher perceived sweetness. The role of the level of bitterness and astringency was mentioned earlier but the type of bitterness is also important. Thus, a beer bittered with tetrahydro-iso-α-acids might have a higher perceived sweetness compared to a beer bittered with iso-α-acids. Other compounds, such as iso-amyl acetate, glycerol, vanillin, furaneol, maltol and other heterocyclic oxygen-containing compounds, might also play a part in perceived sweetness.

29.3.26
Bitterness

This attribute is purely a taste sensation, the receptors for which are concentrated towards the back of the tongue and throat. It is sometimes confused with acidic/sour, but it can be distinguished fairly easily from this attribute due to its slow response and concentration of perception at the back of the tongue and throat. Bitterness score represents the maximum perceived bitterness sensation. In addition, the bitterness taste lingers after the sample has been swallowed. This can be referred to as after-bitterness. After-bitterness score thus represents the strength and duration of the bitterness sensation that lingers in the mouth.

Sweetness tends to mask bitterness, resulting in diminished perceived bitterness.

Panelists need to be consistent in their tasting technique (sample size and how long it is held in the mouth) as this can influence their bitterness and after-bitterness scores. Since the perception of bitterness is fairly slow and the sensation lingers, panelists should take their time when assessing a beer for bitterness. Owing to the lingering nature of bitterness panelists should take a sip of water before each sample to help prevent carryover effects.

Bitterness comes mainly from the iso-α-acids in beer. Hops contain α-acids which are extracted and then isomerized (more technically the reaction is a rearrangement) during kettle boil to give iso-α-acids. Iso-α-acids in beer can also result from addition of hop extracts.

The α-acids in hops can be extracted with liquid CO_2 and then isomerized to produce isomerized hop extracts. Isomerized hop extracts can be added after fermentation, and can give a smoother and less harsh bitterness as they are a higher purity source of iso-α-acids with all the polyphenols removed.

Another type of hop product is reduced hop extract. The variants in this group are produced by liquid CO_2 extraction of the α-acids from the hops, followed by an isomerization and further chemical reaction. These materials occur in a number of forms (e.g. the dihydro, tetrahydro and hexahydro forms). These forms have different bittering capabilities, but the exact relative level of bitterness of the various forms is still being debated. These reduced hop extract materials can be used in a similar manner and for similar reasons to isomerized hop extracts. They also have the advantage of improved light stability in beer.

29.3.27
After-bitterness

This attribute refers to the bitter taste that lingers after the sample has been swallowed. After-bitterness score thus represents the strength and duration of the bitterness sensation that lingers in the mouth. If excessive levels of after-bitterness are present then this may give rise to a coarse/harsh after-taste. Coarse/harsh after-taste can also arise for other reasons.

29.3.28
Astringency

Astringency is a mouthfeel sensation characterized by 'mouth-puckering' as experienced when drinking strong black tea or young red wine. At excessive levels it is considered to be a defect. Astringency tends to be masked by sweetness.

29.3.29
Metallic

This is mainly a taste sensation. Descriptors include rusty water and tinny. Increased iron content is generally associated with older plant as it is caused by excessive corrosion of tank linings. Copper content can arise from traditional copper brewhouses.

29.4
Sensory Evaluation Environment

A professional sensory panel requires a professional standard environment in which to assess samples. Sensory testing is not usually a full-time role and in breweries the panelists are required to allocate time away from their functional work to contribute to the evaluations. The test room should therefore be centrally located to ensure that it is easily accessible to the panelists.

The design should enhance panelist sensitivity and eliminate distracting factors. A basic layout should include a booth area with a number of individual booths for panelists to assess samples in quiet privacy with full concentration. The booths should have individual hatch doors connected to a sample preparation area. Samples can be presented according to individual need without disturbing other panelists. It is highly desirable to have an adjacent room that can be used for post-session discussion, training sessions and special tests. Complete booth design details and other considerations are available [6].

The sample container is best made from glass. Glasses should be deeply colored to mask color, foam and clarity. Ruby red, dark smoke or opaque glasses are recommended.

The sensory environment extends beyond the physical boundaries of the test room. Sensory assessment is an important function and one that provides useful output for critical decision making. Thus, a senior management sponsor or custodian for the sensory panel is required. In addition, it is important that this value-adding function be recognized by senior management and, in particular, that it be recognized in a way that enhances the efforts and motivation of the sensory panel. A well-functioning and professional sensory panel requires coaching and leadership. There is an effort required that is directed to a clear goal. A full-time panel leader will play a pivotal role in this.

29.4.1
Panel Leader

Developing sensory capability is a challenge for many breweries, particularly when many of the employees are not recruited for their knowledge of sensory evaluation. Sensory evaluation involves the application of knowledge and skills from several areas including physiology, psychology, chemistry and statistics. It is not easy to find individuals well educated across such a diverse group of topics. A professional panel leader should be sufficiently knowledgeable in these areas or capable of learning about these over time.

The panel leader needs to have a plan and goals to work towards. The panel leader will have a range of responsibilities which ultimately support the delivery and development of sensory capability in the brewery. These responsibilities can include:

- Plan and develop appropriate resources.
- Train and validate panelists.

- Select appropriate test procedures and research new methods.
- Communicate and advise on application of sensory tools.
- Manage testing processes and activities.
- Data processing and statistical analysis.
- Documentation, reporting and communication of results to stakeholders [7].
- Communicate the value of sensory work undertaken to management.
- Ensure the sensory panel is adequately rewarded for their efforts.

Clearly, management skills are required in addition to sensory expertise.

29.4.2
Sensory Manual

A key responsibility of the panel leader is the development of a 'sensory manual'. A sensory manual should be available in every sensory department. This will ensure a consistent approach over the longer term despite movement in staff and participants. This manual should describe the scope and operations of the sensory programme. It should include the details of how tests are conducted and reported, examples of sensory forms and reports, training modules, sample preparation procedures, scoring system, and other technical information. It will also assist the orderly introduction of improvements and allow the sign off by stakeholders (e.g. technical, brewing and marketing departments).

29.4.3
Panel Motivation

While there is an obvious need for emphasis on panel training and monitoring, there is a less obvious but nevertheless important need for paying due attention to panel motivation. This in many ways contributes to panelist and panel performance, and thus the worth of the output.

With sensory sessions often once or twice a day, such frequency can represent a problem for some participants. This can be aided to a certain degree by enlarging the pool of participants. This reduces the frequency and hence the risks associated with boredom or absence from other work. Maintaining a panel's motivation over an extended period requires some effort. There does not appear to be a simple answer to this issue. What is successful for one company may not be appropriate or fit with the culture of another. There are some general guidelines that may be helpful to consider:

- Participants should be rewarded for their participation not for accuracy.
- Participation should be recognized by the panel leader on a regular basis.
- Management should recognize the contributions of the panel.
- Panel leader and support team should be positive and encouraging.
- Panel leader should avoid reference to right or wrong answers.
- Participants should have access to information of their performance and receive coaching.

29.5
Types of Sensory Tests

Owing to the subjective nature of sensory evaluation and the complex way in which the various flavor attributes interact to affect the sensory receptors, sensory evaluation must be carried out under standard conditions using standard methods. In this way physiological and psychological differences between panelists can be minimized. These standard methods are based on the appropriate International Organization for Standardization methods. Furthermore, it is necessary to select the correct method of testing to provide the information required [8–11].

There are three main types of sensory tests: difference tests, descriptive analysis (also called flavor profiling) and preference tests. A brief description of each test follows:

29.5.1
Paired Comparison Test [12]

This test is used to determine:

- Directional difference – to determine in what way a particular flavor attribute differs between two beers, for example, more sweet or less sweet. Thus, trained panelists must be used.

- Preference – to establish whether there is a preference between two beers in consumer tests. Thus, an untrained consumer panel MUST be used.

29.5.2
Duo–Trio Test [13]

This test is similar to the Triangle Test (below) except one sample is identified as the reference. Panelists require no attribute recognition training.

29.5.3
Triangle Test [14]

This test is used to detect a difference in overall flavor between two beers. Panelists require no attribute recognition training.

29.5.4
Flavor Profile and Rating Test [15]

This test uses a formal, structured quantitative descriptive analysis approach to obtain a flavor profile of the beer(s) in which the intensity of each attribute is assessed. Overall quality scores are used to rate the overall quality of the beer(s).

29.5.5
Fresh and Aged Test

This special form of a profile test estimates the flavor stability (i.e. shelf-life) of the beer(s) by quantitative assessment of samples of various ages for the intensity of oxidized attribute and of overall quality.

29.6
Selection, Training and Validation of Panelists

A successful sensory panel can be likened to a successful sporting team. Selection of appropriate people is important to the panel's success. The selection and development of panelists is a challenge that should not be underestimated in a brewery [16, 17].

The size of a panel will ultimately depend on the number and types of tests to be undertaken. When designing a new sensory programme it is useful to prepare best- and worst-case estimates.

It is important when recruiting a panel to choose people who have a desire to be involved and are committed to giving the time it takes. It can also be a benefit to have someone who knows about the brewing process, but this is not essential. It can be helpful if panelists like or are interested in beer. This is more likely to result in sustained motivation. Health should also be considered. Some people have allergies or persistent health problems.

Training is very important in establishing a sensory panel. The panelists must be trained in the various beer flavors (Section 29.3). It is paramount that all panelists recognize an aroma or taste and call it by the same name. This is where practice makes perfect or close to it. By using beers dosed with known flavors at known concentrations the panelists can assess, discuss and learn. Regular training and feedback is required. Experienced members need to mentor and share their knowledge with the newer members.

It is useful to set assessment exercises for panelists. In this way competency is tested and currency maintained. This also establishes an aspirational standard for newcomers. Only those panelists who have passed the appropriate assessment exercises are eligible to attend the corresponding sensory evaluation sessions. Testing twice a year for flavor recognition (particularly for those flavors that are critical to the range of beers produced in the brewery) is suggested. It is important to recognize individual skills. Some people are blind to certain aromas or tastes while others have notable sensitivities. These variations are allowable and if well understood can be used to good effect when arranging a panel for a specific test or investigation.

Panelists should be aiming for accuracy and consistency. The panel leader should monitor this aspect of panel performance by occasionally inserting duplicate samples. Test beers with added flavor attributes can be inserted and information provided in a discrete way so as not to embarrass or intimidate a panelist.

This learning can provide reinforcement and grow the confidence of panelists. Frequent attendance at sensory sessions maintains and improves a panelist's sensory ability. Reliability of attendance underpins the performance of the panel.

29.7
Building a Sensory Capability

There can be many reasons and applications for a sensory evaluation programme. In a brewery there are three applications that have wide use and provide a focus for training and panel selection. These are assessment of difference, beer quality acceptance assessment and beer flavor profiling. These three applications require different levels of training. Table 29.2 illustrates the types of sensory test a trained panelist can attend:

- Level 1, the most basic level, is the 'Difference Panel' which assesses whether there is an overall flavor difference between two or more beers.

- Level 2, a more complex level, is the 'Beer Quality Acceptance Panel'. This panel would be focused on assessing the quality of samples for acceptance to the next stage of processing in the brewery or into the market.

- Level 3 involves a 'Profile Panel', which is required to give a detailed description of the flavor of beer samples. This level is the most challenging and so requires the most significant training.

29.7.1
Training Programme Overview

A training programme represents the foundation stone of a company's sensory evaluation capability. It is important to develop a competent panel that can perform reliably and in a standardized way so as to maximize the confidence in decision making. What follows then is a detailed description of one possible approach to training.

Table 29.2 Level of training required for test attendance.

Test type	Level 1	Level 2	Level 3
Triangle	×	×	×
Paired comparison (preference only)	×		
Paired comparison (attribute intensity)		×	×
Beer quality acceptance		×	×
Profile/fresh and aged			×

The training programme begins with a description of the technique for flavor assessment (see Section 29.2). The training programme then takes the panelists through the various stages of Level 1, Level 2 and Level 3 referred to earlier.

Generally a training programme is conducted with at least two prospective panelists so that best use can be made of the time involved. Larger numbers are more efficient but once a panel is established new recruits are usually introduced in smaller numbers. Current panelists should be encouraged to attend these sessions to re train but this is not essential.

29.7.2
Level 1 – Difference Panelist

This is the basic level in sensory evaluation programmes. No training in the various flavor attributes is given at this level.

Level 1 training could begin with a discussion session. This could commence with the general requirements for attendance at sensory evaluation sessions. Next, the type of sensory tests Level 1 panelists can attend and the uses of the information gained could be discussed.

The technique used to assess the flavor of a beer should be discussed in detail. Once this is done the panelists should then assess as wide as possible range of beer types applicable to the local brewery market to demonstrate to the panelists the variety of beer flavors and to give them some confidence that they can detect different beer flavors.

An assessment exercise for Level 1 consists of the panelists carrying out a three-beer-type 'Matching Test'. In this test the panelists are presented with two sets of three samples. The panelists have to correctly match the samples of the same beer type. If the panelist can not pass this assessment after three attempts their further participation should be discouraged.

29.7.3
Level 2 – 'Beer Quality Acceptance' Panelist

This level is designed for panelists who are to be accepted as trained for the quality acceptance of beer. Samples are assessed to determine absence of any gross flavor defects and trueness to type or beer type. In the case of finished beer tank samples, assessment of trueness to type is the main concern. In the case of packaging samples (i.e. samples ex filler) beer type is the main concern to ensure that a beer is not packaged in bottles/cans with the wrong beer type label. In assessing trueness to type or beer type only on the job training can be given as this is the only way panelists can learn what is the acceptable batch to batch variation in the flavor of the beer types. For example, a lager may have a characteristic estery aroma, but the intensity may vary.

As Level 2 panelists are used to assess the quality of beer, the attributes chosen must be relevant to this task. The attributes might be: sulfur dioxide, hydrogen sulfide, clove/4-VG, diacetyl, acidic/sour, oxidized, low body and bitterness. Lightstruck may also included in the training as it may be seen during assessment of consumer-complaint beers. Lightstruck is not included as an assessment attribute as it can contaminate the aroma of the sensory room.

The initial training programme could be spread over four sessions. The first training session could commence with a discussion of the type of sensory tests Level 2 panelists can attend and the uses of the information gained. The first demonstration of flavor attributes is then conducted. This should cover attributes that can usually be recognized by untrained panelists to give them some confidence in their sensory evaluation ability. The second training session could cover the remaining flavor attributes. It is commonly found that a single exposure to the training samples is not sufficient for the panelists to remember the flavor attributes. The two final sessions therefore give the panelists two additional exposures to these flavor attributes.

As part of the actual training programme the panelists should assess beers treated to illustrate the various flavor attributes. Generally, a relatively mild flavored beer is used as the base beer. Panelists are asked to smell and taste the reference beer and the spiked sample, and then comment on the difference (nature and degree) between the samples. The panelists are then told the flavor attribute being demonstrated and then this attribute is discussed. Much stronger samples of the same flavor could be used to assist panelists who cannot detect the flavor in the spiked beer sample.

Level 2 assessment consists of two sections. Level 2A covers detection of gross flavor defects, while Level 2B covers the detection of gross flavor defects *and* the recognition of the flavor attribute(s) causing the problem.

Level 2A assessment involves an exercise to test the panelist's ability to recognize beer with acceptable flavor and beer with gross flavor defects. The panelist should receive two sets (presented at two sessions) of eight samples. Thus, each panelist receives a total of 16 samples. Each set of eight samples should consist of three controls (i.e. unspiked) and five spiked beers (one attribute per sample). The panelist is asked to make a 'go/no go' decision for each sample. The panelists are not told how many control or spiked samples are included, but the fact that there might be more than one control sample is mentioned (but not the fact that there are three control samples). After the first session the panelists should *not* be informed of the answers. In the second session the second set of eight samples covers the remaining five attributes. The panelist passes this assessment if they correctly identify 7/10 of the spiked samples as 'no go' *and* 4/6 of the control samples as 'go'.

Level 2B assessment should involve an exercise to test the panelist's ability to correctly identify the attribute responsible for gross flavor defects. The panelist receives two sets of six samples. One sample is the control (clearly indicated as the control) and five spiked beers (one attribute per sample). The panelist is asked to identify the flavor attribute present in the five spiked beers. After the first session the panelist should *not* be informed of the answers. The second set of six samples covers the remaining five attributes. The panelist passes this assessment if 7/10 of the attributes are correctly identified.

29.7.4
Level 3 – 'Profile' Panelist

Level 3 training and assessment should include the following steps:

- Training in recognizing all the flavor attributes listed on the company's Flavor Profile form and instruction in the use of this form.
- An assessment exercise to test panelist's ability to recognize the attributes.
- Training in attribute intensity scoring.
- A ranking exercise to assess panelist's intensity scoring ability.

During the initial training sessions the Level 3 panelists need to be informed that the panelists should:

- Score attributes in a manner that reflects different intensity levels of an attribute in samples and so contributes to the panel's ability to discriminate between beers.

- Score attributes in a manner that positively correlates with appropriate analytical data.

- Score attributes in a manner that positively correlates with the panel average score.

- Score attributes in a repeatable manner.

- Use the full scale of attribute scores.

These requirements ensure that the panel can adequately identify the flavor differences between beer samples.

After training in all the flavor attributes the Level 3A assessment exercise can be held. Level 3A assessment involves testing the panelist's ability to recognize some of the 'new' attributes. Level 3A could cover the attributes DMS/cooked vegetable, acetaldehyde, solvent, estery/fruity, hoppy, butyric, fatty acid, alcoholic, sweetness and metallic.

In the actual assessment the panelist should receive two sets of six samples. One sample is the control (clearly indicated as the control) and five spiked beers (one attribute per sample). The panelist is asked to identify the flavor attribute present in the five spiked beers. After the first session the panelist should *not* be informed of the answers. The second set of six samples covers the remaining five attributes. The panelist passes this assessment if 7/10 of the attributes are correctly identified.

As discussed during the initial stages of Level 3 training, panelists are required to score the intensity of each attribute. Guidelines on the scale used to assess attribute intensity need to be developed. These may vary from brewery to brewery, and may be influenced by the style of the beers brewed and the flavor intensity of the beers.

If the panelist passes the Level 3A assessment they then proceed to training in the use of the attribute intensity scale. These cover the most important attributes, from a flavor profile point of view, at low-, mid- and high-point intensity. A special session may be devoted to oxidized flavor as this flavor is most difficult to assess and score. Samples of various beers and ages should be assessed and discussed.

The panelist then proceeds to Level 3B assessment, which is an exercise to test their ability to detect different levels of attribute intensity. This is a ranking exercise [18] using commercially available or spiked beers.

29.7.5
On-going Level 3 Performance Monitoring

There are a number of methods to monitor the performance of Level 3 panelists.

One means of assessing performance is by examining the repeatability of flavor profile scores. This involves presenting the same beer sample at different profile sessions and determining which attributes show unacceptable variation. Any which show excessive variability will require further re-training.

Another important component of panelist's performance is discrimination ability. This involves measuring the ability of the panelist to detect differences in attribute intensity (for a range of brewery relevant attributes) and to give attribute scores, which are in agreement with expected attribute intensity trends. For example, if it is generally agreed that Brand A is more hoppy than Brand B, the average hoppy score for a number of Brand A samples should be higher than the average hoppy score for a number of Brand B samples.

The agreement of a panelist's scores with those given by the rest of the panel, for each attribute, is also important and is best assessed using an appropriate graph. However absolute agreement is not necessary. Rather what is sought is if the panel average score increases then the individual panelists' scores also increase. Similarly, if the panel average score decreases then the individual panelists' scores also decrease. The graph can also identify those panelists who always give the same attribute score and so are not adding value to the panel. The graph can also identify panelists who are consistently giving higher (or lower) scores compared to the panel average.

29.8
Applications for Flavor Assessment

29.8.1
In-Process Sample Testing

Building the flavor assessment capability is the first step in bringing this important and value-adding benefit to the business. However, unless this capability is applied effectively the benefits can be missed. While the focus of this chapter has been on the flavor assessment of the finished beer, further value can be achieved by extending the reach of flavor assessment. By choosing critical process points and inputs and conducting flavor assessment, further risk management benefit as well as a quality benefit can be achieved. Using the same principles as described earlier attention could be paid to brewing raw materials such as water, wort, kieselguhr slurry, carbonated water and even the containers used to package the beer. Some ingredients are used quickly, but others may be stored for periods of time. With storage comes the heightened risk of contamination. For example, kieselguhr may absorb odors. Empty cans and even bottles can absorb or adsorb musty/moldy taints, particularly trichloroanisole and tribromophenol, which have flavor thresholds measured in the parts per trillion.

This sensory expertise can also be applied as part of 'in-process' control. Such samples include end of fermentation beer, end of maturation beer, bright beer and beer after packaging. The assurance of quality can be achieved by using panelists trained and experienced in these specific samples. 'Beer quality acceptance' assessment will usually be sufficient for these samples. However, more detailed flavor profiles or chemical analysis may be required when corrective and/or preventative actions are necessary.

29.8.2
Final Beer Testing

Final beer testing provides the bulk of a brewery's sensory panel activities. This will cover a number of areas. First, assessment of 'trial' samples, which initially could be done by difference testing and then full flavor profiling. Trials could be either on the pilot scale or full-scale production. The trials could cover the use of alternative raw materials, such as a hops from an alternative source, or the initial testing of new season's malt and hops. Trials could assess the flavor impact of brewing using a new yeast strain. The impact of changes to the mashing temperature regime, wort oxidation level or fermentation temperature could be determined.

Final beer testing should cover both quality control (samples taken from the brewery warehouse) and quality assurance (samples collected from retail outlets). This will usually involve full flavor profiling. The data obtained would be compared to the desired flavor profile (i.e. trueness to type assessment). Trending of profile data is another absolute must to determine if there are any long term flavor changes.

Flavor profiling can also be used to monitor competitor products. The data could then be analyzed by multivariate data analysis methods such as principal components analysis. This technique produces 'maps' showing the relationship between the flavor of several beer samples and the main flavor characteristics of the samples.

29.8.3
Consumer Research

Consumers are the final judges of beer quality and so their opinion is vital. Consumer research would normally be done by a specialized consumer research company. The consumer data mainly identifies the 'liking' aspects.

Consumer research is important for both new and existing products. Consumer research would typically involve a more limited number of trial samples to determine which is the most suitable for full scale production. Consumer research on existing and competitor products might help to identify the reason for the success (or failure) of a product and could identify new market opportunities.

References

1. Bell, G.A. and Watson, A.J. (eds) (1999) *Tastes and Aromas – the Chemical Senses in Science and Industry* University of New South Wales Press, Sydney.
2. Bartoshuk, L.M. (1993) The biological basis of food perception and acceptance, *Food Quality and Preference*, **4**, 21–32.
3. European Brewery Convention (1997) Analytica-EBC Section 13 Sensory Analysis Method 13.12, Fachverlag Hans Carl, Nürnberg.
4. Meilgaard, M.C. (1975) Flavor chemistry of beer. Part II. Flavor and threshold of 239 aroma volatiles, *Master Brewers Association of the Americas Technology Quarterly*, **12**, 151–68.
5. Hill, P. (2003) Keeping the flavour wheel turning – The development of a flavour stability, *Proceedings of the European Brewery Convention Congress, Dublin*, paper 75.
6. European Brewery Convention (1997) Analytica-EBC Section 13 Sensory Analysis Method 13.2, Fachverlag Hans Carl, Nürnberg.
7. European Brewery Convention (1997) Analytica-EBC Section 13 Sensory Analysis Method 13.5, Fachverlag Hans Carl, Nürnberg.
8. European Brewery Convention (1997) Analytica-EBC Section 13 Sensory Analysis Method 13.3, Fachverlag Hans Carl, Nürnberg.
9. Institute of Brewing Flavor Sub-Committee (1995) *Sensory Analysis Manual*, Institute of Brewing, London.
10. Lawless, H.T. and Heymann, H. (eds) (1998) *Sensory Evaluation of Food – Principles and Practices*, Chapman & Hall, London.
11. Meilgaard, M. Civille, G.V. and Carr, B.T. (eds) (1991) *Sensory Evaluation Techniques*, 2nd edn, CRC Press, Boca Raton, FL.
12. European Brewery Convention (1997) Analytica-EBC Section 13 Sensory Analysis Method 13.6, Fachverlag Hans Carl, Nürnberg.
13. European Brewery Convention (1997) Analytica-EBC Section 13 Sensory Analysis Method 13.8, Fachverlag Hans Carl, Nürnberg.
14. European Brewery Convention (1997) Analytica-EBC Section 13 Sensory Analysis Method 13.7, Fachverlag Hans Carl, Nürnberg.
15. European Brewery Convention (1997) Analytica-EBC Section 13 Sensory Analysis Method 13.10, Fachverlag Hans Carl, Nürnberg.
16. European Brewery Convention (2004) Analytica-EBC Section 13 Sensory Analysis Method 13.0, Fachverlag Hans Carl, Nürnberg.
17. European Brewery Convention (1997) Analytica-EBC Section 13 Sensory Analysis Method 13.4, Fachverlag Hans Carl, Nürnberg.
18. European Brewery Convention (1997) Analytica-EBC Section 13 Sensory Analysis Method 13.11, Fachverlag Hans Carl, Nürnberg.
19. Meilgaard, M.C., Dalgliesh, C.E. and Clapperton, J.F. (1979) Beer flavor terminology, *Journal of the American Society of Brewing Chemists*, **37**, 47–52.

30
Technical Approval of Equipment
Walter Flad and Hans Michael Eßlinger

30.1
Generalities

In civil engineering it is common to engage an experienced architect who is responsible for proper planning and tendering of the building structure. Thus, the investor obtains comparable offers of different construction companies, covering the same content of work. In doing this the user can be sure of getting well-defined benefits for the requested services at an adequate price.

This approved practice is also recommend for all brewing equipment and should be performed by engineers with good experience in this field. The tendering for brewing equipment must contain guidelines for technical and technological data, which should be integrated in the sales agreements as measurable guaranteed values.

After installation and initial operation, the guaranteed values should be controlled by an independent acceptance test and a technical approval validation.

30.2
Technical Approval of Brewhouses

The German DIN Standard 8777 can be taken as the basis for technical and technological approval of brewhouses. This standard is worked out from the Normenausschuss Maschinenbau in the Deutsches Institut für Normung based in Frankfurt/Main.

The supplier and brewery have to define what guidelines the control brews have to fulfill [1, 2]. Analytical quality features of the raw materials have to be agreed from both sides. DIN 8777 refers to brews with 100% barley malt for bottom-fermenting beers in a range between 11.5 and 12.5% original gravity. Different processing needs adaptation of the guidelines. If the two parties do not find agreeable results in an internal test, an expert institute should be hired for neutral approval.

Handbook of Brewing: Processes, Technology, Markets. Edited by H. M. Eßlinger
Copyright © 2009 WILEY-VCH Verlag GmbH & Co. KGaA, Weinheim
ISBN: 978-3-527-31674-8

Table 30.1 Grist composition.

	Lauter tun	Mash filter
Husks (%)	18–25	<1
Coarse grits (%)	<10	<3
Fine grits I (%)	35	5–12
Fine grits II (%)	21	28–36
Grits flour (%)	7	18–25
Powder flour (%)	<15	30–41
Husk volume (sieve 1) (ml/100 g)	>700	–

The quality of ground malt depends on the lautering system and should maintain the parameters shown in Table 30.1.

Mash pH must be in the range 5.4–5.8. A mashing intensity of above 104 has to be reached (nitrogen content of brewhouse wort in relation to nitrogen content of the congress wort). The turbidity during lautering (below 40 European Brewing Convention 90° units during 60% of the lautering time) gives an indication of the solids in the full kettle wort before boiling (below 100 mg/l). Oxygen uptake to the hot wort must be prevented and iodine control by photometry has to be below 0.30 $\Delta E_{578\,nm}$ in the full kettle wort.

Extract losses are mostly measured by a spent grain analysis. The values are:

- Total available extract: 1.3–1.8 wt% (<4% dry malt).
- Soluble extract: ≤0.8 wt%.
- Iodine value: <3.0 $\Delta E_{578\,nm}$.

After boiling, the quality of the wort is fixed. Measurement of the volumes of full kettle and cast wort indicates the intensity of evaporation; extract in the cold wort indicates the extract yield, calculating the extract in the wort to the extract coming from the raw materials. The parameters of color, total bound nitrogen (below 45 in the cast wort), heat coagulable nitrogen (15–25 mg/l) and dimethylsulfide (DMS) plus DMS precursor (below 100 µg/l) indicate the effectiveness of boiling. The content of free DMS has to be below 100 µg/l in the cold wort.

30.3
Technical Approval of Filling Lines

Special methods need to be considered for filling lines in order to obtain proper results. Acceptance tests of bottle filling lines should be operated accordingly to the regulations of the German DIN standards (see Table 30.2) that have been constituted specially for beverage filling equipment:

Table 30.2 German DIN standards.

	Title	Original German title
DIN 8782	Terms for Filling Lines and Individual Aggregates	Begriffe für Abfüllanlagen und einzelne Aggregate
DIN 8783	Investigations at Filling Equipment	Untersuchungen an abfülltechnischen Anlagen
DIN 8784	Minimum Specifications and Order-related Specifications	Mindestangaben und auftragsbezogene Angaben

- DIN 8782 [3]: contains definitions of the most common terms in filling lines, such as general terms (filling line, package, etc.), time terms (operating time, downtime, etc.) and efficiency terms for single machines and complete filling lines.
- DIN 8783 [4]: defines important measuring and acceptance test conditions.
- DIN 8784 [5]: contains a collection of particular standard values that should be declared for single filling machines (e.g. palletizer, depalletizer, crating machine, decrating machine, bottle cleaning machine, filler, etc.). By using the forms of this DIN standard, all important technical data are present and ready to check after installation.

Unfortunately these three DIN standards are not available in an official English version. Most suppliers, however, use these standards as self-translated form sheets according to the DIN.

30.3.1
Efficiency tests according to DIN 8782

For efficiency tests of filling lines the following (English language) terms are evident:

- Real runtime: runtime of single machine or of whole line without any breakdown times [s].
- Machine-related downtime: downtime of a machine caused by failures of the machine itself [s].
- Non-machine-related downtime: downtime of a machine caused by external reasons (not of the machine itself) [s].
- Operating time: sum of real runtime and machine-related downtime [s].
- Nominal output (Q_n): nominal output of pieces or bottles per time unit [bottles/h].

- Setting output (Q_s): setting output of pieces or bottles per time unit [bottles/h].
- Real output (Q_r): correctly filled pieces/operating time [bottles/h].
- Supply rate (λ): real output/nominal output [%]. The supply rate may be greater than 100%, if the setting output exceeds the nominal output.
- Efficiency factor (η): real output/setting output [%].

The 'efficiency factor' is the most suitable value to evaluate the performance of a new machine or a new filling line. In this regard performance acceptance tests of filling lines according to DIN 8782 need to determine the real output of the filler in relation to the setting output. Its determination is complex and laborious due to the necessity of differentiation between machine- and non-machine-related stoppages.

In a filling line, the bottle filler usually is the machine that should run at 100% of its setting speed at all times. All downtimes of the filler cannot be regained. Due to this, all other machines of a line are dimensioned to a certain over-capacity with buffer conveyers between to avoid filler stops at every little stoppage within the line.

Hence, in daily practice it is of little use to appraise the efficiency factor of a line. Instead, it is recommended to use the so-called 'utilization factor':

- Utilization factor (φ): correctly filled pieces/(loading time × nominal output) [%]. The utilization factor includes all production losses that are in the responsibility of the workers and management. It is easy to determine and gives a mapping about how well the line capacity is actually used.

30.3.2
Reporting of Efficiency Tests

Table 30.3 gives an example of the correct reporting of an acceptance test. The efficiency for new filling lines (returnable bottles) should be guaranteed with values between 85–90%, depending on the complexity of the line. In general, each machine contributes to line downtimes and this can be estimated at about 1% per machine. The utilization factor varies from 50 to 65% if it is related to the complete loading time of 8 h/shift.

30.4
Other Key Figures

In contrast to the DIN standard figures – special for bottle filling lines – it is also common to use various key figures like OEE (Overall Equipment Effectiveness) or OPI (Operational Performance Indicator) that are well known from other industries.

The **OEE factor** is based on the ideas of Nakajiama [6] and has been enhanced by different industry sectors as summarized by Voigt [7].

Table 30.3 Example of the correct reporting of an acceptance test (data of a test run over 6 h filling time).

General data	
nominal output	50 000 bottles/h
setting output	50 400 bottles/h
Time measurements	
loading time	21 600 s
−non-line-related downtime	−347 s
operating time (loading time−non-line-related downtime)	21 253 s
−line-related downtime	−1 279 s
real runtime (operating time−line-related downtime)	19 974 s
bottle counting	
bottles to the filler	279 636 bottles
−incorrect fillings	−71 bottles
correctly filled bottles (bottles to the filler−% incorrect fillings)	279 565 bottles
Results	
real output (correctly filled bottles/operating time)	47 355 bottles/h
supply rate (real output/nominal output)	94.71%
efficiency (real output/setting output)	93.96%
utilization factor [correctly filled bottles/(loading time × nominal output)]	93.19%

Overall equipment efficiency:

OEE = Availiability x Performance efficiency x Rate of quality

Rate of availiability:

$$\text{Availiability} = \frac{\text{Operating time}}{\text{Loading time}} = \frac{\text{Loading time - Downtime}}{\text{Loading time}}$$

Performance index:

$$\text{Performance efficiency} = \frac{\text{Net operating time}}{\text{Operating time}} = \frac{\text{Processed amount}}{\text{Operating time} \cdot \text{theoretical output}}$$

Rate of quality:

$$\text{Rate of quality} = \frac{\text{Valuable operating time}}{\text{Net operating time}} = \frac{\text{Processed amount - defect amount}}{\text{Processed amount}}$$

Figure 30.1 Terms of OEE [7].

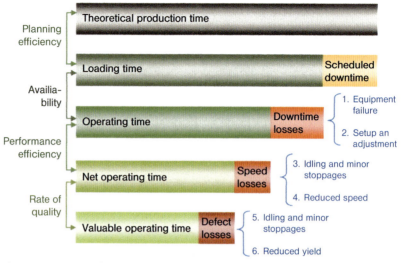

Figure 30.2 Terms and interrelations of Enhanced OEE [7].

The **Enhanced OEE-Factor** is defined as follows:

Planning efficiency × Availability × Performance efficiency × Rate of Quality

The interrelation between these terms are shown in Fig. 30.2.

All methods described can be used accordingly for other filling lines such as keg plants, packaging plants a. s. o.

References

1 DIN 8777 (1996) *Sudhausanlagen in Brauereien*, Deutsches Institut für Normung, Berlin (new edition will be published at the end of 2009).
2 MEBAK (2002) Sudwerkskontrolle, in *Brautechnische Analysenmethoden*. Band II der MEBAK, Selbstverlag der MEBAK, Freising-Weihenstephan, pp. 1–31.
3 DIN 8782 (1984) *Begriffe für Abfüllanlagen und einzelne Aggregate*, Deutsches Institut für Normung eV, Berlin.
4 DIN 8783 (1986) *Untersuchungen an abfülltechnischen Anlagen*, Deutsches Institut für Normung eV, Berlin.
5 DIN 8784 (1993) *Mindestangaben und auftragsbezogene Angaben*, Deutsches Institut für Normung eV, Berlin.
6 Nakajiama, S. (1988) *Introduction to TPM*, Productivity Press, Cambridge.
7 Voigt, T. (2006) Kennzahlen zur Bewertung der Abfülltechnik, Vortrag beim 13. Flaschenkellerseminar, Freising-Weihenstephan.

Index

a

AAS (atomic absorption spectrometry) 447, 466, 467
abscisic acid (ABA) 46, 52, 54
accelerated fermentation 219
– under carbon dioxide pressure 217, 218
Accum, Frederic 32
acetaldehyde 131, 132, 211
– as flavor attribute 680, 682
– as off-flavor 267
acetaldehyde dehydrogenase 132
acetate (acetyl-CoA) 134
acetic acid 252, 483, 484, 688
– in bottom-fermented beer 359
– as off-flavor 267
acetic acid bacteria 478
Acetobacter 268
Acetobacter pasteurianus 477
acetogenesis 638
α-aceto-hydroxybutyrate 132
acetoin 359, 484
acetolactate 128
2-acetolactate 132, 133, 417
α-acetolactate 132
acetyl-CoA 131, 132, 134, 207
acetyl-CoA synthetase 132
2-acetyl-pyrazine 285
2-acetyl-pyridine 385
2-acetyl-pyrrole 383, 385, 386
2-acetyl-pyrroline 177
2-acetyltetrahydropyridine 177
2-acetyl-thiazole 385, 386
4-acetyl-thiazole 385
2-acetyl-thiazoline 385
acid-caustic titration 449, 450
acidic cleaning agents 599
– for keg cleaning 618
acidic/sour flavor attribute 680, 687, 688

acidification phase, of lambic beer production 252
acidity 108, 109
acid malt 164
acidogenesis 638
acrospire length 153
activated-carbon filtration 112
activated-sludge plant 635, 636
active chlorine 601, 612, 618
active noise abatement 672
active transport 126
adaptive control algorithms 543, 544
additive cleaning substances 598
adenosine diphosphate-glucose 49
ADF (apparent degree of fermentation) 444
adhumulone 87
adjunct mashing 176–181
Advant OCS 551
aerated waste water ponds 637
aeration
– during fermentation 214
– in fermentation 209
– of water 112
aerobic metabolism 131
aerobic waste water treatment 635–637, 668
– combined with anaerobic waste water treatment 640
Africa
– beer markets 506, 509
– top brewers 512
after-bitterness, assessing 680, 690
agarose gels 232, 234
agrarian societies, beer and 2–5
AI (aniline index) 422, 423
air, beer transport with 419
air pollution. *see also* emissions
– control regulations 675
albumins 50, 51

Handbook of Brewing: Processes, Technology, Markets. Edited by H. M. Eßlinger
Copyright © 2009 WILEY-VCH Verlag GmbH & Co. KGaA, Weinheim
ISBN: 978-3-527-31674-8

alcohol acyltransferase 134
alcohol chilling test 457
alcohol dehydrogenase 131
alcohol-free beers 235–240
– biological methods 238–240
– bioreactors (immobilized yeast fermentation) 240
– combination physical-biological processes 240
– dialysis 238
– interrupted fermentation (batch processing) 239
– physical methods 236–238
– production techniques 236–240
– reverse osmosis 237, 238
– thermal dealcoholization 236, 237
alcoholic flavor attribute 680, 688
alcohols
– aliphatic 210
– aromatic 210
– as cleaning additives 598
– formation of higher 132, 133
– oxidations of 402
aldehydes 132, 211
– aldol condensation of 402
– in bottom-fermented beer 359
– for disinfection 600–602
– staling process and 399
– Strecker 187, 239, 408, 423
alehouses
– identification of 21
– licensing of 22
ales 23, 248–250, 504, 505, 507
– bitter 248
– in Britain 21, 22
– British pale 28
– British sea power and 22, 23
– brown 249
– cask-conditioned 248–250
– dispensing 348
– India pale 249
– Irish 249
– mild 248
– 'normal' 249
– old 249
– pale 248
– 'real' 249
– Scottish 249
– types of 25
– types of English 248, 249
aleurone cells, cell death of 54
aleurone layer 44, 45, 51, 153, 154
– wheat 47
Alexis barley 404, 406

ale yeast 121–123
algorithms
– advanced 543–545
– classic 542–545
aliphatic alcohols 210
alkaline cleaning agents 597, 598, 609
– aluminum and 612
– in bottle-washing machines 616
– for keg cleaning 618
– temperature and 613
alkalinity 109, 111
allergens, plant food 52
α-acids
– bitter flavor and 690
– cohumulone fraction of 368
– determination of 461
– in hops 87, 91, 97, 99
– lightstruck flavor and 426, 428
Alster 258
Alsterwasser 258
alt beer 223, 224
Altbier 259
Althammer, A. 6
Altbierbowle (Altbier punch) 258
Alt-Schuss (Alt-Shot) 258, 259
alu 2
aluminum, cleaning 597, 612
Alzheimer's disease 517
amaranth *(Amaranthus cruenteus, A. hypochondriacus, A. caudatus)* 56, 57, 69
– for gluten-free beer 246
amaretto, beer and 259
AmBev 511
American lager 28
American Society of Brewing Chemists (ASBC) 437, 439
American Society of Brewing Chemists units 161
Amici, Giovanni Battista 31
amino acids 525
– concentration in beer 453
– foam and 362
– formation of *N*-heterocycles from 385
– in grains 51
– metabolism of by yeast 139
– removal from wort 128, 129
– Strecker's degradation of 402
α-amino compounds 408
1-amino-1-desoxy 408
ammonia 448
ammonia cooling systems 671
ammonia sulfate 457
amylase 31, 54, 55, 170
α-amylase 60, 170, 173, 180, 247, 411, 447

Index | 711

β-amylase 170, 173, 180, 181
α-amylase/trypsin inhibitors 51
α-amylasis 160, 161
amylolysis 160, 161, 171, 458
amylopectin 49, 64, 170, 173
amyloplasts 46, 49
amylose 49
anabolic processes 46–53
– regulation of grain filling 52, 53
– starch formation 47–50
– storage proteins 50–52
anaerobic metabolism 131
anaerobic waste water treatment 637–639, 668
– combined with aerobic waste water treatment 640
analysis, methods for analyzing hop and hop products 92
analytics. see quality control
Anheuser-Busch 36, 511
aniline index (AI) 422, 423
anion exchanger 114, 115
annual capacity, of malting plant 560
anthocyanogens 89, 373, 374, 379, 380
– analysis of total 447
– solubility of 380
– value of 380
antifreeze proteins 53
antimicrobial properties of beer 524
antioxidants
– applied to beer-based mixed drinks 269
– in beer 518
– comparative effects 374
AOX (absorbable organic halogen substances) problem 666, 667
apparent degree of fermentation (ADF) 444
apparent density 440
apparent extract 441
apparent specific density 440, 441
apparent specific gravity 443
Apple Radler 264, 265
applied gauge pressure, calculating 344–346
arabinoxylans 45, 46
arginine 160
Arnhold, Max and Georg 30
aroma 361
aroma hops 89, 94, 95
aroma profile, of alcohol-free beer 237, 238
aroma substances in hops 88, 89, 371
aromatic alcohols 210
aromatic compounds, staling and changes in 402, 403
arteries, beer and protection of 522
articulated-arm robot 583

artificial sweeteners, in beer-based mixed drinks 261, 262
ASBC (American Society of Brewing Chemists) 437, 439
ascorbic acid, applied to beer-based mixed drinks 269
Asia/Pacific
– beer markets 506, 509
– malt and 502
– top brewers 512
asset management systems 552
astringency, assessing 680, 690
atherosclerosis, beer and 516–519
atomic absorption spectrometry (AAS) 447, 466, 467
attenuation 31
A-type amyloplasts 49
Aubry test 151
Aureobasidium 477
Australia
– beer market 509, 510
– malting barley varieties 500
Austria
– beer market 506
– dietetic beer in 241
automated yeast propagation 138, 139
automatic reel changing feature, on pressure-sensitive labelers 333, 334
automation 531–553
– control strategies 542–549
– information technologies 551, 552
– measurement technology 533–542
– objectives of 531
– overview 531–533
– PCS 550, 551
availability, calculating 707
Avena fatua 60

b

Babylonian beer 6
Bacillus spp. 602
back-of-label coding/dating 337
bacteria. *see also individual species*
– acetic acid 478
– beer-spoilage 481–484
– capsular slime-forming acidic acid 477
– diacetyl and 215
– lactic acid 478
balanced alkalinity 109
ballast resistors 655
Balling, Carl Josef Napoleon 31
Balling formula 442, 443
bar cooling 342
bar counter 343

Baretz, Leonhard 33
Barke barley 406
barley adjunct, mashing with 177, 179
barley flakes 177
barley *(Hordeum vulgare)* 3, 14, 43, 59, 60
– acreage and production of 59
– amino acids 51
– brewing 147
– as brewing cereal 17
– catabolic processes in 53
– cleaning 149, 150
– conversion to malt 560, 561
– DMS and 387
– endosperm 45, 46
– genome 44
– grain ripening 48
– intake of 149–163
– malting varieties 500
– organoleptic stability and variety of 404
– pre-cleaning 149
– six-row 498
– starch content 46
– starch granules 50
– steeping 150, 151
– storage of 149–163, 555
– two-row 498
– varieties of 147
– world market in 498–501
barley malts, gelatinization temperatures of 172
barley wine 688
barrels, cleaning 617, 618
base, inspecting 316, 318
basis regulation 495
batch archiving 552
Bauer, M. 32
Bavaria 19
– beer export 29
– lager beer and 25–27
Bavarian lager 28
Bavarian Purity Law (Reinheitgebot) 26, 55, 90, 97, 119, 241
– hot holding and 186
Baverstock, James 31
Baviarian Purity Law (Reinheitgebot) 260
B complex vitamins 525, 526
beauty, beer and 525, 526
beer 22. *see also* beer-based mixed drinks; *individual types*
– alcohol and 521
– antimicrobial properties of 524
– archaeological evidence for 2
– assessing flavor of 676–679
– cancer and 521, 522

– defined 1, 26
– effect of addition of hops on character of 94
– enriched with xanthohumol 97
– global production of 505, 506
– health benefits of 515–528
– with hop flavor 94–97
– polyphenol contents in 376
– production process 531–533
– role in early societies 5
– styles of 27
beer additives 15, 32
beer bar 343
– dispensing direct from 346, 348
– dispensing from above 346, 347
– dispensing from below 346, 347
beer-based mixed drinks 257–273
– antioxidant application 269
– assessment by German Agricultural Society 264–266
– constituent beer 260
– development of 258, 259
– filling 272
– filtration 271, 272
– flavorings and juices 263
– food acids 262, 263
– ingredients and mixing formulations 260–263
– microbiology of 268
– mixing 270, 271
– off-flavors 266–268
– packaging 272
– pasteurization 269, 270
– preservation of 269, 270
– production equipment 273
– quality control of 263–268
– sensory assessment of 264
– sweetening 261, 262
– water quality 261
– wet chemical analysis of 263, 264
beer bath 526
beer belly 527
beer-bread 4, 5
beer color 212, 360, 361
beer consumption 497, 503, 506
beer dispensing. *see* draft beer system
beer-dispensing tanks 350
beer foam 361–366. *see also* gushing
– basis of 362
– beer filtration 365
– cold break removal 364
– influence of brewing water 362
– influence of foam stability 362
– influence of gas 362

– influence of hop products   363
– influence of malt   363
– influence of malt filtration   363
– main fermentation   364
– precocious indicators for foam image   365
– storage conditions   364
– wort boiling and   363, 364
Beer-Lambert law   444, 445
beer lines   342
beer markets   505–510
– Africa   509
– America   507, 508
– Asia   507, 509
– Australia and Pacific   507, 509, 510
– Europe   507, 508
– profitability   510
beer miracles   10
beer pests   478
– detection of   484–489
beer pumps   346–348
Beer Quality Acceptance Panel   695
– training for   696, 697
beer recovery, from yeast   221
beer-spoilage bacteria   481–484
beer staling
– cohumulones and   369, 370
– flavor attributes of   687, 688
beer-strainers   6
beer styles   503–505, 507
beer taxes   23, 27, 497
Belgian lambic beer   250–253
Belgian wheat beer   69
bentonite   232, 271
beor   2, 21
Berliner Weisse   235, 253, 254, 258, 259
– foreign yeast in   481
– production method   254
berries, in brewing   43
Berthelot, P.   31
Berzelius, J.   31
β-acids   87, 426, 428, 461
β-fraction   461
beverage parts   351
biguanides   600, 601
biochemical basics, of anaerobic waste water treatment   637–639
biofilms   478, 479
biogas   637, 648, 669
biogenic amines   525
biological acidification   411, 412
biological degradation, of waste water   624, 666
biological methods, of alcohol-free beer production   238–240

biological oxygen demand (BOD)   623–625, 627, 629, 635, 667, 668
biomass   669
bioreactors   240
biotin   127, 130
Bismarck   259
bitter   676
bitter acids of hops   87, 88, 525
bitter ale   248
bittering hops   89, 94, 95
bitterness, assessing   680, 689, 690
bitter principles, yield of   97, 98
bitter substances   366–370
– aging and appraisal of   409
– cohumulone and   368
– precipitation of, during fermentation   212
bitter units   222, 446
black beer   6
black malt   60
black rice   56, 57
blood alcohol levels, beer and   521
blow-off   648
blue maize   56, 57
Bock beer   237, 688
Bockbierbowle (Bock punch)   259
BOD (biological oxygen demand)   623–625, 627, 629, 635, 667, 668
– in aerobic waste water treatment   636
body, assessing   680, 688, 689
boiled wort color   162, 163
boiler house, heat requirements   647, 648
boiling. *see* wort boiling
Bolivia, brewing in   36
Boorde, Andrew   22
Botec   551
bottle-cleaning machine   614–616
bottled beer   505, 507
– metal ions in   389, 390
bottle infeed and discharge   307, 308
bottle jetting   310
bottle pockets and bottle transport   308
bottle rinsing machines, heat requirements   646
bottles, inspecting   316, 318
bottle washer   305–313
– components   307–311
bottling hall   419, 420
bottling line   305–319
– bottle washer   305–313
– inspection and monitoring units   313–319
bottling plant   565
bottling waste water   625, 630
bottom fermentation   19, 26, 55, 119, 216–219

bottom-fermented beer 222
– composition of finished 359, 360
– qualities of 360
bottom-fermented lager beer, concentration of staling components in 405
bottom-fermenting yeasts 121, 122, 207, 479, 480
– determination of foreign yeasts in 488, 489
bouza 7
Boyle, Robert 12
bragot 21
brain lesions, alcohol and 519
bran 47
branched amylopectins 49
branching enzymes 49
branding in global brewing industry 511, 513
brands, top 20 brands worldwide 513
brassatores 13
Braumat 551
Brautechnische Analysenmethoden von MEBAK 438
braxatores 13
Brazil
– beer market 506, 509, 510
– brewing in 36
– as malt importer 501
BRC (British Retail Consortium) 493, 494, 496
bread, as staple food 18, 19
bread wheat
– genome 44
bread wheat *(Triticum aestivum)* 14
breakthrough 155
Brettanomyces 252, 255, 480, 481
Brettanomyces anomalus 481
Brettanomyces bruxellensis 69, 123
Brettanomyces lambicus 69, 123
brewers, world's top 510, 511
Brewers' Academy of the United States 33
Brewers Gold hops 368
brewers guilds 13, 14
brewery planning 562–576. see also malthouse planning
– capacity calculations 564
– documentation and specifications 575, 576
– investment costs 573–575
– machinery dimensioning 562, 563
– new plant costs 574
– required land area 569–572
– showcase brewery characteristics 562
– utilities and power supply 568, 569

– wort/beer losses 563
brewhouse 431, 563, 564
– development of reductions in 413, 414
– DMS and 388, 389
– as electric power consumer 654
– heat requirements 644, 645
– technical approval of 703, 704
– waste water 625, 630
brewing
– industrialization of 21–25
– internationalization of 33–36
– legislative restrictions on 19
– modules of 4
– prerequisites for 3
– science of 30–33
Brewing Convention of Japan 437, 439
brewing curriculum 33
brewing ordinances 14
brewing privilege 11–13
brewing sciences 32, 33
brewmaxx 551
brew size 564
brew water 105–117
– beer foam and 362
– quality criteria 108–111
– water hardness 105–107
– water treatment 110–117
bright beer tank 418–420, 566, 567
– cleaning 607, 609, 610
British Retail Consortium (BRC) 493, 494, 496
bromoacetic acid 602
brown ale 249
Brown Bavarian lager 28
brown bottles, lightstruck flavor and 424, 425
brown butt-beer 23
Brownian motion 401
brush 45
B-type amyloplasts 49
Buchner, Eduard 120
buckwheat *(Fagopyrum esculentum)* 56, 57, 70, 246
Budweiser 511
building costs, new brewery 574
Bull, John 23
business management 491
butanediol 252
2,3-butanediol 132
2,3-butanedione (diacetyl) 132, 133, 686
Buttes, Henry 22
butyric acid 210, 483
n-butyric acid 686
butyric flavor attribute 680, 686

c

Cab 259
caelia/cerea 8
Cagniard-Latour, C. 31
cake filtration 225, 226
calcium
– flocculation and 123
– gushing and 390
– as yeast nutrient 129, 130
calcium alginate 240
calcium carbonate 112
calcium chloride 110
calcium hardness of water 105, 106
calcium hydrogen carbonate 108
calcium ions 108, 109
– molar mass and conversion factors for 109
calcium oxalate
– precipitation of 390–392
– turbidity and 431
calcium pectate 240
calcium silicate 597
calcium sulfate 110
calcofluor 460, 462
calibration curve 443
callose 50
caloric value of beer, determination of 466
camera inspection technology 319
Caminant barley 404, 406
Canadian beer market 508, 510
cancer, beer and 521, 522
Candida 252
cans 277, 278, 505, 507
– filling 296–300
– in plastic carrier multipacks 593, 594
Canterbury Tales, The (Chaucer) 21
capacitative level sensor 535
capacity
– of brewery 563
– of malting plant 557–561
capillary gas chromatography 91
caprylic acid 237
caramelized flavor 680, 685, 687, 688
caramel malts 163, 175, 224, 685
carbohydrate metabolism 131, 132
carbohydrates
– in beer 523
– usable 466
– in waste water 623
– as yeast nutrients 127, 128
carbonate hardness 105, 106, 110
carbon balance, in waste water treatment 639, 640

carbon dioxide
– accelerated fermentation under carbon dioxide pressure 217, 218
– in beer 525
– beer foam and 361, 362
– beer pumps and 348
– beer transport and 419
– calculating applied gauge pressure 344–346
– as cooling agent 671, 672
– degassing from yeast 141, 142
– in draft beer system 339–341
– in draft beer systems 350
– emission of fermentation 671
– in fermentation 209, 212
– filling and 283, 284
– measurement of 453, 454
– micro-split at inner bottle surface as source for 394
– oxygen content and 283
– plastic bottles and permeability of 280
– reaction with sodium hydroxide 609
– titration curve 450
carbon dioxide extraction, fractionating 372
carbon dioxide gas alert units 356
carbon dioxide lines, in draft beer system 343
carbon dioxide recovery 221, 569
carbonic acid 106, 632
– neutralization of 632–634
carbonyl compounds, staling process and 399–404
carbonyl groups, oxidative degradation of 402
carbonyls 423
cardboard flavor 404
cardiovascular benefits of beer drinking 516
Carlsberg 511
Carlsberg flasks 137, 140
Carlsberg test 461
Carlsberg 1 yeast 120
Carlsberg 2 yeast 120
carpronic acid 483
Cartesian robots 583
caryophyllene 683
caryopsis 44
cask-conditioned ale 248–250
– production methods 249, 250
catabolic pathway 132
catabolic processes 53–55
catechin factors in malt extract 404, 406
cation exchanger 114, 115
catty flavor 404
caustic potash 597

caustic soaker bath and spray 615
CBV (conductometric bittering value) 91
CCD (charge-coupled display) 319
CCV (cylindroconical vessel) 217, 220
celiac sprue 52, 244, 468
cellar beer 222
cellar mold 477
cell membrane 124–126
cellulose 45, 50, 230
cell wall
– flocculation and 123
– yeast 124, 125
Celosia 69
Celts, brewing and 8, 9
Cerabar M flansch 537
cereals 55–70
– acreage and production of brewing 59
– amaranth 56, 57, 69
– barley (*see* barley)
– black rice 56, 57
– blue maize 56, 57
– buckwheat 56, 57, 70
– DNA content 44
– domestication of 43
– einkorn 56, 57, 67
– emmer 56, 57, 67
– grain structure 44–46
– kamut 56, 57
– maize 63, 64
– millets 61–63
– oats 56, 57, 60, 61
– proso millet 56, 57, 62, 63
– pseudo- 69, 70
– quinoa 56, 57, 70
– rice 64
– rye 56, 57, 65
– sorghum 56, 57, 65, 66
– as source for alcoholic drinks 3, 4
– spelt 56, 57, 66, 67
– triticale 56, 57, 68
– tritodeum 56, 57, 68
– wheat 56, 57, 68, 69
certification 491–496
– to ISO standards 495
– legal requirements 495
– management systems and business management 491
– management systems standards 491–494
– principles and similarities 494
– through HACCP 496
– through IFS and BRC 496
cervesia 8
cervisarii 13
CFA (continuous flow analysis) 461, 462

charcoal, gushing and 392, 393
charcoal filtration 199
charge-coupled display (CCD) 319
Charlemagne 11, 13
Chaucer, Geoffrey 21
checklist of in-house measures of waste water 632
cheesy flavor attribute 680, 686
chemical/mechanical cleaning 353
chemical oxygen demand (COD) 199, 623–625, 627, 630, 636, 666, 667
chemicals
– in bottle-washing machine 616, 617
– causing gushing 394, 395
chemiluminescence detector 463
chicha 70
chill haze 428, 430, 433
China
– beer market 506, 509
– brewing in 3
– hops in 10
chitin 125
chit malt 164
chloride 111, 448
chloride ions, ratio to sulfate ions 689
chlorine 447, 598, 600–602, 666
chlorine dioxide 447, 455, 600, 601
chloroacetic acid 602, 667
chloroform 666
CHP (combined heat and power) 657
Christian Middle Ages, brewing in 9–16
chromatographic analysis 461–464
chrome-nickel-steel, cleaning 611, 612
chromosomes 126
chymotrypsin 51
Cicero 8
CIES (Comité International d'Enterprises à Succursales) 493
CIP. *see* cleaning in place (CIP)
cis-acting element 52
cities, brewing and 13
– rise and decline of central European 18–20
citric acid 267, 598
Citrobacter 681
Citrobacter freundii 252
Cladosporium 477
Clara 259
clarification, of beer 213
Clark electrodes 455
Clark sensor 538
cleaning 595–599. *see also* cleaning in place (CIP)
– barrels 617, 618

– cleaning agents   596–599
– disinfection   600–603
– draft beer system   353, 354
– foam   619
– glass bottles   613–617
– goal of   595
– loss   603
– material compatibility   611–613
– methods   603–611
– PET bottles   617
– stack   603
– work safety and environmental protection   620
cleaning agents
– acidic   598, 599
– alkaline   597, 598
cleaning in place (CIP)   596
– centralized CIP station   605
– for closed cleaning circuit   608, 609
– combined   605–611
– heat requirements   646, 647
– measurement record   612
– non-recovery   603, 604
– for open cleaning circuit   608
– program variants   610
– recovery tank   604, 605
– reduction of water usage and   665
climatic zone   658
closed cleaning circuits   608, 609
closed-loop control   533, 542, 543
Clostridium spp.   602
closure inspection   317, 318
clove/4-VG flavor attribute   680, 684
coagulable nitrogen   187, 188
coagulation, of fractions of nitrogenous compounds   452
coal, as fuel source for brewing   20, 24, 648
Cocculus indicus (fishberry)   32
COD (chemical oxygen demand)   199, 623–625, 627, 630, 636, 666, 667
code detection   319
coefficient of performance (COP)   659, 660
cofactors   130
cognitive function, beer drinking and   516, 517
cohumulones   87
– beer aging and   369, 370
– bitter quality and   368
– foam stability and   369
cola
– beer and   258, 264
– wheat beer and   259
Colabier (cola-beer)   259
Colahefe   259

Colaweizen   259
cold break removal   364, 414
cold fermentation
– with conventional storage   216, 217
– with integrated maturation at 12 °C   218, 219
– with programmed maturation at 20 °C   219, 220
– with well-directed maturation in cylindroconical vessel   217
cold room, draft beer system   340
cold storage   215
cold supply   657–662
– cold production   659–661
– cooling requirements   657, 658
– goals for optimal   661, 662
cold trub   201, 202
cold-water zone, in bottle-cleaning machine   615, 616
colloidal silica   232
colloidal stability of beer   213, 428–432
– raw materials and auxiliary materials and   431
– turbidity composition   430
– turbidity formation   430, 431
color
– beer   212, 360, 361
– of malt   162, 163
– photometric measurement of   445, 446
colorless bottles, lightstruck flavor and   424, 425
color malt   175
color spectrometers   319
Columbia, beer market in   509
combined CIP cleaning method   605–611
combined heat and power (CHP)   657
combined packers   582, 583
Combrune, Michael   24, 31
combustion method, of determining nitrogenous compounds   451
compensator tap   345
competitive ELISA   468
complexing agents   598
complexometric titration   450
compressed air   569, 662, 663
– waste heat from   650, 651
compressor number, calculated   659–661
computerized beer dispensing   349, 350
concentrate (retentate)   117
concentration, beer and improved   523, 524
conductivity   449
– measurement of   537
conductometric bittering value (CBV)   91
conductometric values (CVs)   91

congress mash, analyses of   458
consumer protection   495
consumer research   700, 701
consumption data, malthouse   561, 562
contaminants   437, 438, 479
contamination   595
– reduction of   4
continuous flow analysis (CFA)   461, 462
continuous packer   580–582
control strategies   542–549
– advanced algorithms   543–545
– classic algorithms   542, 543
– control of lauter tun   545–550
conversion
– of barley to malt   560, 561
– of water hardness units   107
cooling agents, emissions of   671, 672
Coors   511
COP (coefficient of performance)   659, 660
copper
– cleaning   612
– contributing to beer taste   690
– turbidity and   430
– yeast fermentation and   129, 130
copper adjuncts   176
copper kettles   17, 18
Coriolis-type mass flow meters   537, 538
corn adjuncts, mashing with   177, 179
corn grits   58
corn syrup   261
coumaric acid   135
counter-flow cleaning   618
Crabtree effect   120, 126, 129, 131
cropped yeast   415
cross-flow filtration   230, 231, 233, 234
cross-flow microfiltration   116
crown cork   35, 36
crowners   303, 304
crushed cell layer   46
Cryptococcus   252
crystal malt   163, 175, 224, 685
cultivation, of yeast strains   136–138
curing   557
CVs (conductometric values)   91
cylindroconical vessel (CCV)   217, 220
cytolysis   158–160, 171
cytoplasm   124, 126
Czech Republic   506

d

daily quality control   469–473
dark beers   175, 222
– color of   360
– residual alkalinity for   109

dark malt
– in alt beer   224
– drying of   156
– pH   163
date coding   337
DCS (distributed control system)   550, 551
dead-end filtration   230
DEAE-cellulose   240
trans-2-trans-4-decadienal   400
n-decanal   400
decanoic acid   210
n-decanoic acid   686
trans-2-decenal   400
decoction method   169, 175
deferrization   110
defoamers   597, 598, 618
degassing   648
degradation tests   624
degree of fermentation   444
dehydrin   52
dehydrohumulinic acid   424, 426
Dekkera spp.   123
demanganization   110, 111
dementia, beer and   516, 517
density, extract and   441
density analysis   440, 441
deoxynivalenol   438
6-deoxytetrahydro-β-acids   428
depth filtration   226
dermatitis herpetiformis   244
descriptive analysis   693
desiccation   46, 52
Detergent and Cleanser Law   666
dextrinizing units   447
dextrins   123, 127, 446, 466
diacetyl (2,3-butanedione)
– in bottom-fermented beer   359
– as flavor attribute   680, 686
– formation of   132, 133
– in lambic beer   252
– off-flavors and   211
– potential   215
dialysis, in alcohol-free beer production   238
diastatic malt   164
diastatic power   450, 451
diauxie   161
dicarbonyl α-compound   408
2–6-dichlorophenol   267
diesel   258
dietetic beer   235, 240–242
– production methods   241, 242
Difference Panel   695
– training for   696
difference tests   693

diffusion 126
diffusion channels 126
dihydrohumulone 427
cis-dihydro-iso-humulone 427
trans-dihydro-iso-humulone 427
dimethyl disulfide 682
2,3-dimethyl-pyrazine 385
2,5-dimethyl-pyrazine 385
2,6-dimethyl-pyrazine 385
2-ET-3,5-dimethyl-pyrazine 385
2-ET-3,6-dimethyl-pyrazine 385
dimethylsulfide. *see* DMS (dimethylsulfide)
dimethylsulfoxide (DMSO) 386
dimethyl trisulfide 682
DIN EN ISO 9000 438, 492
DIN EN ISO 9001 491, 492
DIN EN ISO 9004 492
DIN EN ISO 14001 492
DIN EN ISO 17025 494
DIN EN ISO 22000 492, 493
DIN Standard 8777 189, 703
DIN Standard 8782 705, 706
DIN Standard 8783 705
DIN Standard 8784 705
directional difference test 693
discrimination ability 699
disinfecting 600–603. *see also* cleaning
– disinfecting substances 601
– for draft beer systems 353, 354
– goal of 595
– head section 311
dispensing 346–350
– from above the beer bar 346, 347
– beer-dispensing tanks 350
– with beer pumps 346–348
– computerized 349, 350
– direct from beer bar 346, 348
– with pre-mixed gas 348, 349
– types of 346
– from underneath beer bar 346, 347
– use of gas blenders 349
dispensing head 352
dispensing tap 349
dispergators 598
dispersing mill 167
dispersion agents 599
dissolved organic carbon (DOC) 623, 627
dissolved oxygen 538
distillation method, for sulfur dioxide 467, 468
distributed control system (DCS) 550, 551
DMS (dimethylsulfide)
– barley and malt and 387
– brewhouse 388, 389

– fermentation and 211
– formation of 386, 387
– in lambic beer 252
– malt cleaner 388
– temperature and 387
– withering and kilning and 387, 388
– wort boiling and 187–189
DMS (dimethylsulfide)/cooked vegetable flavor attribute 680–682
DMS (dimethylsulfide) precursor 155, 161, 187–189, 386, 388, 409
DMSO (dimethylsulfoxide) 386
DNA content, of cereals 44
DOC (dissolved organic carbon) 623, 627
documentation and specifications, new brewery 575, 576
n-dodecanal 400
dormancy of barley 151
Dortmunder beer 222, 360
dosing 541
double-end bottle washing machine 306, 313, 614
downstream products 93
– addition of 99, 100
Dr. Pepper 259
draft beer system 339–357
– applied gauge pressure calculation 344–346
– beer bars/bar counter requirements 343
– beer line requirements 342
– beer quality in 339–346
– carbon dioxide content 339–341
– carbon dioxide line requirements 343
– cold room 340
– design of 341–346
– dispensing 346–350
– foamhead 341
– gas-pressurized parts 350, 351
– glass-washing equipment 343, 344
– hygiene requirements 352–355
– keg-tapping equipment 351, 352
– pouring beer 341
– refrigeration requirements 342
– room requirements 341, 342
– safety precautions 356
– temperature 339
– testing 355, 356
– time on tap 339
draught beer 505, 507
Dreckiges 259
Drecksack 259
Dreher, Anton 30
drive system 654
drought, grain filling and 52, 53

dry milling 165, 167
dull flavor 687, 688
Dumas, determination of nitrogenous compounds according to 451
duo-trio test 693
dust emissions 672
Dutch brewing 20, 21
dynamic low-pressure boiling 192–195
dynamic viscosity 457

e

EBC (European Brewery Convention) 159, 437
EBC Analytica 438, 440, 461
EBC/ASBC 447
EBC bitterness units 95, 96, 222
EBC color units 212, 360
EBC turbidity units 411, 456, 540
ECD (electron capture detector) 463
Ecotherm 190
EDTA 598
efficiency tests, reporting 706, 707
Egypt, brewing in 6, 7
Ehrlich mechanism 132
einkorn *(Triticum monococcum* L.) 3, 56, 57, 67
elderly, alcohol abuse in 520
electrical energy 643. *see also* power supply
electron capture detector (ECD) 463
electronic density meter 441
electronic level-controlled filling systems 292–296
– Sensomatic VPL-PET 295, 296
– Sensomatic VPVI 294, 295
electronic spin resonance spectroscopy (ESR) 469
electronic volumetric filling systems 296–300
– Volumetric VOC 296, 297–300
– Volumetric VODM-PET 296, 297, 300, 301
ELISA (enzyme-linked immunosorbent assay) 462, 468, 469
emblems
– brewers 15, 16
– guild 14
embryo 44, 45
emissions 670–673
– cooling agents 671
– defined 670
– dust 672
– fermentation carbon dioxide 670
– gaseous 670–672
– noise 672, 673
– sewage disposal plant 669, 670

emmer *(Triticum dicoccum)* 7, 56, 57, 67
Emsgold, Pils and 259
end degradation 624
endoplasmic reticulum 124, 126
endosperm 44, 45
– starchy 45, 46
– wheat 47
energy 643–663, 668–670
– cold supply 657–662
– compressed air supply 662, 663
– conservation of 669
– heat requirement of brewery 643–651
– power supply 651–657
– renewable 669, 670
English ales 21, 22, 28, 248, 249
English degree of water hardness 107
English infusion mashing method 248
English pale ale 28
Enterobacter 268
Enterobacter aerogenes 252
Enterobacter cloacai 252
Enterobacter phase, of lambic beer production 251, 252
environmental factors, in grain filling regulation 52, 53
environmental management systems 492
environmental protection 665–675
– cleaning and 620
– emissions 670–673
– energy 668–670
– legal basis of 673, 674
– waste 673, 674
– waste water 665–668
environmental regulations 673, 674
enzymatic analyses 464–466
enzyme digestibility 49
enzyme-linked immunosorbent assay (ELISA) 462, 468, 469
enzymes
– in barley germination 153
– barley malt 171
– malt 161, 162
– modification of grain for gluten-free beer by 245
enzymological studies 31
Epic of Gilgamesh 5
epoxides 89
equalizing tanks, aeration of waste water 634, 635
equipment, technical approval of 703–707
ergosterol 125, 129
Escherichia coli 355
ESR (electronic spin resonance spectroscopy) 469

Essay on Brewing (Combrune) 31
Esterel 147
ester fractions, in hops 372
esters 134
– in alcohol-free beer 237
– in bottom-fermented beer 359
– fermentation and 210
– flavor and 683
– formation of 134
– in hops 89
– lightstruck flavor and 423
– in wheat beer 174
estery/fruity flavor attribute 680, 682, 683
ethanol 209, 210
– content analysis 441–443
– determination of 465
– formation of 131
ethyl acetate 210, 252, 682
ethyl alcohol, foam and 362
ethylcaproate 210
ethylcaprylate 210
ethylene vinyl alcohol polymer (EVOH) 279
ethyl fenchol 267
ethyl lactate 252
ethyl mercaptan 267
ethyl *n*-butyrate 683
ethyl *n*-hexanoate 682, 683
ethyl *n*-octanoate 683
Études sur la Bière (Pasteur) 32
EurepGAP (Good Agricultural Practices) 493
Europe
– beer consumption in 503, 506
– beer markets in 506–508
– brewing in 33
– hop farms 502
– malt and 502
– malting barley varieties 500
– top brewers 512
European Brewery Convention. *see under* EBC
European Sankey 'S' system 351
eutrophication 667
evacuation 285–287
evaporation 186–189
– flash 194, 197
– thin film 199, 200
– vacuum 197, 198
evaporation condensers 659
evaporation efficiency 187
EVOH (ethylene vinyl alcohol polymer) 279
cis-3-exanal 400
exhaust emission heat exchanger 648
export beers 222, 360

external boilers 192, 193
extinction 444
extract
– analysis of 440–444
– apparent 441
– density and 441
– measuring losses 704
– real 441, 442
extremely high rice addition, mashing with 180, 181

f

facilitated diffusion 126
famines, brewing and 14, 15
FAN (free amino nitrogen) 152, 160, 211, 453
– analysis of 446
– in dark beers 175
– in raw grain worts 63
FAO (Food and Agriculture Organization) 58
farnesene 88
fatty acids
– auto-oxidation of 402
– in fermentation 210
– as flavor attribute 680, 685, 686
– in hops 89
– thermal degradation of 405
– as vitamin 130
– yeast management and 129
Fenton reaction 687
fermentation 4, 207–215. *see also* bottom fermentation; top fermentation
– accelerated 217–219
– aeration 209
– appearance during 213
– beer color and 212
– beer foam and 364
– carbon dioxide content 212
– changes during 209–213
– changes in composition of nitrogen compounds 211
– changes in redox properties of beer 212
– characteristics 564
– clarification and colloidal stabilization and 213
– cold 216–219
– control of 214
– degree of 444
– discovery of 119, 120
– fermenters 214, 215
– flavor stability and 417, 418
– flowchart of 208
– goals of 213

– immobilized yeast   240
– interrupted   238, 239
– metabolic pathways   131–135
– parameters   213, 214
– pH drop and   211, 212
– pitching   207, 208
– precipitation of bitter substances and polyphenols   212
– pressureless warm   217, 218
– science of   31, 32
– topping-up   209
– turbidity and   432
fermentation cellar
– design of   657, 658
– waste water and   625, 630
fermentation pathway   120
fermentation tank   531–534
– cleaning   607–610
fermenters   214, 215
fermentum   12
Fertile Crescent   3, 43
ferula acid   89
ferulic acid   135
festival beers   222
fiber, beer as source of   523–525
Ficaria   12
FID (flame ionization detector)   463
filling   275–319
– of beer-based mixed drinks   272
– bottle washer   305–313
– bottling line parts   305–319
– cans   277, 278
– carbon dioxide content   283, 284
– crowners   303, 304
– electronic level-controlled systems   292–296
– electronic volumetric systems   296–300
– evacuation   285–287
– filling phase   288, 289
– filling pressure   284
– flavor stability and   420–422
– flushing with ring-bowl or pure gas   287, 288
– fobbing   301–303
– framework conditions   281–285
– gases and   282–284
– glass bottles   275–277
– hot   270
– inspection and monitoring units   313–319
– kegs   280, 281
– mechanical level-controlled systems   290–292
– oxygen content   282, 283
– packaging choice   275–281

– plastic bottles   278–280
– pressurization   284, 288
– process steps   285–290
– screw-cappers   304
– settling and snifting   290
– temperature and   285
– volumetric VODM-PET   300, 301
filling area, electric power requirement in   654, 655
filling lines, technical approval of   704–706
filling station   531, 532, 534
fill-level inspection   316, 317
film thickness, in shrink-wrap packaging   590
filter   531, 532
filter media, gushing and   392
filtration   566
– of beer-based mixed drinks   271, 272
– beer foam and   365
– cake   225, 226
– characteristics   566, 567
– cross-flow   230, 231, 233, 234
– dead-end   230
– depth   226
– filter aids for pre-coating   230
– flavor stability and   418–420
– kieselguhr   226–228, 233
– kieselguhr pre-coating   230
– mash   413, 431
– measurements when transported with air   419
– membrane   230, 231
– processes   115–117
– purpose of   225
– record of in filter line   233
– surface   226
– techniques   226–231
– turbidity and   432
– variables influencing   231
filtration and stabilization plant design   232–234
filtration cellar, waste water COD concentrations   630
final beer testing   700
finger millet (*Eleusine caracana* (L.) Gaertn.)   62
fixed-bed reactors   240
flakes   58
– barley   177
flame AAS   467
flame ionization detector (FID)   463
flame photometric detector (FPD)   463
flash evaporation 'Varioboil'   197, 199
flash pasteurization   270, 422, 646, 661

flavanols 377, 378
flavor assessment 676–679, 699, 700
– attendance requirements 678
– final beer testing 700
– in-process sample testing 699, 700
– overall impression 678, 679
– technique 677
flavor attributes 679–690
– acetaldehyde 680, 682
– acidic/sour 680, 687, 688
– after-bitterness 680, 690
– alcoholic 680, 688
– astringency 680, 690
– bitterness 680, 689, 690
– body 680, 688, 689
– butyric 680, 686
– caramelized 680, 685
– cheesy 680, 686
– clove/4-VG 680, 684
– diacetyl 680, 686
– dimethylsulfide/cooked vegetable 680–682
– estery/fruity 680, 682, 683
– fatty acid 680, 685, 686
– floral 680, 684
– fresh grass 680, 684
– grainy/straw 680, 685
– hoppy 680, 683, 684
– hydrogen sulfide/mercaptan 680, 681
– malty 680, 685
– metallic 680, 690
– oxidized 680, 687
– procedures for illustrating 680
– roasted 680, 685
– solvent 680, 682
– spicy 680, 684
– sulfur dioxide 679, 680
– sweetness 680, 689
– yeasty 680, 686
flavorings, in beer-based mixed drinks 263
flavor profile and rating test 693
flavor profiling 693, 700
flavor stability 399–423
– analytical control of 422, 423
– barley variety and 404
– changes in aromatic compounds and 402, 403
– fermentation and maturation and 417, 418
– filling and 420–422
– filtration and 418–420
– flotation and 414, 416
– germination and 404–409
– hot break removal and 414, 415

– indicator substances 403, 404
– malt quality and 409
– mash filtration and 413
– preserving organoleptic stability and 404–423
– reasons for beer aging 401, 402
– staling processes 399, 400
– wort boiling and 413
– wort preparation and 409–413
– yeast handling and 414–417
flavor wheel 681, 683
Flieger 259
flocculation 123
floor area, calculation of malthouse 558, 559
floral flavor attribute 680, 684
flotation 414, 416
flow injection analysis 462
flow measurement 536–538
flue gas, neutralization with 634
fluorescence detector 462
flushing
– with pure gas 287, 288
– with ring-bowl 287, 288
foam 98, 341
– retention of 455, 456
foam cleaning 619
foam image 365
foam-positive malts 363
foam stability 433
– cohumulone and 369
fobbing 301–303
foiling 326, 327
folic acid 519
fonio millet (Digitaria exiliz) 62
food acids, in beer-based mixed drinks 262, 263
food adulteration 32
Food and Agriculture Organization (FAO) 58
food hygiene regulation 495
food safety standards 492, 493
forcing test 456, 457
foreign bodies, inspecting for 316
foreign substances, inspecting for 316
foreign yeasts 480, 481
– determination of 485–489
formaldehyde 151
formalin test 457
formazin 456
formic acid, in bottom-fermented beer 359
forward-scatter detectors 540
fossil fuel combustion emissions 670
fouling 117
4-VG flavor attribute 680, 684

foxtail millet (*Setaria italica* (L.) P. Beauv.) 44, 61, 62
FPD (flame photometric detector) 463
frambozen lambic beer 251–253
frame filter 27
France
– beer market 506
– malting companies 501, 503
Franck, Hans 16
Franconian breweries 28, 29
free aggressive carbon dioxide 111
free amino nitrogen. *see* FAN (free amino nitrogen)
free diffusion 126
French degree of water hardness 107
French paradox 517
fresh and aged test 694, 695
fresh grass flavor attribute 680, 684
fresh hop aroma 683
freshness, cleanness, crispness of flavor 687, 688
friabilimeter 159, 459
Friabilimeter Calibration Network 459
Frohberg yeast strain 120
front-of-label coding/dating 337
fructose 127, 483
fruit beers 504
fruit lambic beers 250–253
– production method 252, 253
Fuchs, J. 31
fuels 648
full packs, space requirements of 567
fumes 671
fumigation 411, 412
fuminosin 438
fungi, gushing and 394
furan 383, 408
furaneol 689
2-furfural 381, 423
furfural derivatives 381
furfurylalcohol 381
Fusarium culmorum 394, 477
Fusarium graminearum 394, 477
Fusarium spp. 151
fuzzy control algorithms 543–545
fuzzy controllers 533
fuzzy PID control loop 545
fuzzy set theory 543

g

GA (gibberellic acid) 54, 153, 154
α-galactosidase 122
Galant 379, 381
galvanic-amperometric sensors 539
γ rays 319
gantry robots 582, 583
gas, as heat source 648
gas blenders 349
gas chromatography (GC) 91, 463, 464
gas chromatography-mass spectroscopy (GC-MS) 91
Gaseosa 259
gaseous emissions 670–672
gases
– in bottom-fermented beer 359, 360
– extractor devices 311
– filling and 282–284
– plastic bottle permeability to 279, 280
gas flashing during mashing 411, 412
gas-pressurized parts, of draft beer systems 350, 351
Gay-Lussac, Joseph Louis 31, 119
GC (gas chromatography) 91, 463, 464
GC-MS (gas chromatography-mass spectroscopy) 91
gelatinization 49, 170, 172
gelatinization temperature (GT) 170, 172
– of adjunct 176–178
geraniol 88
germ 46
– wheat 47
German Agricultural Society 264–266
German degree of water hardness 107
German Purity Law 98
German slider 'A' system 352
Germany
– beer market 506, 508, 510
– brewing in 9, 27–30
– dietetic beer in 241
– hop farms 502
germination 53, 152–155, 405–409
– germination box design 554, 555
– infection with mold spores during 390
– kilning 407–409
– malt quality 409
– process scheme 557–559
– withering 405–407
germination/kilning floor design 557, 558
Gespritztes 259
Gestreiftes 259
geuze beer 5, 481
GFSI (Global Food Safety Initiative) 493
gibberellic acid (GA) 54, 153, 154
gibberellin hormones 153
glass bottles 275–277
– cleaning 613–617
– lightstruck flavor and 424, 425

glass splinters, detection of 317
glass-washing equipment 343, 344
– glass cleaning system 343, 344
– glass-washing machines 344
– two-sink installation 343
glassy kernels 459
gliadins 52
Global Food Safety Initiative (GFSI) 493
globoid bodies 45
globulins 50
β-glucan 61, 158–160, 177, 465
β-1-6-glucan 125
endo-β-glucanase 158
glucanases 54
β-glucanases 60, 177
β-1-3-glucane 125
β-glucan gel 231
(1-3, 1-4)-β-D-glucans 50
α-glucans
– reaction with polyphenols 379
– turbidity and 430
β-glucans
– analysis of 462
– beer foam and 363
– metabolism of 46
– reaction with polyphenols 379
– turbidity and 430
– viscosity and 457
α-glucan tarnishing 173
gluconate 598
Gluconobacter 268
Gluconobacter frateurii 477
glucoronoarabinoxylans 50
glucose 58, 689
– aerobic and anaerobic fate of 131, 132
– as yeast nutrient 127, 128
glucose syrup 261, 262
glucosidases 55
α-1-4-glucosyl starch chains 49
gluten 437
– determination of 468, 469
gluten-free beer 244–246
– conventional 'gluten-containing' raw material 245
– production methods 244–246
– sources of gluten-free sugars and starch 245, 246
glutenins 52
gluten-sensitive enteropathy 52
glycerol 211, 689
glycogen 126, 128
glycols 598
glycolysis 209

glycoproteins 170, 362
Golgi apparatus 124, 126
Good Agricultural Practices 493
Good Manufacturing Practice 493
Goodwin, Henry 24
grading of malt 460
grain combustors 648
grain enlargement 46, 47
grain filling 47
– regulation of 52, 53
grain ripening, timeline of processes 48
grains
– bred for gluten-free beer 245
– ensuring supply of 14, 15
– rich in carbohydrates 246
– spent 673
grain starches 49
grain starch modification 4
grain structure 44–46
grainy/straw flavor attribute 680, 685
graphite tube AAS 467
grass seeds, in brewing 43
gravity figure 440
Great Britain
– ales in 21–23, 28, 248, 249
– beer market 506, 508, 510
– growth in brewing industry 30
Greeks, brewing and 7, 8
green bottles, lightstruck flavor and 424, 425
green malt, infection of 390
grinding technology 165–168
gripper cups/heads 579–584
grist load 564
grits 58
Groll, Josef 30
gross tank space (GTS) 562
gruit 11, 12, 20
GT (gelatinization temperature) 170, 172
– of adjunct 176–178
GTS (gross tank space) 562
gueuze beers 250–253
– production methods 251, 252
guild emblems 14
guilds, brewing 8, 13, 14
Guinness 259
Guinness, Arthur 25
gushing 389–395, 461
– chemical components causing 394, 395
– determination of induced by raw materials 389
– filter media 392
– malt-induced 392–394

– metal ions in bottled beer 389, 390
– precipitation of calcium oxalate crystals 390–392

h

HACCP (Hazard Analysis of Critical Control Points) 493, 496
Haffmans formula 454
Hafnia alvei 252
Haithabu 11, 16
Hall, John 22
Hallertauer hops 377
Hallertauer Magnum hops 368
Hallertau Hersbruck hops 368
Halltertau Magnum hops 377
Halltertau Tradition hops 377
halogenated acetic acids 667
halogenated carboxylic acids 600, 602
halogenated hydrocarbons 666
Hamburg, as brewing center 16, 17
hammer mills 165, 167
hand assessment 460
Hanseatic League, brewing and 16–18
Hansen, C. E. 32
Hansenula 252
happir 5
Harden, Arthur 120
hardness, water 105–107
hard resins 87
Hassall, Arthur Hill 32
HAT vapor condensers 650
Hazard Analysis of Critical Control Points (HACCP) 493, 496
hazardous substances 674
haze 373, 377, 428, 430, 433
health benefits of beer 515–528
hearing, sense of 676
heart attack, beer and 515, 516, 518
heat
– combined heat and power 657
– grain filling and 52, 53
– malthouse system 561, 562
– recovery possibilities 648–651
– staling and 405, 406
heat carrier systems 647, 648
heating plant 568
heat requirements of brewery 643–651
– boiler house 647, 648
– bottle rinsing machines 646
– brewhouse 644, 645
– heat consumers 644–647
– heat recovery possibilities 648–651
– keg cleaning 646
– optimization possibilities 648

– pasteurization, CIP 646, 647
– room heating and system losses 647
– service warm water 645, 646
heat supply 656
heat transfer date coding 337
Heineken 511
Helicobacter pylori infection, beer and 522, 524
hemicelluloses 45
hemp 10, 11
Henry-Dalton formula 453, 454
Henry's law 283
herbal additives 12
Hermbstädt, S. 31
N-heterocycles 381–386
– in malting process 383
– mashing conditions and 383
– presence of 381, 382
– structures of 385
– of wort 384
– wort boiling 386
O-heterocycles 381, 383
S-heterocycles 381
heterofermentative beers 235
hexagram, brewing and 15, 16
hexahydro-iso-α-acids 426
trans-hexahydro-iso-humulones 427
n-hexanal 400
trans-2-hexanal 400
hexanoic acid 210
1-hexanol 64
hexose 209
higher alcohols
– in bottom-fermented beer 359
– formation of 132, 133
high-frequency inspection technology 319
high-gravity beer 237
high original wort
– beer production methods for 247, 248
– brewing with 246–248
high-performance liquid chromatography (HPLC)
– analysis of bitter substances in hops 366–368
– hop addition and 91
– quality control and 462, 463
high-performance reactors 636
high-temperature wort boiling 192, 194
histidine 160
history of brewing 1–36
– advent of lager 25–36
– agrarian societies 2–5
– Celts and Germans 8, 9
– Christian Middle Ages 9–16

– Dutch brewing 20, 21
– Hanseatic League 16–18
– Hellenistic period 7, 8
– hopped beer 16–21
– industrialization of brewing 21–25
– Mesopotamia and Egypt 5–7
homocysteine 519
homofermentative beers 235
homogeneity, of malt kernels 460
hop addition 91–94
hop aroma 97
hop extracts 371, 541
Hopf, G. 32
hop gardens 17
hopped beer 6
– central European cities and 18–20
– Dutch brewing and 20, 21
– Hanseatic League 16–18
hop pellets 371, 541
– nitrate reduction and addition of 93
hopping technology 91–100
– addition of 'downstream products' 99, 100
– beer enriched with xanthohumol 97
– beer with hop flavor 94–97
– foam 98
– hop addition 91–94
– microbiology 99
– yield of bitter principles 97, 98
hop polyphenols 374–378
hop products
– beer foam and 363, 366
– classification of 90
– global use of 502, 503, 505
hoppy 680, 683, 684
hops (*Humulus lupulus*) 85–100
– analytics 91
– aroma substances in 88, 89, 371
– bitter acids 87, 88
– bitter substances in 366–370, 461
– cohumulone contents in 370
– components 87–89
– cultivation of 85–87
– ester fractions in 372
– first use in brewing 10, 11
– global market 501, 502
– global use of 502, 503, 505
– harvesting 86, 87
– main hop-growing areas 504
– polyphenolic-related reactions and 379
– polyphenols in 89, 375
– storage of 100
– as tithe 11
– top hop companies 505

– trading networks for 24
hordeins 52
Hordeum chilense 68
hordine 156
horizontal integration 552
horizontal leaf filter 227–229
hot break 431
hot break removal 414, 415
hot date stamping 337
hot filling 270
hot holding 186
hotmelt labeling 327–332
– with pre-cut labeling 328–330
– reel-fed 330, 331
hot-water zone, in bottle-cleaning machine 615, 616
HPCL (high-performance liquid chromatography) 91, 462, 463
hull 44, 45
humulene 683
humulone 87
hupulones 426
hydrochloric acid 446, 447, 598
hydrocinnamic acid 89
hydrogels 432
hydrogen carbonate ions 108
hydrogen peroxide 598, 601
hydrogen sulfide 211
hydrogen sulfide/mercaptan, as flavor attribute 680, 681
hydrohumulones 426
hydrolases 54, 55, 162
hydrolysis 53–55, 638
hydrometer 31, 441
hydrophile 395
hydrophobins 437
hydrophone 395
hydrosulfide 267
hydroxybenzoic acids 89, 377, 378
hydroxycinnamic acids 377, 378
hydroxymethylfurfural 381, 386, 446
hygiene requirements
– cleaning and disinfecting procedures 353, 354
– in draft beer systems 352–355
– problem areas 354, 355
hygiene target 352
hygiene testing 355
hypertension, beer and 519
hypochlorite 612

i
ICP-OES (inductively coupled plasma optical emission spectrometry) 466, 467

IFS (International Food Standard)   438, 493, 496
immissions   670
immobilized yeast fermentation   240
immunoassays   468
Imperial stout   223
InBev   511
India pale ale   249
indicator substances, for organoleptic beer stability   404, 405
indirect beer pests   478, 487
indirect beer spoilage bacteria   482
indirect kilning   24
indole   381
inductively coupled plasma optical emission spectrometry (ICP-OES)   466, 467
inductive sensor   537
industrialization of brewing   21–25
– lager beer and German   29
information technologies   551–553
infrared inspection technology   319
infusion method, of mashing   169
in-house measures of waste water   631–634
inkjet date coding   337
in-line inspection machine   313–315
in-line measurement   541, 542
inliner, in beer-dispensing tanks   350
inorganic contamination   595
inorganic substances, removal from water   110, 111
inositol   130
in-process sample testing   699, 700
inspection   313–319
– machine types   313–316
– reliability of   319
– tasks   316–318
– technology   318, 319
Institute of Brewing and Distilling   437, 439
Institute of Brewing (London)   33
intelligence, beer and   518
Interbrew   511
intermediate-spray, in bottle-cleaning machine   615
intermittent packers   579, 580
internal boilers   189–192
– layout of   190
– optimized 'Stromboli'   190, 191
– optimized 'Subjet'   190–192
International Food Standard (IFS)   438, 493, 496
International Organization for Standardization   693

International Organization of Legal Metrology   442
interrupted fermentation method   238, 239
invertase   31
invert sugar   261, 262
investment costs, brewery   573–575
iodine
– mash and   173
– photometric iodine reaction   446
iodometric titration   450, 451
ion balance   107
ion chromatography   447
ion exchange   113–115
ions, determination of   447, 448
Ireland
– beer market   506
– brewing in   9, 10, 24, 25
– Irish ale   249
Iris   12
iron
– in brew water   111
– gushing and   389–391
– measurement of   448
– as off-flavor   267
– turbidity and   430
– as yeast nutrient   129, 130
ISFET pH measurement sensor   536
iso-α-acid homologs   369
iso-α-acids   91, 93, 98, 99, 365, 424, 446, 689, 690
– lightstruck flavor and   426
iso-acids, lightstruck flavor and   429
iso-adhumulones   368–370
iso-amyl acetate   210, 682, 683, 689
iso-amyl alcohol   132, 688
iso-butanal   420
iso-butanol   132, 688
iso-butyl acetate   210, 683
iso-cohumulones   369
iso-extract   93, 95–97
– calculating dosage of   99, 100
iso-glucose   261
iso-humulones   368–370, 402, 424, 427
isoleucine   132
isomaltol   383
cis, trans-isomers   369, 370
isooctane   446
iso-pellets   95, 96
ISO standards, certification according to   495
isothermal mashing process   458
trans-iso-umulone   427
iso-valeric acid   210, 686
iso-xanthohumol   379

j

Japan
- beer market 506, 509, 510
- brewing in 36
- as malt importer 501
Jefferson, Thomas 31
jet spray heads 606, 607
Joseph and his Brothers (Mann) 3
juices, in beer-based mixed drinks 263
jump-mash process 239

k

KAC (knowledge-based analytical controller) 545, 546, 548–550
Kaiser, Cajetan 33
kamut 56, 57
kegging lines 577
kegging plant 567
kegs 280, 281
- cleaning 617, 618, 646
keg-tapping equipment 351, 352
ketoacid 132
α-ketoacid 215
ketones 211
α-ketones 408
kettle extracts 93
kettle hop aroma 683, 684
kidney stones, beer and risk of developing 522, 523
kieselguhr filtration 226–228, 233, 566
- beer foam and 365
- flavor stability and 418
- gushing and 392, 393
- horizontal leaf filter 227–229
- metal candle leaf filter 229
- plate and frame filter 227
- waste from 674
kieselguhr-free filtration methods 566
kieselsols 232
kilning 8, 155, 156
- diagram 561
- DMS and 387, 388
- influence on flavor stability 407–409
- planning for 557
kilning tower design 557, 558
kinematic viscosity 457
Kjeldahl, determination of nitrogenous compounds according to 451, 452
Kjeldahl nitrogen (TKN) 627
Klebsiella 268
Klebsiella aerogenes 252
Klencke, H. 32
Klevi, Lesli 524
Kloeckera apiculata 252

Kloeckera spp. 69
knowledge-based analytical controller (KAC) 545, 546, 548–550
knowledge-based controllers 533
Koch, Robert 120
Kolbach index 160, 387
Kölsch beer 224
kräusen 208, 254
kräusenring 213
Krefelder 258, 259
kriek lambic beer 251, 252
Kulmbach 30
Kützing, F. 31

l

labeling 321–337
- date coding and identification 337
- foiling 326, 327
- hotmelt 327–332
- machine construction 321–324
- options 322
- pressure-sensitive 332–334
- roll on/shrink on 331, 332
- sleeving 334–336
- tamper-evident sealing 327, 332
- wet-glue 324–327
labels
- bottle-washing machines and 616, 617
- inspecting position/type 318
- removal of 309
lactic acid 180, 252, 483, 688
lactic acid bacteria 478
Lactobacillus 69, 239, 252, 268, 687
Lactobacillus amylolyticus 99, 480
Lactobacillus backii 483
Lactobacillus brevis 479, 483
Lactobacillus brevisimilis 483
Lactobacillus buchneri 483
Lactobacillus casei 479, 483
Lactobacillus coryniformis 483
Lactobacillus lindneri 479, 483
Lactococcus 268
lager beer 222
- dispensing 348
- popularity of 503, 504, 507
- rise of 25–27
- spread of lager brewing 27–30
- types of 28
lager yeast 121–123
lambic beer 5, 235, 250–253
- foreign yeast in 481
- fruit 252, 253
- production methods 251, 252
land area requirements, brewery 569–572

lanosterol 129
laser date coding 337
lautering 181–186
– grinding and 165
– lauter tun (see lauter tun)
– mash filter 182, 184, 186
– strainmaster 186
lauter tun 167, 168, 182–186, 531, 532, 534
– advanced control of 545–550
– controller structure 548
– control results 548–550
– description of 545–547
– mash transfer to 411
– process characteristics 547, 548
Lavoisier, A. 31
leak tests 355
LEA (late embryogenesis abundant) genes 52
leathery/whiskey flavor 687, 688
lectin hypothesis 123
legislation
– of beer 14
– consumer protection 495
– environmental regulations 675
– quality control and 438
lemonade
– beer and 258, 259, 264
– Pils and 258
Leuconostoc 268
Levant nut 32
level measurement 534, 535
Lg foam tester 456
lichenase 465
Liebig, J. 31
life expectancy, beer and 516
light
– beer-based mixed drinks and 272
– cans and protection from 278
– glass bottles and protection from 277
lighting 655, 656
light (lite) beers 109, 240
lightstruck flavor 277, 399, 423–428, 681
lignin 45, 430
lime softening 112, 113
limit dextrinase 173
limit monitors 541
linalool 88, 97
lining materials, cleaning agents and 613
lipases 54
lipids, oxidation of 402
lipoxygenases 405
Liquicap FMI 51 535
load management, of electrical power supply 656, 657

load values, of waste water 626, 627
London, brewing in 23
Long, J. 31
Long saccharometer 31
loop reactors 240
loss cleaning and disinfecting 603
lotus effect 141
low-carb beers 240, 241
low-pressure boiling 194, 196
low-temperature vapro condensers 649, 650
Lundin fractions 453
lupolone 100
lupuline 378
lupulin gland 87, 88
lupulones 87
lysine 69, 160

m

machine configurations
– for non-returnables 578, 579
– for returnables 577, 578
machine drive 310, 311
machinery dimensioning 562, 563
magnesium
– in beer 526
– gushing and 390
– as yeast nutrient 129, 130
magnesium hardness of water 105, 106
magnesium hydrogen carbonate 108
magnesium hydroxide 112
magnesium ions 108, 109
magnesium sulfate precipitation 452
Maillard products
– high-temperature wort boiling and 194, 195
– kilning and 407, 408
– mashing process and 410
– thiobarbituric acid index and 161, 446
– wort boiling and 189, 193–195
Maillard reactions
– color and flavor production and 155
– in mash 168
– melanoidins and 175
– wort boiling and 187
maize grits 63
maize syrup 58
maize *(Zea mays)* 43, 63, 64
– acreage and production of 59
– aleurone layer 45
– amino acids in 51
– for gluten-free beer 246
– molecular diversity in 53
– protein and starch in 47

– starch content   46
– starch granules   50
malt   5
– acid   164
– beer foam and   363, 366
– caramel   163
– chit   164
– color of   162, 163
– conversion from barley   556
– dark   156, 163, 224
– diastatic   164
– DMS and   387
– enzymes   161, 162
– flavor stability and quality of   409
– global production   501–503
– for gluten-free beer   246
– grading   460
– Hanseatic League trade in   17
– quality of   157, 158, 704
– roasted   164
– smoked   164
– sorghum   164
– special   163, 164
– storage of   555
– top malting companies   503
– trading networks for   24
– wheat   164
– world market   498–501
Malta   505, 507
maltase process   174, 175
maltase rest   174, 175
malt cleaner   388
Malteurop Group   501, 503
malthouse planning   555–559. *see also* brewery planning
– barley and malt storage   555
– calculation example for malting plant   560
– consumption data   561, 562
– floor area calculation   561
– germination   555
– kilning   556
– steeping   556
– steeping, germination and kilning tower design   557
malthouse waste water   625
malt-induced gushing   392–394
malting   4, 147–164
– barley cleaning   149, 150
– barley intake and storage   149–163
– brewing barley   147, 148
– cleaning, storage, and polishing of malt   156, 157
– germination and   53, 54, 152–155

– kilning   155, 156
– malt quality   157, 158
– malt yield   157
– *N*-heterocycles in   383
– quality criteria of barley malt   157–163
– steeping   150, 151
– volume and mass change during   157
malting losses   560
malt mill   531, 532
malt milling   409, 410
– turbidity and   431
maltol   381, 383, 689
maltooxazine   385
maltose   58, 127, 209, 440, 689
maltose rest   239
maltotriose   58, 127, 209, 689
maltoxazine pyrroline   381
malt polyphenols   373, 374
malt replacements   176–181
malt silos   531, 532, 534, 564
maltster's guilds   14
malt treatment   564
malty flavor attribute   680, 685
Malzbier   242
management systems   491
– standards   491–494
manganese   129, 130, 448
manganometric titration   450
Mann, Thomas   3
mannoproteins   125
manufacturing cultures   479, 480
marketing strategies   24
Märzen beers   222, 360
mash filter   167, 182, 184, 186
mash filtration   413
– influence on beer foam   363
– turbidity and   431
mashing   168–181
– adjunct   176–181
– American infusion method   35
– with barley adjunct   177, 179
– biological acidification   411
– with corn adjuncts   177, 179
– dark beer varieties   175, 176
– with extremely high rice addition   180, 181
– flavor stability and   410–413
– maltase process   174
– mashing parameters   169–173
– *N*-heterocycles and   383
– oxygen during   411–413
– with sorghum adjunct   181, 182
– step-mashing process   173, 174
– temperature during   410, 411
– turbidity and   431

– with very high rice adjunct addition 177–180
mash pH 704
mash tun 531, 532, 534
mash tun adjuncts 176
mass balance 442
mass selective detector 463
mass spectrometers 319
material compatibility, cleaning and 611–613
maturation 214, 215–220, 565
– characteristics 565
– flavor stability and 417, 418
Mazout 258
MBT (3-methyl-2-butene-1-thiol) 424–426
mealiness, measuring 459
measurement technology 533–542. see also photometric measurements
– conductivity measurement 537
– dosing 541
– flow measurement 536–538
– 'in-line' measurement 541, 542
– level measurement 534, 535
– limit monitors 541
– oxygen measurement 538, 539
– pH value measurement 536
– pressure measurement 536, 537
– temperature measurement 534, 535
– turbidity measurement 539, 540
measuring devices 31
MEBAK (Mitteleuropäische Brautechnische Analysenkommission)
– analytical methods 437, 438
– boiling system guidelines 190
– malt quality analyses 157, 159
– on quality of beer-based mixed drinks 263, 264
Mecafill VKPV 292
mechanical level-controlled filling systems 290–292
Megasphaera 268, 478, 479, 481
Megasphaera cerevisiae 483, 484
melanoidins
– flavor and 408
– foam stability and 362
– formation of 382, 408
– kilning and 155
– redox potential and 360
– staling process and 401, 402
– turbidity and 430
melibiase 122
membrane filtration 230
membrane processes 115–117
memory effect 335

mercaptans
– carbonyl formation and 403
– as flavor attribute 386, 680, 681
– lightstruck flavor and 211, 424
Merkur hops 368
Merlin thin film evaporator 199, 200
Mesopotamia, brewing in 5, 6
metabolic pathways 131–135
– carbohydrate 131, 132
– formation of esters 134
– formation of higher alcohols 132, 133
– formation of sulfur dioxide 135
– formation of vicinal diketones 132, 133
– phenolic compounds 134, 135
metabolism of cereals 46–55
– anabolic processes 46–53
– catabolic processes 53–55
metal candle leaf filter 229
metal ions in bottled beer 389–391
metallic flavor attribute 680, 690
metals, beer and removal of 524, 525
methane 670
methanogenesis 638
methional 400
methionine 130
Methods of Analysis (ASBC; Institute of Brewing and Distilling; Brewery Convention of Japan) 439
methylbutanal 404
3-methylbutanal 420
2-methylbutanol 132
2-methyl-1-butanol 210
3-methyl-1-butanol 210
3-methyl-2-butene-1-thiol (MBT) 424–426
3-methyl-2-buten-1-thiol 681
methylene chloride 666
S-methyl esters 682
5-methylfurfural 400
3-methyl-3-mercaptobutylformiate 404
S-methylmethionine (DMS precursor) 188–190, 682
methylpropanal 404
2-methyl-1-propanol 210
2-methyl-pyrazine 383, 385, 386
3-methylthio-1-propanol 211
Mexican beer market 506, 508, 510
MIAC (model identification adaptive control) 543, 544
microbiology 31, 32, 477–489
– of beer-based mixed drinks 268
– detection of beer pests 484–489
– foreign yeasts 480, 481
– hop 99
– manufacturing cultures 479, 480

microorganisms
- beer-spoilage bacteria   481–484
- brewery microflora   477–479
- disinfecting   600–603
- preventing growth of   151
microscope   31
microtiter plates   468
mild ale   248
Miller   511
millets *(Panicum miliaceum, Setaria italica)*   61–63
- acreage and production of   59
- amino acids   51
  for gluten-free beer   246
mineral deposits, removing   599
minerals
- in beer   525, 526, 527
- as yeast nutrients   129, 130
mitochondria   124, 126, 127
Mitteleuropäische Brautechnische Analysenkommission. *see* MEBAK
mixed gas, dispensing beer with   348, 349
mixing, of beer-based mixed drinks   270, 271
mixing tanks, aeration of waste water in   634, 635
mobile CIP systems   609–611
model identification adaptive control (MIAC)   543, 544
model reference adaptive control (MRAC)   543, 544
modification of malt kernels   460
modular labeling system   321, 323
Mohren   259
molasses   58
mold amylases   4
mold fungi   477, 478
mold spores, infection of green malt by   390
molecular diversity   53
Molson   511
monasteries, brewing and   9–12
Moniliella   477
mono bromoacetic acid   667
Moorwasser   259
Morgenstern system   113
mPas, milli Pascal seconds (unit of viscosity)   457
MRAC (model reference adaptive control)   543, 544
multidimensional GC-MS   91
multipacks
- paperboard   592, 593
- with plastic carrier   593, 594
Munich, lager brewing in   27–30

Munich beer   222
munu   5
m value   105, 106, 107, 111, 113
mycotoxins   437, 438, 462, 468
myocardial infarction, beer and   515, 516, 518
myrcene   88, 187
Myrica gale   12

n
Nährbier   235, 242
nanofiltration   116
Narziss, L.   555
National Institute of Health   518
National Institute on Alcohol Abuse and Alcoholism   518
NBB medium, for detection of beer pests   484, 485
NDMA (*N*-nitrosodimethylamine)   156, 525
near-IR absorption   442, 443
near-IR transmission spectroscopy   451
near 'real' beer   246
neck finish, inspecting   316, 318
neger   259
nephelometric turbidity units (NTU)   456
Netherlands   506
net tank space (NTS)   562
neutral cleaning agents   597
neutralization, of waste water   632–634, 668
niacin   526
Nigerian beer market   509
ninhydrin   446
nitrate
- in beer   377, 525
- in brew water   111
- determination of   448
- reduction in water   111
- in wort   93
nitric acid   598, 599, 612, 618
nitrite   448
nitrogen
- beer dispensing and   349
- beer foam and   362
- sources for yeast   128, 129
- using during mashing   411
- in waste water   623, 627
nitrogenous compounds
- changes in composition of   211
- determination of   451–453
- fractions of   452, 453
nitrosamines   156
N-nitrosoamine   437
N-nitrosodimethylamine (NDMA)   156, 525
noble hop aroma   683

Index

noise emission 672
– regulations 673
trans-2-nonadienal 400
trans-2-cis-6-nonadienal 400
n-nonal 400
γ-nonalactone 64
non-alcohol beers 504, 505, 507
non-biological stability, measurement of 456, 457
noncarbonate hardness of water 105, 106, 110, 111
(E)-2-nonenal 404
trans-2-nonenal 687
non-ionic surfactants 598
non-recovery CIP cleaning method 603, 604
non-returnables, machine configurations for 578, 579
'normal' ale 249
North America
– beer consumption in 506
– beer markets 506, 508
– malt and 502
– malting barley varieties 500
– top brewers 512
nozzle head, in CIP system 609, 610
NTA 598
NTS (net tank space) 561
NTU (nephelometric turbidity units) 456
nucleus (cell) 124, 126
Nugget hops 368
nursing tissue 44
nutriceutical 526

o

oat malt 164
oats (Avena spp.) 14, 56, 57, 60, 61
– acreage and production of 59
– as brewing grain 15, 19
– proteins in 51
– starch content 46
obligate beer pests 478, 487
obligate beer spoilage bacteria 482
ochratoxin A 438
n-octanoic acid 686
octaoic acid 210
OECD screening test 624
OEE (overall equipment efficiency) 706, 707
off-flavors in beer-based mixed drinks 266–268
oil 648
old ale 249
Omnigrad TR 45 535
one-step cleaners 609
one-step wort cooling 649, 650

opaque beer 62, 65
open cleaning circuits 608
open-loop control 533, 542
operational load analysis 656
Operational Performance Indicators 706
opium 32
OPP (oriented polypropylene) 324
optical oxygen sensors 539
optimized internal boiler
– Stromboli 190, 191
– Subjet 190–192
optodes 455
organic contamination 595
organic loads, in waste water 627
organic substances, removal of problematic 112
organoleptic stability
– indicator substances 404, 405
– measurements to preserve 404–423
oriented polypropylene (OPP) 324
original gravity 442
ortho-phosphate 627
Oryza glaberrima Steud. 64
Oryza japonica 64
Osiris 6
osteoporosis, beer and 526, 527
overall equipment efficiency (OEE) 706, 707
oxalic acid 392
oxazole 381, 385, 408
oxidation reactions, staling and 399, 402
oxidation/UV irradiation, combination processes using 112
oxidized flavor attribute 680, 687
oxidizing cleaning agents 600, 613
oxidizing cleaning booster 598
oxidoreductases 162
oxygen
– aerobic waste water treatment and 635
– in beer 212
– beer foam and 362
– filling and 282, 283, 420, 421, 423
– germination and 407
– mashing process and 411–413
– measurement of 454, 455, 538, 539
– plastic bottles and permeability of 280
– regulation of 648
– staling and 402, 403, 405, 406
– for steeping 150, 151
– as yeast nutrient 129, 139
ozone, measurement of 447

p

packaging 275–281, 505, 507, 534, 577–594
– of beer-based mixed drinks 272

- cans 277, 278
- glass bottles 275–277
- kegs 280, 281
- packing into packs open at top 578–584
- paperboard multipacks 592, 593
- plastic bottles 278–280
- plastic carrier multipacks 593, 594
- selecting machine configuration 577, 578
- shrink-wrap 587–591
- wrap-around 584–587
packs open at top 578–584
- machine design 578–582
- robot technology 582–584
Painter, W. 35, 36
paired comparison test 693, 695
pale ale 248
Pale American lager 28
pale beer 222
pale malt 224
- drying of 155, 156
- pH 163
- TBN 161, 162
pantothenate 130
pantothenic acid 489, 526
paperboard multipacks 592, 593
papery/cardboard flavor 687, 688
Parkinson's disease, beer and 517
Partigyle system 255
passive noise abatement 673
Pasteur, Louis 31, 32, 119, 120
pasteurization 422
- of alcohol-free beer 240
- of beer-based mixed drinks 268–270
- flash 270, 422, 646, 661
- heat requirements 646
- microflora and 477
pathothenate agar 489
Payen, A. 31
PCS (process control system) 533, 550, 551
pearl millet (*Pennisetum glaucum* (L.) R. Br.) 61
peas 246
peat, as heat source 20
Pectinatus 478, 479, 481
Pectinatus cerevisiiphilus 484
Pectinatus frisingensis 484
Pediococcus 69, 99, 252, 254, 268
Pediococcus Claussenii 483
Pediococcus damnosus 479, 483
Pediococcus inopinatus 483
Penn, William 33
PEN (polyethylene naphthalate) bottles 278, 280, 422
2,3-pentandione 211

2,3-pentanediol 132
2,3-pentanedione 132, 133, 686
pentosans 362, 430
PE (population equivalent) 629
pepsin 31
peptidases 54
Pepys, Samuel 22
perforation date coding 337
performance, beer and better 523, 524
performance efficiency, calculating 707
pericarp 44, 45
periplasm 124, 125
perlite 230
- gushing and 392, 393
permanent haze 428
permease 128
permeate 115
Peronospora warning service 86
peroxidases 405
peroxyacetic acid 601, 602, 618
Perzoz, J. 31
pesticides, hop cultivation and 86
PET (polyethylene terephthalate) bottles 278, 279. *see also* plastic bottles
- for beer-based mixed drinks 272
- cleaning 617
- hotmelt labeling of 327, 328
- machine configurations for 577, 578
- in paperboard multipacks 593
- percent of market 505, 507
- in plastic carrier multipacks 593
- pressure-sensitive labeling of 332, 333
- reel-fed hotmelt labeling of 330, 331
- screw-cappers for 304
- shrink-sleeve process and 336
- staling and 421, 422
- washing 306, 307
pH
- acidic/sour flavor and 687
- anaerobic waste water treatment and 639
- of barley malt enzymes in mash 171
- of brew water 111, 113
- during fermentation 209–212
- influence of mash pH on flavor stability 411, 412
- of malt 163
- of mash 383, 704
- measurement of 448, 449, 536
- taste and 361
- of waste water 632–634, 639
phenolic compounds 134
phenylacetaldehyde 404
β-phenylacetate 210
phenylethanal 420

phenylethanol 132
2-phenylethanol 688
2-phenyl-1-ethanol 210
2-phenylethyl acetate 683
phosphatases 168
phosphate 448, 598
– acidic-acting primary converted to basic-acting secondary 108
– in waste water 627, 667
phospholipids 125
phosphoric acid 598, 599, 612, 618
phosphorus 130, 623
phosphorus molybdenum acid precipitation, of nitrogenous compound fractions 452, 453
photodetectors 540
photometric iodine reaction 446
photometric measurements 444–448
– α-amylases 447
– bitter units 446
– of color 445, 446
– free amino nitrogen 446
– ions 447, 448
– photometric iodine reaction 446
– thiobarbituric acid index 446, 447
– total polyphenols and anthocyanogens 447
physical methods, of alcohol-free beer production 236–238
physiology and toxicology 515–528
– beauty benefits of beer 525, 526
– beer and alcohol 521
– beer and bacteria 524
– beer and cancer 521, 522
– beer and concentration, performance, and reaction time 523, 524
– beer and kidney stones 522, 523
– beer and protection of stomach and arteries 522
– beer and removal of metals 524, 525
– beer as clean food 525
– beer as sports drink 523
– beer belly 527
– beer prescription 527, 528
– beneficial minerals in beer 526, 527
– health benefits of beer 515–521
Pichia 252
picric acid 32
picrinic acid plus citric acid test 457
picrotoxin 32
PID controllers 533, 542, 543
Pilsner malt, friabilimeter results 459
Pilsner (Pilsener, Pils) 222
– absorption spectrum of 445

– color 360
– dispensing 348
– Emsgold and 259
– introduction of 30
– lemonade and 258
– residual alkalinity for 109
pipelines, cleaning 610
pirrazoline 408
pitching 207, 208
planning efficiency, calculating 707
plant food allergens 52
plastic bottles 278–280. *see also* PET (polyethylene terephthalate) bottles
– PEN 278, 280, 422
– staling and 421, 422
plastic carrier multipacks 593, 594
plastic screw caps 304
plate and frame filter 227
PLC (programmable logic controller) 550
pleasure, alcohol consumption and 520
Pliny 8, 12, 525
plumule 46
Poland, beer market 506, 508
polarographic-amperometric sensors 538
polarographic sensor 455
pollutant load
– brewery side products 631
– waste water 625, 626, 628
polyamide 279
polycarbonates 598
polyethylene naphthalate (PEN) bottles 278, 280, 422
polyethylene terephthalate (PET) bottles. *see* PET (polyethylene terephthalate) bottles
polypeptides, turbidity and 430
polyphenols 89
– analysis of total 447
– in beer production 371–381
– defined 373
– foam and 362
– health benefits of 437
– hop 374–378
– malt 373, 374
– origin of 373–378
– oxidation of 168
– polyphenolic-related reactions during brewing 379, 380
– precipitation of, during fermentation 212
– reaction path during brewing 380
– redox potential and 360
– staling process and 401, 402
– turbidity and 430
– value of anthocyanogens 380

Index | 737

polysaccharides, turbidity and 430
polyvinylpolypyrrolidone (PVPP) 89, 365, 432
– for anthocyanogens 381
– removal of polyphenol by 374
polyvinylpyrrolidone (PVPP) 232–234, 240
population equivalent (PE) 629
porter 235, 254, 255
– foreign yeast in 481
– introduction of 23–25
– production method 255
post-fermentation treatment of yeast 140
Postgate, John 32
posthumulone 87
potassium
– in beer 526
– determination of 447, 448
– to prevent microbial growth during steeping 151
– as yeast nutrient 129, 130
potassium dichromate 448
potassium permanganate 151
potato starch granules 50
potential beer pests 478, 487
potential beer spoilage bacteria 482
potential diacetyl 215
potentiometer 455
potentiostatic sensors 539
Potsdamer 258
Powell, Jonathan 527
power factor correction 653
power supply 568, 651–657
– combined heat and power 657
– drive system 654
– electric power consumption 653–656
– external 652, 653
– heat supply 656
– lighting 655, 656
– malthouse 560, 561
– measurement at release point 652
– optimization of 656, 657
– requirement figures 651, 652
– requirement for compressed air production 662
– supply contracts 653
pre-coating 226
– filter aids for 230
– methods 230
preference tests 693
prehumulone 87
Preliminary Beer Law 90
Preliminary Purity Law 93
prenylflavonoids 378

prepared cleaners 597
preserving agents 269
pre-soak with pre-spray, in bottle-cleaning machine 615
pressure, measurement of 536, 537
pressureless warm fermentation 217, 218
pressure-sensitive labeling (PSL) 332–334
pressurization 288
Prewmaister, Jorg 16
primary degradation 624
principles and similarities 494
proanthocyanidins 377, 378, 406
– colloidal stability and 428
– turbidity and 89, 433
proantocianidine 432
process bus 551
process control system (PCS) 533, 550, 551
production planning, flexible 552
production process 531–533
professionalism 12–16
Profile Panel 695
– training for 698, 699
profitability, in global beer markets 510
programmable logic controller (PLC) 550
prohibition, in United States 36
prolamin box (P-box) 52
prolamines 50
prolamins 51, 52
1-propanol 210, 688
n-propanol 132
properties and quality 359–395
– aroma substances in hops 371
– beer color 360, 361
– beer foam 361–366
– bitter substances in hops 366–370
– composition of finished, bottom-fermented beer 359, 360
– DMS 386–389
– gushing 389–395
– N-heterocycles 381–386
– polyphenols in beer production 371–381
– qualities of bottom-fermented beer 360
– rate of 707
– redox potential 360
– taste 361
propionic acid 483, 484
proso millet malt 246
proso millet (Panicum miliaceum) 56, 57, 62, 63
proteases 54, 60
proteinases 417, 418
protein-carbohydrate bodies 45
protein coagulation, in wort boiling 187

protein electrophoresis 460
proteins
– barley 46
– cell membrane 125
– reactions with polyphenols 379
– starch 49
– storage 50–52
– turbidity and 231, 232
– waste water 623
– wheat 46
proteolysis 160, 171, 175, 458
Proud, Robert 35
Provisional Beer Law 55, 58, 97
prozyanidine 406
pseudocereals 69, 70
pseudomycels 481
PSL (pressure-sensitive labeling) 332, 333
puranone 383
pure gas flushing 287, 288
purines 130
purity laws 14, 26
p values 106, 107, 111, 113
PVPP. see polyvinylpolypyrrolidone (PVPP); polyvinylpyrrolidone (PVPP)
pycnometer 441
γ-pyranone 408
pyrazine 175, 383, 385, 408
pyrazole 383
pyridoxine 130, 526
pyrimidines 130
pyrridine 408
pyrrol 381, 408
pyrrole 383
pyrroline 383
pyrrolizine 385
pyruvate 120, 131, 132
pyruvate decarboxylase 131
pyruvate dehydrogenase 131, 132
pyruvic acid 483, 484
Python beer line 351
Python water cooling system 342

q

QACs (quaternary ammonium compounds) 600–602
quality. see properties and quality
Quality and Security 438
quality assurance 700
quality control 437–473, 700
– AAS and ICP-OES 466, 467
– acid-caustic titration 449, 450
– α-amylases 447
– analyses 439–469
– bitter units 446
– carbon dioxide measurement 453, 454
– chlorine dioxide measurement 455
– chromatographic analyses 461–464
– color 445, 446
– complexometric titration 450
– conductivity 449
– congress mash 458
– continuous flavor analysis 461
– daily 469–473
– density, extract, alcohol content, original gravity, and degree of fermentation 440–444
– determination of caloric value of beer 466
– determination of nitrogenous compounds 451–453
– diastatic power 450, 451
– electronic spin resonance spectroscopy 469
– ELISHA 468, 469
– enzymatic analyses 464–466
– free amino nitrogen 446
– friabilimeter 459
– grading 460
– gushing 461
– hand assessment 460
– head retention 455, 456
– homogeneity and modification 460
– hop bitter substances 461
– ions 447, 448
– manganometric titration 450
– overview 437–439
– oxygen measurement 454, 455
– pH measurements 448, 449
– photometric iodine reaction 446
– photometric measurements 444–448
– protein electrophoresis 460
– spent grain analysis 459
– sulfur dioxide 467, 468
– thiobarbituric acid index 446, 447
– titration methods 449–451
– total polyphenols and anthocyanogens 447
– turbidity and nonbiological stability 456, 457
– viscosity 457, 458
quality management systems 492, 494
quaternary ammonium compounds (QACs) 600–602
quinoa *(Chenopodium quinoa)* 56, 57, 70, 246

r

radiation inspection technology 319
radicle 46
Radler 258, 271, 272

raffinose 122
raffinose test 122
raspberry
– Berliner Weisse and 253, 259
– lambic beer and 252, 253
raw materials, influence on colloidal stability 431
RDF (real degree of fermentation) 444
reaction time, beer and 523, 524
'real' ale 249
real degree of fermentation (RDF) 444
real extract 441, 442, 689
recipe-controlled processes 551, 552
recirculating cleaning 353, 354
recovery beer, diacetyl and 686
recovery tank CIP cleaning method 604, 605
Recycling and Waste Management Act 673
redox potential 360
redox properties, changes in 212
redox reactions, staling process and 399–402
reduction capacity, polyphenols and 375
reductones 360
reel-fed hotmelt labeling 330, 331
refermentation phase, of gueuze beer production 252
refrigeration plant 568
refrigeration requirements, for draft beer system 342
refrigeration system. *see also* cold supply
– waste heat from 651
Regina 259
regulations
– air pollution control 670
– environmental 670, 671
– food safety 32
– noise emission 672
– Regulation 178/2002 495
– Regulation 852/2004 495
– waste management 673
– water balance 673
reliability of inspection 319
religion, brewing and 5, 6, 10
renewable energy 669, 670
Renewable Energy Source Act 670
repeatability *(r)* 439
reproducibility *(R)* 439
residual alkalinity 109, 111
residual draining, in bottle-cleaning machine 614
residual liquid inspection 316, 318
restriction pressure 345
returnable glass bottles 577

returnables, machine configurations for 577, 578
returned empties, space requirements of 567
reverse osmosis
– in alcohol-free beer production 237, 238
– brew water treatment and 115–117
– for vapor condensate 199
rho-iso-α-acids 93
trans-rho-iso-humulones 427
rH value 212
ribes flavor 404
riboflavin 526
ribosomes 124, 126
rice adjuncts 177–181
rice grits 58
rice *(Oryza sativa)* 43, 44, 64
– acreage and production of 59
– aleurone layer 45
– for gluten-free beer 246
– proteins in 47, 51
– starch granules 50
– starch in 47, 50
Richardson, John 31
right of precinct 13
ring-bowl flushing 287, 288
roasted flavor attribute 680, 685
roasted malt 164, 175
roasted malt beer 224
robot technology, for packaging 582–584
roller mills 165, 167
roll on/shrink on labeling 331, 332
Romans, brewing and 7, 8
room heating 647
rooms in draft beer system 340–342
rotary fillers 290
rotary inspection machine 314, 315
rotary labelers 321, 323
– for hotmelt labeling 330
rotary packers 580–582
Russ 259
Russia 501
– beer market 506, 508, 510
Russ'n Mass 258
rye malt 164
rye *(Secale cereale)* 65
– acreage and production of 59
– amino acids 51
– as dominant grain 14
– properties of 56, 57

s

saacharose 261
Saatz yeast strain 120

SAB Miller 511
saccharification 247, 446
saccharometer 31
Saccharomyces 69, 119, 120, 121
Saccharomyces bayanus 207, 252
Saccharomyces carlsbergensis 121, 479, 480
Saccharomyces cerevisiae 207, 222, 252, 479, 480
– var. *bayanus* 480
– var. *diastaticus* 480
– var. *pastorianus* 480
Saccharomyces dairensis 252
Saccharomyces globosus 252
Saccharomyces inusitatus 252
Saccharomyces pastorianus 121
Saccharomyces phase, of lambic beer production 252
Saccharomyces uvarum 252
– var. *carlsbergensis* 207
safety precautions, draft beer systems 356
Saladin boxes 152
Saladin tithe 21
salty 676
sandwich ELISA 468
saturation pressure 344
sauergut cultures 480
SBR (sequencing batch reactor) processes 635–637
scaling 117
SCARA (Selective Compliance Assembly Robot Arm) 582, 583
Schmutiges 259
Schmutz 259
SchoKo 194–196
Schönfeld test 151
Schussbier 259
Schwan, Theodor 119
Schwann, Theodor 31
Schwarz, Anton 33
Schweinebier 259
Scotland, brewing in 24
Scottish ale 249
screw-cappers 304
scurvy 22
scutellum 44–46, 153
– wheat 47
sealing materials, cleaning agents and 613
seasonal beers 504
secalins 52
seed coat 44, 45
semipermeable membrane 115
senses 675, 676
Sensomatic VPL-PET 295, 296
Sensomatic VPVI 294, 295

sensors 552, 553
sensory evaluation 675–701
– of beer-based mixed drinks 264, 265
– building sensory capability 695–699
– consumer research 700, 701
– environment for 691, 692
– final beer testing 700
– five senses 675, 676
– flavor attributes 679, 690
– how to assess beer flavor 676–679
– in-process sample testing 699, 700
– manual 692
– panel leader 691, 692
– panel motivation 692
– performance monitoring 699
– selection, training, and validation of panelists 694, 695
– sensory tests 693, 694
sensory fatigue 678
sequencing batch reactor (SBR) processes 635–637
sequestering agents 598
serpin superfamily 51
service warm water 645, 646
sesquiterpene hydrocarbons 89
settling 290
sewage disposal plant 670
– air emissions from 671, 672
shekar 6
shikaru 6
shock cooling 432
short-grain millet 61–63
Shot'Malzbier 258
shrink-sleeve process 334, 336
shrink-wrap packaging 587–591
sicera 6, 21
side scatter detectors 540
sidewall, inspecting 316, 318
Siebel, John E. 33, 35
sight 676
signpost formations 480
sikaru 6
silica 233, 271
silica gel 240
silicates 597
silicon 527
silo capacity 561
simulation tests 624
single-end bottle washing machine 306, 312
Sinner's Circle 305
sintered glass 240
six-row barley 498
sixteenth-century price revolution 18

sleeving   334–336
– shrink-sleeve   334, 336
– stretch-sleeve   334, 335
slider 'M' system   352
slime linings   478
smell, sense of   675, 676
smoked malt   164
smoky-flavor beer   222
snifting   290
soaking   354
soda alkalinity   105
sodium, in beer   526
sodium carbonate   599
sodium hydrogen carbonate   108
sodium hydroxide
– for cleaning   597, 609, 615–618
– to prevent microbial growth during steeping   151
sodium hypochlorite   601
sodium metabisulfite   151
sodium metasilicate   597
sodium polyphosphate   599
soft boiling method 'SchoKo'   194
soft resins   87
soils, for hop cultivation   86
solid electrolyte sensors   539
solid separation, of waste water   668
solubilizers   598
soluble nitrogen   160
solvent, as flavor attribute   680, 682
sorghum adjunct   181, 182
sorghum grits   65, 66
sorghum malts   65, 66, 164
sorghum millet   61
sorghum *(Sorghum bicolor)*   43, 65, 66
– acreage and production of   59
– for gluten-free beer   246
– properties of   56, 57
– starch granules   50
Soufflet Group   501, 503
sour flavor   676
South African beer market   509
South America
– beer consumption in   506
– malt and   502
– malting barley varieties   500
space requirements, brewery   570
Spain, beer market   506, 508, 510
special beers   222
specifications, new brewery   575
specific gravity analysis   440, 441, 443
spectral photometers   464
spectrum line photometer   464

spelt *(Triticum spelta)*   14, 56, 57, 66, 67
spent grain analysis   459
spent grain and yeast   674
sphingolipids   125
spicy flavor attribute   680, 684
split streams   629–631
sports drink, beer as   523
spray balls   606, 607
squalene   129
stability of beer   399–433
– colloidal stability   428–432
– flavor stability   399–423
– foam   362, 369
– hop aroma   371
– lightstruck flavor   423–428
– stabilization systems   432, 433
stabilization   231, 232
– filtration and stabilization plant design   232–234
stack cleaning and disinfecting   603
staling process. *see* flavor stability
starch
– breakdown to sugar   170
– formation of   47–50
– gluten-free   246
starch crystallinity   49
starch granules   49
– properties of   50
starch synthases   49
starch syrup   261
starchy endosperm   45, 46
starchy raw materials   43–71
– brewing cereals   55–70
– structure and metabolism   44–55
Statistical Estimates of the Materials of Brewing (Richardson)   31
steeping   150, 151, 554
– process scheme   558, 559
Steinfurth foam stability tester   455, 456
Steinheil, C.   31
step-mashing process   173, 174
sterilization, steam   618
sterols   125, 129
stomach, beer and protection of   522
storage
– of barley and malt   555
– beer foam and   364
– characteristics   564
– of hops   100
– of malt   156, 157
– maturation and   215
– staling and   421, 422
– turbidity and   432
– of yeast   141, 142, 220, 221

storage/cabinet cooling, in draft beer system   342
storage cellar
– design of   657, 658
– waste water and   625, 630
storage proteins   50–52, 55
storage tank   418, 419, 531–533, 565
– cleaning   607–610
stout   95, 254, 504, 505, 507. *see also* porter
– barley flakes in   177
– dispensing   348
– Imperial   223
– Irish   60
strainmaster   186
straw, yeast preservation and   12
Strecker aldehydes   187, 239, 408, 423
stretch-sleeve process   334, 335
strip gluing   324–326
stroke, beer and   516, 518, 519
Stromboli internal boiler   190, 191
strong beers   222, 223
strongly hopped beer   95
strychnine   32
Subjet internal boiler   191, 192
sucrose   49, 127, 209, 440, 689
sucrose test   422
sugars
– in beer-based mixed drinks   261, 262
– fermentable   127, 128
– gluten-free   245
sulfate   111, 448
– determination of   447
sulfate ions, ratio to chloride ions   689
sulfhydryls   360
sulfides   89
sulfur   130
sulfur compounds   211
sulfur dioxide   214
– as allergen   211, 437, 525
– determination of   465, 467, 468
– as flavor attribute   211, 679, 680
– formation of   135
sulfur heterocycles   89
sulfuric acid   598, 599, 612
supply contracts   653
surface filtration   226
surfactants   598, 599, 618
swabbing method, to detect beer pests   485–488
sweet   676
sweetening in beer-based mixed drinks   261, 262
sweetness, assessing   680, 689
Switzerland, dietetic beer in   241

t

Tabarié relationship   442, 443
Tacitus   9
tamper-evident sealing   327, 332
Tango   259
tanning agents   525
tannins, turbidity and   231, 232
Tasmania, brewing in   36
taste   361, 676
taste testing   675
Taurus hops   368
taverns   15
taxes
– on ale   21
– on beer   23, 27, 497
– on brewing   19
TBI (thiobarbituric acid index)   386, 409, 414, 422, 446, 447
TBN (thiobarbituric acid number)   161, 162, 187, 188
TBN (total bound nitrogen)   627
TC-Tank Jet   611
Technical Instruction for Air Pollution Prevention   670
technology
– grinding   165–168
– hopping   91–100
– inspection   318, 319
– lautering   181–186
– mashing   168–181
– measurement   533–542
– robot   582–584
– Xan   243
teff (*Eragrostis tef* (Zucc.) Trotter)   62
temperature
– alkaline cleaning agents and   613
– anaerobic waste water treatment and   638
– of barley malt enzymes in mash   171
– condensation   659
– DMS and   387
– filling   285
– gelatinization   170, 172
– germination and   152, 154
– for hop storage   100
– keg storage   339
– kilning   155, 407–409
– during mashing process   410, 411
– measurement of   534, 535
– pasteurization and   271
– steeping   150
– step-mashing and   174
– withering   405, 406
– yeast propagation and   139
– for yeast storage   142

temperature tests, for draft beer systems 355, 356
Temporary Beer Law 241, 242
testing
– draft beer systems 355, 356
– sensory 693, 694
testing and calibration laboratories 494
tetraden formation 483
tetrahydro- α-acids 428
tetrahydro-iso- α-acids 93, 94, 98, 426, 428, 689
tetrahydro-iso extract 503
trans-tetrahydro-iso-humulone 427
tetraploid durum (*Triticum durum* L., *T. turgidum* L., *T. polonicum* L.) 67, 68
Tettnang Tettnanger hops 377
theoretical production time 706, 707
thermal conductivity detector 463
thermal dealcoholization, in alcohol-free beer production 236, 237
thermal energy 643
thermal energy analyzer (TEA-detector) 463
thermal power stations 669
thermal treatment of beer 422, 423
thermo-analytical measurement 443
thermometer 31
thiamin 130
thiazoles 381
thin film evaporator 199, 200
thiobarbituric acid index (TBI) 386, 409, 414, 422, 446, 447
thiobarbituric acid number (TBN) 161, 162, 187–189
thioesters 89, 682
thiols 211, 681
Thomson, T. 31
tiazole 408
time on tap 339
time-related archiving 552
(TIN) total inorganic nitrogen 627
titah 5
titration methods 449–451
TKN (Kjeldahl nitrogen) 627
tocopherol, applied to beer-based mixed drinks 269
top fermentation 22, 26, 55, 58, 119, 222, 223
top-fermented beers 223, 224
top-fermenting yeasts 121, 122, 207, 479, 480
topping-up 209, 214
Torulopsis 252
total bound nitrogen (TBN) 627
total inorganic nitrogen (TIN) 627

total waste water 627–629
touch, sense of 676
toxic constituents, of waste water 624
trace elements, as yeast nutrients 129, 130
training, for sensory evaluation 694–697
transaminase system 128
transgenic plants 53
transglutaminase 245
transmission 444
tray packer 589, 590
Treatise on the Adulteration of Food and Culinary Poisons, A (Accum) 32
treatment zones, in bottle washing 306, 307
triangle test 693, 695
tricarboxylic acid cycle 131
trichloroanisole 267
2,3,5-trimethyl-pyrazine 385
triticale (*xT. riticosecale* Wittmack) 50, 56, 57, 68
triticins 51
tritordeum 56, 57, 68
trub 140, 142
– cold 202
true density 440
true to type 679
tunnel pasteurizers 646
turbidity 377, 428
– composition of 430
– formation of 430, 431
– measurement of 456, 457, 539, 540
– stabilization 231, 232
turbidity photometers 456
turbulent cleaning 618
two-mash procedure 169, 170
twopenny 23
two-row barley 498
two-sink glass washing installation 343
two-step wort cooling 649
tyrosol 132

u

UB (unfiltered beer) 562
ultrafiltration 116
umami 676
umbels 87, 88
9-undecanal 400
n-undecanal 400
unfiltered beer (UB) 562
United States
– beer consumption in 503, 506
– beer market 506, 508, 510
– brewing in 33–36
– dietetic beer in 240, 241

– hop farms 502
– top brewers 512
U.S. Department of Agriculture 518
U.S. Department of Health and Human Services 518
utilities and power supply 568, 571, 574
– carbon dioxide recovery 568
– with coldness 568
– with compressed air 568
– with electrical power 569
– with fresh water 569
– with heat 568
utilization factor 706
UV irradiation, labeling using 331–333
UV irradiation/oxidation, combination processes using 112

v

vacuoles 124, 126
vacuum evaporation 197, 198
valeric acid 483
valine 128, 129, 132
vanillin 689
van Leeuwenhoek, Antonie 119
vapor condensate 191, 202
– re-use of 201
– waste heat from 650
vapor condenser, heat recovery and 649, 650
vapors 670
Varioboil 197, 199
vats 24
Venezuelan beer market 509
vertical integration 552
very high rice adjunct addition, mashing with 177–180
vicinal diketones 211, 686
– formation of 132, 133
4-vinylguaiacol 89, 134, 223
4-vinylphenol 134, 223
viscosity
– dynamic 457
– kinematic 457
– measurement of 457, 458
vitamins
– in beer 525, 526
– as yeast nutrients 130
Volumetric VOC filler 296–300
Volumetric VODM-PET filler 296, 297, 300, 301

w

Wagner, John 35
Wakley, Thomas 32
Wanderhaufen vessel 152

warm maturation 215
waste 673, 674
– kieselguhr 674
– production 673
– spent grains and yeast 674
waste heat
– from compressed air 650, 651
– from refrigerating system 651
– from vapor condensate 650
waste management regulations 674
waste water 621–640
– aerobic treatment 635–637, 640
– anaerobic treatment 637–640
– analysis of 623–625
– biological degradation of 666
– composition of 666
– constituents 622–624
– direct discharge of 666
– environmental protection and 665–668
– indirect discharge of 666–668
– in-house measures 631–634
– measurement 626, 627
– mixing and equalizing tanks 634, 635
– neutralization 632–634
– neutralization of 668
– parameters of 627–631
– pollutant load 625, 626
– split streams 629–631
– total 627–629
– treatment of 632–640
– types of 621, 622
– volume and concentrations 626, 627
– waste water plant planning 625, 626
water. *see also* brew water; waste water
– in beer-based mixed drinks 261
– beer quality and 105
– in bottle washing 306
– fresh water requirements in brewery 621, 622
– reduction of usage 665
– for steeping 150
– in yeast cell 124
water balance regulations 674
water supply 569
water treatment 110–117
– activated-carbon filtration 112
– aeration 112
– combination processes using oxidation/UV irradiation 112
– deferrization and demanganization 110, 111
– heat requirements 648
– ion exchange 113–115
– lime softening 112, 113

– membrane processes   115–117
– nitrate reduction   111
– removal of problematic inorganic substances   110, 111
– removal of problematic organic substances   112
– waste water   632–640
weakly hopped beer   94
Weiba   259
Weizenbock beer   223
Wendes   16
Western Brewer   35
wet chemical analysis, of beer-based mixed drinks   263, 264
wet end of the line   305
wet-glue labeling   324–327
wet milling   165
wheat beer mixtures   259
wheat beers   68, 69, 223, 235, 504
– Belgian   69
– dispensing   348
– phenolic compounds in   134, 135
– white   27
– Xan   243
wheat group (*Triticum* L.)   66–69
– einkorn   67
– emmer   67
– spelt   66, 67
– tetraploid durum   67, 68
– triticale   68
– tritordeum   68
– wheat   68, 69
wheat kernel
– nutrients in   47
wheat malt   17, 164, 224
wheat *(Triticum aestivum)*   3, 43, 68, 69
– acreage and production of   59
– aleurone layer   45
– amino acids   51
– endosperm   45, 46
– grain ripening   48
– properties of   56, 57
– proteins in   47, 51
– starch granules   50
– starch in   46, 47
Whewell, William   2
whirlpool   531, 532, 534
Whitbread, Samuel   24
white barley beer   27
white wheat beer   27
widgets   278, 299
Wiener beer   360
WinCoS   551
winter barley malts, beer foam and   363

Winthrop, John   35
Wismar   17
withering   387, 388, 405–407, 557
wood, as fuel source for brewing   20, 27, 648
woodruff   253, 259
working pressure   344
work safety, cleaning and   620
world beer production (1900, 2003)   34
world market   497–513
– Africa   509
– America   508
– Asia   509
– Australia and Pacific   509, 510
– barley and malt market   498–501
– beer consumption   503
– beer markets   505–510
– beer styles   503–505
– branding   511, 513
– Europe   507, 508
– global malt production   501
– hop market   501, 502
– packaging   505
– profitability   510
– raw materials   498–503
– statistics   498–505
– top brewers   510–512
– use of hop and hop products   502, 503
wort   4
– cooling   27
– fermentation and composition of   213
– flavor stability and preparation of   409–413
– high original   246–248
– N-heterocycles of   384
– as nutrient for yeast   127
– from oat malt   61
– quality of   704
– yeast cultivation in   136, 137
wort/beer losses, calculating   563
wort boiling   186–202, 386, 413, 414
– beer foam and   363, 364
– boiling systems   190–200
– cold trub   202
– dynamic low-pressure boiling   194, 196
– evaporation   186–189
– external boiler   192, 193
– flash evaporation 'Varioboil'   197–199
– high temperature   194, 195
– hot holding   186
– internal boilers   190–192
– optimized internal boiler 'Stromboli'   190, 191
– optimized internal boiler 'Subjet'   191, 192
– soft boiling method 'SchoKo'   194
– thin film evaporator 'Merlin'   199, 200

– turbidity and   431
– vacuum evaporation   197, 198
– vapor condensate   201, 202
– wort stripping   196
wort cooler   531–534
wort cooling   564
– heat recovery and   648, 649
– optimization of   661
wort kettle   531, 532, 534
wort production   165–202
– flowchart   166
– grinding technology   165–168
– for lambic and gueuze beers   251
– lautering technology   181–186
– mashing technology   168–181
– wort boiling   186–202
– wort stripping   196
wort sugar content   31
wrap-around packaging   584–587

x

Xan beer   235, 243
Xan technology   243
xanthohumol   89, 97, 243, 377–379, 437, 522
Xenophon   6
xerogels   432
X-rays   319

y

yeast   119–142
– ale *vs.* lager   121–123
– beer recovery from   221
– brewing   119–127
– brewing privilege and   12
– chemical composition of   124
– classification of   121
– dried   137
– flocculation   123
– foreign   480, 481, 485–489
– history of yeast research   119, 120
– morphology and chemical composition   124–127
– repitching   140
– spent   674
– strains of brewing   120, 121
yeast cellar   138–140
yeast cold-contact process   238, 239
yeast conservation starter culture   4
yeast crop   140, 141, 220, 221
yeast management   127–142, 566
– cultivation   135–138
– esters, formation of   134
– flavor stability and   414–417
– higher alcohols, formation of   132, 133
– metabolic pathways during propagation and fermentation   131–135
– nutrient requirements and intake   127–130
– phenolic compounds   134, 135
– post-fermentation treatment   140
– propagation in yeast cellar   138–140
– sulfur dioxide, formation of   135
– treatment after cropping and storage   141, 142
– vicinal diketones, formation of   132
– vitamins and other growth factors   130
– yeast crop   140, 141, 220, 221
yeast storage   220, 221
yeasty flavor attribute   680, 686
Young, William   120
young beer   207

z

zearalenon   438
zinc   127, 129, 130
Zosimus of Panopolis   7
Z protein   51
zymolectin   123
zymosterol   125
Zymotechnic Institute   33
zythos   7, 8